SATELLITE COMMUNICATIONS SYSTEMS

Second Edition

WILEY SERIES IN COMMUNICATION AND DISTRIBUTED SYSTEMS

Integrated Digital Communications Networks (Volume 1 and 2)
G. Pujolle, D. Seret, D. Dromard and E. Horlait

Security for Computer Networks, Second Edition
D.W. Davies and W.L. Price

Elements of Digital Communication
J.C. Bic, D. Duponteil and J.C. Imbeaux

Satellite Communications Systems, Second Edition (Systems, Techniques and Technology)
G. Maral and M. Bousquet

Using Formal Description Techniques (An Introduction to ESTELLE, LOTOS and SDL)
Edited by Kenneth J. Turner

Security Architecture for Open Distributed Systems
S. Muftic, A. Patel, P. Sanders, R. Colon, J. Heijnsdijk and U. Pulkinnen

Future Trends in Telecommunications
R.J. Horrocks and R.W.A Scarr

SATELLITE COMMUNICATIONS SYSTEMS

Systems, Techniques and Technology

Second Edition

G. Maral

Ecole Nationale Supérieure des Télécommunications
Toulouse, France

M. Bousquet

Ecole Nationale Supérieure de l'Aéronautique et de l'Espace
Toulouse, France

Translated by J.C.C. Nelson

University of Leeds
UK

JOHN WILEY & SONS
Chichester • New York • Brisbane • Toronto • Singapore

Other Wiley Editorial Offices

John Wiley & Sons, Inc., 605 Third Avenue,
New York, NY 10158-0012, USA

Jacaranda Wiley Ltd, G.P.O. Box 859, Brisbane,
Queensland 4001, Australia

John Wiley & Sons (Canada) Ltd, 22 Worcester Road,
Rexdale, Ontario M9W 1L1, Canada

John Wiley & Sons (SEA) Pte Ltd, 37 Jalan Pemimpin #05-04,
Block B, Union Industrial Building, Singapore 2057

British Library Cataloguing in Publication Data

A catalogue record for this book is available from the British Library

ISBN 0 471 93032 6; 0 471 93582 4 (pbk)

Typeset in $10\frac{1}{2}/12$ pt Palatino by Thomson Press (India) Ltd., New Delhi
Printed and bound in Great Britain by the Bath Press Ltd, Bath, Avon

CONTENTS

ACKNOWLEDGEMENT xii
ACRONYMS xiii
NOTATION xv

1 INTRODUCTION 1
- 1.1 The birth of satellite communications systems 1
- 1.2 Development 1
- 1.3 The architecture of a satellite communications system 2
  - 1.3.1 The space segment 2
  - 1.3.2 The ground segment 4
- 1.4 Types of orbit 5
- 1.5 Technological progress 9
- 1.6 The development of services 11
- 1.7 Conclusions and prospects 12
- References 12

2 LINK ANALYSIS 15
- 2.1 The characteristic parameters of an antenna 15
  - 2.1.1 Gain 15
  - 2.1.2 The radiation pattern 16
  - 2.1.3 The angular beamwidth 17
  - 2.1.4 Polarisation 18
- 2.2 The power emitted in a given direction 21
  - 2.2.1 Equivalent isotropic radiated power (EIRP) 21
  - 2.2.2 Power flux density 22
- 2.3 Received signal power 22
  - 2.3.1 Power received by the receiving antenna 22
  - 2.3.2 Example 1: The uplink 23
  - 2.3.3 Example 2: The downlink 24
  - 2.3.4 The practical case 25
  - 2.3.5 Conclusion 27
- 2.4 Noise power at the receiver input 27
  - 2.4.1 The origins of noise 27
  - 2.4.2 Characterisation and definition of noise 28
  - 2.4.3 Noise temperature of an antenna 29
  - 2.4.4 Noise temperature of an attenuator 30
  - 2.4.5 Noise temperature of a device containing several elements in cascade 30
  - 2.4.6 Application to the noise temperature of receiving equipment 30
  - 2.4.7 Example 31
  - 2.4.8 Conclusion 32
- 2.5 Signal-to-noise ratio at the receiver input 32
  - 2.5.1 Definitions 32
  - 2.5.2 Expression 32
  - 2.5.3 Figure of merit for receiving equipment 33

2.5.4 The antenna noise temperature 33
2.5.5 Noise temperature of the receiver 37
2.5.6 Example 1: Uplink (clear sky) 38
2.5.7 Example 2: Uplink (with rain) 41
2.5.8 Example 3: Downlink (clear sky) 42
2.5.9 Example 4: Downlink (with rain) 44
2.5.10 Conclusion 45
2.6 Influence of the propagating medium 45
2.6.1 The effects of precipitation 46
2.6.2 Other effects 50
2.6.3 Conclusion 51
2.7 Compensation for the effects of the propagating medium 52
2.7.1 Cross-polarisation 52
2.7.2 Attenuation 52
2.7.3 Site diversity 53
2.7.4 Adaptivity 53
2.7.5 Conclusion: the availability–cost relation 55
2.8 Constraints 55
2.8.1 Regulatory aspects 56
2.8.2 Operational constraints 60
2.9 Signal-to-noise ratio for a station-to-station link (single access) 61
2.9.1 Repeater modelling 62
2.9.2 Expression for $(C/N_0)_T$ 64
2.10 Example 66
2.10.1 Calculation of the repeater gain at saturation $(G_{sat})_{SL}$ 66
2.10.2 Calculation of C/N_0 for the up- and downlinks and the total link when the repeater operates in saturation 67
2.10.3 Calculation of the input back-off IBO and the output back-off OBO to achieve $(C/N_0)_T = 80$ dB(Hz) and the corresponding values of $(C/N_0)_u$ and $(C/N_0)_D$ 67
2.10.4 Value of $(C/N_0)_T$ when produced under the effect of rain causing an attenuation of 6 dB on the uplink 68
2.10.5 Value of $(C/N_0)_T$ when produced under the effect of rain causing an attenuation of 6 dB on the downlink with a reduction of 2 dB in the figure of merit of the earth station due to the increase of antenna noise temperature 68
References 69

3 TRANSMISSION TECHNIQUES FOR A SATELLITE CHANNEL 71
3.1 Signal characteristics 71
3.1.1 Telephone channel signals 71
3.1.2 Television signals 72
3.1.3 Radio broadcast programmes 76
3.1.4 Data 77
3.2 A model of the channel 77
3.3 Performance objectives 78
3.3.1 Telephone 80
3.3.2 Television 80
3.3.3 Radio broadcast programmes 81
3.3.4 Data 81
3.4 Availability objectives 82
3.5 Propagation time 82
3.5.1 Propagation time on the space link 82
3.5.2 Propagation time on the network 83
3.5.3 The case of the telephone 83
3.5.4 The case of data transmission 84
3.6 Analogue transmission 85
3.6.1 Baseband processing 85
3.6.2 Multiplexing 87

3.6.3 Amplitude modulation (AM) 89
3.6.4 Demodulation of an amplitude modulated wave 90
3.6.5 Frequency modulation (FM) 90
3.6.6 Demodulation of a frequency modulated wave 91
3.6.7 Telephone transmission on SCPC/FM 92
3.6.8 Telephone transmission on FDM/FM 94
3.6.9 Telephone transmission in FDM/SSB-AM 96
3.6.10 Television transmission in SCPC/FM 96
3.6.11 Energy dispersion 98
3.7 Digital transmission 99
3.7.1 Digitisation of analogue signals 99
3.7.2 Time division multiplexing (TDM) 100
3.7.3 Digital speech concentration and channel multiplication 101
3.7.4 Synchronism between networks 102
3.7.5 Encryption 102
3.7.6 Channel encoding 103
3.7.7 Digital modulation 104
3.7.8 Demodulation 109
3.7.9 Decoding and error correction 114
3.7.10 Energy dispersion 118
3.7.11 Conclusion—comparison between analogue and digital transmission 120
References 124

4 MULTIPLE ACCESS 125
4.1 Traffic laws 125
4.2 Traffic routing 126
4.2.1 One carrier per link 126
4.2.2 One carrier per transmitting station 127
4.2.3 Comparison 127
4.3 The principle of multiple access 128
4.3.1 Access to a particular channel 128
4.3.2 Multiple access to the satellite repeater 130
4.4 Frequency division multiple access (FDMA) 130
4.4.1 Transmission schemes 130
4.4.2 Adjacent channel interference 131
4.4.3 Intermodulation 132
4.4.4 Carrier power-to-noise power spectral density ratio for a station-to-station link 137
4.4.5 Throughput of FDMA 138
4.4.6 Intelligible crosstalk 139
4.4.7 Conclusion 139
4.5 Time division multiple access (TDMA) 140
4.5.1 Burst generation 141
4.5.2 Frame structure 142
4.5.3 Burst reception 144
4.5.4 Synchronisation 145
4.5.5 Throughput of TDMA 151
4.5.6 Conclusion 153
4.6 Code division multiple access (CDMA) 154
4.6.1 Direct sequence transmission (DS-CSMA) 154
4.6.2 Transmission by frequency hopping (FH-CDMA) 157
4.6.3 Code generation 160
4.6.4 Synchronisation 160
4.6.5 The throughput of CDMA 162
4.6.6 Conclusion 164
4.7 Fixed and on-demand assignment 165
4.7.1 The principle 165
4.7.2 Comparison between fixed and on-demand assignment 166

4.7.3 Example of an on-demand assignment system in FDMA: the SPADE system 167
4.7.4 On-demand assignment with TDMA 168
4.7.5 Centralised or decentralised management of demand assignment 171
4.7.6 Conclusion 172
4.8 Random access 172
4.8.1 Totally asynchronous protocols 173
4.8.2 Protocols with synchronisation 177
4.8.3 Protocols with assignment on demand (DAMA) 178
4.9 Conclusion 178
References 179

5 MULTIBEAM SATELLITE NETWORKS 181
5.1 Advantages and disadvantages of multibeam satellites 181
5.1.1 Advantages 182
5.1.2 Disadvantages 186
5.2 Interconnection by transponder hopping 188
5.3 Interconnection by on-board switching (SS/TDMA) 189
5.3.1 The principle 189
5.3.2 Frame structure 191
5.3.3 Window structure 192
5.3.4 Assignment of packets in the frame (burst time plan) 193
5.3.5 Synchronisation 194
5.3.6 Frame throughput 195
5.4 Interconnection by beam scanning 197
5.5 Intersatellite links (ISL) 197
5.5.1 Links between geostationary and low orbit satellites (GEO-LEO) 197
5.5.2 Links between geostationary satellites (GEO–GEO) 198
5.5.3 Links between low orbit satellites (LEO–LEO) 202
5.5.4 Frequency bands 203
5.5.5 Radio-frequency links 203
5.5.6 Optical links 204
5.5.7 Conclusion 207
5.6 Conclusion 207
References 208

6 REGENERATIVE SATELLITE NETWORKS 211
6.1 Conventional and regenerative transponders 211
6.2 Comparison of link budgets 211
6.2.1 Linear channel without interference 211
6.2.2 Non-linear channel without interference 214
6.2.3 Non-linear channel with interference 215
6.3 On-board processing 216
6.3.1 Downlink coding 216
6.3.2 Information storage 218
6.4 Impact on the earth segment 218
6.4.1 The uplink 220
6.4.2 The downlink 220
6.5 Conclusion 222
References 222

7 ORBITS 223
7.1 Keplerian orbits 223
7.1.1 Kepler's laws 223
7.1.2 Newton's law 223

7.1.3 Relative movement of two point bodies 224
7.1.4 Orbital parameters 227
7.1.5 The earth's orbit 232
7.1.6 Earth–satellite geometry 240
7.1.7 Eclipses of the sun 246
7.1.8 Sun–satellite conjunction 247
7.2 Useful orbits for space communication 248
7.2.1 Elliptical orbits with non-zero inclination 248
7.2.2 Geosynchronous elliptic orbits with zero inclination 262
7.2.3 Circular geosynchronous orbits with non-zero inclination 263
7.2.4 Sub-synchronous circular orbits with zero inclination 265
7.2.5 Geostationary satellite orbits 265
7.3 Perturbations of the orbit 276
7.3.1 The nature of the perturbations 277
7.3.2 The effect of perturbations; orbit perturbation 280
7.3.3 Perturbations of the orbit of geostationary satellites 282
7.3.4 Orbit corrections: station keeping of geostationary satellites 290
7.4 Conclusion 307
References 307

8 EARTH STATIONS 311
8.1 Station organisation 311
8.2 Radio-frequency characteristics 312
8.2.1 Effective isotropic radiated power 312
8.2.2 Figure of merit of the station 314
8.2.3 Standards defined by international organisations 315
8.3 The antenna subsystem 321
8.3.1 Radiation characteristics (major lobe) 321
8.3.2 Side-lobe radiation 321
8.3.3 Antenna noise temperature 323
8.3.4 Types of antenna 329
8.3.5 Pointing angles of an earth station antenna 334
8.3.6 Mountings to permit antenna pointing 337
8.3.7 Tracking 344
8.4 The radio-frequency sub-system 355
8.4.1 Receiving equipment 355
8.4.2 Transmission equipment 359
8.4.3 Redundancy 366
8.5 Communication sub-systems 366
8.5.1 Frequency translation 367
8.5.2 Amplification, filtering and equalisation 369
8.5.3 Modulation and demodulation 372
8.5.4 Additional functions 373
8.5.5 Time division multiple access terminals 374
8.6 The network interface sub-system 377
8.6.1 Multiplexing and demultiplexing 377
8.6.2 Digital speech interpolation (DSI) 378
8.6.3 Digital circuit multiplication equipment (DCME) 380
8.6.4 Echo suppression and cancellation 383
8.6.5 Equipment specific to SCPC transmission 384
8.7 Monitoring and control; auxiliary equipment 384
8.7.1 Monitoring alarms and control (MAC) 384
8.7.2 Electrical power 385
8.8 Conclusion 386
References 386

9 THE COMMUNICATION PAYLOAD 391
9.1 Mission and characteristics of the payload 391
9.1.1 Functions of the payload 391
9.1.2 Characterisation of the payload 392
9.1.3 The relationship between the radio-frequency characteristics 393
9.2 Conventional transponders 394
9.2.1 Characterisation of non-linearities 394
9.2.2 Repeater architecture 403
9.2.3 Equipment characteristics 408
9.3 Transponders for multibeam satellites 420
9.3.1 Fixed interconnection 421
9.3.2 Reconfigurable (semi-fixed) interconnection 421
9.3.3 On-board switching 422
9.4 Regenerative transponders 427
9.4.1 Examples of satellites with on-board regeneration 428
9.4.2 Equipment for regenerative transponders 432
9.4.3 Equipment technology 436
9.5 Antenna coverage 437
9.5.1 Geographical coverage 438
9.5.2 Geometric coverage 441
9.5.3 Global coverage 442
9.5.4 Reduced or spot coverage 444
9.5.5 Evaluation of antenna pointing error 446
9.5.6 Conclusion 455
9.6 Antenna characteristics 455
9.6.1 Antenna functions and characteristics 455
9.6.2 The radio-frequency footprint 458
9.6.3 Circular beam 458
9.6.4 Elliptical beams 461
9.6.5 The influence of depointing 463
9.6.6 Shaped beams 465
9.6.7 Multiple beams 469
9.6.8 Types of antenna 473
9.6.9 Antenna technologies 474
9.7 Conclusion 483
References 484

10 ORGANISATION OF TELECOMMUNICATION PLATFORMS 497
10.1 Subsystems 497
10.2 Attitude control 499
10.2.1 Attitude control functions 500
10.2.2 Attitude sensors 502
10.2.3 Attitude determination 503
10.2.4 Actuators 507
10.2.5 The principle of gyroscopic stabilisation 509
10.2.6 Stabilisation by rotation 511
10.2.7 'Three-axis' stabilisation 514
10.3 The propulsion subsystem 520
10.3.1 Characteristics of thrusters 520
10.3.2 Chemical propulsion 523
10.3.3 Electric propulsion 527
10.3.4 Organisation of the propulsion subsystem 531
10.3.5 Example calculation 535
10.4 The electric power supply 536
10.4.1 Primary energy sources 536
10.4.2 Secondary energy sources 542
10.4.3 Conditioning and protection circuits 547
10.4.4 Example calculations 554

10.5 Telecontrol, tracking and command (TTC) 555
10.5.1 Frequencies used 556
10.5.2 The command link 557
10.5.3 Telemetry links 559
10.5.4 Location 559
10.5.5 On-board data handling 563
10.6 Thermal control and structure 567
10.6.1 Thermal control specifications 567
10.6.2 Passive control 569
10.6.3 Active control 572
10.6.4 Structure 573
10.6.5 Conclusion 576
10.7 Developments and trends 576
References 577

11 SATELLITE INSTALLATION AND LAUNCHERS 581
11.1 Installation in orbit 581
11.1.1 General principle 581
11.1.2 Calculation of the required velocity increments 583
11.1.3 Inclination correction and circularisation 584
11.1.4 The apogee (or perigee) motor 593
11.1.5 Injection into orbit with a conventional launcher 597
11.1.6 Injection into orbit from a quasi-circular low altitude orbit 602
11.1.7 Operations during installation (station acquisition) 604
11.1.8 Injection into orbits other than geostationary 608
11.1.9 The launch window 609
11.2 Launchers 610
11.2.1 China 611
11.2.2 Europe (Ariane) 612
11.2.3 Europe other than Ariane 623
11.2.4 The United States 624
11.2.5 India 638
11.2.6 Israel 638
11.2.7 Japan 639
11.2.8 USSR 641
11.2.9 Cost of installation in orbit 643
References 643

12 THE SPACE ENVIRONMENT 647
12.1 Vacuum 647
12.1.1 Characterisation 647
12.1.2 Effects 647
12.2 The mechanical environment 648
12.2.1 The gravitational field 648
12.2.2 The earth's magnetic field 649
12.2.3 Solar radiation pressure 650
12.2.4 Meteorites and material particles 651
12.2.5 Torques of internal origin 651
12.2.6 The effect of communication transmissions 652
12.2.7 Conclusions 652
12.3 Radiation 653
12.3.1 Solar radiation 653
12.3.2 Terrestrial radiation 654
12.3.3 Thermal effects 654
12.3.4 Effects on materials 656
12.4 Flux of high energy particles 657
12.4.1 Cosmic particles 657
12.4.2 Effects on materials 658

12.5 The environment during installation 661
12.5.1 The environment during launching 662
12.5.2 Environment in the transfer orbit 662
References 662

13 RELIABILITY OF SATELLITE COMMUNICATIONS SYSTEMS 663
13.1 Introduction of reliability 663
13.1.1 Failure rate 663
13.1.2 The probability of survival or reliability 664
13.1.3 Probability of death or unreliability 665
13.1.4 MTTF—mean lifetime 666
13.1.5 Reliability during the wear-out period 667
13.2 Satellite system availability 667
13.2.1 No back-up satellite in orbit 668
13.2.2 Back-up satellite in orbit 668
13.2.3 Conclusion 668
13.3 Sub-system reliability 669
13.3.1 Elements in series 669
13.3.2 Elements in parallel (static redundancy) 670
13.3.3 Dynamic redundancy (with switching) 671
13.3.4 Equipment having several failure modes 675
13.4 Component reliability 676
13.4.1 Component reliability 676
13.4.2 Component selection 676
13.4.3 Manufacture 678
13.4.4 Quality assurance 678
References 680

INDEX 683

ACKNOWLEDGEMENT

Reproduction of figures extracted from the 1990 Edition of CCIR Volumes (XVIIth Plenary Assembly, Dusseldorf, 1990) and from the "Handbook on Satellite Communications (ITU Geneva, 1988)" is made with the authorisation of the International Telecommunication Union (ITU) as copyright holder.

The choice of the excerpts reproduced remains under the sole responsibility of the authors and does not engage in any way the ITU.

The complete CCIR Volumes can be obtained from:

International Telecommunication Union
General Secretariat-Sales Section
Place des Nations, 1211 GENEVA 20 (Switzerland)

Tel: + 41 22 730 51 11 **Tg:** Burinterna Geneva
Telefax: 2/m + 41 22 730 51 94 **Tlx:** 421 000 uit ch

ACRONYMS

ABM	apogee boost motor
ACI	adjacent channel interference
ACK	acknowledgement
ADPCM	adaptative pulse code modulation
A/D	analog-to-digital conversion
AKM	apogee kick motor
AM	amplitude modulation
AMP	amplifier
AOR	Atlantic Ocean region
APC	adaptive predictive coding
AR	axial ratio
ARQ	automatic repeat request
ARQ-SW	automatic repeat request—stop and wait
ARQ-GB(N)	automatic repeat request—go back N
ARQ-SR	automatic repeat request—selective repeat
ATA	auto-tracking antenna
ATC	adaptive transform coding
BEP	bit error probability
BER	bit error rate
BFN	beam forming network
BPF	band pass filter
BPSK	binary phase shift keying
BSS	broadcasting satellite service
BW	bandwidth
CATV	cable television
CCI	cochannel interference
CDMA	code division multiple access
CFM	companded frequency modulation
CCIR	International Radio Consultative Committee
CCITT	The International Telegraph and Telephone Consultative Committee
CONUS	continental US
DASS	demand assignment signaling and switching unit
DCME	digital circuit multiplication equipment
DE	differentially encoded
dB	decibel
dBm	unit for expression of power level in dB with reference to 1 mW
dBm0	unit for expression of power level in dBm at a point of zero relative level (a point of a telephone channel where the 800 Hz test signal has a power of 1 mW)
DBS	direct broadcasting satellite
DC	direct current
D/C	down-converter
DCME	digital circuit multiplication equipment
DCU	distribution control unit
DEMOD	demodulator
DE-M-PSK	differentially encoded M-ary phase shift keying
DEMUX	demultiplexer
DM	delta modulation
D-M-PSK	differential M-ary phase shift keying
DSI	digital speech interpolation
DTE	data terminal equipment
DTTL	data transition tracking loop
DUT	device under test
EBU	European Broadcasting Union
EIRP	effective isotropically radiated power
ENR	excess noise ratio
EPC	electric power conditioner
ERL	echo return loss
ES	earth sation
ESA	European Space Agency
EUTELSAT	European Telecommunications Satellite Organisation
EMC	electromagnetic compatibility
EMI	electromagnetic interference
FEC	forward error correction
FET	field effect transistor
FETA	field effect transistor amplifier
FDM	frequency division multiplex
FDMA	frequency division multiple access
FM	frequency modulation
FMA	fixed-mount antenna
FSS	fixed satellite service
GaAs	gallium arsenide
GCE	ground communication equipment
GDE	group delay equalizer
GEO	geostationary orbit
GPS	Global Positioning System
GTO	geostationary transfer orbit
HPA	high power amplifier
HPB	half power beamwidth
IAU	international astronomical unit
IBA	Independent Broadcasting Authority
IBO	input back-off
IBS	international business service
IF	intermediate frequency
IFRB	International Frequency Registration Board

IM intermodulation
IMUX input multiplexer
INMARSAT International Maritime Satellite Organisation
INTELSAT International Telecommunications Satellite Consortium
IOR Indian Ocean region
IPA intermediate power amplifier
IPE initial pointing error
ISDN integrated services digital network
ISL intersatellite link
ITU International Telecommunication Union

LEO low earth orbit
LHCP left hand circular polarisation
LNA low noise amplifier
LO local oscillator
LPC linear predictive coding
LPF low pass filter
LRE low rate encoding

MAC multiplexed analog components (also monitoring, alarm and control)
MOD modulator
MOS metal-oxide semiconductor
M-PSK M-ray phase shift keying
MSK minimum shift keying
MUX multiplexer
MX mixer

NACQ no acknowledgement
NH northern hemisphere
NRZ non return to zero

OMUX output multiplexer
OBO output back-off

PAM payload assist module
PABX private branch automatic exchange
PB primary body (orbits)
PCM pulse code modulation
pdf probability density function
PFD power flux density
PKM perigee kick motor
PM phase modulation
PODA priority oriented demand assignment
POL polarisation
POR Pacific Ocean region
psd power spectral density
PSK phase shift keying
PTA programme tracking antenna

QPSK quaternary phase shift keying

RAM random access memory
RARC Regional Administrative Radio Conference
RCVO receive only
RCVR receiver
Rec recommendation
Rep report
RF radio frequency
RFI radiofrequency interference
RHCP right-hand circular polarisation
RL return loss
Rx receiver

SB secondary body (orbits)
SBC sub-band coding
SC suppressed carrier
SCPC single channel per carrier
S/C spacecraft
SEP symbol error probability
SH southern hemisphere
SHF super high frequency (3 GHz to 30 GHz)
SKW satellite-keeping window
SL satellite
SMATV satellite based master antenna
SMS satellite multi-services
SNR signal-noise ratio
SORF start of receive frame
SOTF start of transmit frame
SPADE single-channel-per-carrier PCM multiple access demand assignment equipment
SPDT single-pole double-throw (switch)
SPMT single-pole multiple-throw (switch)
SPU satellite position uncertainty
SS satellite switch
SSB single sideband
SS-TDMA satellite switched TDMA
SSMA spread spectrum multiple access
SSPA solid state power amplifier
STS space transportation system
SW switch
SWR standing wave ratio
SYNC synchronization

TDM time division multiplex
TDMA time division multiple access
TEM transverse electromagnetic
TIE terrestrial interface equipment
T/R transmit/receive
Tx transmitter
TTC telemetry, tracking and command
TTY telegraphy
TV television
TWT travelling wave tube
TWTA travelling wave tube amplifier

U/C up-converter
UHF ultra high frequency (300 MHz to 3 GHz)
UW Unique Word

VHF very high frequency (30 MHz to 300 MHz)
VOW voice order wire
VSAT very small aperture terminal
VPA variable power divider
VPD variable phase shifter
VSWR voltage standing wave ratio

WARC World Administrative Radio Conference

Xponder transponder

NOTATION

a	orbit semi-major axis
A	azimuth angle (also attenuation, area, availability, traffic density and carrier amplitude)
A_{eff}	effective aperture area of an antenna
A_{AG}	attenuation by atmospheric gases
A_{RAIN}	attenuation due to precipitation and clouds
A_p	attenuation of radiowave by rain for percentage p of an average year
B	bandwidth
b	voice channel bandwidth (3100 Hz from 300 to 3400 Hz)
B_n	noise measurement bandwidth at baseband (receiver output)
B_N	noise equivalent bandwidth of receiver
c	velocity of light $= 3 \times 10^8$ m/s
C	carrier power
C/N_0	carrier power-to-noise power spectral density ratio (W/Hz)
$(C/N_0)_U$	uplink carrier power-to-noise power spectral density ratio
$(C/N_0)_D$	downlink carrier power-to-noise power spectral density ratio
$(C/N_0)_{IM}$	carrier power-to-intermodulation noise power spectral density ratio
$(C/N_0)_I$	carrier power-to-interference noise power spectral density ratio
$(C/N_0)_{I,U}$	uplink carrier power-to-interference noise power spectral density ratio
$(C/N_0)_{I,D}$	downlink carrier power-to-interference noise power spectral density ratio
$(C/N_0)_T$	carrier power-to-noise power spectral density ratio for total link
D	diameter of a reflector antenna (also used as a subscript for 'downlink'
e	orbit eccentricity
E	elevation angle (also energy and electric field strength)
E_b	energy per information bit
E_c	energy per channel bit
f	frequency (Hz)
f_c	carrier frequency
f_m	frequency of a modulating sine wave
f_{max}	maximum frequency of the modulating baseband signal spectrum
f_D	downlink frequency
f_U	uplink frequency
F	noise figure
ΔF_{max}	peak frequency deviation of a frequency modulated carrier
F_s	sampling frequency
g	peak factor
G	power gain (also gravitational constant)
G_{sat}	gain at saturation
G_R	receiving antenna gain
G_T	transmitting antenna gain
G_{Rmax}	maximum receiving antenna gain
G_{Tmax}	maximum transmitting antenna gain
G_{SL}	satellite repeater gain
$(G_{sat})_{SL}$	saturation gain of satellite repeater
G/T	gain to equivalent noise temperature ratio of a receiving equipment
i	inclination of the orbital plane
k	Boltzmann's constant $= 1.379 \times 10^{-23}$ W/K Hz
k_{FM}	FM modulation frequency deviation constant (MHz/V)
k_{PM}	PM phase deviation constant (rad/V)
K_P	AM/PM conversion coefficient
K_T	AM/PM transfer coefficient
l	earth sation latitude
L	earth station-to-satellite relative longitude (also loss in link budget calculations, and loading factor of FDM/FM multiplex)
L_e	effective path length of radiowave through rain (km)
L_{FRX}	receiver feeder loss
L_{FTX}	transmitter feeder loss
L_{FS}	free space loss
L_{POINT}	depointing loss
L_{POl}	antenna polarisation mismatch loss
L_R	receiving antenna depointing loss
L_T	transmitting antenna depointing loss
m	satellite mass

M	mass of the Earth (kg) (also number of possible states of a digital signal)
N_o	noise power spectral density (W/Hz)
$(N_o)_U$	uplink noise power spectral density (W/Hz)
$(N_o)_D$	downlink noise power spectral density (W/Hz)
$(N_o)_T$	total link noise power spectral density (W/Hz)
$(N_o)_I$	interference power spectral density (W/Hz)
N	noise power (W) (also number of stations in a network)
p	pre-emphasis/companding improvement factor (also rainfall annual percentage)
p_w	rainfall worst month time percentage
P	power (also number of bursts in a TDMA frame)
P_b	information bit error rate
P_c	channel bit error rate
P_T	power fed to the antenna (W)
P_{Tx}	transmitter power (W)
P_R	received power (W)
P_{Rx}	power at receiver input (W)
(P_i^1)	input power in a single carrier operation mode
(P_o^1)	output power in a single carrier operation mode
$(P_i^1)_{sat}$	input power in a single carrier operation mode at saturation
$(P_o^1)_{sat}$	saturation output power in a single carrier operation mode
(P_i^n)	input power in a multiple carrier operation mode (*n* carrier)
(P_o^n)	output power in a multiple carrier operation mode (*n* carriers)
P_o^{IMX}	power of intermodulation product of order X at output of a non linear device in a multicarrier operation mode
Q	quality factor
r	distance between centre of mass (orbits)
R	slant range from earth station to satellite (km) (also symbol or bit rate)
R_b	information bit rate (s^{-1})
R_c	channel bit rate (s^{-1})
R_{call}	mean number of calls per unit time
R_e	earth radius = 6378 km
R_o	geostationary satellite altitude = 35 786 km
R_p	rainfall rate (mm/h)
R_S	symbol (or signalling) rate (s^{-1})
S	user signal power (W)
S/N	signal-to-noise power ratio at user's end
T	period of revolution (orbits) (s) (also noise temperature (K))
T_A	antenna noise temperature (K)
T_{AMB}	ambient temperature (K)
T_b	information bit duration (s)
T_B	burst duration (s)
T_c	channel bit duration (s)
T_e	effective input noise temperature of a four port element system (K)
T_F	frame duration (s) (also feeder temperature)
T_m	effective medium temperature (K)
T_o	reference temperature (290 K)
T_R	receiver input noise temperature (K)
T_S	symbol duration (s)
T_{SKY}	clear key contribution to antenna noise temperature (K)
T_{GROUND}	ground contribution to antenna noise temperature (K)
U	subscript for 'uplink'
v	true anomaly (orbits)
V_s	satellite velocity (m/s)
$V_{Lp/p}$	peak-to-peak luminance voltage (V)
$V_{Tp/p}$	peak-to-peak total video signal voltage (including synchronization pulses)
V_{Nrms}	root-mean-square noise voltage (V)
w	psophometric weighting factor
X	intermodulation product order (IMX)
α	angle from boresight of antenna (also nadir angle)
γ	vernal point
Γ	spectral efficiency (bit/s Hz)
δ	declination angle (also delay)
η	antenna aperture efficiency
θ_{3dB}	half power beamwidth of an antenna
λ	wavelength $= c/f$
μ	$= GM$ (G: gravitational constant, M: mass of earth)
ρ	code rate
σ	Stefan–Boltzmann constant $= 5.67 \times 10^{-8}$ $Wm^{-2} K^{-4}$
ϕ	satellite–earth station angle from the Earth's centre
ψ	polarisation angle
ω	argument of perigee
Ω	right ascension of the ascending node

1 INTRODUCTION

This chapter consists of a descriptive approach to the characteristics of satellite communication systems. Its aim is to satisfy the curiosity of an impatient reader and facilitate a deeper understanding by directing him or her to the appropriate chapter without imposing the need to read the whole work from start to finish.

1.1 THE BIRTH OF SATELLITE COMMUNICATIONS SYSTEMS

Radio communication by satellite is the outcome of research in the area of communications whose objective is to achieve ever increasing ranges and capacities with the lowest possible costs.

The Second World War favoured the expansion of two very distinct technologies—missiles and microwaves. The expertise eventually gained in the combined use of these two techniques opened up the era of satellite communication. The service provided in this way usefully complements that previously provided exclusively by ground networks using radio and cables.

The space era started in 1957 with the launching of the first artificial satellite (sputnik). Subsequent years have been marked by various experiments including the following: Christmas greetings from President Eisenhower broadcast by SCORE (1958), the reflecting satellite ECHO (1960), store-and-forward transmission by the COURIER satellite (1960), wideband repeater satellites (TELSTAR and RELAY in 1962) and the first geostationary satellite SYNCOM (1963).

In 1965, the first commercial geostationary satellite INTELSAT I (or Early Bird) inaugurated the long series of INTELSATs; in the same year, the first Soviet communication satellite of the MOLNYA series was launched.

1.2 DEVELOPMENT

The first satellite systems provided a low capacity at a relatively high cost; for example, INTELSAT I weighed 68 kg at launch for a capacity of 480 telephone channels and an annual cost of $32 500 per channel at the time. This cost resulted from a combination of the cost of the launcher, that of the satellite, the short lifetime of the satellite (1.5 years) and its low capacity. The reduction in cost is the result of much effort which has led to the production of reliable launchers which can put heavier and heavier satellites into orbit (3750 kg at launch for INTELSAT VI). In addition, increasing expertise in microwave techniques has enabled realisation of contoured multibeam antennas whose beams adapt to the shape of continents, re-use of the same band of frequencies from one beam to the other and incorporation of higher power transmission amplifiers. Increased satellite capacity has led to a reduced cost per telephone channel (80 000 channels on INTELSAT VI for an estimated annual cost per channel of $380 in 1989).

In addition to the reduction in the cost of communication, the most outstanding feature is the diversity of services offered by satellite telecommunication systems. Originally these were designed to carry communications from one point to another, as with cables, and the extended coverage of the satellite was used to advantage to establish long distance links; hence Early Bird enabled stations on opposite sides of the Atlantic Ocean to be connected. As a consequence of the limited performance of the satellite, it was necessary to use earth stations equipped with large antennas and therefore of high cost (around $10 million for a station equipped with a 30 m diameter antenna). The increasing size and power of satellites has permitted a consequent reduction in the size of earth stations, and hence their cost, with a consequent increase in number. In this way it has been possible to exploit another feature of the satellite which is its ability to collect or broadcast signals from or to several locations. Instead of transmitting signals from one point to another, transmission can be from a single transmitter to a large number of receivers distributed over a wide area or, conversely, transmission can be from a large number of stations to a single central station often called a hub. In this way, multipoint data transmission networks, satellite broadcast networks and data collection networks have been developed. Broadcasting can be either to relay transmitters (or cable heads) or directly to the private consumer (the latter are commonly called direct broadcast by satellite television systems). These networks operate with small earth stations having antennas with a diameter between 0.6 and 3.5 m at a cost between $500 and $50 000.

1.3 THE ARCHITECTURE OF A SATELLITE COMMUNICATIONS SYSTEM

Figure 1.1 shows the various components of a satellite communication system. It comprises a ground segment and a space segment.

1.3.1 The space segment

The space segment contains the satellite and all terrestrial facilities for the control and monitoring of the satellite. This includes the tracking, telemetry and command stations (TT&C) together with the satellite control centre where all the operations associated with station-keeping and checking the vital functions of the satellite are performed.

The radio waves transmitted by the earth stations are received by the satellite; this is called the *uplink*. The statellite in turn transmits to the receiving earth stations; this is the *downlink*. The quality of a radio link is specified by its carrier-to-noise ratio. The parameters involved are discussed in Chapter 2 which is devoted to link analysis. The important factor is the quality of the total link, from station to station, and this is determined by the quality of the uplink and that of the downlink. The quality of the total link determines the quality of the signals delivered to the end user in accordance with the type of modulation and coding used. These aspects are discussed in Chapter 3 which deals with transmission techniques over a satellite channel.

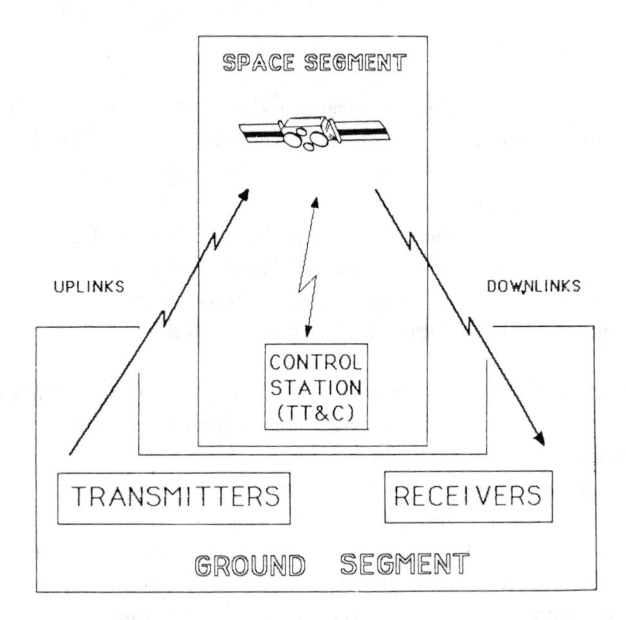

Figure 1.1 The elements of a satellite communication system.

The satellite forms a mandatory point of passage for a group of simultaneous links. In this sense, it can be considered as the nodal point of a network. Access to the satellite, and to a satellite transponder, by several carriers implies the use of specific techniques, called multiple access techniques. The mode of operation of these techniques differs between a satellite with only a single beam (monobeam satellite) and one with several beams (multibeam satellite). The problems posed by multiple access of uplinks to the satellite and their interconnection with the downlinks are presented in Chapter 4 in the context of a monobeam satellite and in Chapter 5 in the context of a multibeam satellite. If the system contains several satellites, these satellites can be connected by radio or optical links; these are intersatellite links. The details of these links are treated in Chapter 5.

The satellite consists of a payload and a platform. The payload consists of the receiving and transmitting antennas and all the electronic equipment which supports the transmission of carriers. The platform consists of all the subsystems which permit the payload to operate. These include:

- Structure.
- Electric power supply.
- Temperature control.
- Attitude and orbit control.
- Propulsion equipment.
- Tracking, telemetry and command (TT&C) equipment.

The detailed architecture and technology of the payload equipment are explained in Chapter 9. The architecture and technologies of the platform are considered in Chapter 10. The operations of orbit injection and the various types of launcher are the subject of Chapter 11. The space environment and its effects on the satellite are presented in Chapter 12.

The satellite has a dual role:

- To amplify the received carriers for retransmission on the downlink. The carrier power at the input of the satellite receiver is of the order of 100 pW to 1 nW. The carrier power at the output of the transmission amplifier is of the order of 10 to 100 W. The power gain is thus of the order of 100 to 130 dB.
- To change the frequency of the carrier to avoid re-injection of a fraction of the transmitted power into the receiver; the rejection capability of the input filters at the downlink frequency combines with the low antenna gains between the transmitting output and the receiving input to ensure isolation of the order of 150 dB.

To fulfil its function the satellite can operate as a simple relay. The change in frequency is achieved by means of a frequency converter. This is the case with all currently operational commercial satellites. One speaks of 'transparent' or 'conventional' satellites. However a new generation of satellites (starting with ACTS and ITALSAT) is emerging. They are called 'regenerative' satellites and are equipped with demodulators; baseband signals are, therefore, available on board. The change in frequency is achieved by modulating a new carrier for the downlink. The dual operation of modulation and demodulation can be accompanied by processing of the baseband signal with varying levels of complexity. The performance of conventional and regenerative satellites is compared in Chapter 6.

To ensure a service with a specified availability, a satellite communication system must make use of several satellites in order to ensure redundancy. A satellite can cease to be available due to a failure or because it has reached the end of its lifetime. In this respect it is necessary to distinguish between the reliability and the lifetime of a satellite. Reliability is a measure of the probability of a breakdown and depends on the reliability of the equipment and any schemes to provide redundancy. The lifetime is conditioned by the ability to maintain the satellite on station in the nominal attitude, that is the quantity of fuel available for the propulsion system and attitude and orbit control [ARN-82]. In a system provision is generally made for an operational satellite, a backup satellite in orbit and a backup satellite on the ground. The reliability of the system will involve not only the reliability of each of the satellites but also the reliability of launching. An approach to these problems is treated in Chapter 13.

1.3.2 The ground segment

The ground segment consists of all the earth stations; these are most often connected to the end-user's equipment by a terrestrial network or, in the case of small stations (Very Small Aperture Terminal, VSAT), directly connected to the end-user's equipment. Stations are distinguished by their size which varies according to the volume of traffic to be carried on the space link and the type of traffic (telephone, television or data). The largest are equipped with antennas of 30 m diameter (Standard A of the INTELSAT network). The smallest have 0.6 m antennas (direct television receiving stations). Fixed, transportable and mobile stations can also be distinguished. Some stations are both transmitters and receivers. Others are only receivers; this is the case, for example with receiving stations for a satellite broadcast system or a distribution system for television or data signals. Figure 1.2 shows the typical architecture of an earth station for both transmission and reception. Chapter 2 introduces the characteristic parameters of the earth station which appear in the link budget calculations. Chapter 3 presents

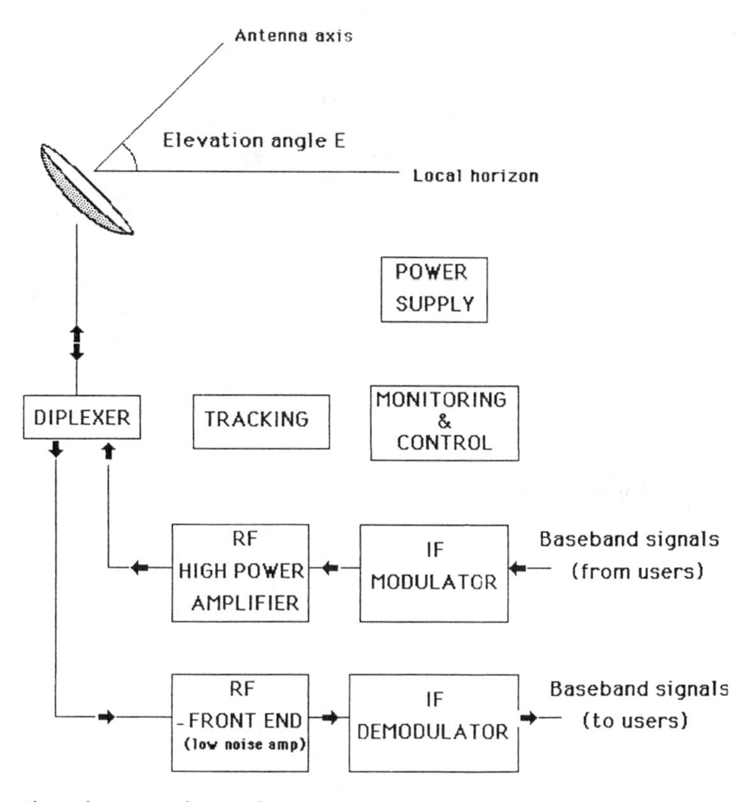

Figure 1.2 The architecture of an earth station.

the characteristics of signals supplied to earth stations by the user either directly or through a terrestrial network, the signal processing at the station (such as multiplexing, pre-emphasis, de-emphasis, companding, digital concentration, coding, scrambling, encryption, and so on), transmission and reception (including modulation and demodulation). Chapter 8 treats the architecture and equipment of an earth station.

1.4 TYPES OF ORBIT

The orbit is the trajectory followed by the satellite in equilibrium between two opposing forces (Figure 1.3). These are the force of attraction, due to the earth's gravitation, directed towards the centre of the earth and the centrifugal force associated with the curvature of the satellite's trajectory. The trajectory is within a plane and shaped as an ellipse with a maximum extension at the apogee and a minimum at the perigee. The satellite moves more slowly in its trajectory as the distance from the earth increases. Chapter 7 provides a definition of the orbital parameters.

The most favourable orbits are as follows:

- Elliptical orbits inclined at an angle of 64° with respect to the equatorial plane. This type of orbit is particularly stable with respect to irregularities in terrestrial gravitational

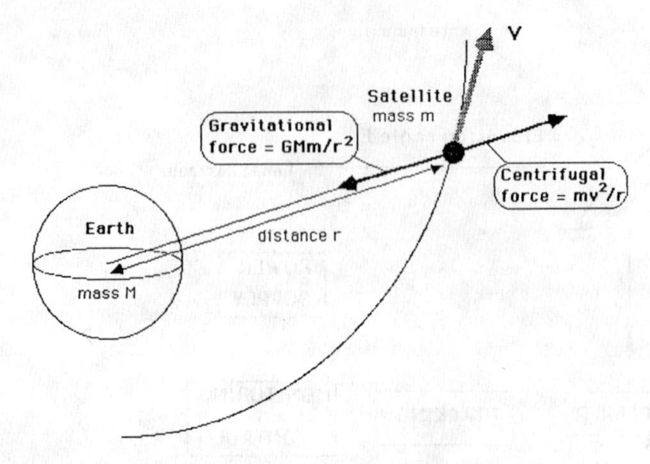

Figure 1.3 The forces which determine the trajectory of a satellite.

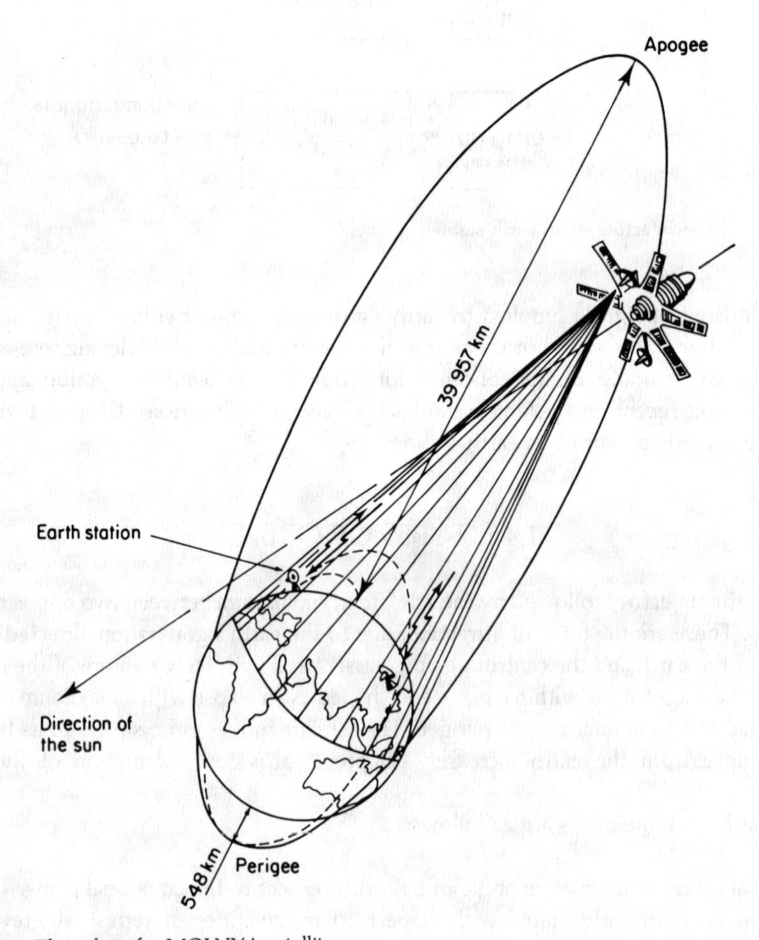

Figure 1.4 The orbit of a MOLNYA satellite.

potential and, owing to its inclination, enables the satellite to cover regions of high latitude for a large fraction of the orbital period as it passes to the apogee. This type of orbit has been adopted by the USSR for the satellites of the MOLNYA system with a period of 12 hours [JOH-88]. Figure 1.4 shows the geometry of the orbit and Figure 1.5 shows the track of the satellite on the surface of the earth. Notice that the satellite remains above the regions located under the apogee for a period of the order of 8 hours. Continuous coverage can be ensured with three phased satellites on different orbits. Several studies relate to elliptical orbits with a period of 24 h (TUNDRA orbits) or a multiple of 24 h [ABA-88], [DON-84]. These orbits are particularly useful for satellite systems for communication with mobiles where the masking effects caused by surrounding obstacles such as buildings and trees and multiple path effects are pronounced at low elevation angles (say less than 30°). In fact, inclined elliptic orbits can provide the possibility of links at medium latitudes when the satellite is close to the apogee with elevation angles close to 90°; these favourable conditions cannot be provided at the same latitudes by geostationary satellites. An operational system, called ELLIPSAT, comprising 24 satellites in two different orbits inclined at 64° (2903 km/426 km), has been proposed to offer permanent coverage of the USA [ELL-91].

- Circular inclined orbits. The altitude of the satellite is constant and equal to several hundreds of kilometres. The period is of the order of one and a half hours. With near 90° inclination, this type of orbit guarantees that the satellite will pass over every region of the earth. This is the reason for choosing this type of orbit for observation satellites (for example the SPOT satellite; altitude 830 km, orbit inclination 98.7°, period 101 minutes). One can envisage the establishment of store-and-forward communications if the satellite is equipped with a means of storing information. Several systems with worldwide coverage using constellations of satellite carriers in low altitude circular orbits (of the order of 1000 km) have been proposed recently (IRIDIUM, GLOBAL STAR, ODYSSEY, ARIES, LEOSAT, STARNET, etc) [ANA-92].

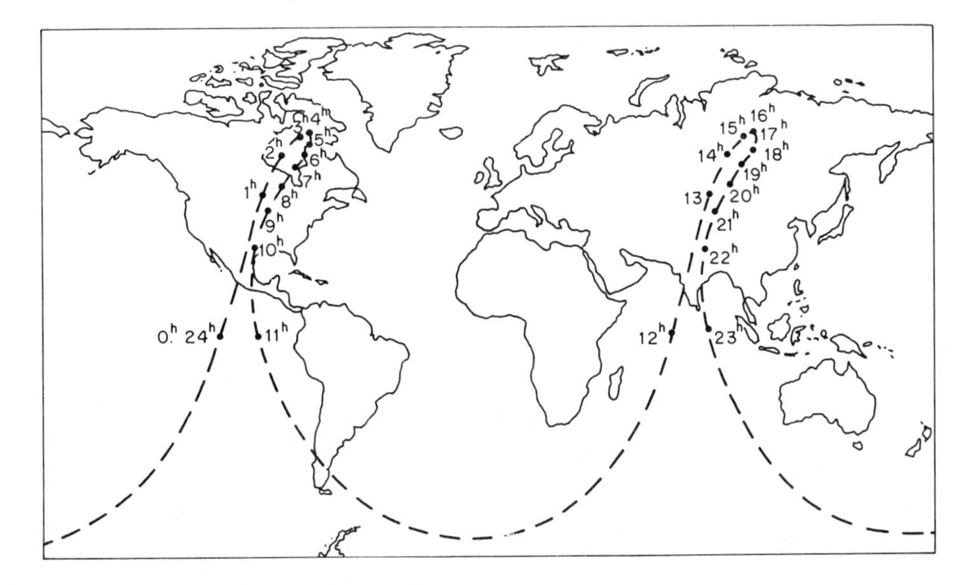

Figure 1.5 The track of a MOLNYA satellite.

- Circular orbits with zero inclination (equatorial orbits). The most popular is the geostationary satellite orbit; the satellite orbits around the earth at an altitude of 35 786 km, and in the same direction as the earth. The period is equal to that of the rotation of the earth and in the same direction. The satellite thus appears as a point fixed in the sky and ensures continuous operation as a radio relay in real time for the area of visibility of the satellite (43% of the earth's surface).

The choice of orbit depends on the nature of the mission, the acceptable interference and the performance of the launchers:

- The extent and latitude of the area to be covered; contrary to widespread opinion, the altitude of the satellite is not a determining factor in the link budget for a given earth coverage. Chapter 2 shows that the propagation attenuation varies as the inverse square of the distance and this favours a satellite following a low orbit on account of its low altitude; however, this disregards the fact that the area to be covered is then seen through a larger solid angle. The result is a reduction in the gain of the satellite antenna which offsets the distance advantage. Now a satellite following a low orbit provides only limited space coverage at a given time and limited time at a given location. Unless low gain antennas (of the order of a few dB) which provide low directivity and hence almost omnidirectional radiation are installed, earth stations must be equipped with satellite tracking devices which increase the cost. The geostationary satellite thus appears to be particularly useful for continuous coverage of extensive regions. However, it does not permit coverage of the polar regions which are accessible by satellites in inclined elliptical orbits or polar orbits.

- The elevation angle of earth stations; a satellite in an inclined or polar elliptical orbit can appear overhead at certain times which enables communication to be established in urban areas without encountering the obstacles which large buildings constitute for elevation angles between 0° and approximately 70°. With a geostationary satellite, the angle of elevation decreases as the difference in latitude or longitude between the earth station and the satellite increases.

- Transmission duration and delay; the geostationary satellite provides a continuous relay for stations within visibility but the propagation time of the waves from one station to the other is of the order of 0.25 s. This requires the use of echo control devices on telephone channels or special protocols for data transmission. A satellite moving in a low orbit confers a reduced propagation time. The transmission time is thus low between stations which are close and simultaneously visible to the satellite, but it can become long (several hours) for distant stations if only store-and-forward transmission is considered.

- Interference; geostationary satellites occupy fixed positions in the sky with respect to the stations with which they communicate. Protection againt interference between systems is ensured by planning the frequency bands and orbital positions. The small orbital spacing between adjacent satellites operating at the same frequencies leads to an increase in the level of interference and this impedes thc installation of new satellites. Different systems could use different frequencies but this is restricted by the limited number of frequency bands assigned for space radiocommunication by the radiocommunication regulations. In this context one can refer to an 'orbit-spectrum' resource which is limited. With orbiting satellites, the geometry of each system changes with time and the relative geometries of one system with respect to another are variable and difficult to synchronise. The probability of interference is thus high.

- The performance of launchers; the mass which can be launched decreases as the altitude increases.

The geostationary satellite is certainly the most popular. At the present time there are around 150 geostationary satellites in operation within the 360° of the whole orbital arc. Some parts of this orbital arc, however, tend to be highly congested (for example above the American continent and the Atlantic).

1.5 TECHNOLOGICAL PROGRESS

Figure 1.6 shows the technological developments since the start of the satellite communication era.

The start of commercial satellite telecommunications can be traced back to the commissioning of INTELSAT I (Early Bird) in 1965. Until the beginning of the 1970s, the services provided

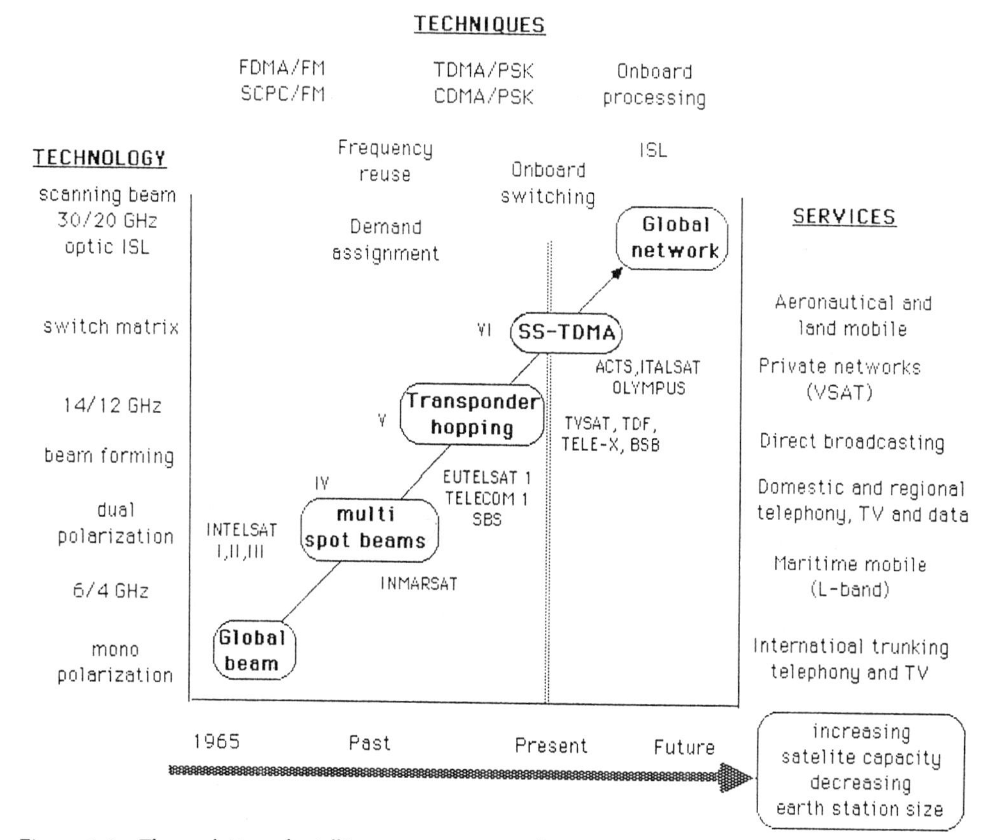

Figure 1.6 The evolution of satellite communication techniques.

were telephone and television (TV) signal transmission between continents. The satellite was designed to complement the submarine cable and played essentially the role of a telephone trunk connection. The goal of increased capacity has led rapidly to the institution of multibeam satellites and the re-use of frequencies first by orthogonal polarisation and subsequently by spatial separation (see Chapter 5). The transmission techniques were analogue and each carrier conveyed either a TV signal or frequency division multiplexed telephone channels (see Chapter 3). Multiple access to the satellite (see Chapter 4) was resolved by frequency division multiple access (FDMA). The increasing demand for a large number of low capacity links, for example for national requirements or for communication with ships led in 1980 to the introduction of demand assignment (see Chapter 4) first using FDMA with single channel per carrier/frequency modulation (SCPC/FM) or phase shift keying (PSK) and subsequently using time division multiple access/phase shift keying (TDMA/PSK) in order to profit from the flexibility of digital techniques (see Chapter 3). Simultaneously, the progress of antenna technology (see Chapter 9) enabled the beams to conform to the coverage of the service area; in this way the performance of the link was improved while reducing the interference between systems. However, the increasing number of beams made interconnection of the network more and more difficult; until this time it had been achieved by hops between transponders. From this arose the institution, at the present stage of development, of onboard switching (satellite switched time division multiple access (SS-TDMA), see Chapter 5).

The future will bring rapid developments in the following areas:

- On-board processing (see Chapter 6), by taking advantage of the possibilities provided by on-board carrier demodulation (regenerative satellites).
- Intersatellite links (see Chapter 5).
- Scanning or hopping beams (see Chapters 6 and 9).
- The use of higher frequencies (30/20 GHz and 50/40 GHz) in spite of the propagation problems caused by rain (see Chapter 2).

Ambitious programmes are in progress in these areas as follows:

- In Europe the European Space Agency (ESA) is financing a development plan containing three sections [REU-88].

  (a) Payload and Spacecraft Development and Experimentation (PSDE) which concerns the operation of experimental missions prior to operational systems for the 1990 decade. Three missions are presently programmed; these are an advanced aeronautical and mobile maritime payload (Aramis) on SAT-1, an advanced technological communications satellite (Artemis, previously SAT-2) and a communication mission for ground mobile systems using inclined orbits (Archimedes) on SAT-3.

  (b) Advanced Systems and Technology Programme (ASTP) which contributes technological support to PSDE.

  (c) European Data and Relay Satellite (EDRS) which is a geostationary relay satellite between satellites in low orbit and ground control stations in the context of the European orbital infrastructure programme.

- In the United States, NASA is completing work on the Advanced Communications Technology Satellite (ACTS) whose launch is expected in 1993. This experimental system is preparing for commercial exploitation of the 30/20 GHz bands and will enable operational experience of on-board regeneration with beam scanning to be acquired [INU-88]. After the ACTS project, NASA will continue its advanced technology programme in the area of both payloads and platforms [BAG-86], [FAY-86]

- In Japan, the telecommunication satellite programme CS (Sakura) comprising the satellites CS-1 (1977), CS-2 (1983) and CS-3 (1988) is a success. The launching of an experimental satellite ETS-VI in 1993 will extend the experience acquired in the 30/20 GHz bands/to the 50/40 GHz bands [MOR-87].

Together, these programmes will enable the increase of capacity provided by satellite systems to be continued and will respond to the increasing demand for telecommunication. They will also permit the size, and consequently cost, of earth stations to be further reduced. This is a fundamental aspect of the promotion of satellite telecommunication.

Previous developments, driven essentially by the constraints of the dominant telephone and data exchange markets will be of benefit to systems for satellite communication with mobile vehicles. Hence the present expertise in multibeam techniques, or hopping beams in the short term, are important advantages for the development of such systems in which small station size is a fundamental constraint.

Finally, for completeness, the technological developments associated with direct broadcasting by satellite must be mentioned; this does not pose major network problems but does imply the operation of satellites with a high transmission power (amplifiers of 200 to 300 W) or medium power (amplifiers of around 100 W) according to the extent of the area to be covered.

1.6 THE DEVELOPMENT OF SERVICES

Initially designed as 'trunks' which duplicate long-distance terrestrial links, satellite links have rapidly conquered specific markets. A satellite telecommunication system has three properties which are not, or only to a lesser extent, found in terrestrial networks; these are:

- The possibility of broadcasting.
- A wide bandwidth.
- Rapid set-up and ease of reconfiguration.

The preceding section describes the state of technical development and shows the development of the ground segment in respect of a reduction in the size of stations and a decreasing station cost. Initially a satellite system contained a small number of earth stations (several stations per country equipped with antennas of 15 to 30 m diameter collecting the traffic from an extensive area by means of a ground network). Subsequently, the number of earth stations has increased with a reduction in size (antennas of 1 to 4 m) and a greater geographical dispersion. The stations have become closer to the user. In some cases they support only reception (RCVO, receive only). The potential of the services offered by satellite telecommunications has thus diversified. Three classes of service can now be distinguished as follows:

- Trunking telephony and television programme exchange; this is a continuation of the original service. The traffic concerned is part of a country's international traffic. It is collected and distributed by the ground network on a scale appropriate to the country concerned. Examples are INTELSAT and EUTELSAT (TDMA network); the earth stations are equipped with 15 to 30 m diameter antennas.

- 'Multiservice' systems; telephone and data for user groups who are geographically dispersed. Each group shares an earth station and accesses it through a ground network whose extent is limited to one district of a town or an industrial area. Examples are TELECOM 1, SBS, EUTELSAT 1, TELE-X and INTELSAT (IBS network); the earth stations are equipped with antennas of 3 to 10 m diameter.

- Very small aperture terminal (VSAT) systems; low capacity data transmission (uni- or bi-directional), television or digital sound programme broadcasting [CRE-89]. Most often, the user is directly connected to the station. Examples of such networks are: EQUATORIAL, INTELNET or INTELSAT and so on; the earth stations are equipped with antennas of 0.6 to 1.2 m diameter. Mobile users can also be included in this category.

1.7 CONCLUSIONS AND PROSPECTS

Satellite telecommunications are now part of our environment. Every day we receive and transmit information by satellite, often without knowing it. The availability of the service is high and can be as high as 99.5%. The difficulty of the enterprise should not be forgotten; failure of the launcher and failure of the satellite are formidable and much feared dangers in this context.

Satellite telecommunication will have to face the increasing competition of fibre optic ground networks in the next 10 to 20 years. Installation of these networks has started and, in time, the most industrialised countries will be entirely cabled. Such networks offer both wide bandwidth and high capacity; these features have so far been characteristic of satellites, hence the competition mentioned previously. In this context, one could imagine that satellites will be integrated into such networks and will appear as supporting elements to improve security within the ground network of wide area links. However, one can also imagine that competition will force the operators of satellite systems to offer specialised services which will use the characteristics of satellite communication more specifically; examples are broadcasting and data collection, access to mobile vehicles, radiolocation and so on.

Whatever the assumption, one can be assured that satellites will continue to occupy an important place as a means of communication.

REFERENCES

[ABA-88] M. Abadie (1988) La solution Archimède, *Télématique Magazine*, No. 23, pp. 71–72.
[ANA-92] F. Ananasso, G. Rondinelli, P. Palmucci, B. Pavesi (1992) Small satellites applications: a new

perspective in satellite communications, *AIAA 14th International Communication Satellite Systems Conference, Washington DC, March 22–26, 1992*, pp. 911–915.

[ARN-82] J.F. Arnaud (1982) *Les moissons de l'espace* Editions Plon.

[BAG-86] J.W. Bagwell (1988) Technology achievements and projections for communication satellites of the future, *AIAA 11th International Conference, San Diego, CA*, pp 289–297.

[CRE-89] E. Crespo, J.M. Freixe, A. Nelson, N. Pham (1989) High quality digital sound distribution vai satellite, *8th International Conference on Digital Satellite Communications, Pointre à Pitre, Guadeloupe, April 24–28, 1989* pp. 399–406.

[DON-84] P. Dondl (1984) Loopus opens a new dimension in satellite communications, *International Journal of Satellite Communications*, **2**, No. 4, pp. 241–250.

[ELL-91] ELLIPSAT Corporation (1991) Application of authority to construct ELLIPSO 2, Dossier déposé auprès de la FCC, June 3, 1991.

[FAY-86] K.A. Faymon (1988) Spacecraft 2000, *AIAA 11th International Conference, San Diego, CA* pp. 88–91.

[INU-88] T. Inukai, D. Jupin, R. Lindstrom, D. Meadows (1988) ACTS TDMA network control architecture, *AIAA 12th International Conference, Arlington, VA*, pp. 225–239.

[JOH-88] N.L. Johnson (1988) Satcom in the Soviet Union, *Satellite Communications*, pp. 21–24, June.

[MOR-87] T. Mori, T. Iida (1987) Japan's space development programs for communications: an overview, *IEEE Journal on Selected Areas in Communications*, **SAC-5**, No. 4, pp. 624–629.

[REU-88] K.E. Reuter (1988) The European long term space plan, *ESA Bulletin*, No. 54, pp. 14–29, May.

2 LINK ANALYSIS

This chapter deals with the transmission of radio waves between two earth stations, one transmitting and one receiving, via a satellite. In this context, the link consists of two sections, the uplink from the earth station to the satellite and the downlink from the satellite to the receiving earth station.

The aim of the chapter is to determine the signal-to-noise ratio at the receiver input. This ratio depends on the characteristics of the transmitter, the transmission medium and the receiver. The uplink and downlink are first considered separately. Then the expression for the signal-to-noise ratio for the complete link between the two earth stations is established.

In this chapter, the term 'signal' relates to the carrier modulated by the information content. Techniques which permit the information content to be processed and superimposed on the carrier are described in Chapter 3.

Although the principles described are valid in any context, the examples will treat the case of a geostationary satellite. It will be assumed that the frequency of the carrier is between 1 GHz and 30 GHz since this corresponds to the majority of present and future applications to the end of this century.

The only case to be considered in this chapter will be that of a single station-to-station link (with single access to a transponder). The specific aspects of the case of a network, which implies the establishment of several simultaneous links (with multiple access to a transponder), are presented in Chapter 4.

2.1 THE CHARACTERISTIC PARAMETERS OF AN ANTENNA

2.1.1 Gain

The gain of an antenna is the ratio of the power radiated (or received) per unit solid angle by the antenna in a given direction to the power radiated (or received) per unit solid angle by an isotropic antenna fed with the same power. The gain is maximum in the direction of maximum radiation (the electromagnetic axis of the antenna, also called the boresight) and has a value given by:

$$G_{max} = (4\pi/\lambda^2)A_{eff} \tag{2.1}$$

where $\lambda = c/f$ and c is the velocity of light $= 3 \times 10^8$ m/s and f is the frequency of the electromagnetic wave. A_{eff} is the equivalent electromagnetic surface area of the antenna. For an antenna with a circular aperture or reflector of diameter D and geometric surface $A = \pi D^2/4$,

$A_{\text{eff}} = \eta A$, where η is the efficiency of the antenna. Hence:

$$G_{\text{max}} = \eta(\pi D/\lambda)^2 = \eta(\pi Df/c)^2 \tag{2.2}$$

Expressed in dBi (the gain relative to an isotropic antenna), the actual antenna gain is:

$$G_{\text{max,dBi}} = 10 \log \eta(\pi D/\lambda)^2 = 10 \log \eta(\pi Df/c)^2 \qquad \text{(dB)}$$

The global efficiency η of the antenna is the product of several factors which take account of the illumination law, spill-over loss, surface impairments, resistive and mismatch losses etc.:

$$\eta = \eta_{\text{i}} \times \eta_{\text{s}} \times \eta_{\text{f}} \times \eta_{\text{z}} \cdots \tag{2.3}$$

The *illumination efficiency* η_{i} specifies the illumination law of the reflector with respect to uniform illumination. Uniform illumination ($\eta_{\text{i}} = 1$) leads to a high level of secondary lobes. A compromise is achieved by attenuating the illumination at the reflector boundaries (aperture edge taper). In the case of a Cassegrain antenna (see Section 8.3.4.3), the best compromise is obtained for an illumination attenuation at the boundaries of 10 to 12 dB which leads to an illumination efficiency η_{i} of the order of 91%.

The *spill-over efficiency* η_{s} is defined as the ratio of the energy radiated by the primary source which is intercepted by the reflector to the total energy radiated by the primary source. The difference constitutes the spill-over energy. The larger the angle under which the reflector is viewed from the source, the greater the spill-over efficiency. However, for a given source radiation pattern, the illumination level at the boundaries becomes less with large values of view angle and the illumination efficiency collapses. A compromise leads to a spill-over efficiency of the order of 80%.

The *surface finish efficiency* η_{f} takes account of the effect of surface impairments on the gain of the antenna. The actual parabolic profile differs from the theoretical one. In practice, a compromise must be found between the effect on the antenna characteristics and the cost of fabrication. The effect on the on-axis gain is of the form:

$$\Delta G = \exp[-B(4\pi\varepsilon/\lambda)^2]$$

where ε is the fabrication tolerance, the deviation between the actual and theoretical profiles measured perpendicularly to the concave face, and B is a factor, less than or equal to 1, whose value depends on the radius of curvature of the reflector. This factor increases as the radius of curvature decreases. For parabolic antennas of focal distance f, it varies as a function of the ratio f/D, where D is the diameter of the antenna. In practice, B is the order of 0.7 and ε of the order of $\lambda/16$; this leads to a surface finish efficiency of the order of 90%.

The other losses, including ohmic and impedance mismatch losses η_{z} are of less importance.

In total, the global efficiency η, the product of the various efficiencies, is typically between 55 and 75%.

2.1.2 The radiation pattern

The radiation pattern indicates the variations of gain with direction. For an antenna with a circular aperture or reflector this pattern has rotational symmetry and is completely represented

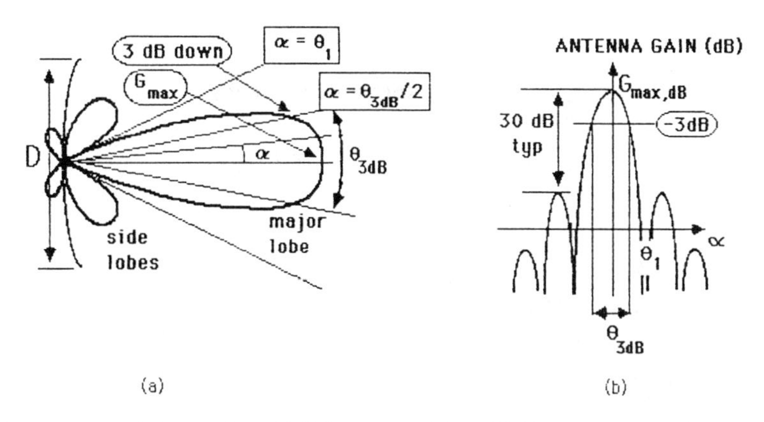

Figure 2.1 Antenna radiation pattern. (a) Polar representation. (b) Cartesian representation.

within a plane in polar co-ordinate form (Figure 2.1a) or cartesian co-ordinate form (Figure 2.1b). The main lobe which contains the direction of maximum radiation and the side lobes can be identified.

2.1.3 The angular beamwidth

This is the angle defined by the directions corresponding to a given gain pattern with respect to the maximum value. The 3 dB beamwidth, indicated on Figure 2.1a by θ_{3dB}, is very often used. The 3 dB beamwidth corresponds to the angle between the directions in which the gain falls to half its maximum value. The 3 dB beamwidth is related to the ratio λ/D by a coefficient whose value depends on the chosen illumination law. For uniform illumination, the coefficient has a value of 58.5°. With non-uniform illumination laws, which lead to attenuation at the reflector boundaries, the 3 dB beamwidth increases and the value of the coefficient depends on the particular characteristics of the law. The value currently used is 70° which leads to the following expression:

$$\theta_{3dB} = 70(\lambda/D) = 70(c/fD) \qquad \text{(degrees)} \tag{2.3a}$$

In a direction α with respect to the boresight, the value of gain is given by:

$$G(\alpha)_{dB} = G_{max,dB} - 12(\alpha/\theta_{3dB})^2 \qquad \text{(dB)} \tag{2.4}$$

This expression is valid only for sufficiently small angles (between 0 and $\theta_{3dB}/2$).

Combining expressions (2.2) and (2.3a) it can be seen that the maximum gain of an antenna is a function of the 3-dB beamwidth and this relation is independent of frequency:

$$G_{max} = \eta(\pi Df/c)^2 = \eta(\pi 70/\theta_{3dB})^2 \tag{2.5}$$

If a value $\eta = 0.6$ is considered, this gives:

$$G_{max} = 29\,000/(\theta_{3dB})^2 \tag{2.6}$$

in which θ_{3dB} is expressed in degress.

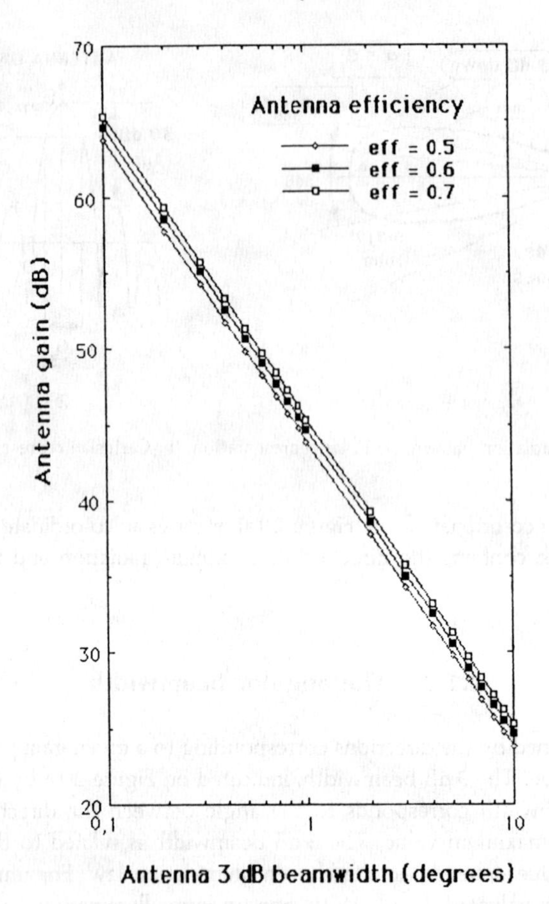

Figure 2.2 Antenna gain in the direction of maximum radiation as a function of the angular beamwidth θ_{3dB} for three values of efficienty ($\eta = 0.5$, $\eta = 0.6$ and $\eta = 0.7$).

Figure 2.2 shows the relationship between 3 dB beamwidth and maximum gain for three values of antenna efficiency. The gain is expressed in dBi and the 3 dB beamwidth in degrees.

$$G_{\max,\mathrm{dBi}} = 44.6 - 20 \log \theta_{3\mathrm{dB}}$$

$$\theta_{3\mathrm{dB}} = 170/10^{(G_{\max,\mathrm{dBi}}/20)} \quad \text{(degrees)}$$

2.1.4 Polarisation

The wave radiated by an antenna consists of an electric field component and a magnetic field component. These two components are orthogonal and perpendicular to the direction of propagation of the wave; they vary with the frequency of the wave. By convention, the polarisation of the wave is defined by the direction of the electric field. In general, the direction of the electric field is not fixed and its amplitude is not constant. During one period, the projection of the extremity of the vector representing the electric field onto a plane

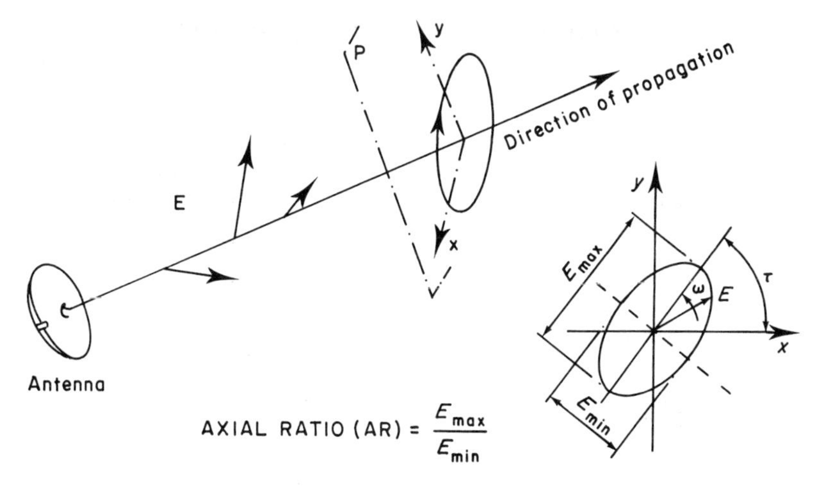

Figure 2.3 Characterisation of the polarisation of an electromagnetic wave.

perpendicular to the direction of propagation of the wave describes an ellipse; the polarisation is said to be elliptical (Figure 2.3).

Polarisation is characterised by the following parameters:

—Direction of rotation (with respect to the direction of propagation): right-hand (clockwise) or left-hand (counter-clockwise).
—Axial ratio (AR): $AR = E_{max}/E_{min}$, that is the ratio of the major and minor axes of the ellipse. When the ellipse is a circle (axial ratio $= 1 = 0$ dB), the polarisation is said to be circular. When the ellipse reduces to one axis (infinite axial ratio, the electric field maintains a fixed direction) the polarisation is said to be linear.
—Inclination τ of the ellipse.

Two waves are in orthogonal polarisation if their electric fields describe identical ellipses in opposite directions. In particular, the following can be obtained:

—Two orthogonal circular polarisations described as right-hand circular and left-hand circular (the direction of rotation is for an observer looking in the direction of propagation).
—Two orthogonal linear polarisations described as horizontal and vertical (relative to a local reference).

An antenna designed to transmit or receive a wave of given polarisation can neither transmit nor receive in the orthogonal polarisations. This property enables two simultaneous links to be established at the same frequency between the same two locations; this is described as frequency re-use by orthogonal polarisation. To achieve this either two polarised antennas must be provided at each end or, preferably, one antenna which operates with the two specified polarisations may be used. This practice must, however, take account of imperfections of the antennas and the possible depolarisation of the waves by the transmission medium. These effects lead to mutual interference of the two links.

This situation is illustrated in Figure 2.4 which relates to the case of two orthogonal linear

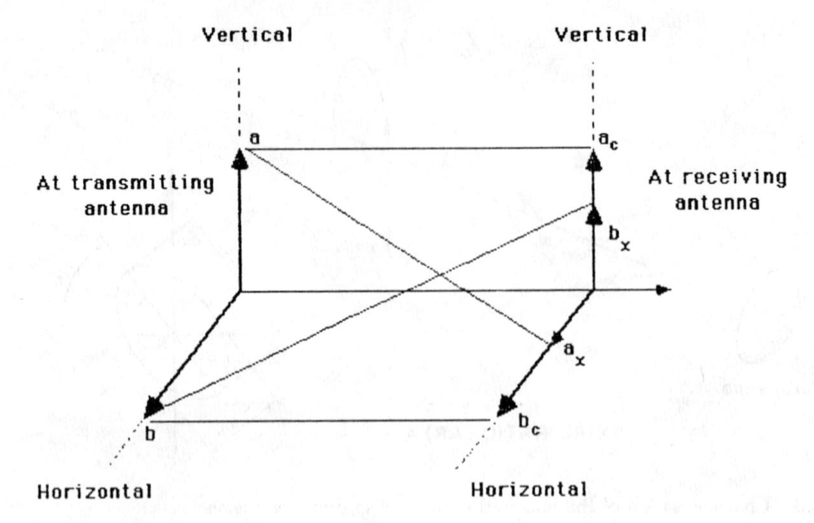

Figure 2.4 Amplitude of the transmitted and received electric field for the case of two orthogonal linear polarisations.

polarisations (but the illustration is equally valid for any two orthogonal polarisations). Let a and b be the amplitudes, assumed to be equal, of the electric field of the two waves transmitted simultaneously with linear polarisation, a_C and b_C be the amplitudes received with the same polarisation and a_X and b_X be the amplitudes received with orthogonal polarisations. The following are defined:

—The *cross-polarisation isolation* $XPI = a_C/b_X$ or b_C/a_X, hence:

$$\text{XPI (dB)} = 20 \log (a_C/b_X) \text{ or } 20 \log (b_C/a_X) \qquad \text{(dB)}$$

—The *cross-polarisation discrimination* (when a single polarisation is transmitted) $\text{XPD} = a_C/a_X$, hence:

$$\text{XPD (dB)} = 20 \log (a_C/a_X) \qquad \text{(dB)}$$

In practice, XPI and XPD are comparable and are often confused within the term isolation.

For a quasi-circular polarisation characterised by its value of axial ratio AR, the cross-polarisation discrimination is defined as:

$$\text{XPD} = 20 \log [(\text{AR} + 1)/(\text{AR} - 1)] \qquad \text{(dB)}$$

Conversely, the axial ratio AR can be expressed as a function of XPD by:

$$\text{AR} = (10^{\text{XPD}/20} + 1)/(10^{\text{XPD}/20} - 1)$$

The values and relative values of the components vary as a function of direction with respect to the antenna boresight. The antenna is thus characterised for a given polarisation by a radiation pattern for nominal polarisation (co-polar) and a radiation pattern for orthogonal polarisation (cross-polar). Cross-polarisation discrimination is generally maximum on the antenna axis and degrades for directions other than those corresponding to the direction of maximum gain.

2.2 THE POWER EMITTED IN A GIVEN DIRECTION

2.2.1 Equivalent isotropic radiated power (EIRP)

The power radiated per unit solid angle by an isotropic antenna fed from a radio-frequency source of power P_T is given by:

$$P_T/4\pi \qquad \text{(W/steradian)}$$

In a direction where the value of transmission gain is G_T, any antenna radiates a power per unit solid angle equal to:

$$G_T P_T/4\pi \qquad \text{(W/steradian)}$$

The product $P_T G_T$ is called the 'equivalent isotropic radiated power' (EIRP). It is expressed in W.

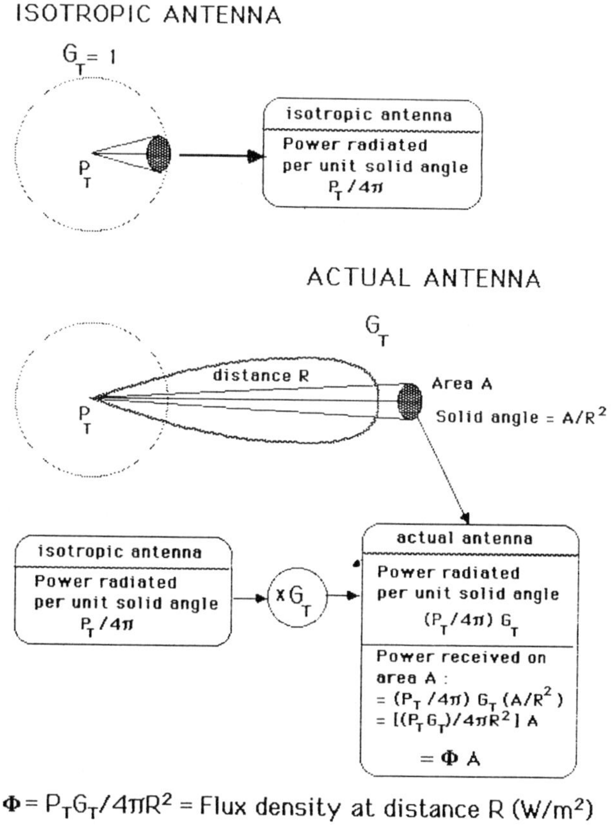

$\Phi = P_T G_T/4\pi R^2$ = Flux density at distance R (W/m^2)

Figure 2.5 Power flux density.

2.2.2 Power flux density (Figure 2.5)

A surface of effective area A situated at a distance R from the transmitting antenna subtends a solid angle A/R^2 at the transmitting antenna. It receives a power equal to:

$$P_R = (P_T G_T/4\pi)(A/R^2) = \Phi A \qquad \text{(W)} \tag{2.7}$$

The magnitude $\Phi = P_T G_T/4\pi R^2$ is called the 'power flux density'. It is expressed in W/m^2.

2.3 RECEIVED SIGNAL POWER

2.3.1 Power received by the receiving antenna (Figure 2.6)

A receiving antenna of effective area A_{Reff} located at a distance R from the transmitting antenna receives a power equal to:

$$P_R = \Phi A_{Reff} = (P_T G_T/\pi R^2) A_{Reff} \qquad \text{(W)} \tag{2.8}$$

The equivalent area of an antenna is expressed as a function of its receiving gain G_R by the expression:

$$A_{Reff} = G_R/(4\pi/\lambda^2) \qquad (m^2) \tag{2.9}$$

Hence an expression for the received power:

$$\begin{aligned} P_R &= (P_T G_T/4\pi R^2)(\lambda^2/4\pi)G_R \\ &= (P_T G_T)(\lambda/4\pi R)^2 G_R \\ &= (P_T G_T)(1/L_{FS})G_R \qquad \text{(W)} \end{aligned} \tag{2.10}$$

where $L_{FS} = (4\pi R/\lambda)^2$ is called the free space loss and represents the ratio of the received and transmitted powers in a link between two isotropic antennas. Figure 2.7 gives the value of $L_{FS}(R_0)$ as a function of frequency for a geostationary satellite and a station situated exactly under the satellite at a distance $R = R_0 = 35\,786$ km equal to the altitude of the satellite. Notice that L_{FS} is of the order of 200 dB. For any station whose position is represented by its relative latitude and longitude l and L with respect to the geostationary satellite (since the satellite is situated in the equatorial plane, l is the geographical latitude of the station),

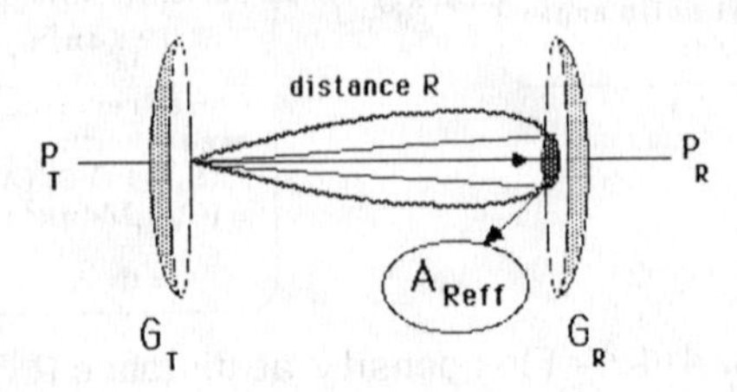

Figure 2.6 The power received by a receiving antenna.

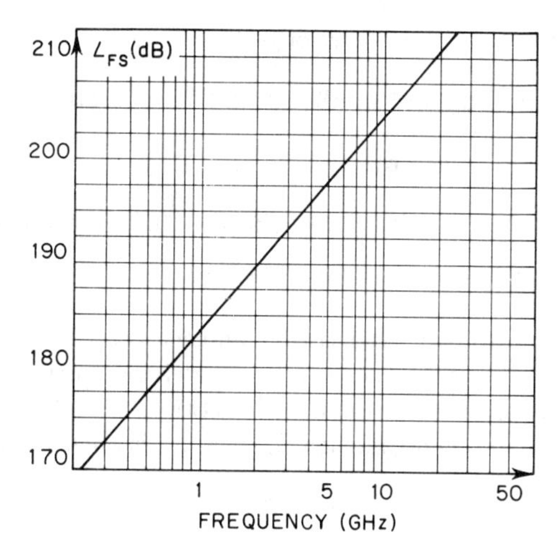

Figure 2.7 The attenuation of free space.

the value of $L_{FS}(R_0)$ provided by Figure 2.7 must be corrected by the term $(R/R_0)^2$, hence:

$$L_{FS} = (4\pi R/\lambda)^2 = (4\pi R_0/\lambda)^2 (R/R_0)^2 = L_{FS}(R_0)(R/R_0)^2$$

where $(R/R_0)^2 = 0.42(1 - \cos I \cos L)$ (see Chapter 7, equation (7.58)). The value of $(R/R_0)^2$ is between 1 and 1.356 (0 to 1.3 dB).

2.3.2 Example 1: The uplink

Consider the transmitting antenna of an earth station equipped with an antenna of diameter $D = 4$ m. This antenna is fed with a power P_T of 100 W, that is 20 dB(W), at a frequency $f_U = 14$ GHz. It radiates this power towards a geostationary satellite situated at a distance of 40 000 km from the station on the axis of the antenna. The beam of the satellite receiving antenna has a width $\theta_{3dB} = 2°$. It is assumed that the earth station is at the centre of the region covered by the satellite antenna and consequently benefits from the maximum gain of this antenna. The efficiency of the satellite antenna is assumed to be $\eta = 0.55$ and that of the earth station to be $\eta = 0.6$. The power flux density at the satellite and the power received by it will be calculated.

—Power flux density at the satellite:

$$\Phi = P_T G_{Tmax}/4\pi R^2 \qquad (\text{W/m}^2)$$

Gain of the earth station antenna:

from equation (2.2),

$$G_{Tmax} = \eta(\pi D/\lambda_U)^2 = \eta(\pi D f_U/c)^2$$
$$= 0.6(\pi \times 4 \times 14 \times 10^9/3 \times 10^8)^2 = 206\,340 = 53.1 \text{ dB}$$

The equivalent isotropic radiated power of the earth station (on the axis) is given by:

$$P_T G_{Tmax} = 53.1 \text{ dB} + 20 \text{ dB(W)} = 73.1 \text{ dB(W)}.$$

The power flux density is given by:

$$\Phi = P_T G_{Tmax}/4\pi R^2 = 73.1 \text{ dB(W)} - 10 \log(4\pi(4 \times 10^7)^2)$$
$$= 73.1 - 163 = -89.9 \text{ dB(W/m}^2).$$

—The power recieved (in dBW) by the satellite antenna is obtained using equation (2.10):
P_R = EIRP − attenuation of free space + gain of receiving antenna.
The attenuation of free space $L_{FS} = (4\pi R/\lambda_U)^2 = (4\pi R f_U/c)^2 = 207.4$ dB.
The gain of the satellite receiving antenna $G_R = G_{Rmax}$ is obtained using equation (2.2):

$$G_{Rmax} = \eta(\pi D/\lambda_U)^2$$

The value of D/λ_U is obtained using (2.3), hence $\theta_{3dB} = 70(\lambda_U/D)$ from which $D/\lambda_U = 70/\theta_{3dB}$ and $G_{Rmax} = \eta(70\pi/\theta_{3dB})^2 = 6650 = 38.2$ dB.

Notice that the antenna gain does not depend on frequency when the beamwidth, and hence the area covered by the satellite antenna, is imposed.
In total:

$$P_R = 73.1 - 207.4 + 38.2 = -96.1 \text{ dB(W)}, \qquad \text{that is } 0.25 \text{ nW or } 250 \text{ pW}.$$

2.3.3 Example 2: The downlink

Consider the transmitting antenna of a geostationary satellite fed with a power P_T of 10 W, that is 10 dB(W) at a frequency $f_D = 12$ GHz and radiating this power in a beam of width θ_{3dB} equal to 2°. An earth station equipped with a 4 m diameter antenna is located on the axis of the antenna at a distance of 40,000 km from the satellite. The efficiency of the satellite antenna is assumed to be $\eta = 0.55$ and that of the earth station to be $\eta = 0.6$. The power flux density at the earth station and the power received by it will be calculated.

—Power flux density at the earth station:

$$\Phi = P_T G_{Tmax}/4\pi R^2 \qquad (\text{W/m}^2)$$

The gain of the satellite antenna is the same in transmission as in reception since the beamwidths are made the same (notice that this requires two separate antennas on the satellite since the diameters cannot be the same and are in the ratio $f_u/f_D = 14/12 = 1.17$).
Hence:

$$(\text{EIRP})_{SL} = P_T G_{Tmax} = 38.2 \text{ dB} + 10 \text{ dB(W)} = 48.2 \text{ dB(W)}$$

The power flux density is:

$$\Phi = P_T G_{Tmax}/4\pi R^2 = 48.2 \text{ dB(W)} - 10 \log(4\pi(4 \times 10^7)^2) = 48.2 - 163$$
$$= -114.8 \text{ dB(W/m}^2)$$

—The power (in dBW) received by the antenna of the earth station is obtained using expression (2.10):

$$P_R = \text{EIRP} - \text{attenuation of free space} + \text{gain of the receiving antenna.}$$

The attenuation of free space is $L_{FS} = (4\pi R/\lambda_D)^2 = 206.1$ dB.

The gain $G_R = G_{R\max}$ of the ground station receiving antenna is obtained using expression (2.2), hence:

$$G_{R\max} = \eta(\pi D/\lambda_D)^2 = 0.6(\pi \times 4/0.025)^2 = 151\ 597 = 51.8\ \text{dB}$$

In total:

$$P_R = 48.2 - 206.1 + 51.8 = -106.1\ \text{dB(W)}, \qquad \text{that is } 25\ \text{pW}.$$

2.3.4 The practical case

In practice, it is necessary to take account of additional losses due to various causes:

—losses associated with attenuation of waves as they propagate through the atmosphere;
—losses in the transmitting and receiving equipment;
—losses due to imperfect alignment of the antennas;
—polarisation mismatch losses.

2.3.4.1 Attenuation in the atmosphere

The attenuation of waves in the atmosphere, denoted by L_A is due to the presence of gaseous components in the troposphere, water (rain, clouds, snow and ice) and the ionosphere. A quantitative presentation of these effects is given in Section 2.6. The overall effect on the power of the received signal can be taken into account by replacing L_{FS} in expression (2.10) by L, called the path loss, where:

$$L = L_{FS} L_A \tag{2.11}$$

2.3.4.2 Losses in the transmitting and receiving equipment

Figure 2.8 clarifies these losses:

—The loss L_{FTX} between the transmitter and the antenna; to feed the antenna with a power P_T it is necessary to provide a power P_{TX} at the output of the transmission amplifier such that:

$$P_{TX} = P_T L_{FTX} \qquad \text{(W)} \tag{2.12}$$

Expressed as a function of the power of the transmission amplifier the EIRP can be written:

$$\text{EIRP} = P_T G_T = (P_{TX} G_T)/L_{FTX} \qquad \text{(W)} \tag{2.13}$$

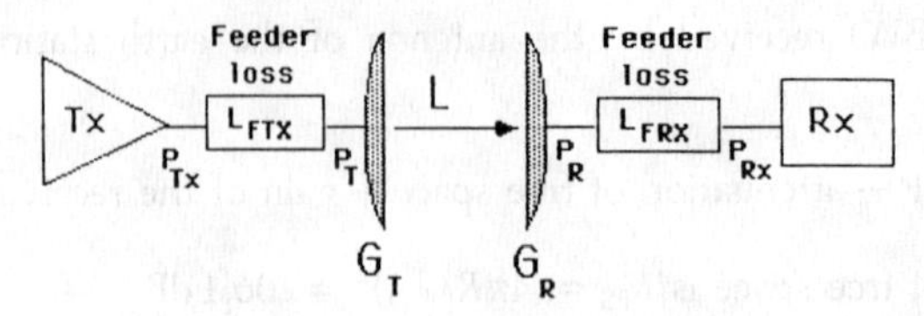

Figure 2.8 Losses in the terminal equipment.

—The loss L_{FRX} between the antenna and the receiver; the signal power P_{RX} at the input of the receiver is equal to:

$$P_{RX} = P_R / L_{FRX} \qquad \text{(W)} \tag{2.14}$$

2.3.4.3 *Losses due to imperfect alignment of the antennas*

Figure 2.9 shows the geometry of the link for the case of imperfect alignment of the transmitting and receiving antennas. The result is a loss of antenna gain which can be inserted in expression (2.10) in the form of a misalignment loss L_T on transmission and a misalignment loss L_R on reception. These losses are a function of the misalignment angles of transmission (α_T) and reception (α_R) and are evaluated using expression (2.4). Their value in dB is given by:

$$L_T = 12(\alpha_T/\theta_{3DB})^2 \qquad \text{(dB)}$$

$$L_R = 12(\alpha_R/\theta_{3DB})^2 \qquad \text{(dB)} \tag{2.15}$$

2.3.4.4 *Losses due to polarisation mismatch*

It is also necessary to consider the polarisation mismatch loss L_{POL} observed when the receiving antenna is not oriented with the polarisation of the received wave. In a link with circular polarisation, the transmitted wave is circularly polarised only on the axis of the antenna and becomes elliptical off this axis. Propagation through the atmosphere can also change circular into elliptical polarisation (see Section 2.6). In a linearly polarised link, the wave can be subjected to a rotation of its plane of polarisation as it propagates through the atmosphere.

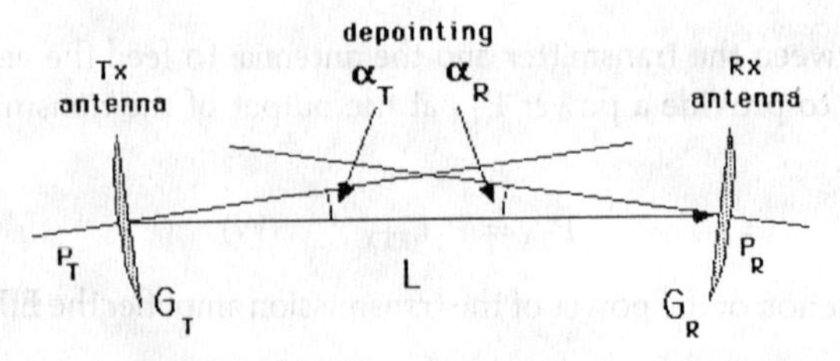

Figure 2.9 Geometry of the link.

Finally, the receiving antenna may not have its plane of polarisation aligned with that of the incident wave. If γ is the angle between the two planes, the polarisation mismatch loss L_{POL} (in dB) is equal to $20 \log \cos \gamma$. In the case where a circularly polarised antenna receives a plane polarised wave, L_{POL} has a value of 3 dB.

Considering all sources of loss, the defining expression (2.10) for the signal power at the receiver input becomes:

$$P_{RX} = (P_{TX} G_{Tmax} / L_T L_{FTX})(1/L_{FS} L_A)(G_{Rmax} / L_R L_{FRX} L_{POL}) \qquad \text{(W)} \qquad (2.16)$$

2.3.5 Conclusion

Equations (2.10) and (2.16) which express the signal power at the input to the receiver are of the same form; they are the product of three factors:

—The first (EIRP) characterises the transmitting equipment,
—The second ($1/L$) characterises the transmission medium,
—The third (the gain of the receiver) characterises the receiving equipment.

In its most complete form, as in (2.16), the expression for these factors is:

—EIRP $= (P_{TX} G_{Tmax} / L_T L_{FTX})$ (W)

This expression takes account of the losses L_{FTX} between the transmission amplifier and the antenna and the reduction in antenna gain L_T due to misalignment of the transmitting antenna.

—$1/L = 1/L_{FS} L_A$

The path loss L takes account of the attenuation of free space L_{FS} and the attenuation in the atmosphere L_A.

—The gain of the receiving equipment: $G = G_{Rmax} / L_R L_{FRX} L_{POL}$

This expression takes account of the losses L_{FRX} between the antenna and the receiver, the loss of antenna gain L_R due to misalignment of the receiving antenna and the polarisation mismatch losses L_{POL}.

2.4 NOISE POWER AT THE RECEIVER INPUT

2.4.1 The origins of noise

Noise is a signal without information content which adds itself to the useful signal. It reduces the ability of the receiver to reproduce the information content of the useful signal correctly.

The origins of noise are as follows:

—The noise emitted by natural sources of radiation located within the antenna reception area,
—The noise generated by electronic components in the equipment.

Signals from transmitters other than those which it is wished to receive are also classed as noise. This noise is described as interference.

2.4.2 Characterisation and definition of noise

Harmful noise power is that which occurs in the bandwidth of the useful signal. Normally this is that of the receiver. A very much used noise model is that of white noise for which the power spectral density N_0 (W/Hz) is constant in the frequency band involved (Figure 2.10). The equivalent noise power N(W) measured in a bandwidth B_N (Hz) has a value:

$$N = N_0 B_N \qquad \text{(W)} \tag{2.17}$$

Real noise sources do not always have a constant power spectral density, but the model is convenient for representation of real noise observed over a limited bandwidth.

2.4.2.1 *Noise temperature of a two-port noise source*

The noise temperature of a noise source delivering an available noise power N is given by:

$$T = N/kB = N_0/k \qquad \text{(K)} \tag{2.18}$$

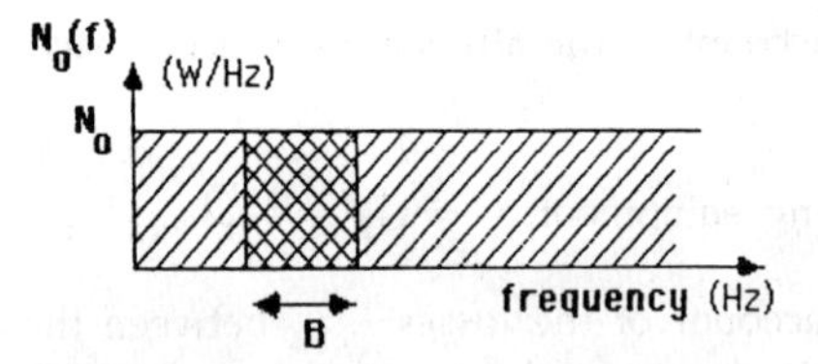

Figure 2.10 Spectral density of white noise.

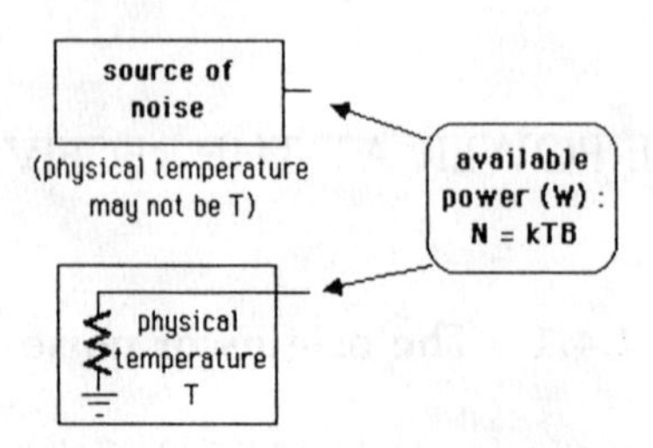

Figure 2.11 Definition of the noise temperature of a noise source.

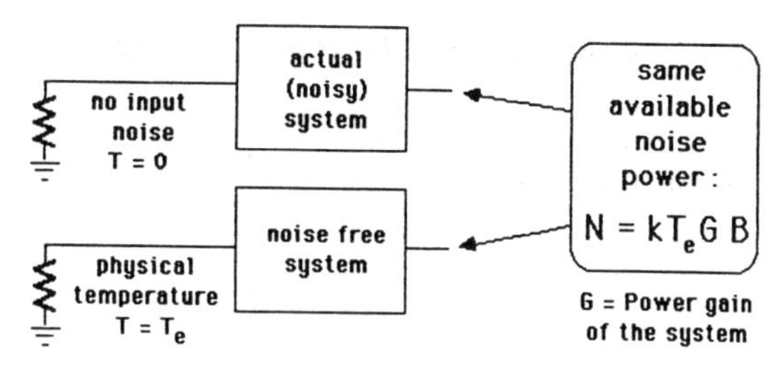

Figure 2.12 Equivalent input noise temperature of a system.

where k is Boltzmann's constant $= 1.379 \times 10^{-23} = -228.6$ dB (W/Hz K), T represents the thermodynamic temperature of a resistance which delivers the same available noise power as the source under consideration (Figure 2.11). Available noise power is the power delivered by the source to a device which is impedance matched to the source.

2.4.2.2 *Noise temperature of a four-port element*

The noise temperature T_e of a four-port element is the thermodynamic temperature of a resistance which, placed at the input of the element assumed to be noise-free, establishes the same available noise power at the ouput of the element as the actual element without the noise source at the input (Figure 2.12). T_e is thus a measure of the noise generated by the internal components of the four-port element.

The noise figure of this four-port element is the ratio of the total available noise power at the output of the element to the component of this power engendered by a source at the input of the element with a noise temperature equal to the reference temperature $T_0 = 290$ K.

Assume that the element has a power gain G, a bandwidth B and is driven by a source of noise temperature T_0; the total power at the output is $Gk(T_e + T_0)B$. The component of this power originating from the source is GkT_0B. The noise figure is thus:

$$F = [Gk(T_e + T_0)B]/[GkT_0B] = (T_e + T_0) = 1 + T_e/T_0 \tag{2.19}$$

2.4.3 Noise temperature of an antenna

An antenna picks up noise from radiating bodies within the radiation pattern of the antenna. The noise output from the antenna is a function of the direction in which it is pointing, its radiation pattern and the state of the surrounding environment. The antenna is assumed to be a noise source characterised by a noise temperature called the noise temperature of the antenna T_A (K).

Let $T_b(\theta, \varphi)$ be the brightness temperature of a radiating body located in a direction (θ, φ), where the gain of the antenna has a value $G(\theta, \varphi)$. The noise temperature of the antenna is obtained by integrating the contributions of all the radiating bodies within the radiation pattern of the antenna. The noise temperature of the antenna is thus:

$$T_A = (1/4\pi) \iint T_b(\theta, \varphi)\, G(\theta, \varphi)\, d\Omega \tag{2.20}$$

2.4.4 Noise temperature of an attenuator

An attenuator is a four-port element containing only passive components (which can be classed as resistances) all at temperature T_F which is generally the ambient temperature. If L_F is the attenuation caused by the attenuator, the noise temperature of the attenuator is:

$$T_e = (L_F - 1)T_F \qquad \text{(K)} \tag{2.21}$$

If $T_F = T_0$, the noise figure of the attenuator, by comparison of (2.17) and (2.18) is:

$$F_F = L_F.$$

2.4.5 Noise temperature of a device containing several elements in cascade

Consider a device containing a chain of N four-port elements in cascade, each element j having a power gain G_j ($j = 1,2,\ldots,N$) and a noise temperature T_{ej}.
The noise temperaturre of the device is:

$$T_e = T_{e1} + T_{e2}/G_1 + T_{e3}/G_1G_2 + \cdots + T_{iN}/G_1G_2,\ldots,G_{N-1} \qquad \text{(K)} \tag{2.22}$$

The noise factor is obtained from (2.19):

$$F = F_1 + (F_2 - 1)/G_1 + (F_3 - 1)/G_1G_2 + \cdots + (F_N - 1)/G_1G_2,\ldots,G_{N-1} \tag{2.23}$$

2.4.6 Application to the noise temperature of receiving equipment

Consider the receiving equipment shown in Figure 2.13. This consists of an antenna connected to a receiver. The connection is a lossy one and is at a thermodynamic temperature T_F (which is close to $T_0 = 290\,\text{K}$). It introduces an attenuation L_{FRX} which corresponds to a gain

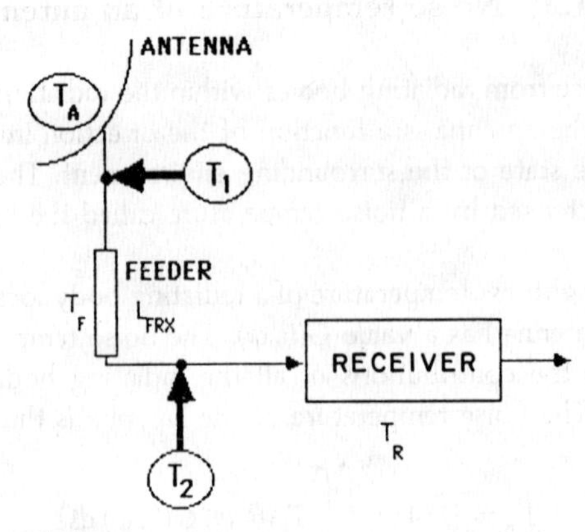

Figure 2.13 A receiving system.

$G_{FRX} = 1/L_{FRX}$ and is less than 1. The noise temperature of the system will be determined at two points as follows:

—At the antenna output, before the connection losses, temperature T_1,
—At the receiver input, after the losses, temperature T_2.

The noise temperature T_1 at the antenna output is the sum of the noise temperature of the antenna T_A and the noise temperature of the subsystem consisting of the connection and the receiver in cascade. The noise temperature of the connection is given by (2.21). From (2.22) the noise temperature of the sybsystem is $(L_{FRX} - 1)T_F + T_R/G_{FRX}$. Adding the contribution of the antenna, considered as a noise source, this becomes:

$$T_1 = T_A + (L_{FRX} - 1)T_F + T_R/G_{FRX} \qquad \text{(K)} \tag{2.24}$$

Now consider the receiver input, this noise must be attenuated by a factor L_{FRX}. Replacing G_{FRX} by $1/L_{FRX}$, one obtains the noise temperature T_2 at the receiver input:

$$T_2 = T_1/L_{FRX} = T_A/L_{FRX} + T_F(1 - 1/L_{FRX}) + T_R \qquad \text{(K)} \tag{2.25}$$

This noise temperature, which takes account of the noise generated by the antenna and the connection together with the receiver noise is called the system noise temperature. Notice that measurement of noise at this point would reflect only the noise contribution of the antenna and the connection.

2.4.7 Example

Consider the receiving system of Figure 2.13 with the following values:

—Antenna noise temperature: $T_A = 50\,\text{K}$
—Thermodynamic temperature of the connection: $T_F = 290\,\text{K}$
—Noise temperature of the receiver: $T_R = 50\,\text{K}$.

The system noise temperature will be calculated for two cases: (1) no losses in the connection between the antenna and the receiver and (2) connection loss $L_{FRX} = 1\,\text{dB}$. The system noise temperature will be calculated at the receiver input.

Using equation (2.25), $T_2 = T_A/L_{FRX} + T_F(1 - 1/L_{FRX}) + T_R$.
For case (1), $T_2 = 50 + 50 = 100\,\text{K}$ and for case (2)

$$T_2 = 50/10^{0.1} + 290(1 - 1/10^{0.1}) + 50 = 39.7 + 59.6 + 50 = 149.3\,\text{K}.$$

Notice the influence of the connection with losses; it reduces the antenna noise but it makes its own contribution to the noise and this finally causes an increase in the system noise temperture. The contribution of an attenuation to the noise can be quickly estimated using the following rule: every attenuation of 0.1 dB upstream of the receiver makes a contribution to the system noise temperature at the receiver input of $290(1 - 1/10^{0.01}) = 6.6\,\text{K}$ or around 7 K. To realise a receiving system with a low noise temperature it is imperative to avoid losses upstream of the receiver.

2.4.8 Conclusion

The noise contribution in a receiving system is determined by the noise temperature at a given point in the system, most often the receiver input; this noise temperature is called the system noise temperature. It is obtained by summing, at this point, all noise temperatures corresponding to noise generated upstream and all noise temperatures equivalent to the noise generated downstream of the point considered.

2.5 SIGNAL-TO-NOISE RATIO AT THE RECEIVER INPUT

2.5.1 Definitions

The signal-to-noise ratio enables the relative magnitude of the received signal to be specified with respect to the noise present at the receiver input. Several ratios for specifying this relative magnitude can be envisaged:

—The ratio of signal power to noise power; this approach seems to be the most natural since two magnitudes of the same kind are being compared. It is usual to designate the power of the modulated carrier by C. As the noise power is N, the ratio is written C/N.
—The ratio of the signal power to the spectral density of the noise; this is written C/N_0 and is expressed in Hz. It has the advantage, with respect to the ratio C/N, of not in any way presupposing the bandwidth used. In fact, the latter implies knowledge of the equivalent noise bandwidth B_N of the receiver which is adjusted to the bandwidth B occupied by the modulated carrier. In the course of the design, one can require to evaluate the link quality before the nature of the transmitted signals is specified. The bandwidth occupied by the carrier is then unknown and this prevents further insight into the C/N value.
—The ratio of the signal power to the noise temperature; this ratio is derived from the ratio C/N_0 by multiplication by Boltzmann's constant k. It is written C/T and is expressed in W/K.

The ratio C/N_0 corresponds to the most widespread practice and is adopted in the following.

2.5.2 Expression

The power of the signal received at the receiver input is given by (2.16).
Hence

$$C = P_{RX}.$$

The noise spectral density at the same point is $N_0 = kT$, where T is given by (2.24).
Hence:

$$C/N_0 = [(P_{TX}G_{T\max}/L_T L_{FTX})(1/L_{FS}L_A)(G_{R\max}/L_R L_{FRX} L_{POL})] \\ /[T_A/L_{FRX} + T_F(1 - 1/L_{FRX}) + T_R](1/k) \quad \text{(Hz)} \qquad (2.25)$$

This expression can be interpreted as follows:

$$C/N_0 = \text{(transmitter EIRP) (1/path loss) (receiver gain/noise temperature)} (1/k) \quad \text{(Hz)} \quad (2.26)$$

C/N_0 can also be expressed as a function of the power flux density Φ:

$$C/N_0 = \Phi(\lambda^2/4\pi) \text{ (receiver gain/noise temperature) } (1/k) \quad \text{(Hz)} \quad (2.27)$$

where

$$\Phi = \text{(transmitter EIRP)}/(4\pi R^2) \quad (\text{W/m}^2)$$

Finally, it can be verified that evaluation of C/N_0 is independent of the point chosen in the receiving chain as long as the signal power and the noise power spectral density are calculated at the same point.

2.5.3 Figure of merit for receiving equipment

The expression (2.26) for C/N_0 introduces three factors:

—The first (EIRP) characterises the transmitting equipment,
—The second ($1/L = 1/L_{FS}L_A$) characterises the transmission medium,
—The third (receiver gain/noise temperature) characterises the receiving equipment. It is called the figure of merit, or G/T, of the receiving equipment.

By examining (2.25) it can be seen that the figure of merit G/T of the receiving equipment is a function of the antenna noise temperature T_A and the equivalent noise temperature T_R of the receiver. These magnitudes will now be quantified.

2.5.4 The antenna noise temperature

There are two cases to be considered:

—That of a satellite antenna (the uplink).
—That of an earth station antenna (the downlink).

2.5.4.1 *The satellite antenna (the uplink)*

The noise captured by the antenna is noise from the earth and from outer space. The beamwidth of a satellite antenna is equal to or less than the angle of view of the earth from the satellite, that is 17.5° for a geostationary satellite. Under these conditions, the major contribution is that from the earth. For a beamwidth θ_{3dB} of 17.5°, the antenna noise temperature is given by Figure 2.14 [NJO-85]. It depends on frequency and the orbital position of the satellite. For a smaller width (a narrow beam) the temperature depends on the frequency and the area covered; the continents radiate more noise than the oceans. In the absence of precise figures 290 K can be taken.

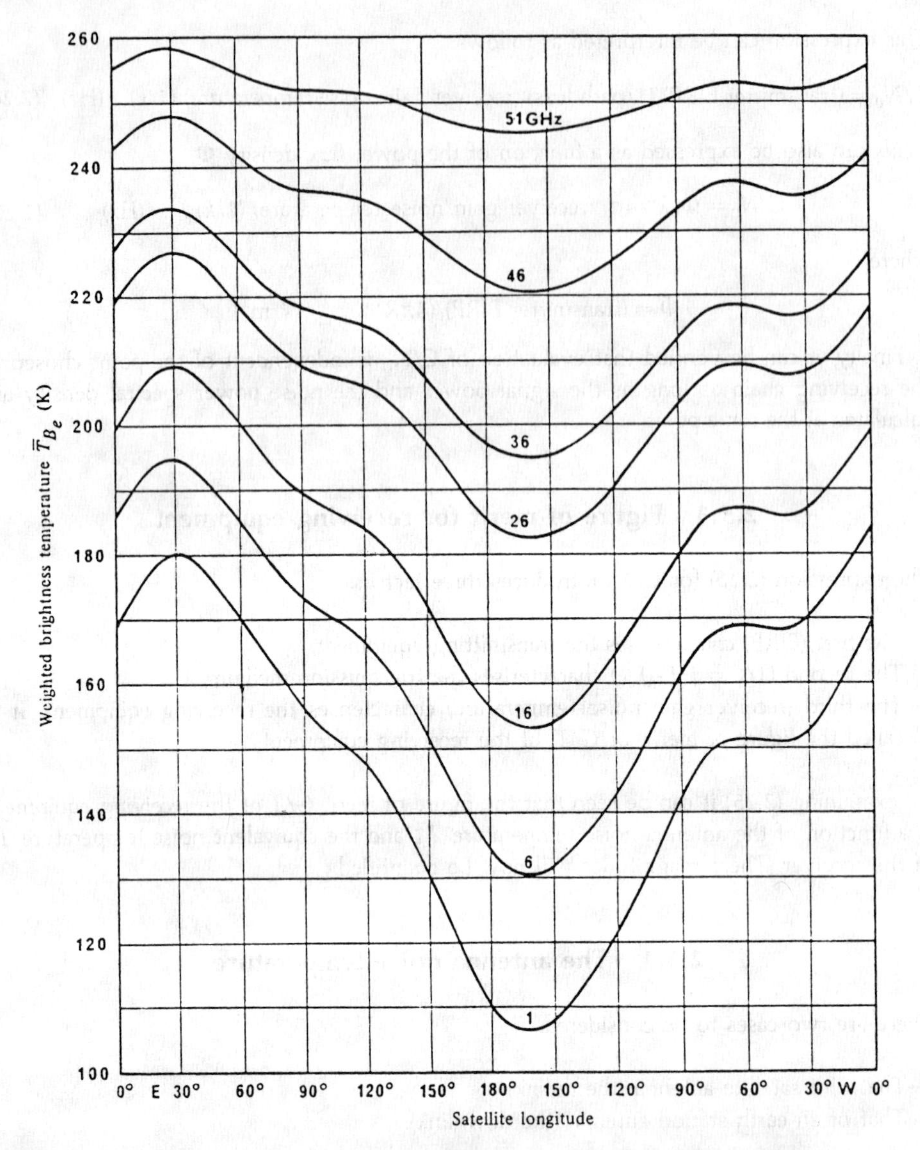

Figure 2.14 Antenna temperature of a geostationary satellite as a function of orbital position for frequencies between 1 and 51 GHz [NJO-85]. (Reproduced by permission of the American Geophysical Union).

2.5.4.2 *The earth station antenna (the downlink)*

The noise captured by the antenna consists of noise from the sky and noise due to radiation from the earth. Figure 2.15 shows the situation.

(a) 'Clear sky' conditions. At frequencies greater than 2 GHz the greatest contribution is that of the non-ionised region of the atmosphere which, being an absorbent medium, is a noise source. In the absence of meteorological formations (conditions described as 'clear sky'),

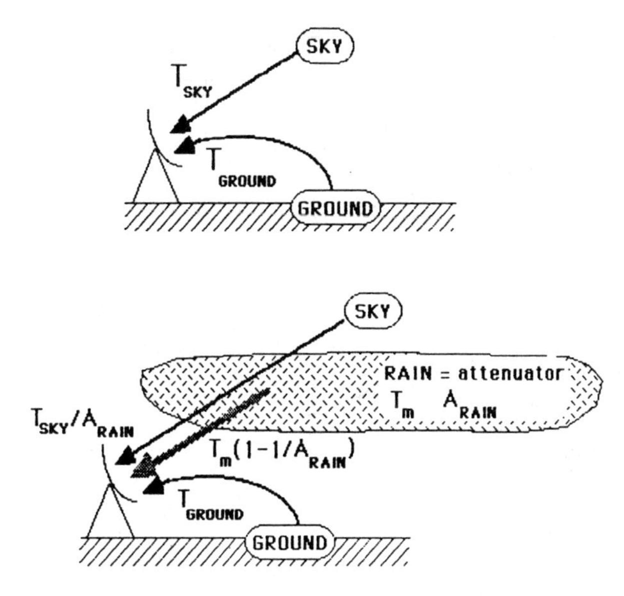

Figure 2.15 Contributions to the noise temperature of an earth station.

the antenna noise temperature contains contributions due to the sky and the surrounding ground.

The sky noise contribution is determined from expression (2.20), where $T_b(\theta, \varphi)$ is the brightness temperature of the sky in the direction (θ, φ). In practice, only that part of the sky in the direction of the antenna boresight contributes to the integral as the gain has a high value only in that direction. As a consequence, the noise contribution of the clear sky T_{SKY} can be assimilated with the brightness temperature for the angle of elevation of the antenna. Figure 2.16 shows the clear sky noise temperature as a function of frequency and elevation angle [CCIR, Report 720].

Radiation from the ground in the vicinity of the earth station is captured by the side lobes of the antenna radiation pattern and partly by the main lobe when the elevation angle is small. The contribution of each lobe is determined by $T_i = G_i(\Omega_i/4\pi)T_G$, where G_i is the mean gain of the lobe of solid angle Ω_i and T_G the brightness temperature of the ground. The sum of these contributions yields the value T_{GROUND}. The following can be taken as a first approximation [CCIR, Report 390]:

—$T_G = 290$ K for lateral lobes whose elevation angle E is less than $-10°$
—$T_G = 150$ K for $-10° < E < 0°$
—$T_G = 50$ K for $0° < E < 10°$
—$T_G = 10$ K for $10° < E < 90°$

The antenna noise temperature is thus given by:

$$T_A = T_{SKY} + T_{GROUND} \qquad \text{(K)} \tag{2.28}$$

To this noise may be added that of individual sources which are located in the vicinity of

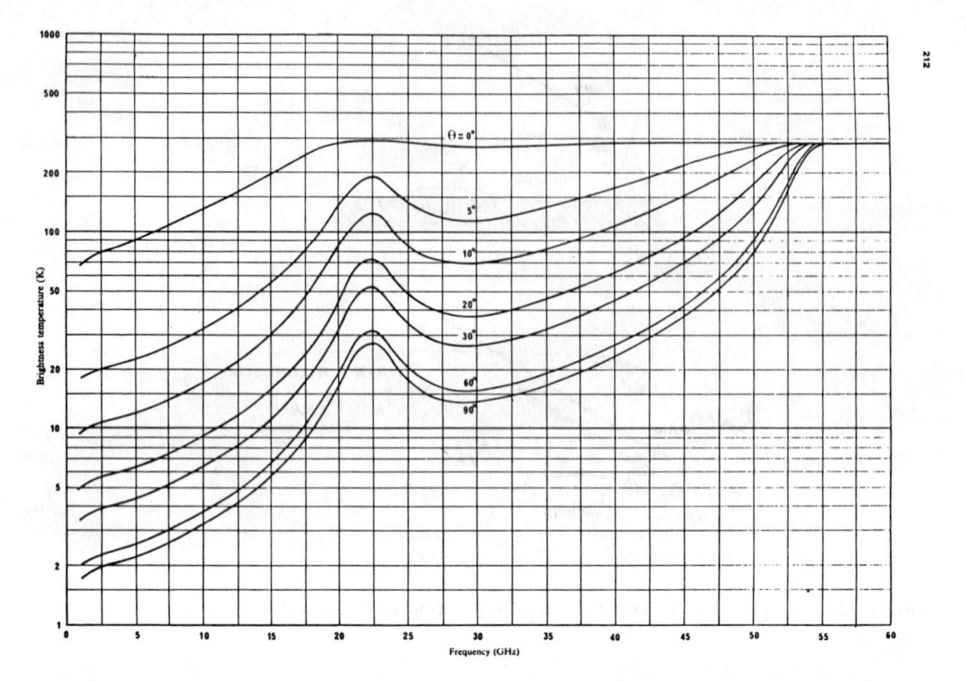

Figure 2.16 Noise temperature of clear sky as a function of frequency and elevation angle E for mean atmospheric humidity (7.5 g/cm^3 humidity at ground level) and standard temperature and pressure conditions at ground level. (From CCIR Rep 720–2. Reproduced by permission of the ITU.)

the antenna boresight. For a radio source of apparent angular diameter α and noise temperature T_n at the frequency considered and measured at ground level after attenuation by the atmosphere, the additional noise temperature ΔT_A for an antenna of beamwidth θ_{3dB} is given by:

$$\Delta T_A = T_n(\alpha/\theta_{3dB})^2 \qquad \text{if } \theta_{3dB} > \alpha \qquad \text{(K)}$$
$$\Delta T_A = T_n \qquad \text{if } \theta_{3dB} < \alpha \qquad \text{(K)} \qquad (2.29)$$

For earth stations pointing towards a geostationary satellite, only the sun and the moon need to be considered. The sun and the moon have an apparent angular diameter of 0.5°. There is an increase of noise temperature when these heavenly bodies are aligned with the earth station pointing towards the satellite. This particular geometrical configuration can be foreseen. To be more specific, at 12 GHz a 13 m antenna undergoes a noise temperature increase due to the sun, at a time of steady sun, by an amount ΔT_A equal to 12,000 K. The conditions of occurrence and the value of ΔT_A as a function of the antenna diameter and frequency are discussed in detail in Chapters 7 and 8. For the moon, the increase is at most 250 K at 4 GHz [CCIR Report 390].

Figure 2.17 shows the variation of antenna noise temperature T_A as a function of elevation angle E for various types of antenna at different frequencies in a clear sky [CCIR, Report 868] [CCIR Handbook]. It can be seen that the antenna noise temperature decreases as the elevation angle increases.

(b) Conditions of rain. The antenna noise temperature increases during the presence of

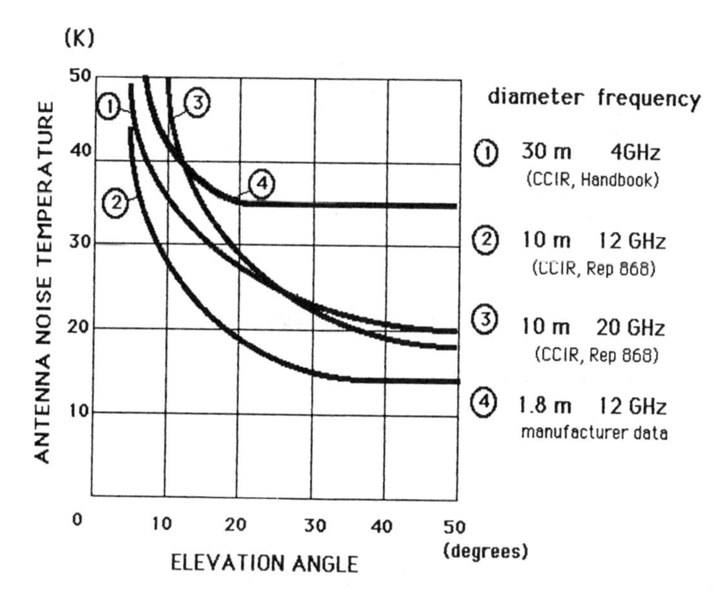

Figure 2.17 Typical values of antenna noise temperature T_A as a function of elevation angle E. Curve 1: diameter = 30 m, frequency = 4 GHz. Curve 2: diameter = 10 m, frequency = 12 GHz.Curve 3: diameter = 10 m, frequency = 20 GHz. Curve 4: diameter = 1.8 m, frequency = 12 GHz [Curve 1: *CCIR Handbook on Satellite Communications*, p. 209, 1985] [Curves 2 and 3: CCIR Report 868] [Curve 4: manufacturer's data (Accatel Telspace)].

meteorological formations, such as clouds and rain, which constitute an absorbent, and consequently emissive, medium. Using expression (2.24), the antenna noise temperature becomes:

$$T_A = T_{SKY}/A_{RAIN} + T_m(1 - 1/A_{RAIN}) + T_{GROUND} \quad \text{(K)} \qquad (2.30)$$

where A_{RAIN} is the attenuation and T_m the mean thermodynamic temperature of the formations in question. For T_m, values between 260 K and 280 K can be assumed [THO-83] [CCIR, Report 564].

In conclusion, the antenna noise temperature T_A, is a function of:

—the frequency,
—the elevation angle,
—the atmospheric conditions (clear sky or rain).

Consequently, the figure of merit of an earth station must be specified for particular conditions of frequency, elevation angle and atmospheric conditions.

2.5.5 Noise temperature of the receiver

Figure 2.18 shows the arrangement of the receiver. By using (2.21), the noise temperature T_R of the receiver can be expressed as:

$$T_R = T_{LNA} + T_{MX}/G_{LNA} + T_{IF}/G_{LNA}G_{MX} \quad \text{(K)} \qquad (2.31)$$

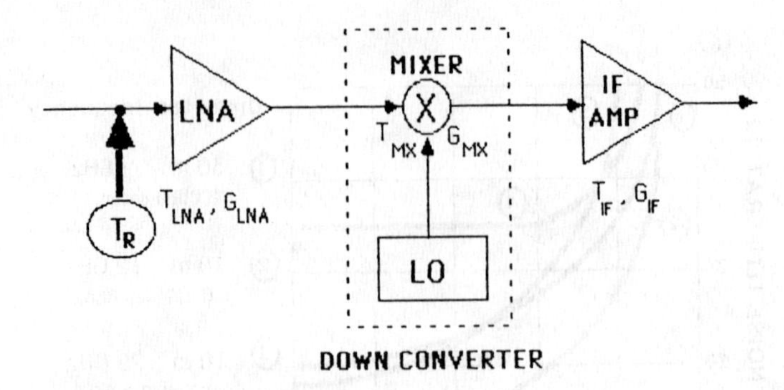

Figure 2.18 The architecture of a receiver.

Example:

Low noise amplifier (LNA): $T_{LNA} = 150\,K$, $G_{LNA} = 50\,dB$

Mixer: $T_{MX} = 850\,K$, $G_{MX} = -10\,dB$ ($L_{MX} = 10\,dB$)

IF amplifier: $T_{IF} = 400\,K$, $G_{IF} = 30\,dB$

Hence:

$$\begin{aligned} T_R &= 150 + 850/10^5 + 400/10^5 10^{-1} \\ &= 150\,K \end{aligned}$$

Notice the benefit of the high gain of the low noise amplifier which limits the noise temperature T_R of the receiver to that of the low noise amplifier T_{LNA}.

2.5.6 Example 1: Uplink (clear sky)

Figure 2.19 shows the geometry of the uplink. It is assumed that the transmitting earth station is on the edge of the 3 dB coverage of the satellite receiving antenna. The data are as follows:

—Frequency: $f_U = 14\,GHz$
—For the earth station (ES):
Transmitting amplifier power: $P_{TX} = 100\,W$
Loss between amplifier and antenna: $L_{FTX} = 0.5\,dB$
Antenna diameter: $D = 4\,m$
Antenna efficiency: $\eta = 0.6$
Maximum pointing error: $\alpha_T = 0.1°$
—Earth station–satellite distance: $R = 40\,000\,km$
—Atmospheric wave attenuation: $L_A = 0.3\,dB$ (typical value for attenuation by atmospheric gases at this frequency for an elevation angle of 10°)

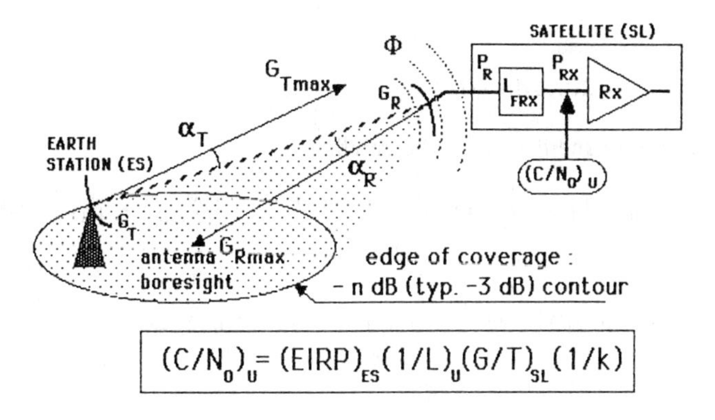

Figure 2.19 The geometry of an uplink.

—For the satellite (SL):
Receiving beam aperture: $\theta_{3dB} = 2°$
Antenna efficiency: $\eta = 0.55$
Receiver noise figure: $F = 3$ dB
Losses between antenna and receiver: $L_{FRX} = 1$ dB
Thermodynamic temperature of the connection: $T_F = 290$ K
Antenna noise temperature: $T_A = 290$ K

To calculate the EIRP of the earth station:

$$(EIRP)_{ES} = (P_{TX}G_{Tmax}/L_T L_{FTX}) \qquad (W) \tag{2.32}$$

with:

$$P_{TX} = 100 \text{ W} = 20 \text{ dB(W)}$$

$$G_{Tmax} = \eta(\pi D/\lambda_U)^2 = \eta(\pi D f_U/c)^2 = 0.6[\pi \times 4 \times (14 \times 10^9)/(3 \times 10^8)]^2 = 206\,340$$
$$= 53.1 \text{ dBi}$$

$$L_T(\text{dB}) = 12(\alpha_T/\theta_{3dB})^2 = 12(\alpha_T D f_U/70c)^2 = 0.9 \text{ dB}$$

$$L_{FTX} = 0.5 \text{ dB}$$

Hence:

$$(EIRP)_{ES} = 20 \text{ dB(W)} + 53.1 \text{ dB} - 0.9 \text{ dB} - 0.5 \text{ dB} = 71.7 \text{ dB(W)}$$

To calculate the attenuation on the upward path (U):

$$L_U = L_{FS} L_A \tag{2.33}$$

with:

$$L_{FS} = (4\pi R/\lambda_U)^2 = (4\pi R f_U/c)^2 = 5.5 \times 10^{20} = 207.4 \text{ dB}$$

$$L_A = 0.3 \text{ dB}$$

Hence:

$$L_U = 207.4\ \text{dB} + 0.3\ \text{dB} = 207.7\ \text{dB}$$

To calculate the figure of merit G/T of the satellite:

$$(G/T)_{SL} = (G_{R\max}/L_R L_{FRX} L_{POL})/[T_A/L_{FRX} + T_F(1 - 1/L_{FRX}) + T_R] \qquad (K^{-1}) \quad (2.34)$$

with:

$$G_{R\max} = \eta(\pi D/\lambda_U)^2 = \eta(\pi.70/\theta_{3dB})^2 = 0.55(\pi.70/2)^2 = 6650 = 38.2\ \text{dBi}$$

$$L_R = 12(\alpha_R/\theta_{3dB})^2$$

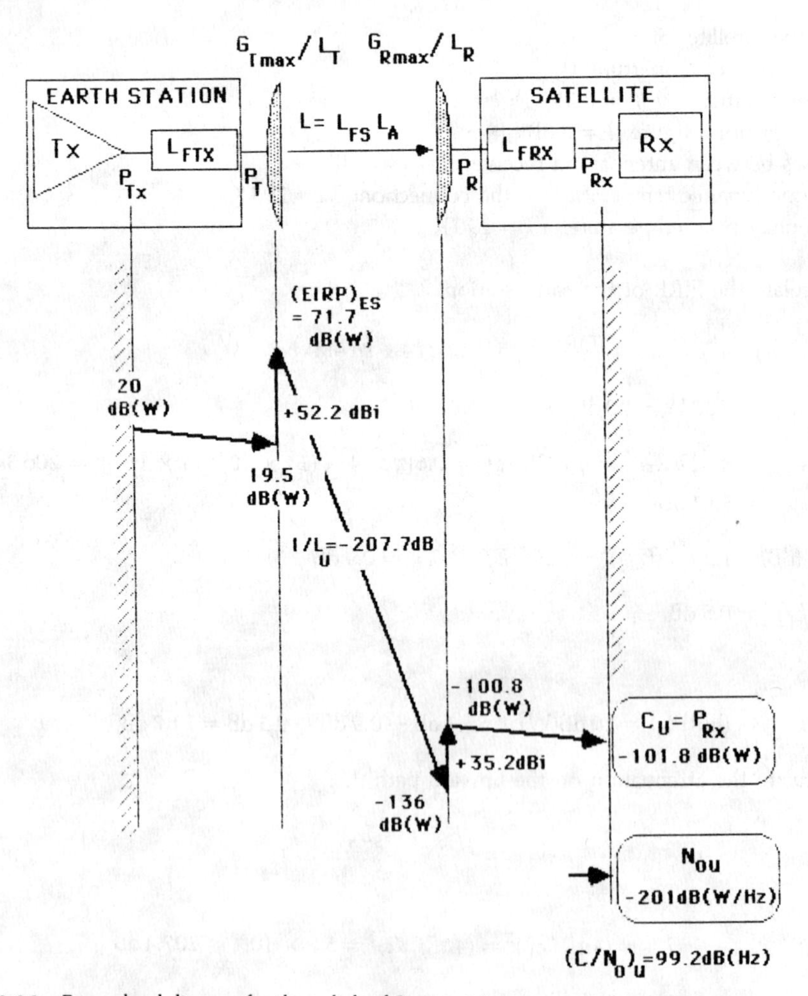

Figure 2.20 Power level diagram for the uplink of Section 2.5.6.

As the earth station is on the edge of the 3 dB coverage area, $\alpha_R = \theta_{3dB}/2$ and

$$L_R = 3\,\text{dB}$$

$$L_{FRX} = 1\,\text{dB}$$

$$L_{POL} = 0\,\text{dB}$$

$$T_A = 290\,\text{K}$$

$$T_F = 290\,\text{K}$$

$$T_R = (F-1)T_O = (10^{0.3}-1)290 = 290\,\text{K}$$

Hence:

$$(G/T)_{SL} = 38.2 - 3 - 1 - 10\log[290/10^{0.1} + 290(1 - 1/10^{0.1}) + 290]$$
$$= 6.6\,\text{dB K}^{-1}$$

Notice that when the thermodynamic temperature of the connection between the antenna and the satellite receiver is close to the antenna noise temperature, which is the case in practice, the equivalent noise temperature at the receiver input is $T \approx T_F + T_R \approx 290 + T_R$. It is, therefore, needlessly costly to install a receiver with a very low noise figure on-board a satellite.

To calculate the ratio C/N_0 for the uplink:

$$(C/N_0)_U = (\text{EIRP})_{ES}(1/L_U)(G/T)_{SL}(1/k) \qquad (\text{Hz}) \tag{2.35}$$

Hence:

$$(C/N_0)_U = 71.7\,\text{dBW} - 207.7\,\text{dB} + 6.6\,\text{dB K}^{-1} + 228.6\,\text{dBW/Hz K} = 99.2\,\text{dB Hz}$$

Figure 2.20 summarises the variations in signal power level throughout the trajectory.

2.5.7 Example 2: Uplink (with rain)

In the presence of rain, propagation attenuation is greater due to the attenuation A_{RAIN} caused by rain in the atmosphere. This is in addition to the attenuation due to gases in the atmosphere (0.3 dB). A typical value of attenuation due to rain for an earth station situated in a temperate climate (for example in Europe) can be considered to be $A_{RAIN} = 10\,\text{dB}$. Such an attenuation would typically be exceeded, at a frequency of 14 GHz, for 0.01% of the time for an average year. This gives $L_A = 0.3\,\text{dB} + 10\,\text{dB} = 10.3\,\text{dB}$.

Hence:

$$L_U = 207.4\,\text{dB} + 10.3\,\text{dB} = 217.7\,\text{dB}$$

Other things being equal, this gives:

$$(C/N_0)_U = 71.7\,\text{dB(W)} - 217.7\,\text{dB} + 6.6\,\text{dB(K}^{-1}) + 228.6\,\text{dB(W/Hz K)} = 89.2\,\text{dB(Hz)}$$

The ratio $(C/N_0)_U$ for the uplink would be greater than the value calculated in this way for 99.99% of the time for an average year.

2.5.8 Example 3: Downlink (clear sky)

Figure 2.21 shows the geometry of the downlink. It is assumed that the receiving earth station is located at the edge of the 3 dB coverage area of the satellite receiving antenna. The data are as follows:

—Frequency: $f_D = 12$ GHz
—For the satellite (SL):

Transmitting amplifier power: $P_{TX} = 10$ W
Loss between amplifier and antenna: $L_{FTX} = 1$ dB
Transmitting beam aperture: $\theta_{3dB} = 2°$
Antenna efficiency: $\eta = 0.55$

—Earth station–satellite distance: $R = 40\,000$ km
—Atmospheric attenuation: $L_A = 0.3$ dB (typical attenuation by atmospheric gases at this frequency for an elevation angle of 10°)
—For the earth station (ES):
Receiver noise figure: $F = 2.2$ dB
Loss between antenna and receiver: $L_{FRX} = 0.5$ dB
Thermodynamic temperature of the connection: $T_F = 290$ K
Antenna diameter: $D = 4$ m
Antenna efficiency: $\eta = 0.6$
Maximum misalignment: $\alpha_R = 0.1°$
Ground noise temperature: $T_{GROUND} = 45$ K

To calculate the EIRP of the satellite:

$$(\text{EIRP})_{SL} = (P_{TX}G_{T\max}/L_T L_{FTX}) \quad (\text{W}) \tag{2.36}$$

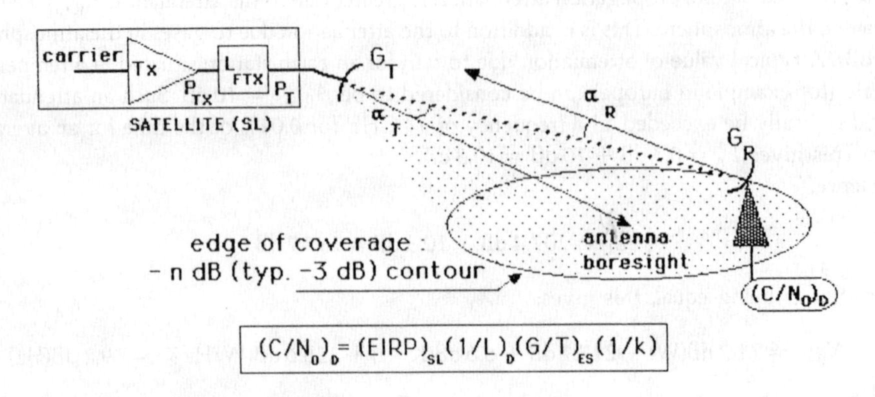

Figure 2.21 The geometry of a downlink.

with:

$$P_{TX} = 10\,W = 10\,dB(W)$$

$$G_{Tmax} = \eta(\pi D/\lambda_D)^2 = \eta(\pi 70/\theta_{3dB})^2 = 0.55(\pi 70/2)^2 = 6650 = 38.2\,dBi$$

$$L_T(dB) = 3\,dB$$

$$L_{FTX} = 1\,dB$$

Hence:

$$(EIRP)_{SL} = 10\,dB(W) + 38.2\,dB - 3\,dB - 1\,dB = 44.2\,dB(W)$$

To calculate the attenuation on the downlink (D):

$$L_D = L_{FS}L_A \tag{2.37}$$

with:

$$L_{FS} = (4\pi R/\lambda_D)^2 = (4\pi R f_D/c)^2 = 4.04 \times 10^{20} = 206.1\,dB$$

$$L_A = 0.3\,dB$$

Hence:

$$L_D = 206.1\,dB + 0.3\,dB = 206.4\,dB$$

To calculate the figure of merit G/T of the earth station:

$$(G/T)_{ES} = (G_{Rmax}/L_R L_{FRX} L_{POL})/[T_A/L_{FRX} + T_F(1 - 1/L_{FRX}) + T_R] \qquad (K^{-1})$$

with:

$$G_{Rmax} = \eta(\pi D/\lambda_D)^2 = \eta(\pi D f_D/c)^2 = 0.6(\pi \times 4 \times 12 \times 10^9/3 \times 10^8)^2$$
$$= 151\,597 = 51.8\,dBi$$

$$L_R\,(dB) = 12(\alpha_R/\theta_{3dB})^2 = 12(\alpha_R D f_D/70c)^2 = 0.6\,dB$$

$$L_{FRX} = 0.5\,dB$$

$$L_{POL} = 0\,dB$$

$T_A = T_{SKY} + T_{GROUND}$ with $T_{SKY} = 20\,K$ (see Figure 2.14 for $f = 12\,GHz$ and $E = 10°$) and $T_{GROUND} = 45\,K$ from which $T_A = 65\,K$

$$T_F = 290\,K$$

$$T_R = (F - 1)T_O = (10^{0.22} - 1)290 = 191.3\,K$$

Hence:

$$(G/T)_{ES} = 51.8 - 0.6 - 0.5 - 10\log[65/10^{0.05} + 290(1 - 1/10^{0.05}) + 191.3]$$

$$= 26.2\,dB(K^{-1})$$

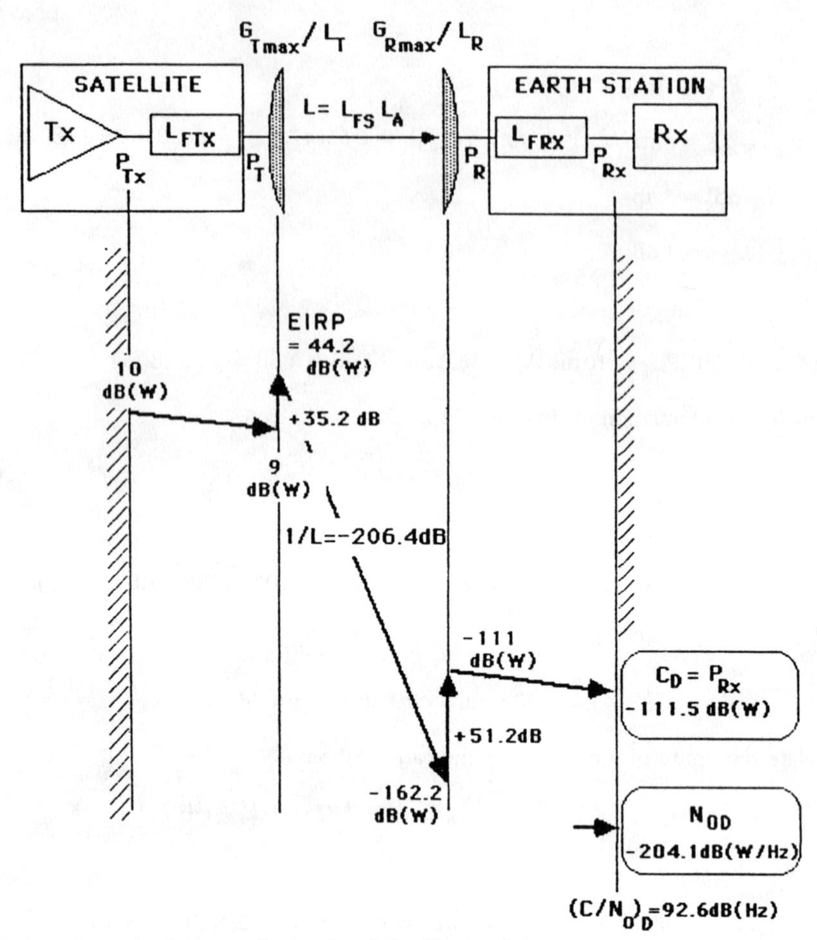

Figure 2.22 Power level diagram for the downlink of Section 2.5.8.

To calculate the ratio C/N_0 for the downlink:

$$(C/N_0)_D = (\text{EIRP})_{SL}(1/L_D)(G/T)_{ES}(1/k) \qquad \text{(Hz)} \tag{2.38}$$

Hence:

$$(C/N_0)_D = 44.2\ \text{dB(W)} - 206.4\ \text{dB} + 26.2\ \text{dB(K}^{-1}) + 228.6\ \text{dB(W/Hz.K)}$$

$$= 92.6\ \text{dB(Hz)}$$

Figure 2.22 summarises the variations of power signal level throughout the trajectory.

2.5.9 Example 4: Downlink (with rain)

$A_{RAIN} = 7$ dB will be taken as a typical value of attenuation due to rain for an earth station situated in a temperate climate (for example in Europe) which will be exceeded, at a frequency of 12 GHz, for more than 0.01% of the time during an average year; this gives $L_A = 0.3$ dB + 7 dB = 7.3 dB. Hence: $L_D = 206.1 + 7.3$ dB = 213.4 dB. The antenna noise temperature is given by:

$$T_A = T_{SKY}/A_{RAIN} + T_m(1 - 1/A_{RAIN}) + T_{GROUND} \qquad \text{(K)} \tag{2.39}$$

Taking

$$T_m = 275 \text{ K}$$

$$T_A = 20/10^{0.7} + 275(1 - 1/10^{0.7}) + 45 = 265 \text{ K}$$

Hence

$$(G/T)_{ES} = 51.8 - 0.6 - 0.5 - 10 \log [265/10^{0.05} + 290(1 - 1/10^{0.05}) + 191.3]$$
$$= 24.1 \text{ dB(K}^{-1})$$

To calculate the ratio C/N_0 for the downlink:

$$(C/N_0)_D = (\text{EIRP})_{SL}(1/L_D)(G/T)_{ES}(1/k) \qquad \text{(Hz)}$$

Hence:

$$(C/N_0)_D = 44.2 \text{ dB(W)} - 213.4 \text{ dB} + 24.1 \text{ dB(K}^{-1}) + 228.6 \text{ dB(W/Hz K)} = 83.5 \text{ dB(Hz)}$$

The ratio $(C/N_0)_D$ for the downlink will be greater than the value calculated in this way for 99.99% of the time during an average year.

2.5.10 Conclusion

The quality of the link between a transmitter and a receiver can be characterised by the ratio of the signal power to the noise power spectral density C/N_0. This is a function of the characteristics of the terminal equipment of the link, the transmitter EIRP and the receiver figure of merit G/T and the properties of the transmission medium. In a satellite link between two earth stations, two links must be considered—the uplink characterised by the ratio $(C/N_0)_U$ and the downlink characterised by the ratio $(C/N_0)_D$. The propagation conditions in the atmosphere affect the uplink and downlink differently; rain reduces the value of the ratio $(C/N_0)_U$ by decreasing the value of received power C_U while it reduces the value of $(C/N_0)_D$ by reducing the value of received power C_D and increasing the downlink system noise temperature. Denoting the resulting degradation by $\Delta(C/N_0)$ gives:

$$\Delta(C/N_0)_U = \Delta C_U = (A_{RAIN})_U \tag{2.40}$$

$$\Delta(C/N_0)_D = \Delta C_D - \Delta(G/T) = (A_{RAIN})_D + \Delta T \tag{2.41}$$

2.6 INFLUENCE OF THE PROPAGATING MEDIUM

On both the up- and downlinks, the carrier passes through the atmosphere. Recall that the range of frequencies concerned is from 1 to 30 GHz. From the point of view of wave propagation at these frequencies, only two regions of the atmosphere have an influence—the troposphere and the ionosphere. The troposphere extends practically from the ground to an altitude of 15 km. The ionosphere is situated between around 70 and 1000 km. The regions where their influence is maximum are in the vicinity of the ground for the troposphere and at an altitude of the order of 400 km for the ionosphere [CNES-CNET 1982].

The influence of the atmosphere has been mentioned previously in order to introduce the losses L_A due to atmospheric attenuation into expression (2.11) and in connection with antenna noise temperature. However, other phenomena can occur. Their nature and significance will now be explained.

The predominant effects are those caused by absorption and depolarisation due to tropospheric precipitation (rain and snow). These are particularly significant for frequencies greater than 10 GHz. The occurrence of precipitation is defined by the percentage of time during which a given intensity level is exceeded. Low intensities with negligible effects correspond to high percentages of time (typically 20%); these are described as 'clear sky' conditions. High intensities, with significant effects, correspond to small percentages of time (typically 0.01%); these are described as 'rain' conditions. These effects can degrade the quality of the link below an acceptable threshold. The availability of a link is thus directly related to the temporal precipitation statistics. In view of their importance, the effects of precipitation will be presented first. The effects of other phenomena will be examined later.

2.6.1 The effects of precipitation

The intensity of precipitation is measured by the rainfall rate R expressed in mm/h. The temporal precipitation statistic is given by the cumulative probability distribution which indicates the annual percentage p (%) during which a given value of rainfall rate R_p (mm/h) is exceeded. In the absence of precise precipitation data for the location of the earth station involved in the link, the data of CCIR Report 563 can be used. To be more specific, in Europe a rainfall rate of $R_{0.01}$ ($p = 0.01\%$ is the annual percentage most used to analyse systems, it corresponds to 53 minutes per year) is around 30 mm/h with the exception of some Mediterranean regions where the frequency of storms (heavy precipitation for a short time interval) leads to a value of $R_{0.01} = 50$ mm/h. In equatorial regions, $R_{0.01} = 120$ mm/h (Florida in the USA for example) or even 160 mm/h (Central America).

Precipitation causes two effects:

—Attenuation.
—Cross polarisation.

2.6.1.1 *Attenuation [CCIR Report 564]*

The value of attenuation due to rain A_{RAIN} is given by the product of the specific attenuation γ_R (dB/km) and the effective path length of the wave in the rain L_e (km), that is:

$$A_{RAIN} = \gamma_R L_e \quad \text{(dB)} \tag{2.42}$$

The value of γ_R depends on the frequency and intensity R_p (mm/h) of the rain. The result is a value of attenuation which is exceeded during a percentage of time p. Determination of A_{RAIN} proceeds in several steps:

(1) Calculation of the height of the rain h_R (km):

$$h_R\,(\text{km}) = 3 + 0.028 \qquad \text{if } 0 < \text{latitude} < 56^\circ$$
$$= 4 - 0.075(\text{latitude} - 36) \text{ if latitude} \geqslant 36^\circ$$

(2) Calculation of the slant path length in the rain:
$L_S = (h_R - h_S)/\sin E$ (valid for an elevation angle $E > 5^\circ$)
h_s = height of the earth station above mean sea level (km)

(3) Calculation of $r_{0.01}$, the reduction factor for 0.01% of the time, which takes account of the inhomogeneity of the rain:

$$r_{0.01} = 1/(1 + (L_S/L_0)\cos E)$$

where $L_0\,(\text{km}) = 35\exp(-0.015\,R_{0.01})$.

(4) Calculation of L_e:

$$L_e = L_S r_{0.01} \qquad (\text{km})$$

(5) Determination of $R_{0.01}$, exceeded for 0.01% of an average year, where the earth station is located:

(6) Determination of γ_R using the nomogram of Figure 2.23 as a function of $R_{0.01}$ and frequency. In the case of circular polarisation of the wave, the mean value of the attenuations obtained for each plane polarisation is taken.

(7) The value of attenuation due to rain exceeded for 0.01% of an average year is thus:

$$A_{\text{RAIN}}(p = 0.01) = \gamma_R L_e \qquad (\text{dB})$$

The value of attenuation exceeded for a percentage p between 0.001% and 1% is:

$$A_{\text{RAIN}}(p) = A_{\text{RAIN}}(p = 0.01) \times 0.12 p^{-(0.546 + 0.043 \log p)} \qquad (\text{dB})$$

It is sometimes required to estimate the attenuation exceeded during a *percentage p_w of any month* (that is the least favourable month). The corresponding annual percentage is given by

$$p = 0.3\,(p_w)^{1.15} \qquad (\%) \tag{2.43}$$

Typical values of attenuation due to rain exceeded for 0.01% of an average year can be deduced from the previous procedure for regions where the rainfall rate $R_{0.01}$ is exceeded for 0.01% of an average year with a value of 30 to 50 mm/h. This gives around 0.1 dB at 4 GHz, 5 to 10 dB at 12 GHz, 10 to 20 dB at 20 GHz and 25 to 40 dB at 30 GHz.

Attenuation due to rain clouds or fog can be calculated using the same procedure [CCIR Report 721]. The specific attenuation γ_C is calculated as:

$$\gamma_C = KM \qquad (\text{dB/km}) \tag{2.44}$$

where $K = 1.1 \times 10^{-3} f^{1.8}$.

f is expressed in GHz from 1 GHz to 30 GHz, K in (dB/km)/(g/m^3) and M = water concentration of the cloud (g/m^3).

Attenuation due to rain clouds and fog is small compared with that due to precipitation except for clouds and fog with a high water concentration. For an elevation angle $E = 20^\circ$,

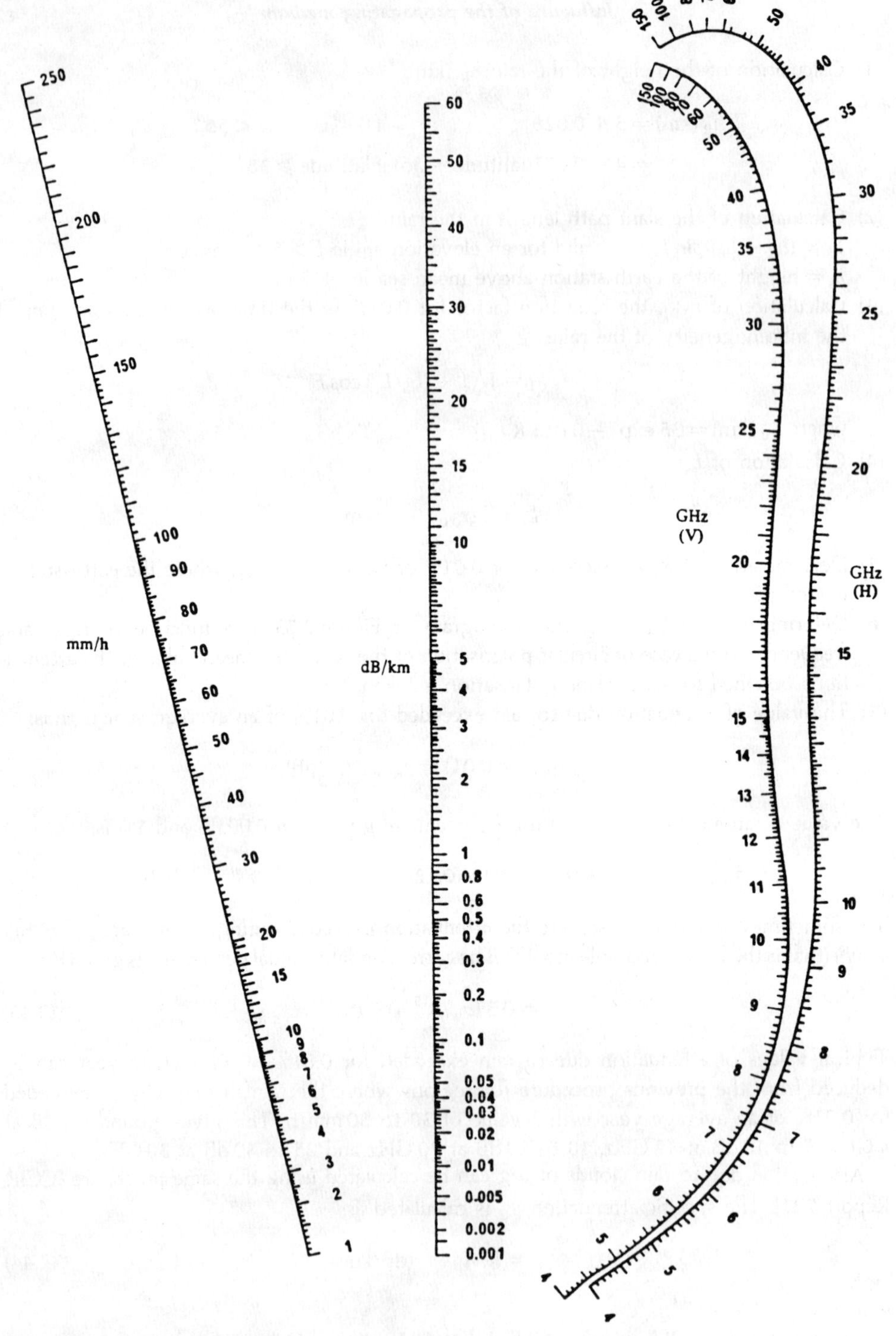

☞ with circular polarization use the arithmetic mean of attenuation with horizontal and vertical polarization

Figure 2.23 Nomogram for determination of the specific attenuation γ_R as a function of the frequency (GHz) and rain density R (mm/h) [CCIR Report 721]. (Reproduced by permission of the ITU.)

one can expect 0.5 to 1.5 dB at 15 GHz and 2 to 4.5 dB at 30 GHz. This attenuation, however, is observed for a greater percentage of the time.

Attenuation due to ice clouds is smaller still. Dry snow has little effect. Although wet snowfalls can cause greater attenuation than the equivalent rainfall rate, this situation is very rare and has little effect on attenuation statistics. The degradation of antenna characteristics due to accumulation of snow and ice may be more significant than the effect of snow along the path.

2.6.1.2 *Cross-polarisation [CCIR Report 722]*

Part of the energy transmitted in one polarisation is transferred to the orthogonal polarisation state. Cross-polarisation occurs as a result of differential attenuation and differential phase shift between two orthogonal characteristic polarisations. These effects originate in the non-spherical shape of raindrops. A commonly accepted model for a falling raindrop is an oblate spheroid with its major axis canted to the horizontal and with deformation dependent upon the radius of a sphere of equal volume. It is commonly accepted that canting angles vary randomly in space and time. The angle of the characteristic polarisations to the horizontal and 'vertical' (i.e. the direction perpendicular to both the horizontal and the propagation path) is often termed the effective canting angle.

The relationship between cross-polarised discrimination XPD and the copolarised path attenuation A_{RAIN} is of importance for predictions based on attenuation statistics. The following relationship is in approximate agreement with long-term measurements in the frequency range between about 3 and 37 GHz:

$$\mathrm{XPD} = U - 20\log(A_{RAIN}) \qquad \mathrm{dB}) \tag{2.45}$$

where:

$$U = 30\log(f) - D(E) + \kappa^2 + I(\tau) \qquad (\mathrm{dB})$$

f is the frequency in GHz, E the elevation in degrees and τ the polarisation tilt angle (for linear polarisation) relative to the horizontal.

The term $D(E)$ varies approximately with elevation angle E as given by:

$$D(E) = 40\log(\cos E) \qquad (\mathrm{dB})$$

However it is recognised that this term does not predict the elevation dependence for elevation angles close to 90°.

The term κ^2 is believed to depend primarily on the degree of random spread of the raindrop canting angles averaged over the path. For a Gaussian model of the raindrop canting angle distribution, $\kappa^2 = 0.0053\,\sigma^2$, where σ (in degrees) has been termed the effective standard deviation of the inclination of the raindrop canting angle distribution. Because κ^2 depends on several factors, σ cannot necessarily be interpreted solely in terms of the canting angle distribution.

The factor $I(\tau)$ can be omitted for circular polarisation. It represents approximately the improvement of linear polarisation with respect to circular polarisation. If the effective canting angle is assumed to vary randomly within a rainstorm and from storm to storm and to have a Gaussian distribution with zero mean and standard deviation σ_m, then $I(\tau)$ can be expressed by:

$$I(\tau) = -10\log\{0.5[1 - \cos(4\tau)\exp(-\kappa_m^2)]\} \qquad \text{(dB)}$$

where $\kappa_m^2 = 0.0024\,\sigma_m$ (σ_m in degrees).

Values of σ can be taken as 0°, 5°, 10° and 15° for time percentages of 1, 0.1, 0.01, and 0.001 respectively at 14/11 GHz [ROG-86]. A value of $\sigma_m = 5°$ would appear to give a sufficiently conservative maximum improvement of $I = 15$ dB for $\tau = 0°$ or 90°.

Typically one can expect a value of XPD less than 20 dB for 0.01% of the time [STR-83], [ROG-86], [YOS-86].

Snow (dry or wet) causes similar phenomena.

2.6.2 Other effects

2.6.2.1 *Attenuation by atmospheric gases [CCIR Report 564]*

Attenuation due to gas in the atmosphere depends on the frequency, the elevation angle, the altitude of the station and the water vapour concentration. It is negligible at frequencies less than 10 GHz and does not exceed 1 to 2 dB at 22 GHz (the frequency corresponding to a water vapour absorption band) for mean atmospheric humidity and elevation angles greater than 10°.

2.6.2.2 *Attenuation by sandstorms [CCIR Report 721]*

The specific attenuation (dB/km) is inversely proportional to the visibility and depends strongly on the humidity of the particles. At 14 GHz it is of the order of 0.03 dB/km for dry particles and 0.65 dB/km for particles of 20% humidity. If the path length is 3 km, the attenuation can reach 1 to 2 dB.

2.6.2.3 *Refraction [CCIR Report 718], [CCIR Report 564], [CCIR Report 263]*

The troposphere and the ionosphere have different refractive indices. The refractive index of the troposphere decreases with altitude, is a function of meteorological conditions and is independent of frequency. That of the ionosphere depends on frequency and the electronic content of the ionosphere. Both are subject to rapid local fluctuations. The effect of refraction is to cause curvature of the trajectory of the wave, variation of wave velocity and hence propagation time. Variations of refractive index, called 'scintillation', cause variations in the angle of arrival, amplitude and phase of the transmitted wave. The most troublesome phenomenon is ionospheric scintillation; it is greater when the frequency is low and the earth station is close to the equator. The received signal varies in amplitude and the peak-to-peak amplitude of these variations, at a frequency of 11 GHz and medium latitudes, can exceed 1 dB for 0.01% of the time. The other phenomena are significant only for low elevation angles (less than 10°) or when the carrier is used for precise distance measurements.

2.6.2.4 *The Faraday effect [CCIR Report 263]*

The ionosphere introduces a rotation of the plane of polarisation of a linearly polarised wave. The angle of rotation is inversely proportional to the square of the frequency. It is a function of the electronic content of the ionosphere and consequently varies with time, the season and the solar cycle. The order of magnitude is several degrees at 4 GHz. Since cyclic variations can be predicted, this effect can be compensated by a consequent rotation of the antenna polarisation. However, some perturbations (geomagnetic storms, for example) are sudden and unpredictable. The result, for a small percentage of the time, is an attenuation $L_{POL}(\text{dB}) = 20\log(\cos\gamma)$ of the received signal (see Section 2.3.4.4 above) and the appearance of a cross-polarised component which reduces the value of cross-polarisation discrimination XPD. For an angle of rotation γ, the value of XPD is given by XPD (dB) $= -20\log(\tan\gamma)$. For the case of a rotation of 9° at a frequency of 4 GHz [CCIR Report 555], this gives $L_{POL} = 0.1$ dB and XPD = 16 dB.

2.6.2.5 *Cross-polarisation due to ice crystals*

Ice clouds, where high altitude ice crystals are in a region close to the 0°C isotherm, are also the cause of cross-polarisation. However, in contrast to rain and other hydrometeors, this effect is not accompanied by attenuation. It causes a reduction in the value of XPD given by (2.45) by an amount C_{ice} (dB) $= (0.3 + 0.1\log p)$, XPD, where p is the percentage of time [ROG-86]. This reduction is around 2 dB for a percentage of time equal to 0.01%.

2.6.2.6 *Influence of the ground—multipath effects*

When the earth station antenna is small, and hence has a beam with a large angular width, the received signal can be the result of one wave received directly and a wave of equivalent amplitude received after reflection on the ground or environmental obstacles. In the case of destructive superposition (phase opposition), a large attenuation is observed. This effect does not exist when the earth station is equipped with an antenna which is sufficiently directional to eliminate the reflected wave.

2.6.3 Conclusion

At low frequencies (less than 10 GHz), the attenuation L_A is generally small and the principal cause of degradation of the link is cross-polarisation. This is caused by the ionosphere and by high altitude ice crystals in the troposphere. At higher frequencies, the phenomena of attenuation and cross-polarisation are both observed. These are caused essentially by atmospheric gases, rainfall and other hydrometeors.

Statistically, these phenomena become greater when a short percentage of time is considered. The availability of the link increases when these effects can be compensated. Compensation techniques are available and will now be discussed.

2.7 COMPENSATION FOR THE EFFECTS OF THE PROPAGATING MEDIUM

2.7.1 Cross polarisation [GHO-88]

The method of compensation relies on modification of the polarisation characteristics of the earth station (see Chapter 8). Compensation is achieved as follows:

—For the uplink, by correcting the polarisation of the transmitting antenna by anticipation so that the wave arrives matched to the satellite antenna,
—For the downlink, by matching the antenna polarisation to that of the received wave.

Compensation can be automatic; the signals transmitted by the satellite must be made available (as beacons) so that the effects of the propagating medium can be detected and the required control signal deduced [CCIR Report 555].

2.7.2 Attenuation

The mission specifies a value of the ratio C/N_0 greater than or equal to $(C/N_0)_{required}$ during a given percentage of the time, equal to $(100-p)\%$. For example, 99.99% of the time implies $p=0.01\%$. As seen in the examples of Sections 2.5.6 to 2.5.9, the attenuation A_{RAIN} due to rain causes a reduction of the ratio C/N_0 given by:

$$(C/N_0)_{rain} = (C/N_0)_{clear\ sky} - A_{RAIN}\,(\text{dB}) \qquad (\text{dB})(\text{Hz})) \qquad (2.46)$$

for an uplink and:

$$(C/N_0)_{rain} = (C/N_0)_{clear\ sky} - A_{RAIN}\,(\text{dB}) - \Delta(G/T)(\text{dB}(\text{Hz})) \qquad (2.47)$$

for a downlink.

$\Delta(G/T) = (G/T)_{clear\ sky} - (G/T)_{rain}$ represents the reduction (in dB) of the figure of merit of the earth station due to the increase of noise temperature.

For a successful mission, one must have $(C/N_0)_{rain} = (C/N_0)_{required}$; this can be achieved by including a margin $M(p)$ in the clear sky link budget with $M(p)$ defined by:

$$M(p) = (C/N_0)_{clear\ sky} - (C/N_0)_{required}$$
$$= (C/N_0)_{clear\ sky} - (C/N_0)_{rain} \qquad (\text{dB}) \qquad (2.48)$$

The value of A_{RAIN} to be used is a function of the time percentage p. It increases as p decreases.

Making provision for a margin $M(p)$ in the clear sky link requirement implies an increase of the EIRP which requires a higher transmitting power. For high attenuations which are encountered for a small percentage of the time and at the highest frequencies (see Section 2.6.1.1), the extra power necessary can exceed the capabilities of the transmitting equipment. Other solutions must then be considered as follows:

—site diversity,
—adaptivity.

2.7.3 Site diversity

High attenuations are due to regions of rain of small geographical extent. Two earth stations at two distinct locations 1 and 2 can establish links with the satellite which, at a given time t, suffer attenuations of $A_1(t)$ and $A_2(t)$ respectively; $A_1(t)$ is different from $A_2(t)$ as long as the geographical separation is sufficient. The signals are thus routed to the link less affected by attenuation. On this link the attenuation is $A_D(t) = \min[A_1(t), A_2(t)]$. The mean attenuation for a single location is defined as $A_M(t) = [A_1(t) + A_2(t)]/2$; all values are in dB.

Two concepts are useful to quantify the improvement provided by location diversity [CCIR Report 564], [PRA-86] as follows:

—The diversity gain.
—The diversity improvement factor.

2.7.3.1 *Diversity gain $G_D(p)$*

This is the difference (in dB) betweeen the mean attenuation at a single location $A_M(p)$, exceeded for a time percentage p, and the attenuation with diversity $A_D(p)$ exceeded for the same time percentage p. Hence, for a downlink for example, the required margin $M(p)$ at a given location is obtained from (2.48) and (2.47):

$$M(p) = A_{RAIN} + \Delta(G/T) \qquad \text{(dB)} \tag{2.49}$$

With site diversity, the required margin becomes:

$$M(p) = A_{RAIN} + \Delta(G/T) - G_D(p) \qquad \text{(dB)} \tag{2.50}$$

2.7.3.2 *Diversity improvement factor F_D*

This is the ratio between the percentage of time p_1 during which the mean attenuation at a single site exceeds the value A dB and the percentage of time p_2 during which the attenuation with diversity exceeds the same value A dB.

Figure 2.24 shows the relation between p_2 and p_1 as a function of the distance between the two locations. These curves can be modelled by the following relations:

$$p_2 = (p_1)^2(1 + \beta^2)/(p_1 + 100\beta^2) \tag{2.51}$$

with $\beta^2 = 2 \times 10^{-4} d$, when the distance $d > 5$ km.

Site diversity also provides protection against scintillation and cross-polarisation.

2.7.4 Adaptivity

Adaptivity involves variation of certain parameters of the link for the duration of the attenuation in such a way as to maintain the required value for the ratio C/N_0.

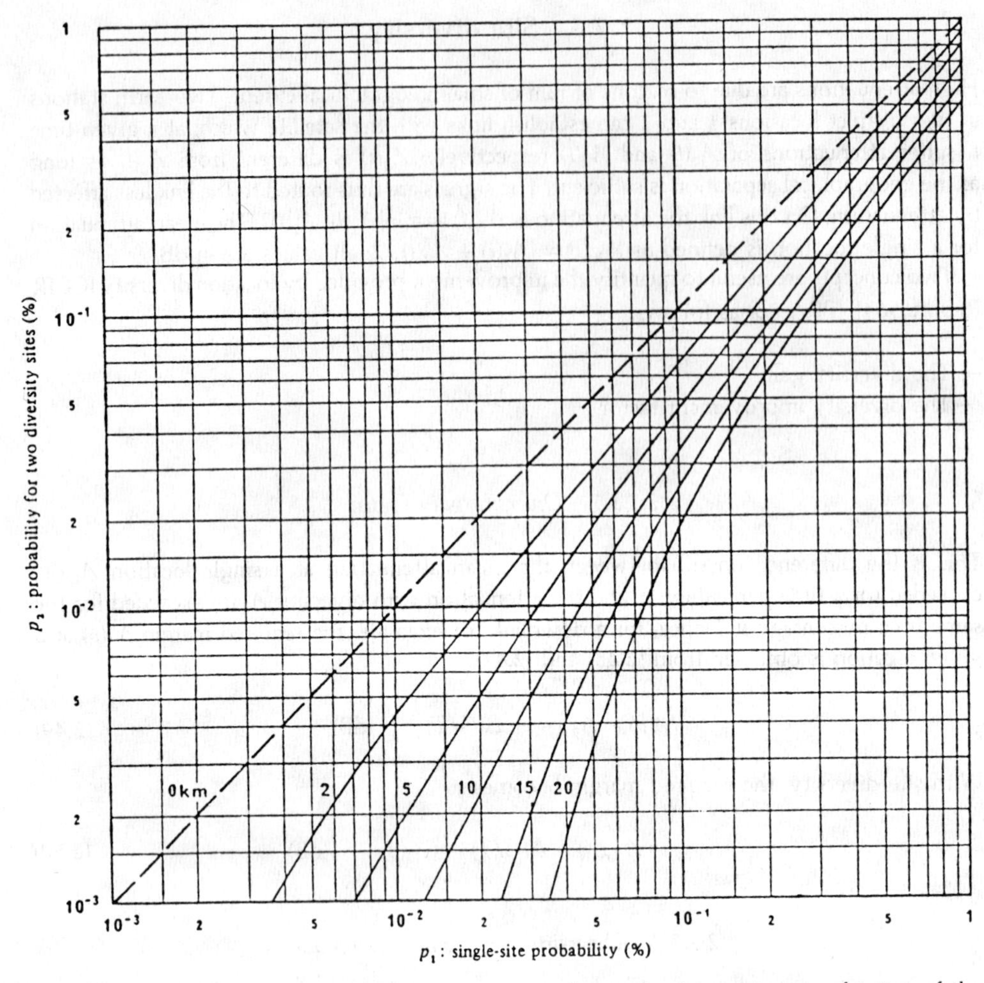

Figure 2.24 Probability of exceeding attenuation A simultaneously at two sites as a function of the distance between the two sites [CCIR Rep 564]. (Reproduced by permission of the ITU.)

Several approaches can be envisaged as follows:

—Assignment of an additional resource, which is normally kept in reserve, to the link affected by attenuation. This additional resource can be:

- An increase of transmission time (such as an unoccupied frame time slot in the case of TDMA multiple access, see Chapter 4) with or without the use of error correcting codes [MCM-86], [BAR-88];
- Use of a frequency band at a lower frequency which is less affected by the attenuation [WIL-88], [CAR-88];
- Use of higher EIRP on the uplink [HOR-88].

—Reduction of capacity; the link affected by the attenuation has its capacity reduced. In the

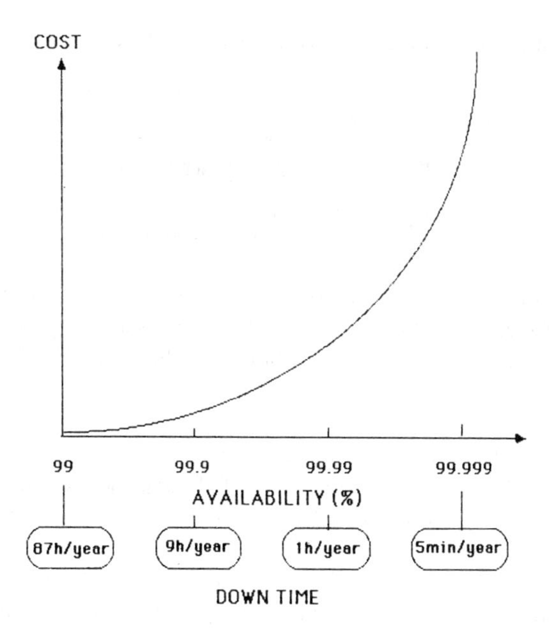

Figure 2.25 Link cost as a function of availability.

case of digital transmission (Chapter 3), the reduction in information rate enables an error correcting code to be used for a constant transmission rate. The combined effect of the reduction in information rate and the decoding gain enables a margin to be provided. An example is given in Section 6.3 for the case of a regenerative satellite downlink.

2.7.5 Conclusion: the availability–cost relation

A low unavailability (0.01% of the time for example) corresponds to a high availability (99.99% for the example considered). If only the effects of the propagating medium are considered as the cause of unavailability, the accepted value of unavailability represents the percentage of time p during which a given attenuation can be exceeded. The value of this attenuation is high when p is small, that is the required availability is high. Since the methods used to compensate for attenuation become more costly as the attenuation increases, the specified availability has a marked effect on system cost. This is shown in Figure 2.25.

2.8 CONSTRAINTS

The constraints in choosing the parameters of the link are in three categories:

—Regulations.
—Operational constraints.
—Propagation conditions.

As propagation conditions have been discussed above, the first two items will be considered below.

2.8.1 Regulatory aspects

2.8.1.1 *Administrative organisation*

The International Telecommunication Union (ITU) is the United Nations organisation for telecommunications. One of the objectives of the ITU is to ensure compatible radio networking by avoiding 'harmful interference' between different systems [THU-87]. To this end it has established the following:

—Committees to examine technical and operational matters; examples are the International Radiocommunication Consultative Committee (CCIR) and the International Telegraph and Telephone Consultative Committee (CCITT). These committees produce Reports and Recommendations.
—World and Regional Administrative Radio Conferences which are convened to discuss particular telecommunications topics and to carry out total or partial revision of the administrative regulations, particularly the Radiocommunication Regulations (RR).
—The International Frequency Registration Board (IFRB) which is responsible for registration of the frequency assignments made by countries to their radio stations and verification that they conform to the assignment rules.

Every radiocommunication system must conform to the provisions contained in the Radiocommunications Regulations. These regulations divide satellite telecommunication into various space radiocommunication services.

2.8.1.2 *Space radiocommunication services*

The Radiocommunication Regulations distinguish the following services:

—Fixed Satellite Service (FSS).
—Mobile Satellite Service (MSS) with three particular services:
- Maritime Mobile Satellite Service (MMS),
- Aeronautical Mobile Satellite Service (AMS),
- Land Mobile Satellite Service (LMS).

—Broadcasting Satellite Service (BSS).
—Earth Exploration Satellite Service (EES).
—Space Research Service (SRS).
—Space Operation Service (SOS).
—Inter-Satellite Service (ISS).
—Amateur Satellite Service (ASS).

Appendix 2 from [MAR-86], pp. 366–385, extracted from Chapter 1 of the *Radio Regulations*, specifies the definition of these services. Table 2.1 presents examples of the functions provided within these services [THU-87].

Table 2.1 Examples of the functions provided by the various space communication services.

Type of Link (↑ = uplink, ↓ = downlink)		Function provided	Services
Constant frequency transmission (radio beacon)	↓	Tracking,	SOS
		Navigation	RNS
		Frequency standards	SFS
		Observation of the earth and ionosphere	SRS, EES
Connecting downlink	↓	Maintenance telemetry	SOS
		Operational telemetry	SRS, EES
		Transmission of acquired data	SRS, EES
Broadcasting links	↓	Time signals	TSS
		Data*	BSS
		Sound programs*	BSS
		Television programmes*	BSS
Fixed links	↓	Point-to-zone and broadcast	FXS
	↑↓	Point-to-multipoint (with multiple access)	FXS
		Point-to-point	FXS
		Transportable stations	FXS or MSS
Links with mobiles	↑↓	Land	MSS (LMS)
		Maritime	MSS (MMS)
		Aeronautical	MSS (AMS)
Bidirectional connecting link	↑↓	With mobile service satellites	FXS or MSS
Connecting uplink	↑	With broadcast satellites	FXS
		Feeder	FXS
		Data collection	SOS or FXS
		[Search and rescue]	[AMS]
		Telecommand	SOS
Constant frequency transmission	↑	Accurate antenna pointing	SOS
Radiometer	↑	Observation by passive detectors	SRS, EES
Radio identification	↑↓	Navigation [Search and rescue]	SRS [MSS]
		Direction finding, Surveying	RIS
		Observation by sounding (active detectors)	SRS, EES
		Traffic management	MSS

*Separate or combined reception.
MSS, Mobile satellite service; MMS, Maritime mobile satellite service; LMS, Land mobile satellite service; AMS, Aeronautical mobile satellite service; RNS, Radio navigation satellite service; SFS, Standard frequencies satellite service; BSS, Broadcasting satellite service; EES, Earth Exploration satellite service; SRS, Space research service; SOS, Space operation service; TSS, Time signals satellite service; RIS, Radio identification satellite service.

2.8.1.3 *Frequency allocation*

The concept of a radiocommunication service is applied to the allocation of frequency bands and analysis of the conditions for sharing a given band among compatible services. To this end the world has been divided into three regions as follows:

—Region 1: Europe, Africa, the Middle East and the USSR.
—Region 2: The Americas.
—Region 3: Asia except the Middle East and USSR, Oceania.

Frequency allocations to a given service can depend on the region. According to region, the bands allocated can be exclusive (bands allocated uniquely to the service) or shared (bands shared among several services). To be more precise:

—Fixed satellite service links use the following bands:

(a) Around 6 GHz for the uplink and around 4 GHz for the downlink (systems described as 6/4 GHz or C Band). These bands are occupied by the oldest systems (such as INTELSAT, American domestic systems etc.) and tend to be saturated.
(b) Around 8 GHz for the uplink and around 7 GHz for the downlink (systems described as 8/7 GHz or X Band). These bands are reserved, by agreement between administrations, for government use.
(c) Around 14 GHz for the uplink and around 12 GHz for the downlink (systems described as 14/12 GHz or Ku band). This corresponds to current operational developments (such as EUTELSAT, TELECOM I and II etc.)
(d) Around 30 GHz for the uplink and around 20 GHz for the downlink (systems described as 30/20 GHz or Ka band); this is currently used for experimental and pre-operational purposes.

The bands above 30 GHz will be used eventually in accordance with developing requirements and technology.

—Mobile satellite service links currently use the bands around 1.6 GHz for the uplink and 1.5 GHz for the downlink (systems described as 1.6/1.5 GHz or L band)
—Broadcasting Satellite Service links contain only downlinks using bands around 12 GHz. The uplink appertains to the Fixed-Satellite Service and is called a feeder link. For further details refer to the Radiocommunication Regulations (RR).

2.8.1.4 Fixed-Satellite Service allotment plan

The World Radio Administrative Conferences 1985 and 1988 (WARC-ORB-85 and WARC-ORB-88) have retained the principle of a plan which guarantees every country equal access to the geostationary satellite orbit and to the frequency bands allotted to space services using this orbit. The plan adopted concerns fixed satellite services in the 6/4, 14/12 and 30/20 GHz bands. This plan contains two parts as follows:

—An allotment plan which enables each administration to satisfy the hardware requirements of national services for at least one orbital position on an arc and in one or more predetermined bands. The allotment plan relates to the following bands:
4500–4800 MHz (downlink),
6725–7025 MHz (uplink),
10.7–10.95 GHz and 11.2–11.45 GHz (downlink),
12.15–13.25 GHz (uplink).

—Procedures which permit requirements other than those appearing in the allotment plan to be satisfied.

Each allotment consists of:

—An orbital position on a predetermined arc.
—A bandwidth in the bands mentioned.
—A service area.

The procedures associated with the plan permit an orbital position to be modified within the limits of the predetermined arc in such a way as to give more flexibility to the plan. They include the possibility of using a national allotment for a regional system (serving several neighbouring countries).

2.8.1.5 *Interference with terrestrial systems*

Most of the frequency bands allocated to space radiocommunication are also allocated on a shared basis to terrestrial radiocommunication. To facilitate this sharing, a number of provisions has been introduced into the *Radiocommunication Regulations* [RR, Articles 27 and 28] and a co-ordination procedure has been instituted between earth and terrestrial stations [RR, Article 11].

Four types of interference between systems can be distinguished:

—A satellite interfering with a terrestrial station.
—A terrestrial station interfering with a satellite.
—An earth station interfering with a terrestrial station.
—A terrestrial station interfering with an earth station.

Figure 2.26 illustrates the geometry associated with these forms of interference. The provisions intended to reduce them are numerous. The most evident are as follows:

(a) Limitation of the power flux density produced on the Earth's surface by satellites [RR, Article 28, Section IV]; see also [CCIR Report 387 and Recommendation 358]
(b) Limitation of the EIRP emitted by terrestrial stations in the direction of the orbit of geostationary satellites [RR, Article 27, Section II]; see also [CCIR Recommendation 406 and Reports 393 and 790]
(c) Limitation of the minimum elevation angle of an earth station antenna [RR, Article 28, Section III]
(d) Limitation of the EIRP of the earth station on the horizon [RR, Article 28, Section II]; see also [CCIR, Report 386]
(e) Limitation of off-axis EIRP density levels from earth stations [CCIR Recommendation 524]
(f) Use of energy dispersion techniques for analogue transmission using angular modulation and digital transmission in the fixed satellite service [CCIR Recommendation 446]
(g) The co-ordination procedure which governs every earth and terrestrial station installation [RR, Article 11, Appendix 28]; see also [CCIR Recommendation 359 and Report 382]
(h) The station-keeping conditions [RR, Article 29, Section III] [CCIR Recommendation 484] and the orbital spacing between satellites [CCIR Report 453 and 559]

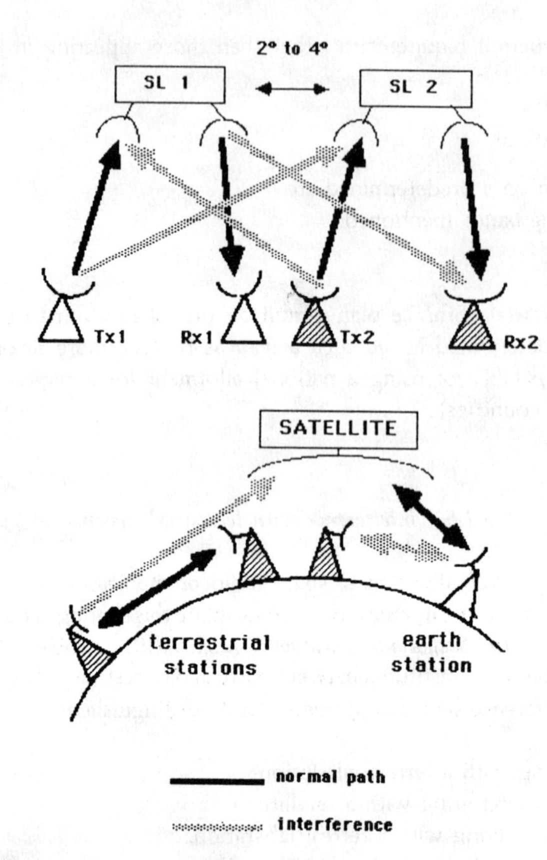

Figure 2.26 The geometry of interference between systems.

(i) The specifications for the radiation diagram of earth station antennas [CCIR Recommendations 465 and 580, Reports 390 and 391] and those of the satellite [CCIR Report 558 (fixed satellite service) and 810 (broadcast satellite)].

For more complete information it is useful to consult Volumes IV and IX (second part) of the CCIR.

2.8.2 Operational constraints

These constraints relate to:

—Realisation of a C/N_0 ratio greater than or equal to a specified value for a given percentage of the time.

—Provision of an adequate satellite antenna beam for coverage of the service area (uplink and downlink); this imposes the value of satellite antenna gain.

—The level of interference between satellite systems; orbital separation between satellites operating in identical frequency bands may be as low as a few degrees. Under these conditions it is important that the earth station antenna produces a beam of sufficiently small angular width and with sufficiently small secondary lobes. This avoids emission of excessively large signals towards an adjacent satellite or reception of signals from this satellite which interfere excessively with the required signal. However, the size of the antenna should not be too large; otherwise, considering the station-keeping tolerances of the satellite, the satellite will move significantly within the principal lobe. In the absence of a costly tracking system this would involve large variations in antenna gain.

—The total cost should be minimal.

The first of these constraints implies a minimum value of the product EIRP × G/T for each link (up and down). The two following constraints limit the degree of exchange between EIRP and G/T for all pairs of values giving the minimum value of their product:

—On the uplink, the noise temperature of the satellite is influenced to a large extent by the high noise temperature of the earth and, taking account of the constraint on coverage, the G/T of the satellite can hardly be significant (it is of the order of plus or minus a few dB(K^{-1})). It is up to the ground station to ensure a sufficient EIRP and, taking account of antenna constraints, design flexibility resides above all in the output power of the transmitting amplifier.

—On the downlink, the output power of the amplifier used is generally limited by amplifier technology and by the size of the platform which limits primary power generation. Taking account of the coverage constraint on the satellite antenna, the EIRP of the satellite is limited (of the order of a few tens of dB(W)). It is necessary to compensate with a high ground station G/T and, taking account of the antenna constraint, design flexibility resides above all in the receiver noise temperature.

At this stage the various constraints remain sufficiently flexible for several technical solutions to be envisaged. An attempt is made to choose the most economic solution in order to satisfy the last constraint to be stated, that of cost.

2.9 SIGNAL-TO-NOISE RATIO FOR A STATION-TO-STATION LINK (SINGLE ACCESS)

Section 2.5 presents an expression for the signal-to-noise ratio C/N_0 for each of the two links (up and down) which make up the station-to-station link. It now remains to establish the expression for the ratio C/N_0 for the total station-to-station link (Figure 2.27).

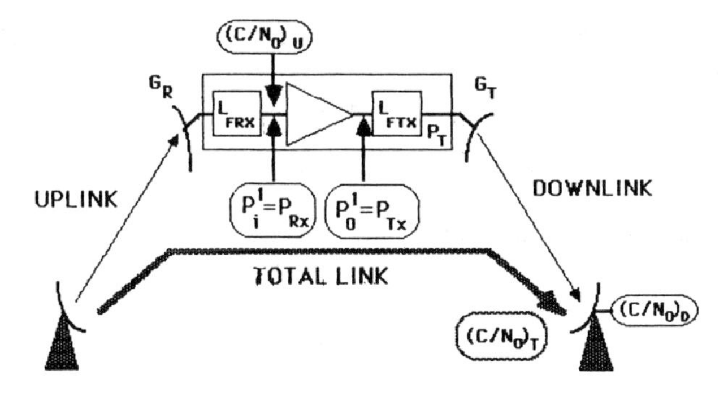

Figure 2.27 Complete station-to-station link.

The following notation is used:

—$(C/N_0)_U$ for the value of signal-to-noise ratio on the uplink.
—$(C/N_0)_D$ for the value of signal-to-noise ratio on the downlink.
—$(C/N_0)_T$ for the value of signal-to-noise ratio on the total link.

2.9.1 Repeater modelling

The repeater has two functions:

—to amplify the power of the received signal,
—to change the frequency of the carrier.

A conventional repeater is considered here, without demodulation of the carrier. The regenerative satellite case is examined in Chapter 6. Only the case of a single link between two earth stations is considered here and consequently the repeater supports only a single carrier (single carrier operation). The case of multicarrier operation is examined in Chapter 4.

2.9.1.1 *The transfer characteristic*

From an energy point of view, which is that of the link budget, frequency changing is transparent. Modelling the repeater (Figure 2.28), therefore, involves consideration only of the transfer characteristic between the power at the input of the satellite receiver, denoted by (P_i^1), with $(P_i^1) = P_{RX}$, and the power at the output of the transmitting amplifier, denoted by (P_o^1), with $(P_o^1) = P_{TX}$. The index *i* signifies the input, the index o signifies the output and the index 1 indicates single carrier operation. Observe the nonlinear nature of this characteristic; the output power increases with input power until it reaches a value $(P_i^1)_{sat}$ for which the output power passes through a maximum called the 'saturation output power', denoted by $(P_o^1)_{sat}$. The notation 'sat' stands for saturation.

2.9.1.2 *Input and output back-off*

The output power at saturation is that commonly referred to in connection with the power of the transmitting amplifier. The input and output powers are normalised to their saturation values; hence:

$$\text{IBO} = (P_i^1)/(P_i^1)_{sat}, \qquad \text{OBO} = (P_o^1)/(P_o^1)_{sat}$$

IBO is the input back-off and OBO is the output back-off.

2.9.1.3 *Power gain*

As the repeater is non-linear, the power gain G_{SL} (SL for satellite) depends on the operating

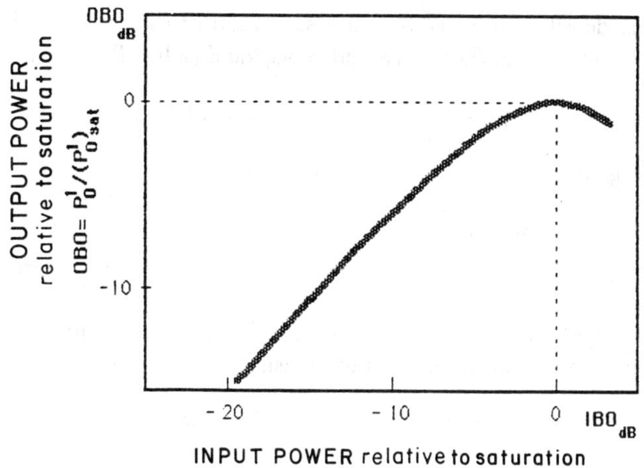

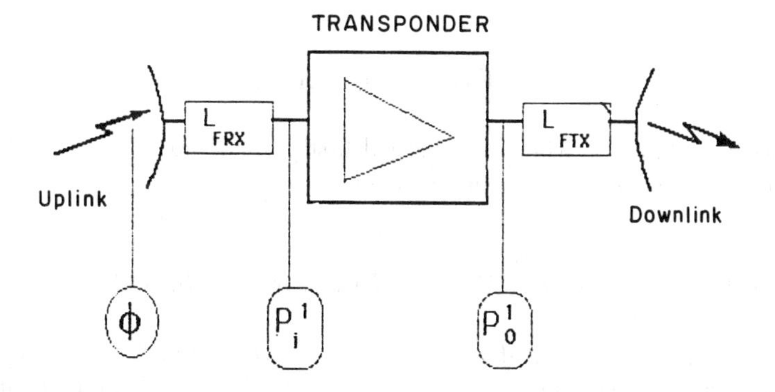

Figure 2.28 Power transfer characteristic of a satellite transponder (single carrier operation).

point. The saturation gain $(G_{sat})_{SL}$ of the repeater is defined as:

$$(G_{sat})_{SL} = (P_o^1)_{sat}/(P_i^1)_{sat} \tag{2.52}$$

At any operating point the power gain G_{SL} of the repeater has a value:

$$G_{SL} = (P_o^1)/(P_i^1) = \mathrm{OBO}(P_o^1)_{sat}/\mathrm{IBO}(P_i^1)_{sat} = \mathrm{OBO}/\mathrm{IBO}(G_{sat})_{SL} \tag{2.53}$$

2.9.1.4 *The relation between gain, EIRP and power flux density at saturation*

The repeater is characterised by two quantities as follows:

—The saturation EIRP in single carrier operation $(\mathrm{EIRP}_{sat})_{SL}$. Using the notation of Section 2.3 and Figure 2.6 and replacing P_{TX} by $(P_o^1)_{sat}$, it is defined as:

$$(\mathrm{EIRP}_{sat})_{SL} = (P_o^1)_{sat} G_{Tmax}/L_T L_{FTX} \qquad \mathrm{(W)} \tag{2.54}$$

—The power flux density necessary to cause saturation of the repeater $(\Phi_{sat})_{SL}$. Using the notation of Section 2.3 and Figure 2.8 and replacing P_{RX} by $(P_i^1)_{sat}$, it is defined as:

$$(\Phi_{sat})_{SL} = P_R/A_{Reff} = (P_i^1)_{sat} L_{FRX} L_R L_{POL} (4\pi/\lambda_U^2)/G_{Rmax} \quad (W/m^2) \quad (2.55)$$

The saturation gain of the repeater is then given by:

$$\begin{aligned}(G_{sat})_{SL} &= (P_o^1)_{sat}/(P_i^1)_{sat} \\ &= (L_T L_{FTX} L_{FRX} L_R L_{POL}/G_{Tmax} G_{Rmax})(4\pi/\lambda_U^2)[(\text{EIRP}_{sat})_{SL}/(\Phi_{sat})_{SL}] \quad (2.56)\end{aligned}$$

Notice that the input back-off IBO can also be expressed as the ratio of the actual power flux density and the saturation power flux density:

$$\text{IBO} = (P_i^1)/(P_i^1)_{sat} = (\Phi)_{SL}/(\Phi_{sat})_{SL} \quad (2.57)$$

Hence the transfer characteristic for single carrier operation given in Figure 2.28 shows both the relationship between normalised input and output power and between normalised input and output power flux density.

2.9.2 Expression for $(C/N_0)_T$

The expression for $(C/N_0)_T$ will be established firstly in the absence of interference from other systems and then allowing for interference.

2.9.2.1 *Expression for $(C/N_0)_T$ without interference from other systems*

The power of the signal received at the input of the earth station receiver is C_D. The noise at the input of the earth station receiver corresponds to the sum of the following:

—The downlink noise considered in isolation (the noise temperature of the receiving system T_D as in expression 2.24) which defines the ratio C/N_0 for the downlink $(C/N_0)_D$ and can be calculated as in the example of Section 2.5.8 with $(N_0)_D = kT_D$.
—The uplink noise retransmitted by the satellite.
Hence:

$$(N_0)_T = (N_0)_D + G(N_0)_U \quad (W/Hz) \quad (2.58)$$

where $G = G_{SL} G_T G_R / L_{FTX} L_D L_{FRX}$ is the total power gain between the satellite receiver input and the earth station receiver input. This is formed by the contribution of the repeater gain G_{SL}, the gain of the satellite transmitter G_T/L_{FTX}, the attenuation on the downlink L_D and the gain of the earth station receiver G_R/L_{FRX}.
This gives:

$$\begin{aligned}(C/N_0)_T^{-1} &= (N_0)_T/C_D \\ &= [(N_0)_D + G(N_0)_U]/C_D = (N_0)_D/C_D + (N_0)_U/G^{-1}C_D \quad (Hz^{-1}) \quad (2.59)\end{aligned}$$

In the above expression the term $G^{-1}C_D$ represents the signal power at the satellite receiver input. Hence $(N_0)_U/G^{-1}C_D = (C/N_0)_U^{-1}$. Finally:

$$(C/N_0)_T^{-1} = (C/N_0)_U^{-1} + (C/N_0)_D^{-1} \qquad (\mathrm{Hz}^{-1}) \tag{2.60}$$

In this expression:

$$\begin{aligned}(C/N_0)_U &= (P_i^1)/(N_0)_U = \mathrm{IBO}(P_i^1)_{sat}/(N_0)_U \\ &= \mathrm{IBO}(P_o^1)_{sat}/(G_{sat})_{SL}(N_0)_U \\ &= \mathrm{IBO}(C/N_0)_{U,sat} \qquad (\mathrm{Hz}) \\ (C/N_0)_D &= \mathrm{OBO}(\mathrm{EIRP}_{sat})_{SL}(1/L_D)(G/T)_{ES}(1/k) \\ &= \mathrm{OBO}(C/N_0)_{D,sat} \qquad (\mathrm{Hz})\end{aligned}$$

$(C/N_0)_{U,sat}$ and $(C/N_0)_{D,sat}$ are the values of C/N_0 for the uplink and downlink when the repeater operates at saturation. L_D represents the attenuation on the downlink and is given by (2.11) and $(G/T)_{ES}$ the figure of merit of the earth station.

2.9.2.2 *Expression for $(C/N_0)_T$ allowing for interference created by other systems*

The geometry relating to interference has been given in Figure 2.26. The signals emitted by other systems are superimposed on the useful signals of the station-to-station link at two levels:

—At the input of the repeater on the uplink.
—At the input of the receiver on the downlink.

The effect of interference is similar to an increase of the thermal noise on the link affected by interference. It is allowed for in the equations in the form of an increase of the spectral density:

$$N_0 = (N_0)_{\text{without interference}} + (N_0)_I \qquad (\mathrm{W/Hz}) \tag{2.61}$$

where $(N_0)_I$ represents the increase of the noise power spectral density due to interference. A ratio $(C/N_0)_I$ which expresses the signal power in relation to the spectral density of the interference can be associated with $(N_0)_I$; these are $(C/N_0)_{I,U}$ for the uplink and $(C/N_0)_{I,D}$ for the downlink. This leads to modification of equation (2.60) by replacing $(C/N_0)_U$ and $(C/N_0)_D$ by the following expressions:

$$\begin{aligned}(C/N_0)_U^{-1} &= [(C/N_0)_U^{-1}]_{\text{without interference}} + (C/N_0)_{I,U}^{-1} \qquad (\mathrm{Hz}^{-1}) \\ (C/N_0)_D^{-1} &= [(C/N_0)_D^{-1}]_{\text{without interference}} + (C/N_0)_{I,D}^{-1} \qquad (\mathrm{Hz}^{-1})\end{aligned} \tag{2.62}$$

The total expression becomes:

$$(C/N_0)_T^{-1} = (C/N_0)_U^{-1} + (C/N_0)_D^{-1} + (C/N_0)_I^{-1} \qquad (\mathrm{Hz}^{-1}) \tag{2.63}$$

where $(C/N_0)_U$ and $(C/N_0)_D$ are the values appearing in expression (2.60) and:

$$(C/N_0)_I^{-1} = (C/N_0)_{I,U}^{-1} + (C/N_0)_{I,D}^{-1} \qquad (\mathrm{Hz}^{-1}) \tag{2.64}$$

2.10 EXAMPLE

It is required to establish a link between two earth stations by means of a satellite (Figure 2.27). The data are as follows:

—Uplink frequency: $f_U = 14\,\text{GHz}$.
—Downlink frequency: $f_D = 12\,\text{GHz}$.
—Attenuation of the downward path: $L_D = 206\,\text{dB}$.
—For the satellite:

—Power flux density required to saturate the repeater:

$$(\Phi_{sat})_{SL} = -90\,\text{dB(W/m}^2)$$

—Gain on the axis of the receiving antenna: $G_{Rmax} = 30\,\text{dB}$
—Figure of merit: $(G/T)_{SL} = 3.4\,\text{dB(K}^{-1})$
—Repeater characteristic (single carrier operation):

$$\text{OBO(dB)} = \text{IBO(dB)} + 6 - 6\exp[\text{IBO(dB)}/6]$$

—Equivalent isotropic radiated power at saturation:

$$(\text{EIRP}_{sat})_{SL} = 50\,\text{dB(W)}$$

—Gain on the axis of the transmitting antenna: $G_{Tmax} = 40\,\text{dB}$

The following losses are neglected:
—Transmission and reception: $L_{FRX} = L_{FTX} = 0\,\text{dB}$
—Polarisation mismatching $L_{POL} = 0\,\text{dB}$
—Unfavourable geography and misalignment: $L_R = L_T = 0\,\text{dB}$
—For the earth station:
—Figure of merit: $(G/T)_{ES} = 25\,\text{dB(K}^{-1})$

It is assumed that there is no interference.

2.10.1 Calculation of the repeater gain at saturation $(G_{sat})_{SL}$

$$(G_{sat})_{SL} = (P_o^{\,1})_{sat}/(P_i^{\,1})_{sat}$$

From (2.54):

$$(P_o^{\,1})_{sat} = (\text{EIRP}_{sat})_{SL} L_T L_{FTX}/G_{Tmax} \qquad \text{(W)}$$

Hence:

$$(P_o^{\,1})_{sat} = 50\,\text{dB(W)} - 40\,\text{dB} = 10\,\text{dB(W)} = 10\,\text{W}$$

From (2.55)

$$(P_i^{\,1})_{sat} = (\Phi_{sat})_{SL} G_{Rmax}/L_{FRX} L_R L_{POL}(4\pi/\lambda_U^2) \qquad \text{(W)}$$

hence:

$$(P_i^1)_{sat} = -90\,\text{dB(W/m}^2) + 30\,\text{dB} - 44.4\,\text{dB (m}^2) = -104.4\,\text{dB (W)} = 36\,\text{pW}$$

$$(G_{sat})_{SL} = (P_o^1)_{sat}/(P_i^1)_{sat} = 10\,\text{dB (W)} - (-104.4\,\text{dB(W)}) = 114.4\,\text{dB}$$

2.10.2 Calculation of C/N_0 for the up- and downlinks and the total link when the repeater operates in saturation

From (2.60):

$$(C/N_0)_{\mathrm{U,sat}} = (P_{\mathrm{i}}^{1})_{\mathrm{sat}}/k\,T_{\mathrm{SL}} = (P_{\mathrm{i}}^{1})_{\mathrm{sat}}(G/T)_{\mathrm{SL}}/kG_{\mathrm{Rmax}}$$

$$(C/N_0)_{\mathrm{U,sat}} = -104.4 + 3.4 - (-228.6) - 30 = 97.6\ \mathrm{dB\,(Hz)}$$

$$(C/N_0)_{\mathrm{D,sat}} = (\mathrm{EIRP}_{\mathrm{sat}})_{\mathrm{SL}}(1/L_{\mathrm{D}})(G/T)_{\mathrm{ES}}(1/k) \qquad (\mathrm{Hz})$$

$$(C/N_0)_{\mathrm{D,sat}} = 50 - 206 + 25 - (-228.6) = 97.6\ \mathrm{dB\,(Hz)}$$

$$(C/N_0)^{-1}_{\mathrm{T,sat}} = (C/N_0)^{-1}_{\mathrm{U,sat}} + (C/N_0)^{-1}_{\mathrm{D,sat}} \qquad (\mathrm{Hz}^{-1})$$

$$(C/N_0)_{\mathrm{T,sat}} = 94.6\ \mathrm{dB\,(Hz)}$$

2.10.3 Calculation of the input back-off IBO and the output back-off OBO to achieve $(C/N_0)_{\mathrm{T}} = 80$ dB (Hz) and the corresponding values of $(C/N_0)_{\mathrm{U}}$ and $(C/N_0)_{\mathrm{D}}$

One must have:

$$(C/N_0)^{-1}_{\mathrm{U}} + (C/N_0)^{-1}_{\mathrm{D}} = 10^{-8}\,\mathrm{Hz}^{-1}$$

Hence:

$$\mathrm{IBO}^{-1}(C/N_0)^{-1}_{\mathrm{U,sat}} + \mathrm{OBO}^{-1}(C/N_0)^{-1}_{\mathrm{D,sat}} = 10^{-8}\,\mathrm{Hz}^{-1}$$

This gives:

$$10^{-\mathrm{IBO(dB)}/10} + 10^{-\mathrm{OBO(dB)}/10} = 10^{1.76}$$

with:

$$\mathrm{OBO\,(dB)} = \mathrm{IBO\,(dB)} + 6 - 6\exp(\mathrm{IBO(dB)}/6)$$

Numerical solution gives:

$$\mathrm{IBO} = -16.4\ \mathrm{dB}$$

$$\mathrm{OBO} = -10.8\ \mathrm{dB}$$

Hence:

$$(C/N_0)_{\mathrm{U}} = \mathrm{IBO}(C/N_0)_{\mathrm{U,sat}} = -16.4\ \mathrm{dB} + 97.6\ \mathrm{dB\,(Hz)} = 81.2\ \mathrm{dB\,(Hz)}$$

$$(C/N_0)_{\mathrm{D}} = \mathrm{OBO}(C/N_0)_{\mathrm{D,sat}} = -10.8\ \mathrm{dB} + 97.6\ \mathrm{dB\,(Hz)} = 86.8\ \mathrm{dB\,(Hz)}$$

2.10.4 Value of $(C/N_0)_T$ when produced under the effect of rain causing an attenuation of 6 dB on the uplink

The attenuation of 6 dB on the uplink increases the input back-off by 6 dB. The new value of IBO becomes:

$$\text{IBO (dB)} = -16.4\text{ dB} - 6\text{ dB} = -22.4\text{ dB}$$

The new value of output back-off corresponding to this is:

$$\text{OBO (dB)} = \text{IBO (dB)} + 6 - 6\exp(\text{IBO(dB)}/6) = -16.5\text{ dB}$$

Hence:

$$(C/N_0)_U = \text{IBO}(C/N_0)_{U,sat} = -22.4\text{ dB} + 97.6\text{ dB (Hz)} = 75.2\text{ dB (Hz)}$$

$$(C/N_0)_D = \text{OBO}(C/N_0)_{D,sat} = -16.5\text{ dB} + 94.6\text{ dB (Hz)} = 81.1\text{ dB (Hz)}$$

and, from (2.60)

$$(C/N_0)_T = 74.2\text{ dB (Hz)}$$

To regain the required value $(C/N_0)_T = 80$ dB (Hz), it is necessary to increase the $(\text{EIRP})_{ES}$ of the transmitting earth station by 6 dB.

2.10.5 Value of $(C/N_0)_T$ when produced under the effect of rain causing an attenuation of 6 dB on the downlink with a reduction of 2 dB in the figure of merit of the earth station due to the increase of antenna noise temperature

The value of $(C/N_0)_U$ remains unchanged, hence: $(C/N_0)_U = 81.2$ dB (Hz).

The value of $(C/N_0)_D$ loses 8 dB, hence: $(C/N_0)_D = 86.8$ dB (Hz) $-$ 8 dB $= 78.8$ dB (Hz). From which: $(C/N_0)_T = 76.8$ dB (Hz).

To regain the required value $(C/N_0)_T = 80$ dB (Hz), it is necessary to increase the $(\text{EIRP})_{ES}$ of the transmitting earth station in such a way that the value of IBO satisfies the equation:

$$\text{IBO}^{-1}(C/N_0)^{-1}_{U,sat} + \text{OBO}^{-1}(C/N_0)^{-1}_{D,sat} = 10^{-8}\text{ Hz}^{-1}$$

in which:

$$(C/N_0)_{U,sat} = 97.6\text{ dB (Hz)}$$

$$(C/N_0)_{D,sat} = 97.6\text{ dB (Hz)} - 8\text{ dB} = 89.6\text{ dB (Hz)}$$

This gives:

$$\text{IBO} = -13\text{ dB}$$

$$\text{OBO} = -7.7\text{ dB}$$

It is necessary to increase the $(\text{EIRP})_{ES}$ of the earth station transmission by -13 dB $-$ $(-16.4\text{ dB}) = 3.4$ dB.

REFERENCES

[BAR-88] S.K. Barton, S.E. Dinwiddy (1988) A technique for estimating the throughput of adaptative TDMA fade countermeasure systems, *International Journal of Satellite Communications*, **6**, No. 3, pp. 331–341.

[CAR-88] F. Carassa, G. Tartara (1988) Frequency diversity and its applications, *International Journal of Satellite Communications*, **6**, No. 3, pp. 313–322.

[GHO-88] A. Ghorbani, N.J. McEwan (1988) Propagation theory in adaptative cancellation of cross-polarization, *International Journal of Satellite Communications*, **6**, No. 1, pp. 25–28.

[HOR-88] J. Horle (1988) Uplink power control of satellite earth station as a fade countermeasure of 20/30 GHz communications systems, *International Journal of Satellite Communications*, **6**, No. 3, pp. 323–330.

[MAR-86] G. Maral, M. Bousquet (1986) *Satellite Communications Systems*, Wiley.

[MCM-86] G.R. McMillen, B.A. Mazur, T. Abdel-Nabi (1986) Design of a selective FEC subsystem to counteract rain fading in Ku-Band TDMA systems, *International Journal of Satellite Communications*, **4**, No. 2, pp. 75–82.

[NJO-85] E.G. Njoku, E.K. Smith (1985) Microwave antenna temperature of the Earth from geostationary orbit, *Radio Science*, **20**, No. 3, pp. 591–599.

[PRA-86] T. Pratt, C.W. Bostian (1986) *Satellite Communications*, Wiley.

[ROG-86] D.V. Rogers, J.E. Allnutt (1986) System implications of 14/11 GHz path depolarization. Part 1: predicting the impairments, *International Journal of Satellite Communications*, **4**, No. 1, pp. 1–11.

[STR-83] S.J. Struharik (1983) Rain and ice depolarization measurements at 4 GHz in Sitka Alaska, *COMSAT Technical Review*, **13**, No. 2, pp. 403–435, Fall.

[THO-83] R.W. Thorn, J. Thirlwell, D.J. Emerson (1983) Slant path radiometer measurements in the range 11–30 GHz at Martlesham Heath, England, *3rd International Conference of Antennas and Propagation, ICAP 83*, pp. 156–161.

[THU-87] M. Thue (1987) Organisations Internationales dans le domaine des Télécommunications, *Note Technique NT/DICET/AIN/102*, Centre National D'etudes des Télécommunications, Nov.

[WIL-88] M.J. Willis, B.G. Evans (1988) Fade countermeasures at Ka band for Olympus, *International Journal of Satellite Communications*, **6**, No. 3, pp. 301–311.

[WIT-88] D.J. Withers (1988) Space radio services and WARC-MOB-87, *International Journal of Satellite Communications*, **6**, No. 1, pp. 25–28.

[YOS-86] Y. Hosoya, H. Fukushi, K. Satoh, S. Tsuchiya, Y. Otsu (1986) Propagation characteristics, *IEEE Transactions on Aerospace and Electronic Systems, Special Issue on Japan's CS Communications Satellite Experiments*, **AES-22**, No. 3, pp. 255–263.

3 TRANSMISSION TECHNIQUES FOR A SATELLITE CHANNEL

This chapter deals with techniques which enable signals to be sent from one user to another. In this chapter, the term 'signal' relates to the voltage representing the information transmitted from one user to another (such as telephone, television, telex, data, and so on). Such a signal is called a 'baseband' signal. If the baseband signal is analogue, the voltage which represents it can take any value within a given range. If the signal is digital, the voltage takes discrete values, of which there are a finite number, within a given range.

In all cases, the baseband signal modulates the carrier discussed in the previous chapter in order to access the radio-frequency channel for routing via the satellite. Before modulating the carrier, the signal is generally subjected to specific processing.

The following will be examined in succession:

—Analogue transmission of telephone and television signals by satellite.
—Digital transmission of telephone, radio broadcast and data signals by satellite.

3.1 SIGNAL CHARACTERISTICS

The following baseband signals will be considered:

—Speech on a telephone channel.
—Television.
—Broadcast radio programmes.
—Data.

These are the most common. Telex and facsimile signals transmitted on telephone channels have different characteristics from those of the voice and the results presented below do not strictly apply.

3.1.1 Telephone channel signals

A telephone channel signal occupies a band from 300 Hz to 3400 Hz. The test tone for the channel is a pure sinusoid at a frequency of 800 Hz (CCITT) or 1000 Hz (USA). Its power at the point of zero relative level in any telephone channel with a reference impedance of 600 Ω is 1 mW or 0 dBm0 (the suffix 0 indicates that the value expressed in dBm is referenced to the zero relative level point). The maximum energy of a signal representing speech is in the region of 800 Hz and 99% of the energy is situated below 3000 Hz. The signal power of an 'average talker' (with the averaging performed over a sufficient number of talkers) relative to the zero relative level point is given by [JON-84], [LAB-84]:

$$P_m = P_a + 0.115\,\sigma^2 + 10 \log \tau \qquad (\text{dBm}_0) \tag{3.1}$$

where $P_a = -12.9$ dBm0 represents the average power of the speech signal, $\sigma = 5.8$ dB is the standard deviation of the normal distribution of active speech power, $\tau = 0.25$ is the activity factor of a talker (this factor takes account of the periods of silence reserved for listening to the correspondent and pauses in the discussion). In total $P_m = -15$ dBm0 [CCITT Recommendation G 223].

3.1.2 Television signals [CCIR, Report 624]

Colour television standards are as follows: NTSC (Japan, USA, Canada, Mexico, some South American countries and Asia), PAL (Europe except France, Australia, other South American countries and some African countries) and SECAM (France, USSR, Eastern countries and other African countries). Recently, a new standard called MAC (Multiplexed Analogue Components) has been proposed for satellite broadcasting (direct broadcast television) [CCIR Reports 1073, 1074].

3.1.2.1 *Luminance and chrominance components*

The television signal contains three components: the luminance signal which represents the image in black and white, the chrominance signal which represents the colour and the sound. Figure 3.1 shows how the luminance and chrominance signals are generated; the television camera produces three voltages E_R, E_B and E_G representing the red (R), blue (B) and green (G) components of the colour at one point of the scanned image (525 lines per frame and 60 fields per second—that is 30 images per second—in the NTSC standard, 625 lines per frame and 50 fields per second—that is 25 images per second—in the PAL and SECAM standards). These signals are filtered by the gamma filters in order to compensate for the non-linear response of the receiving cathode-ray tube and then combined to generate the luminance signal E_{Y^*} defined by:

$$E_{Y^*} = 0.3E_{R^*} + 0.59E_{G^*} + 0.11E_{B^*} \tag{3.2}$$

and the two components of the chrominance signal $E_{R^*} - E_{Y^*}$ and $E_{B^*} - E_{Y^*}$ which contain the information required to reconstruct the components of the original colour signal.

3.1.2.2 *NTSC, PAL and SECAM colour television signals*

The composite video signal is formed by summing the E_{Y^*} signal and a sub-carrier modulated by the two components of the chrominance signal (Figure 3.1). A sub-carrier modulated by

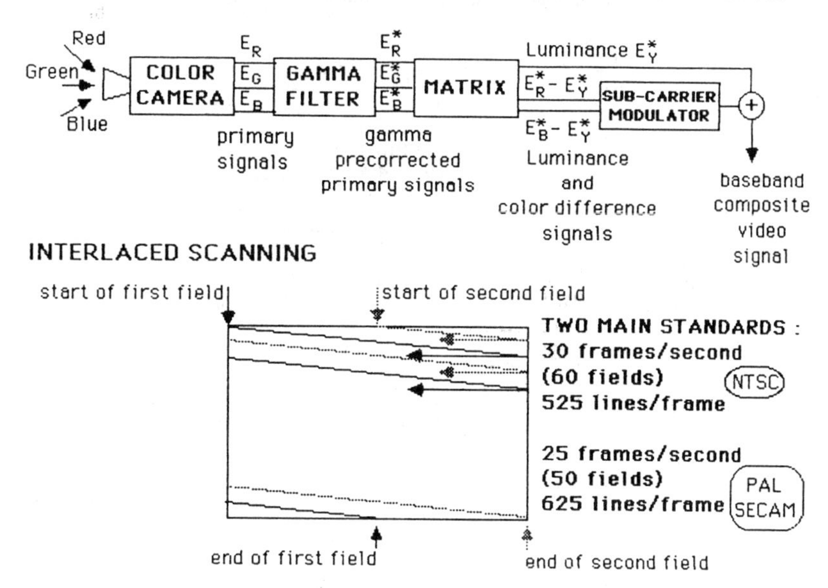

Figure 3.1 Generation of television signals.

the sound is added to the composite video signal. The modulation techniques depend on the particular system. Figure 3.2 shows the spectrum of a television signal.

3.1.2.3 *The MAC colour television signal [CCIR Reports 1073, 1074]*

The composite video signal, described above, has the advantage of compatibility with a monochrome video signal but also certain disadvantages as follows:

—The receiver cannot separate the luminance and chrominance components perfectly. Hence, for example, rapid variations of luminance are interpreted as variations in colour and striations are introduced into the image. Reduction of these effects inevitably causes a loss of horizontal resolution.

—With satellite transmission using frequency modulation (see Section 3.6.6) the chrominance component is affected by noise to a greater extent.

—The sound does not have the quality of digital sound to which the public is now accustomed (as with digital recording and compact discs for example). The multiplexed analogue components (MAC) standard has the objective of remedying the above disadvantages while supporting extension to simultaneous broadcasting of several services and compatibility with future high definition television standards.

Figure 3.3 shows different MAC standards and indicates the formation of the signal which is transmitted at radio frequency:

—A-MAC: The radio-frequency carrier is modulated by a frequency division multiplex signal;

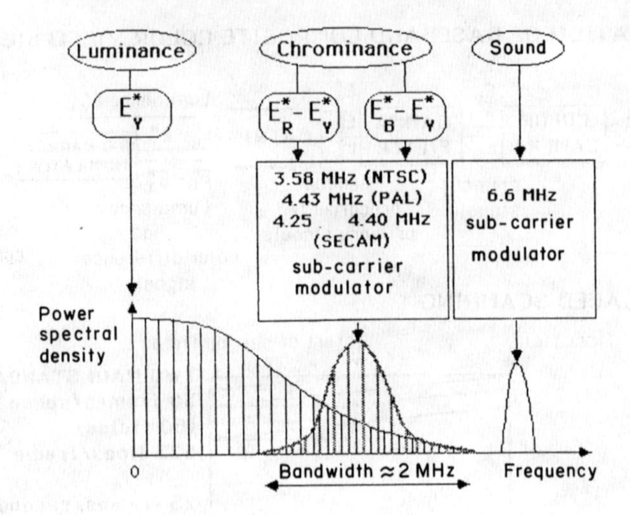

Figure 3.2 The spectrum of the composite television signal.

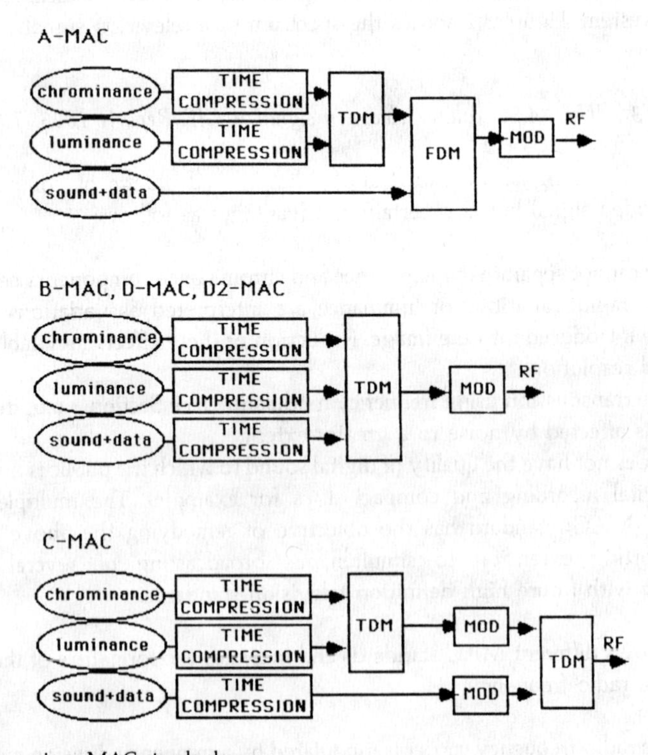

Figure 3.3 The various MAC standards.

this contains a digital signal (sound or data) and a signal formed by time division multiplexing the analogue luminance and chrominance signals compressed in time.

—B-MAC, D-MAC, D2-MAC: The radio-frequency carrier is modulated by a time division multiplexed signal consisting of a digital signal (time compressed sound or data) and time compressed analogue luminance and chrominance signals. The three standards differ in the digital coding format.

—C-MAC: The transmission is a time division multiplex of two carriers, one modulated by a time division multiplex of the time compressed analogue luminance and chrominance signals and the other by a digital signal (time compressed sound or data).

Figure 3.4 shows the time distribution of the components of a MAC signal.

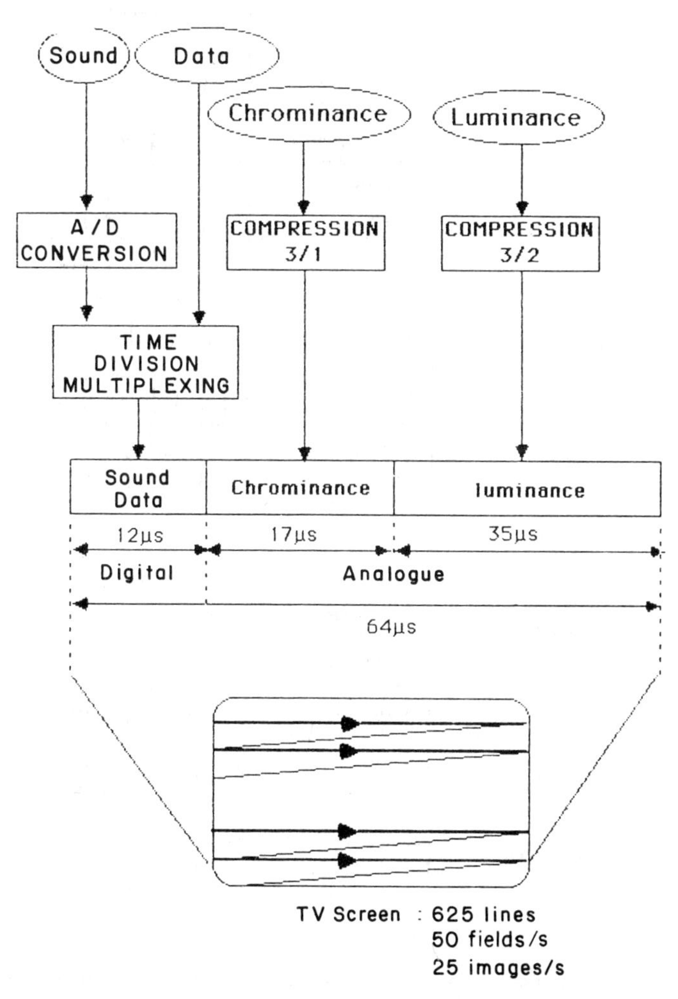

Figure 3.4 Allocation of one line period to the components of a MAC signal.

3.1.3 Radio broadcast programmes

A high quality radio broadcast programme occupies a band from 40 Hz to 15 kHz. The test signal is a pure sinusoid at a frequency of 1 kHz. Its power relative to the zero reference level for an impedance of 600 Ω is 1 mW or 0 dBm0s (the 's' suffix indicates that the value relates to the broadcast test signal). The mean power of an acoustic programme is −4 dBm0 and the peak power (exceeded for a fraction less than 10^{-5} of the time) is equal to 12 dBm0 [CCIR Report 491].

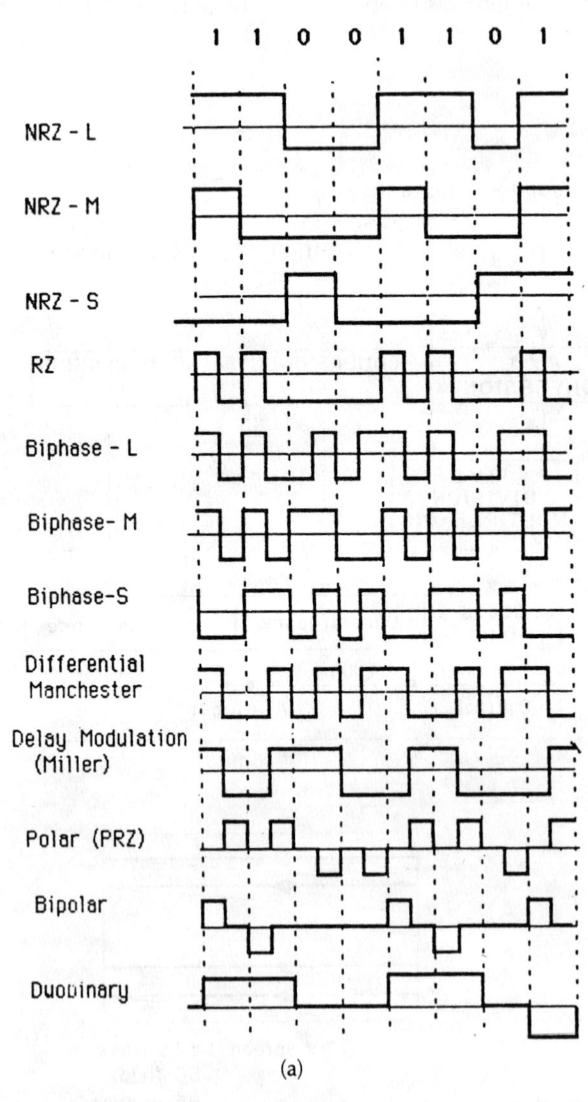

Figure 3.5 (a) Representation of a binary baseband message according to code. (b) Power spectral density according to the representation of a binary message.

3.1.4 Data

Different waveforms can be chosen for the voltage representing the binary element, or bit, to be transmitted. The choice of a particular representation depends on the spectral characteristics to be given to the baseband signal which consists of a time sequence of these waveforms. Figure 3.5a shows examples which correspond to the same binary message according to the chosen representation. Table 3.1a defines the representations of Figure 3.5a. Figure 3.5b shows the associated power spectral densities for a message containing an equal number of logic 1 and 0 states [STE-87], [TUG-82], [BIC-86].

3.2 A MODEL OF THE CHANNEL

Figure 3.6 shows the transmission channel from one user terminal to another. If the terminal is located at some distance from the station, it will be connected to it through a terrestrial network (Figure 3.6a). This is the case for large stations which are connected to the terrestrial network by means of a station/network interface. For small stations (VSAT), it can be expected that the station and terminal will be at the same location. There is, therefore, only a station/terminal interface (Figure 3.6b). Between the station/network or station/terminal interface and the transmitting antenna is the earth station equipment which provides the baseband signal processing functions, intermediate frequency (IF) modulation and conversion to radio frequency (IF/RF). The inverse operations take place at the receiving station.

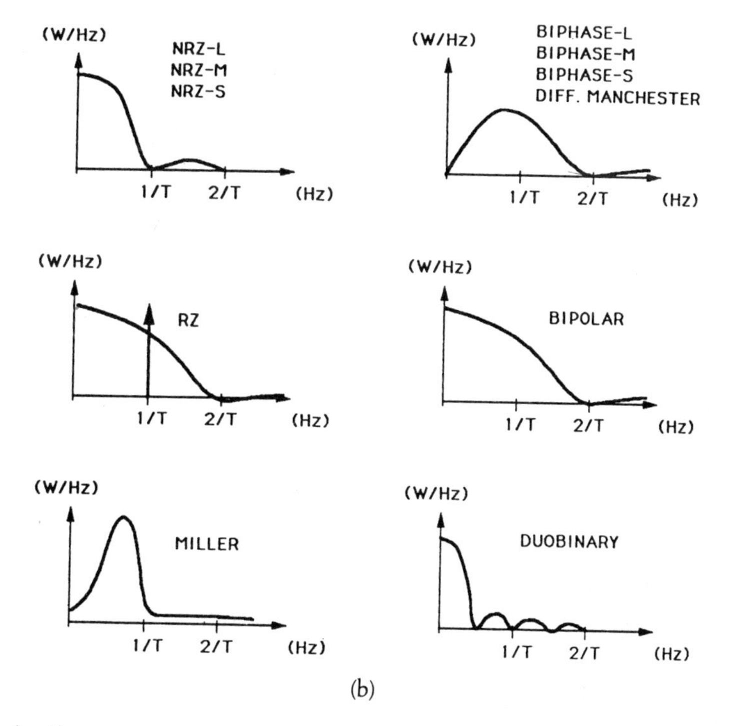

Figure 3.5 *(cont.)*

Table 3.1a Definition of the representations of a binary message

Designation	Definition
NON-RETURN TO ZERO (NRZ-L)	1 = High level 0 = Low level
NON-RETURN TO ZERO-MARK (NRZ-M)	1 = Transition at the start of the interval 0 = No transition
NON-RETURN TO ZERO-SPACE (NRZ-S)	1 = No transition 0 = Transition at the start of the interval
RETURN TO ZERO (RZ)	1 = Pulse in the first half of the interval 0 = No pulse
BIPHASE-LEVEL (MANCHESTER)	1 = Falling transition at the centre of the interval 0 = Rising transition at the centre of the interval
BIPHASE-MARK	Always a transition at the start of the interval 1 = Transition at the centre of the interval 0 = No transition at the centre of the interval
BIPHASE-SPACE	Always a transition at the start of the interval 1 = No transition at the centre of the interval 0 = Transition at the centre of the interval
DIFFERENTIAL MANCHESTER	Always a transition at the centre of the interval 1 = No transition at the start of the interval 0 = Transition at the start of the interval
DELAY MODULATION (MILLER)	1 = Transition at the centre of the interval 0 = No transition if following a 1 No transition at the end of the interval if followed by a 1
POLAR (PRZ)	Always a transition at the centre of the interval 1 = Pulse of positive polarity 0 = Pulse of negative polarity
BIPOLAR	1 = pulse in the first part of the interval with alternate polarities from one pulse to the next 0 = No pulse
DUOBINARY	1 = Positive or negative level, opposite to the previous 1 if an odd number of 0s between the two 0 = No pulse

3.3 PERFORMANCE OBJECTIVES

According to the nature of the transmitted signal, the performance objectives have been fixed by the CCIR. The quality of the signals delivered to the user is defined at the station/network interface level (Figure 3.6a) or the station/terminal interface (Figure 3.6b) by:

—The ratio S/N = baseband signal power/baseband noise power when the signal is analogue.
—The bit error rate (BER) when the signal is digital.

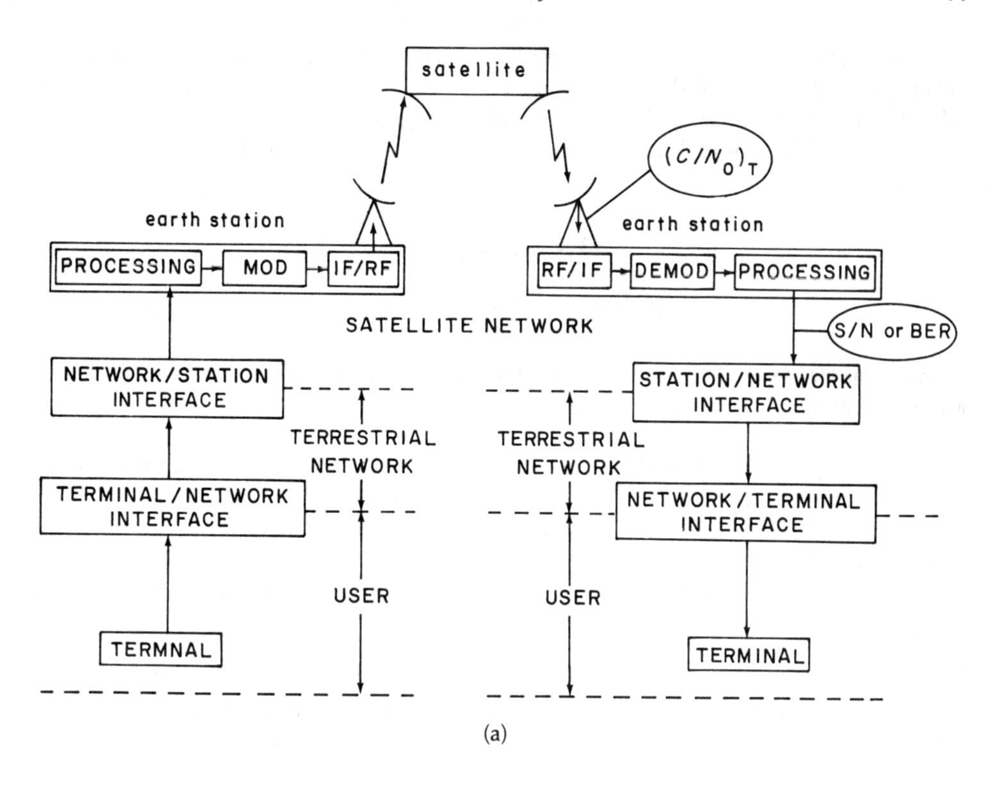

(a)

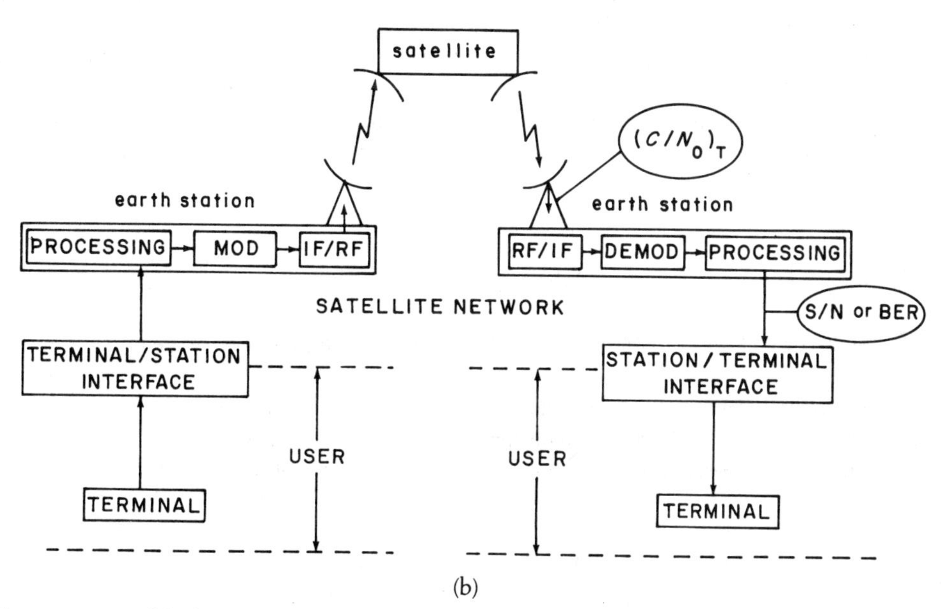

(b)

Figure 3.6 Model of a transmission channel from one terminal to another. (a) Connection by terrestrial network. (b) Direct connection (VSAT).

These two quantities S/N and BER depend on the value of $(C/N_0)_T$ as will be seen in Sections 3.6 and 3.7. Recall that $(C/N_0)_T$ quantifies the quality of the station-to-station link and is defined at the input of the earth station receiver. An expression for it as a function of the quality of the up- and downlinks is given in (2.63) of Chapter 2.

3.3.1 Telephone

3.3.1.1 *Analogue transmission*

CCIR Recommendation 353 stipulates that the noise power at a zero relative level point in any telephone channel must not exceed:

—10 000 pWOp psophometrically weighted one minute mean power, for more than 20% of any month,
—50 000 pWOp psophometrically weighted one minute mean power, for more than 0.3% of any month,
—1000 000 pWOp unweighted (with an integration time of 5 ms) for more than 0.01% of any year.

The noise power mentioned above is defined at a zero transmission level point where the test signal power has a value of 1 mW. The ratio S/N of the test signal power to the noise power is thus given by:

$$S/N = 10^9/N_{\mathrm{pWOp}} \tag{3.3}$$

where N_{pWOp} is the power mentioned in the above Recommendation. Consequently, the value 10 000 $\mathrm{pW_{Op}}$ corresponds to a S/N ratio = 50 dB.

The psophometrically weighted power (identified by the suffix p in pWOp) is that measured at the output of a psophometric filter whose gain is intended to reproduce the curve of ear sensitivity as a function of frequency [CCITT Recommendation G 223]. Figure 3.7 shows the gain of the psophometric filter and its effect on the noise. The established improvement in the S/N ratio is $w = 2.5$ dB.

3.3.1.2 *Digital transmission*

CCIR Recommendation 522 stipulates that the bit error rate must not exceed:

—one part in 10^6, 10-minute mean value for more than 20% of any month.
—one part in 10^4, 1-minute mean value for more than 0.3% of any month.
—one part in 10^3, 1-second mean value for more than 0.05% of any month.

3.3.2 Television

CCIR Recommendation 567 stipulates that the ratio of the nominal luminance signal amplitude to the mean square weighted noise in the 0.01 to 5 MHz band must be not less than 53 dB

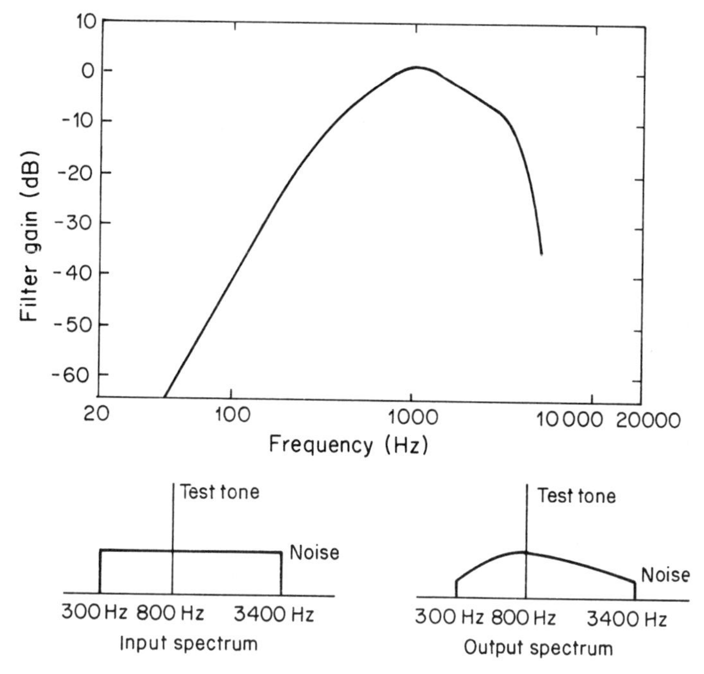

Figure 3.7 The gain of a psophometric filter and its effect on noise.

during more than 1% of any month and not less than 45 dB during more than 0.1% of any month. The weighting used (video measurement weighting) takes account of the sensitivity of the average viewer to the noise frequency components [CCIR Recommendation 567].

3.3.3 Radio broadcast programmes

CCIR Recommendation 505 stipulates that the noise power must not exceed $-42\,\mathrm{dB_{q0ps}}$ during more than 20% of any month (in the notation $\mathrm{dB_{q0ps}}$, the suffix p signifies that the noise measurement is weighted and the suffix q indicates the use of quasi-peak measuring equipment conforming to the specifications of CCIR Recommendation 468). An increase of 4 dB is accepted for 1% of the time and 12 dB for 0.1% of the time. When the broadcast programme is transmitted by satellite in digital form, the performance objective is stipulated in terms of error probability. These errors have the effect of generating audible 'clicks'. To limit their frequency to about one per hour, a binary error rate of the order of 10^{-9} is required [CCIR Report 648].

3.3.4 Data

CCIR Recommendation 614 stipulates that the bit error rate for satellite transmission of data at 64 kbit/s at a frequency below 15 GHz on a link which is part of an Integrated Services Digital Network (ISDN) must not exceed:

—10^{-7} during more than 10% of any month,

—10^{-6} during more than 2% of any month,
—10^{-3} during more than 0.03% of any month.

No standard has yet been produced for data rates other than 64 kbit/s.

3.4 AVAILABILITY OBJECTIVES

Availability is the fraction of time during which a service conforming to the specifications is provided. It is affected by both equipment breakdowns and propagation phenomena.

CCIR Recommendation 579 stipulates that the unavailability must not exceed:

—0.2% of the year in the case of breakdown (interruption of the service must be less than 18 hours per year)
—0.2% of any month if the service interruption is due to propagation.

The effects of propagation on the quality of the link have been examined in Chapter 2. Breakdown involves the earth station and satellite equipment. For earth stations, service interruption due to conjunction of the station, the satellite and the sun is regarded as a breakdown. As far as the satellite is concerned, it is necessary to consider its reliability, which is determined by breakdowns of on-board equipment, breaks during an eclipse when the only source of on-board energy is solar and the life time of the satellite. In general an operational satellite, a back-up satellite in orbit and a back-up satellite on the ground are provided. Finally, availability depends also on the reliability of the launchers which are indispensable for replacement of satellites at the end of their life. The approach to these problems is treated in more detail in Chapter 13.

3.5 PROPAGATION TIME

Returning to Figure 3.6, it can be seen that the propagation time of signals from one terminal to another is the sum of the propagation times on the space link (station-to-station) t_{SS} and the propagation times t_N on the network at departure and arrival.

3.5.1 Propagation time on the space link

The propagation time on the space link depends on the satellite orbit. It is least with satellites in low orbit, but the relative variations of propagation time are much greater than in the case of a geostationary satellite for which the variations are small although the propagation time is relatively long.

The propagation time on a space link is given by the following relation

$$t_{SS} = (R_U + R_D)/c \tag{3.4}$$

where R_U and R_D are the distances from the earth station to the satellite on the up- and downlinks respectively and c is the velocity of light ($c = 3 \times 10^8$ m/s). The distances R_U and

R_D are given by expression (7.40) of Chapter 7 and depend on the location of the stations with respect to the satellite at the time considered. For a geostationary satellite, a range of propagation times can be determined by considering two extreme cases:

(1) A vertical trajectory: $R_U = R_D = R_0$, where R_0 is the altitude of the satellite ($R_0 = 35\,786$ km). In this case $t_{SS} = 238$ ms.
(2) An oblique trajectory corresponding to stations at the limit of visibility (elevation angle $= 0°$): $R_U = R_D = (R_0 + R_E)\cos(17.4°/2)/c$, where R_E is the radius of the earth ($R_E = 6738$ km). In this case $t_{SS} = 278$ ms.

3.5.2 Propagation time on the network

The total propagation time on the network can be calculated using the following expression [CCITT Recommendation G114]:

$$t_N = 12 + (0.004 \times \text{distance in km}) \quad \text{(ms)} \tag{3.5}$$

It is reasonable to take a mean value of 30 ms for the sum of the propagation times on each network [CCIR Report 383].

3.5.3 The case of the telephone

Figure 3.8a shows the path followed by telephone signals during a conversation. A high value of propagation time causes an unpleasant effect known as an echo in telephone circuits. The origin of this phenomenon is the interface between the subscriber lines which are bidirectional (2-wire circuit) to reduce the cost and the long distance links between switching centres where the two directions of transmission are generally separate (4-wire circuit). This interface is realised with a differential coupler which can be seen in Figure 3.8a and in more detailed form in Figure 3.8b. When the impedance Z_1 of the subscriber line is equal to the load impedance Z_2, every signal received on the link from B to A is tansmitted to subscriber A and every signal emitted by subscriber A is transmitted on the link from A to B. In general the impedances Z_1 and Z_2 are not equal, since Z_2 is fixed and Z_1 depends on the line in service. Under these conditions, part of the received signal power passes directly from the B to A link to the A to B link. Subscriber B thus hears his own voice with a delay equal to the round trip return time, this is the phenomenon of echo. It is more noticeable when the propagation time is high.

CCITT Recommendation G114 stipulates that the propagation time $t_{SS} + t_N$ between subscribers must not exceed 400 ms. It recommends the use of echo suppressors [CCITT Recommendation 164] or echo cancellers [CCITT Recommendation 165] when this time is between 150 ms and 400 ms. No particular precaution is to be taken when it is less than 150 ms.

For links between subscribers established with geostationary satellites this leads to the following:

—The need to install echo suppressors or cancellers (see Chapter 10).
—The need to avoid a 'double hop', that is establishment of a link through two satellites

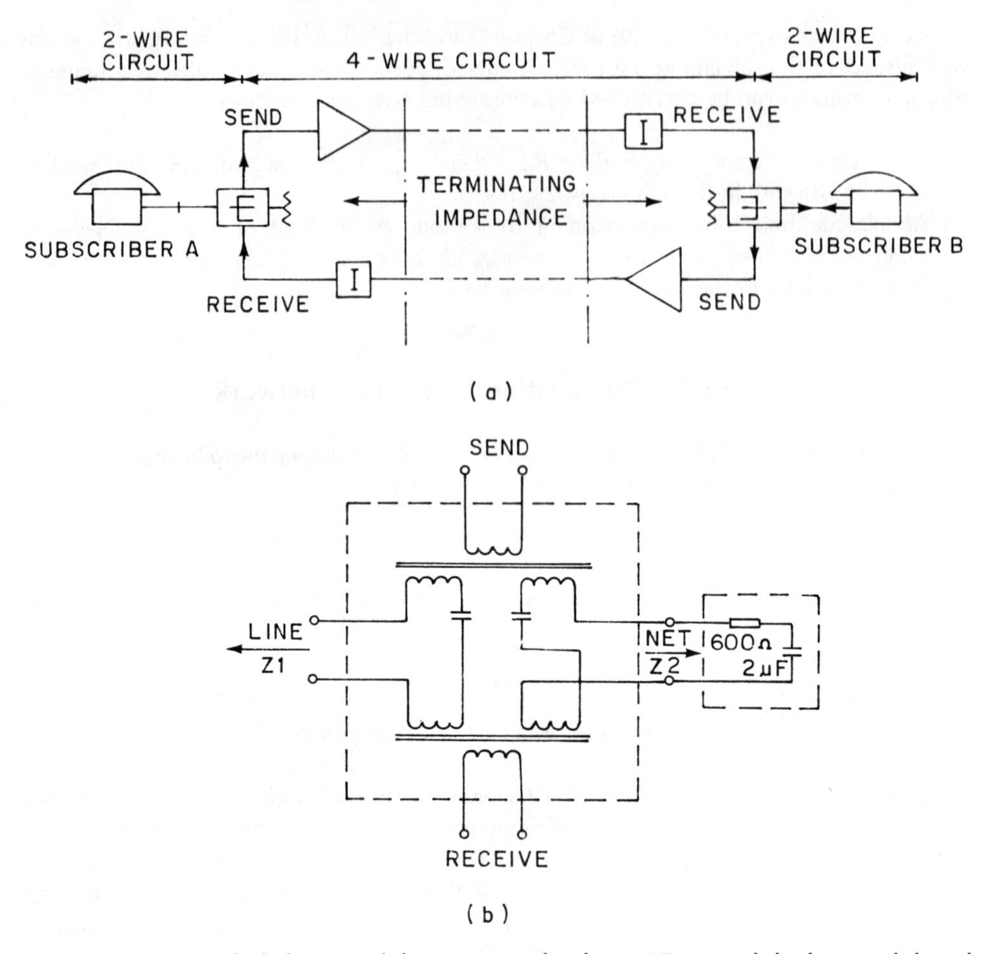

Figure 3.8 Routing of telephone signals between two subscribers. (a) Two-wire links, four-wire links and the location of differential couplers. (b) The principle of the differential coupler.

without an intersatellite link. If the system contains an intersatellite link, the propagation time t_{ISL} between the two satellites must remain less than 90 ms. The orbital separation between the two satellites for a propagation time $t_{ISL} = R_{ISL}/c$ is given by the following expression:

$$\theta = 2 \arcsin ct_{ISL}/2(R_E + R_0) \tag{3.6}$$

With $R_E = 6378$ km, $R_0 = 35\,786$ km and $t_{ISL} < 90$ ms, this gives $\theta < 37°$.

3.5.4 The case of data transmission

Terminals which use ARQ procedures for automatic repetition of data packets received in error (see Section 3.7.9) must be equipped with buffer memories of sufficient size to store all the packets transmitted during the duration of the round trip trajectory.

3.6 ANALOGUE TRANSMISSION

Analogue transmission is characterised by the following:

—Processing performed on the baseband signal (before modulation and after demodulation) in order to improve the quality of the link.
—The number of channels supported by the carrier. In the case of a single channel one refers to single channel per carrier (SCPC) transmission.
In the case of several channels transmitted by frequency division multiplexing, one refers to FDM transmission.
—The type of modulation used. The most widely used is frequency modulation (FM). As the modulation envelope is constant (the carrier amplitude is not affected by the modulating signal), it is robust with respect to the non-linearities of the satellite channel (see Section 2.9). On the other hand, for a given quality of link, it offers the useful possibility of a trade-off between the signal-to-noise ratio and the bandwidth occupied by the carrier. In view of the use on some recent satellites of field effect transistors which are more linear than the travelling wave tubes which they replace, experiments have been performed in the realisation of single sideband amplitude modulation (SSB-AM) [LAB-84], [BRA-84], [JON-84].

The station-to-station space link is generally identified by the multiplexing/modulation combination. After reviewing baseband processing, frequency division multiplexing and modulation techniques, the following will be examined successively:

—Telephone transmission using SCPC/FM, FDM/FM and FDM/SSB-AM.
—Television transmission using SCPC/FM.

3.6.1 Baseband processing

The purpose is to improve the quality of the space link using methods whose realisation cost is less than that arising from modification of one of the parameters involved in the link budget. The principal methods used are:

(a) For telephony:
—Speech activation.
—Pre- and de-emphasis.
—Compression and expansion (companding).
(b) For television:
—Pre- and de-emphasis.

3.6.1.1 Speech activation

The principle is to establish the space link only when the subscriber is actually speaking. As the activity factor is 0.25 (Section 3.1.1), its application to multicarrier SCPC systems (see Section 4.4) should permit a reduction of the power required on the satellite by about 6 dB.

In practice, allowing guard times for activation and deactivation of the carrier and the sensitivity of the system-to-noise spikes, the reduction is only of the order of 4 dB. The parameters to be taken into account are:

—The activation threshold (− 30 to − 40 dBm0).
—The carrier activation time (6 to 10 ms).
—The carrier deactivation time (150 to 200 ms).

3.6.1.2 *Pre- and de-emphasis*

The noise at the output of the demodulator of a frequency modulation transmission has a parabolic spectral density; the high frequency components of the signal are more affected by noise than the low frequencies. Figure 3.9 shows the principle of pre- and de-emphasis. The pre-emphasis filter before modulation increases the amplitude of the high frequency components. The cross-over frequency is the frequency for which the gain of the pre-emphasis filter is 0 dB. After demodulation, the de-emphasis filter, whose gain follows a law which is the inverse of that of pre-emphasis, reduces the amplitude of the high frequency signal components and the noise. The signal is restored without spectral distortion. The noise power in the band is reduced. In this way the signal-to-noise ratio S/N is improved. The improvement is of the order of 4 to 5 dB for telephony and around 13 dB for television.

3.6.1.3 *Compression and expansion (companding) [CCITT Recommendations G162 and G166]*

An improvement in the signal-to-noise ratio S/N at the output of the demodulator is obtained by reducing the dynamic range of the signal before modulation (compression) and performing

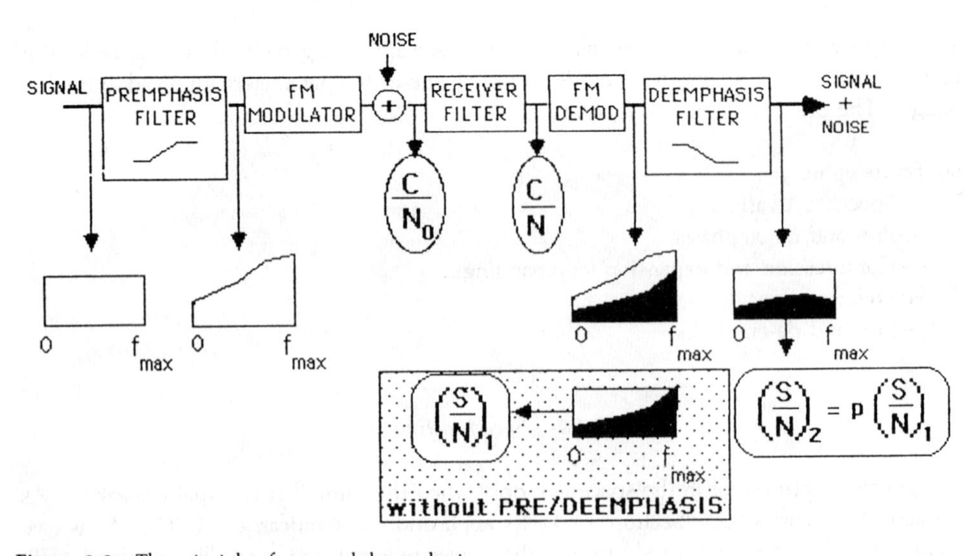

Figure 3.9 The principle of pre- and de-emphasis.

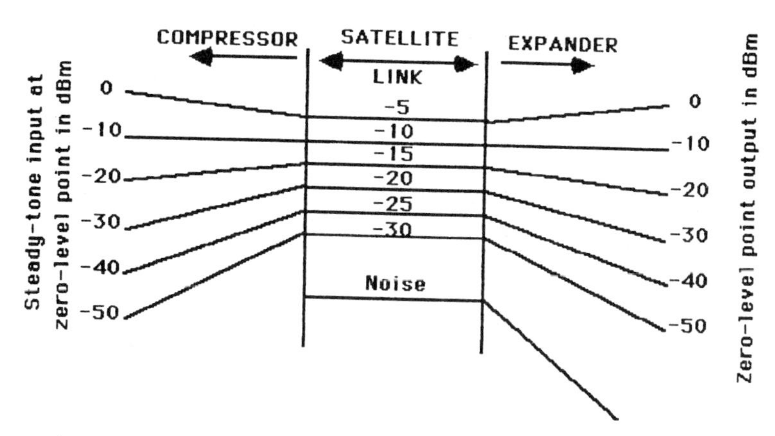

Figure 3.10 The principle of companding.

the inverse operation after demodulation. Figure 3.10 illustrates the principle. When control of the device gain is adapted to the power of syllables, the technique is called syllabic companding. Compression in general reduces the dynamic range by half and expansion by a factor of two restores the original dynamic range. On expansion, the noise at the demodulator output is subjected to attenuation since it is at a low power level. For example, if the noise level at the receiver input between syllables, words and phrases, is -25 dB with respect to the zero reference level, the corresponding noise level after expansion is -50 dB. The improvement in signal-to-noise ratio is the result of this attenuation. The improvement is subjective since it is associated with the absence of perceived noise during silences in the conversation. It is considered to be of the order of 15 dB.

3.6.2 Multiplexing

Multiplexing consists of combining the signals from several users into a single signal which then forms the signal which modulates the carrier. After demodulation the individual signals are separated by an operation called demultiplexing. For analogue telephone transmission, multiplexing is by frequency division (FDM); the subscribers' signal spectra are translated in frequency and located adjacent to each other in the frequency spectrum with a space of 4 kHz reserved for each channel. Figure 3.11 illustrates the principle of multiplexing and demultiplexing.

3.6.2.1 Capacity and spectral occupation of FDM telephony [CCIR Recommendations 481 and 482]

Table 3.1b indicates the capacities and frequency bands of multiplexed telephony used in satellite communication.

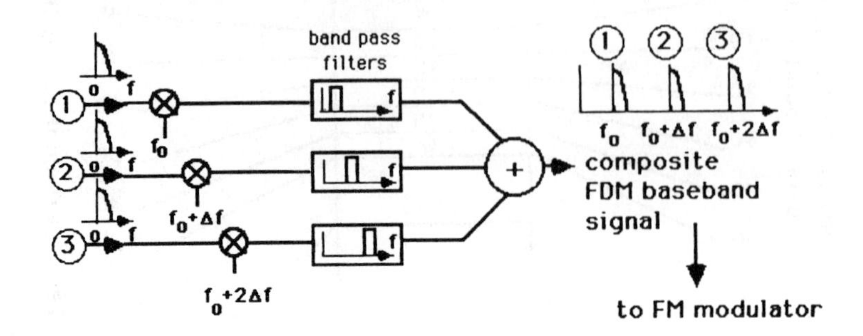

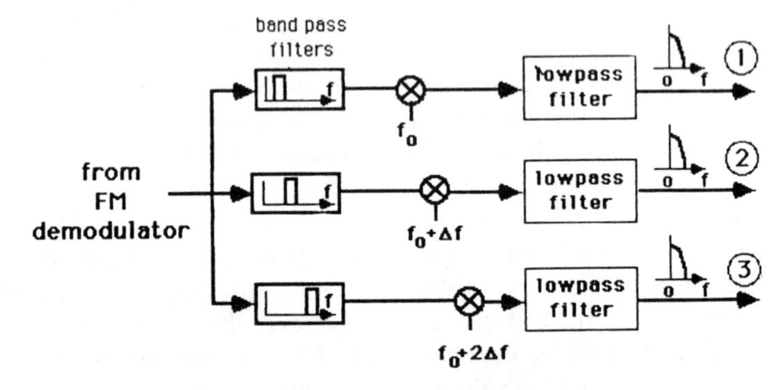

Figure 3.11 The principle of FDM multiplexing and demultiplexing.

3.6.2.2 *Mean power of FDM telephony*

The telephone channels which constitute the multiplex are not correlated. When the number n of channels is sufficiently high ($n > 240$), the multiplexed signal can be considered as white Gaussian noise in the band occupied. Its mean power S_m is then the product of the number of channels n and the mean power in each channel which is $P_m = -15$ dBm0 (see Section 3.1.1); hence $S_m = nP_m$. Expressed in dB this gives:

$$S_m(\text{dB}) = 10 \log S_m = -15 + 10 \log n(\text{dBm0}) \qquad \text{for } n > 240 \tag{3.7}$$

For a number of channels n less than 240, it is assumed that:

$$S_m(\text{dB}) = 10 \log S_m = -1 + 4 \log n(\text{dBm0}) \qquad \text{for } n < 240 \tag{3.8}$$

3.6.2.3 *Peak power of a multiplexed signal*

Let S_{max} be the power which is exceeded with a probability of 10^{-3}. It can be shown that:

$$S_{max}/S_m = 10 \tag{3.9}$$

Table 3.1b Capacities and frequency bands of analogue multiplexes used in satellite communication. [CCIR Rec 481]

System capacity (number of channels)	Limits of band occupied by telephone channels (kHz)
12	12–60
24	12–108
36	12–156
48	12–204
60	12–252
72	12–300
96	12–408
132	12–552
192	12–804
252	12–1052
312	12–1300
372	12–1548
432	12–1796
492	12–2044
552	12–2292
612	12–2540
792	12–3284
972	12–4028
1092	12–4892
1200	12–5340
1332	12–5884
1872	12–8120

3.6.3 Amplitude modulation (AM)

3.6.3.1 *The principle*

Let $v(t)$ be the voltage representing the modulating signal and F_c the carrier frequency. Amplitude modulation associates a variation of carrier amplitude with the voltage $v(t)$ whose peak amplitude is V_{max}. The expression for the carrier is:

$$c(t) = A[1 + m_{AM}v(t)/V_{max}] \cos(\omega_c t) \tag{3.10}$$

where m_{AM} is the amplitude modulation index, assumed to be between 0 and 1.

3.6.3.2 *Spectral occupation*

In the form of (3.10) the spectrum of the modulated carrier contains a component at frequency f_c (that of the unmodulated carrier) and two sidebands on each side of this component which replicate the spectrum of the modulating signal. If f_{max} is the maximum frequency of the spectrum of the modulating signal, the bandwidth occupied by the carrier is:

$$B = 2f_{max} \qquad \text{(Hz)} \tag{3.11}$$

Transmission of one of the two sidebands can be suppressed by filtering. A single sideband amplitude modulated signal is thus generated (SSB-AM). Under these conditions, spectral occupancy is reduced by half and becomes:

$$B = f_{max} \qquad \text{(Hz)} \tag{3.12}$$

3.6.4 Demodulation of an amplitude modulated wave

3.6.4.1 *The principle*

For amplitude modulation with two sidebands, demodulation can be non-coherent. In satellite transmission, amplitude modulation is considered only in cases where bandwidth constraints are severe. It is then better to use single sideband modulation and coherent demodulation. The principle of coherent demodulation is to multiply the received modulated carrier by an unmodulated carrier, generated locally, which has the same frequency and phase as the received carrier. Low-pass filtering at the output of the multiplier restores the modulating signal.

3.6.4.2 *Signal-to-noise ratio at the demodulator output*

With single sideband amplitude modulation, the signal-to-noise ratio at the demodulator output is given by

$$S/N = (C/N_0)_T / f_{max} \tag{3.13}$$

3.6.5 Frequency modulation (FM)

3.6.5.1 *The principle*

Let $v(t)$ be the voltage representing the modulating signal and f_C be the nominal carrier frequency. Frequency modulation associates a frequency deviation of the carrier $\Delta F(t) = f(t) - f_C$, which is proportional to $v(t)$, with the voltage $v(t)$:

$$\Delta F(t) = f(t) - f_C = k_{FM} v(t) \qquad \text{(Hz)} \tag{3.14}$$

k_{FM}(Hz/V) characterises the modulator.

3.6.5.2 *Modulation index*

If the modulating signal is sinusoidal of frequency f_m and amplitude A, it causes a peak frequency deviation of the carrier of value $\Delta F_{max} = k_{FM} A$. The modulation index m_{FM} is defined as:

$$m_{FM} = \Delta F_{max}/f_m \quad (3.15)$$

3.6.5.3 *Spectral occupation*

The spectrum of a carrier modulated by a sinusoidal signal of frequency f_m occupies a bandwidth given by Carson's formula:

$$B = 2(m_{FM} + 1)f_m \qquad \text{(Hz)} \quad (3.16)$$

In practice, Carson's formula is used even when the modulating signal is not sinusoidal. Hence, assuming that the non-sinusoidal modulating signal occupies a band $(0 - f_{max})$, f_m in (3.15) and (3.16) is replaced by the maximum frequency f_{max} of the modulating signal:

$$B = 2(m_{FM} + 1)f_{max} \qquad \text{(Hz)} \quad (3.17)$$

with $m_{FM} = \Delta F_{max}/f_{max}$.

3.6.6 Demodulation of a frequency modulated wave

3.6.6.1 *The principle*

The carrier at the demodulator input has a carrier-to-noise ratio $(C/N_0)_T$. The demodulator identifies the instantaneous frequency deviation $\Delta F(t)$ of the carrier and recovers a voltage $u(t)$ such that:

$$u(t) = \sigma_{FM}\Delta F(t) \qquad \text{(V)} \quad (3.18)$$

σ_{FM}(V/Hz) characterises the demodulator.

3.6.6.2 *Noise spectral density at the demodulator output*

The noise spectral density at the demodulator output is given by [DUP-73 p. 212], [SHA-79 p. 354]:

$$N_0(f) = N_0(\sigma_{FM}/A)^2(2\pi f)^2 \qquad \text{(W/Hz)} \quad (3.19)$$

where A is the amplitude of the carrier and N_0 the noise spectral density at the demodulator input, assumed to be constant. The noise spectral density at the demodulator output is not constant; it increases parabolically with frequency.

3.6.6.3 *Signal-to-noise ratio at the demodulator output*

For a modulating signal with a spectral width $(0 - f_{max})$ the signal-to-noise ratio S/N at the demodulator output, when the noise power is measured in a bandwidth $B_N = f_{max}$, is given by:

$$S/N = [3/(2f_{max})](\Delta F_{max}/f_{max})^2(C/N_0)_T \tag{3.20}$$

3.6.6.4 *Demodulation gain*

The modulated carrier occupies a bandwidth B at the demodulator input where B is given by Carson's formula. It is assumed that the receiver has an equivalent noise bandwidth B_N equal to the spectral occupation of the carrier, hence $B_N = B$. The noise power at the demodulator input is $N = N_0 B_N$ and the carrier power-to-noise power ratio has a value $C/N = (C/N_0)_T B_N$.

Equation (3.20) becomes:

$$S/N = (3/2)(B_N/f_{max})(\Delta F_{max}/f_{max})^2 C/N = 3(1 + m_{FM})m_{FM}^2 C/N \tag{3.21}$$

When m_{FM} is sufficiently large, $S/N = 3m_{FM}^3 C/N$. The value of S/N is greater than that of C/N; demodulation of a frequency modulated wave provides a 'demodulation gain' in terms of signal-to-noise ratio. Notice that this advantage implies providing the satellite link with a bandwidth greater than that of the modulating signal. Since demodulation gain increases with occupied bandwidth, a low value of C/N can be compensated by an increase in the bandwidth used in the case where the satellite link is limited in power. It is this principle of 'bandwidth C/N exchange' (also called 'bandwidth-power exchange'), for a given S/N, which makes frequency modulation very well suited to analogue signal transmission by satellite.

3.6.6.5 *Demodulator threshold*

Equation (3.21) is valid only for values of C/N greater than a minimum value called the 'demodulator threshold'. Below this threshold, noise of an impulse type appears at the demodulator output and this degrades the signal-to-noise ratio S/N with respect to the value given by (3.21). The demodulator threshold depends on the type of demodulator. It is close to 10 dB for conventional demodulators and can be 5 to 6 dB with 'improved threshold' demodulators.

3.6.7 Telephone transmission on SCPC/FM

The carrier is modulated by the signal of a single telephone channel (SCPC = Single Channel Per Carrier). There is, therefore, no multiplexing.

3.6.7.1 *Signal-to-noise ratio at the demodulator output [FER-75]*

The ratio of test signal power-to-noise power at the demodulator output is given by:

$$S/N = 3[(\Delta F_r)^2/(f_{max})^3]pw(C/N_0)_T \tag{3.22}$$

with:

ΔF_r = r.m.s. frequency deviation due to the test signal
f_{max} = maximum frequency of the telephone channel = 3400 Hz
p = improvement factor due to pre-emphasis and de-emphasis (6.3 dB) and companding (17 dB), if any
w = psophometric weighting factor (2.5 dB)
$(C/N_0)_T$ = carrier power-to-noise spectral density ratio at the receiver input (Hz).

3.6.7.2 *Required bandwidth*

The receiver must have a bandwidth equal to the bandwidth B occupied by the carrier. The equivalent noise bandwidth of the receiver B_N is taken equal to B:

$$B_N = B = 2(\Delta F_p + f_{max}) \qquad \text{(Hz)} \tag{3.23}$$

where:

$$\Delta F_p = \text{peak frequency deviation (Hz)} = gL\Delta F_r$$

The expression for ΔF_p is derived as follows; the peak frequency deviation due to an active speech signal is related to the effective frequency deviation by a factor g which is a function of the acceptable clipping. If clipping is defined by the r.m.s. deviation due to an active speech signal exceeding a level 15 dB below the peak deviation, then $g = 12.6$. Furthermore, the r.m.s. frequency deviation due to an active speech signal and the r.m.s. frequency deviation due to the test signal are in a ratio, called the load factor L, which is that of the r.m.s. signal amplitudes. The ratio of the amplitudes is equal to the square root of the powers. Hence, from expression (3.1) where the activity factor is omitted since an active speech signal is considered:

$$L = 10^{(P_a + 0.115\sigma^2)/20} = 0.35 \qquad \text{without companding}$$

$$L = 10^{(P_a/2 + 0.002875\sigma^2)/20} = 0.53 \qquad \text{with companding} \tag{3.24}$$

3.6.7.3 *Example 1: SCPC/FM transmission (without companding)*

A satellite link between two stations occupying a bandwidth of 25 kHz is considered. The quality target is $S/N = 50$ dB. Taking account of (3.23), the peak frequency deviation must not exceed $\Delta F_p = B_N/2 - f_{max} = 12{,}500 - 3400 = 9100$ Hz. From this the value of $\Delta F_r = \Delta F_p/gL = 9100/(12.6 \times 0.35) = 2063.5$ Hz is deduced. The required value of $(C/N_0)_T$ is calculated from (3.22), hence:

$$(C/N_0)_T = (S/N)/\{3[(\Delta F_r)^2/(f_{max})^3]pw\} = 76.1 \text{ dB (Hz)}.$$

The corresponding value required for $(C/N)_T$ is:

$$(C/N)_T = (C/N_0)_T/B_N = 32.1 \text{ dB}.$$

3.6.7.4 *Example 2: SCPC/CFM transmission (with companding; CFM = Companded FM)*

The above example is reconsidered with $L = 0.53$ and $p\,(\mathrm{dB}) = 17\,\mathrm{dB} + 6.3\,\mathrm{dB} = 23.3\,\mathrm{dB}$. This gives:

$$\Delta F_r = \Delta F_p/gL = 9100/(12.6 \times 0.53) = 1362.7\,\mathrm{Hz}.$$

$$(C/N_0)_T = 62.7\,\mathrm{dB(Hz)}.$$

The corresponding value required for $(C/N)_T$ is:

$$(C/N)_T = (C/N_0)_T/B_N = 18.7\,\mathrm{dB}.$$

The gain due to companding is thus 13.4 dB.

3.6.8 Telephone transmission on FDM/FM

The carrier is frequency modulated by an FDM telephone multiplex of n channels. The pre-emphasis characteristics are given in CCIR [Recommendation 464].

3.6.8.1 *Signal-to-noise ratio at the demodulator output*

The worst signal-to-noise ratio is that of the upper voice channel in the multiplexed spectrum. The S/N ratio of the test signal power and the noise power in this channel is given by:

$$(S/N) = (\Delta F_r/f_{max})^2(1/b)pw(C/N_0)_T \qquad (3.25)$$

where:

ΔF_r = r.m.s. frequency deviation due to the test signal.
f_{max} = maximum frequency of the multiplex signal (a function of the number of multiplexed telephone channels, see Table 3.1).
b = bandwidth of one telephone channel (3100 Hz).
p = improvement factor due to pre- and de-emphasis (4 dB) and companding, if used (17 dB).
w = psophometric weighting factor (2.5 dB).
$(C/N_0)_T$ = carrier power-to-noise spectral density at the receiver input (Hz).

Recall that $S/N = 50\,\mathrm{dB}$ for a weighted noise power of 10 000 pW0p (see Section 3.3.1.1, equation (3.3)).

3.6.8.2 *Required bandwidth*

The receiver must have a bandwidth equal to the bandwidth B occupied by the carrier. The equivalent noise bandwidth of the receiver B_N is taken equal to

$$B_N = B = 2(\Delta F_p + f_{max}) \qquad \text{(Hz)} \tag{3.26}$$

where ΔF_p = peak frequency deviation (Hz) $= gL\Delta F_r$

The expression for ΔF_p is derived as follows; the peak frequency deviation due to the multiplex signal relates to the r.m.s. frequency deviation by a factor g equal to the peak to r.m.s. amplitude ratio of the multiplex signal.

The ratio of amplitudes is equal to the square root of the powers. From (3.9):

$$g = \sqrt{(10)} = 3.16 \tag{3.27}$$

Furthermore, the r.m.s. frequency deviation due to the multiplex signal and the r.m.s. frequency deviation due to the test signal are in a ratio, called the load factor L, which is that of the r.m.s. amplitudes of these signals. It follows, therefore, from equations (3.1), (3.7) and (3.8) [SZA-81], [JON-84], [CCIR Report 708–11] that:

—without companding:

$$\begin{aligned} L &= 10^{(-15 + 10 \log n)/20} && \text{for } n > 240 \\ L &= 10^{(-1 + 4 \log n)/20} && \text{for } n < 240 \end{aligned} \tag{3.28}$$

—with companding:

$$\begin{aligned} L &= 10^{(-11.2 + 10 \log n)/20} && \text{for } n > 240 \\ L &= 10^{(2.8 + 4 \log n)/20} && \text{for } n < 240 \end{aligned} \tag{3.29}$$

When each multiplexed channel is subjected to companding (FDM/CDM = FDM/Companded FM), with a constant required S/N ratio, a given value of $(C/N_0)_T$ and a constant spectral occupancy, an increase in the capacity of the multiplexed signal is obtained. An example is given in [SZA-81] of an increase by a factor of 1.7.

3.6.8.3 *Example 1: FDM/FM transmission (without companding)*

Consider a satellite repeater of 36 MHz bandwidth. The value of $(C/N_0)_T$ required to transmit a multiplex of 972 channels will be calculated.

The maximum frequency of the multiplex is, from Table 3.1, $f_{max} = 4028$ kHz. From (3.28) $L = 5.54$. From (3.26) $\Delta F_p = B_N/2 - f_{max} = 13.97$ MHz and $\Delta F_r = \Delta F_p/gL = 13.97/(3.16 \times 5.54) = 0.798$ MHz. Substituting into (3.25) for $S/N = 50$ dB with pw (dB) = 4 dB + 2.5 dB = 6.5 dB gives $(C/N_0)_T = 92.5$ dB(Hz). The corresponding value of $(C/N)_T$ is $(C/N)_T = (C/N_0)_T(1/B_N) = 16.9$ dB.

3.6.8.4 *Example 2: FDM/CFM transmission (with companding)*

The same bandwidth of 36 MHz is used to transmit the same multiplex. With companding the load factor is modified according to equation (3.29) from which, with $n = 972$, $L = 8.6$. This gives $\Delta F_r = \Delta F_p/gL = 13.97/(3.16 \times 8.6) = 0.514$ MHz. Substituting into (3.25) for

$S/N = 50$ dB with pw (dB) = 17 dB + 4 dB + 2.5 dB = 23.5 dB, the required value for $(C/N_0)_T$ becomes $(C/N_0)_T = 79.3$ dB (Hz). The corresponding value of $(C/N)_T$ is $(C/N)_T = (C/N_0)_T (1/B_N) = 3.7$ dB. The gain provided by companding is thus 13.2 dB. In practice the full benefit of this gain cannot be obtained since it is necessary to ensure a value of $(C/N)_T$ at least equal to the demodulation threshold (see Section 3.6.6.5).

FDM/CFM transmission can also be compared with FDM/FM transmission in terms of capacity for the same bandwidth of 36 MHz and the same ratio $(C/N_0)_T = 92.5$ dB (Hz). Considering that the maximum frequency f_{max} of the multiplexed signal varies more or less as 4500 n, where n is the number of multiplexed channels, equations (3.25), (3.26) and (3.29) indicate an FDM/CFM capacity of 2550 telephone channels compared with the capacity of 972 channels given by Example 1 in FDM/FM.

3.6.9 Telephone transmission in FDM/SSB-AM [JON-84]

The carrier is frequency modulated by a frequency division multiplex (FDM) of n telephone channels. Considering that the maximum frequency f_{max} of the multiplex varies more or less as 4500 n, where n is the number of multiplexed channels, the spectrum occupied by the carrier from (3.12) is 4500 n Hz. The carrier power C is $C = nP_m$, where P_m is given by (3.1). A ratio P_m/N for the voice of 35 dB corresponds to a quality target on each channel determined by $S/N = 50$ dB for the test signal. The required ratio $(C/N)_T$, expressed in dB, is thus $(C/N)_T = 35$ dB + $10 \log n$. From the bandwidth point of view, a satellite repeater of 36 MHz bandwidth permits transmission of 8000 telephone channels, in the absence of power limitation. But, in order to achieve the required carrier-to-noise power ratio of 74 dB, a link budget is required which is all the more difficult to realise since the non-linearity of the repeater introduces intermodulation noise. This constraint can be reduced by introducing companding. In the study cited in the reference concerning the use of an RCA satellite repeater equipped with a solid state amplifier, the possibility is noted of transmitting, with companding and in the absence of interference from adjacent systems, 6500 channels with a $(C/N)_T$ ratio = 57.45 dB.

3.6.10 Television transmission in SCPC/FM

The carrier is modulated by the television signal after passing through the preemphasis filter. After demodulation the signal is de-emphasised. The pre- and de-emphasis characteristics are given in CCIR [Report 212, Recommendation 405].

3.6.10.1 *Signal-to-noise ratio at the demodulator output*

The quality target for a television transmission is stipulated in terms of the weighted signal-to-noise ratio (see Section 3.2.2). This is given by:

$$S/N = (3/2)(\Delta F_{Tpp}/B_n)^2(1/B_n)pw(C/N_0)_T \tag{3.30}$$

where ΔF_{Tpp} is the peak-to-peak frequency deviation for a signal at the input of the transmission chain of frequency equal to the cross-over frequency f_r of pre- and de-emphasis

Table 3.2 Improvement in S/N ratio provided by pre- and de-emphasis and videometric weighting [CCIR Report 637].

System type (lines per frame/frames per second)	System name	Noise measurement bandwidth B_n (MHz) at receiver output	Combined effect pre- and de-emphasis plus weighting pw (dB)
525/60	M	4.2	12.8
	unified	5.0	14.8
625/50	B, G, H	5.0	16.3
	I	5.0	12.9
	Unified	5.0	13.2
	D, K, L	6.0	18.1

($f_r = 1.512$ MHz for 625/50 systems and 0.762 MHz for 525/60 systems) and peak-to-peak amplitude V_{Tpp} equal to the peak-to-peak amplitude of the video signal including the synchronising pulses. B_n is the bandwidth of noise measurement at the receiver output and is equal to the maximum spectral frequency of the video signal f_{max} or 5 MHz when the noise measurement is made with the unified weighting filter recommended in CCIR Recommendation 568. The product pw represents the combined effect of pre- and de-emphasis and the weighting. Table 3.2 gives the values of pw for various systems [CCIR Report 637].

3.6.10.2 *Required bandwidth*

The receiver must have a bandwidth equal to the bandwidth B occupied by the carrier. The equivalent noise bandwidth of the receiver B_N must be equal to [CCIR Report 215]:

$$B_N = B = \Delta F_{Tpp} + 2f_{max} \qquad \text{(Hz)} \tag{3.31}$$

ΔF_{Tpp} is defined as previously and f_{max} is the maximum frequency in the spectrum of the video signal.

3.6.10.3 *Example 1: Transmission of 625/50 television by INTELSAT*

Data: $\Delta F_{Tpp} = 15$ MHz, $f_{max} = 6$ MHz, $B_n = 5$ MHz.

From (3.30): $S/N = (3/2)\,(15\text{ MHz}/5\text{ MHz})^2(1/5\text{ MHz})\,(C/N_0)_T\,pw$, where pw is given by Table 3.2 as $pw = 13.2$ dB. This gives $S/N = 5.6 \times 10^{-5}(C/N_0)_T$.
Expressed in dB:

$$S/N = -42.5 + (C/N_0)_T \qquad \text{(dB)}$$

To obtain a value of 45 dB, it is necessary to have $(C/N_0)_T = 87.5$ dB (Hz). With 89.6 dB (Hz) INTELSAT provides a S/N of 47.1 dB.

From (3.31) the bandwidth occupied by the modulated carrier is:

$$B = 15\text{ MHz} + 2 \times 6\text{ MHz} = 27\text{ MHz}.$$

However, INTELSAT uses only 15.75 MHz of bandwidth in order to be able to transmit two television carriers simultaneously on the same 36 MHz bandwidth repeater. Therefore frequency overdeviation occurs. Taking $B_N = 15.75$ MHz, this gives:

$$S/N = 5.6\,10^{-5}(C/N)_T \times B_N = 882(C/N)_T$$

Expressed in dB:

$$S/N = 29.5 \text{ dB} + (C/N)_T \qquad \text{(dB)}$$

To obtain a value of 45 dB, it is necessary to have $(C/N)_T = 15.5$ dB.

3.6.10.4 *Example 2: Transmission of 625/50 television by the ASTRA satellite*

Data: $\Delta F_{Tpp} = 13.5$ MHz, $f_{max} = 6$ MHz, $B_n = 5$ MHz.
From (3.30): $S/N = (3/2)\,(13.5 \text{ MHz}/5 \text{ MHz})^2\,(1/5 \text{ MHz})\,(C/N_0)_T\,pw$, with $pw = 13.2$ dB. This gives:

$$S/N = 4.6\,10^{-5}(C/N_0)_T$$

Expressed in dB:

$$S/N = -43.4 + (C/N_0)_T \qquad \text{(dB)}$$

To obtain a value of 45 dB, it is necessary to have $(C/N_0)_T = 88.4$ dB(Hz). From (3.31), the bandwidth occupied by the modulated carrier is:

$$B = 13.5 \text{ MHz} + 2 \times 6 \text{ MHz} = 25.5 \text{ MHz}$$

Considering $B_N = B$:

$$S/N = 4.6\,10^{-5}(C/N)_T \times B_N = 1173\,(C/N)_T$$

Expressed in dB:

$$S/N = 30.7 \text{ dB} + (C/N)_T \qquad \text{(dB)}$$

To obtain a value of 45 dB it is necessary to have $(C/N)_T = 14.3$ dB. With a C/N of 13 dB, ASTRA provides a S/N of 43.7 dB.

3.6.11 Energy dispersion

The CCIR recommendation [CCIR Recommendation 446] for the use of energy dispersion techniques in radio-frequency transmissions in order to limit interference between radio communication systems sharing the same frequency bands is mentioned in Section 2.8.1.4.

When the modulation index of a frequency modulated carrier is low, the power of the modulated carrier is concentrated in a narrow band in the vicinity of the carrier and the risk of interference in increased. This is the case, for example, for an FDM/FM transmission with a low load or an SCPC/FM television transmission when the image contains large portions of constant luminance.

The principle of energy dispersion is to superimpose a low frequency triangular signal on the modulating signal before modulation. This signal is subtracted from the demodulated signal on reception. For telephone transmission, a dispersion signal with a frequency between 20 and 150 Hz is used. For television transmission, the dispersion signal must be synchronised to the field frequency (50 or 60 Hz according to the system) [CCIR Report 384].

3.7 DIGITAL TRANSMISSION

Digital transmission relates to links for which the user's terminals produce digital signals (computers for example). But it is also possible to transmit signals of analogue origin (telephone or sound broadcasting for example) in digital form. Although this choice implies an increased baseband, it permits signals from diverse origins to be transmitted on the same satellite channel and the satellite link to be incorporated in the Integrated Services Digital Network (ISDN, see Section 3.3.3). This implies the use of Time Division Multiplex (TDM) techniques.

Figure 3.12 shows the elements of the digital transmission chain. These elements will now be described in succession.

3.7.1 Digitisation of analogue signals

Digitisation of analogue signals implies three stages [DUP-73], [SHA-79], [MAR-87]:

—Sampling
—Quantisation
—Source coding

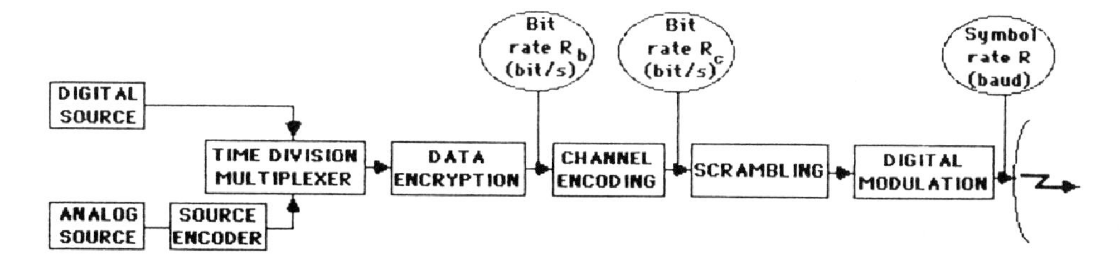

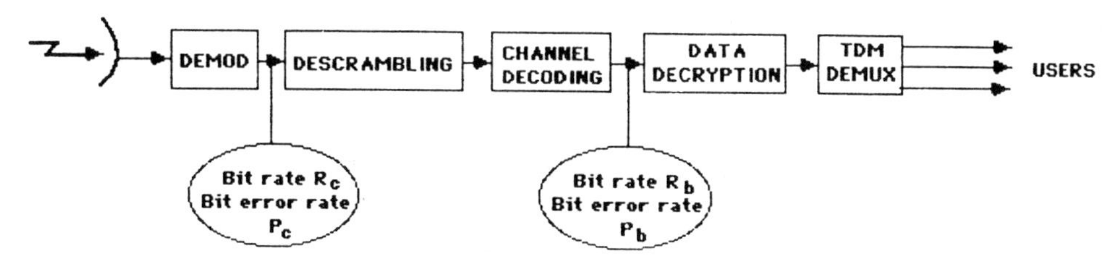

Figure 3.12 The elements of a digital satellite transmission chain.

3.7.1.1 *Sampling*

Sampling must be performed at a frequency F_S equal to at least twice the maximum frequency f_{max} of the spectrum of the analogue signal. The signal at the sampler output is a sequence of amplitude modulated pulses (pulse amplitude modulation (PAM). For the voice on a telephone channel $f_{max} = 3400$ Hz and $F_S = 8$ kHz. For a radio programme, $f_{max} = 15$ kHz and $F_S = 32$ kHz [CCIR Recommendation 606].

3.7.1.2 *Quantisation*

Each sample is then quantised into a finite number M of discrete levels. This quantisation introduces an error which is described as quantisation noise. Quantisation can be either uniform or non-uniform according to the quantisation step which may be independent of, or a function of, the sample magnitude. In the case of non-uniform quantisation it is possible to adapt the quantisation law to the amplitude distribution of the samples in order to maintain a constant signal-to-quantisation noise ratio for all sample amplitudes. This operation is called compression. For speech samples, two types of compression are currently used; these are 'μ law' and 'A law' compression [SCH-80].

3.7.1.3 *Source encoding*

Quantised samples have a finite number M of levels which can be represented by a finite alphabet of signals which will be transmitted on the link. This operation is called source encoding (PCM; pulse code modulation) in order to distinguish it from channel encoding which provides protection against transmission errors (see Sections 3.7.6 and 3.7.9). Most often the element of the alphabet is a binary signal and it is thus necessary to transmit $m = \log_2 M$ bits per sample and this determines the bit rate:

$$R_q = F_S \log_2 M \tag{3.32}$$

For example, for the telephone, if $M = 2^8 = 256$ (Europe), 8 bits per sample are required. With $F_S = 8$ kHz, this gives $R_q = 64$ kbit/s. For a radio programme, source encoding with compression is used which provide a bit rate of 384 kbit/s [CCIR Recommendation 660]. Various techniques have been used to reduce the bit rate. These techniques take advantage of the existence of redundancy between successive samples. In this way a bit rate $R_b \leqslant R_q$ is achieved and this is the information rate to be transmitted. These techniques, called low rate encoding (LRE), are described for example in [JAY-76]. They are applicable for speech and vision. For telephony the most widely used equipment uses adaptive differential encoding (ADPCM, Adaptive Differential PCM) which provides a value $R_q = 32$ kbit/s [CCITT-G721].

3.7.2 Time division multiplexing (TDM)

Time division multiplexing (TDM) consists of interleaving in time the bits relating to different signals. For multiplexing digital telephone channels two channels recommended by the CCITT [CCITT Recommendations G732 and G733] are widespread—the European standard of the

CEPT (European Conference on Post and Telecommunications) and the 'T-carrier' standard used in Japan and North America (USA and Canada).

3.7.2.1 *The CEPT hierarchy*

The CEPT standard is based on a frame of 256 binary elements. The frame duration is 125 microseconds. The bit rate is 2.048 Mbit/s. The multiplex capacity is 30 telephone channels, 16 bits per frame being used for signalling and the frame synchronisation signal. The highest capacities are obtained by successive multiplexing of multiplexers of equal capacity. In this way a multiplexing hierarchy is established which contains several levels; each level is constructed by multiplexing 4 multiplexed channels with a capacity equal to the capacity of the immediately lower level.

3.7.2.2 *The 'T-carrier' hierarchy*

This standard is based on a frame of 192 bits obtained by multiplexing 24 samples each of 8 bits to which one frame alignment bit is added. Each frame thus contains 193 bits. The frame duration is 125 microseconds. The bit rate is 1.544 Mbit/s. The multiplex capacity is 24 channels (23 + 1 for signalling). The multiplexing hierarchy differs between Japan and North America.

Table 3.3 summarises the characteristics of these multiplexing techniques.

3.7.3 Digital speech concentration and channel multiplication

Systems for digital speech concentration (Digital Speech Interpolation (DSI)) use the activity factor of telephone channels in order to reduce the number of satellite channels required to transmit a given number of terrestrial channels. The speech interpolation technique is based on the fact that in a normal telephone conversation each participant monopolises the circuit for only around half the time. As the silences between syllables, words and phrases increase so does the unoccupied time. Hence on average the activity time of a circuit is from 35 to 40% of the connection time. By making use of the actual activity of the channels, several

Table 3.3 Characteristics of CEPT and T-carrier multiplexes

	CEPT		USA/Canada		Japan	
Hierarchy level	Throughput (Mbit/s)	Capacity (channels)	Throughput (Mbit/s)	Capacity (channels)	Throughput (Mbit/s)	Capacity (channels)
1	2 048	30	1 544	24	1 544	24
2	8 448	120	6 312	96	6 312	96
3	34 368	480	44 736	672	32 064	480
4	139 264	1 920	274 176	4 032	97 728	1 440
5	557 056	7 680			400 352	5 760

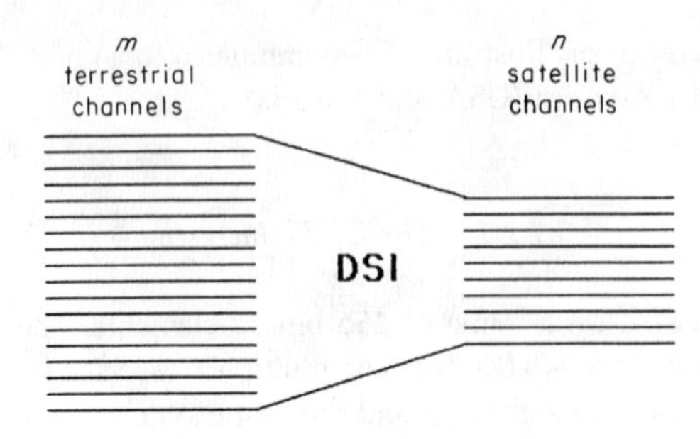

Figure 3.13 Digital speech interpolation (DSI).

users can be permitted to share the same telephone circuit. Figure 3.13 shows this principle. The gain of the digital concentrator is given by the ratio m/n. In the INTELSAT/EUTELSAT system, 240 terrestrial channels require only 127 satellite channels plus one assignment channel and the gain is $240/127 = 1.9$. The preceding gain assumes satellite telephone channels at 64 kbit/s. By adding a low rate encoder (LRE) to the digital speech concentrator, the gain can be further increased. For example, with encoding at 32 kbit/s a gain increase by a factor of 2 can be obtained. These techniques are used in digital circuit multiplication equipment (DCME). More details on the operation of these systems are given in Section 8.6.

3.7.4 Synchronism between networks

Due to movement of the satellite in its orbit, even if it is a geostationary satellite for which this movement is small but not zero (see Chapter 7), a Doppler effect is observed and the received binary rate is not always equal to the transmitted binary rate. Furthermore, the terrestrial networks at the end of the satellite link (see Figure 3.6a), when digital, do not always have strictly synchronous clocks. To compensate for these variations, buffer memories are provided at the station/network interfaces. In choosing the size of these memories account must be taken of the station-keeping specifications and use can be made of CCITT Recommendation G811 which recommends plesiochronism between digital interfaces. Plesiochronism exists when the clocks of each network have a discrepancy of at most $\pm 10^{-11}$. This leads to a frame slip for a multiplex frame of 125 microseconds once every 72 days. For more details consult [FEH-81], [CCIR Report 707].

3.7.5 Encryption

Encryption is used when it is wished to prevent exploitation of, or tampering with, transmitted messages by unauthorised users. It consists of performing an algorithmic operation in real time bit-by-bit on the binary stream. The set of parameters which defines the transformation is called the 'key'. Although the use of encryption is often associated with military communications, commercial satellite systems are increasingly induced by their customers to propose encrypted links particularly for commercial and administrative networks. In fact, due to the extended coverage of satellites and the easy access to them by small stations,

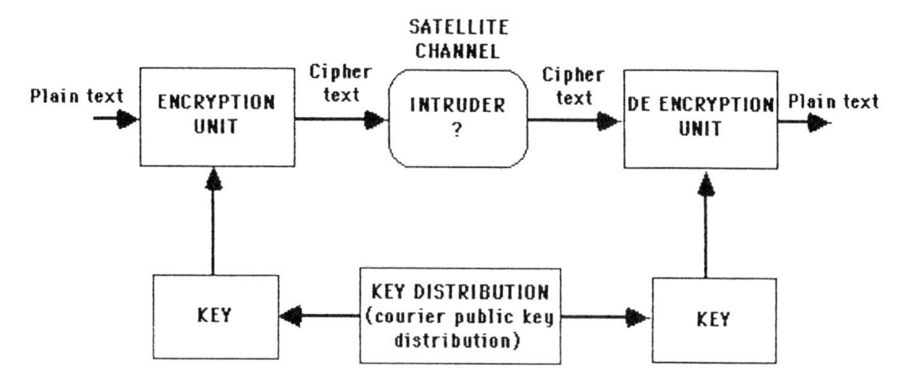

Figure 3.14 The principle of encrypted transmission.

eavesdropping and message falsification are potentially within the reach of a large number of agents of reduced means [BIC-82].

Figure 3.14 illustrates the principle of encrypted transmission. The encryption and de-encryption units operate with a key provided by the key generation units. Acquisition of a common key implies a secure method for key distribution.

Encryption consists of two aspects:

—That of confidentiality—avoiding exploitation of the message by unauthorised persons.
—That of authenticity—providing protection against any modification of the message by an intruder.

Two techniques are used [TOR-81]:

—On-line encryption (stream cipher)—each bit of the original binary stream (plain text) is combined using a simple operation (for example modulo 2 addition) with each bit of a binary stream (the keystream) generated by a key device. The latter, for example, could be a pseudorandom sequence generator whose structure is defined by the key.
—Encryption by block (block ciphering)—the transformation of the original binary stream into an encrypted stream is performed block-by-block according to logic defined by the key.

3.7.6 Channel encoding

Figure 3.15 illustrates the principle of channel encoding. It has the objective of adding redundant bits to the information bits; the former will be used at the receiver to detect and correct errors. The addition of these bits is performed in blocks or by convolution [MAR-87], [SHA-79], [CLA-81], [BHA-81], [BIC-86], [MIC-85]. The code rate is defined as:

$$\rho = n/(n + r) \tag{3.33a}$$

where r is the number of bits added for n information bits.

The bit rate at the encoder input is R_b. At the output it is greater, and equal to R_c.

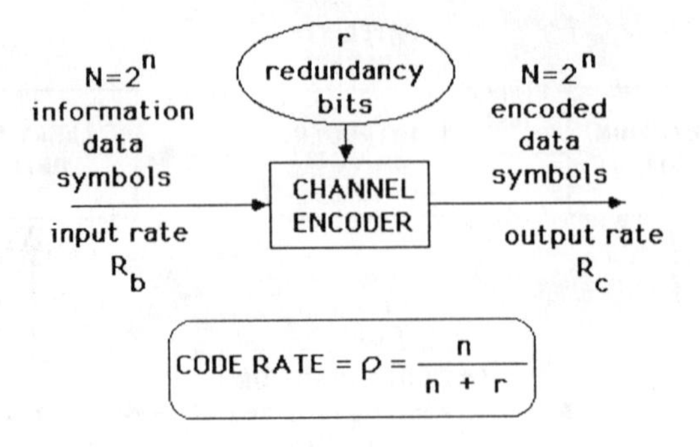

Figure 3.15 The principle of channel encoding.

Hence:

$$R_c = R_b/\rho \qquad \text{(bit/s)} \tag{3.33b}$$

3.7.7 Digital modulation

Figure 3.16 shows the principle of a modulator. It consists of:

—A symbol generator.
—An encoder.
—A radio-frequency signal (carrier) generator.

The symbol generator generates symbols with M states, where $M = 2^m$, from m consecutive bits of the binary input stream. The encoder establishes a correspondence between the M states of these symbols and M possible states of the transmitted carrier. Two types of coding are practised:

—Direct encoding—one state of the symbol defines one state of the carrier.
—Encoding of transitions (differential encoding)—one state of the symbol defines a transition between two consecutive states of the carrier.

For a bit rate R_c (bit/s) at the modulator input, the modulation rate R_s at the demodulator

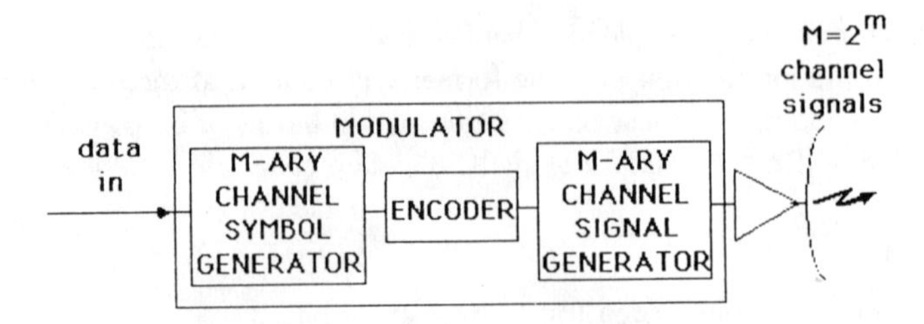

Figure 3.16 The principle of a modulator for digital transmission.

output (the number of changes of state of the carrier per second) is given by:

$$R_s = R_c/m = R_c/\log_2 M \qquad \text{(baud)} \tag{3.34}$$

Phase modulation (phase shift keying (PSK) is particularly well suited to satellite links. In fact, it has the advantage of a constant envelope and in comparison with frequency shift keying (FSK) it provides better spectral efficiency (number of bits/s transmitted per unit of radio-frequency bandwidth—see Section 3.7.7.4). Two types of phase shift keying will be considered:

(a) Two-state modulatiom ($M = 2$):
 —With direct encoding—Binary Phase Shift Keying (BPSK).
 —With differential encoding—Differentially Encoded BPSK (DE-BPSK).
(b) Four-state modulation ($M = 4$):
 —With direct encoding—Quadrature Phase Shift Keying (QPSK).
 —With differential encoding—Differentially Encoded QPSK (DE-QPSK).

3.7.7.1 *Two state modulation—BPSK and DE-BPSK*

Figure 3.17 shows the structure of a two-state phase modulator. There is no symbol generator since the binary symbol is identified by the input bit. Let b_k be the logical value of a bit at the modulator input in the time interval $[kT_c, (k+1)T_c]$. The encoder transforms the input bit b_k into a bit m_k such that:

—For direct encoding (BPSK): $m_k = b_k$.
—For differential encoding (DE-BPSK): $m_k = b_k \oplus m_{k-1}$, where $\oplus$ represents the 'exclusive' OR logical operation.

The radio-frequency signal generator is controlled by the bit m_k which is represented in the time interval $[kT_c, (k+1)T_c]$ by a voltage $v(kT_c) = \pm V$. The carrier of frequency $f_c = \omega_c/2\pi$ can be expressed during this interval:

$$C(t) = A\cos(\omega_c t + \theta_k) = \mathrm{v}(kT_c)A\cos(\omega_c t) \qquad \text{(V)} \tag{3.35}$$

where $\theta_k = \bar{m}_k \pi$ and $\bar{m}_k$ is the logical complement of m_k. $\theta_k = 0$ if $m_k = 1$ and $\theta_k = \pi$ if

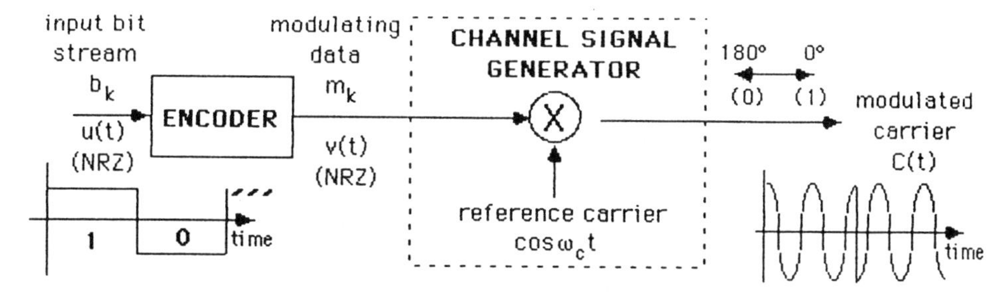

Figure 3.17 Two-state phase modulator (BPSK).

Table 3.4 Relation between bit and carrier phase in BPSK.

(a) Direct encoding

b_k	Phase
0	π
1	0

(b) Differential encoding

	Previous state		New State		
b_k	m_{k-1}	Phase	m_k	Phase	
0	0	π	0	π	No phase change
	1	0	1	0	
1	0	π	1	0	Phase change
	1	0	0	π	

$m_k = 0$. During this period the carrier thus exhibits the same phase state from the two phase states of 0 and π. The last term of expression (3.35) shows that this phase modulation can be regarded as suppressed carrier amplitude modulation with two amplitude states $\pm V$ (notice that the envelope remains constant). This modulation can be realised simply, as shown in Figure 3.17, by multiplication of the carrier by the voltage $v(t)$. Table 3.4 shows the relation between b_k and the carrier phase for the two types of encoding.

3.7.7.2 *Four state modulation: QPSK and DE-QPSK*

Figure 3.18 shows the configuration of a four-state phase modulator. The symbol generator is a serial-parallel converter which generates two binary streams A_k and B_k, each of bit rate $R_C/2$, from the input stream of bit rate R_c. The symbol A_kB_k is a 'dibit' which occupies the time interval $[kT_s, (k+1)T_s]$ equal to $T_s = 2T_c$ or the duration of two bits. The encoder

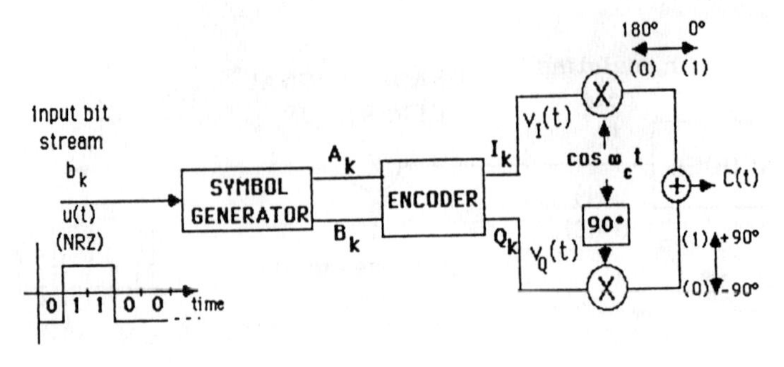

Figure 3.18 Four-state phase modulator (QPSK).

transforms dibit A_kB_k into dibit I_kQ_k. The encoder realises the following function:

—Direct encoding (QPSK):

$$I_k = A_k \qquad Q_k = B_k \tag{3.36}$$

—Differential encoding (DE-QPSK):

$$\begin{aligned} I_k &= I_{k-1}\bar{A}_k\bar{B}_k + \bar{I}_{k-1}A_kB_k + Q_{k-1}A_k\bar{B}_k + \bar{Q}_{k-1}\bar{A}_kB_k \\ Q_k &= I_{k-1}\bar{A}_kB_k + \bar{I}_{k-1}A_k\bar{B}_k + Q_{k-1}\bar{A}_k\bar{B}_k + \bar{Q}_{k-1}A_kB_k \end{aligned} \tag{3.37}$$

The radio-frequency signal generator superposes two carriers in quadrature. These two carriers are amplitude modulated (with suppressed carrier) by bits I_k and Q_k which are represented in the time interval $[kT_s, (k+1)T_s]$ by voltages $v_I(kT_s) = \pm V$ and $v_Q(kT_s) = \pm V$. The expression for the carrier during the interval $[kT_s, (k+1)T_s]$ is:

$$\begin{aligned} C(t) &= v_I(kT_s)A\cos(\omega_c t) + v_Q(kT_s)A\sin(\omega_c t) \\ &= A\sqrt{2}V\cos(\omega_c t - \Phi) \qquad \text{(V)} \end{aligned} \tag{3.38}$$

where $\Phi = 45°$, $135°$, $225°$ or $315°$ according to the values of the voltages $v_I(kT_s)$ and $v_Q(kT_s)$. It can be seen that the carrier can take one of four phase states, each state being associated with one value of the symbol I_kQ_k. In general two phase states separated by 90° are associated with two dibits I_kQ_k which differ by a single bit (Gray code). Hence an error at the receiver in recognising the phase between two adjacent phases leads to an error in a single bit. Table 3.5 shows the correspondence between dibit A_kB_k and carrier phase for the two types of coding.

Table 3.5 Relation between the bit pair A_kB_k and carrier phase in QPSK.

(a) Direct encoding

A_kB_k	Phase
00	$5\pi/4$
01	$3\pi/4$
10	$7\pi/4$
11	$\pi/4$

(b) Differential encoding

A_kB_k	Phase change
00	0
01	$\pi/2$
10	π
11	$3\pi/2$

3.7.7.3 *Variants of QPSK*

In QPSK modulation, the voltages which modulate the two carriers in quadrature change simultaneously and the carrier can be subjected to a phase change of 180°. In a satellite link which includes filter components, large phase shifts cause amplitude modulation of the carrier. The non-linearity of the channel transforms these amplitude variations into phase variations (see Chapter 9) which degrade the performance of the demodulator. Several variants of QPSK modulation have been proposed to limit the amplitude of the phase shift. These are:

—Offset QPSK (OQPSK), also called staggered QPSK (SQPSK)—the voltage is shifted by half the symbol duration. Only one of the two carriers in quadrature is subjected to a phase change at one time and hence the resulting carrier phase can vary only in steps of 90°.

—Minimum shift keying modulation (MSK), also called fast frequency shift keying (FFSK)—the voltage representing bits I_k and Q_k is no longer constant for the duration of the symbol but varies according to a sine law. It can be shown that this modulation is a particular case of modulation by frequency shift keying (FSK) [GRO-76]. The carrier

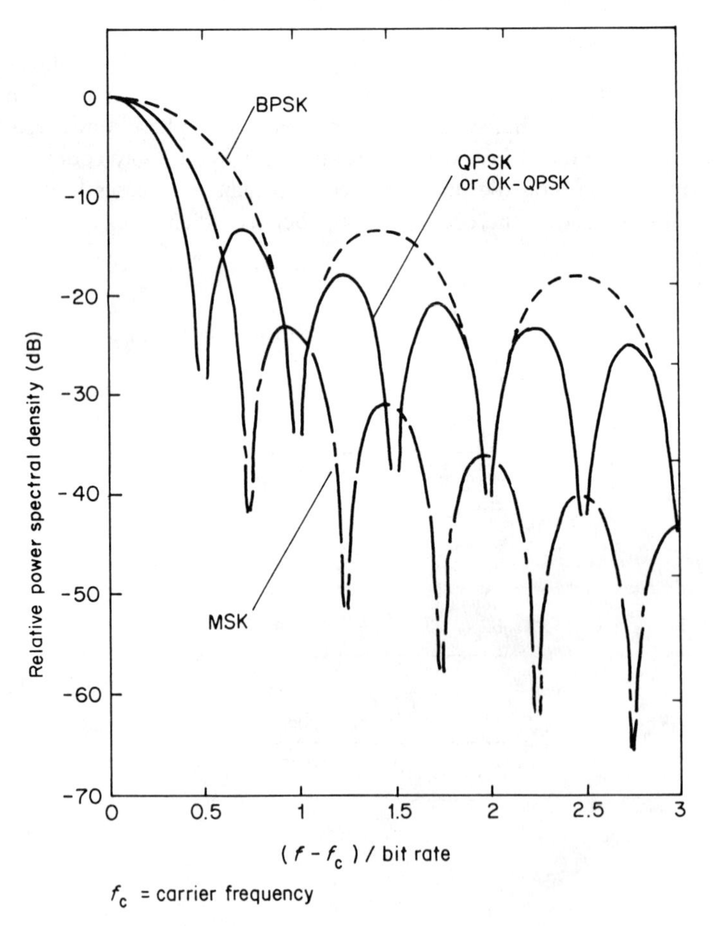

Figure 3.19 The spectrum of digital carriers.

phase changes linearly for the duration of the symbol to reach the required value of phase shift at the end.

In spite of their advantages these types of modulation are not used in operational systems. The advantages are not sufficient to justify the high cost of development of new modulators and demodulators and the replacement of numerous QPSK modulators and demodulators already in operation. An excellent presentation of these types of modulation will be found in [FEH-81].

3.7.7.4 *Spectral efficiency*

Figure 3.19 shows the form of the spectrum of the digital carriers presented in this section. An important parameter in the choice of modulation type for a space link is the spectral occupancy of the carrier in the satellite repeater. In fact, the link operator pays for the bandwidth occupied and is reimbursed at the user's expense. His benefit increases as spectral occupation decreases and the number of paying users increases, that is the throughput is high. This leads to the concept of spectral efficiency which is defined as the ratio of the capacity R_c (bit/s) of a carrier to the bandwidth occupied B (Hz), hence:

$$\Gamma = \mathrm{R}_c/B \qquad \text{(bit/s Hz)} \tag{3.39}$$

It can be shown that for BPSK modulation, the theoretical spectral efficiency is 1 bit/s Hz. For QPSK modulation, it is 2 bit/s Hz [PRO-83]. In practice, taking account of the imperfections of the transmission channel (such as non-optimal filtering and non-linearities), the spectral efficiency is of the order of 0.7–0.8 bit/s Hz for BPSK and 1.4–1.6 bit/s Hz for QPSK.

3.7.8 Demodulation

The role of the demodulator is to identify the phase (or phase shift) of the received carrier and to deduce from it the value of the bits of the transmitted binary stream. Demodulation can be:

—Coherent: the demodulator makes use of a local sinusoidal reference signal having the same frequency and phase as the modulated wave at the transmitter. The demodulator interprets the phase of the received carrier by comparing it with the phase of the reference signal. Coherent demodulation enables the binary stream to be reconstructed for both cases of transmission encoding—direct (BPSK and QPSK) and differential (DE-BPSK and DE-QPSK).

—Differential: the demodulator compares the phase of the received carrier for the duration of transmission of a symbol and its phase for the duration of the preceding symbol. The demodulator thus detects phase changes. The transmitted information can be recovered only if it is contained in phase changes; differential demodulation is always associated with differential encoding on transmission. This type of modulation and demodulation is identified by the initials D-BPSK (differential demodulation—BPSK) or D-QPSK (differential demodulation— QPSK).

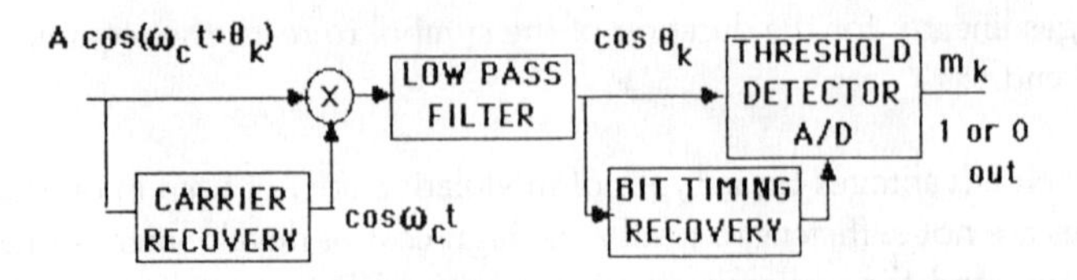

Figure 3.20 The structure of a coherent demodulator for two state phase modulation.

The structure of BPSK and QPSK demodulators will first be examined then the performance of the various types of modulation and demodulation will be compared.

3.7.8.1 *BPSK and DE-BPSK demodulators*

(a) Coherent demodulation (Figure 3.20): the received carrier is multiplied by the reference signal $\cos \omega_c t$ provided by a carrier recovery circuit. The output of the multiplier is:

$$\begin{aligned} V(t) &= C(t) \times \cos \omega_c t \\ &= A \cos (\omega_c t + \theta_k) \cos \omega_c t \\ &= (A \cos \theta_k + A \cos (2\omega_c t + \theta_k)/2 \end{aligned} \tag{3.40}$$

The low-pass filter which follows the multiplier eliminates the component of frequency $2f_c$. At the filter output the voltage is $v \cos \theta_k = \pm v$. This is compared with the 0 threshold by the threshold detector selected at the middle of the bit interval by a bit synchronisation circuit. The detector is in error if noise changes the sign of the voltage.

(b) Differential demodulation (Figure 3.21): the received DE-BPSK carrier is fed into a delay line (with a delay equal to the duration of one symbol) and is multiplied by the delayed output of the line. The result of the multiplication, $v^2 \cos (\omega_c t + \theta_k) \cos (\omega_c t + \theta_{k-1})$, is filtered by a low-pass filter the output of which is $v^2 \cos (\theta_k - \theta_{k-1})$; the value of m_k is deduced from the sign of $v^2 \cos (\theta_k - \theta_{k-1})$.

3.7.8.2 *QPSK and DE-QPSK demodulators*

(a) Coherent demodulation (Figure 3.22): the demodulator is an extension of coherent demodulation of a BPSK carrier to channels in phase and quadrature.

(b) Differential demodulation (Figure 3.23): the demodulator is an extension of differential demodulation of a DE-BPSK carrier to channels in phase and quadrature.

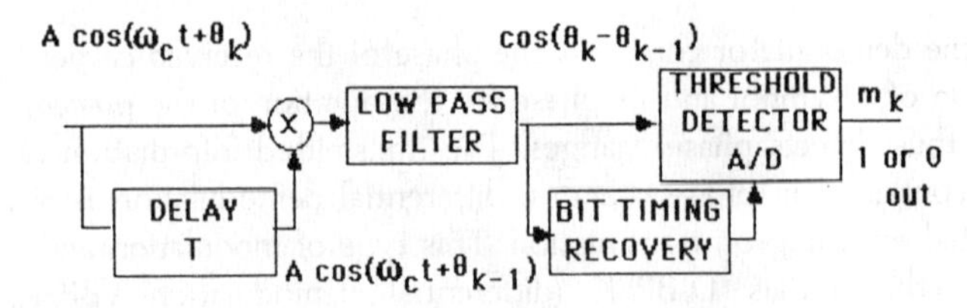

Figure 3.21 The structure of a differential demodulator for two state phase modulation.

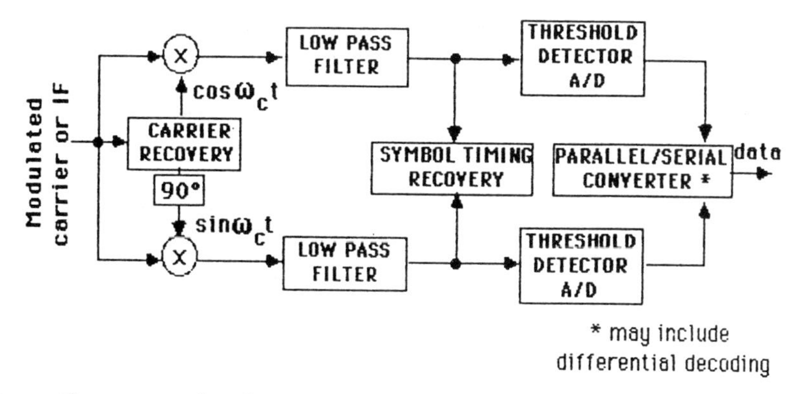

Figure 3.22 The structure of a coherent demodulator for four state phase modulation.

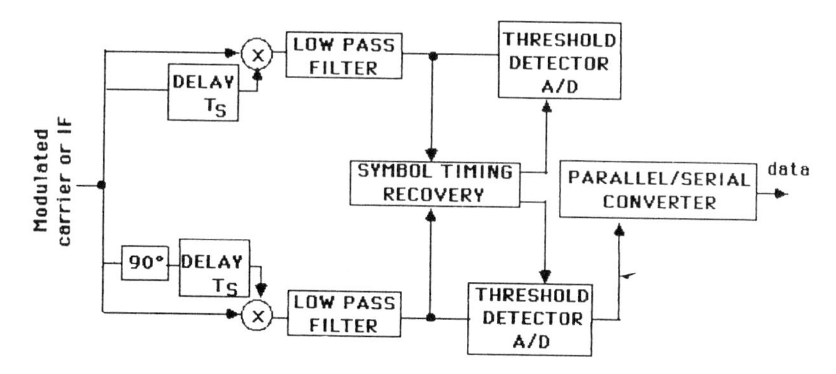

Figure 3.23 The structure of a differential demodulator for four state phase modulation.

3.7.8.3 *Performance*

Phase (or phase shift) identification errors under the influence of noise lead to errors in identification of the received symbol and hence the received bit. The measure of demodulator performance is the bit error rate (BER). The error rate can be estimated from the theoretical error probability (EP). This error probability is always less than the measured error rate. Nevertheless it permits the various demodulation techniques to be compared and the degradation of a real system to be quantified.

For two-state modulation, the symbol is identified as the bit. The symbol error probability SEP represents the bit error probability BEP:

$$\text{BEP} = \text{SEP} \tag{3.41}$$

For four-state modulation where association of the symbols $I_k Q_k$ with the phase states follows a Gray code, the bit error probability BEP is given by:

$$\text{BEP} = \text{SEP}/2 \tag{3.42}$$

Table 3.6 Expressions for binary error probabilities (BEP)

Type of modulation–demodulation	Binary error probability
Coherent demodulation:	
Direct encoding:	
BPSK	$(1/2)\,\mathrm{erfc}\sqrt{(E_c/N_0)}$
QPSK	$(1/2)\,\mathrm{erfc}\sqrt{(E_c/N_0)}$
Differential encoding:	
DE-BPSK	$\mathrm{erfc}\sqrt{(E_c/N_0)}$
DE-QPSK	$\mathrm{erfc}\sqrt{(E_c/N_0)}$
Differential demodulation:	
(Differential encoding only)	
D-BPSK	$(1/2)\exp(-E_c/N_0)$

More generally:

$$\mathrm{BEP} = \mathrm{SEP}/\log_2 M \qquad \text{for } M \geqslant 2 \tag{3.43}$$

Table 3.6 gives the expressions for the bit error probabilities BEP for the demodulators considered previously [PRO-83], [FEH-81]. The function erfc is the complementary error function defined by:

$$\mathrm{erfc}(x) = (2/\sqrt{\pi})\int_x^{\infty} \mathrm{e}^{-u^2}\mathrm{d}u \tag{3.44}$$

This function is tabulated with some useful approximate formulae in [BIC-86, Volume 2, p. 276].

The ratio E_c/N_0 arises in the expression for error probability, where E_c is the energy per bit. This is the product of the power of the received carrier for the duration of one bit, namely $E_c = CT_c = C/R_c$. Hence:

$$E_c/N_0 = (C/R_c)N_0 = (C/N_0)/R_c \tag{3.45}$$

For a conventional satellite link the value of C/N_0 to be used is $(C/N_0)_T$. For a regenerative satellite link the value of C/N_0 to be used is either that of the uplink or that of the downlink.

Figure 3.24 shows the corresponding error probability curves.

3.7.8.4 *Comparison*

In practice, the project specification stipulates a given error rate (see quality targets, Section 3.3). The error probability is thus fixed and the required value of E_c/N_0 is to be determined. The types of modulation and demodulation are compared on the basis of the values of E_c/N_0 obtained for each type. Table 3.7 indicates the theoretical values of E_c/N_0 necessary to achieve a given bit error probability BEP for each type of modulation and demodulation. The figures in brackets indicate the difference between the value of E_c/N_0 for the modulation/demodula-

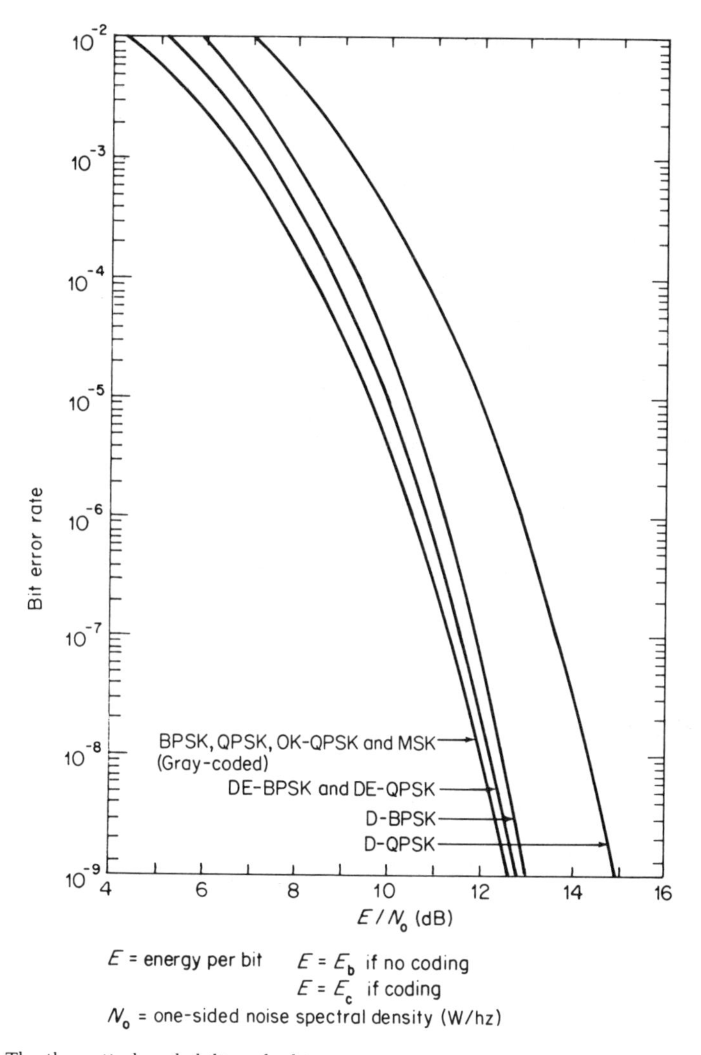

Figure 3.24 The theoretical probability of a bit error.

tion type considered and the value obtained with QPSK. This quantity is called the 'degradation in E/N_0'. This degradation is, in this case, associated with a choice of modulation/demodulation type of lower performance, in terms of bit error probability, than BPSK or QPSK. It expresses the increase in E_c/N_0 required to regain the same binary error probability and consequently the increase in C/N_0 on the satellite link for the same throughput.

In practice the value of E_c/N_0 required to achieve a given BEP is higher than the value given in Table 3.7 as a result of demodulator implementation degradation. For instance, the value of E/N_0 required to achieve BEP $= 10^{-6}$ using a coherent demodulator for demodulation of a differentially encoded modulation (DE-BPSK or DE-QPSK) could be as high as 12.3 dB, considering 0.3 dB for theoretical degradation from the theoretical performance of coherent demodulation using direct encoding plus 1.5 dB for the implementation degradation of the actual demodulator.

Table 3.7 Theoretical values of E_c/N_0 to achieve a given bit error probability for transmitted binary elements (E_c = energy per transmitted bit, N_0 = noise spectral density). Δ = degradation in E_c/N_0 relative to B-PSK and Q-PSK

BEP	BPSK	DE-BPSK (Δ) DE-QPSK	D-BPSK (Δ)	D-QPSK (Δ)
10^{-3}	6.8 dB	7.4 dB (0.6 dB)	7.9 dB (1.1 dB)	9.2 dB (2.4 dB)
10^{-4}	8.4 dB	8.8 dB (0.4 dB)	9.3 dB (0.9 dB)	10.7 dB (2.3 dB)
10^{-5}	9.6 dB	9.9 dB (0.3 dB)	10.3 dB (0.7 dB)	11.9 dB (2.3 dB)
10^{-6}	10.5 dB	10.8 dB (0.3 dB)	11.2 dB (0.7 dB)	12.8 dB (2.3 dB)
10^{-7}	11.3 dB	11.5 dB (0.2 dB)	11.9 dB (0.6 dB)	13.6 dB (2.3 dB)
10^{-8}	12.0 dB	12.2 dB (0.2 dB)	12.5 dB (0.5 dB)	14.3 dB (2.3 dB)
10^{-9}	12.6 dB	12.8 dB (0.2 dB)	13.0 dB (0.4 dB)	14.9 dB (2.3 dB)

3.7.9 Decoding and error correction

The decoder uses the redundancy introduced at the encoder in order to detect and correct errors. In this respect, two techniques can be distinguished which can be used independently or simultaneously:

—Forward-acting error correction (FEC).
—Automatic repeat request (ARQ).

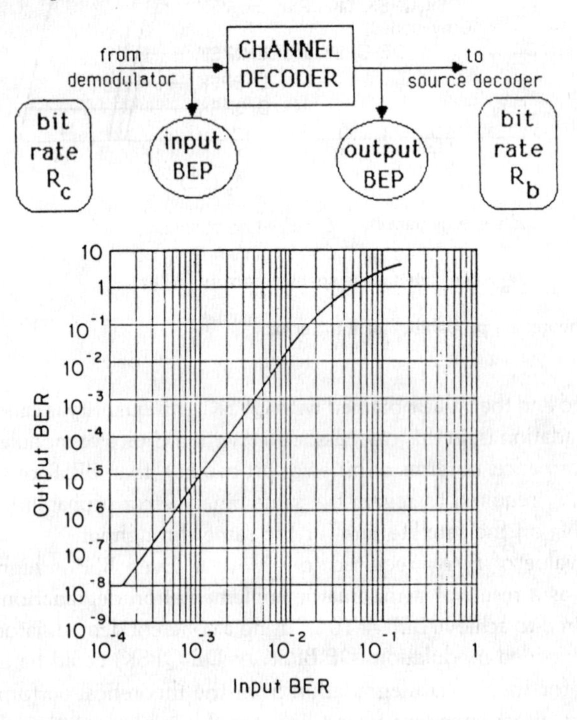

Figure 3.25 The relation between bit error probability $(BEP)_{out}$ at the output of an error correcting decoder and bit error probability at the input $(BEP)_{in}$.

3.7.9.1 *Forward-acting error correction*

At the decoder input, the bit rate is R_c and the bit error probability is $(BEP)_{in}$. At the output, the information rate is again R_b, which was that at the encoder input (see Section 3.7.5). Because of the error correction provided by the decoder, the bit error probability $(BEP)_{out}$ is lower than at the input. Figure 3.25 depicts an example of the relation between $(BEP)_{out}$ and $(BEP)_{in}$. The value of $(BEP)_{in}$ is given as a function of E_c/N_0, according to the modulation/demodulation type, by one of the curves of Figure 3.24. By combining this curve with the curve of Figure 3.25 for the encoding/decoding scheme considered, the curves of Figure 3.26 can be established. These curves establish the performance of the modulation and encoding system. Notice that the bit error probability is expressed as a function of E_b/N_0 which is related to E_c/N_0 by the expression:

$$E_b/N_0 = E_c/N_0 - 10 \log \rho \tag{3.46}$$

where ρ is the code rate defined by (3.32). The 'decoding gain' is defined as the difference in E_b/N_0, at a given value of BEP, between transmissions with and without encoding. Table 3.8 indicates several typical values of decoding gain. The advantage to the link budget provided by the decoding gain is paid for by an increase in the bandwidth used on the space link. In fact, it is necessary to transmit a bit rate R_c which is greater than the bit rate R_b, and, according to (3.39), the bandwidth used is $B = R_c\Gamma = R_b/\rho\Gamma$.

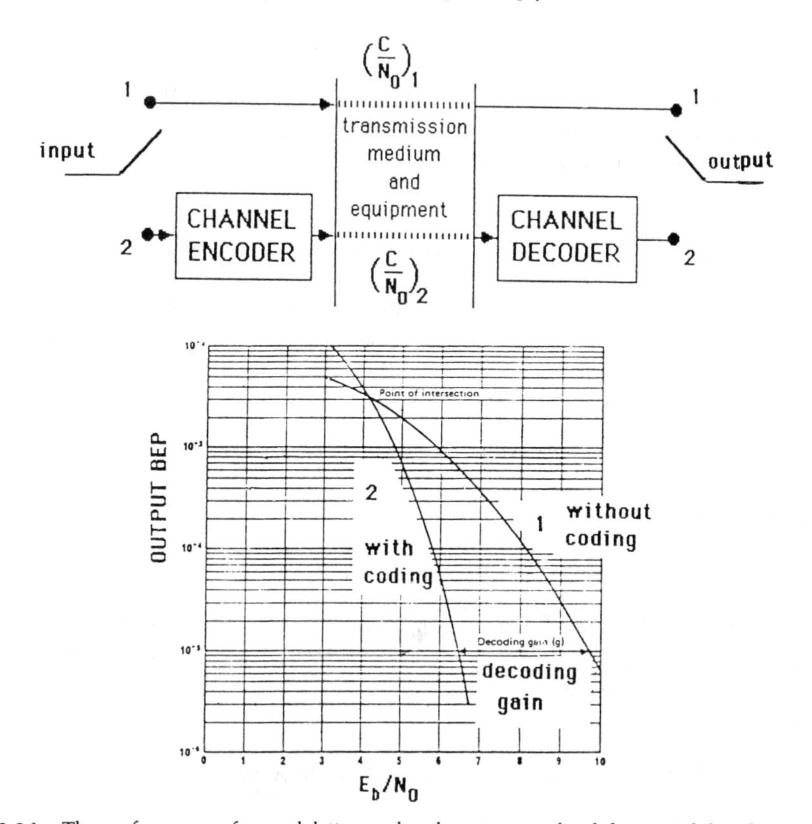

Figure 3.26 The performance of a modulation and coding system; the definition of decoding gain.

Table 3.8 Typical values of decoding gain

Coding ratio ρ	E_b/N_0 required for BEP $= 10^{-6}$	Decoding gain
1	10.5 dB	0 dB
7/8	8.0 dB	2.5 dB
3/4	6.2 dB	4.3 dB
2/3	5.7 dB	4.8 dB
1/2	5.3 dB	5.2 dB

3.7.9.2 *Example*

Consider a satellite link which provides a ratio $(C/N_0)_T = 85$ dB(Hz). The bandwidth of the repeater is 36 MHz. It is required to ensure transmission at an information rate $R_b = 36$ Mbit/s with a bit error probability BEP $= 10^{-6}$. The modulation used is QPSK with a spectral efficiency $\Gamma = 1.5$ bit/s Hz.

(a) In the absence of forward-acting error correction ($\rho = 1$). The transmitted bit rate is $R_c = R_b = 36$ Mbit/s. The bandwidth used is: $B = 36$ Mbit/s/1.5 bit/s Hz $=$ 24 MHz. Hence a fraction of the repeater bandwidth is used. The required value of $(C/N_0)_T$ is:

$$(C/N_0)_1 = (E_b/N_0)_1 R_b$$

As there is no coding $E_b/N_0 = R_c/N_0$. To ensure the BEP $= 10^{-6}$, from Table 3.8, $(E_b/N_0)_1 = 10.5$ dB. A demodulator implementation degradation of 1.5 dB will be taken. Hence $(E_b/N_0)_1 = 10.5\text{ dB} + 1.5\text{ dB} = 12$ dB.
Hence:

$$(C/N_0)_1 = (E_b/N_0)_1 R_b = 12\text{ dB} + 10 \log 36 \times 10^6 = 87.6\text{ dB(Hz)}$$

The value required for $(C/N_0)_T$ is greater than the value available; the link is power limited. In contrast, the bandwidth used is less than the bandwidth of the repeater.

(b) With forward-acting error correction ($\rho = 2/3$). The binary transmission rate is $R_c = R_b/\rho = 36\text{ Mbit/s} \times 3/2 = 54$ Mbit/s. The bandwidth used is:

$$B = 54\text{ Mbit/s}/1.5\text{ bit/s Hz} = 36\text{ MHz}.$$

The required value of $(C/N_0)_T$ is:

$$(C/N_0)_2 = (E_b/N_0)_2 R_b$$

To ensure BEP $= 10^{-6}$, from Table 3.8, $(E_b/N_0)_2 = 5.7$ dB. There is thus a benefit of a decoding gain of 4.8 dB. The demodulator implementation degradation remains at 1.5 dB. Hence, $(E_b/N_0)_2 = 5.7\text{ dB} + 1.5\text{ dB} = 7.2$ dB.

Hence:

$$(C/N_0)_2 = (E_b/N_0)_2 R_b = 7.2\text{ dB} + 10 \log 36 \times 10^6 = 82.8\text{ dB(Hz)}$$

that is, 4.8 dB less than previously due to the decoding gain. This time the required value of $(C/N_0)_T$ is less than the available value. By occupying all of the repeater bandwidth, an economy of power is achieved which leads to a power margin of 85 dB (Hz) − 82.8 dB (Hz) = 2.2 dB.

3.7.9.3 *Automatic repeat request*

This technique applies particularly in the case of data packet transmission. The decoder detects errors but does not correct them. In the case of error detection, a retransmission request is sent to the transmitter. It is, therefore, necessary to provide a return channel. This can be a satellite or terrestrial channel. The use of error detecting codes requires the ability to control the source throughput and allow a variable decoding delay. These disadvantages are compensated by the simplicity of decoder realisation, the possibility of adapting to varying error statistics and obtaining low error rates.

Three basic techniques are employed [LIN-84], [BIC-86], [BHA-81]:

—Retransmission with stop and wait or reception acknowledgement (ARQ Stop-and-wait).
—Continuous retransmission (Go-Back-N ARQ).
—Selective retransmission (Selective-repeat ARQ).

Figure 3.27 illustrates the principle of these three techniques.

The performance is measured in terms of efficiency, expressed as the ratio of the mean number of information bits transmitted to the total number of bits which could be transmitted during the same time.

3.7.9.4 *Example*

Consider a digital satellite link with a capacity $R = 48$ kbit/s. The round-trip return time is taken to be $T = 600$ ms. The bit error probability is BEP $= 10^{-4}$. Transmission is in blocks of $n = 1000$ bits.

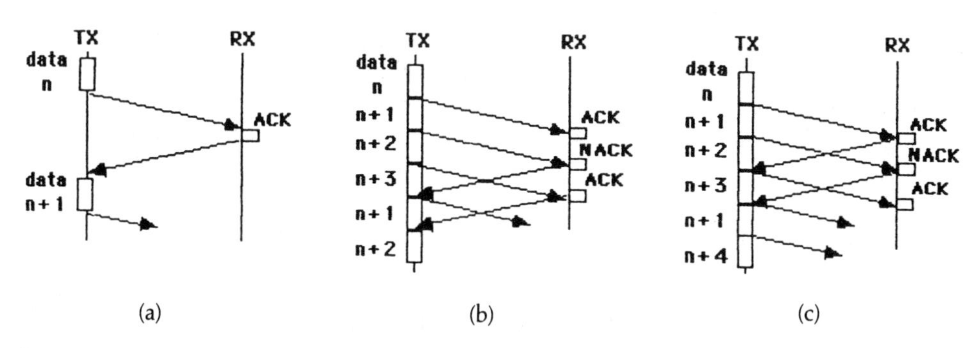

Figure 3.27 Error detection with retransmission. (a) Retransmission with stop and wait or reception acknowledgement (ARQ Stop-and-wait). (b) Continuous retransmission (Go-Back-N ARQ). (c) Selective retransmission (Selective-repeat ARQ).

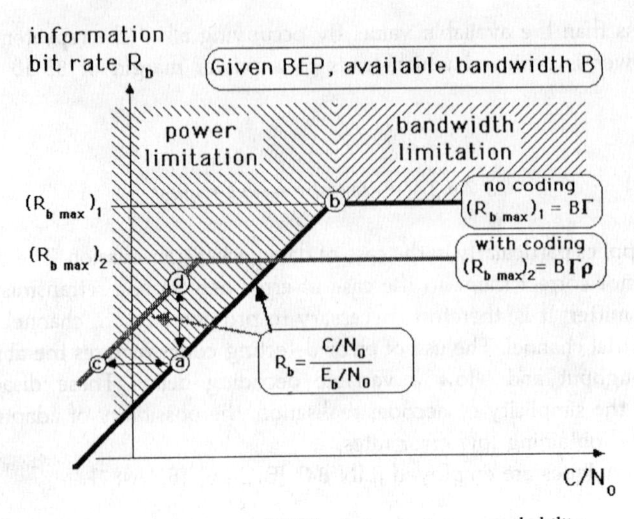

Figure 3.28 Information rate as a function of C/N_0 at constant error probability.

The error probability in a block is $P_B = 1-(1-\text{BEP})^n = 1-\exp(-n\text{BEP})$ for $n\text{BEP} \ll 1$, hence $P_B = 0.1$.

—Efficiency in ARQ-SW: $\eta = k(1-P_B)/(n+RT) = 0.03$
—Efficiency in ARQ-GB (N): $\eta = k(1-P_B)/(n(1-P_B)+RTP_B) = 0.23$
—Efficiency in ARQ-SR: $\eta = k(1-P_B)/n = 0.84$.

The increase in efficiency from one technique to another is accompanied by an increase in the complexity of the equipment.

3.7.9.5 *Conclusion: bandwidth-power interchange*

For a given error probability, error correcting coding permits the required value of $(C/N_0)_T$ to be reduced on the satellite link on condition that a greater radio-frequency bandwidth occupation is accepted. As $(C/N_0)_T$ is related to the carrier power, it is customary to call this practice 'power-bandwidth exchange'. Notice that, in this exchange, the code rate plays a role analogous to that of modulation index in frequency modulation (see Section 3.6.6.4); reducing the code rate for a constant information bit rate R_b widens the bandwidth and enables power to be economised by reducing C/N_0. This is illustrated in Figure 3.28 which shows the various possible combinations of information bit rate R_b and required value of $C/N_0 = (C/N_0)_T$ at constant error probability according to the code rate.

3.7.10 Energy dispersion

The recommendation made by the CCIR [CCIR Recommendation 446] for the use of energy dispersion techniques in order to limit interference between radio communication systems sharing the same frequency bands has been recalled in Section 3.6.11. In digital transmission when the binary stream is random the carrier energy is spread throughout the spectrum of the modulating signal. By limiting the transmitted EIRP of the satellite, one can remain below the limit on surface power density at ground level (see Section 2.8.1.4). In contrast, if the binary stream contains a repeated fixed pattern, lines appear in the spectrum of the modulated carrier and their amplitude can lead to the limit on surface power density at ground level being exceeded. The principle of energy dispersion is to generate a modulating binary stream which has random properties regardless of the structure of the binary stream containing the information [CCIR Report 384]. This operation which is performed at the transmitter before modulation is called scrambling. On reception the inverse operation, performed after demodulation, is called descrambling. Figure 3.29 shows an example of scrambler and descrambler realisation. Each bit of the binary stream carrying information is combined by modulo 2 addition with each bit generated by a pseudorandom sequence generator (this operation resembles on-line encryption, see Section 3.7.5). The pseudorandom sequence generator consists of a shift register with various feedback paths. The descrambler contains the same pseudorandom sequence generator and, by virtue of the properties of modulo 2 addition, the combination by modulo 2 addition of the bits of the demodulated binary stream

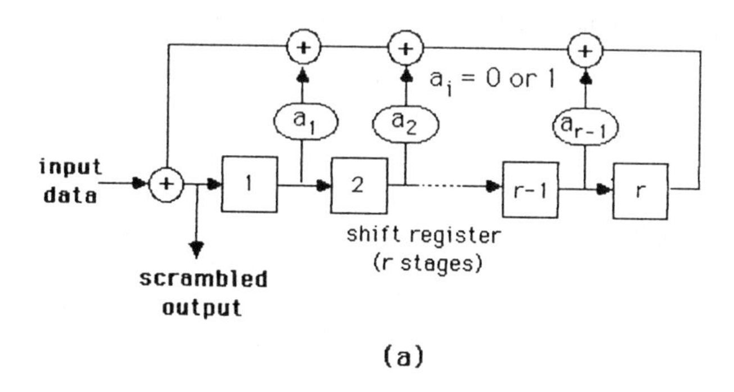

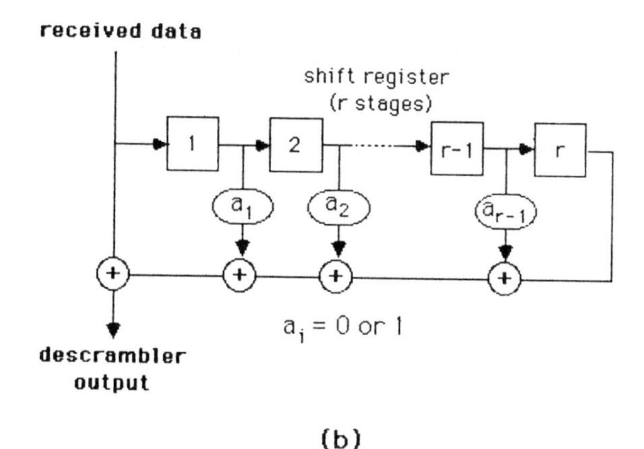

Figure 3.29 (a) A scrambler. (b) A descrambler.

with those of the random sequence provides recovery of the information content. This implies synchronism of the two pseudorandom sequence generators. The arrangement of Figure 3.29 automatically ensures synchronism; after r bits transmitted without error the r stages of the scrambling and descrambling shift registers are in the same state. However, an error in one bit produces as many errors in an interval of r bits as there are non-zero coefficients a_i in the feedback paths [FEH-81], [DES-85]. An additional advantage provided by scrambling is suppression of sequences of logical 0's or 1's which, in NRZ-L coding, can lead to a loss of synchronisation of the bit timing recovery circuit and introduce detection errors at the demodulator output as a result of a timing error in the instant of decision (see Figures 3.21 to 3.24).

3.7.11 Conclusion—comparison between analogue and digital transmission

Consider a satellite repeater providing a bandwidth of 36 MHz. It is assumed that the carrier occupies all the repeater bandwidth (single access). Two approaches will be compared (Figure 3.30):

—The carrier is frequency modulated by a multiplex of analogue telephone signals (FDM/FM).
—The carrier is QPSK modulated by a multiplex of digital telephone signals (TDM/QPSK).

In each case the multiplex capacity, as a number of telephone channels, will be determined as a function of the required $(C/N_0)_T$. The quality target will be taken as that of the CCIR, namely:

—For analogue transmission: the noise power should not exceed 10 000 pW0p, psophometrically weighted 1-minute mean power for more than 20% of any month (CCIR Recommendation 353). Under these conditions it has been seen (Section 3.3.1.1) that the minimum S/N ratio of test signal power to noise power is $S/N = 50$ dB.
—For digital transmission: the bit error rate must not exceed 10^{-6}, 10-minute mean value, for more than 20% of any month [CCIR Recommendation 522].

3.7.11.1 *Analogue transmission*

According to Table 3.1, the multiplex capacity is a function of the maximum frequency f_{max} of the multiplex. The spectral extent of the carrier modulated by the multiplexed telephone signals is limited by the repeater bandwidth, that is 36 MHz. This bandwidth is also that of the receiver, $B_{IF} = 36$ MHz. By using equation (3.26) the value of peak frequency deviation ΔF_p is given by

$$\Delta F_p = B_{IF}/2 - f_{max} \quad \text{(Hz)} \tag{3.47}$$

The expression for the S/N ratio, given by (3.25), is recalled below:

$$(S/N) = (\Delta F_r/f_{max})^2(1/b)pw(C/N_0)_T$$

The r.m.s. frequency deviation ΔF_r due to the test signal occurs in this expression; it is known

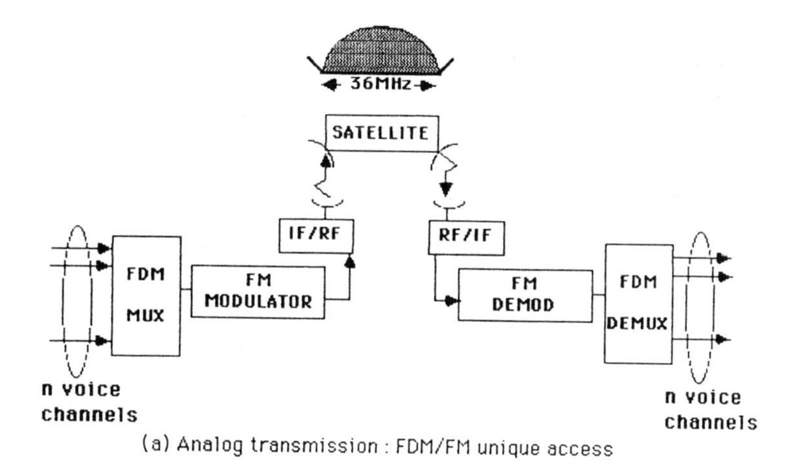

(a) Analog transmission : FDM/FM unique access

(b) Digital transmission : TDM/QPSK unique access

Figure 3.30 Link architecture for comparison of analogue and digital transmission.

(Section 3.6.8.2) to be equal to:

$$\Delta F_r = \Delta F_p / gL \qquad \text{(Hz)} \tag{3.48}$$

with $g = \sqrt{(10)} = 3.16$

$$L = 10^{(-15 + 10\log n)/20} \qquad \text{for } n > 240$$

$$L = 10^{(-1 + 4\log n)/20} \qquad \text{for } n < 240$$

On this basis, Table 3.9 can be constructed; it indicates the required value of $(C/N_0)_T$ for a capacity of n multiplexed telephone channels. In addition, the value of C/N, to verify that operation is above the demodulator threshold is also given.

Table 3.9 Required values of $(C/N_0)_T$ and C/N in accordance with the capacity of an analogue telephone multiplex with single carrier transmission occupying all the bandwidth of a 36 MHz transponder.

Number of telephone channels n	Max. frequency of multiplex f_{max}(kHz)	Peak frequency deviation ΔF_p kHz)	Effective frequency deviation ΔF_r (kHz)	Required $(C/N_0)_T$ dB (Hz)	C/N (dB)
1 092	4 892	13 108	705	95.3	19.7
972	4 028	13 972	797	92.5	16.9
792	3 284	14 716	930	89.4	13.8
612	2 540	15 460	1 111	85.6	10.0

3.7.11.2 *Digital transmission*

Let R_b be the bit rate associated with a telephone channel. It will be assumed for simplicity that transmission of n telephone channels at a rate R_b also requires transmision of signals occupying 5% of the multiplex capacity. This multiplex is transmitted after encoding at a rate R_c and this rate modulates the QPSK carrier which thus occupies a bandwidth $B = 36$ MHz.

Hence:

—Bit rate of one telephone channel: R_b (bit/s)
—Multiplex capacity: R (bit/s)
—Number of telephone channels: $n = R/(1.05\ R_b)$
—Bit rate of modulating binary stream: $R_c = R/\rho$, where ρ is the code rate
—Bandwidth used: $B = R_c/\Gamma$, where Γ is the spectral efficiency of QPSK modulation ($\Gamma = 1.5$ bit/s Hz).

In total, the number of telephone channels is given by:

$$n = B\rho\Gamma/(1.05R_b) \tag{3.49}$$

Table 3.10 Required values of $(C/N_0)_T$ and C/N in accordance with the capacity of a digital telephone multiplex with single carrier transmission occupying all the bandwidth of a 36 MHz transponder

Coding ratio ρ	Number of telephone channels n	$(C/N_0)_T$ dB (Hz)	C/N dB
1	803	87.9	12.3
7/8	703	84.8	9.2
3/4	602	82.2	6.7
2/3	535	81.3	5.7
1/2	401	79.6	4.0

Furthermore, the required value of (C/N_0) is given by:

$$(C/N_0)_T = (E_b/N_0)R$$

where the value of E_b/N_0 is obtained from Table 3.8 in accordance with the chosen coding scheme. Table 3.10 indicates the results.

The results of Tables 3.9 and 3.10 are shown in Figure 3.31 and represented by the

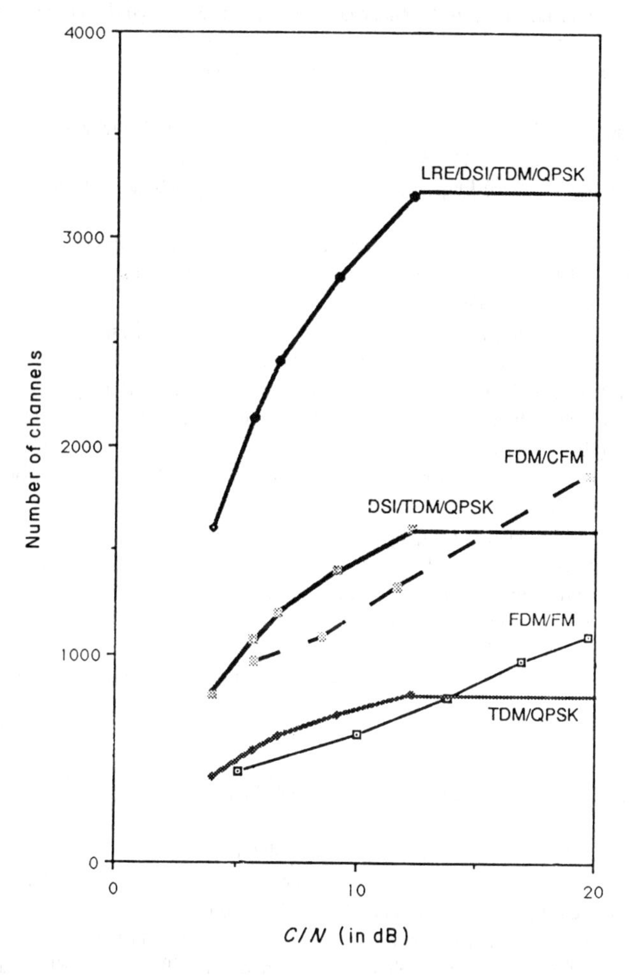

Figure 3.31 Comparison of analogue and digital transmission of a telephone multiplex by a single carrier occupying the total bandwidth of a transponder of 36 MHz bandwidth. TDM/QPSK: source coding at 64 kbit/s, digital time division multiplexing, four-state phase modulation with direct encoding and coherent demodulation. FDM/FM: frequency division multiplexing, frequency modulation. DSI/TDM/QPSK: source coding at 64 kbit/s, digital speech interpolation, digital time division multiplexing, four-state phase modulation with direct encoding and coherent demodulation. FDM/CFM: companding, frequency division multiplexing, frequency modulation. LRE/DSI/TDM/QPSK: source coding at 32 kbit/s, digital speech interpolation, digital time division multiplexing, four-state phase modulation with direct encoding and coherent demodulation.

FDM/FM and TDM/QPSK curves. This figure extends the comparison to other transmission schemes.

REFERENCES

[BIC-82] J.C. Bic, J.C. Bousquet, M. Oberle (1982) Privacy over digital satellite links, *5th International Conference on Digital Satellite Communications, Genoa, Italy, March 23–26*, pp. 243–249.

[BIC-86] J.C. Bic, D. Duponteil, J.C. Imbeaux (1986) *Eléments de communications numériques.* Dunod.

[BHA-81] V.K. Bhargava, D. Hacoun, R. Matyas, P. Nuspl (1981) *Digital Communications by Satellite*, Wiley.

[BOU-87] M. Bousquet, G. Maral (1987) 'Digital communications: satellite systems', *Systems and Control Encyclopedia*, pp. 1050–1057.

[BRA-84] W.H. Braun, J.E. Kleiger (1984) RCA satellite networks: high technology and low user cost, *Proceedings of the IEEE*, **72**, No. 11, pp. 1483–1505, Nov.

[CAM-76] S.J. Campanella (1976) Digital speech interpolation, *COMSAT. Technical Review*, **6**, No. 1, Spring.

[CLA-81] G.C. Clark, J.B. Cain (1981) *Error Correction Coding for Digital Communications*, Plenum Press.

[DES-85] K. Dessouky (1985) Effect of digital scrambling on satellite communications links, *International Conference on Communications, Paper 16.1.*

[DUP-73] J. Dupraz (1973) *Théorie de la Communication*, Editions Eyrolles.

[FEH-81] K. Feher (1981) *Digital Communications: Satellite/Earth Station Engineering*, Prentice-Hall.

[FER-75] M.E. Ferguson (1979) Design of FM single channel per carrier systems, *IEEE International Conference on Communications, June 16–18, 1975*, Vol. 1, pp 12–11/12–16, also in *'Satellite Communications'*, IEEE Press, pp. 336–341.

[GRO-76] S. Gronomeyer, A. McBride (1976) MSK and offset QPSK modulation, *IEEE T rans. on Communications*, **COM-24**, No. 8, pp. 809–820.

[JAY-76] N.S. Jayant (1976) *Waveform Quantization and Coding*, IEEE Press.

[JON-84] K. Jonnalagadda, L. Schiff (1984) Improvements in capacityof analog voice multiplex systems carried by satellite, *Proceedings of the IEEE*, 72, No. 11, Nov., pp. 1537–1547.

[LAB-84] E. Laborde, P.J. Freedenberg (1984) Analytical comparisons of CSSB and TDMA/DSI satellite transmission and techniques, *Proceedings of the IEEE*, **72**, No 11, Nov, pp.1548–1555.

[LIN-84] S. Lin, D.J. Costello, M.J. Mikller (1984) Automatic Repeat Request error control schemes, *IEEE Communications Magazine*, **22**, No. 12, pp. 5–16, Dec.

[MAR-87] G. Maral (1987) Digital communications: fundamentals, *Systems and Control Encyclopedia*, pp. 1043–1050.

[Mic-85] A.M. Michelson, A.H. Levesque (1985) *Error Control Techniques for Digital Communications*, Wiley.

[PRO-83] J.G. Proakis (1983) *Digital Communications*, McGraw-Hill.

[SCH-80] M. Schwartz (1980) Information, *Transmission, Modulation and Noise.* McGraw-Hill.

[SHA-79] K.S. Shanmugam (1979) *Digital and Analog Communication Systems*, Wiley.

[STE-87] M. Stein, (1987) *Les Modems pour transmission de données.* Masson.

[SZA-81] G. Szarvas, H. Suyderhound (1981) Enhancement of FDM-FM satellite capacity by use of compandors, *COMSAT Technical Review*, **11**, No. 1. Spring, pp. 1–58.

[TOR-81] D.J. Torrier (1981) *Principles of Military Communications Systems*, Artech House.

[TUG-82] D. Tugal, O. Tugal (1982) *Data Transmission*, McGraw-Hill.

4 MULTIPLE ACCESS

This chapter deals with techniques which permit several stations on the same network to interchange information via the nodal point which the satellite represents. Between the transmitting and receiving antennas, the satellite contains a repeater which usually consists of one or more channels, called transponders, operating in parallel on different sub-bands of the total bandwidth used (Chapter 9). Information transfer between several earth stations implies the establishment of several simultaneous station-to-station links on the same satellite channel. Depending on the chosen solution, the satellite channel amplifies one or more carriers. In the latter case, it is necessary to modify the link budget expressions presented in Chapter 2 to take account of the effects of intermodulation resulting from the non-linearity of the satellite channel transfer characteristics. It is also necessary to consider the problem of sharing the output power of the satellite channel among all the carriers.

Only single beam satellite networks will be considered in this chapter. The more complex case of multibeam satellite networks is treated in Chapter 5. In the context of a single beam satellite, the carriers transmitted by all the network stations access the same satellite receiving antenna and these same stations can receive all the carriers retransmitted by the satellite antenna. If the beam is sufficiently large (beam width of the order of 17°) a satellite channel provides coverage of all regions of the Earth visible to the satellite (for example, the global beam of INTELSAT). With a smaller angular aperture the satellite channel provides coverage of only part of the Earth (a limited geographical region or country; examples are regional satellites such as EUTELSAT and national ones such as TELECOM I and II).

After a brief review of the statistical laws governing the traffic, this chapter successively analyses the problems of routing information and multiple access by presenting the following three fundamental techniques:

—Frequency Division Multiple Access (FDMA).
—Time Division Multiple Access (TDMA).
—Code Division Multiple Access (CDMA).

4.1 TRAFFIC LAWS

The traffic intensity A is defined as

$$A = R_{\text{call}} T_{\text{call}} \qquad \text{(Erlang)} \tag{4.1}$$

where

R_{call} = mean number of calls per unit time
T_{call} = mean duration of a communication.

It is assumed that the number of users generating calls is much greater than the number of communication channels C provided and blocked calls are not stored. Under these conditions, Erlang's formula indicates the probability that n channels are occupied ($n \leqslant C$):

$$E_n(A) = (A^n/n!) \Big/ \sum_{k=0}^{k=C} (A^k/k!) \tag{4.2}$$

The probability of blocking is given by:

$$B(C, A) = E_C(A) \tag{4.3}$$

4.2 TRAFFIC ROUTING

The problem is posed in the following terms: given a demand for traffic in a network of N stations, how is this traffic to be routed? For this it is necessary to establish an adequate information transfer capacity between each pair of stations. This capacity is defined as a function of demand and acceptable blocking probability (a typical value is 0.5 to 1%). Let C_{XY} be the capacity, expressed as a number of telephone channels or bit/s, for information transfer demand t_{XY} from station X to station Y. The set of capacities available for exchanges between the N stations is described by a matrix of dimension N with a zero leading diagonal ($\dot{C}_{XX} = 0$). For example, for a network containing three stations (X = A, B, C; Y = A, B, C):

From station	To station A	B	C
A	—	C_{AB}	C_{AC}
B	C_{BA}	—	C_{BC}
C	C_{CA}	C_{BC}	—

Information transfer occurs in accordance with the techniques described in Chapter 3 and implies modulation of the radio-frequency carrier relayed by the satellite channel. Two approaches are possible:

—Establish one carrier per link.
—Establish one carrier per transmitting station.

These two approaches are shown in Figure 4.1 for the case of a network with three stations A, B and C.

4.2.1 One carrier per link (Figure 4.1a)

One carrier carries the traffic t_{XY} information from station X to station Y. The number of carriers is equal to the number of non-zero coefficients in the above matrix, that is $N(N-1)$. The matrix coefficients define the capacity of each carrier.

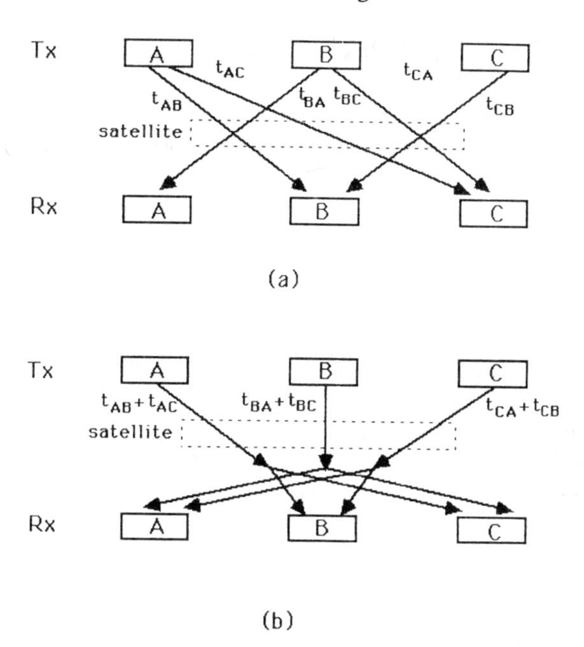

Figure 4.1 Traffic routing. (a) One carrier per link. (b) One carrier per station.

4.2.2 One carrier per transmitting station (Figure 4.1b)

The broadcasting property of the satellite is used; this enables every station to receive all the carriers transmitted to the satellite (for a single beam satellite). Under these conditions, it can be seen that the task of carrying all the traffic from station X to all other stations can be assigned to a single carrier. The number of carriers is equal to the number of stations N. The capacity of each carrier is given by the sum of the coefficients of the row of the above matrix which corresponds to the transmitting station.

4.2.3 Comparison

It can be observed that the 'one carrier per link' approach leads to a greater number of carriers than the 'one carrier per transmitting station' approach and each carrier has a smaller capacity. However, the receiving station receives only traffic which is intended for it while in the case of one carrier per transmitting station, the receiving station Y must extract 'X to Y' traffic from the carrier received from station X from the total traffic transmitted by station X on this carrier.

The choice between these two approaches is an economic one. It depends on other considerations such as the number of satellite channels, the bandwidth of the satellite channel and the multiple access technique used. In general, the fact that a large number of carriers is relayed by the satellite is a greater penalty than having to transmit carriers of higher capacity. The 'one carrier per transmitting station' approach is the most often used.

4.3 THE PRINCIPLE OF MULTIPLE ACCESS

The problem of multiple access arises when several carriers are handled simultaneously by a satellite repeater which is a nodal point of the network. There are two aspects to be considered as follows:

—Multiple access to a particular repeater channel.
—Multiple access to a satellite repeater.

4.3.1 Access to a particular channel

Each channel (transponder) amplifies every carrier whose spectrum falls within its passband at a time when the channel is in an operational state. The resource offered by each channel can thus be represented in the form of a rectangle in the time/frequency plane. This rectangle represents the bandwidth of the channel and its duration of operation (Figure 4.2). In the

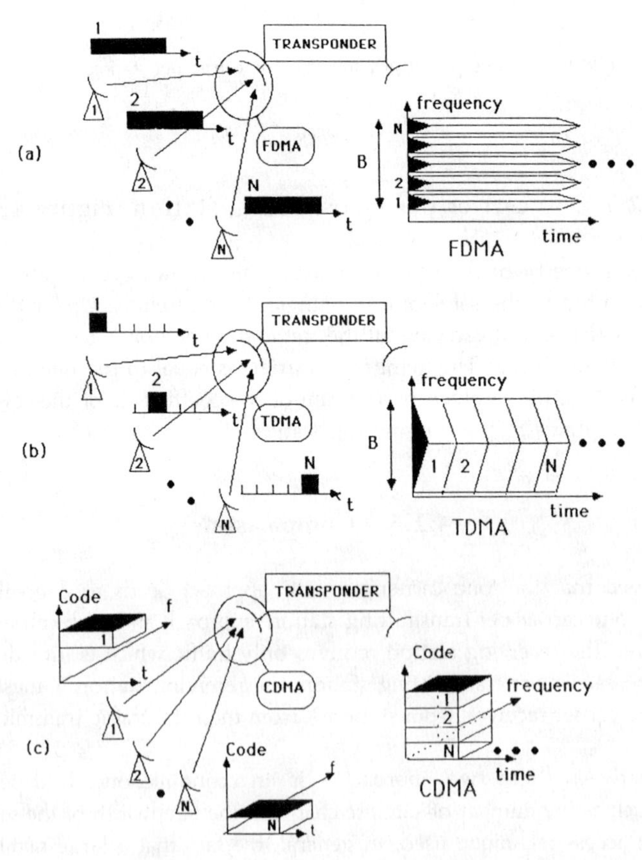

Figure 4.2 The principle of multiple access. (a) Frequency division multiple access (FDMA). (b) Time division multiple access (TDMA). (c) Code division multiple access (CDMA). (B = channel (transponder) bandwidth).

absence of special precautions, carriers occupy this rectangle simultaneously and mutually interfere. To avoid this interference, it is necessary for earth station receivers to be able to discriminate between the received carriers. This discrimination can be achieved:

—As a function of the location of the carrier energies in the frequency domain. If the spectra of the carriers each occupy a different sub-band, the receiver can discriminate between carriers by filtering. This is the principle of multiple access by frequency division (FDMA, Figure 4.2a)

—As a function of the temporal location of the carrier energies; several carriers received sequentially by the receiver can be discriminated by temporal gating even if they occupy the same frequency band. This is the principle of multiple access by time division (TDMA, Figure 4.2b).

—By the addition of a 'signature' which is known to the receiver and is specific to each carrier. This ensures identification of the carrier even when all the carriers occupy the same frequency band simultaneously. The signature is most often realised by means of pseudo random codes (Pseudo Noise (PN) codes), hence the name Code Division Multiple Access (CDMA, Figure 4.2c). The use of such codes has the effect of considerably broadening the carrier spectrum in comparison with that which it would have if modulated only by the useful information. This is why CDMA is also called Spread Spectrum Multiple Access (SSMA).

Several types of multiple access as defined above can be combined; Figure 4.3 illustrates the range of combinations.

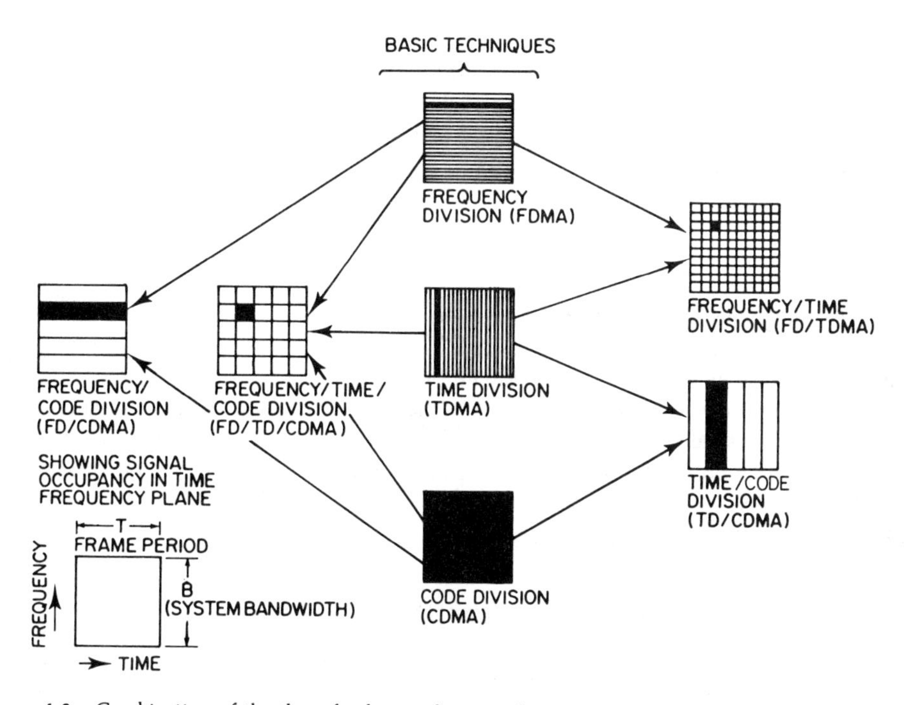

Figure 4.3 Combination of the three fundamental types of multiple access into hybrid access types.

4.3.2 Multiple access to the satellite repeater

Multiple access to a particular repeater channel (transponder) implies prior multiple access to the satellite repeater. Access to a satellite repeater is achieved as a function of frequency (or polarisation) of the carrier. For every carrrier combination there is, therefore, an obligatory FDMA access to the repeater (this includes polarisation discrimination as equivalent to frequency discrimination) together with FDMA, TDMA or CDMA access to each channel. The corresponding combinations of Figure 4.3 can thus be considered as representative of multiple access to a satellite repeater. In all cases, the spectral occupation of a carrier must not exceed the channel bandwidth.

4.4 FREQUENCY DIVISION MULTIPLE ACCESS (FDMA)

The bandwidth of a repeater channel is divided into sub-bands; each sub-band is assigned to the carriers transmitted by an earth station. With this type of access, the earth stations transmit continuously and the channel transmits several carriers simultaneously at different frequencies. It is necessary to provide guard intervals between each band occupied by a carrier to allow for imperfections of oscillators and filters. The downlink receiver selects the required carrier in accordance with the appropriate frequency. The Intermediate Frequency (IF) amplifier provides the filtering.

Depending on the multiplexing and modulation techniques used, several transmission schemes can be considered. In each case, the channel carries several carriers simultaneously and the non-linear transfer characteristic of the channel is the cause of a major problem—that of intermodulation between the carriers.

4.4.1 Transmission schemes

The various transmission schemes correspond to the different combinations of multiplexing and modulation examined in Chapter 3. Figure 4.4 illustrates the most common situations.

4.4.1.1 *FDM/FM/FDMA (Figure 4.4a)*

The baseband signals from the network or users are analogue. They are combined to form a frequency division multiplex (FDM) signal. This multiplexed analogue signal frequency modulates a carrier which accesses the satellite on a particular frequency at the same time as other carriers on other frequencies from other stations. To minimise intermodulation products, and consequently the number of carriers (see Section 4.3.2), traffic routing is performed according to the 'one carrier per transmitting station' principle. The FDM multiplex signal thus consists of all the signal frequencies destined for the other stations. Figure 4.5 shows an example of a network of three stations.

4.4.1.2 *TDM/PSK/FDMA (Figure 4.4b)*

The baseband signals from the network or users are digital. They are combined to form a time division multiplex (TDM) signal. The binary stream representing this multiplexed signal

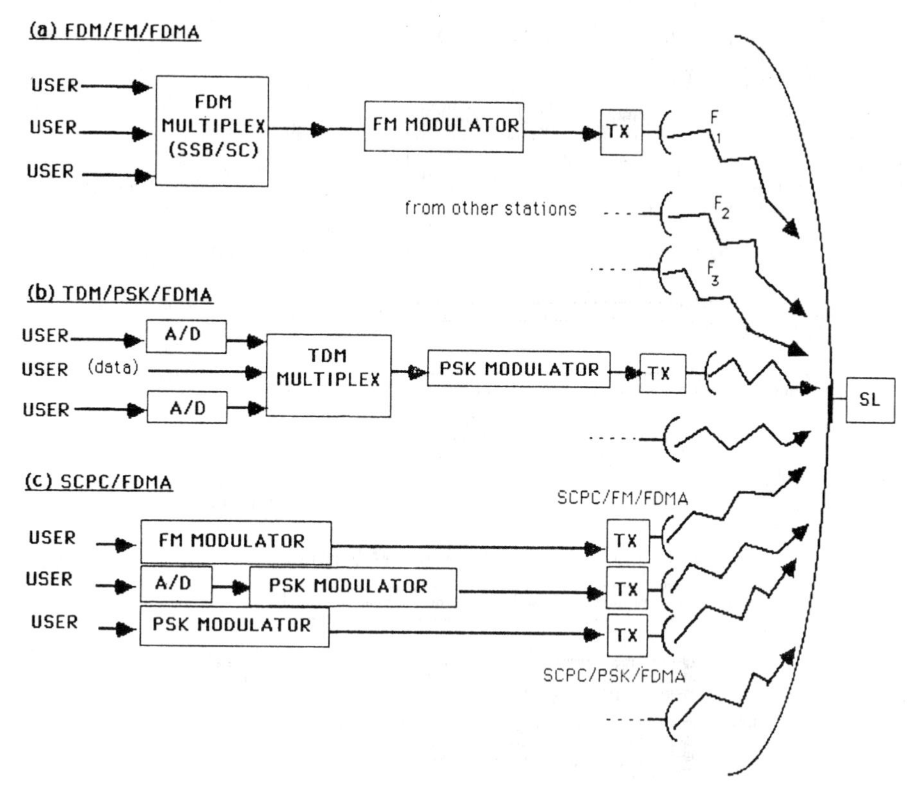

Figure 4.4 FDMA transmission configurations. (a) FDM/FM/FDMA. (b) TDM/PSK/FDMA. (c) SCPC/FDMA.

modulates a carrier by phase-shift keying (PSK) which accessses the satellite at a particular frequency at the same time as other carriers on other frequencies from other stations. To minimise intermodulation products, and consequently the number of carriers (see Section 4.3.2), traffic routing is performed according to the 'one carrier per transmitting station' principle. The TDM multiplex signal thus consists of all the temporal signals destined for other stations. Figure 4.5 shows an example of a network of three stations.

4.4.1.3 *SCPC/FDMA (Figure 4.4c)*

The baseband signals from the network or users each modulate a carrier directly, in either analogue or digital form according to the nature of the signal considered (SCPC). Each carrier accesses the satellite on its particular frequency at the same time as other carriers on other frequencies from the same or other stations. Information routing is thus performed according to the 'one carrier per link' principle.

4.4.2 Adjacent channel interference

As shown in Figure 4.6, the channel bandwidth is occupied by several carriers at different frequencies. The channel transmits these to all the earth stations situated in the coverage

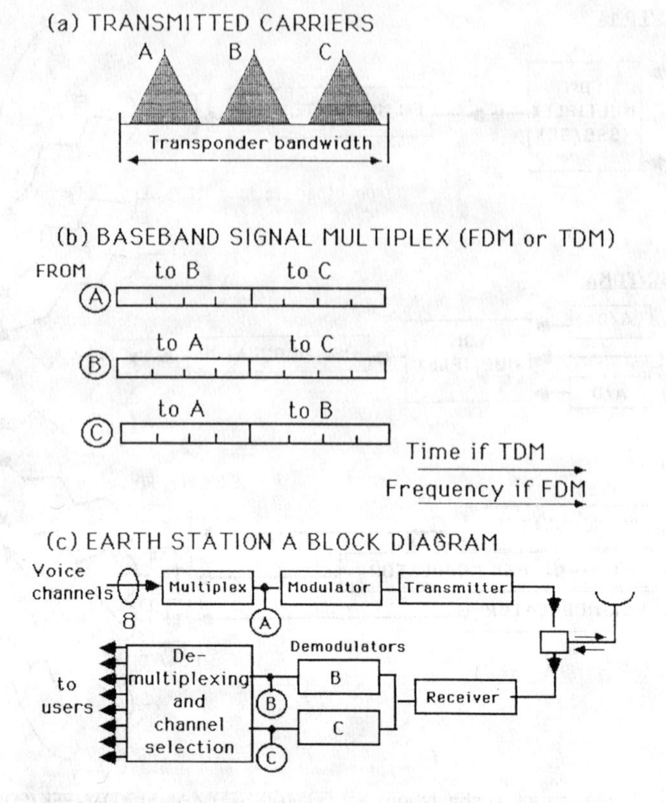

Figure 4.5 Example of a three-station FDMA system with 'one carrier per station' routing.

area of the satellite antenna. The carriers must be filtered by the receiver at each earth station and this filtering is easier to realise when the carrier spectra are separated from each other by a wide guard band. However, the use of wide guard bands leads to inefficient use of the channel bandwidth and a higher operating cost, per carrier, of the space segment. There is, therefore, a technical and economic compromise to be made. Whatever the compromise chosen, part of the power of a carrier adjacent to a given carrier will be captured by the receiver tuned to the frequency of the carrier considered. This causes noise due to interference, called adjacent channel interference (ACI). This interference is additional to the interference between systems analysed in Section 2.9.2 and can be included in the term $(C/N_0)_I$ which appears in expression (2.63) for $(C/N_0)_T$.

4.4.3 Intermodulation [SHA-65], [SEV-66], [SPO-67]

4.4.3.1 Definition of intermodulation products

It was seen in Section 2.9.1 that a satellite repeater channel has a nonlinear transfer characteristic. By the nature of frequency division multiple access, this amplifier simultaneously amplifies several carriers at different frequencies. The earth station itself has a non-linear power amplifier

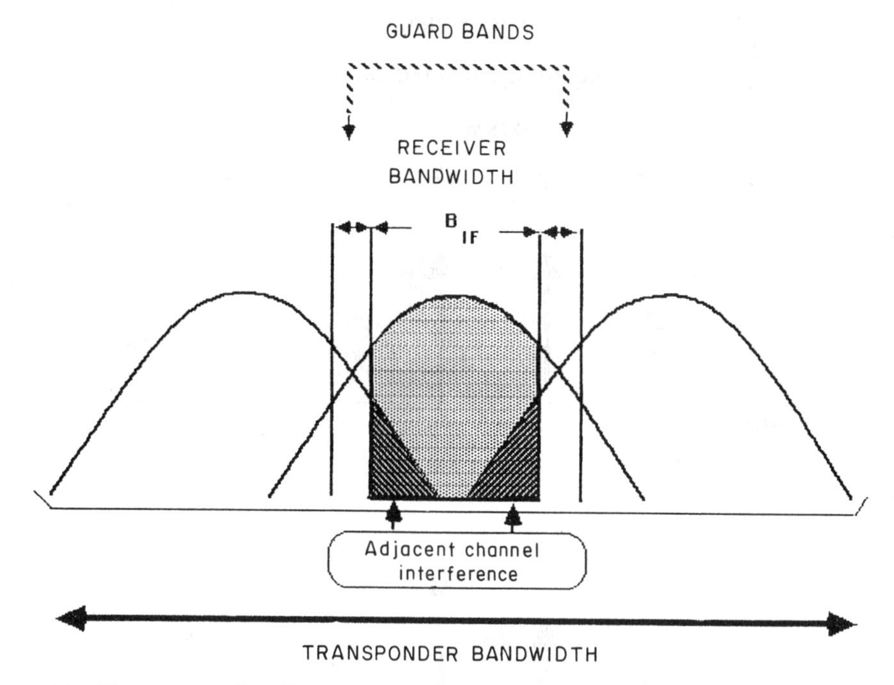

Figure 4.6 The spectrum of an FDMA transponder and adjacent channel interference.

and this amplifier can be fed with several carriers at different frequencies. In general, when N sinusoidal signals at frequencies $f_1, f_2, \ldots, f_N$ pass through a non-linear amplifier, the output contains not only the N signals at the original frequencies but also undesirable signals called intermodulation products. These appear at frequencies f_{IM} which are linear combinations of the input frequencies thus:

$$f_{IM} = m_1 f_1 + m_2 f_2 + \cdots + m_N f_N \qquad \text{(Hz)} \tag{4.4}$$

where $m_1, m_2, \ldots, m_N$ are positive or negative integers.

The quantity X is called the order of an intermodulation product such that:

$$X = Im_1I + Im_2I + \cdots + Im_NI \tag{4.5}$$

When the centre frequency of the amplifier passband is large compared with the bandwidth, which is the case for a satellite repeater channel, only the odd order intermodulation products fall within the channel bandwidth. Moreover, the amplitude of the intermodulation products decreases with the order of the product. Hence, in practice, only products of order 3, and to a lesser extent 5, are significant. Figure 4.7 shows the generation of intermodulation products from two unmodulated carriers at frequencies f_1 and f_2. It can be seen that, in the case of unmodulated carriers of unequal amplitude, the intermodulation products are greater at high frequencies if the carrier of greater amplitude is that which has the higher frequency and at low frequencies if the carrier of greater amplitude is that which has the lower frequency. This shows the advantage of locating the most powerful carriers at the extremities of the channel bandwidth.

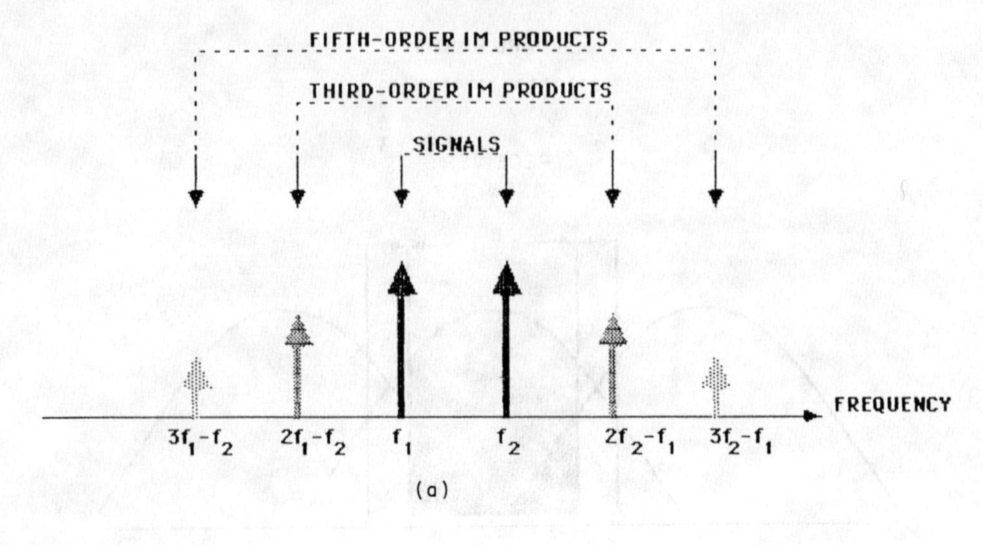

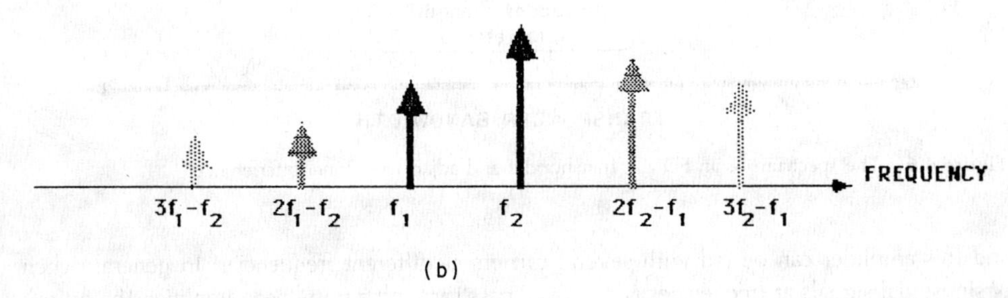

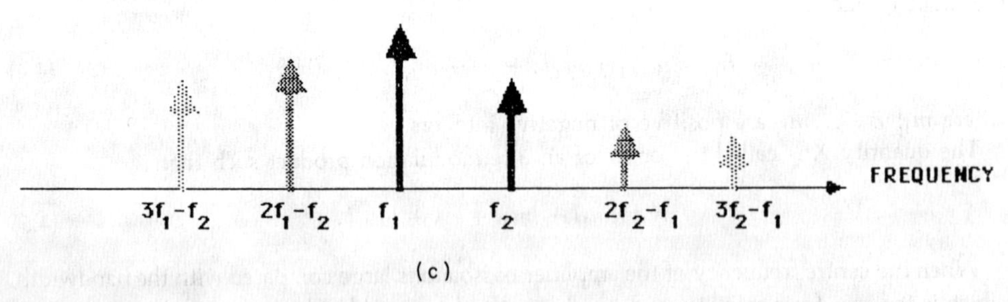

Figure 4.7 Intermodulation products for the case of two sinusoidal signals (unmodulated carriers). (a) Equal amplitudes, (b) and (c) unequal amplitudes.

4.4.3.2 *Transfer characteristic of a non-linear amplifier in multicarrier operation*

Figure 4.7b shows the power transfer characteristic of a satellite repeater channel in single carrier operation. In general, the form of this characteristic is valid for every non-linear amplifier. It is now necessary to extend this model to the case of multi-carrier operation. For

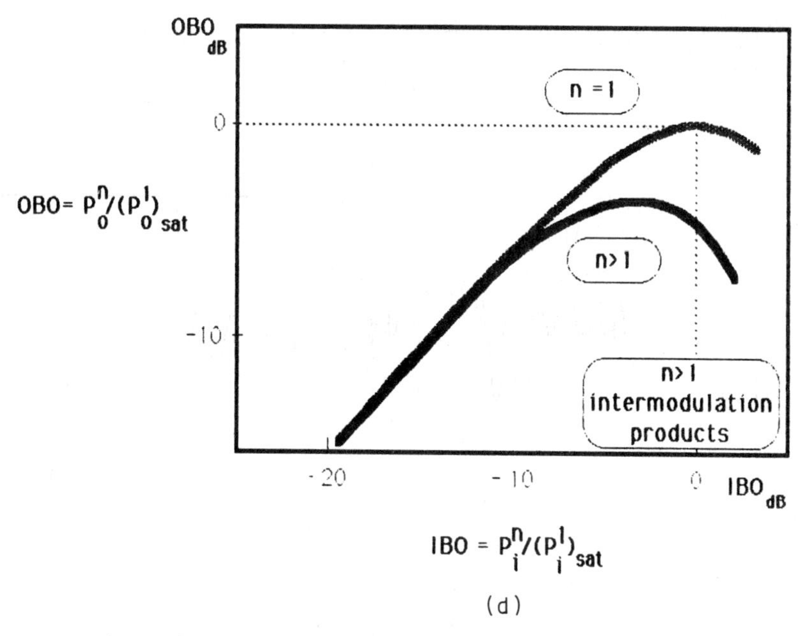

Figure 4.7(*cont.*) (d) Transfer characteristics of a non-linear amplifier in multicarrier operation.

this, the following notation will be used:

(P_i^1) = Carrier power at the amplifier input (i = input) in single carrier operation,
(P_i^n) = Power of one carrier (from n) at the amplifier input in multicarrier operation,
(P_o^1) = Carrier power at the amplifier output (o = output) in single carrier operation,
(P_o^n) = Power of one carrier (from n) at the amplifier output in multicarrier operation,
(P_o^{IMX}) = Power of intermodulation product of order X at the amplifier output in multicarrier operation.

The definition of input and output back-off, given in Section 2.9.1 for the case of single carrier operation, is generalised to the case of multicarrier operation as follows:

$$\text{IBO} = (P_i^n)/(P_i^1)_{sat}$$
$$\text{OBO} = (P_i^n)/(P_o^1)_{sat}$$

In the above expressions, the subscript 'sat' indicates the value of the quantity considered at saturation.

Figure 4.7d shows the form of the variation of OBO as a function of IBO together with the variation of the ratio $(P_o^{IMX})/(P_o^1)_{sat}$.

4.4.3.3 *Intermodulation noise*

When the carriers are modulated, the intermodulation products are no longer spectral lines since their power is dispersed over a spectrum which extends over a band of frequencies

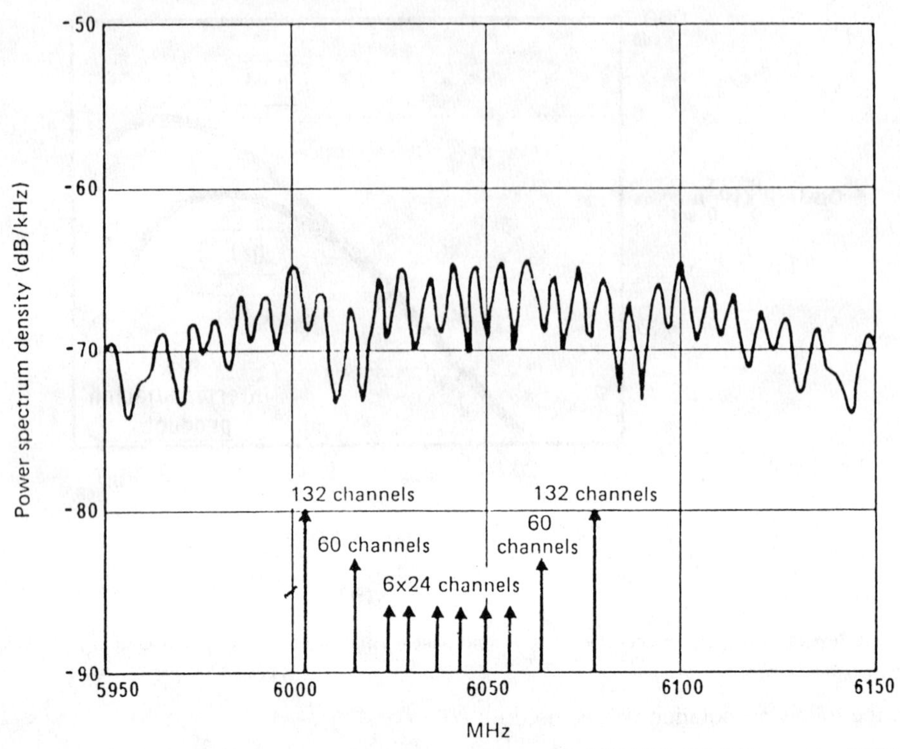

Figure 4.8 Intermodulation noise spectrum generated by several FDM/FM/FDMA carriers (CCIR-88, p. 84.]. (Reproduced by permission of the ITU.)

[GAG-84]. If the number of carriers is sufficiently high, superposition of the spectra of the intermodulation products leads to a spectral density which is sensibly constant over the whole of the amplifier bandwidth and this justifies treatment of intermodulation products as white noise. Figure 4.8 shows an example of the intermodulation noise spectrum in a channel carrying 10 FDM/FM/FDMA carriers [CCIR-88, p. 84].

4.4.3.4 *The carrier to intermodulation noise spectral density power ratio* $(C/N_0)_{IM}$

The spectral density of intermodulation noise power is denoted by $(N_0)_{IM}$. Its value depends on the transfer characteristic of the amplifier and the number and type of carriers amplified. A carrier power to intermodulation noise power spectral density ratio $(C/N_0)_{IM}$ can be associated with each carrier at the amplifier output. This ratio can be deduced from an amplifier characteristic of the type given in Figure 4.7d by estimating, for example, $(N_0)_{IM}$ as $(P_o^{IMX})/B$, where B, is the spectral width of the modulated carrier. Hence $(C/N_0)_{IM} = (P_o^{IMX})/B)$. Figure 4.9 shows the form of the variation of $(C/N_0)_{IM}$ as a function of back-off and number of carriers. It can be seen that the ratio $(C/N_0)_{IM}$ becomes smaller as saturation is approached (the nonlinear characteristic is more severe) and the number of carriers increases (an increase of the total power of the intermodulation products).

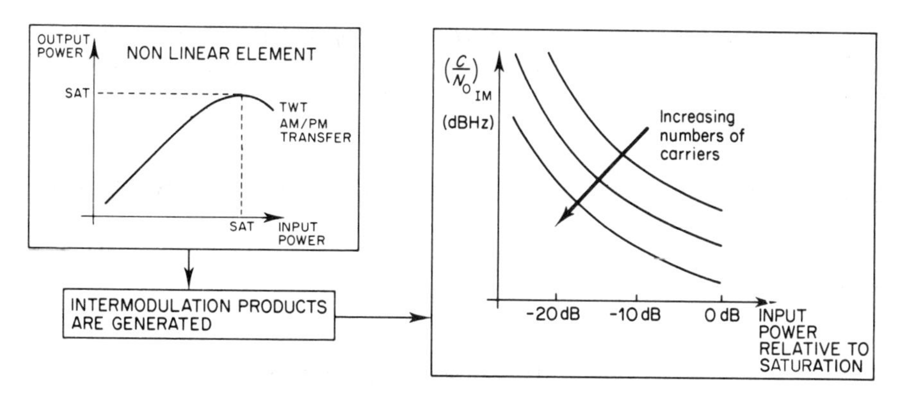

Figure 4.9 Variation of $(C/N_0)_{IM}$ as a function of back-off and number of carriers.

4.4.4 Carrier power-to-noise power spectral density ratio for a station-to-station link

4.4.4.1 *Expression*

Intermodulation noise is added to the other sources of noise analysed in Chapter 2 and equation (2.53) for the carrier power-to-noise power spectral density ratio for the complete station-to-station link $(C/N_0)_T$ is modified as follows:

$$(C/N_0)_T^{-1} = (C/N_0)_U^{-1} + (C/N_0)_D^{-1} + (C/N_0)_I^{-1} + (C/N_0)_{IM}^{-1} \qquad (\text{Hz}^{-1}) \qquad (4.6)$$

with:

$$(C/N_0)_{IM}^{-1} = (C/N_0)_{IM,U}^{-1} + (C/N_0)_{IM,D}^{-1}$$

where $(C/N_0)_{IM,U}$ and $(C/N_0)_{IM,D}$ correspond to the generation of intermodulation noise in the transmitting earth station and the satellite repeater channel respectively.

In this case the expressions for the ratios $(C/N_0)_U$, $(C/N_0)_D$ and $(C/N_0)_{IM}$ are to be used with values of input and output back-off IBO and OBO for operation of the amplifier in multicarrier mode with carriers of equal power. The output power of the amplifier is shared among the carriers, the thermal noise and the intermodulation noise to which the interference noise for the channel is added.

If the carriers at the amplifier input are of unequal power, the power at the amplifier output is shared unequally between carriers and noise. Therefore the amplifier does not have the same power gain for all carriers and a capture effect can arise; carriers of high power acquire more power than carriers of low power [SHA-65]. For carriers of high power the value of the ratio is greater than that given by equation (4.6). For carriers of low power, it is smaller. Generation of intermodulation products is also observed between noise on the uplink and the carriers; this effect can be taken into account in the form of an increase in the noise temperature at the channel input [TAM-87].

4.4.4.2 *The influence of back-off*

Figure 4.10 shows the variation of each of the terms in equation (4.6) as a function of input back-off IBO assuming the equivalent interference noise to be negligible. Because of the opposite direction of variation of the term $(C/N_0)_{IM}$ compared to that of the ratios $(C/N_0)_U$ and $(C/N_0)_D$, the value of $(C/N_0)_T$ passes through a maximum for a non-zero value of back-off. Two effects are, therefore, observed which are consequences of using the same channel to amplify several carriers:

—The total power at the output of the channel is less than that which would exist in the absence of back-off.
—The useful power per carrier is reduced by allocation of part of the total power to intermodulation products.

4.4.5 Throughput of FDMA

It can be seen from Figure 4.10 that the value of $(C/N_0)_T$ is always less than the value obtained in single carrier operation. On the other hand, the maximum value of $(C/N_0)_T$ becomes less as the back-off is increased and this is the case when the number of carriers increases. Figure 4.11 shows the relative variation, with number of telephone channels, of the total capacity of a satellite channel of 36 MHz bandwidth for global coverage by the INTELSAT IV or IVA satellite. The transmission scheme is of the FDM/FM/FDMA type (Section 4.4.1.1). The carriers are modulated by multiplexed signals of equal capacity. As the number of carriers increases, the bandwidth allocated to each carrier must decrease and this leads to a reduction of the capacity of the modulating multiplexed signal. As the total capacity is the product of the capacity of each carrier and the number of carriers, it could be imagined that the total capacity would remain sensibly constant. But it is not; the total capacity decreases

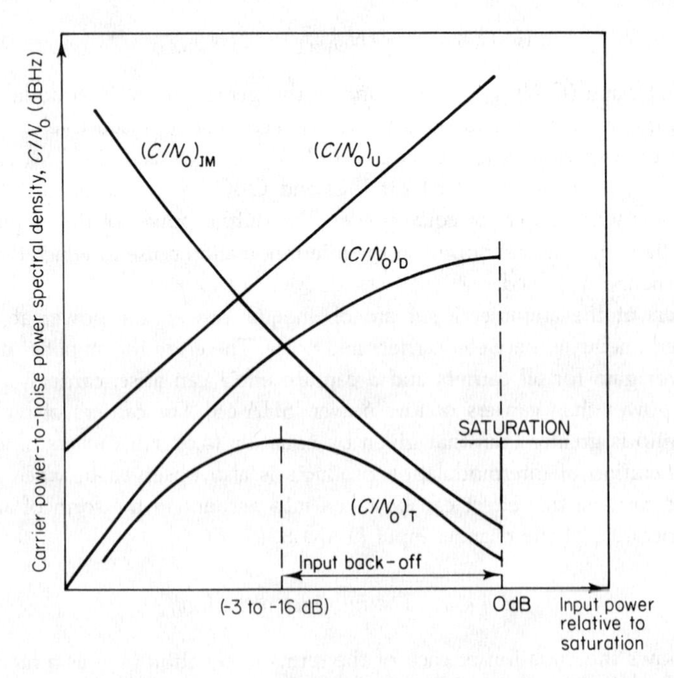

Figure 4.10 Variation of $(C/N_0)_U$, $(C/N_0)_D$, $(C/N_0)_{IM}$ and $(C/N_0)_T$ as a function of input back-off IBO.

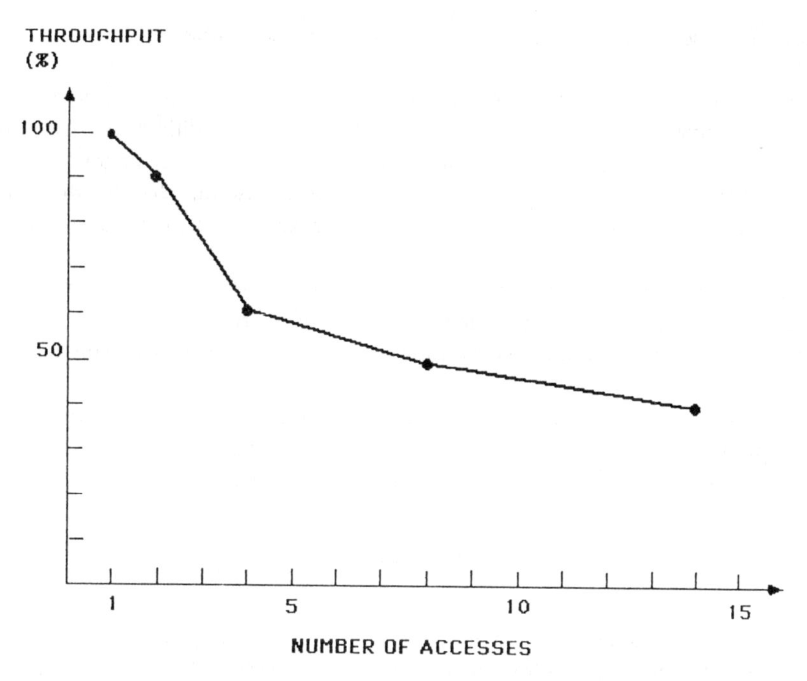

Figure 4.11 Efficiency of an FDMA transmission; the curve indicates the relative variation of the total capacity as a number of telephone channels of an INTELSAT transponder with a bandwidth of 36 MHz as a function of the number of accesses, that is the number of carriers of FDM/FM/FDMA type. The value indicated as 100% represents the total capacity of the multiplex which modulates the carrier for the case of single access to the transponder.

as the number of carriers increases. This results from the fact that each carrier is subjected to a reduction in the value of $(C/N_0)_T$ since the back-off is large when the number of carriers is high. The curve of Figure 4.11 thus represents the throughput of an FDMA system as a function of the number of accesses; it effectively shows the ratio of the total real channel capacity (the source of revenue for the network operator) and the potential capacity of the channel (for which the network operator pays).

4.4.6 Intelligible crosstalk

Intelligible crosstalk arises when an amplitude modulated carrier coexists with frequency modulated carriers. Due to the non-linear amplitude-phase transfer characteristic of the channel (see Chapter 8), the amplitude modulation is transferred to the other carriers in the form of parasitic phase modulation and this is detected as an intelligible signal by the frequency demodulator of the earth station receiver.

4.4.7 Conclusion

Frequency division multiple access (FDMA) is characterised by continuous access to the satellite in a given frequency band. This technique has the advantage of simplicity and relies on the use of proven equipment. However, it has some disadvantages:

—Lack of flexibility in case of reconfiguration; to accommodate capacity variations it is necessary to change the frequency plan and this implies modification of transmitting frequencies, receiving frequencies and filter bandwidths of the earth stations.
—Loss of capacity when the number of accesses increases due to the generation of intermodulation products and the need to operate at a reduced satellite transmitting power (back-off).
—The need to control the transmitting power of earth stations in such a way that the carrier powers at the satellite input are the same in order to avoid the capture effect. This control must be performed in real time and must adapt to attenuation caused by rain on the uplinks.

This is the oldest access technique and it remains the most used despite the disadvantages. It tends to perpetuate itself due to investments made in the past and its known operational advantages which include the absence of synchronisation between earth stations.

4.5 TIME DIVISION MULTIPLE ACCESS (TDMA)

Figure 4.12 shows the operation of a network according to the principle of time division multiple access. The earth stations transmit discontinuously during a time T_B. This transmission is called a burst. A burst transmission is inserted within a longer time structure of duration T_F called a frame period and this corresponds to the periodic time structure within which all stations transmit. Each carrier representing a burst occupies all of the channel bandwidth. Hence the channel carries one carrier at a time.

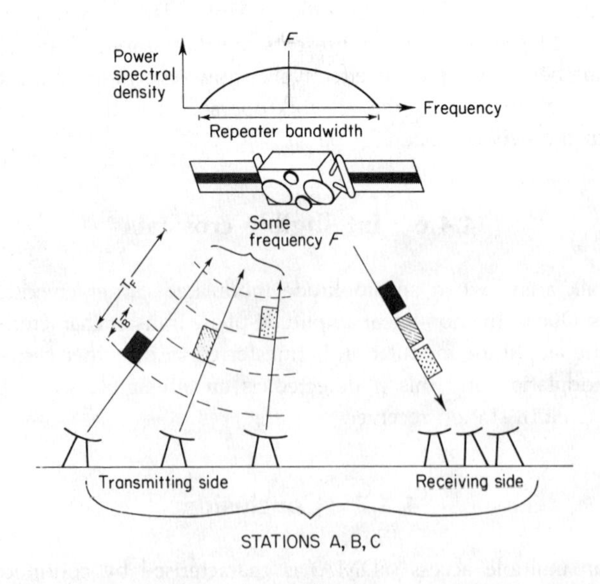

Figure 4.12 Operation of a network according to the principle of time division multiple access (TDMA).

4.5.1 Burst generation

The burst corresponds to the transfer of traffic from the station considered. This transfer can be made in accordance with the 'one carrier per link' method; in this case the station transmits $N-1$ bursts per frame, where N is the number of stations on the network and the number of bursts P in the frame is given by $P = N(N-1)$. With the 'one carrier per station' method the station transmits a single burst per frame and the number of bursts P in the frame is equal to N. Each burst thus travels in the form of sub-bursts of traffic from station to station. Due to the decrease of throughput of the channel as the number of bursts increases (see Section 4.5.5) the 'one carrier per station' approach is generally retained.

Figure 4.13 illustrates burst generation. The earth station receives information in the form of a continuous binary stream of rate R_b from the network or user interface. This information must be stored in a buffer memory while waiting for the burst transmission time. When this time arises, the contents of the memory are transmitted in a time interval equal to T_B. The bit rate R which modulates the carrier is thus given by:

$$R = R_b(T_F/T_B) \qquad \text{(bit/s)} \tag{4.7}$$

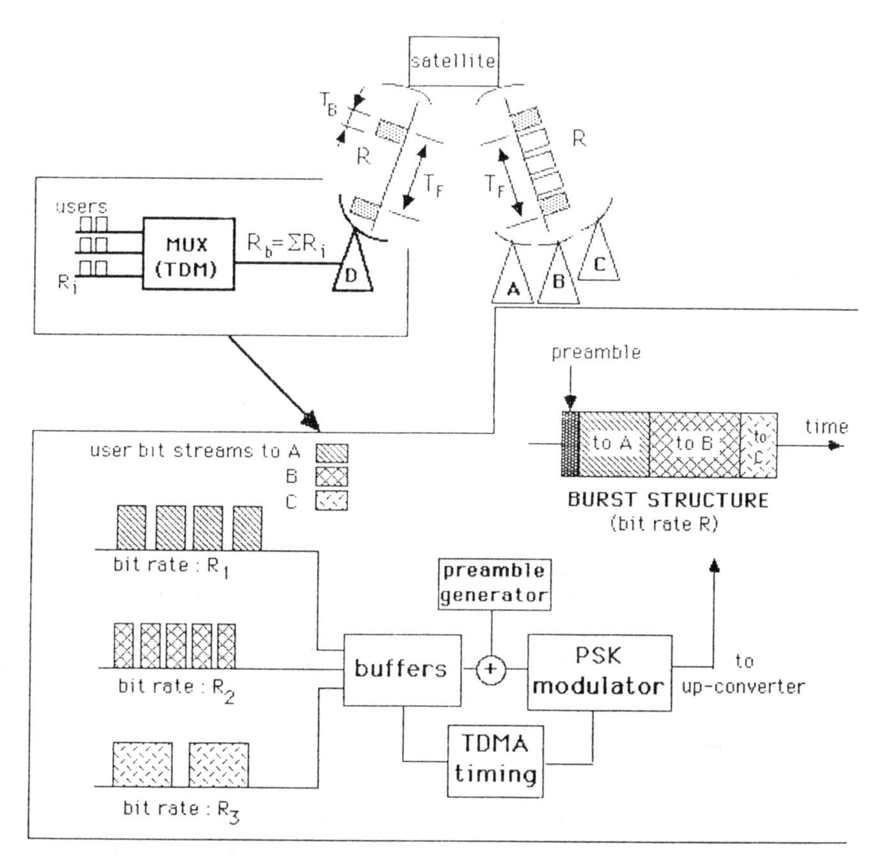

Figure 4.13 Burst generation. R_i = User rate (bit/s), R_b = information rate of the multiplex (bit/s) = $\sum R_i$, R = rate in each burst (bit/s), T_B = burst duration (s), T_F = frame duration (s).

The value of R is high when the burst duration is short and consequently the transmission duty cycle (T_B/T_F) of the station is low. Hence, for example, if $R_b = 2$ Mbit/s and $(T_F/T_B) = 10$, modulation occurs at 20 Mbit/s. Notice that R represents the total capacity of the network; that is the sum of the station capacities in bit/s. If all stations have the same capacity, the duty cycle (T_F/T_B) represents the number of stations on the network.

It can now be seen why this type of access is always associated with digital transmission; it is easy to store bits for a frame period and to empty a digital memory in the shorter period of one burst. Performing this type of processing on analogue information is not easy.

The structure of a burst can be seen in Figure 4.13. This consists of a header, or preamble, and a traffic field. The header has several functions:

—To permit the demodulator of the receiving earth station, in the case of coherent demodulation, to recover the carrier generated by the local oscillator at the transmitter. For this purpose the header contains a bit sequence which provides a constant carrier phase.
—To permit the detector of the receiving earth station to synchronise its bit decision clock to the symbol rate; for this purpose the header contains a bit sequence providing alternating opposite phases.
—To permit the earth station to identify the start of a burst by detecting, by means of a correlator, a group of bits called a 'unique word' (UW). The unique word enables the receiver to resolve carrier phase ambiguity in the case of coherent demodulation. Knowing the start of the burst, the bit rate and having (if required) resolved the phase ambiguity, the receiver can then identify all the bits occurring after the unique word.
—To permit the transfer of service messages between stations (telephone and telex) and signalling.

The traffic field is located at the end of the header and this corresponds to the transmission of useful information. In the case of the 'one carrier per station' method where the burst transmitted by a station carries all the information from this station to the other stations, the traffic field is structured in sub-bursts which correspond to the information transmitted by the station to each of the other stations.

4.5.2 Frame structure

The frame is formed at satellite level. It consists of all the bursts transmitted by the earth stations placed one after the other, if transmission synchronisation of the stations is correct. To take account of synchronisation imperfections, a period without transmission, called a guard time, is provided between each burst. Figure 4.14 shows the frame used in the INTELSAT and EUTELSAT networks. The length of the frame is 2 ms. The guard time occupies 64 symbols or 128 bits and this corresponds to a time interval of 1 microsecond. Notice the presence of two types of burst:

—Those of traffic stations, with a header of 280 symbols, or 560 bits, and a traffic field structured in multiples of 64 symbols in accordance with the capacity of each station.
—Those of reference stations with a header of 288 symbols, or 576 bits, and without a traffic field. The reference station is the station which defines the frame clock by transmitting its reference burst; all the network traffic stations must synchronise themselves to the

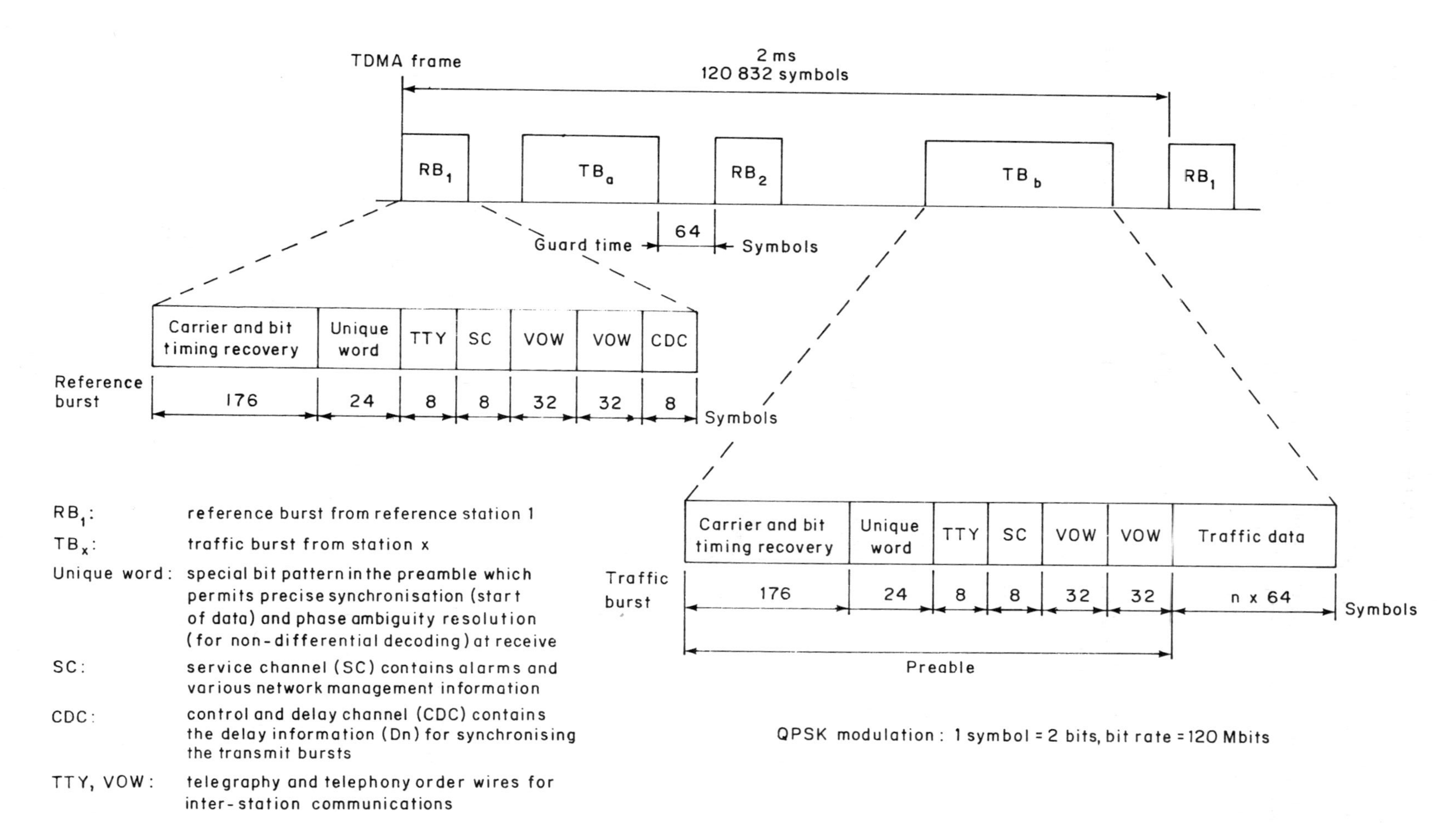

Figure 4.14 Frame structure (INTELSAT/EUTELSAT standard). (From CCIR-88. Reproduced by permission of the ITU.)

reference station by locating their burst with a constant delay with respect to the reference station burst, called the reference burst. Because of its fundamental role in correct operation of the network, the reference station is replicated. This is why there are two reference bursts per frame; one is transmitted by each of the two mutually synchronised reference stations.

4.5.3 Burst reception

On the downlink, each station receives all bursts in the frame. Figure 4.15 illustrates the processing at the receiving station.

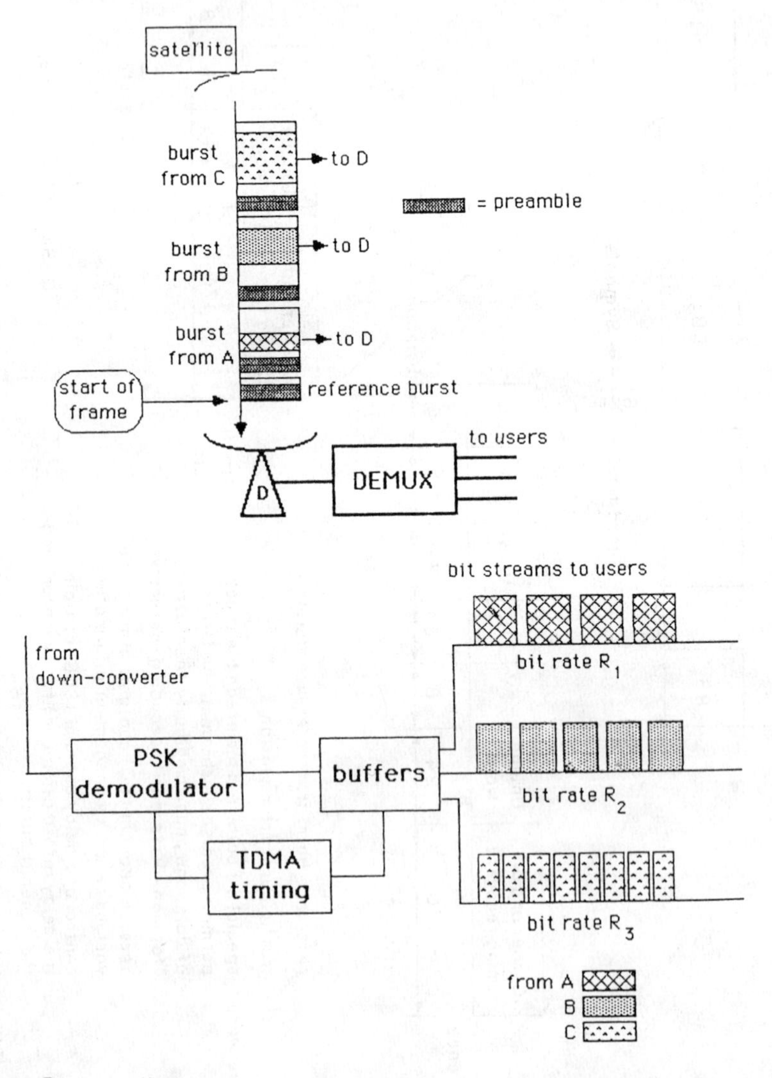

Figure 4.15 Burst reception.

The receiving station identifies the start of each burst of the frame by detection of the unique word; it then extracts the traffic which is intended for it and is contained in a sub-burst of the traffic field of each burst. This traffic is received discontinuously with a bit rate R. To restore the original bit rate R_b in the form of a continuous binary stream, the information is stored in a buffer memory for one frame period and is read out at a rate R_b during the following frame.

It is fundamental for identification of the burst contents that the receiving station must be able to recognise the unique word at the start of each burst. The unique word detector establishes correlation between each bit sequence at the output of the receiver bit detector which is of the same length as the unique word and a replica of the unique word stored in the correlator memory. Only received sequences which produce a correlation peak greater than a threshold are retained as unique words. The performance of the unique word detector is measured by two quantities [FEH-83]:

—The pobability of non-detection, that is the probability of not detecting the presence of a unique word at the start of burst reception.
—The probability of a false alarm, that is the probability of falsely identifying the unique word in any binary sequence, for example in the traffic field.

The probability of non-detection decreases when:

—The bit error rate of the link decreases.
—The length of the unique word decreases.
—The correlation threshold decreases.

The probability of a false alarm is independent of the bit error rate on the link and decreases when:

—The length of the unique word increases.
—The correlation threshold increases.

A compromise must, therefore, be found; in practice, the probability of a false alarm is reduced without increasing the probability of non-detection by taking advantage of a priori knowledge of the frame structure in order to perform correlation only in the time intervals when the unique word is expected.

4.5.4 Synchronisation

Synchronisation of transmissions from different network stations is necessary. Its purpose is to avoid burst recovery from others in the frame. Such a recovery would lead, due to the resulting interference, to the impossibility of the earth station receiver detecting information under typical conditions. Before considering synchronisation, it is important to establish the order of magnitude of the disturbances associated with the imperfections of the geostationary satellite orbit.

4.5.4.1 *Residual movements of a geostationary satellite*

The orbit control of the satellite defines a station-keeping 'window' whose typical dimensions are 0.1° in longitude and latitude. Furthermore the eccentricity of the orbit is limited to a maximum value of the order of 0.001. The satellite thus moves, as indicated in Figure 4.16, in a volume of the order of 75 km × 75 km × 85 km. This introduces an altitude variation of around 85 km with a periodicity of 24 h which has two effects:

—A variation in round-trip propagation time of around 570 μs. This quantifies the magnitude of potential daily displacement of a burst in the frame in the absence of correcting action. This value is to be compared with the frame duration (from 2 ms to 20 ms),

—A Doppler effect which, if the maximum displacement velocity of the satellite is considered to be 10 km/h, causes displacement of the position of a burst in the frame from one station at a rate of around 20 ns/s. With a guard time between two bursts of 1 μs, and assuming the particular case of displacement in opposite directions of two consecutive bursts in the frame, the time for the drift to absorb the guard time between the two bursts is of the order of $(1/2)\,(1 \times 10^{-6}/20 \times 10^{-9})\,\text{s} = 25\,\text{s}$. This determines the timescale for undertaking corrective action. Notice that this time is greater than the round-trip propagation time of the bursts and indicates that control of burst position can be based on observation of position error.

4.5.4.2 *Relation between the start of a frame on transmission and reception*

Any station n (n = 1, 2, ..., N) must transmit its burst in such a way that it arrives at the satellite with a delay d_n with respect to the reference burst. As shown in Figure 4.17, the value of the delay d_n has a particular value for each station. The set of values of d_n determines the arrangement of bursts in the frame (burst time plan). Positioning is correct when station

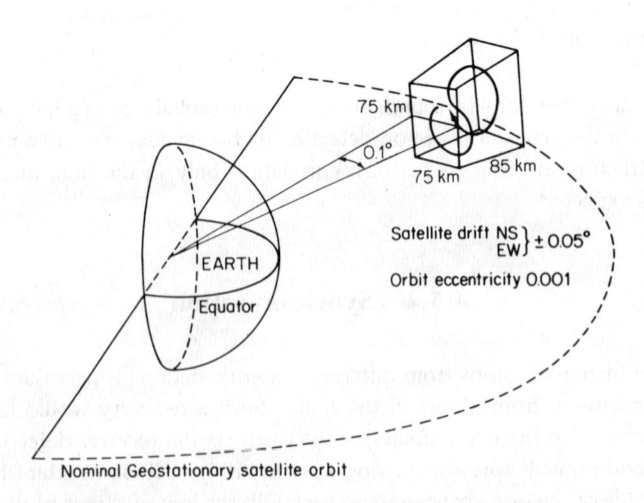

Figure 4.16 Evolution of the volume occupied by a geostationary satellite in the course of an orbital period (24 h).

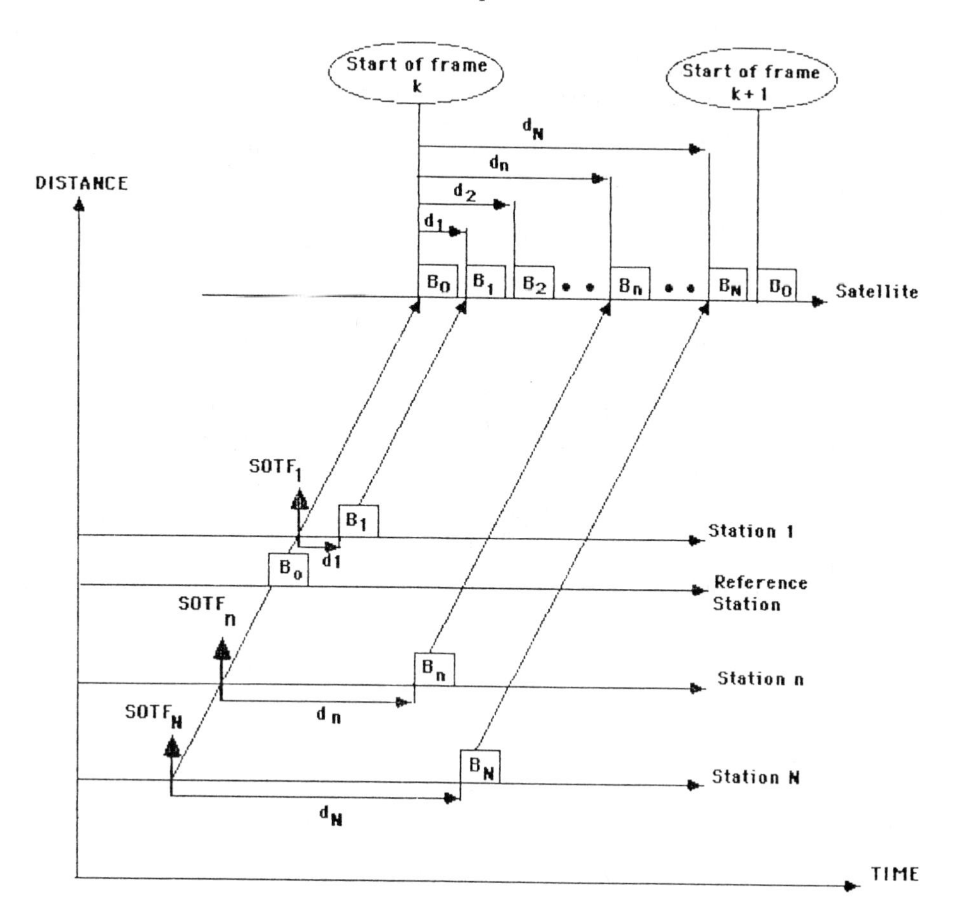

Figure 4.17 Burst assignment within the frame; each station n locates its burst at satellite level with a delay d_n (n = 1, 2, ..., N) with respect to the reference burst which defines the start of the frame. The vertical arrow at station n indicates the start of the transmission frame ($SOTF_n$) for this station.

n transmits with a delay d_n with respect to the start of the frame being transmitted as defined by the time $SOTF_n$ (start of transmit frame). This time $SOTF_n$ is the instant at which the station must transmit in order to position its burst in the frame time slot occupied by the reference burst. The problem of synchronising station n is thus that of determining $SOTF_n$. Once this instant is known, it is merely necessary for station n to transmit with a delay d_n with respect to $SOTF_n$.

With a single beam satellite, station n receives all of the frame on the downlink. Detection of the unique word of the reference burst determines the start time of the received frame $SORF_n$ (start of receive frame). Figure 4.18 shows the time relationship between the start of the transmitted frame $SOTF_n$ and the start of the received frame $SORF_n$; $SORF_n$ is equal to the start time of the frame (k) at the satellite plus the propagation time on the downlink R_n/c, where R_n is the distance of the satellite from ground station n and c is the velocity of light. The start time of frame ($k + m$), where m is an integer, is equal to $SOTF_n$ plus the propagation time on the uplink R_n/c. The time separating the start of frame (k) and frame

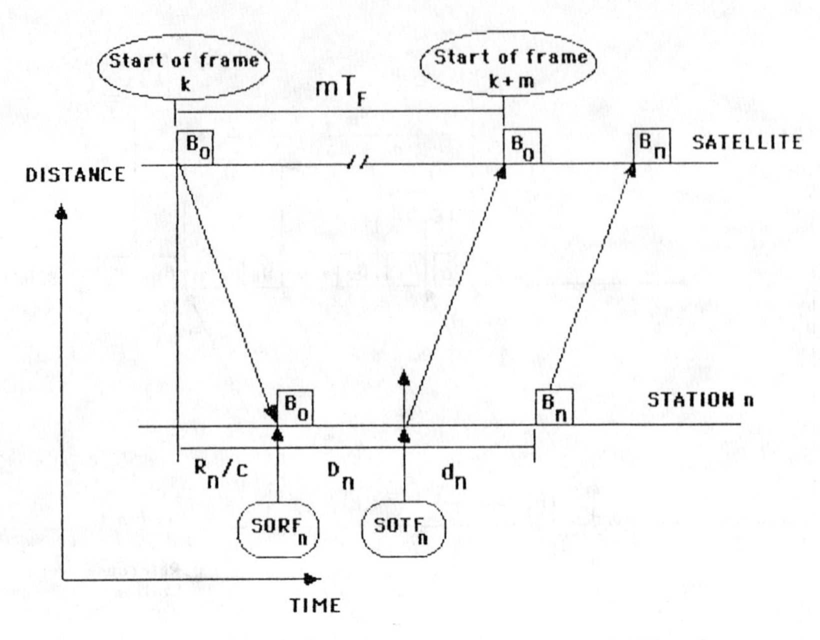

Figure 4.18 The relation between the frame start times on transmission $SOTF_n$ and on reception $SORF_n$ for a station n.

$(k+m)$ at the satellite is by definition mT_F. Hence the relationship:

$$SOTF_n - SORF_n = D_n = mT_F - 2R_n/c \qquad \text{(s)} \qquad (4.8)$$

For this quantity to be positive, it is necessary to choose m such that mT_F is greater than the value of $2R_n/c$ for station n which is furthest from the satellite. For example, for the TELECOM 1 network, the value of m is taken as 14 which corresponds, for a frame duration $T_F = 20$ ms, to a maximum roundtrip propagation time of 280 ms [BOU-81].

In summary, station n identifies $SORF_n$ by detecting the unique word of the reference burst and transmits at an instant $D_n + d_n$ later. Depending on the method of determining the value of D_n, two types of synchronisation can be distinguished:

—Closed loop synchronisation.
—Open loop synchronisation.

4.5.4.3 *Closed loop synchronisation*

Figure 4.19 illustrates this method. Station n observes the position of its burst in the frame relative to the reference burst by measuring the time between detection of the unique word of the reference burst and detection of the unique word of its own burst. Let $d_{on}(j)$ be the value observed on reception of the frame for which the value $D_n(j)$ had been used to determine the transmission time. The difference $e_n(j) = d_{on}(j) - d_n$ is the burst position error. The station

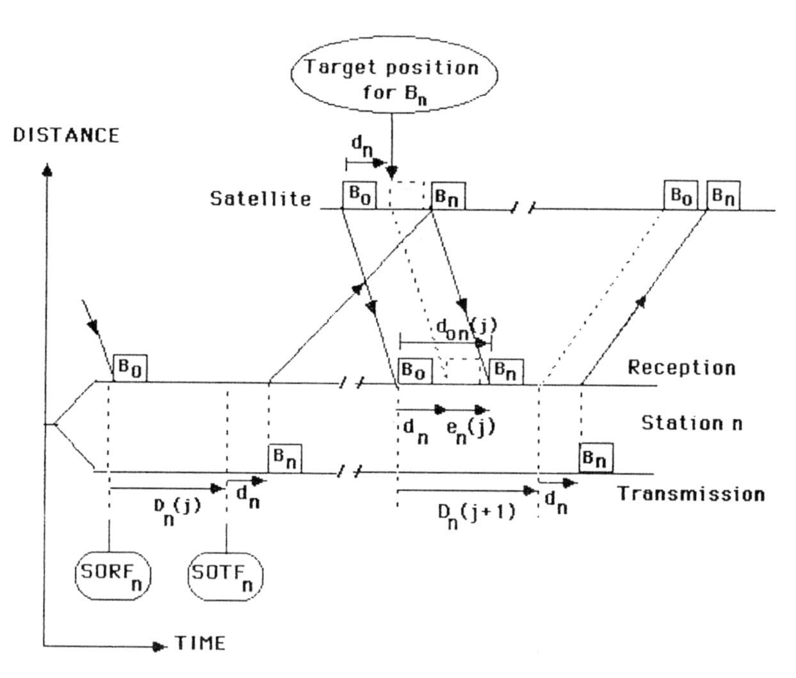

Figure 4.19 Closed loop synchronisation; station n observes the position of its burst and consequently corrects the transmission time.

then increases the value of D_n according to the following algorithm:

$$D_n(j+1) = D_n(j) - e_n(j) \qquad \text{(s)} \tag{4.9}$$

and uses the new value of D_n to determine the transmission time. Notice that the minimum time necessary to make a correction is equal to the round-trip propagation time for the station furthest from the satellite, that is of the order of 280 ms.

4.5.4.4 *Open loop synchronisation*

This is used particularly for networks with assignment on demand where the burst position of traffic stations is controlled by the reference station (see Section 4.7.4); this method relies on knowledge of the satellite position and calculation of the distance R_n between the satellite and each ground station. The satellite position can be provided by the orbit control station (space segment). If decoupling of responsibility between the space segment and the ground segment is required, two auxiliary stations must be provided in addition to the reference station. Figure 4.20 illustrates the approach; the two auxiliary stations and the reference station measure the propagation time of their bursts. The two auxiliary stations communicate these values to the reference station which determines the satellite position by triangulation and calculates the distance of the satellite from each network station. The reference station

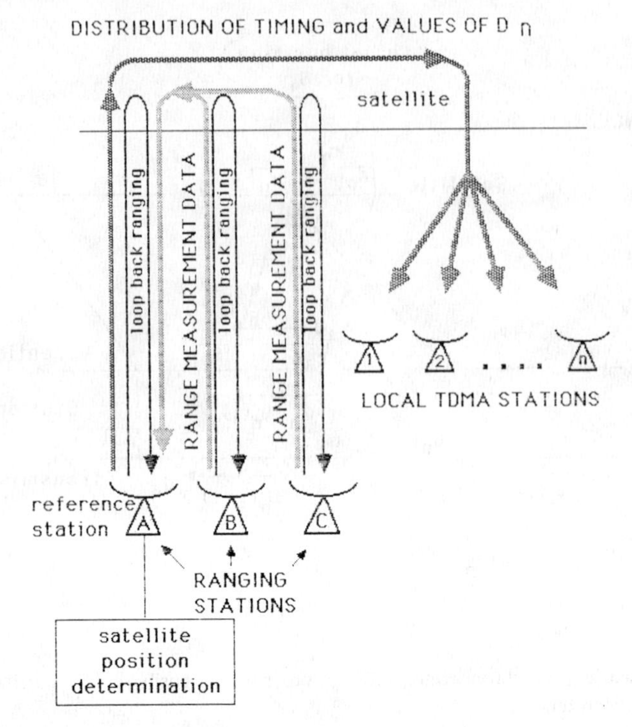

Figure 4.20 Open loop synchronisation.

broadcasts the value of D_n calculated from expression (4.8) by means of the reference burst (field CDC of Figure 4.14). Notice that the time before correction is equal to the time required to measure the propagation time (one round trip) plus the time required for transmission of this information by the two auxiliary stations to the reference station (one round trip) plus the calculation time and finally the time to broadcast the values of D_n. This time can amount to several seconds and consequently implies longer guard times than in the case of closed loop synchronisation.

4.5.4.5 *Acquisition of synchronisation*

Acquisition of synchronisation by a station is achieved each time the station wishes to enter the network. Operation can be in closed or open loop. In closed loop the station transmits a low power burst, generally modulated by a pseudo-random sequence, observes its position, corrects it to give its nominal position and then operates on full power to transmit the useful information. Modulation by a pseudo-random sequence facilitates acquisition by virtue of the autocorrelation properties of the sequence, which permit measurement of the position error, and energy dispersion which limits interference by the entering station with transmissions from stations which are carrying traffic. In open loop, the entering station receives the value of D_n from the reference station and transmits at an instant $D_n + d_n$ after receiving the reference burst.

4.5.5 Throughput of TDMA

4.5.5.1 *Definition*

The throughput of TDMA transmission can be measured by the ratio of the channel capacity in single carrier operation (only one access) and the capacity of the same channel for the multiple access case. It is assumed that the whole bandwidth is occupied in both cases. In single carrier operation, the transfer capacity is $R = B\Gamma$, where B is the channel bandwidth (Hz) and Γ is the spectral efficiency of the modulation (bit/s Hz). In the case of multiple accesses, the capacity is $R(1 - \sum t_i/T_F)$, where $\sum t_i$ represents the sum of the times not devoted to transmission of traffic (guard times plus burst headers). The throughput is thus:

$$\eta = 1 - \sum t_i/T_F \tag{4.10}$$

It expresses the ratio of the time devoted to transmission of traffic (which is the source of revenue for the network operator) and the total channel utilisation time (for which the network operator pays). The throughput is greater when the frame duration T_F is high and when $\sum t_i$ is small.

The throughput depends on the number P of bursts in the frame. Let p be the number of bits in the header and g the equivalent duration in bits of the guard time. Assuming that the frame contains two reference bursts, this gives:

$$\eta = 1 - (P + 2)(p + g)/RT_F \tag{4.11}$$

where R is the bit rate of the frame (bit/s).

The throughput as a function of the number of accesses, that is the number of stations N on the network, depends on the routing arrangement adopted. It is known (Section 4.5.1) that:

—in the case of a 'one carrier per link' arrangement, $P = N(N - 1)$
—in the 'one carrier per station' routing arrangement, $P = N$.

Since throughput is low when P is high, the advantage of adopting a 'one carrier per station' arrangement can be seen.

Throughput is directly involved in calculating the capacity of a network as a number of telephone channels. Let r be the signalling rate associated with one telephone channel and n the number of telephone channels; this gives:

$$n = \eta R/r \tag{4.12}$$

This expresses the number of telephone channels in the frame. The number of terrestrial telephone channels assigned to the network depends on the possible use of concentration by digital speech interpolation (DSI) and the corresponding gain (see Section 3.7.3).

4.5.5.2 *Frame duration considerations*

A long frame duration implies a higher storage capacity in the transmitting and receiving earth station buffer memories. On the other hand, the frame duration is involved in the transmission time of information from one terrestrial network–station interface to another station–terrestrial network interface; this is effectively equal to the round-trip return time increased by the transmission and reception storage time. Since the storage time is at most equal to the duration of one frame, this gives:

$$\text{Transmission time} = \text{Round trip propagation time} + 2T_F \quad \text{(s)} \tag{4.13}$$

For telephone transmission, CCITT Recommendation G 114 stipulates that the propagation time between subscribers must not exceed 400 ms (Section 3.5.3). Accepting that the round-trip propagation time of radio waves cannot exceed 278 ms and it is reasonable to allow 30 ms for the sum of the propagation times in the end networks, the following condition must be observed in order to satisfy Recommendation G 114:

$$T_F \leqslant (i/2)(400 - 278 - 30) = 46 \text{ ms} \tag{4.14}$$

In practice, frame durations vary from 750 μs to 20 ms.

4.5.5.3 *Guard and header time considerations*

For a given frame duration the throughput increases as $\sum t_i$ is decreased. This implies:

—A reduction of the guard times; this approach is limited by the precision of the synchronisation method. A closed loop method is preferable to an open loop one in this respect.

—A reduction of the headers; it is important to provide circuits in the receivers for rapid clock and bit rate recovery. In this context the advantage of differential demodulation with respect to coherent demodulation can be seen. However, transmision throughput is not the only criterion and the worse performance in terms of differential demodulation error probability must not be forgotten. One can attempt to reduce the duration of the unique word but this involves an increase in the probability of a false alarm in detection of the unique word (Section 4.5.3).

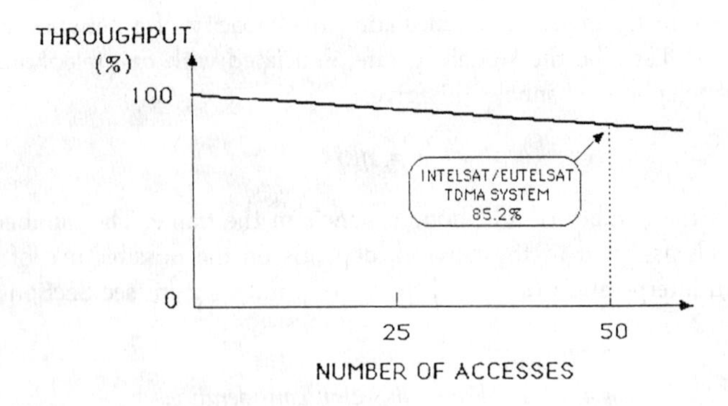

Figure 4.21 The efficiency of the INTELSAT/EUTELSAT TDMA system; the 100% value indicated for a single access corresponds to the capacity of the single carrier which passes through the transponder and is transmitted continuously.

4.5.5.4 *Example*

The variation of throughput as a function of the number of bursts of traffic P, equal to the number N of traffic stations or the number of accesses, can be examined by inserting the values of the INTELSAT/EUTELSAT standard indicated in Figure 4.14 into equation (4.11). Taking $p = 560$, $g = 128$, $R = 120.832$ Mbit/s and $T_F = 2$ ms gives:

$$\eta = 1 - 2.85\ 10^{-3}(P + 2) \tag{4.15}$$

This expression is represented by the curve of Figure 4.21. Notice the relatively slow decrease (with respect to that of the figure for FDMA for example) of throughput as a function of number of accesses. Hence for a number of accesses equal to 50, the throughput is still 85%.

4.5.6 Conclusion

Time division multiple access (TDMA) is characterised by access to the channel during a time slot. This has certain advantages:

—At each instant the channel amplifies only a single carrier which occupies all of the channel bandwidth; there are no intermodulation products and the carrier benefits from the saturation power of the channel. However nonlinearity exists and, combined with the effects of filtering on transmission and reception, introduces a degradation with respect to the ideal digital transmission performance presented in Chapter 3.
—Transmission throughput remains high for a large number of accesses.
—There is no need to control the transmitting power of the stations.
—All stations transmit and receive on the same frequency whatever the origin or destination of the burst; this simplifies tuning.

TDMA, however, has certain disadvantages:

—The need for synchronisation.
—The need to dimension the station for transmission at high throughput.

Consider a station-to-station link. The quality target is specified in terms of error probability. The imposed value determines the required value of the ratio E/N_0. The ratio C/N_0 for the whole link is determined by the relation established in Chapter 3 and recalled here:

$$C/N_0 = (E/N_0)R \tag{4.16}$$

It can be seen that C/N_0 is proportional to R for which the expression is given by equation (4.7). For a capacity R_b, a station must be dimensioned to transmit a throughput R which is high when the duty cycle T_B/T_F is low. (In FDMA the station transmits a throughput R_b at radio frequency and consequently the required C/N_0 is smaller for a duty cycle T_B/T_F.) This disadvantage of TDMA is partly compensated by the higher power provided by the channel on the downlink compared with the FDMA case where back-off is necessary.

Overall, TDMA implies more costly equipment at the earth stations. The cost of this equipment is, however, compensated by better utilisation of the space segment due to the higher transmission throughput in the case of a large number of accesses. Furthermore, digital processing leads to operational simplicity.

4.6 CODE DIVISION MULTIPLE ACCESS (CDMA)

With code division multiple access (CDMA), network stations transmit continuously and together on the same frequency band of the channel. There is, therefore, interference between the transmissions of different stations and this interference is resolved by the receiver which identifies the 'signature' of each transmitter; the signature is presented in the form of a binary sequence, called a code, which is combined with the useful information at each transmitter. The set of codes used must have the following correlation properties:

—Each code must be easily distinguishable from a replica of itself shifted in time.
—Each code must be easily distinguishable regardless of other codes used on the network.

Transmission of the code combined with the useful information requires the availability of a much greater radio-frequency bandwidth than that required to transmit the information alone using the techniques described in Chapter 3. This is the reason why one refers to spread spectrum transmission.

Two techniques are used in CDMA:

—The technique described as direct sequence (DS).
—The technique described as frequency hopping (FH).

4.6.1 Direct sequence transmission (DS-CSMA)

4.6.1.1 *The principle*

Figure 4.22 illustrates the principle; the binary message to be transmitted *m(t)*, of bit rate $R_b = 1/T_b$, is coded in NRZ so that $m(t) = \pm 1$ and is multiplied by a binary sequence *p(t)*, itself coded in NRZ so that $p(t) = \pm 1$, of bit rate $R_c = 1/T_c$ which is much greater (by 10^2 to 10^6) than the bit rate R_b. The binary element of the sequence is called a chip in order to distinguish it from the binary element (bit) of the message. The composite signal then modulates a carrier by phase-shift keying (BPSK for example—see Section 3.7.7) whose frequency is the same for all network stations. The transmitted signal *s(t)* can be expressed by:

$$s(t) = m(t)p(t) \cos \omega_c t \qquad \text{(V)} \tag{4.17}$$

At the receiver the signal is coherently demodulated by multiplying the received signal by a replica of the carrier. Neglecting thermal noise, the signal *r(t)* at the input of the detector low-pass filter (LPF) is given by:

$$\begin{aligned} r(t) &= m(t)p(t) \cos \omega_c t(2 \cos \omega_c t) \\ &= m(t)p(t) + m(t)p(t) \cos 2\omega_c t \qquad \text{(V)} \end{aligned} \tag{4.18}$$

The detector low-pass filter eliminates the high frequency components and retains only the

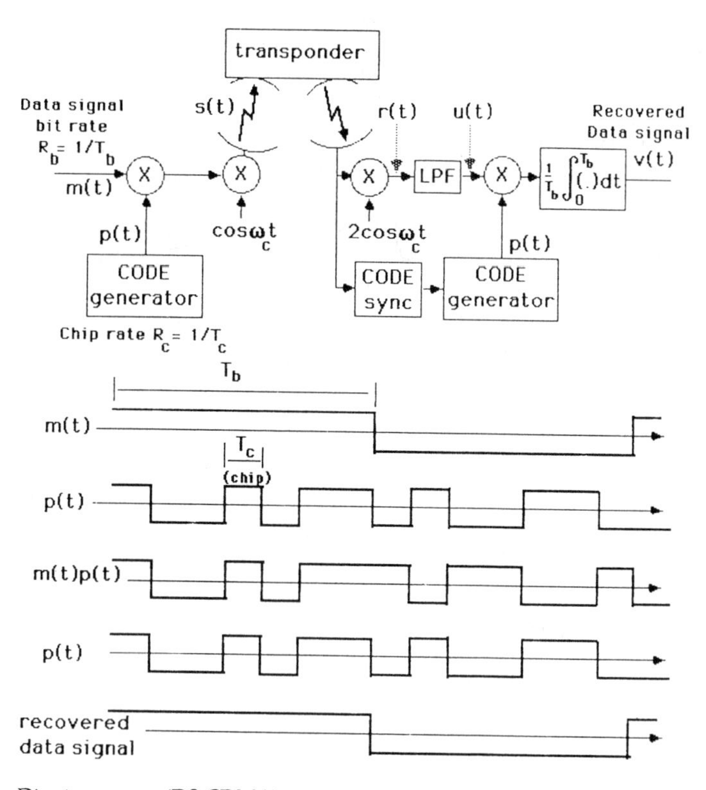

Figure 4.22 Direct sequence (DS-CDMA).

low frequency component $u(t) = m(t)p(t)$. This component is then multiplied by the local code $p(t)$ in phase with the received code. In the product $p(t)^2 = 1$. At the output of the multiplier this gives:

$$x(t) = m(t)p(t)p(t) = m(t)p(t)^2 = m(t) \qquad \text{(V)} \tag{4.19}$$

This signal is then integrated over one bit period to filter the noise. The transmitted message is recovered at the integrator output.

4.6.1.2 *Spectral occupation*

The spectrum of the carrier $s(t)$, of power P, is given by:

$$S(f) = P/R_c[\sin(\pi(f-f_c)/R_c)/(\pi(f-f_c)/R_c)]^2 \qquad \text{(W/Hz)} \tag{4.20}$$

It is represented in Figure 4.23. For comparison purposes, this spectrum is superimposed on that which the carrier would have if modulated by the message $m(t)$ alone. It can be seen that, in CDMA transmission, $s(t)$ has a spectrum which is broadened by the spreading ratio R_c/R_b. This is the result of combining the message with the chip sequence. It will now be shown that this combination permits multiple access.

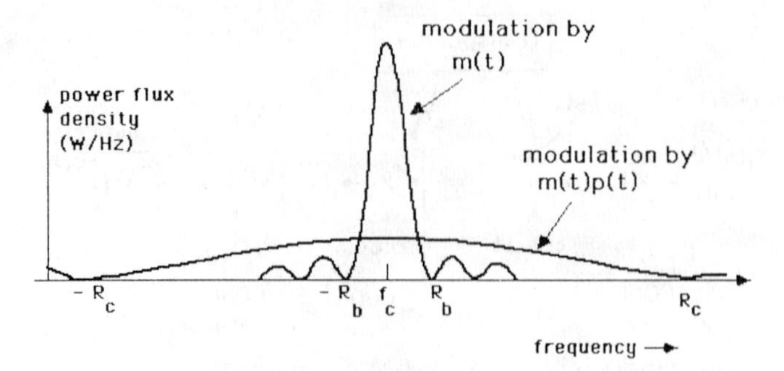

Figure 4.23 The spectrum of the carrier in DS-CDMA together with the spectrum which the carrier would have if modulated by the message $m(t)$.

4.6.1.3 *Realisation of multiple access*

The earth station receives from the channel the useful signal $s(t)$ superimposed on the signals $s_i(t)$ $(i = 1, 2, \ldots, N-1)$ of the $N-1$ other users transmitted on the same frequency; hence:

$$r(t) = s(t) + \sum s_i(t) \qquad \text{(V)} \tag{4.21}$$

with:

$$s(t) = m(t)p(t) \cos \omega_C t$$

$$\sum s_i(t) = \sum m_i(t)p_i(t) \cos \omega_C t$$

The multiplier output signal is given by:

$$x(t) = m(t)p(t)^2 + \sum m_i(t)p_i(t)p(t) = m(t) + \sum m_i(t)p_i(t)p(t) \qquad \text{(V)} \tag{4.22}$$

The message is now superimposed on noise due to interference. If care has been taken to choose codes with a low cross correlation function this noise will be small. Multiplication of $\sum m_i(t)p_i(t)$ by $p(t)$ at the receiver implies spreading the spectrum of each of the messages $m_i(t)$ which have already been spread. The noise spectral density $\sum m_i(t)p_i(t)p(t)$ is consequently low. The interference noise power in the bandwidth of the useful message $m(t)$ is thus low.

In the preceding discussion it has been assumed that multiplication by the chip sequence is performed on the binary message at baseband. It should be noted that expression (4.17) is also obtained by multiplying the carrier by the chip sequence after the carrier has been modulated by the binary message. In the same way, the operations of demodulation and despreading can be reversed at the receiver. If spread spectrum transmission is used to realise the 'multiple access' function, it is preferable on reception to proceed firstly to despreading and then to demodulation. In the other case (described above for reasons of simplicity of explanation), coherent demodulation necessitates recovery of the reference carrier in a spectrum (obtained by non-linear processing of spread and modulated carriers) containing the other reference carriers with high power levels. By proceeding firstly to despreading, the spectra of the unwanted carriers are spread and recovery of the required reference carrier is performed under favourable signal-to-noise conditions. On transmission, technological simplicity tends to a preference for spreading before modulation.

4.6.1.4 *Protection against interference between systems*

The signals transmitted by systems sharing the same frequency band as that used by the network can be narrow band carriers (medium capacity FDM/FM/FDMA carriers, for example). Let $J(t) \cos \omega_C t$ be such a carrier. The signal at the multiplier output is:

$$x(t) = m(t) + J(t)p(t) \qquad \text{(V)} \tag{4.23}$$

The interference noise is spread by the receiver. The interference power in the bandwidth of the useful message $m(t)$ is small.

This property is useful:

—For military applications when one wishes to avoid interference from an enemy transmitting a high power in a narrow band. (Spread spectrum transmission also provides the possibility of transmitting with discretion in view of the low spectral density of the carrier.)
—For civil applications, when one wishes to receive signals with small antennas in congested bands (for example at 4 GHz); due to the wide aperture of the antenna beam, the station reeceives carriers from adjacent satellites with a relatively high power. Spreading the spectrum of the carriers by the receiver limits the interference power in this case.

4.6.1.5 *Protection against multiple trajectories*

A link has multiple trajectories when the radio signal follows paths of different lengths and arrives at the receiver in the form of a useful signal accompanied by replicas delayed in time. This arises, for example, in mobile satellite links where the downlink wave is captured at the same time as its reflections from surrounding objects. The reflected signals thus appear as interference. If the time delay between the direct wave and the reflected waves is greater than the duration T_c of a chip, there is no longer correlation between the received code and the local code for the reflected waves and the spectrum of the reflected signals is spread. In this way there is a benefit of protection from interference from multiple trajectories.

4.6.2 Transmission by frequency hopping (FH-CDMA)

4.6.2.1 *The principle*

Figure 4.24 illustrates the principle. The binary message $m(t)$ to be transmitted is of rate $R_b = 1/T_b$ and is coded in NRZ. It modulates a carrier whose frequency $f_c(t) = \omega_c(t)/2\pi$ is generated by a frequency synthesizer controlled by a binary sequence, or code, generator. This generator delivers chips with a bit rate R_c. The principle will be illustrated by means of modulation by phase-shift keying (BPSK), although other types of modulation can be adopted, particularly frequency shift keying (FSK). The transmitted signal is thus of the form:

$$s(t) = m(t) \cos \omega_c(t)t \tag{4.24}$$

The carrier frequency is determined by a set of $\log_2 N$ chips, where N is the number of

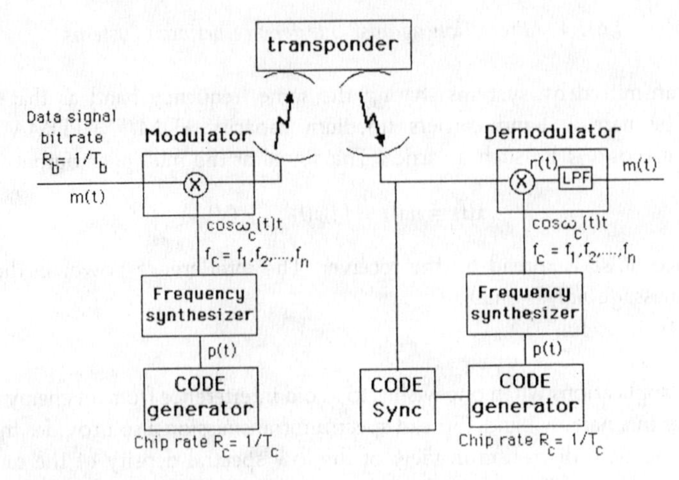

Figure 4.24 Frequency hopping (FH-CDMA).

possible carrier frequencies. It changes each time the code has generated $\log_2 N$ consecutive chips. The carrier frequency thus changes in steps. The frequency step is $R_H = R_c/\log_2 N$.

At the receiver the carrier is multiplied by an unmodulated carrier generated under the same conditions as at the transmitter. If the local code is in phase with the received code, the multiplier output signal is:

$$r(t) = m(t)\cos\omega_c(t)t \times 2\cos\omega_c(t)t = m(t) + m(t)\cos 2\omega_c(t)t \quad (4.25)$$

The second term is eliminated by the low pass filter (LPF) of the demodulator.

4.6.2.2 *Spectral occupation*

Three types of system can be considered:

—those for which there is one frequency step per information bit: $R_H = R_b$;
—those for which there are several frequency steps per bit: $R_H \gg R_b$;
—those for which a frequency step covers several bits: $R_H \ll R_b$.

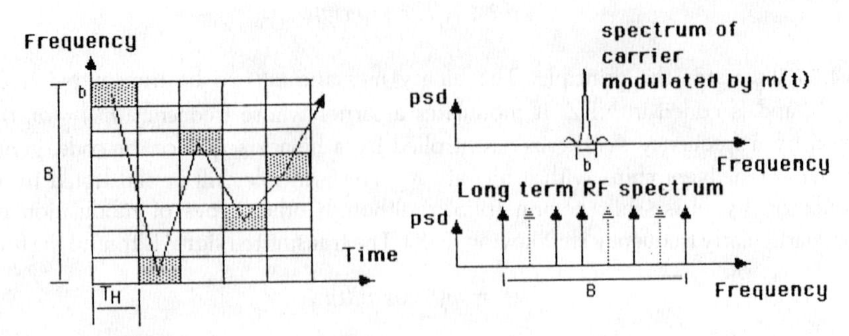

Figure 4.25 Spectral distribution in FH-CDMA for $R_H \ll R_b$.

Figure 4.25 shows an example of transmission with $R_H \ll R_b$. The short-term carrier spectrum (the spectrum for a period $T_H = 1/R_H$) has the characteristics of a BPSK carrier modulated by a binary stream of bit rate R_b and consequently occupies a bandwidth b approximately equal to R_b. The long-term spectrum consists of the superposition of the N carriers of the short-term spectrum. Hence it has a wider spectrum B. The spreading factor is B/b. The progress of the transmission of a carrier can be represented on the frequency–time grid of Figure 4.25 where each case represents one frequency state of the carrier at a given time.

4.6.2.3 *Realisation of multiple access*

The various network carriers follow different trajectories on the grid of Figure 4.25. At the receiver, only the carrier whose trajectory coincides with that of the carrier regenerated by

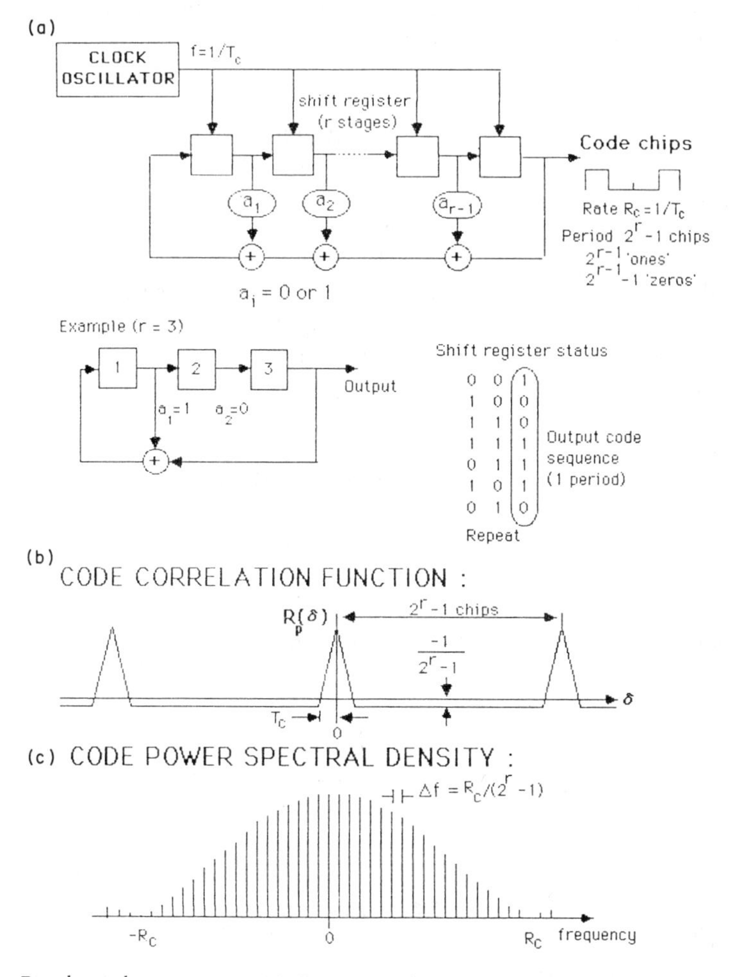

Figure 4.26 Pseudo-random sequence. (a) Generation. (b) Autocorrelation function. (c) Power spectral density.

the local synthesiser is demodulated. Hence the multiplier output signal, during an interval T_H when the synthesizer frequency is constant and equal to $\omega_c/2\pi$, is:

$$r(t) = [m(t) \cos \omega_c t + \sum m_i(t) \cos \omega_{ci} t] \times 2 \cos \omega_c t \quad (4.26)$$

At the output of the low-pass filter one finds $m(t)$ accompanied by noise caused by the possible presence of carriers such that $\omega_{ci} = \omega_c$. The probability of such an event is small when the number of frequency bands on the grid is high and hence the spectrum spreading factor B/b is large. The spectral density of the long-term interference noise spectrum can thus be made small.

4.6.2.4 *Protection against interference*

In a similar manner to the case of the direct sequence, interference caused by fixed frequency carriers is subjected to spectrum spreading at the receiver which limits the noise power in the bandwidth of the useful message $m(t)$.

4.6.3 Code generation

Figure 4.26 shows an example of an arrangement for the generation of a pseudo-random code sequence. The arrangement consists of a set of r flip-flops forming a shift register with a set of feedback paths provided with 'exclusive or' operations. The state of the flip-flops changes at the clock rate R_c. The stream of chips at the output is periodic with a period of $2^r - 1$ and each period contains $2^{r-1} - 1$ chips equal to 0 and 2^{r-1} chips equal to 1. The figure also shows the form of the autocorrelation function of the sequence together with its frequency spectrum.

4.6.4 Synchronisation

Synchronisation of the receiver pseudo-random sequence generator and the pseudo-random sequence which spreads the spectrum of the received carrier is a fundamental condition for realisation of multiple access. It is this condition which enables the receiver to detect the useful message $m(t)$. Synchronisation consists of two phases:

—sequence acquisition;
—tracking.

The principle of acquisition will be illustrated for the case of direct sequence transmission (DS-CDMA).

4.6.4.1 *Acquisition*

Figure 4.27 shows the principle of a possible acquisition scheme (DS-CDMA); the received carrier $s_1(t)$ is multiplied by the locally generated sequence $p(t + \partial)$. This is not in phase with

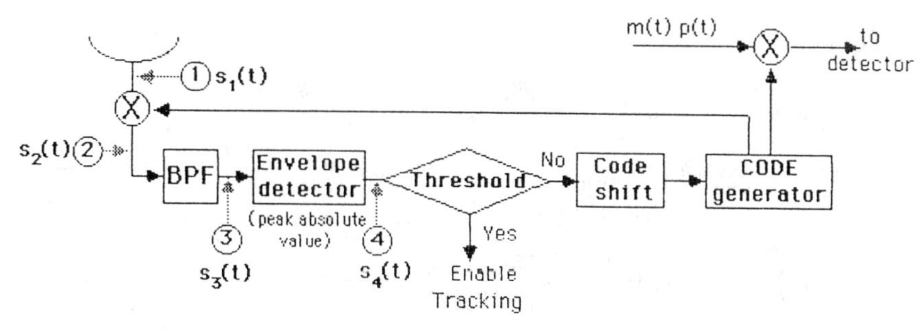

Figure 4.27 The principle of code acquisition in a DS-CDMA system.

the received sequence $p(t)$ and the shift is denoted by ∂. The multiplier output $s_2(t)$ is fed to a band-pass filter which is centered on the carrier frequency ω_c and has a bandwidth wide with respect to the spectrum of $m(t)$ but narrow with respect to the spectrum of $p(t)$. The filter thus has the effect of averaging the product $p(t)p(t+\partial)$ and the filter output signal can be expressed by:

$$s_3(t) = m(t)p(t)p(t+\partial)\cos\omega_c t \tag{4.27}$$

An envelope detector follows which detects the peak value of the filter output signal. As the amplitude of the carrier modulated by $m(t)$ is constant, the signal at the envelope detector output provides the absolute value of the autocorrelation function of $p(t)$, hence:

$$s_4(t) = Ip(t)p(t+\partial)I = IR_p(\partial)I \tag{4.28}$$

It is known (Figure 4.26c) that this function has a pronounced maximum for $\partial = 0$. The amplitude of the output voltage of the envelope detector is measured for a given value of ∂, then, if this voltage is less than a fixed threshold, ∂ is incremented by an amount equal to the duration of a chip T_c. The operation is repeated until the amplitude of the envelope detector output exceeds the fixed threshold indicating that the correlation peak for $\partial = 0$ has been achieved. One then proceeds to the tracking mode.

It is good practice to accumulate the results of several measurements for a given ∂ by placing an integrator, with a time interval equal to several periods of the pseudo-random sequence, between the envelope detector and the threshold detector.

4.6.4.2 *Tracking*

Figure 4.28 shows the principle of the tracking arrangement; the acquisition loop is duplicated with an 'advance' branch and a 'delay' branch. The signal produced by the pseudo-random sequence generator in the 'advance' branch is $p(t+T_c/2)$, that produced in the 'delay' branch is $p(t-T_c/2)$. The two signals at the envelope detector outputs are subtracted to produce an error signal $e(\partial) = |R_p(\partial + T_c/2)| - |R_p(\partial - T_c/2|$ which, after filtering, controls the advance or delay of the sequence generator. The sign of $e(\partial)$ indicates the direction of the correction to be performed and variation of $e(\partial)$ as a function of ∂ has the form characteristic of an error signal in a control loop.

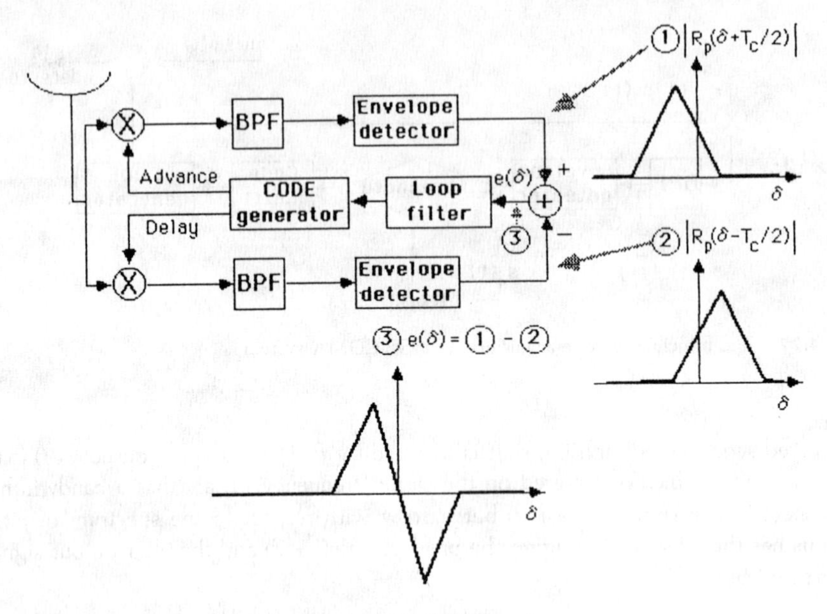

Figure 4.28 The principle of code tracking in a DS-CDMA system.

Some realisations, for acquisition and for tracking, replace the envelope detector by an energy detector (quadratic detector) [SIM-85, Vol. 3, Chap. 2]. This does not modify the principle but it does modify the form of the error signal characteristic. Other possibilities include numerical calculation of the convolution between the received signal and the locally generated code [GUO-90], [GUO-91].

4.6.5 The throughput of CDMA

The throughput of CDMA can be considered as the ratio of the total capacity provided by a channel in the case of single access, that is a single carrier modulated without spectrum spreading, and that of a channel transmitting several carriers modulated in CDMA simultaneously. The total capacity of the channel is then the product of the capacity of one carrier and the number of carriers, that is the number of accesses. The capacity of a carrier is R_b. What is the maximum number of accesses?

4.6.5.1 *Maximum number of accesses*

Consider the case of direct sequence modulation (DS-CDMA). Assume for simplicity that the N received carriers are all of equal power C. The useful carrier power at the receiver input is thus C. As the information rate carried by this carrier is R_b, the energy per information bit is $E_b = C/R_b$. Neglecting thermal noise in the noise power at the receiver input and retaining only the contribution of interference noise, the noise power spectral density N_0 at the receiver input is $N_0 = (N-1)C/B_N$, where B is the equivalent noise bandwidth of the receiver. This gives:

$$E_b/N_0 = B_N/R_b(N-1) \tag{4.29}$$

The spectral efficiency $\Gamma = R_c/B_N$ of the digital modulation used can be introduced into this expression. This then gives:

$$E_b/N_0 = R_C/R_b(N-1)\Gamma \tag{4.30}$$

As the quality of the link is stipulated by a given error rate, the value of E_b/N_0 is imposed. From this the maximum number of accesses N_{max} is deduced and is given by:

$$N_{max} = 1 + (R_c/R_b)/\Gamma(E_b/N_0) \tag{4.31}$$

4.6.5.2 *Throughput expression*

The maximum total capacity of the network is equal to $N_{max}R_b$. The capacity of a single carrier modulated without spectrum spreading and occupying a bandwidth B_N would be R_c. The throughput η of CDMA is thus given by the ratio:

$$\eta = N_{max}R_b/R_c \tag{4.32}$$

4.6.5.3 *Example*

Consider a CDMA network occupying the whole of a 36 MHz channel. The receiving bandwidth is $B_N = 36$ MHz. It will be assumed that each carrier has the capacity of one telephone channel, that is 64 Kbit/s. With BPSK modulation of theoretical spectral efficiency $\Gamma = 1$ bit/sHz the chip rate is $R_c = B_N/\Gamma = 36$ Mbit/s and the spreading ratio is $36 \times 10^6/64 \times 10^3 = 563$. Table 4.1 shows the maximum number of accesses, the maximum total capacity of the network and the throughput for a chosen error probability. The throughput, of the order of 10%, is low compared, for example, with TDMA (Section 4.4.5). The values in the table are optimistic; thermal noise is neglected, user codes are assumed to be orthogonal and no account is taken of degradation due to the demodulator.

Table 4.1 The performance of a CDMA access network using a 36 MHz transponder and binary phase shift keying (BPSK). Each carrier has the capacity of one 64 kbit/s telephone channel

Required error probability	E_b/N_0	Maximum number of accesses N_{max}	Maximum total capacity	Efficiency (%)
10^{-4}	8.4 dB	82	5.3 Mbit/s	15
10^{-5}	9.6 dB	62	4 Mbit/s	11
10^{-6}	10.5 dB	51	3.3 Mbit/s	9

4.6.6 Conclusion

Code division multiple access operates on the principle of spread spectrum transmission, recalled in Figure 4.29. The code sequence which serves to spread the spectrum constitutes the 'signature' of the transmitter. The receiver recovers the useful information by reducing the spectrum of the carrier transmitted in its original bandwidth. This operation simultaneously spreads the spectrum of other users in such a way that these appear as noise of low spectral density.

Code division multiple access has the following advantages:

—It is simple to operate since it does not require any transmission synchronisation between stations. The only synchronisation is that of the receiver to the sequence of the received carrier.
—It offers useful protection properties against interference from other systems and interference due to multiple paths; this makes it attractive for networks of small stations with large antenna beamwidth and for satellite communication with mobiles.

The main disadvantage is the low throughput; a large bandwidth of the space segment is used for a low total network capacity with respect to the capacity of a single unspread carrier.

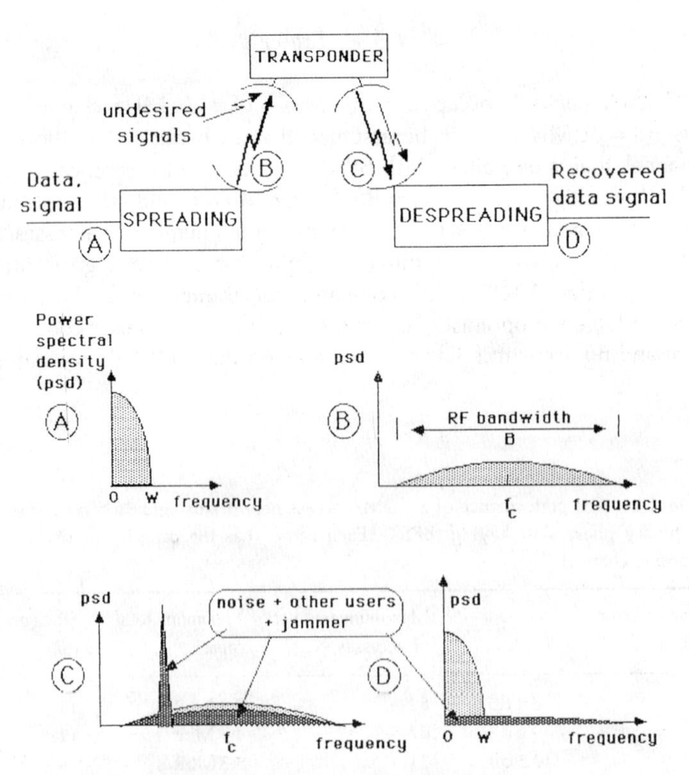

Figure 4.29 Spread spectrum transmission in a code division multiple access system.

4.7 FIXED AND ON-DEMAND ASSIGNMENT

4.7.1 The principle

Traffic routing implies access by each carrier transmitted by the earth stations to a channel. For each of the three fundamental modes (FDMA, TDMA and CDMA) described in the previous sections, each carrier is assigned a portion of the resource offered by the satellite or by a channel (a frequency band, a time slot or a fraction of the total power). This assignment can be defined once and for all (a fixed assignment) or in accordance with requirements (on-demand assignment).

With fixed assignment the capacity of each earth station is fixed independently of the traffic demand from the terrestrial network to which it is connected. An earth station can receive a traffic request from the network to which it is connected greater than the capacity which is allocated to it. It must then refuse some calls; this is a blocking situation in spite of the fact that other stations may have excess capacity available. Because of this, the resource constituted by the satellite network is poorly exploited.

On-demand assignment indicates that the resource offered by the channel can be assigned in a variable manner to the various stations in accordance with demand. There will, therefore, be the possibility of transferring capacity from stations with excess capacity to stations with excess demand.

On-demand assignment has so far found practical applications only for FDMA and TDMA. The description of on-demand assignment will, therefore, be limited to these two cases. As for multiple access, a distinction can be made between:

—on-demand assignment of a channel;
—on-demand assignment of a repeater, which can support several channels.

4.7.1.1 *The channel case*

On-demand assignment can be performed in accordance with one of the following situations [FRE-74], [TIR-83]:

—Variable destination: all or part of the resource provided by the channel is allocated permanently to the channels from one station and assigned as a function of demand. With FDMA, stations transmit at given fixed frequencies and are able to receive at several frequencies which change with the traffic demand. With TDMA, it implies variable organisation, at the transmitting station, of the sub-bursts within a burst as a function of demand without changing the assignment, at channel level, of frame time slots to bursts (bursts are of fixed duration and position).
—Variable origin: all or part of the resource provided by the channel is allocated permanently to the channels received by one station and assigned as a function of demand. With FDMA stations can transmit at several frequencies which change with the traffic demand and receive at given fixed frequencies. With TDMA, it implies that each station

transmits, at channel level, in frame time slots allocated permanently to each receiving station.

4.7.1.2 *The satellite case*

It is necessary to extend the preceding as follows:

—Variable destination: with FDMA, frequency agility on reception extends to all bands of the channels. With TDMA, it is necessary to create frequency agility on reception for all bands of the channels.
—Variable origin: with FDMA frequency agility on transmission extends to all bands of the repeaters. With TDMA, it is necessary to create frequency agility on reception for all bands of the channels.

The concept of totally variable assignment can now be introduced; all of the resource provided by the channels is available for all of the stations. With FDMA, frequency agility must exist for all bands of the channels both on transmission and reception. With TDMA, each station must be equipped with a frequency agile modulator in order to be able to access every channel (it is assumed that the maximum capacity of a station is limited to that of one channel), and as many demodulators as there are channels. The assignment of frame time slots in each channel is variable as a function of demand and hence the duration and position of bursts is variable.

4.7.2 Comparison between fixed and on-demand assignment

Consider a satellite network containing 20 stations. Each station must transmit traffic to the 19 other stations via a satellite repeater channel whose capacity S is 1520 channels. A blocking probability of 0.01 is required. The traffic intensity per channel will be calculated for the case of fixed assignment and for that of on-demand assignment.

4.7.2.1 *Fixed assignment*

The capacity of the channel is shared among 20 stations. Each station thus has $1520/20 = 76$ channels available. These channels are shared among the 19 destinations. There are, therefore, $76/19 = 4$ channels per destination. The maximum traffic intensity A must be determined such that the blocking probability $B(C = 4, A) = E_{C=4}(A)$ remains less than 0.01. It is found, by using expression (4.2) that $A = 0.87$ Erlang, that is 0.217 Erlang per channel.

4.7.2.2 *On-demand assignment*

The total capacity S of the channel can be assigned to any station whatever the destination. One must realise the condition $B(S = 1520, A) = E_{S=1520}(A) < 0.01$ which leads to $A = 1491$ for the 1520 channels and hence an intensity $A/1520 = 0.98$ Erlang per channel.

4.7.3 Example of an on-demand assignment system in FDMA: the SPADE system

4.7.3.1 Description

The SPADE system (Single channel per carrier PCM multiple access demand assigned equipment [PUE-71], [EDE-72]) is an example of an on-demand asssignment system in FDMA. It is designed to use one INTELSAT channel of 36 MHz bandwidth. The channel provides a capacity of 800 digital channels for carriers modulated in QPSK at a rate of 64 kbit/s. Following a call from a user, the station to which this user is connected transmits a message on a signalling channel common to all the destination stations of the station to which the called user is connected. In this message it indicates which channel it intends to use from among the channels available at the particular time. The called station acts similarly and all the stations update their lists of available channels by removing the two newly assigned channels from these lists. At the end of the conversation, the channels are made available for new assignments. The carriers are voice activated, that is the carrier is transmitted only in the presence of speech. This permits the number of carriers amplified by the channel to be reduced in a ratio equal to the activity factor of a telephone channel (Section 3.1.1).

The signalling channel is a common channel of 160 kHz bandwidth. The stations access it in TDMA to exchange messages relating to control of on-demand assignment. This control is decentralised among all network stations.

Table 4.2 indicates the principal characteristics of the SPADE system.

4.7.3.2 Comparison between SPADE and a fixed assignment system

Table 4.3 presents a comparison between the capacity of an INTELSAT channel of 36 MHz bandwidth with permanent assignment for different FDM/FM/FDMA carriers and that of

Table 4.2 Characteristics of the SPADE system

Channel characteristics	
Coding	PCM
Modulation	QPSK
Capacity	64 kbit/s
Channel bandwidth	38 kHz
Channel spacing	45 kHz
Stability	± 2 kHz
Maximum error rate	10^{-4}
Signalling channel characteristics	
Access type	TDMA
Capacity	128 kbit/s
Modulation	BPSK
Frame period	50 ms
Burst period	1 ms
Number of accesses	50 (49 stations + 1 reference)
Maximum error rate	10^{-7}

Table 4.3 Comparison of demand and fixed assignment in FDMA for a transponder of 36 MHz bandwidth

Type of link	Carrier spectral width	Number of channels	Number of accesses	Total capacity
Fixed assignment				
FDM/FM	5 MHz	60	7	420
FDM/FM	2.5 MHz	24	14	336
Demand assignment (SPADE)				
SCPC/FM	0.045 MHz	1	800	800

the SPADE system [PUE-71]. The advantage provided by SPADE is greater when the number of accesses in fixed assignment is high and the capacity per access is small. The penalty of the advantage provided by the SPADE system is the cost of the equipment required in the earth stations to provide control of the on-demand assignment. It has been possible to reduce this cost in some national systems where the control is centralised in a single station (Algeria for example).

4.7.4 On-demand assignment with TDMA

4.7.4.1 *Example 1—The TELECOM 1 system*

The TELECOM 1 satellite carries a payload at 14/12 GHz which consists of six channels of which five provide communication for the TRANSDYN digital multiservice network of France Telecom [BOU-81], [LOM-81], [GUE-83]. This network operates in TDMA with on-demand assignment. Each station transmits on the frequency of one channel and can receive the five channels simultaneously; there is, therefore, frequency agility on reception only. Station transmissions are distributed among the five channels. Satellite accesses are established call by call in accordance with demands from terrestrial switching stations to which the earth stations are connected. Figure 4.30 shows the frame structure. Figure 4.31 shows the various traffic and signalling channels for routing traffic and demand assignment messages respectively. Management of on-demand assignment is centralised at the reference station. Assignment demands from the earth station to the reference station are transmitted by traffic station bursts (in bursts denoted by 'a from *n* to 0' in Figure 4.30) and the return assignments are transmitted by the reference station to the traffic stations by means of the reference burst. Assignment of bursts from different stations in the frame (burst time plan) remains fixed for the duration of one super-frame (256 frames). Change of assignment occurs at the start of a super-frame. The system is capable of processing 10 calls per second and a point-to-point link at 64 kbit/s is established in less than 6 seconds in 95% of cases [CNE-83]. Table 4.4 summarises the characteristics of the demand assignment system of TELECOM 1.

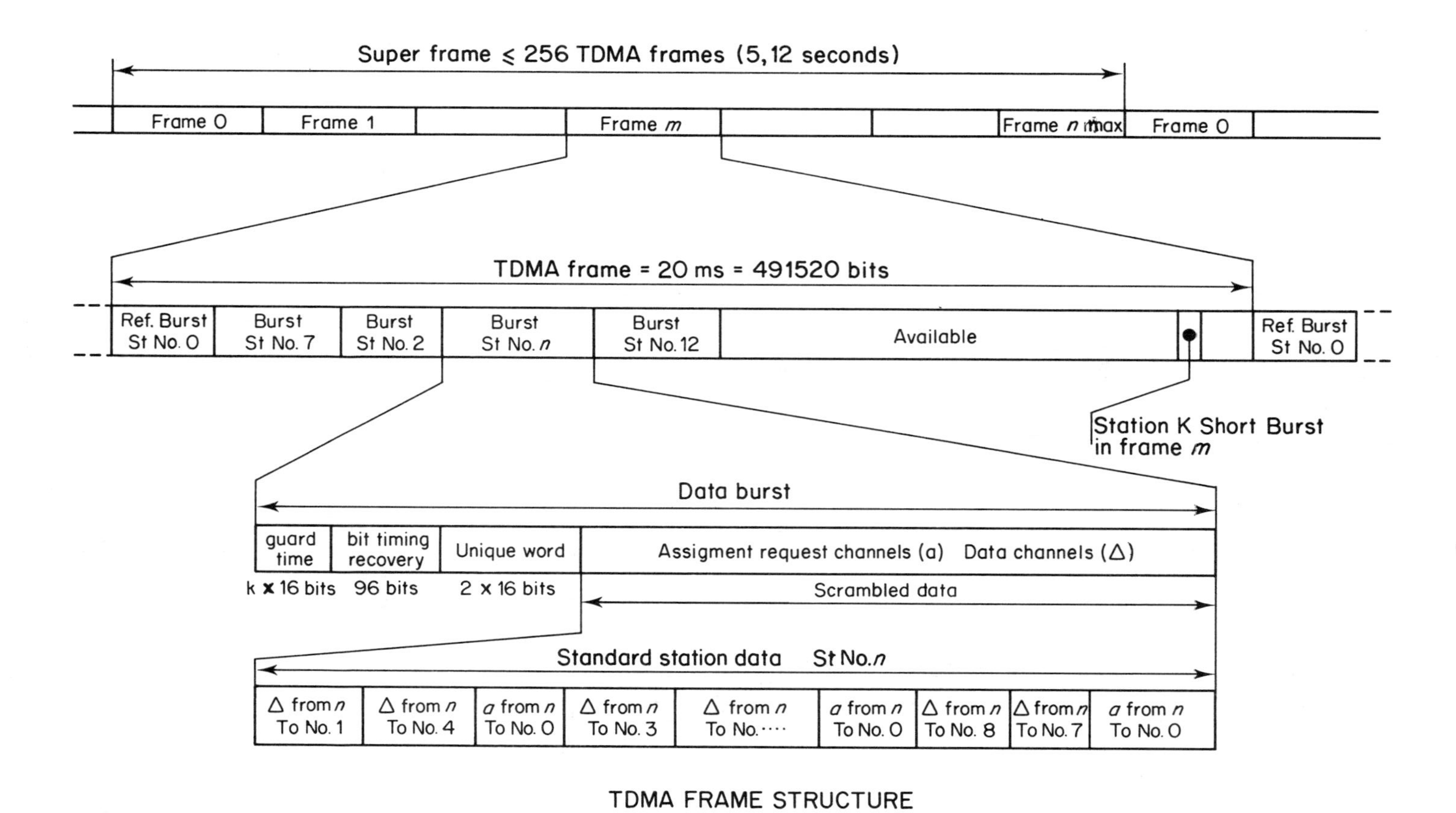

Figure 4.30 Frame structure in a TELECOM 1 transponder system.

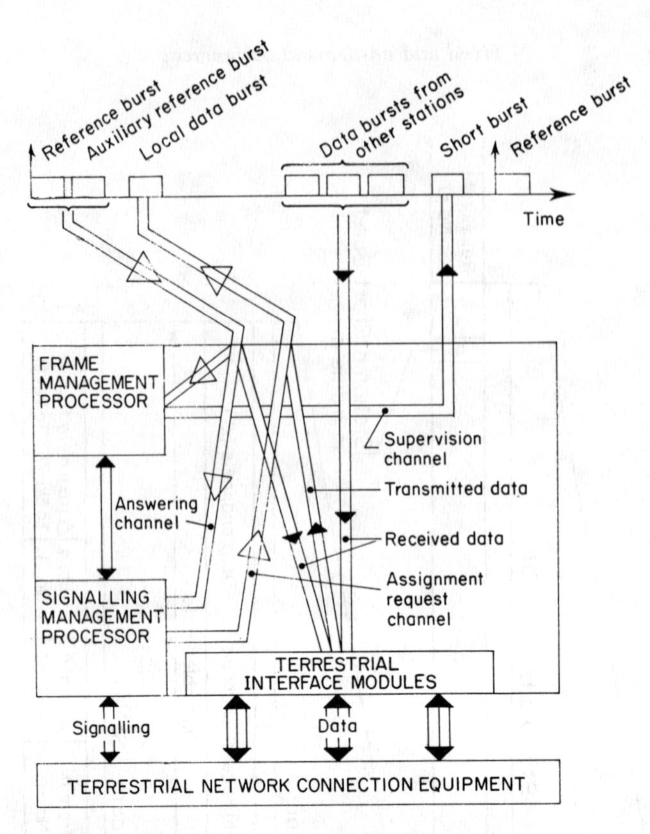

Figure 4.31 Traffic and signalling channels in the TELECOM 1 system.

Table 4.4 Characteristics of the TELECOM 1 TDMA demand assignment system [BOU-81], [LOM-81], [GUE-83]

Modulation	BPSK with differential encoding
Frame period	20 ms
Superframe period	5.12 s (256 frames)
Capacity	24,576 Mbit/s
Error rate	10^{-6} (99% of the time)
Transponder bandwidth	36 MHz
User data rate	2.4 kbit/s, 4.8 kbit/s, 9.6 kbit/s (with reservation), 32 kbit/s (telephony in DPCM) 48 kbit/s, $n \times 64$ kbit/s ($n = 1$, 2, 4, 8, 16 or 30 with reservation or call by call), 2048 Mbit/s (with reservation)
Demand assignment	Centralised control Assignment messages transmitted to earth stations by the reference burst. Assignment requests transmitted by the traffic stations in sub-bursts (CCITT No. 7) with capacities from 12.8 kbit/s to 32 kbit/s and an error correcting code of ratio 4/5. Replies by the reference burst (CCITT No. 7) of capacity 64 kbit/s with error correcting coding of ratio 4/5.
Performance	The system has a processing capacity of 10 calls per second. A point-to-point link is established in less than 6 seconds in 95% of cases.
Traffic station	Antenna diameter: 3.5 m Capacity is 255 bursts for transmission and reception.
Network	62 stations per transponder $\times 5$ transponders = 310 stations.

4.7.4.2 *Example 2—The SBS system [SCH-80], [GOO-84]*

The SBS system is a satellite network intended to provide companies, banks, insurance companies, government organisations and the like with wide bandwidth switched links for telephony, data, television and facsimile by means of earth stations located directly in company establishments. The stations are equipped with 5.5 m or 7.6 m antennas. Data rates offered to users are from 2.4 kbit/s to 6.4 Mbit/s. The error rate is 10^{-4} during 99.5% of the time. The user can obtain 10^{-7} during 99.5% of the time by using an error correcting code. The frame duration is 15 ms and the frame data rate is 48 Mbit/s. Management of demand assignment is centralised.

4.7.5 Centralised or decentralised management of demand assignment

Management is centralised when management of demand assignment is realised in a single station; this implies that ordinary traffic stations send demand messages to the central station and this determines the assignment of resources and transmits this assignment to the whole network. Management is decentralised when stations transmit their demands on a common signalling channel. These demands are taken into account by each station and the state of the resources is updated at each station.

Table 4.5 presents a comparison of the advantages and disadvantages of centralised management with respect to those of decentralised management.

Table 4.5 Comparison of centralised and decentralised control of demand assignment

Principle	Advantages	Disadvantages
Centralised control		
Traffic stations send request message to the central station	Stations do not have to perform the assignment—Low equipment cost	Correct operation of the network depends on that of the control station—Reduced reliability
Determination of resource assignment by the central station and transmission of this assignment to the whole network	Reduced signalling since the whole network does not need to be informed	The need for a redundant control station The establishment time of a link is penalised by the double hop
Decentralised control		
Request transmitted on a common signalling channel	No control station —Better network reliability	More complex equipment at each station —Higher cost
Each station updates the state of the resources	Reduced establishment time	More signalling

4.7.6 Conclusion

Fixed assignment is recommended for networks involved in routing large volumes of traffic between a small number of stations of high capacity. Demand assignment provides better utilisation of the satellite network in the case of a large number of stations of low capacity per access with large variations in demand. Each station can thus benefit occasionally from a greater capacity than that which it would have in the case of a fixed assignment. Management of the assignment implies a link establishment time of the order of a second. When the link is required for several minutes, as is the case for average telephone conversations, this establishment time is of no consequence. The choice of a demand assignment technique must, therefore, take the following aspects into account:

—Specifications on the user side: traffic density, number of destinations, blocking probability.
—The gain resulting from the operation of demand assignment; this involves comparing the increase of revenue resulting from a higher traffic throughput for a given blocking probability with the increased expense involved in the installation of equipment to manage demand assignment.
—The choice between centralised and decentralised management.

The establishment time of the link can, however, be a decisive factor for some types of traffic such as that of data exchanged between data processing systems. The traffic generated in communication between computers, or between computers and computer terminals, is characterised by a large variation in the duration of messages and the intervals between messages. Furthermore, the user often imposes a clause concerning transmission time which can be short compared with the time between messages. Under these conditions, the establishment time of a link can exceed the utilisation time and this corresponds to inefficient use of the network. On the other hand, the establishment time of the link can lead to an unacceptable transmission time. It is thus preferable to resort to random access as described in the following section.

4.8 RANDOM ACCESS

This type of access is well suited to networks containing a large number of stations where each station is required to transmit short randomly generated messages with long dead times between messages. The principle of random access is to permit transmission of messages almost without restriction in the form of limited duration bursts which occupy all the bandwidth of the transmission channel. It is, therefore, multiple access with time division and random transmission. The possibility of collisions between bursts at the satellite is accepted. In the case of collision, the earth station receiver will be confronted with interference noise which can compromise message identification. Retransmission of all or part of the burst will be necessary. Protocol types are distinguished by the means provided to overcome this disadvantage.

The performance of these protocols is measured in terms of the throughput and the mean transmission delay. Throughput is the ratio of the volume of traffic delivered at the destination to the maximum capacity of the transmission channel. The transmission time (delay) is a random variable. Its mean value indicates the mean time between the generation of a message and its correct reception by the destination station.

These protocols have been the object of numerous studies since 1970 without finding commercial application on satellite networks. Recently, their practical application has become of great importance in the context of private networks using small stations (VSAT—very small aperture terminals) which have been widely developed to provide satellite communication between computers and distant terminals [FUJ-86], [RAY-87a], [RAY-87b], [RAY-88], [WOL-87].

4.8.1 Totally asynchronous protocols

4.8.1.1 The 'ALOHA' protocol

Figure 4.32 illustrates the principle of multiple random access in accordance with this protocol [ABR-77], [HAY-81], [ABR-73]. The packets are transmitted by each earth station without any restriction on the time of transmission. It is, therefore, a totally asynchronous protocol. In the absence of collisions (the case of Figure 4.32a), the destination stations transmit an acknowledgement in the form of a short return packet (ACK).

Figure 4.32b illustrates the case of a collision. The destination station receiver is not able to identify the message and does not send an acknowledgement. If an acknowledgement is not received within a fixed time interval after transmission, the transmitting station retransmits the message; the time interval is set to a value slightly greater than twice the round trip propagation time of the carrier wave. This new transmission occurs after a random time which is determined independently at each station in order to avoid a further collision.

In the case where the user population is homogeneous (the packet duration and message generation rate are constant), it can be shown that the traffic carried S (packets correctly interpreted by the receiver) as a function of the total traffic G (original and retransmitted messages) is given by the relation:

$$S = G\exp(-2G) \qquad \text{(packets/time slot)} \tag{4.33}$$

S and G are expressed as a number of packets per time slot equal to the common packet duration. Consequently, S represents the transmission throughput. This curve is shown in Figure 4.33. Figure 4.34 shows the variation of mean transmission time as a function of the throughput S. It can be seen that the ALOHA protocol does not exceed a throughput of 18% and the mean transmission time increases very rapidly as the traffic increases due to the increasing number of collisions and packet retransmissions.

4.8.1.2 The selective reject (SREJ) ALOHA protocol [RAY-87c]

With totally asynchronous transmission, collision between packets is most often partial. With the ALOHA protocol, the coherence of the packet is destroyed even by a partial collision. This leads to retransmission of the contents of the whole packet although only a part has suffered a collision. The SREJ-ALOHA protocol has been designed to avoid a complete

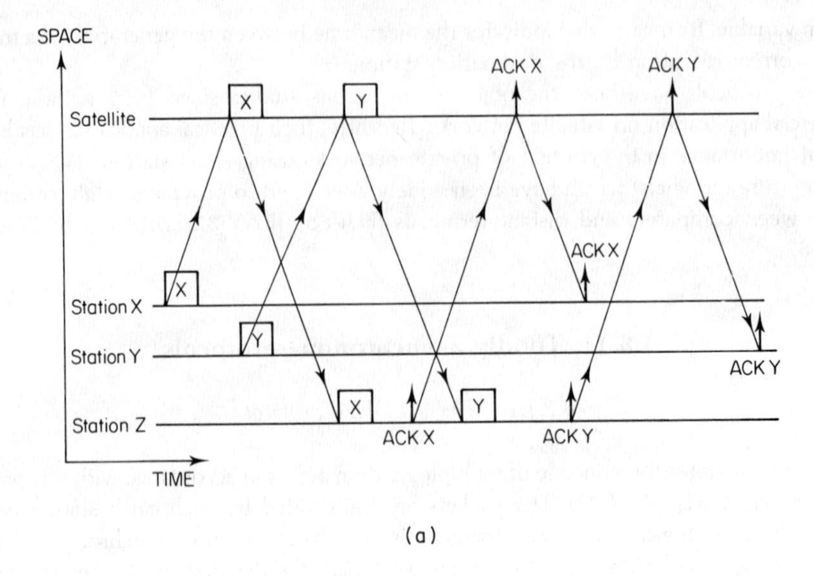

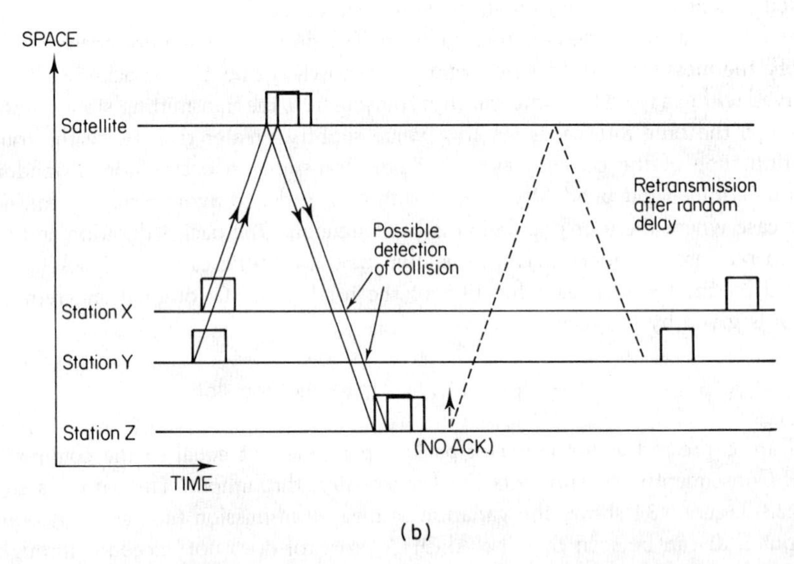

Figure 4.32 Distance–time diagram illustrating the principle of the ALOHA random multiple access protocol. (a) Without collision. (b) With collision.

retransmission. The transmitted packet is divided into sub-packets each having its own header and protocol bits. When a collision occurs, only the sub-packets involved are retransmitted. Figure 4.35 illustrates the principle of the SREJ-ALOHA protocol. The transmission throughput of the protocol is greater than that of the ALOHA protocol. The practical limit, of the order of 30%, is caused by the addition of headers to the sub-packets. The SREJ-ALOHA protocol is well suited to applications in which the messages have variable lengths.

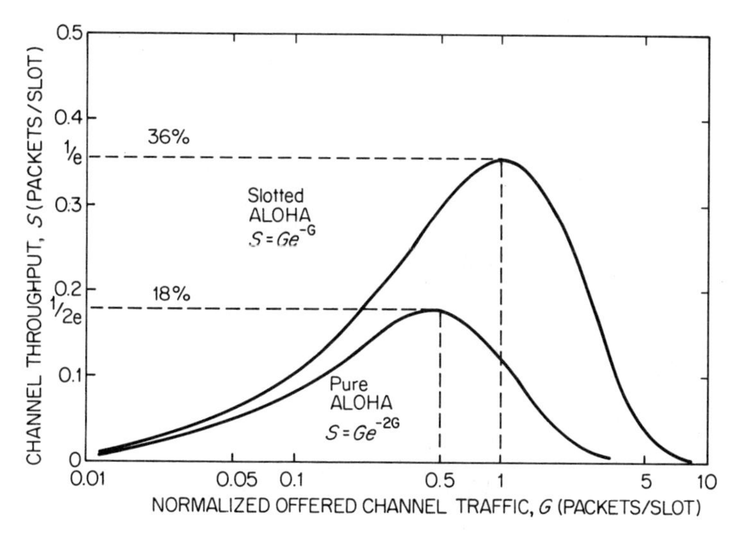

Figure 4.33 Transmission efficiency.

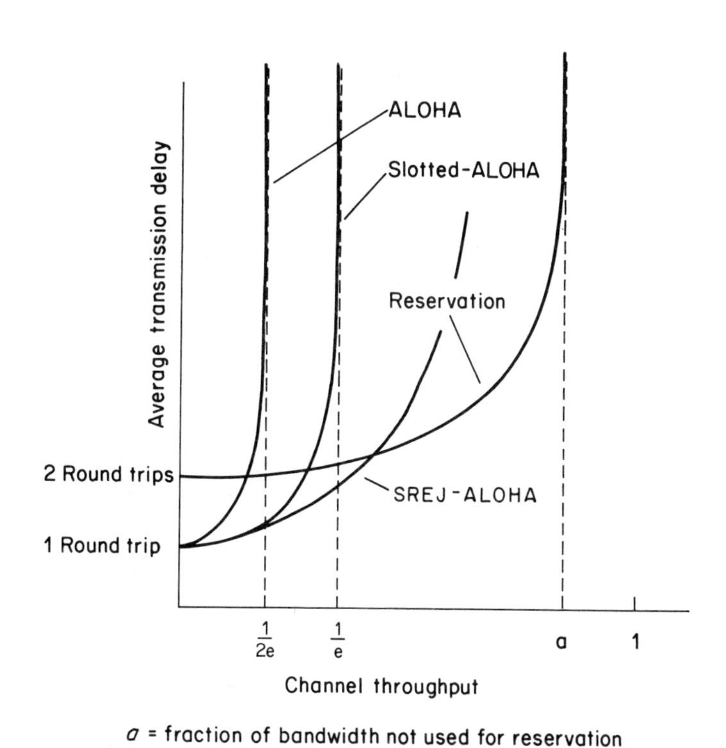

Figure 4.34 Mean transmission time.

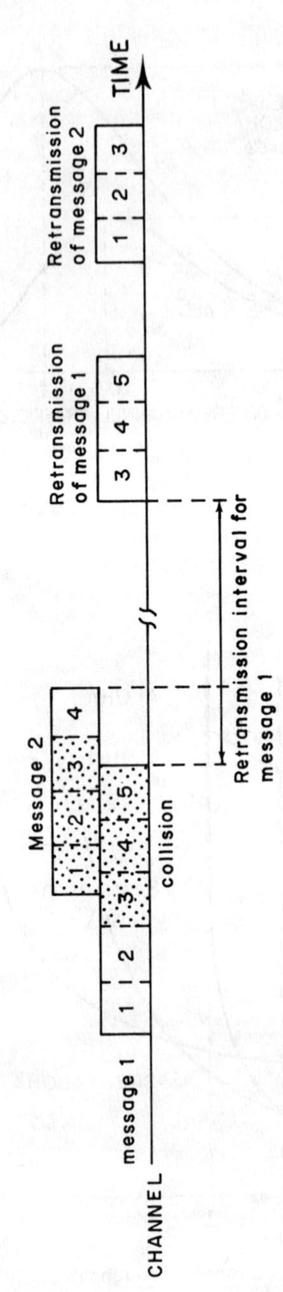

Figure 4.35 The principle of the SREJ-ALOHA protocol.

4.8.1.3 *The time-of arrival collision resolution algorithm (CRA) protocol [CAP-79]*

This protocol provides an improvement to the ALOHA protocol by avoiding the possibility that a packet which has already been subjected to a collision encounters another packet during its retransmission. To achieve this, stations avoid transmitting new packets in the time slots provided for retransmission of packets which have suffered a first collision. This protocol implies a procedure for identifying packets which have suffered a collision and the setting up of temporary co-ordination of transmissions. The throughput is from 40 to 50%. Although very promising, a protocol of this kind tends to be complex to implement and considerable work is still required prior to operational use.

4.8.2 Protocols with synchronisation

4.8.2.1 *The slotted ALOHA (S-ALOHA) protocol*

Transmissions from stations are now synchronised in such a way that packets are located at the satellite in time slots defined by the network clock and equal to the common packet duration. Hence there cannot be partial collisions; every collision arises from complete superposition of packets. The timescale of the collision is thus reduced to the duration of a packet whereas with the ALOHA protocol this timescale is equal to the duration of two packets as shown in Figure 4.36. This divides the probability of collision by two and the throughput becomes:

$$S = G\exp(-G) \qquad \text{(packets/time slot)} \tag{4.34}$$

This curve is represented in Figure 4.33. The increased throughput due to synchronisation can be seen.

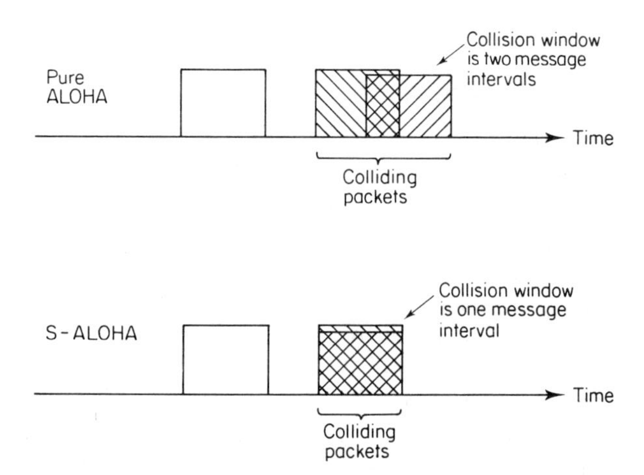

Figure 4.36 Collision diagrams with the ALOHA and S-ALOHA protocols.

4.8.2.2 *The announced retransmission random access (ARRA) protocol*

This protocol increases the throughput of S-ALOHA by introducing a frame structure which permits numbering of time slots. Each packet incorporates additional information indicating the slot number reserved for retransmission in case of collision. This protocol enables collisions between new messages and retransmissions to be avoided. The throughput is of the order of 50 to 60%.

4.8.3 Protocols with assignment on demand (DAMA)

These protocols are intended to increase transmission throughput by an advance capacity reservation procedure (Demand Assignment Multiple Access, DAMA). A station reserves a particular time slot within a frame for its own use.

Reservation can be implicit or explicit [RET-80]:

—Implicit reservation is reservation by occupation, that is every slot occupied once by the packet from a given station remains assigned to this station in the frames which follow. This protocol is called R-ALOHA [CRO-73], [ROB-73]. The disadvantage is that a station is in a position to capture all the time slots of a frame for itself. The advantage is the absence of establishment time for a reservation.

—Explicit reservation involves a station sending a request to occupy certain time slots to a control centre. Two examples of this protocol are R-TDMA and C-PODA (Contention based priority oriented demand assignment) [JAC-78]. The disadvantage of these protocols is the establishment time which can be prohibitive in some interactive applications. Figure 4.34 shows the variation of transmission time for protocols of the DAMA type with explicit reservation as a function of transmission throughput.

4.9 CONCLUSION

There is a large variety of solutions to the problem of mutliple access to a repeater by a group of network stations. The choice of access type depends above all on economic considerations; there are the global cost in terms of investment and operating costs and the benefits in terms of revenues.

General indications can be given according to the type of traffic:

—For traffic characterised by long messages, implying continuous or quasicontinuous transmission of a carrier, FDMA, TDMA and CDMA access techniques are the most appropriate. This involves, for example, telephone traffic, television transmission and videoconferencing. If the volume of traffic per carrier is large and the number of accesses is small (trunking), FDMA has the advantage of operational simplicity. When the traffic per carrier is small and the number of accesses large, FDMA loses much in efficiency of usage of the space segment and TDMA and CDMA are the best candidates. However, TDMA requires relatively costly earth station equipment. For small stations exposed to inter-system interference, CDMA may be preferred despite its low throughput.

Selection of FDMA or TDMA multiple access also implies a choice between fixed and demand assignment. Economic considerations will prevail; the increase in revenue resulting from higher traffic is compared with the increased expense involved in the installation of equipment to control demand assignment.

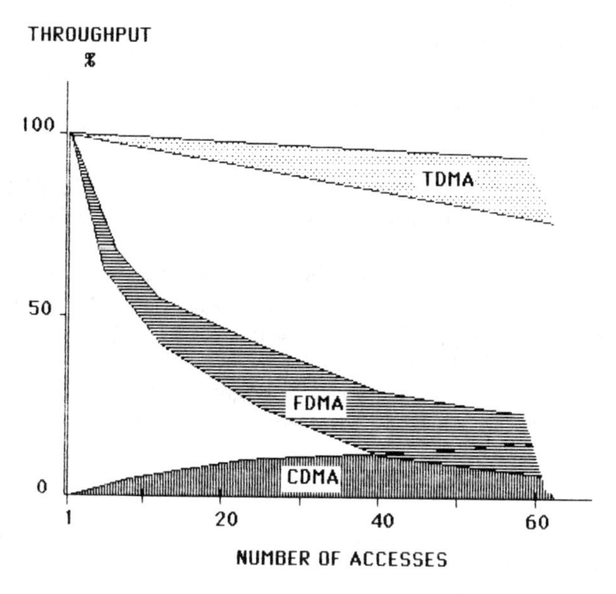

Figure 4.37 Comparison of throughput for different multiple access techniques. A 100% throughput corresponds to the capacity considering *one access* only (one carrier within a single transponder).

Figure 4.37 summarises the results stated previously relating to transmission throughput and can be used in the choice of type of multiple access.

—For traffic characterised by short messages and random generation with long dead times between messages, random access is the most appropriate. Figure 4.34 illustrates the compromise between a short transmission delay with low throughput of the space segment and efficient utilisation of the space segment with a longer transmission delay.

REFERENCES

[ABR-73] N. Abramson (1973) Packet switching with satellites, *NCC AFIPS Conference Proceedings*, **42**, pp. 695–702.

[ABR-77] N. Abramson (1977) The throughput of packet broadcasting channel, *IEEE Tansactions on Communications*, **COM-25**, No. 1, pp. 117–128.

[BOU-81] J.C. Bousquet (1981) Time division multiple access system with demand assignment for intra company network using the satellite TELECOM 1, *5th International Conference on Digital Satellite Communications, Genoa (Italy).*

[CAP-79] J.I. Capetenakis (1979) Tree algorithms for packet broadcast channels. *IEEE Transactions on Information Theory*, pp. 505–513, Sept.

[CCIR-88] CCIR (1988) *Handbook on Satellite Communications*, Geneva.

[CNE-83] CNES-CNET (1983) *Télécommunications Spatiales, ouvrage collectif par des ingénieurs du CNES et du CNET*, Tome III, Masson.

[CRO-73] W. Crowther, R. Rettberg, D. Walden (1973) A system for broadcast communications: reservation ALOHA, *Proceedings of the 6th International System Science Conference*, Hawai, pp. 371–374.

[EDE-72] B.I. Edelson, A.M. Werth (1972) Spade system progress and application, *COMSAT Technical Review*, **2**, No. 1, pp. 221–242.

[FEH-83] K. Fehr (1983) *Digital Communications*, Prentice Hall.

[FRE-74] G. Frenkel (1974) The grade of service in multiple access satellite communications systems, *IEEE Transactions on Communications Technology*, **COM-22**, pp 1681–1685.

[FUJ-86] A. Fujii, Y. Teshigawara, S. Tejima, Y. Matsumoto (1986) AA/TDMA adaptive satellite access method for mini-earth stations for satellite communications network, *Globecom 86*, pp. 1494–1499.

[GAG-84] R.M. Gagliardi (1984) *Satellite Communications*, Lifetime Learning Publications.

[GOO-84] B. Goode (1984) SBS TDMA-DA system with VAC and DAC, *Proceedings of the IEEE*, **72**, No. 11, pp. 1594–1610, Nov.

[GUE-83] J.P. Guenin, J.C. Bernard Dende, Y. Choi, A. Hoang-van (1983) The TELECOM 1 satellite system: architecture of the common signalling network, *6th International Conference on Digital Satellite Communications, Phoenix*, pp. V.17–V.21.

[GUO-90] X.Y. Guo, G. Maral, A. Marguinaud, R. Sauvagnac (1990) A fast algorithm for the pseudonoise sequence acquisition in direct sequence spread spectrum systems, ESA-WPP-019, *Proceedings of the Second International Workshop on Digital Signal Processing Techniques Applied to Space Communications (DSP90), Politecnici di Turino, Turin (Italy, 24–25 September, 1990).*

[GUO-91] X.Y. Guo, G. Maral, A. Marguinaud, R. Sauvagnac (1991) Méthode de calcul rapide de convolution entre deux séquences dont l'une est binaire, *Annales des télécommunications*, Tome 46, No. 3–4, pp. 181–190.

[HAY-81] J.F. Hayes (1981) Local distribution in computer communications, *IEEE Communications Magazine*, pp. 6–14, March.

[JAC-78] I.M. Jacobs, R. Binder, E.V. Hoversten (1978) General purpose packet satellite networks, *Proceedings of the IEEE*, **66**, No. 11, pp. 1448–1467.

[LOM-81] D. Lombard, F. Rancy (1981) TDMA demand assignment operation in TELECOM 1 business service network, *NTC81, New Orleans*, pp. 6.2.2.1–6.2.2.5.

[PUE-71] J.G. Puente, W.G. Schmidt, A.M. Werth (1971) Multiple access techniques for commercial satellites, *Proceedings of the IEEE*, **59**, No. 2, pp. 218–219.

[RAY-87a] D. Raychaudhuri, K. Joseph (1987) Ku-band satellite data networks using very small aperture terminals—Part 1: multi-access protocols, *International Journal on Satellite Communications*, pp. 195–212, April–June.

[RAY-87b] D. Raychaudhuri (1987) Ku-band satellite data networks using very small aperture terminals—Part 2: system design, *International Journal on Satellite Communications*, pp. 195–212, July-Sept.

[RAY-87c] D. Raychaudhuri (1987) Stability, throughput and delay of asynchronous selective reject ALOHA, *IEEE Transactions on Communications*, **COM-35** (7), 767–772.

[RAY-88] D. Raychaudhuri, K. Joseph (1988) Channel access protocols for Ku-band VSAT networks: a comparative evaluation, *IEEE Communications Magazine*.

[RET-80] G. Retnadas (1980) Satellite multiple access protocols, *IEEE Communications Magazine*, **18**, No. 5, pp. 16–22.

[ROB-73] L.G. Roberts (1973) Dynamic allocation of satellite capacity through packet reservation, *NCC73*, pp. 711–716.

[SCH-80] H. Schnipper (1980) Market aspects of satellite business services, *EASCON 80*, pp. 92–94.

[SEV-66] J.L. Sevy (1966) The effect of multiple CW and FM signals passed through a hard limiter or TWT, *IEEE Transactions on Communications*, **COM-14**, No. 5, pp. 568–578.

[SHA-65] P.D. Shaft (1965) Limiting of several signals and its effect on communications system peformance, *IEEE Transactions on Communications*, **COM-13**, No. 4, pp. 504–512.

[SIM-85] M.K. Simon, J. Omura, R.A. Scholtz, B.K. Levitt (1985) *Spread Spectrum Communications*, Computer Science Press.

[SPO-67] J.H. Spoor (1967) Intermodulation noise of FDM/FM communications through a hard limiter, *IEEE Transactions on Communications*, **COM-15**, No. 4, pp. 557–565.

[TAM-87] I. Tamir, Y. Rappaport (1987) Generalized satellite link model and its application to the transmission plan, *International Journal of Satellite Communications*, **5**, pp. 49–56.

[TIR-83] S. Tirro (1983) Satellites and switching, *Space Communications and Broadcasting*, **1**, No. 1, pp. 97–133.

[WOL-87] C.J. Wolejsza, D. Taylor, M. Grossmann, W.P. Osborne (1987) Multiple access protocols for data communications via VSAT networks, *IEEE Communications Magazine*, **25**, no. 7, July.

5 MULTIBEAM SATELLITE NETWORKS

Single beam satellite networks as discussed in the previous chapter have disadvantages in accordance with one of the following:

—The satellite may provide coverage of the whole region of the earth which is visible from the satellite and thus permit long-distance links to be established, for example from one continent to another. In this case, the gain of the satellite antenna is limited by the aperture angle of the beam. Hence, for a geostationary satellite, such coverage implies a 3 dB aperture angle for the antenna beam of 17.5° and consequently an antenna gain of around 20 dB.
—The satellite may provide coverage of only part of the earth (a region or country) by means of a narrow beam (a zone or spot beam). One thus benefits from a higher antenna gain due to a reduction of the aperture angle of the antenna beam, but the system can be connected to stations situated outside its coverage only by terrestrial or intersatellite links.

With a single beam satellite, it is therefore necessary to choose between interconnection of a large number of stations (extended coverage) and provision of a favourable link budget by means of a high satellite antenna gain (reduced coverage).

The multibeam satellite permits these two alternatives to be reconciled; satellite coverage is extended since it results from the juxtaposition of several beams and each beam provides an antenna gain which increases as the beamwidth decreases. The performance improves as the number of beams increases; the limit is provided by the antenna technology and the mass of the satellite whose complexity increases with the number of beams.

This chapter presents the advantages of a system design involving the operation of multibeam satellites. The problems of realisation of a multibeam satellite are treated in Chapter 9.

5.1 ADVANTAGES AND DISADVANTAGES OF MULTIBEAM SATELLITES

The two configurations of Figure 5.1 will be compared; in one the satellite provides global coverage with a single beam of angular width $\theta_{3dB} = 17.5°$ (Figure 5.1a) and in the other the satellite supports narrow beams of angular width $\theta_{3dB} = 1.75°$ with a consequently reduced coverage (Figure 5.1b).

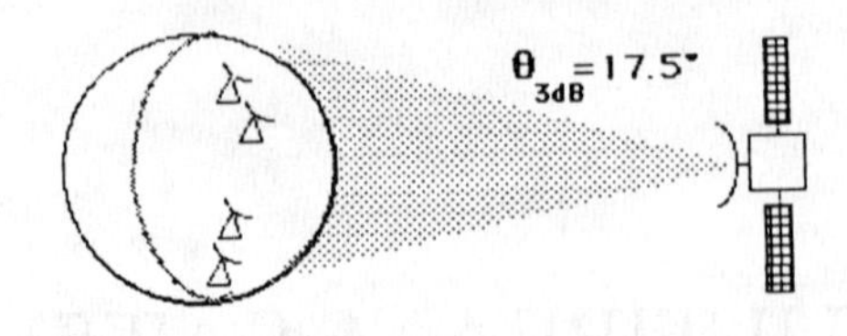

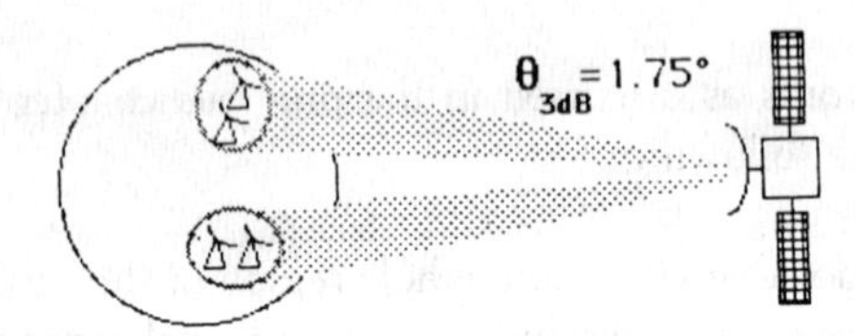

Figure 5.1 Global coverage and coverage by several narrow beams.

5.1.1 Advantages

5.1.1.1 *Impact on the earth segment*

The expression for $(C/N_0)_U$ for the uplink is given by (see Section 2.5.2):

$$(C/N_0)_U = (\text{EIRP})_{\text{station}}(1/L_U)(G/T)_{\text{satellite}}(1/k) \qquad (\text{Hz}) \qquad (5.1)$$

Assuming that the noise temperature at the satellite receiver input is $T_{\text{satellite}} = 800\,\text{K} = 29\,\text{dB(K)}$ and is independent of the beam coverage (this is not rigorously true but satisfies a first approximation). Let $L_U = 200\,\text{dB}$ and neglect the implementation losses. Expression (5.1) becomes (all terms in dB):

$$(C/N_0)_U = (\text{EIRP})_{\text{station}} + (G_R)_{\text{satellite}} - 0.4 \qquad (\text{dB(Hz)}) \qquad (5.2)$$

where $(G_R)_{\text{satellite}}$ is the gain of the satellite receiving antenna. This relation is represented in Figure 5.2 for the two cases considered:

—Global coverage ($\theta_{3\text{dB}} = 17.5°$) which implies $(G_R)_{\text{satellite}} = 29\,000/(\theta_{3\text{dB}})^2 = 20\,\text{dB}$.
—Narrow beam coverage ($\theta_{3\text{dB}} = 1.75°$) which implies $(G_R)_{\text{satellite}} = 29\,000/(\theta_{3\text{dB}})^2 = 40\,\text{dB}$.

The expression for $(C/N_0)_D$ for the downlink is given by:

$$(C/N_0)_D = (\text{EIRP})_{\text{satellite}}(1/L_D)(G/T)_{\text{station}}(1/k) \qquad (\text{Hz}) \qquad (5.3)$$

Assume that the power of the carrier transmitted by the satellite is $P_T = 10\,\text{W} = 10\,\text{dB(W)}$. Let $L_U = 200\,\text{dB}$ and neglect the implementation losses. Expression (5.3) becomes (all terms in dB):

$$(C/N_0)_D = (G_T)_{\text{satellite}} + (G/T)_{\text{station}} + 38.6 \qquad (\text{dB(Hz)}) \qquad (5.4)$$

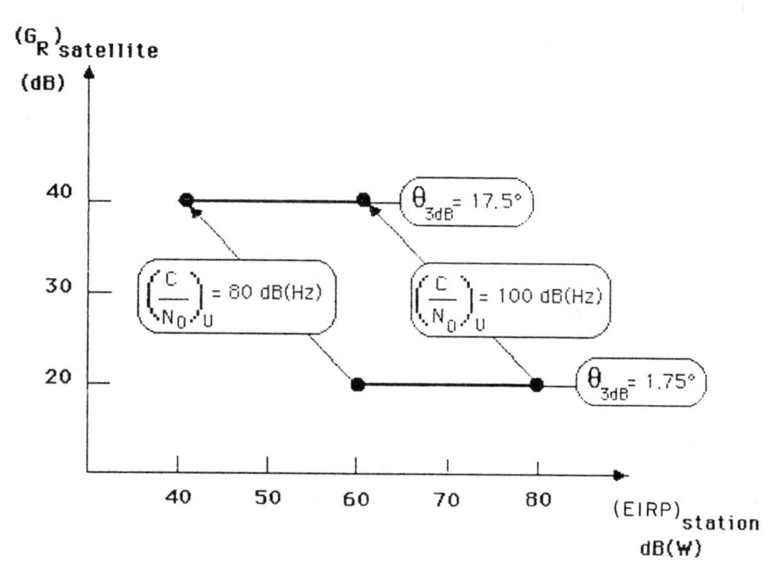

Figure 5.2 Comparison of the EIRP values required for an earth station in the case of global coverage ($\theta_{3dB} = 17.5°$) and in the case of a narrow beam ($\theta_{3dB} = 1.75°$).

This relation is represented in Figure 5.3 for the two cases considered:

—Global coverage ($\theta_{3dB} = 17.5°$) which implies $(G_T)_{satellite} = 29\,000/(\theta_{3dB})^2 = 20$ dB.
—Narrow beam coverage ($\theta_{3dB} = 1.75°$) which implies $(G_T)_{satellite} = 29\,000/(\theta_{3dB})^2 = 40$ dB.

In Figures 5.2 and 5.3 the oblique arrows indicate the reduction in $(EIRP)_{station}$ and $(G/T)_{station}$ when changing from a satellite with global coverage to a multibeam satellite. In this case the multibeam satellite permits an economy of size, and hence cost, of the earth segment. For instance, a 20 dB reduction of $(EIRP)_{station}$ and $(G/T)_{station}$ may result in a tenfold reduction of the antenna size (perhaps from 30 m to 3 m) with a cost reduction for the earth station perhaps from 10 M$ to 50 k$. If an identical earth segment is retained (a vertical displacement towards the top) an increase of C/N_O is achieved which can be transferred to an increase of capacity, if sufficient bandwidth is available, at constant signal quality (in terms of S/N or BER).

5.1.1.2 *Frequency re-use*

Frequency re-use consists of using the same frequency band several times in such a way as to increase the total capacity of the network without increasing the allocated bandwidth. An example has been seen in Chapter 2 (Section 2.1.4) of frequency re-use within the same beam by the use of orthogonal polarisation. In the case of a multibeam satellite the isolation resulting from antenna directivity can be exploited to re-use the same frequency band in different beams. Figure 5.4 compares the principle of frequency re-use by orthogonal polarisation (Figure 5.4a) and the principle of re-use by angular beam separation (Figure 5.4b). In both cases the bandwidth allocated to the system is B. The system uses this bandwidth

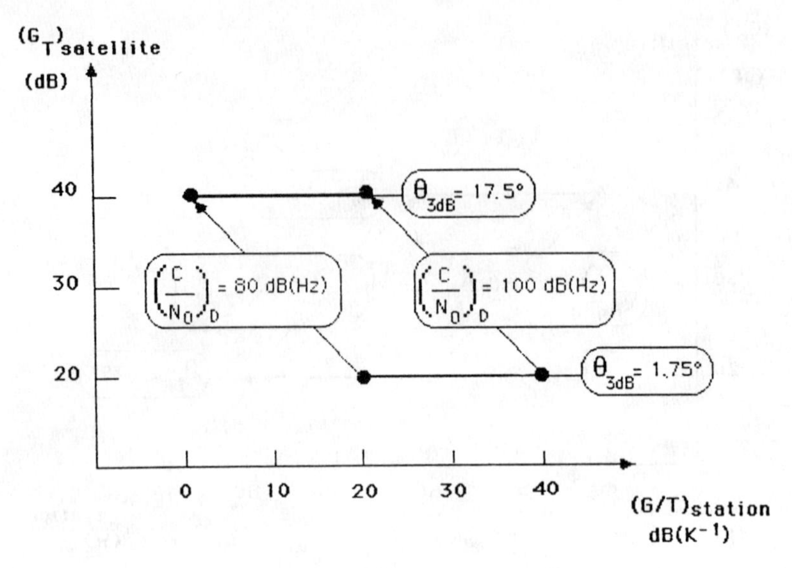

Figure 5.3 Comparison of the required values of factor of merit G/T for an earth station in the case of global coverage ($\theta_{3dB} = 17.5°$) and in the case of a narrow beam ($\theta_{3dB} = 1.75°$).

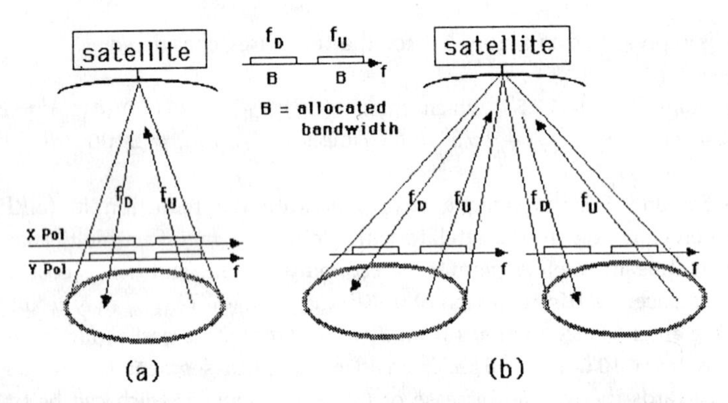

Figure 5.4 Frequency re-use. (a) By orthogonal polarisation, (b) By angular separation of the beams in a multibeam satellite system.

B centred on the frequency f_U for the uplink and on the frequency f_D for the downlink. In the case of re-use by orthogonal polarisation, the bandwidth B is used twice only. In the case of re-use by angular separation, the bandwidth B can be re-used for as many beams as the permissible interference level allows. Both types of frequency re-use can be combined.

The frequency re-use factor is defined as the number of times that the bandwidth B is used. In theory, a multibeam satellite with M beams to which a bandwidth B is allocated and which combines re-use by angular separation and re-use by orthogonal polarisation within each beam has a frequency re-use factor of $2M$. This signifies that it can claim the capacity which would be offered by a single beam satellite with single polarisation using a bandwidth of $2M \times B$. In practice the frequency re-use factor depends on the configuration of the service area which determines the coverage before it is provided by the satellite. If the service area

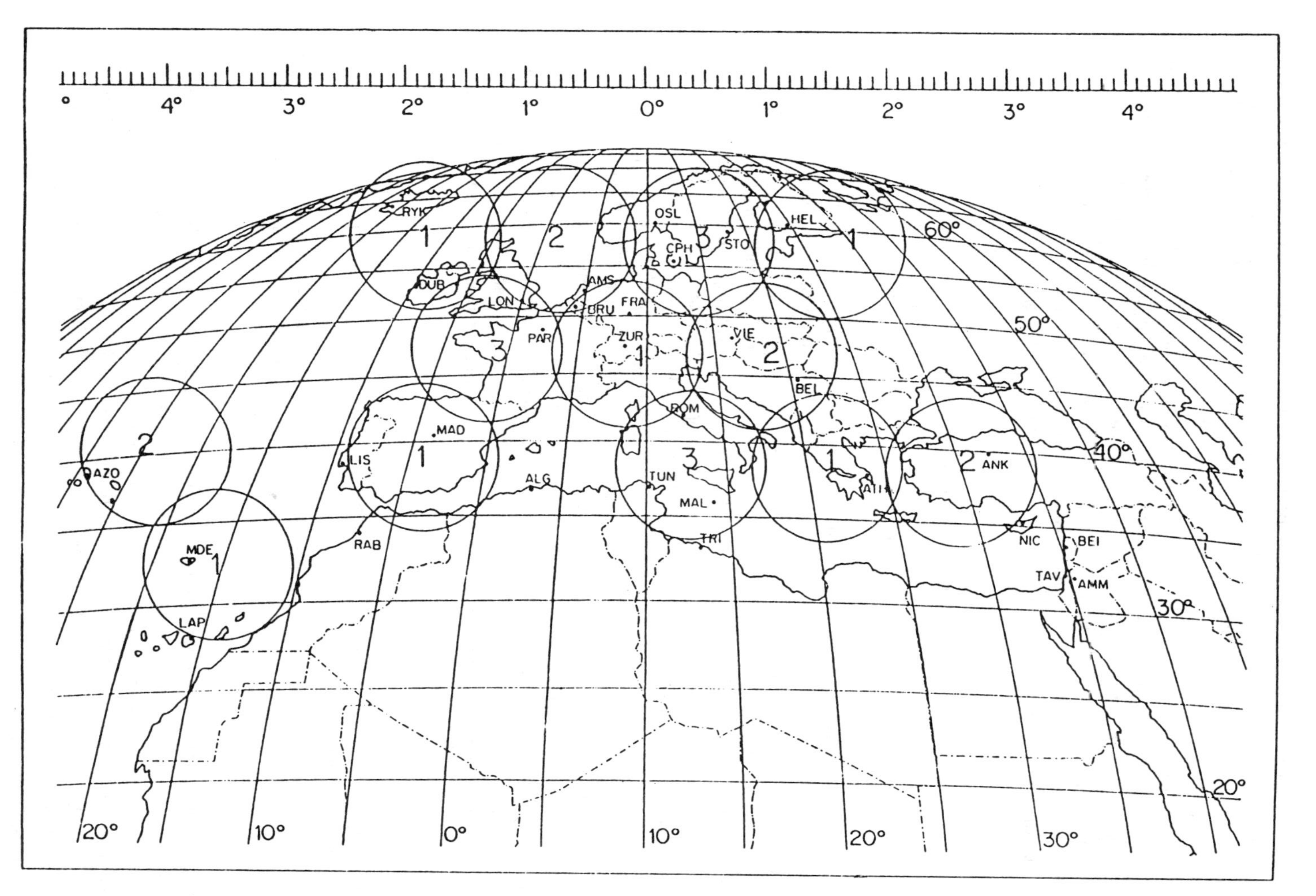

Figure 5.5 Example of European coverage by a multibeam satellite system [LOP-82]. (Reproduced by permission of the European Space Agency.)

consists of several widely separated regions (for example, urban areas separated by extensive rural areas), the angular separation of the beams can be sufficient to permit re-use of the same band in all beams. The frequency re-use factor can then attain the theoretical value of $2M$. Figure 5.5 shows a type of coverage more appropriate to regions with a more homogeneous population distribution [LOP-82]. As the beams are contiguous, if the same frequency band cannot be used from one beam to the other. In this example, the bandwidth allocated is divided into three equal separate sub-bands and each is used in beams (1, 2 and 3) with sufficient angular separation from each other, the equivelent bandwidth, in the absence of re-use by orthogonal polarisation, has a value given by: $6 \times (B/3) + 4 \times (B/3) + 3 \times (B/3) = 4.3\,B$ for $M = 13$ beams. The frequency re-use factor is then 4.3.

5.1.2 Disadvantages

5.1.2.1 *Inteference between beams*

Figure 5.6 illustrates interference generation in a multibeam satellite system. The allocated bandwidth B is divided into two sub-bands B_1 and B_2. The figure shows three beams. Beams 1 and 2 use the same band B_1. Beam 3 uses band B_2.

On the uplink (Figure 5.6a), the carrier at frequency f_{U1} of bandwidth B_1 transmitted by the beam 2 earth station is received by the antenna lobe defining beam 1 with a low but non-zero gain. The spectrum of this carrier superimposes itself on that of the carrier of the same frequency emitted by the beam 1 earth station which is received with the maximum antenna gain. The carrier of beam 2 will therefore appear as interference noise in the spectrum of the carrier of beam 1. This noise is called co-channel interference (CCI). Furthermore, part of the power of the carrier at frequency f_{U2} emitted by the earth station of beam 3 is introduced during filtering to define the satellite channels (see Chapter 8) in the channel occupied by carrier f_{U1}. In this case it consists of adjacent channel interference (ACI) analogous to that encountered in connection with frequency division multiple access in Section 4.3.2.

On the downlink (Figure 5.6b) the beam 1 earth station receives the carrier at frequency f_{D1} emitted with maximum gain in the antenna lobe defining beam 1. The following are superimposed on the spectrum of this carrier:

—The spectra of the uplink adjacent channel and co-channel interference noise retransmitted by the satellite,
—The spectrum of the carrier at the same frequency f_{D1} emitted with maximum gain in beam 2 and with a small but non-zero gain in the direction of the beam 1 station. This represents additional co-channel interference (CCI).

The effect of interference appears as an increase of thermal noise under the same conditions as interference noise between systems analysed in Section 2.9.2. It must be included in the term $(C/N_0)_I$ which appears in expression (2.46). Taking account of the multiplicity of sources of interference, which become more numerous as the number of beams increases, relatively low values of $(C/N_0)_I$ may be achieved and the contribution of this term impairs the performance in terms of $(C/N_0)_T$ of the total link. An estimate of the typical contribution of interference noise in a multibeam satellite link puts this contribution at 40% of the total noise [DAV-81].

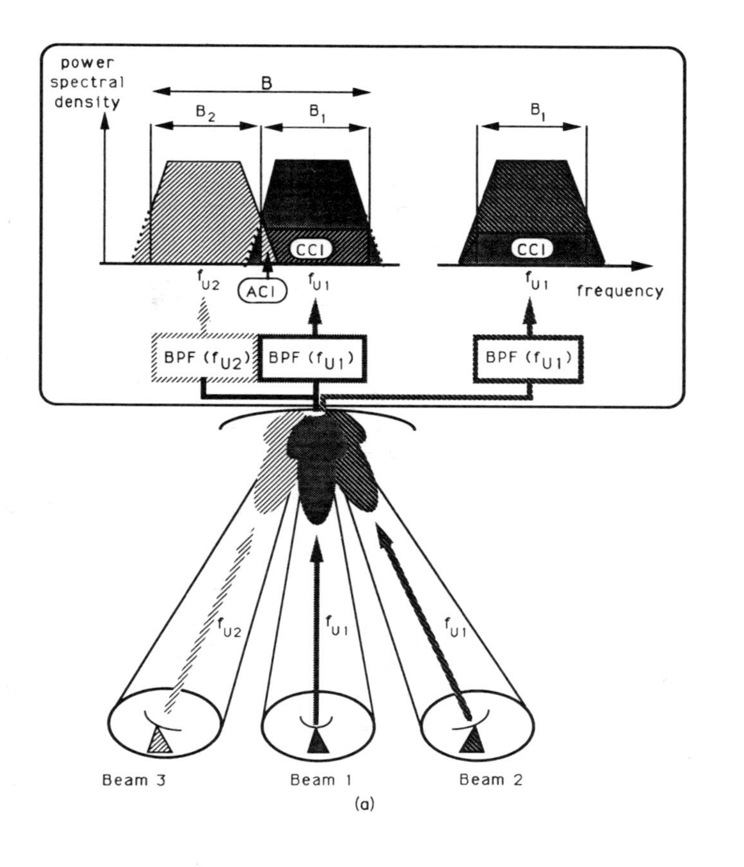

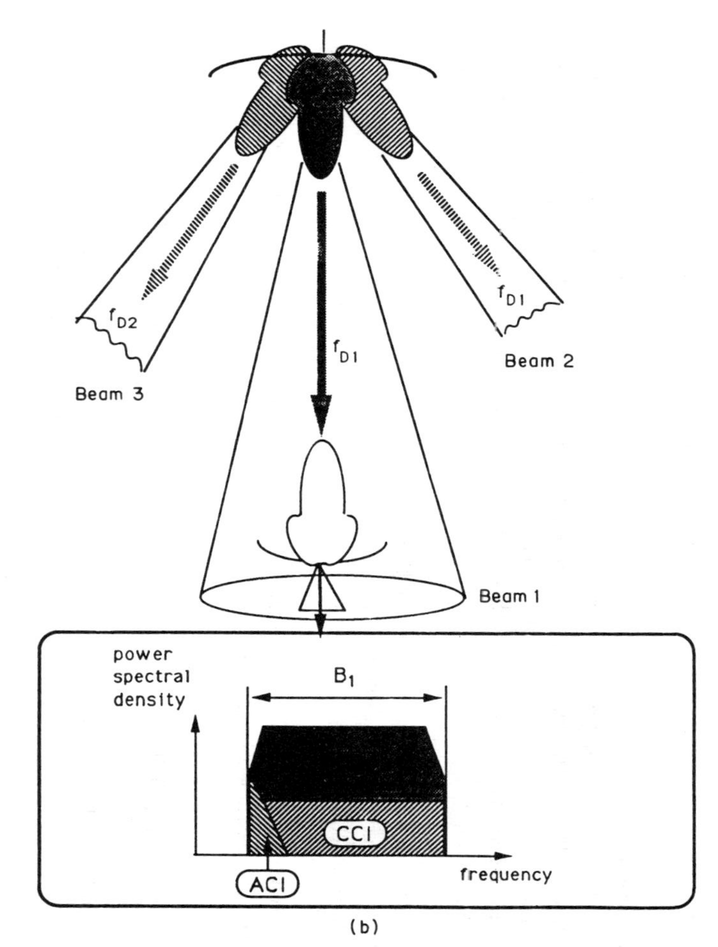

Figure 5.6 Internal interference between beams in a multibeam satellite system: (a) uplink, (b) downlink.

5.1.2.2 *Interconnection between coverage areas*

A multibeam satellite system must be in a position to interconnect all network earth stations and consequently must provide interconnection of coverage areas. The complexity of the payload is added to that of the multibeam satellite antenna subsystem which is already more complex than that of a single beam satellite.

Using transparent transponders, three techniques can be envisaged for interconnection of coverage areas:

—Interconnection by transponder hopping.
—Interconnection by on-board switching (SS/TDMA).
—Interconnection by beam scanning.

These three solutions will be discussed below in succession. The approach using regenerating repeaters is examined in Chapter 6.

5.2 INTERCONNECTION BY TRANSPONDER HOPPING

The band allocated to the system is divided into as many sub-bands as there are beams. A set of filters on board the satellite separates the carriers in accordance with the sub-band

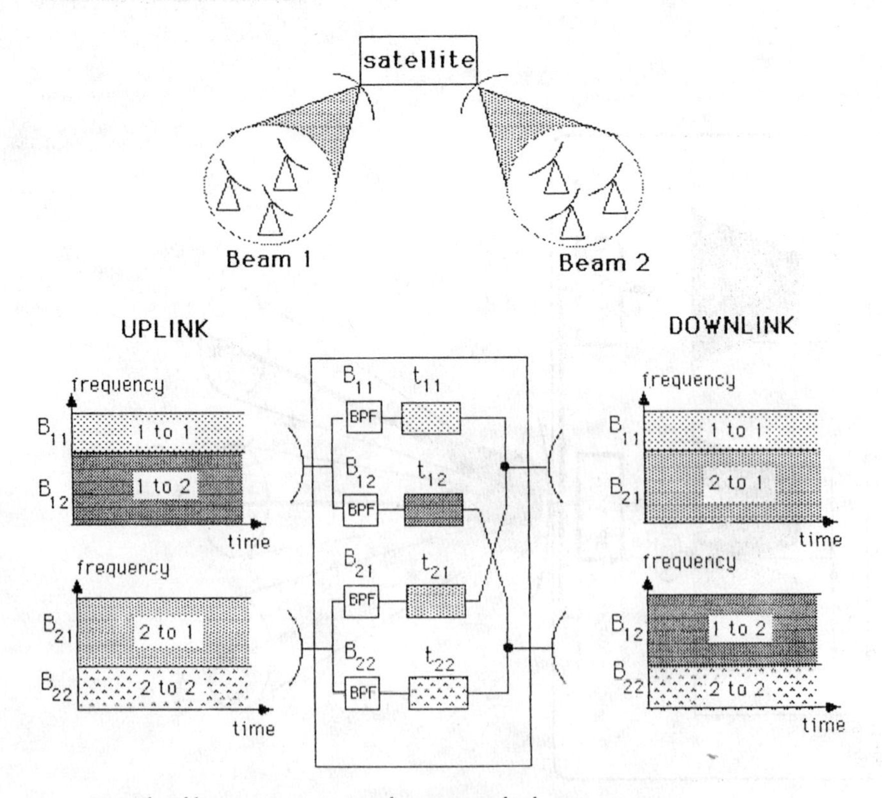

Figure 5.7 Example of beam interconnection by transponder hopping.

Table 5.1 Frequency agility required to ensure beam interconnection in accordance with the type of coverage

Type of coverage		Frequency and polarisation agility	Example
Uplink	Downlink		
Global	Global	On transmission or on reception	TELECOM 1
Spot	Global	On reception	
Global	Spot	On transmission	EUTELSAT 1
Spot	Spot	On transmission and on reception	INTELSAT V

occupied. The output of each filter is connected by a transponder to the antenna of the destination beam. It is necessary to use a number of filters and transponders at least equal to the square of the number of beams. Figure 5.7 illustrates this concept for an example with two beams. According to the type of coverage, the earth stations must be able to transmit and/or receive on several frequencies and polarisations in order to hop from one transponder to another (transponder hopping). Table 5.1 indicates the type of frequency agility required to ensure interconnection between beams according to the type of coverage. The capacity offered to traffic can be varied between beams, within the total capacity defined by the system bandwidth, by modification of the sub-band assignments and hence by modification of the connections between input filters and transponders (INTELSAT, for example [FUE-77], [DIC-78]). This operation is realised by telecommand, from time to time, in accordance with long-term traffic fluctuations.

5.3 INTERCONNECTION BY ON-BOARD SWITCHING (SS/TDMA)

5.3.1 The principle

Beam switching by transponder hopping is a solution when the number of beams is low. The number of transponders increases at least as the square of the number of beams; with a large number of beams the satellite becomes too heavy. It is therefore necessary to consider on-board switching. The principle is illustrated in Figure 5.8. The payload includes a programmable switch matrix having a number of inputs and outputs equal to the number of beams. This matrix connects each received beam to each transmitted beam by way of a receiver and a transmitter. The number of repeaters is thus equal to the number of beams. The distribution control unit (DCU) associated with the matrix establishes the sequence of connection states between each input and the outputs during a frame in such a way that the carriers arriving at the satellite in each beam are routed to the destination beams. Since interconnection between two beams is cyclic, stations must store traffic from users and

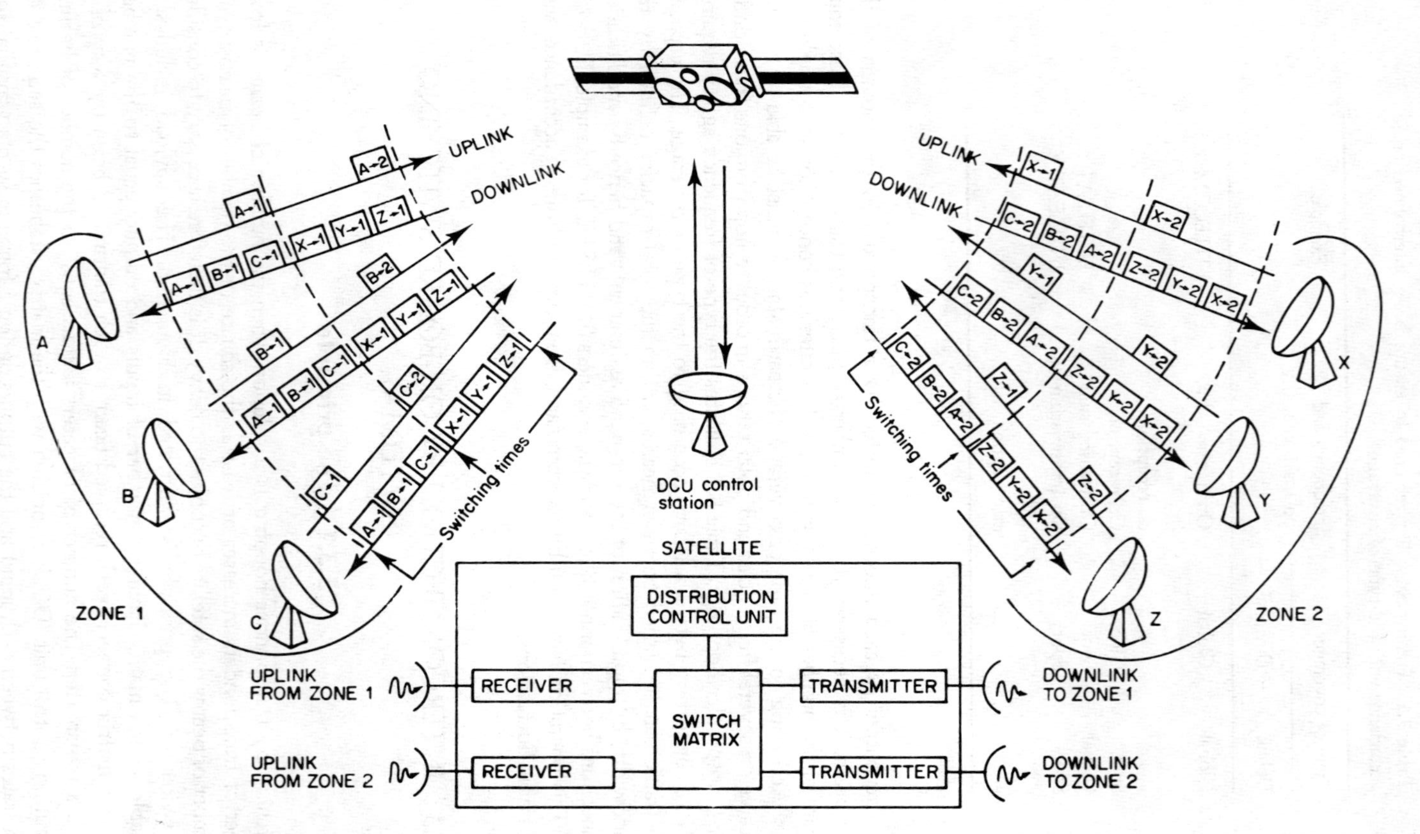

Figure 5.8 The principle of on-board switching (SS-TDMA).

transmit it in the form of bursts when the required interconnection between beams is realised. This technique can thus be used in practice only with digital transmission and access of the TDMA type. This is why it is called satellite switched time division multiple access (SS-TDMA) [SCH-69].

5.3.2 Frame structure

Figure 5.9 shows an example of a switch state sequence for a 3 × 3 matrix. This sequence contains four connection states which are used for the duration of the frame represented in

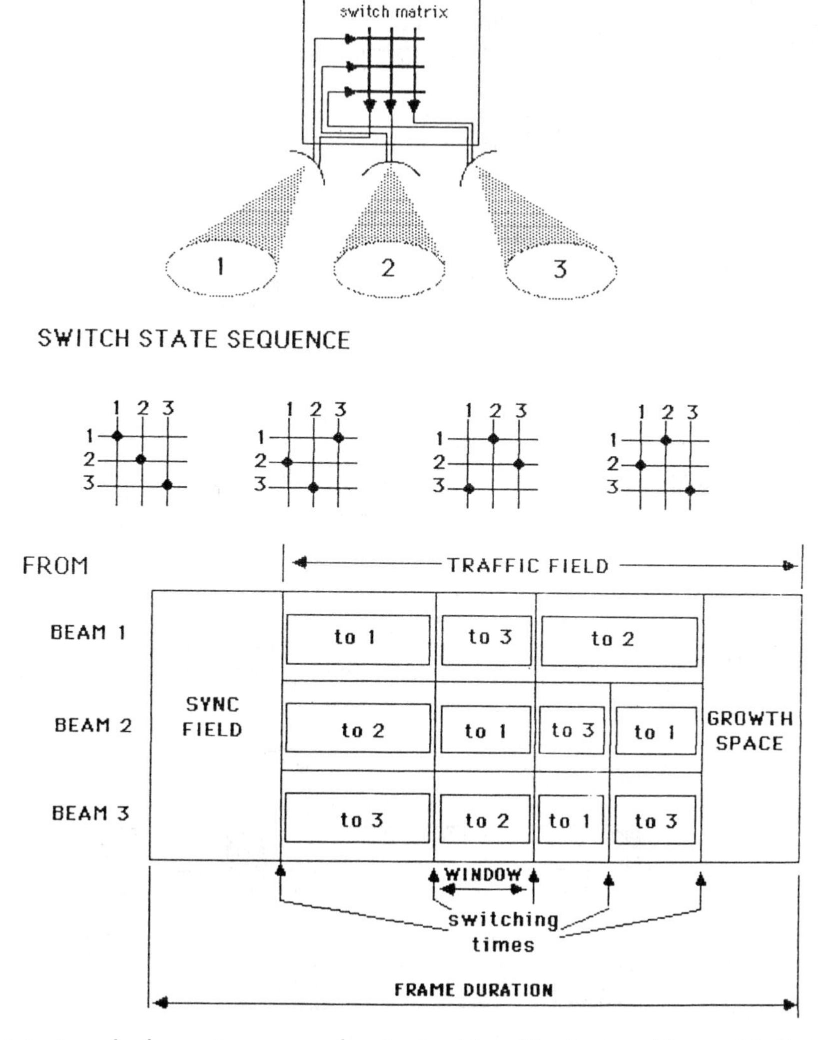

Figure 5.9 Example of connection sequence for a 3 × 3 matrix and the structure of the associated frame.

the lower part of the figure. The frame contains a synchronising field (see Section 5.3.5) and a traffic field. Bursts from stations are routed to their destinations in the traffic field. The traffic field contains a succession of switching modes during which the switching matrix retains the same connection state. The duration of a connection between an up beam and a down beam is called a window [TIR-83]. A window can extend over the duration of several switch modes.

5.3.3 Window structure

Figure 5.10 shows the way in which bursts are arranged in the time interval of one window. The figure shows that bursts are transmitted by stations A, B and C from beam 3 to beam

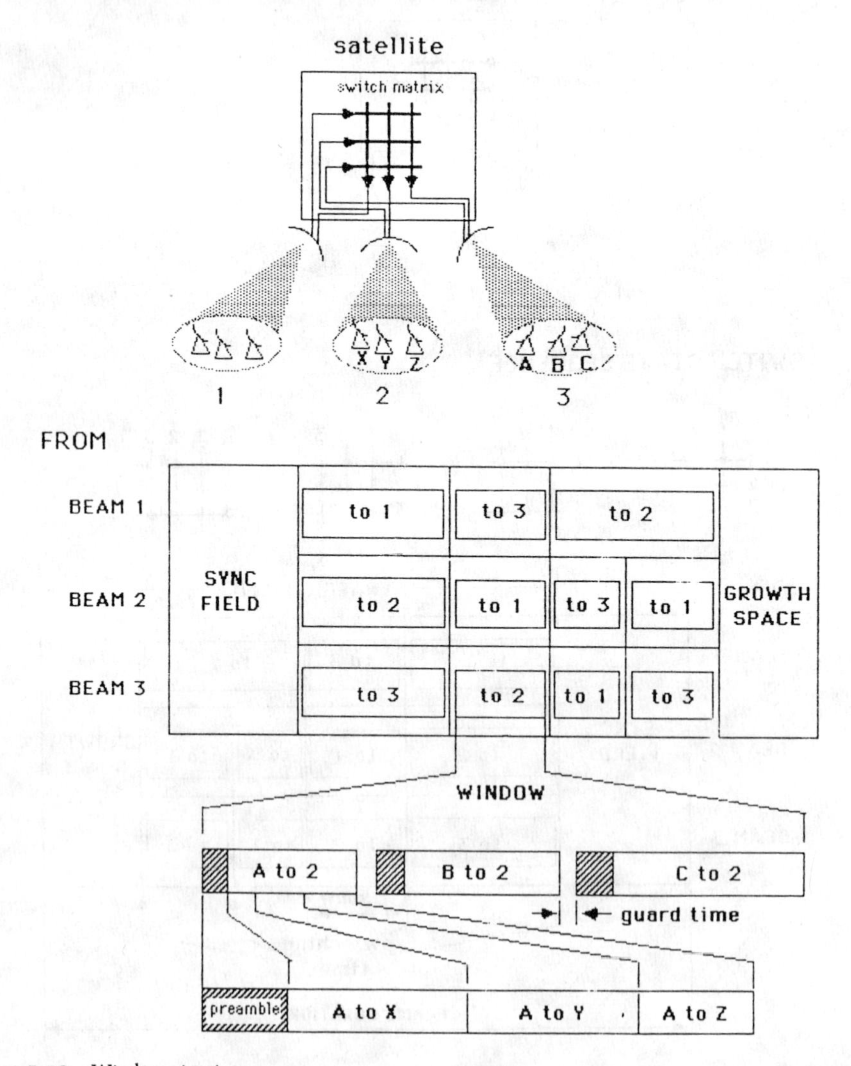

Figure 5.10 Window structure.

2. Each burst transmitted by a station during the window time considered consists of several sub-bursts which contain station to station information.

5.3.4 Assignment of packets in the frame (burst time plan)

The assignment of bursts in the frame must maximise the use of satellite transponders. Transponders are exploited best when the windows are occupied entirely by traffic bursts. This is possible only when traffic distribution between beams is balanced. In practice this is not always the case. A traffic matrix can be established which describes the traffic from one beam to another. For example, for a three-beam satellite (1, 2 and 3), this matrix is as shown below.

	To beam	1	2	3	
From beam	1	t_{11}	t_{12}	t_{13}	S_1
	2	t_{21}	t_{22}	t_{23}	S_2
	3	t_{31}	t_{32}	t_{33}	S_3
		R_1	R_2	R_3	

The sum of each row S_i $(i = 1, 2, 3)$ represents the traffic transmitted by all stations on the same beam. The sum of each column R_j $(j = 1, 2, 3)$ represents the beam traffic received by all stations. In the case of balanced traffic distribution between beams, the sums S_i and R_j are equal. Otherwise, one of these sums is greater than all the others. The corresponding line of the matrix (row or column) is called the critical line.

It can be shown that the minimum frame time to route the traffic bursts transmitted by all stations is the time required to transmit the traffic of the critical line of the traffic matrix at the rate considered [INU-79]. Numerous algorithms permit optimal filling of frames with bursts. A classification of these algorithms is given in [MAR-87].

An SS-TDMA network can operate with fixed or demand assignment. With demand assignment, variations of capacity allocated to stations are obtained by variation of burst length as with TDMA (Section 4.6). Variation of station burst length is accompanied by variation of the position of bursts of other stations, and consequently a change in the assignment of other bursts (burst time plan change). There are three types of burst time plan change:

—burst time plan change without switching mode allocation change.
—burst time plan change accompanied with switching mode allocation change but not accompanied with switch state sequence change.
—burst time plan change accompanied with switch state sequence change.

In the first case, only earth stations are concerned. In the two other cases, it is necessary to load information concerning the new sequence of switch states or the new switching mode allocation into the distribution control unit (DCU) memory by means of a special purpose link (which can be the telecommand link). The assignment changes occur at the start of a super-frame in order to guarantee synchronisation of the change among all stations and the satellite.

5.3.5 Synchronisation

There are two aspects of network synchronisation:

—Synchronisation of earth stations.
—Synchronisation of the earth segment with the satellite.

Synchronisation of earth stations is achieved as in single beam TDMA by one of the methods (closed or open loop) presented in Section 4.4.4. A difficulty appears in closed loop, however, due to the fact that the stations of one beam do not receive their bursts transmitted to other

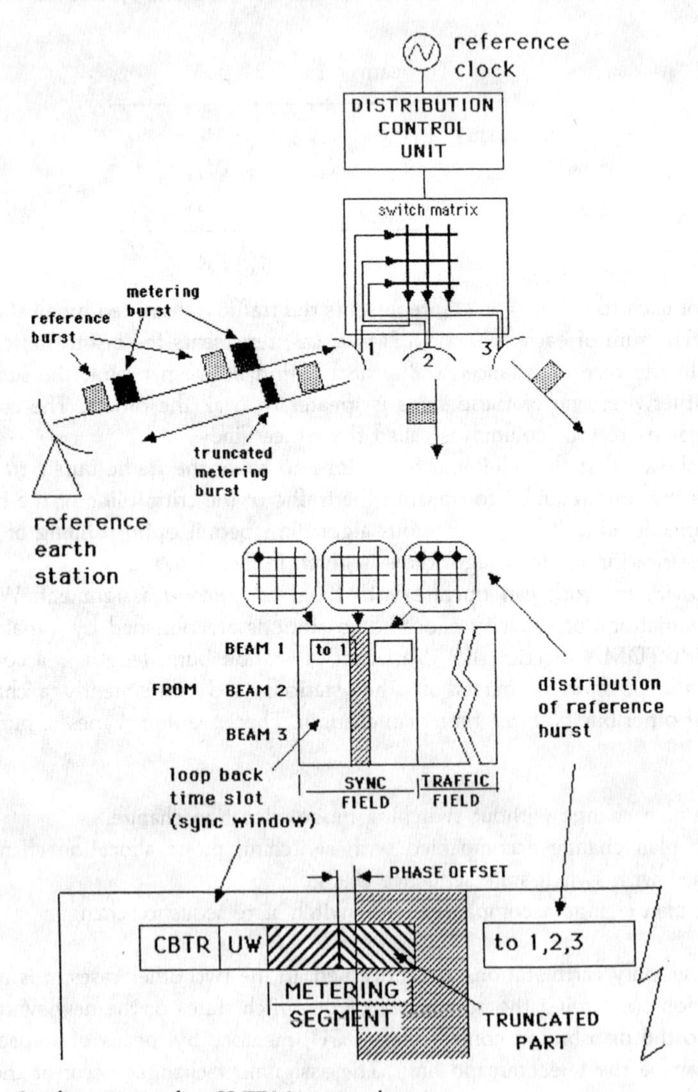

Figure 5.11 Synchronisation of an SS-TDMA network.

beams and thus cannot determine their positioning error. One solution, called co-operative feedback closed loop synchronisation [CAM-81], involves making this measurement at a beam destination station; this station in turn indicates the value of error observed in one of its bursts intended for the station concerned. This method reduces the frame throughput and can be used only if the number of beams is low. In order not to degrade the frame throughput with a large number of beams, it is preferable to use an open loop method.

Synchronisation of the earth segment with the satellite implies a choice in the location of the network clock. Should it be on the ground in the reference station as in TDMA or should it be on board the satellite by virtue of its location at a privileged nodal point? Leaving the clock in the reference station necessitates a demodulator and a unique word detection circuit on board the satellite. Installing the clock on board the satellite implies a modulator to broadcast this clock on all beams. In both cases, additional equipment is required on board the satellite and this can be subject to breakdown. The stability of the clock can also be questioned in the on-board case; is it compatible with CCITT Recommendation G811 (see Section 3.7.4) which recommends plesiosynchronism between networks and hence a stability of 10^{-11}? One solution is to consider the distribution control unit (DCU) clock to be the network clock [SHI-71], [CAM-81], [CAR-80].

The reference station identifies the rate of this clock and synchronises reference burst transmission; the principle is illustrated in Figure 5.11. The reference station transmits another burst called the metering burst before the reference burst. With steady state synchronisation, this burst arrives at the satellite in the frame synchronisation field straddling the first two switch modes; the first mode establishes a return connection to the reference station beam (beam 1 in the figure) and the second mode does not provide any connection between input and output. The metering burst thus returns, truncated, to the reference station and this controls its transmission to maintain a constant truncation and hence remain in synchronism with the control unit clock. Transmission of the reference burst follows that of the metering burst with a constant delay. It is distributed by the satellite on all beams by means of the third switching mode of the synchronisation field which connects beam 1 with all the beams. The control unit clock rate can be compared with that of a more precise clock installed in the reference station. Phase corrections can be sent by telecommand in such a way that the required stability is ensured [INU-81].

5.3.6 Frame throughput

The definition of throughput is given in Section 4.4.5. The expression for it is:

$$\eta = 1 - \Sigma t_i / T_F \qquad (5.5)$$

where Σt_i represents the sum of the times not devoted to information transmission and T_F is the frame duration. There are four components of Σt_i:

—The synchronisation field.
—The packet headers and the guard times including the intervals reserved for switching of the on-board matrix. In comparison with a single beam satellite, a station must transmit

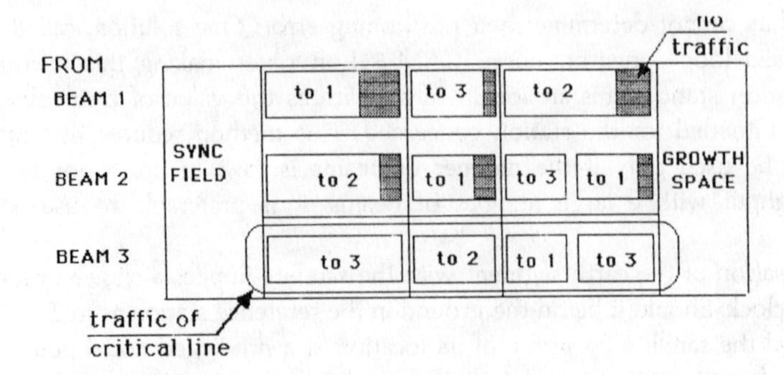

Figure 5.12 Example of SS-TDMA frame packing in the case of non-uniform traffic distribution.

several times within a frame if its packets are destined for several beams. As each packet has a header, there are more dead times.

—In the case of unbalanced traffic distribution among the beams, the critical section determines the minimum duration of the switching modes and some frame windows will not be filled, thereby leaving transponders inactive. This is the situation illustrated in Figure 5.12 where the critical row of the matrix is S_3, that is the traffic transmitted on beam 3 determines the minimum duration of the set of switching modes.

—In the case of demand assignment, burst assignment cannot be optimum at a given instant and dead times are introduced into the windows when a change of assignment occurs without a change of switching mode or sequence of switch states.

Overall, it is difficult to provide values since throughput depends on traffic distribution. Simulations have indicated values of 75 to 80% [TIR-83]. For every assumption, throughput is less than with a single beam satellite.

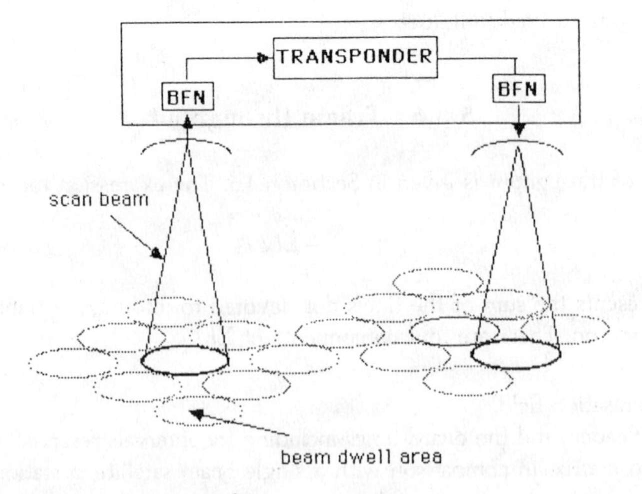

Figure 5.13 Interconnection by scanning beams.

5.4 INTERCONNECTION BY BEAM SCANNING

Each coverage area is illuminated cyclically by an antenna beam whose orientation is controlled by a beam-forming network which is part of the antenna sub-system on board the satellite [REU-77]. The area stations transmit or receive their bursts when the area is illuminated by a beam. In the absence of on-board memory, at least two beams are necessary at a given instant—one to establish the uplink and one to establish the downlink. The illumination duration is proportional to the volume of traffic to be carried between the two areas. Figure 5.13 illustrates this concept.

5.5 INTERSATELLITE LINKS (ISL)

Intersatellite links (ISL) can be considered as particular beams of multibeam satellites; the beams in this case are directed not towards the earth but towards other satellites. For bidirectional communication between satellites, two beams are necessary—one for transmission and one for reception. Network connectivity implies the possibility of interconnecting beams dedicated to intersatellite links and other links at the payload level.

Three classes of intersatellite link can be distinguished:

—Links (GEO-LEO) between geostationary earth orbit (GEO) and low earth orbit (LEO) satellites; also called inter-orbital links (IOL).
—Links between geostationary satellites (GEO-GEO).
—Links between low orbit satellites (LEO-LEO).

The applications of these links will first be examined, then the parameters which dimension an intersatellite link will be identified. The technological aspects are treated in the context of the payload in Chapter 9.

5.5.1 Links between geostationary and low orbit satellites (GEO-LEO)

This type of link serves to establish a permanent relay via a geostationary satellite between one or more earth stations and a group of satellites proceeding in a low orbit at an altitude of the order of 500 to 1000 km. For economic and political reasons, one does not wish to install a network of stations which is so large that at every instant the passing satellites are visible from at least one station. One or more geostationary satellites are therefore used which are permanently and simultaneously visible both from stations and low orbit satellites; they serve to relay communications. This technique also permits possible limitations of the terrestrial network to be avoided.

This concept is presently operated in the NASA tracking network by means of the tracking and data relay satellite (TDRS) which, in particular, provides communication with the American space shuttle [LAN-82], [SCH-84]. A European programme is in progress for the installation of a data relay satellite (DRS) to provide communication between the ground and the European shuttle HERMES or the COLUMBUS station [DIC-87]. Retransmission of the data from a terrestrial observation satellite (SPOT) to the processing centre at Toulouse by means of a geostationary satellite is also considered [ARN-88].

5.5.2 Links between geostationary satellites (GEO–GEO)

5.5.2.1 *Increasing the capacity of a system*

Consider a multibeam satellite network. Figure 5.14 illustrates the case of a three beam satellite (Figure 5.14a). It will be assumed that the traffic demand increases and exceeds the capacity of the satellite. It is therefore necessary to launch a satellite of greater capacity and this implies risks, development costs and the availability of a suitable launcher; alternatively a second satellite identical to the first could be launched with the traffic shared between the two satellites. To avoid interference, the two satellites must be in two sufficiently distant orbital positions. To ensure interconnectivity among all stations, it is necessary to equip all

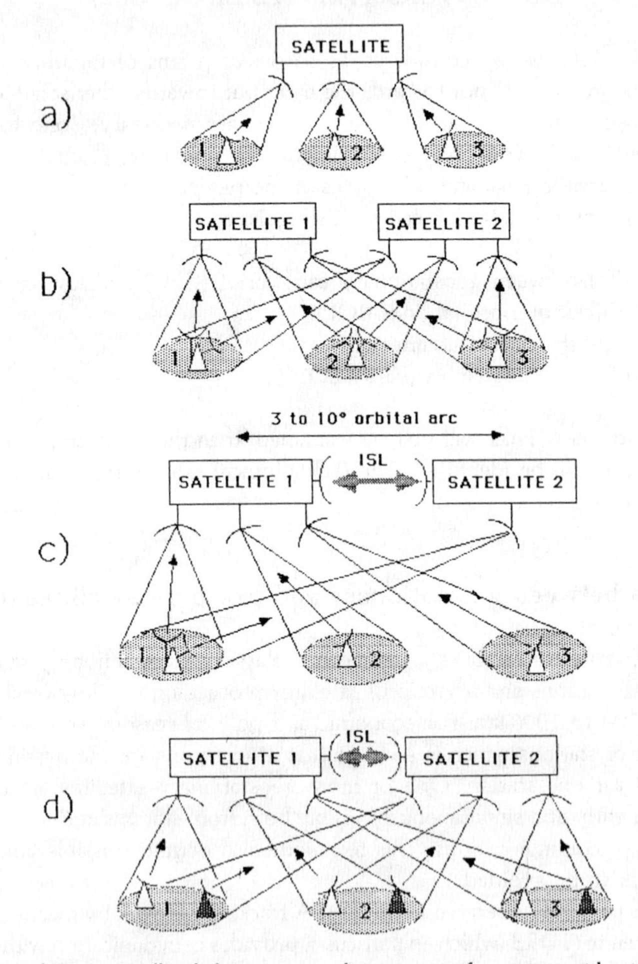

Figure 5.14 Use of an intersatellite link to increase the capacity of a system without heavy investment in the earth segment. (a) Network with a single satellite. (b) A second satellite is launched to increase the capacity of the space segment; the stations must be equipped with two antennas. (c) With an intersatellite link, only the stations of the most heavily loaded region must be equipped with two antennas. (d) The stations are distributed between the two satellites. The intersatellite link carries the traffic between the two groups of stations.

stations with two antennas each pointing towards a different satellite (Figure 5.14b). With satellites provided with intersatellite transponders one can either:

—Equip the stations of region 1, assumed to be generating the excess traffic, with a second antenna and retain the same configuration for the stations of regions 2 and 3 (Figure 5.14c). The instersatellite link carries the excess traffic of region 1.

Or:

—Distribute the stations, each with a single antenna, into two groups, each associated with one satellite (Figure 5.14d). The instersatellite link carries the traffic between the two groups.

The choice is economic and depends on the case considered. A study carried out for a European system shows, for example, that the choice of an intersatellite link is economically

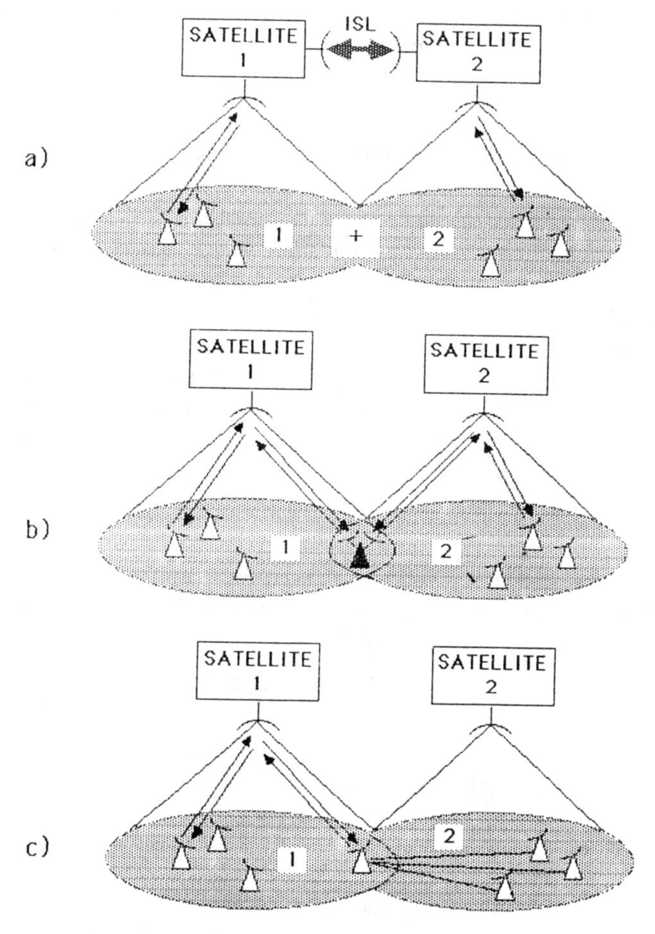

Figure 5.15 Extension of system coverage. (a) Interconnection of the stations of each coverage by an intersatellite link. (b) Interconnection without an intersatellite link by a station common to the two networks. (c) Interconnection without an intersatellite link by a terrestrial network.

justified if the mass of the intersatellite transponders does not exceed 20% of the launch mass of the satellite and if 15 to 20% at most of stations must be equipped with two antennas [PUC-88].

5.5.2.2 *Extension of the coverage of a system*

An intersatellite link permits earth stations of two networks to be interconnected and hence the geographical coverage of the two satellites to be combined (Figure 5.15a). The alternative solutions are either:

—To install an interconnecting earth station equipped with two antennas in the common part of the two coverages, if it exists (Figure 5.15b).

Or:

—To make the connection, by means of the terrestrial network, from the stations of one network to a station of the other network situated on the common border of the two coverages (Figure 5.15c).

5.5.2.3 *Increase of the minimum elevation angle of earth stations*

Long distance links by a single satellite require earth stations with a small elevation angle, sometimes less than 10°. This causes a degradation of G/T for the receiving station (see Section 2.5.4) and increases the risk of interference with terrestrial microwave relays. If the link passes through two geostationary satellites connected by an intersatellite link, the elevation angle increases. Hence, a link by a single satellite with an elevation angle of 5° becomes, with two satellites separated by 30°, a link with an elevation angle of 20° for equatorial stations (Figure 5.16) and 15° for stations at a latitude of 45°. This would be the case between London and Tokyo, for example, with two satellites above the Indian Ocean.

5.5.2.4 *Reduction of the constraints on orbital position*

The orbital position of a satellite is often the result of a conflict, resolved by means of a procedure called co-ordination, between the desire of the satellite operator to ensure coverage

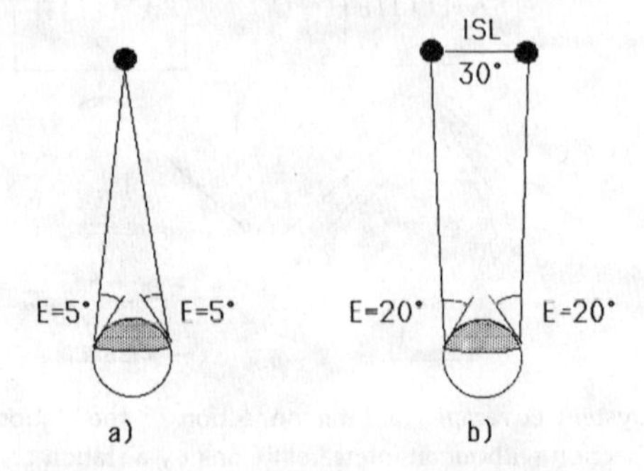

Figure 5.16 Increase of the minimum elevation angle of earth stations.

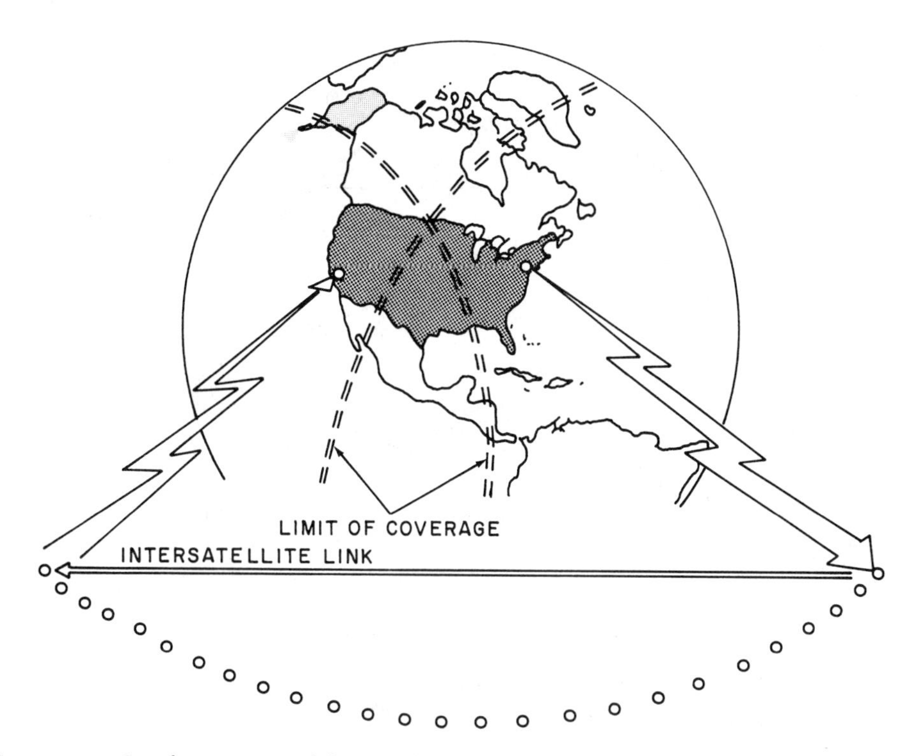

Figure 5.17 Complete coverage of the United States in spite of saturation of the orbital arc [MOR-89]. (Reproduced by permission from Morgan, N.L. and Gordon, G.D. (1989) *Communications Satellite Handbook,* ©1989 John Wiley & Sons Inc.)

of the service area under the best conditions and the need to avoid interference with established systems. The problem becomes acute above continents and particularly for the orbital arc above the American continent. Intersatellite links, when they permit traffic to be shared among several satellites in different orbital positions, provide the operator with some latitude in the positioning of his satellites. Figure 5.17 shows an example of a solution by positioning two satellites at the extremities of the congested arc while guaranteeing coverage of the whole of the United States to the operator [MOR-89].

5.5.2.5 *Satellite clusters*

The principle is to locate several separate satellites in the same orbital position with a separation of around 100 km and interconnection by intersatellite links [VIS-79], [WAD-80], [WAL-82]. The satellites are thus all in the principal lobe of an earth station antenna and appear equivalent to a single large capacity satellite which would be too large to be launched by an existing launcher. The cluster is put in place by successive launching of the satellites which form it. As all the satellites are subjected to the same perturbations, orbital control is simplified. In the case of breakdown of a satellite, it can be replaced in the cluster. Finally, the configuration of the cluster can be modified in accordance with traffic demands.

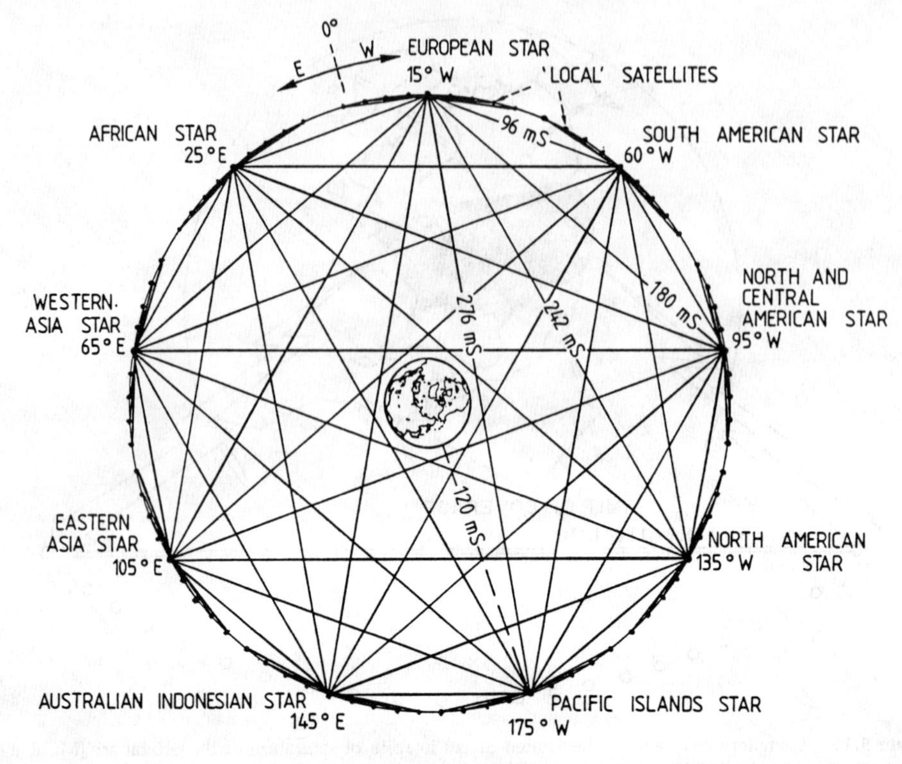

Figure 5.18 A global network [GOL-82]. (Reproduced by permission of the American Institute of Aeronautics and Astronautics.)

5.5.2.6 *A global network*

Figure 5.18 shows a futuristic design of a global network based on nine geostationary STAR satellites, which establish a basis for worldwide communication, and a set of local satellites connected to these by regional intersatellite links [GOL-82].

5.5.3 Links between low orbit satellites (LEO–LEO)

The advantages of low orbit satellites (see Section 1.4) and the increasing congestion of geostationary satellite orbits suggest the future development of orbiting satellites. In fact, the disadvantages of an orbiting satellite (limited duration of communication time and relatively small coverage) can be reduced in a network containing a large number of satellites which are interconnected by intersatellite links and equipped with a means of switching between beams. An example of a network of this type is proposed in [BRA-84] and [BIN-87]. Motorola Inc. has planned to construct, launch and operate such a network, called IRIDIUM, for worldwide cellular personal communication services. The system incorporates a constellation of 77 satellites [LEO-91].

5.5.4 Frequency bands

Table 5.2 indicates the frequency bands allocated to intersatellite links by the Radiocommunication Regulations. These frequencies correspond to strong absorption by the atmosphere and have been chosen to provide protection against interference between intersatellite links and terrestrial systems. However, these bands are shared with other space services and the limitation on interference level is likely to impose constraints on the choice of the defining parameters of intersatellite links [CCIR Reports 451, 465, 874, 951].

Table 5.2 also indicates the wavelengths envisaged for optical links. These result from the transmission characteristics of the components.

5.5.5 Radio-frequency links

The link budget equations presented in Chapter 2 are directly applicable. Propagation losses reduce to free space losses since there is no passage through the atmosphere. Antenna pointing error can be maintained at around a tenth of the beamwidth and this leads to a pointing error loss of the order of 0.5 dB. The antenna temperature in the case of a GEO-GEO link, in the absence of solar conjunction, is of the order of 10 K. Table 5.3 indicates typical values for the terminal equipment. For practical applications antenna dimensions of the order

Table 5.2 Frequency bands for intersatellite links

Radio frequency	22.55–23.55 GHz
(Radio Communications	32–33 GHz
Regulations 1986)	54.25–58.2 GHz
	59–64 GHz
	116–134 GHz
	170–182 GHz
	185–190 GHz
Optical	.8–0.9 micron (AlGaAs laser diode)
	1.06 micron (Nd:YAG laser diode)
	0.532 micron (Nd: YAG laser diode)
	10.6 micron (CO_2 laser)

Table 5.3 Typical values for terminal equipment of a radio-frequency intersatellite link

Frequency	Receiver noise factor	Transmitter power
23–32 GHz	3–4.5 dB	150 W
60 GHz	4.5 dB	75 W
120 GHz	9 dB	30 W

of 1 to 2 m should be considered. Considering a frequency of 60 GHz and transmission and reception losses of 1 dB leads to:

— A receiver factor of merit G/T of the order of 25 to 29 dB(K^{-1}).
—A transmitter EIRP of the order of 72 to 78 dB(W).

Because of the relatively wide lobe of the antenna (0.2° at 60 GHz for a 2 m antenna), major problems in establishing the link would not be expected. Each satellite must be able to orientate its receiving antenna in the direction of the transmitting satellite with a precision of the order of 0.1° to acquire a beacon signal which will subsequently be used for tracking.

The development of high capacity radio-frequency intersatellite links between geostationary satellite systems will imply re-use of frequencies from one beam to another. In view of the small angular separation of the satellites, it would be preferable to use narrow beam antennas with reduced secondary lobes in order to avoid interference between systems. Consequently, and in view of the limited antenna size imposed by the launcher and the technical complexity of the antennas which may be deployed, the use of high frequencies is indicated. The use of optical links may be usefully considered in this context.

5.5.6 Optical links

In comparison with radio links, optical links have specific characteristics which are briefly described here. For a more complete presentation, refer to [KAT-87], [GAG-84, Chapter 9], [IJSC-88].

5.5.6.1 *Establishing a link*

Two aspects should be indicated:

—The small diameter of the telescope which is typically of the order of 0.3 m. In this way one is freed from congestion problems and aperture blocking of other antennas in the payload.
—The narrowness of the optical beam which is typically 5 microradians. Notice that this width is several orders of magnitude less than that of a radio beam and this is an advantage for protection against interference between systems. But it is also a disadvantage since the beamwidth is much less than the precision of satellite attitude control (typically 0.1° or 1.75 mrad). Consequently an advanced pointing device is necessary; this is probably the most difficult technical problem.

Because of this, the intersatellite link must be dimensioned differently in the following three phases:

—Acquisition: the beam must be as wide as possible in order to reduce the acquisition time. But this requires a high power laser transmitter. A laser of lower mean power can be used which emits pulses of high peak power with a low duty cycle. The beam scans the region of space where the receiver is expected to be located. When the receiver receives the signal, it enters a tracking phase and transmits in the direction of the received signal. On receiving the return signal from the receiver, the transmitter also enters the tracking phase. The typical duration of this phase is 10 seconds.

—Tracking: the beams are reduced to their nominal width. Laser transmission becomes continuous. In this phase, which extends throughout the following, the pointing error control device must allow for movements of the platform and relative movements of the two satellites. In particular, allowance must be made for the narrowness of the beams and the expected displacement of the satellite during the round-trip propagation time of the light between the two satellites. Hence it is necessary to transmit information with an angular displacement with respect to the direction of reception in the direction of the expected satellite displacement. This is called 'point ahead correction'. The angular shift can reach 80 mrad and consequently is much greater than the beam width.

—Communications: information is exchanged between the two ends.

5.5.6.2 *Transmission*

Laser sources operate in single and multi-frequency modes. In single frequency mode spectral occupation varies between 10 kHz and 10MHz. In multi-frequency mode it is from 1.5 to 10 nm. The power emitted depends on the type of laser. Table 5.4 gives orders of magnitude.

Modulation can be internal or external. Internal modulation implies direct modification of the operation of the laser. External modulation is a modification of the light beam after its emission by the laser. The intensity, the frequency, the phase and the polarisation can be modulated. Phase and polarisation modulation are external. Intensity and frequency modulation can be internal or external. Polarisation modulation requires the presence of two detectors in the receiver, one for each polarisation. Because of this it is preferable to reserve polarisation for multiplexing of two channels.

The intensity distribution of a laser beam, as a function of angle with respect to the maximum intensity, follows a Gaussian law. The on-axis gain is given by:

$$G_{\mathrm{Tmax}} = 32/(\theta_{\mathrm{T}})^2 \tag{5.6}$$

where θ_{T} is the total beamwidth at $1/e^2$. The choice of θ_{T} depends on the pointing accuracy. With imprecise pointing, a high θ_{T} is better but gain is lost. If θ_{T} is reduced, there is benefit in gain but the pointing error loss increases. It can be shown that, if the pointing error is

Table 5.4 Typical values of transmitted power for lasers

Type of laser	Wavelength	Transmitted power
Solid state (laser diode)		
AlGaAs	0.8–0.9 micron	About 100 mW
InPAaGa	1.3–1.5 micron	About 100 mW
Nd:YAG	1.06 micron	0.5 to 1 W
Nd:YAG	0.532 micron	100 mW
Gas laser		
CO_2	10.6 micron	Several tens of watts

essentially an alignment error, the (maximum gain × pointing error loss) product is maximum when $\theta_T = 2.8 \times$ (pointing error) [KAT-87, p. 51]. In general, for a pointing error of any kind, the beamwidth may be adapted to the pointing error.

In addition to losses due to pointing error, transmission losses and degradation of the wavefront in the emitting optics occur.

5.5.6.3 *Transmission*

Transmission loss reduces to the free space loss:

$$L = (\lambda/4\pi R)^2 \tag{5.7}$$

where λ is the wavelength and R is the distance between transmitter and receiver.

5.5.6.4 *Reception*

The receiving gain of the antenna is given by:

$$G_R = (\pi D_R/\lambda)^2 \tag{5.8}$$

where D_R is the effective diameter of the receiver.

The receiving losses include optical transmission losses and, with coherent detection, losses associated with degradation of the wavefront (the quality of the wavefront is an important characteristic for optimum mixing of the received signal field and that of the local oscillator at the detector surface).

Filtering also introduces losses, since the transmission coefficient of the filter decreases with the bandwidth. Filtering has the purpose of reducing the contribution of noise from external sources. A typical filter width is from 0.1 to 100 nm.

The detector converts incident photons on the surface of the detector into electrons which are detected as an electric current of intensity I_S given by:

$$I_S = (P_S/hf)\eta eG \quad \text{(A)} \tag{5.9}$$

where:
I_S = intensity of signal current (A),
P_S = useful optical power received (W),
h = Planck's constant = 6.6×10^{-34} J/Hz,
f = laser frequency (Hz),
η = quantum efficiency of the photodetector,
e = charge on the electron (C),
G = photodetector gain.
(P_S/hf) represents the number of photons received per second. $S = \eta e/hf$ is the sensitivity of the photodetector (A/W). Hence:

$$I_S = SGP_S \tag{5.10}$$

To this current is added a noise current whose components are the noise associated with the received optical flux, the noise due to the dark current and the noise of the electronic amplifying circuits. The electrical signal-to-noise ratio S/N at the demodulator output depends on the type of demodulation:

—For direct detection:

$$S/N = (SGP_S)^2/(\sigma_{dd})^2 \tag{5.11}$$

—For homodyne detection:

$$S/N = 4(SG)^2 P_S P_{LO} \eta_H L_P / [(\sigma_{dd})^2 + (\sigma_{LO})^2]$$

—For heterodyne detection:

$$S/N = 4(SG)^2 P_S P_{LO} \eta_H L_P / [(\sigma_{dd})^2 + (\sigma_{LO})^2] \tag{5.12}$$

where:

σ_{dd} = effective value of the noise current with direct detection (A).
P_{LO} = local oscillator power at the photodetector input (W).
η_H = mixing efficiency.
L_P = loss due to polarisation mismatching.
σ_{LO} = effective value of noise current due to the local oscillator (A).

In theory, coherent detection (homodyne or heterodyne) confers a higher value of S/N ratio. However, in the case of alignment error between the local oscillator and the beam signal, mixing efficiency is degraded. This type of detection cannot therefore be used for acquisition and tracking. As the receiver used for these functions can also be used for communication purposes, there is no advantage in weight or power in using coherent techniques. However, for high throughputs (greater than 1 Gbit/s), the power required for direct detection is excessive and it is necessary to resort to coherent detection.

5.5.7 Conclusion

The choice between radio and optical links depends on the mass and power consumed. In general terms, it can be said that the advantage is with radio links for low throughputs (less than 1 Mbit/s). For high capacity links (several tens of Mbit/s) optical links command attention.

5.6 CONCLUSION

Multibeam satellite systems make it possible to reduce the size of earth stations and hence the cost of the earth segment. Frequency re-use from one beam to another permits an increase in capacity without increasing the bandwidth allocated to the system. However, interference between adjacent channels, which occurs between beams using the same frequencies, limits the potential capacity increase particularly as interference is greater with earth stations equipped with small antennas.

Intersatellite links permit the following:

—The use of a geostationary satellite as a relay for permanent links between low orbit satellites and a network of a small number of earth stations.

—An increase in system capacity by combining the capacities of several geostationary satellites.
—The planning of systems with a higher degree of flexibility.
—Consideration of systems providing a permanent link and worldwide coverage using low orbit satellites as an alternative to systems using geostationary satellites.

Optical technology is more advantageous in terms of mass and power consumption for high capacity links.

REFERENCES

[ARN-88] M. Arnaud, A. Barumchercyk, E. Sein (1988) An experimental optical link between an earth remote sensing satellite SPOT4 and a European data relay satellite, *International Journal of Satellite Communications*, **6**, No. 2, pp. 127–140.

[BIN-87] R. Binder, S.D. Huffman, I. Guarantz, P.A. Vena (1987) Crosslink architectures for a multiple satellite system, *Proceedings of the IEEE*, **75**, No. 1, pp. 74–82, Jan.

[BRA-84] K. Brayer (1984) Packet switching for mobile earth stations via low orbiting satellite network, *Proceedings of the IEEE*, **72**, No. 11, pp. 1627–1636, Nov.

[CAM-81] S.J. Campanella, R.J. Colby (1981) Network control for TDMA and SS/TDMA in multiple beam satellite systems, *5th International Conference on Digital Satellite Communications*, pp. 335–343, Genoa, March.

[CAR-80] C.R. Carter, Survey of synchronization techniques for a TDMA satellite switched system, *IEEE Transactions on Communications*, **COM-28**, No. 8, pp. 1291–1301, August.

[DAV-81] R.S. Davies (1981) Optimization of SS/TDMA communication satellite payload, *5th International Conference on Digital Satellite Communications*, pp. 435–439, Genoa, March.

[DIC-78] J.L. Dicks, M.P. Brown (1978) INTELSAT V satellite transmission design, *ICC 78*, pp. 2.2.1–2.2.5.

[DIC-87] A. Dickinson, S.E. Dinwiddy, J. Sandberg (1987) The European data relay system as part of the in-orbit infrastructure, *ESA Bulletin*, pp. 47–52.

[FUE-77] J.C. Fuenzalida, P. Rivalan, H.J. Weiss (1977) Summary of the INTELSAT V communications performance specifications, *COMSAT Technical Review*, **7**, No. 1, pp. 311–326, Spring.

[GAG-84] R.M. Gagliardi (1984) *Satellite Communications*, Lifetime Learning Publications.

[GOL-82] E. Golden (1982) The wired sky, *AIAA 9th International Conference*, San Diego, pp. 174–180.

[INU-79] T. Inukai (1979) Efficient SS/TDMA time slot assignment algorithm, *IEEE Transactions on Communications*, **COM-27**, No. 10, pp. 1449–1455.

[INU-81] T. Inukai, S.J. Campanella (1981) On board clock correction for SS/TDMA and baseband processing satellites, *COMSAT Technical Review*, **11**, No. 1, pp. 77–102, Spring.

[IJSC-88] *International Journal of Satellite Communications*, Special Issue on Intersatellite links, *6*, No. 2, 1988.

[KAT-87] M. Kaitzman (Ed.) (1987) *Laser Satellite Communications*, Prentice-Hall.

[LAN-82] R.B. Landon, H.G. Raymond (1982) Ku band satellite communications via TDRSS, *AIAA 9th International Conference, San Diego*, pp. 741–755.

[LEO-91] R.J. Leopold (1991) Low earth orbit global cellular communications network, ICC'91, pp. 1108/1111.

[LOP-82] M. Lopriore, A. Saitto, G.K. Smith (1982) A unifying concept for future fixed satellite service payloads for Europe, *ESA Journal*, *6*, No. 4, pp. 371–396.

[MAR-87] G. Maral, M. Bousquet (1987) Performance of fully variable demand assignment SS-TDMA system, *International Journal of Satellite Communications*, **5**, No. 4, pp. 279–290.

[MOR-89] W.L. Morgan, G.D. Gordon (1989) *Communications Satellite Handbook*, Wiley.

[PUC-88] A. Purio, E. Saggese (1988) Identification of requirements for intersatellate links, *International Journal of Satellite Communications*, **6(2)**, 106–117.

[REU-77] D.O. Reudink, Y.S. Yeh (1977) A scanning spot beam satellite system, *The Bell System Technical Journal*, Oct.

[SCH-69] W.G. Schmidt (1969) An onboard switched multiple access for millimeter wave satellites, *1st International Conference on Digital Satellite Communications*, London, Nov.

[SCH-84] J.J. Schwartz, L. Schuchman (1984) NASA's satellite relay tracking and data acquisition program, *EASCON*, pp. 157–162.

[SHI-71] N. Shimasaki, R.A. Rapuano (1971) Synchronization for a communications distribution center on board a satellite, *International Conference on Communications*, pp. 42.20–42.25, Montreal.

[TIR-83] S. Tirro (1983) Satellites and switching, *Space Communications and Broadcasting*, **1**, No. 1, pp. 97–133.

[VIS-79] P.S. Visher (1979) Satellite clusters, *Satellite Communications*, **3**, No. 9, pp. 22–27.

[WAD-80] D.V.Z. Wadsworth (1980) Satellite cluster provides modular growth of communications functions, *International Telemetering Conference ITC80*, p. 209, San Diego, Oct.

[WAL-82] J.G. Walker (1982) The geometry of satellite clusters, *Journal of the British Interplanetary Society*, **35**, pp. 345–354.

6 REGENERATIVE SATELLITE NETWORKS

The imminent prospect of regenerative satellites is mentioned in Chapter 1. It may seem strange that it has been necessary to wait almost a quarter of a century after the launching of the first operational satellites to see the appearance of the first regenerative satellites with the launching of the ITALSAT satellite in January 1991 and the planned launching of the ACTS satellite in 1993. In fact, the potential advantages, described in this chapter, which can be achieved by signal regeneration on board the satellite are accompanied by several uncertainties which include the effect of increased complexity on reliability, the effect on flexibility of use, the ability to cope with unexpected changes in traffic demand (both volume and nature) and new operational procedures. For an experienced operator, the strategy is to minimise risks rather than acquire a substantial but uncertain advantage. This partly explains the continued use of conventional non-regenerative satellites.

Will the progressive commissioning of regenerative satellite networks lead to a revolution in satellite communication? Or will it be a simple evolution in the continuous operation of existing networks? This chapter attempts to provide the basic answers by indicating firstly the effect of demodulation followed by on-board remodulation on the link budget as established for a conventional satellite in Chapter 2, then by describing the various factors which can be obtained from the presence of baseband signals on board the satellite and finally by examining the consequences for the earth segment.

6.1 CONVENTIONAL AND REGENERATIVE TRANSPONDERS

Figure 6.1 shows the difference between a regenerative satellite link and a conventional one. With the regenerative satellite, baseband signals, which have modulated the uplink carrier, are available at the output of the demodulator and these signals are used (possibly after processing not shown in the figure but discussed below in Section 6.3) to modulate the downlink carrier. Hence the change of frequency from uplink to downlink which is obtained by mixing with the local radio-frequency oscillator in a conventional satellite is obtained in this case by modulation of a new carrier.

6.2 COMPARISON OF LINK BUDGETS

6.2.1 Linear channel without interference

This is the case of digital links, the only case considered in practice. It is assumed that the probability of error at the output of the demodulator is that given by theory (Table 3.6), that is, there is no degradation due to filtering or non-linearities.

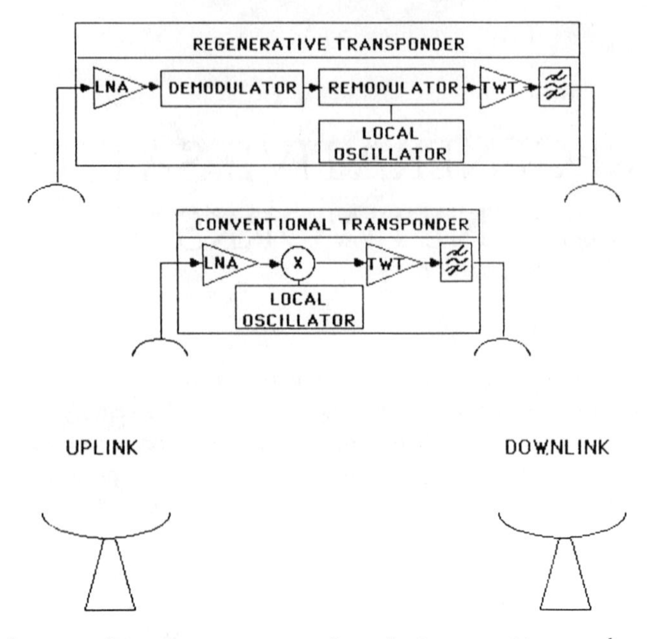

Figure 6.1 Architecture of a regenerative transponder and a transparent transponder.

6.2.1.1 *Link with conventional transponder* (Figure 6.2)

The performance of the link is specified in terms of the error probability at the output of the earth station demodulator. This error probability is a function of the ratio $(E/N_0)_T$ given by expression (3.36) in Section 3.7 and recalled here:

$$(E/N_0)_T = (C/N_0)_T/R_c \tag{6.1}$$

where R_C is the carrier data rate and $(C/N_0)_T$ is the ratio of carrier power to noise spectral density of the station-to-station link given by expression (2.60) of Section 2.9 and recalled here:

$$(C/N_0)_T^{-1} = (C/N_0)_U^{-1} + (C/N_0)_D^{-1} \tag{6.2}$$

Defining $(E/N_0)_U = (C/N_0)_U/R_c$ and $(E/N_0)_D = (C/N_0)_D/R_c$ and using (6.1) and (6.2) gives:

$$(E/N_0)_T^{-1} = (E/N_0)_U^{-1} + (E/N_0)_D^{-1} \tag{6.3}$$

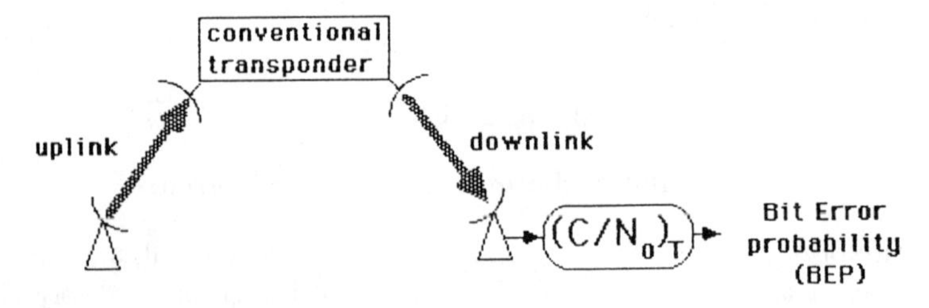

Figure 6.2 Link by transparent satellite.

6.2.1.2 *Link with regenerative transponder* (Figure 6.3)

The error probability is expressed as the probability of having an error on the uplink (BER_U) and no error on the downlink ($1 - BER_D$) or no error on the uplink ($1 - BER_U$) and an error on the downlink (BER_D), hence:

$$BER = BER_U(1 - BER_D) + (1 - BER_U)BER_D \quad (6.4)$$

As BER_U and BER_D are small compared with 1, this becomes:

$$BER = BER_U + BER_D \quad (6.5)$$

BER_U is a function of $(E/N_0)_U$ and BER_D is a function of $(E/N_0)_D$.

6.2.1.3 *Comparison at constant error probability*

The value of the required probability is given as follows:

—For a conventional satellite, the value of $(E/N_0)_T$ is determined by the error probability specified for the link. The required performance is obtained for a set of values of $(E/N_0)_U$ and $(E/N_0)_D$ combined using equation (6.3). The is shown by curve A of Figure 6.4 for an error probability of 10^{-4} and QPSK modulation with coherent demodulation.
—For a regenerative satellite, by calculating BER_U and BER_D from (6.5) with the constraint $BER = \text{constant} = 10^{-4}$ and deducing the corresponding pairs of values of $(E/N_0)_U$ and $(E/N_0)_D$, curves B and C of Figure 6.4 are obtained. These curves correspond to the cases of QPSK modulation with coherent demodulation (curve B) and differential demodulation (curve C) on the uplink. On the downlink demodulation is coherent in both cases. The curves are parameterised as a value of the ratio $\alpha = (E/N_0)_U/(E/N_0)_D$.

Comparing curve A and curve B, it can be seen that the regenerative transponder provides a reduction of 3 dB in the value required for E/N_0 for the uplink and the downlink when the links are identical ($\alpha = 0$ dB). This is explained by the fact that the regenerative transponder does not transmit the amplified uplink noise along with the signal on the downlink, as with a conventional repeater.

However, for very different values of E/N_0 this advantage disappears. For example, for

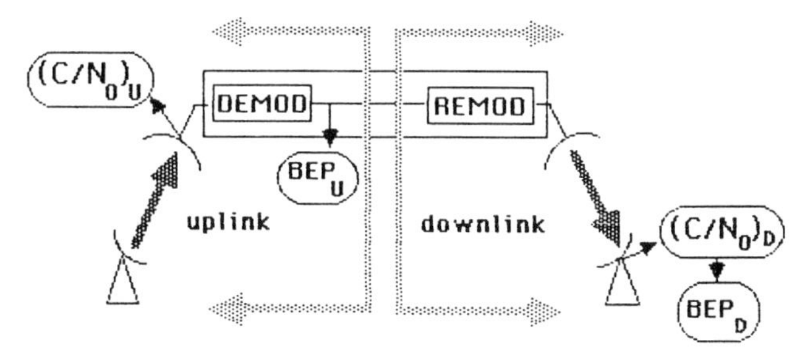

Figure 6.3 Link by regenerative satellite.

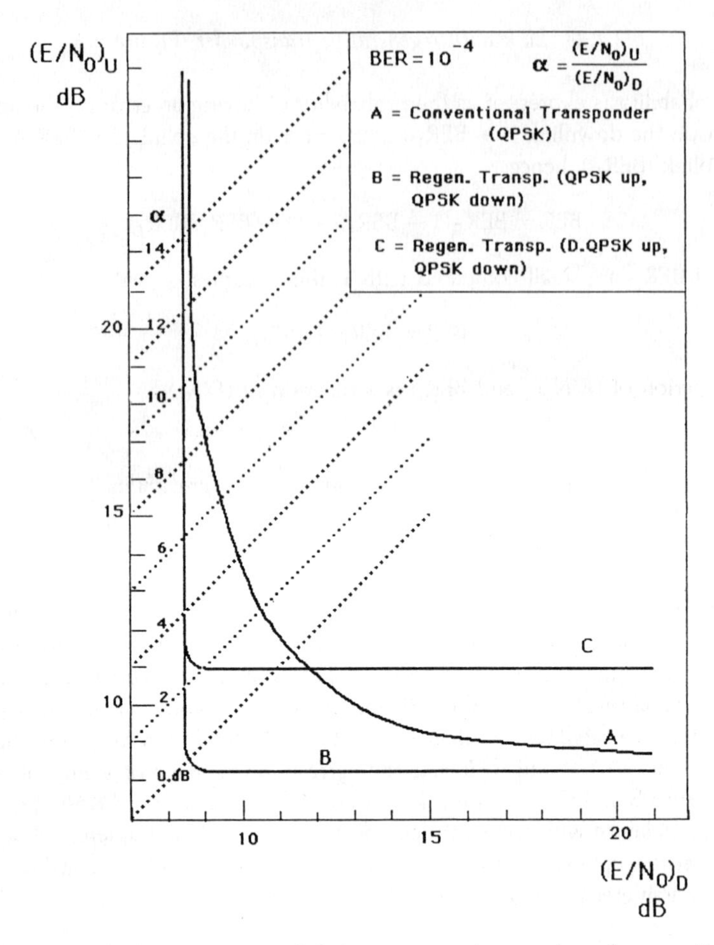

Figure 6.4 Comparison of station-to-station links by transparent transponder and regenerative transponder for the same bit error probability (BEP $= 10^{-4}$) (linear channel). A, transparent transponder. B, regenerative transponder with QPSK modulation and coherent demodulation on the uplink. C, regenerative transponder with QPSK modulation and differential demodulation on the uplink.

α greater than 12 dB the two curves join; in this case the uplink noise is negligible and the performance of the total link reduces in both cases to that of the downlink.

Curve C indicates that by dimensioning the uplink for a value of $(E/N_0)_U$ greater than that of $(E/N_0)_D$ by about 4 dB, on-board differential demodulation can be used without degrading the global performance and this is simpler to realise than coherent demodulation.

6.2.2 Non-linear channel without interference

This corresponds more closely to a real system since a real channel is non-linear and band limited; the combination of non-linearities and filtering introduces a performance degradation of the demodulator which increases as the chain of non-linearities and filters increases. This

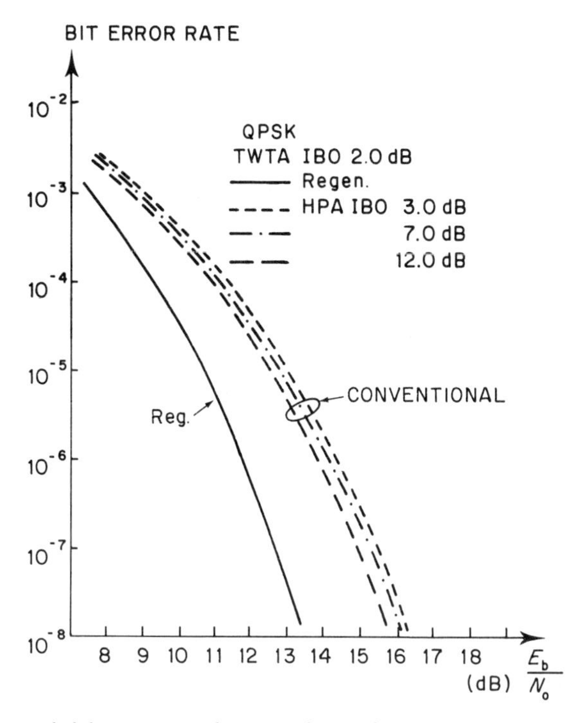

Figure 6.5 Bit error probability BEP as a function of E/N_0 for a link with a transparent transponder and for a link with a regenerative transponder in the absence of interference for the case where $\alpha = (E/N_0)_U/(E/N_0)_D$ is high. TWTA IBO: input back-off of the on-board travelling wave tube. HPA IBO: input back-off of the earth station transmitting amplifier [WAC-81]. (© 1981 IEEE. Reproduced by permission.)

is the case for a link with a conventional transponder (two non-linearities and filters—on transmission at the earth station and at the transponder). With a regenerative transponder, separation of the up and down links means that there is now only one non-linearity and filter per link. Figure 6.5 shows the results obtained by means of computer simulation [WAC-81] for the case where $(E/N_0)_U$ is much greater than $(E/N_0)_D$. This figure and other results show that the regenerative transponder can provide between 2 and 5 dB reduction in E/N_0 with respect to a conventional transponder even when the ratio $\alpha = (E/N_0)_U/(E/N_0)_D$ is large.

6.2.3 Non-linear channel with interference

6.2.3.1 *Link with conventional transponder*

The value of $(C/N_0)_T$ depends on the value of $(C/N_0)_{T\ \text{without interference}}$ in the absence of interference and $(C/N_0)_I$ due to interference on the up and down links (see Section 2.9.2). More precisely:

$$(C/N_0)_T^{-1} = (C/N_0)_T^{-1}{}_{\text{without interference}} + (C/N_0)_I^{-1} \tag{6.6}$$

where:

$$(C/N_0)_T^{-1}{}_{\text{without interference}} = (C/N_0)_U^{-1} + (C/N_0)_D^{-1}$$
$$(C/N_0)_I^{-1} = (C/N_0)_{I,U}^{-1} + (C/N_0)_{I,D}^{-1}$$

By putting $E/N_0 = C/N_0/R_c$, the relations between the values of E/N_0 can be deduced from these equations:

$$(E/N_0)_T^{-1} = (E/N_0)_T^{-1}{}_{\text{without interference}} + (E/N_0)_I^{-1} \tag{6.7}$$

From the curve of Figure 6.5 (which implies that α is large), BER $= 10^{-4}$ with QPSK modulation requires that $(E/N_0)_T = 11$ dB. The upper curve of Figure 6.6 shows the relation between $(E/N_0)_I$ and $(E/N_0)^{-1}_{T\,\text{without interference}}$ obtained from (6.7) for this case.

6.2.3.2 *Link with regenerative transponder*

This is also in the class where $\alpha = (E/N_0)_U/(E/N_0)_D$ is high. The bit error rate of the station-to-station link is defined by the error rate on the downlink and a probability of 10^{-4} requires, from Figure 6.5, $(E/N_0)_D = (E/N_0)_T^{-1} = 9$ dB. The lower curve of Figure 6.6 shows the relation between $(E/N_0)_I$ and $(E/N_0)^{-1}_{T\,\text{without interference}}$ for this case.

Comparison of the two curves of Figure 6.6 shows that, for a given link quality (BER $= 10^{-4}$), the ratio $(E/N_0)_I$ is less for a link with a regenerative transponder. This implies that the link can be established with a higher level of interference. This is a useful advantage in the context of a multibeam satellite network where high levels of interference must be taken into account (see Section 5.1.2).

6.3 ON-BOARD PROCESSING

The availability of binary digits on board the satellite encourages processing of these digits before retransmission.

6.3.1 Downlink coding

Error correcting coding can be used on one of the up- and downlinks. For the downlink, the coder is located on board the satellite and is activated by telecommand. The link thus benefits from the decoding gain, but it must be possible to increase the transmission rate of the link by a factor equal to the inverse of the coding ratio (see Section 3.7.8). This implies that the downlink is limited in power but not in bandwidth. If the link is limited in bandwidth, the transmission rate must be maintained, and consequently the information rate reduced, on both the downlink and the uplinks which feed it. This reduction of throughput provides a margin on $(C/N_0)_D$ which is added to that provided by the decoding gain. Let $(C/N_0)_1$ and $(C/N_0)_2$ be the values of $(C/N_0)_D$ without and with coding respectively. Hence:

$$(C/N_0)_1 = (E_b/N_0)_1 R_{b1} \tag{6.8}$$

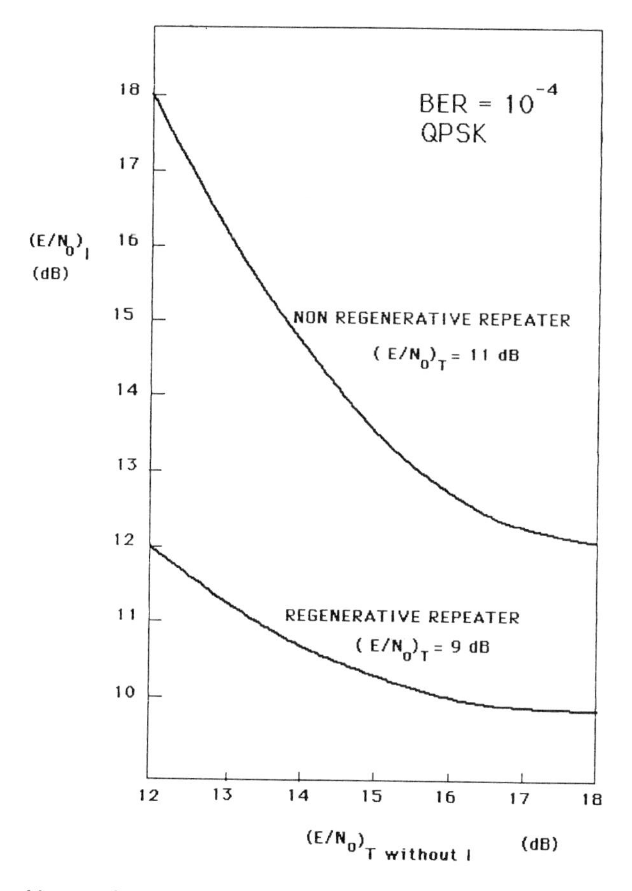

Figure 6.6 Permissible interference level; comparison of links with regenerative and transparent transponders.

where R_{b1}, the information rate, is equal to the rate R_c which modulates the carrier,

$$(C/N_0)_2 = (E_b/N_0)_2 R_{b2} \tag{6.9}$$

where

$$R_{b2} = \rho R_c.$$

The margin realised above (in dB) is thus equal to:

$$\begin{aligned} \text{Margin} &= \Delta(C/N_0)_D \\ &= (C/N_0)_1 - (C/N_0)_2 \\ &= [(E_b/N_0)_1 - (E_b/N_0)_2] - 10 \log \rho \\ &= \text{Decoding gain} + \text{gain provided by rate reduction.} \end{aligned} \tag{6.10}$$

For example, consider the use of a code with coding ratio $\rho = 1/3$ and a decoding gain of 5 dB; at constant bandwidth a margin on the required value of $(C/N_0)_D$ of 10 dB is obtained.

The price to be paid is a reduction of 2/3 in the information rate and hence a reduction in the capacity of this downlink. This margin of 10 dB can be used to compensate for temporary degradation due to rain for links at 20 GHz [FIO-78], [HOL-80].

6.3.2 Information storage

6.3.2.1 *Baseband switching*

The availability of binary code on board the satellite at the output of the uplink carrier demodulators permits switching between receiving and transmitting antennas to be no longer at radio frequency but at baseband. In present designs, the switching device contains a time–space–time (TST) structure that is:

—A set of memories (T) which enable the binary digits from the uplinks to be stored for a limited period (for example one frame duration).
—An interconnection matrix (S).
—A set of memories (T) which permit the binary digits to be stored before transmission in the form of a multiplex of the information grouped by receiving station.

Henceforth the constraint of immediate routing of received information to the destination downlink disappears. The sequence of switching states of a radio-frequency switching matrix is replaced by memory followed by multiplexing of the stored binary digits before their transmission on the various downlinks. This permits earth stations, in particular, to transmit all their information in the same burst and hence to transmit only a single burst per frame. The number of bursts per frame is reduced and the efficiency of the frame increases.

The TST structure is well suited to switching circuits of the telephone type. To make way for integration of services in satellite networks, new architectures are being examined [ESA-88], [CAM-88]. They are based on the principle of packet switching and transfer in asynchronous mode (asynchronous transfer mode, ATM). The destination of a burst is indicated by an address provided in the header. Packet routing is thus achieved by logical memory addressing. This routing must be associated with flow control. Under these conditions, identical processing can be realised for telephone circuits and computer data.

6.3.2.2 *Rate conversion*

A change of rate between the uplink and the downlink is not possible with a conventional satellite. Stations can, therefore, be interconnected only by carriers of the same capacity and this can be restricting. For example, interconnection of a network of large sations carrying intercontinental traffic and small stations (VSAT) in the network of a private organisation implies a terrestrial connection and a double hop as shown in Figure 6.7a. In contrast, by virtue of on-board demodulation, the binary digits of the traffic between networks received on carriers with different data rates can be switched at baseband and combined before transmission on the various downlinks in accordance with their destination and independently of the capacity of the carrier (Figure 6.7b). One example of realisation is shown in the same figure [NUS-86]. To avoid bulk and excessive power consumption, only high rate carriers which

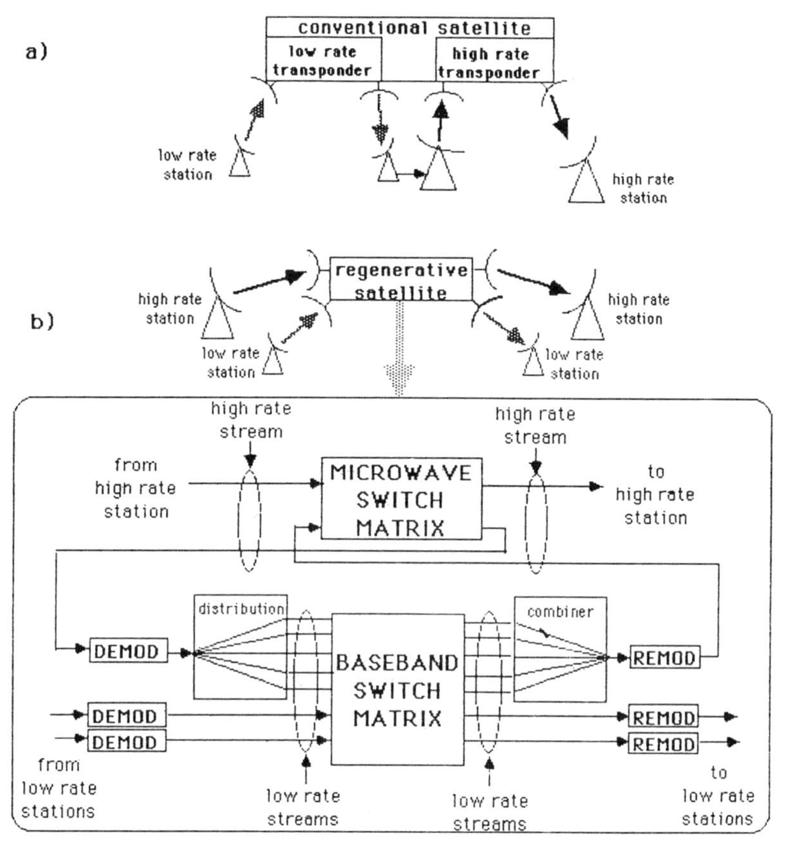

Figure 6.7 Interconnection of two networks with carriers of different capacities. (a) The case of the transparent satellite. (b) The case of the regenerative satellite.

contain traffic destined for a low rate network are routed to a high rate demodulator. The other high rate carriers are switched at radio frequency and are, therefore, not regenerated.

6.3.2.3 *Beam scanning satellites*

Once the techniques of dynamic real-time forming of antenna beams have been mastered, the replacement of multibeam satellites by single beam satellites with a beam which sequentially scans the various regions of the service zone (Figure 6.8) can be considered. The set of dwell areas which are covered sequentially by the beam form the coverage area of the system. When the beam is in a given dwell area, the information destined for stations in the area is extracted from the on-board memory and transmitted in multiplexed form. Simultaneously, regional stations transmit information destined for all other network stations to the satellite. This is stored in the on-board memory for later transmission at the time when the beam passes over the destination area.

An inherent advantage of this type of system is the disappearance of fixed simultaneous

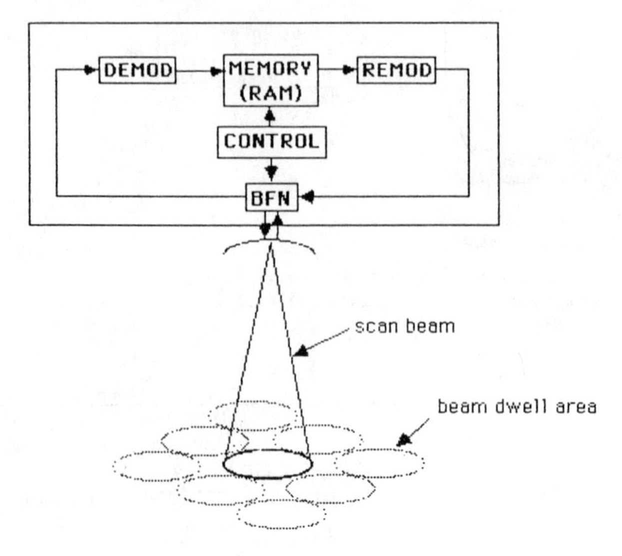

Figure 6.8 Single beam regenerative scanning satellite network.

beams and hence of co-channel interference (CCI). The NASA ACTS satellite uses two scanning beams, one for the uplinks and the other for the downlinks [NAD-88].

6.4 IMPACT ON THE EARTH SEGMENT

Regeneration, compared with a transparent satellite system, permits a reduction in the cost of earth stations by reducing the EIRP of the stations and also, if the transparent system uses FDMA multiple access, the G/T of the stations.

6.4.1 The uplink

The uplink is often overdimensioned so that the performance in terms of C/N_0 of the total link is determined by that of the downlink which is limited by the power available on board the satellite. With a transparent transponder, on account of expression (6.2), it is necessary to provide a ratio $\alpha = (E/N_0)_U/(E/N_0)_D = (C/N_0)_U/C/N(_0)_D$ of the order of 10 dB. With a regenerative transponder, the curves of Figure 3.24 shown that the error probability on the uplink becomes negligible in comparison with that of the downlink when the value of the ratio α is greater than around 2 dB. This reduction of the ratio α leads to a reduction in the EIRP of the earth station and consequently its cost.

Another factor also permits reduction of the EIRP. As a consequence of the storage of binary digits in the baseband switching device, the stations of a given region can transmit continuously on different frequencies (FDMA multiple access) as shown in Figure 6.9. In comparison with time division multiple access (TDMA), the rate transmitted by each station is less. The situation is:

—With TDMA:

$$(C/N_0)_U = (E/N_0)R_{TDMA}$$

—With FDMA:

$$(C/N_0)_U = (E/N_0)R_{FDMA} \tag{6.11}$$

with $R_{FDMA} = R_{TDMA}(T_B/T_F)$, where T_B is the duration of the burst transmitted in TDMA and T_F is the frame duration. As T_B/T_F is less than 1, it can be seen that the value of C/N_0 required with FDMA is less than with TDMA.

6.4.2 The downlink

As shown in Figure 6.9, the satellite amplifier emits a single carrier modulated by the multiplex of the binary digits destined for stations on the beam considered. In comparison with a transparent FDMA system, the transponder operates at saturation. The downlink benefits from the maximum EIRP of the satellite and the figure of merit G/T of the earth stations can be reduced. In comparison with a transparent TDMA system, the received bursts are generated from a single carrier which originates from the satellite oscillator instead of being transmitted by different stations equipped with independent oscillators. It is, therefore,

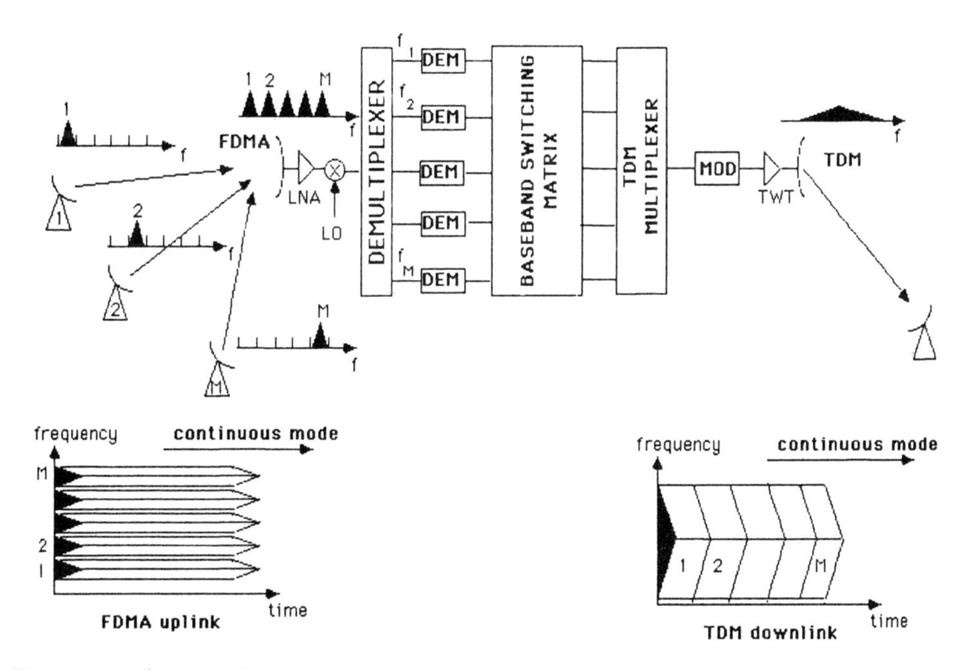

Figure 6.9 The network associated with a regenerative transponder–multiple station access to the transponder using FDMA and TDM transmission from the transponder to the stations. This approach favours the installation of low cost earth stations.

no longer necessary to provide station receivers with fast carrier and bit rate acquisition circuits.

6.5 CONCLUSION

Compared with transparent systems, regenerataive satellite systems tolerate a higher level of interference. They offer the possibility of on-board processing which simplifies the earth station equipment. The more favourable link budget also permits a reduction of the demands on station transmission and reception specifications; all this leads to a reduction in the cost of stations. These systems thus favour the expansion of networks containing a large number of small stations close to be user.

However, we have no experience of such systems. What will be the operational flexibility? What will be the reliability? Questions which arise now concern not only the more complex on-board equipment but also the software. It will be possible to confirm the expected advantages only after a period of experience whose duration is difficult to estimate.

REFERENCES

[CAM-88] S.J. Campanella, B. Pontano, H. Chalmers (1988) Future switching satellites, *12th AIAA International Communications Satellite Systems Conference, Arlington*, pp. 264–273.

[ESA-88] ESA (1988) Study on the applicability of asynchronous time division techniques to satellite communications systems, *Final Study Report, Contract No. 7300/87/F/RD(SC)*, July.

[FIO-78] F. Fiorica (1978) Use of regenerative repeaters in digital communications satellites, *7th AIAA International Communications Satellite Systms Conference, San Diego*, pp. 524–532.

[HOL-80] W.M. Holmes (1980) Multigigabit satellite on board signal processing, *8th AIAA International Communications Satellite Systems Conference, Orlando*, pp. 623–626.

[NAD-88] F.M. Naderi, S.J. Campanella (1988) NASA's advanced communications satellite (ACTS): an overview of the satellite, the network, and the underlying technologies, *12th AIAA International Communications Satellite Systems Conference, Arlington*, pp. 204–224.

[NUS-86] P.P. Nuspl, R. Peters, T. Abdel-Nabi (1986) On-board processing for communications satellite systems, *7th International Conference on Digital Satellite Communications*, pp. 137–148. Munich.

[WAC-81] M. Wachira, V. Arunachalam, K. Feher, G. Lo (1981) Performance of power and bandwidth efficient modulation techniques in regenerative and conventional satellite systems, *ICC 81, Denver*, pp. 37.21–37.2.5.

7 ORBITS

This chapter examines various aspects of the satellite's motion around the earth; these include Keplerian orbits, orbit parameters, perturbations, eclipses and the geometric relationships between satellites and earth stations.

7.1 KEPLERIAN ORBITS

These orbits are named after Kepler who established, at the start of the seventeenth century, that the trajectories of planets around the sun were ellipses and not combinations of circular movements as had been thought since the time of Pythagoras. Keplerian movement is the relative movement of two point bodies under the sole influence of their Newtonian attractions.

7.1.1 Kepler's laws

These laws arise from observation by Kepler of the movement of the planets around the sun:

(a) The planets move in a plane; the orbits described are ellipses with the sun at one focus (1602).
(b) The vector from the sun to the planet sweeps equal areas in equal times (the law of areas, 1605).
(c) The ratio of the square of the period T of revolution of a planet around the sun to the cube of the semi-major axis a of the ellipse is the same for all planets (1618).

7.1.2 Newton's law

Newton extended the work of Kepler and in 1667 discovered the universal law of gravitation. This law states that two bodies of mass m and M attract each other with a force which is proportional to their masses and inversely proportional to the square of the distance r between them:

$$F = GMm/r^2 \tag{7.1}$$

where G is a constant, called the universal gravitation constant, and $G = 6.672 \times 10^{-11}\,\text{m}^3\,\text{kg}^{-1}\,\text{s}^{-2}$.

As the mass of the earth $M = 5.974 \times 10^{24}\,\text{kg}$, the product GM has a value $\mu = GM = 3.986 \times 10^{14}\,\text{m}^3/\text{s}^2$.

From the universal law of gravitation and using the work of Galileo, a contemporary of Kepler, Newton proved Kepler's laws and identified the assumptions (the problem of two spherical and homogeneous bodies). He also modified these laws by introducing the concept of orbit perturbations to take account of actual movements.

7.1.3 Relative movement of two point bodies

The movement of satellites around the earth observes Kepler's laws to a first approximation. The proof results from Newton's law and the following assumptions:

—The mass m of the satellite is small with respect to the mass M of the earth which is assumed to be spherical and homogeneous.
—Movement occurs in free space; the only bodies present are the satellite and the earth.

The actual movement must take account of the fact that the earth is neither spherical nor homogeneous, the attraction of the sun and moon and other perturbing forces.

7.1.3.1 Keplerian potential

Kepler's laws treat the relative movement of two bodies by applying Newton's law. It is convenient to consider the body of greater mass to be fixed with the other moving round it (as the force of attraction is the same for the two bodies, the resulting acceleration is much greater for the body of low mass than for the other).

Consider an orthogonal reference (Figure 7.1) whose origin is at the centre of the earth and whose z axis coincides with the line of the poles (assumed fixed in space). The satellite SL of mass m ($m \ll M$) is at a distance r from the centre of the earth 0 ($\mathbf{r}$ is the vector O–SL).

The force of gravitation F acting on the satellite can be written:

$$\mathbf{F} = -GMm\mathbf{r}/r^3 \qquad \text{(N)} \tag{7.2}$$

(F is a vector centred on SL along SL–O)

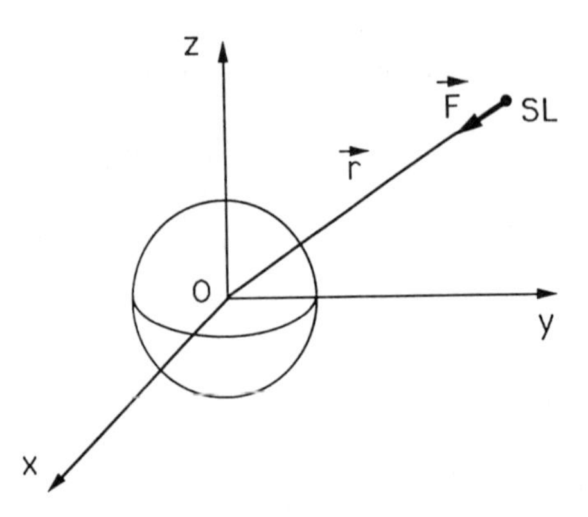

Figure 7.1 Geocentric reference system.

This force always applies to the centre of gravity of the two bodies and, in particular, the centre of the earth O. It is a central force. It derives from a potential U such that $U = GM/r = \mu/r$.

$$\mathbf{F} = \mathrm{d}/\mathrm{d}r[\mu/\mathbf{r}] = \mathrm{grad}\ U \qquad (\mathrm{N}) \tag{7.3}$$

7.1.3.2 *The angular momentum of the system*

The angular momentum **H** of the system with respect to the point O can be written:

$$\mathbf{H} = \mathbf{r} \wedge m\mathbf{V} \qquad (\mathrm{Nm\,s}) \tag{7.4}$$

where v is the velocity of the satellite. The momentum theorem states that the vector differential with respect to time of the instantaneous angular momentum is equal to the moment **M** of the external forces about the origin of the angular momentum:

$$\mathrm{d}\mathbf{H}/\mathrm{d}t = \mathbf{M} \qquad (\mathrm{Nm}) \tag{7.5}$$

In the system under consideration, the only external force **F** passes through the origin. The moment **M** is therefore zero, $\mathrm{d}\mathbf{H}/\mathrm{d}t$ is also equal to zero. The result is that the angular momentum **H** is of constant magnitude, direction and sign. As the angular momentum is always perpendicular to **r** and **V**, movement of the satellite occurs in a plane which passes through the centre of the earth and has a fixed orientation in space perpendicular to the angular momentum vector.

In this plane, the satellite is identified by its polar co-ordinates r and θ (Figure 7.2). Hence:

$$\mathbf{H} = \mathbf{r} \wedge \mathrm{m}\mathbf{V} = \mathbf{r} \wedge m(\mathbf{V}_\mathrm{R} + \mathbf{V}_\mathrm{T}) = \mathbf{r} \wedge m\mathbf{V}_\mathrm{R} + \mathbf{r} \wedge m\mathbf{V}_\mathrm{T}$$

As $\mathbf{V}_\mathrm{R}$ passes through O, the vector product $\mathbf{r} \wedge m\mathbf{V}_\mathrm{R} = 0$.

Hence: $\mathbf{H} = \mathbf{r} \wedge m\mathbf{V}_\mathrm{T}$, that is $|\mathbf{H}| = H = r \times mr\,\mathrm{d}\theta/\mathrm{d}t$. From which:

$$H = r^2\,\mathrm{d}\theta/\mathrm{d}t = C \qquad (\mathrm{Nm\,s}) \tag{7.6}$$

As the angular momentum is constant, C is constant.

The expression $r^2\mathrm{d}\theta/\mathrm{d}t$ represents twice the area swept by the radius vector r during $\mathrm{d}t$. This area is thus constant and Kepler's area law is verified.

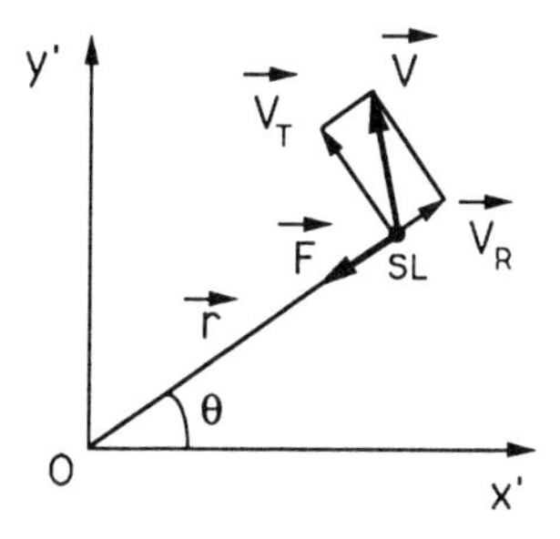

Figure 7.2 Location of the satellite in polar co-ordinates.

7.1.3.3 *Equation of motion*

The satellite describes a trajectory such that at every point it is in equilibrium between the inertial force $m\gamma$ and the force of attraction (Figure 7.2):

$$\mathbf{F} = m\gamma = -m\mathbf{r}/r^3 \qquad \text{(N)}$$

Equilibrium of the radial components of forces leads to:

$$\mathrm{d}^2r/\mathrm{d}t^2 - r(\mathrm{d}\theta/\mathrm{d}t)^2 = -\mu/r^2 \qquad (\mathrm{m\,s^{-2}}) \tag{7.7}$$

where: $\mathrm{d}^2r/\mathrm{d}t^2$ represents the variation of radial velocity, and $r(\mathrm{d}\theta/\mathrm{d}t)^2$ is the centripetal acceleration.

Taking account of (7.6): $r^2\,\mathrm{d}\theta/\mathrm{d}t = H$ gives:

$$\mathrm{d}^2r/\mathrm{d}t^2 - H^2/r^3 = -\mu/r^2 \qquad (\mathrm{m\,s^{-2}}) \tag{7.8}$$

The equation of the orbit is obtained by eliminating time from this equation. The derivative of r with respect to time is put in the form:

$$\mathrm{d}r/\mathrm{d}t = (\mathrm{d}r/\mathrm{d}\theta)\,(\mathrm{d}\theta/\mathrm{d}t)$$

with $\mathrm{d}r/\mathrm{d}\theta = -(1/\rho^2)\,\mathrm{d}\rho/\mathrm{d}\theta$, by putting $\rho = 1/r$ and

$$\mathrm{d}\theta/\mathrm{d}t = H/r^2 \tag{7.6}$$

Hence:

$$\mathrm{d}r/\mathrm{d}t = -H\,\mathrm{d}\rho/\mathrm{d}\theta \quad \text{and} \quad \mathrm{d}^2r/\mathrm{d}t^2 = -H^2\rho^2\,(\mathrm{d}^2\rho/\mathrm{d}\theta^2)$$

Equation (7.8) becomes:

$$\mathrm{d}^2\rho/\mathrm{d}\theta^2 + \rho = \mu/H^2 \tag{7.9}$$

This equation is integrated to give: $\rho = \rho_0\cos(\theta - \theta_0) + \mu/H^2$ (ρ_0 and θ_0 are the constants of integration) which, replacing ρ by $1/r$, can be written:

$$r = (\mu/H^2)/[1 + \rho_0(H^2/\mu)\cos(\theta - \theta_0)],$$

hence:

$$r = p/[1 + e\cos(\theta - \theta_0)] \qquad \text{(m)} \tag{7.10}$$

with $p = (H^2/\mu)$ and $e = (\rho_0 H^2/\mu)$.

This is the equation in polar co-ordinates of a conic section with focus at the origin O, radius vector r and argument θ with respect to an axis making an angle θ_0 with the radius vector r_0 of minimum length along an axis which is an axis of symmetry of the conic section.

7.1.3.4 *Trajectories*

Equation (7.10), for the value $(\theta - \theta_0) = 0$, leads to: $r_0 = p/(1+e)$

from which: $e = (p/r_0) - 1 = (H^2/\mu r_0) - 1$

with $H = r_0 V_0$, since the velocity V_0 is perpendicular to the minimum radius vector.
The quantity e can thus be written:

$$e = (r_0 V_0^2/\mu) - 1 \tag{7.11}$$

The type of conic section depends on the value of e:

For $e = 0$, $V_0 = \sqrt{(\mu/r_0)}$, the trajectory is a circle.
For $e < 1$, $V_0 = \sqrt{(2\mu/r_0)}$, the trajectory is a ellipse.
For $e = 1$, $V_0 = \sqrt{(2\mu/r_0)}$, the trajectory is a parabola.
For $e > 1$, $V_0 = \sqrt{(2\mu/r_0)}$, the trajectory is a hyperbola.

Only values of $e < 1$ correspond to a closed trajectory around the earth and are thus of use for communication satellites. Values of $e \geqslant 1$ correspond to trajectories which lead to the satellite freeing itself from terrestrial attraction (probes).

7.1.3.5 *Energy of the satellite in the trajectory*

The concept of the energy of the satellite in the trajectory is introduced by setting the variation of potential energy between the current point on the trajectory and the point chosen as the origin (such as the extremity of the minimum length radius vector) equal to the variation of kinetic energy between these two points:

$$(1/2)m(V^2 - V_0^2) = m\mu[(1/r) - (1/r_0)], \qquad \text{hence:}$$

$$(V_0^2/2) - \mu/r_0 = (V^2/2) - \mu/r = E \qquad \text{(J)} \tag{7.12}$$

E is a constant which is equal, for unit mass ($m = 1$), to the sum of the kinetic energy $V^2/2$ and the potential energy $-\mu/r$ (equal to the potential μ/r with a change of sign). This is the total energy of the system.

7.1.4 Orbital parameters

The orbits of communication satellites are thus in general ellipses defined in the orbital plane by the equation:

$$r = p/[1 + e\cos(\theta - \theta_0)] \qquad \text{with } e < 1$$

7.1.4.1 *Shape parameters: semi-major axis and eccentricity*

The radius vector r is maximum for $\theta - \theta_0 = \pi$ and corresponds to the apogee of the orbit:

$$r_A = p/(1 - e) \qquad \text{(m)} \tag{7.13}$$

The radius vector r_p corresponding to the perigee of the orbit is that of minimum length r_0 ($r_p = r_0$).

The sum $r_p + r_A$ represents the *major axis* of the ellipse of length $2a$. Adding (7.11) and (7.13) gives:

$$a = 1/2(r_p + r_A) = p/(1 - e^2) \qquad \text{(m)} \qquad (7.14)$$

from which:

$$H^2/\mu = p = a\,(1 - e^2).$$

By putting $\theta - \theta_0$ equal to v, the equation of the ellipse becomes:

$$r = a(1 - e^2)/(1 + e \cos v) \qquad \text{(m)} \qquad (7.15)$$

The *eccentricity* (e) and *semi-major axis* (a) parameters appear and these define the shape of the orbit.

The eccentricity e can be written:

$$e = (r_A - r_p)/(r_A + r_p) \qquad (7.16a)$$

Also:

$$r_p = a(1 - e) \qquad (7.16b)$$

$$r_A = a(1 + e) \qquad (7.16c)$$

7.1.4.2 *Energy and velocity of the satellite*

The energy $E = (V_0^2/2) - \mu/r_0$ can be written $E = (H^2 - 2\mu r_0)/2r_0^2$ since $H = V_0 r_0$.

From (7.14), the semi-major axis can be written in the form:

$$a = \mu r_0^2/(2\mu r_0 - H^2)$$

and hence the *energy* E has a value:

$$E = -\mu/2a \qquad \text{(J)} \qquad (7.17)$$

Introducing the expression for E into (7.12) gives $(V^2/2) - \mu/r = -\mu/2a$, which leads to an expression for the *velocity* V of the satellite:

$$V = \sqrt{\mu[(2/r) - (1/a)]} \qquad (\text{m s}^{-1}) \qquad (7.18a)$$

where: $\mu = GM = 3.986 \times 10^{14}4\ \text{m}^3\,\text{s}^{-2}$ and r is the distance from the satellite to the centre of the earth.

In the case of a circular orbit ($r = a$), the velocity is constant:

$$V = \sqrt{(\mu/a)} \qquad (\text{m s}^{-1}) \qquad (7.18b)$$

7.1.4.3 *Period of the orbit*

The duration of rotation of the satellite in the orbit, or *period* (T) is related to the area Σ of the ellipse by the law of areas which leads to $\Sigma = HT/2$. From (7.14), $H = \sqrt{[a\mu(1 - e^2)]}$. The area

Table 7.1 Altitude, radius, period and velocity for some examples of circular orbits

Altitude (km)	Radius (km)	Period (s)	Velocity ($m\,s^{-1}$)
200	6 578	5 309	7 784
290	6 668	5 419	7 732
800	7 178	6 052	7 450
20 000	26 378	42 636	3 887
35 786	42 164	86 164	3 075

of the ellipse is also given by $\pi a^2\sqrt{(1-e^2)}$. Hence:

$$T = 2\pi\sqrt{(a^3/\mu)} \qquad \text{(s)} \tag{7.19}$$

Some examples of values of the period T and velocity V for a circular orbit as a function of satellite altitude are given in Table 7.1 (the radius of the earth R_E is taken as 6378 km).

7.1.4.4 *Position of the satellite in the orbit—anomalies*

In the plane of the orbit, using the notation of Figure 7.3, the equation of the orbit in polar co-ordinates is given by equation (7.15):

$$r = a(1-e^2)/(1+e\cos v) \qquad \text{(m)}$$

True anomaly (v). The position of the satellite is determined by the angle v called the true anomaly, an angle counted positively in the direction of movement of the satellite from 0 to 360°, between the direction of the perigee and the direction of the satellite.

Eccentric anomaly (E). The position of the satellite can also be defined by the eccentric anomaly E, which is the argument of the image in the mapping which transforms the elliptical trajectory into its principal circle (see Figure 7.3). The true anomaly v is related to the eccentric anomaly E by:

$$\cos v = (\cos E - e)/(1 - e\cos E) \tag{7.20a}$$

and by:

$$\tan(v/2) = \sqrt{[(1+e)/(1-e)]}\tan(E/2) \tag{7.20b}$$

Conversely, the eccentric anomaly e is related to the true anomaly v by:

$$\tan(E/2) = \sqrt{[(1-e)/(1+e)]}\tan(v/2) \tag{7.20c}$$

and by:

$$\cos E = (\cos v + e)/(1 + e\cos v) \tag{7.20d}$$

Finally, the following relation avoids singularities in the calculations:

$$\tan[(v-E)/2)] = (A\sin E)/(1 - A\cos E) = (A\sin v)/(1 + A\cos v) \tag{7.20e}$$

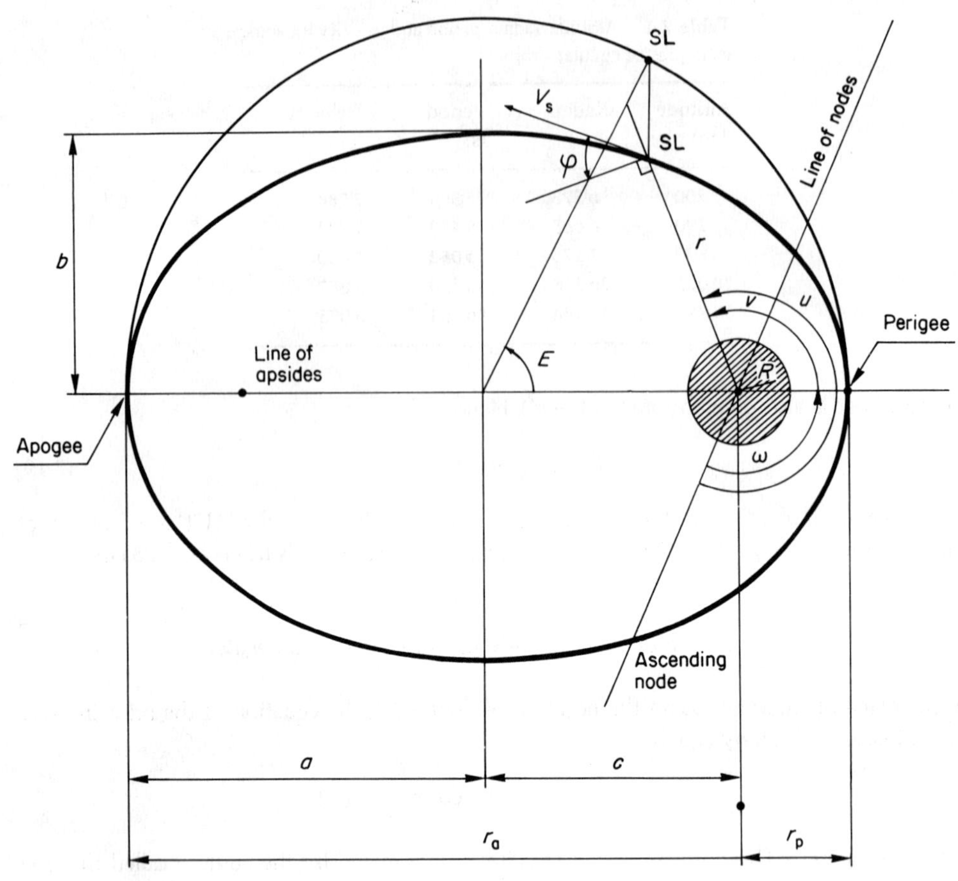

Figure 7.3 Parameters which define the form of the orbit. $b = a\sqrt{(1-e)^2}$, $c = \sqrt{(a^2 - b^2)}$, $e = c/a$.

with:

$$A = e/[1 + \sqrt{(1 - e^2)}]$$

The distance r of the satellite from the centre of the earth can be written:

$$r = a(1 - e \cos E) \qquad \text{(m)} \tag{7.21}$$

Mean movement (n) It is permissible to define the mean movement of the satellite n as the mean angular velocity of the satellite of period T in its orbit:

$$n = 2\pi/T \qquad \text{(rad/s)} \tag{7.22}$$

Mean anomaly (M). The position of the satellite can thus be defined by the mean anomaly M which will be the true anomaly of a satellite in a circular orbit of the same period T. The mean anomaly is expressed by:

$$M = (2\pi/T)(t - t_p) = nt - M_0 \qquad \text{(rad)} \tag{7.23}$$

where t_p is the instant of passing through the perigee. The mean anomaly is related to the eccentric anomaly by Kepler's equation:

$$M = E - e \sin E \qquad \text{(rad)} \tag{7.24}$$

7.1.4.5 *Position of the orbital plane in space*

The position of the orbital plane in space is specified by means of two parameters—the *inclination i* and the *longitude of the ascending node* Ω. These parameters are defined as shown in Figure 7.4 with respect to a reference frame whose origin is the centre of mass of the earth, whose Oz axis is in the direction of terrestrial angular momentum (the axis of rotation normal to the equatorial plane), whose Ox axis (normal to Oz) in the orbital plane is oriented in the direction of a reference defined below and whose Oy axis in the orbital plane is such that the reference system is regular.

Inclination at the plane of the orbit (i). This is the angle of the ascending node, counted positively in the forward direction between 0 and 180°, between the normal (directed towards the east) to the line of nodes in the equatorial plane and the normal (in the direction of the velocity) to the line of nodes in the orbital plane. This is also the angle at the centre of the reference frame between the angular momentum **H** of the orbit and the Oz axis (the direction of the pole).

Right ascension of the ascending node (Ω). This is the angle taken positively from 0 to 360° in the forward direction, between the reference direction and that of the ascending node of the orbit (the intersection of the orbit with the plane of the equator, the satellite crossing this plane from south to north).

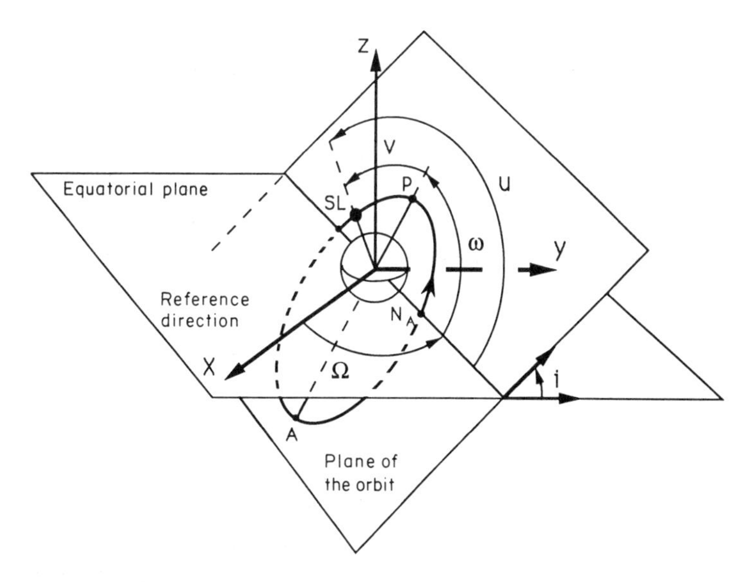

Figure 7.4 Positioning of the orbit in space.

The reference direction is given by the line of intersection of the equatorial plane and the plane of the ecliptic, oriented positively towards the sun (see Section 7.1.5.2 and Figure 7.5). With the Keplerian assumptions for the orbit of the earth around the sun, this line (which is contained in the equatorial plane) maintains a fixed orientation in space with time and passes through the sun at the spring equinox thereby defining the direction of the vernal point (γ).

In reality, the irregularities of terrestrial rotation (Section 7.1.5.3) cause the direction of intersection of the planes to very somewhat; the reference defined in this way is therefore not inertial and does not permit orbital motions to be integrated. Also, arbitrary axes may be defined, for example the position of the reference frame at a particular date. The date usually adopted is noon on 1 January 2000. On this date, the track of the line considered on the celestial arch (a sphere of infinite radius centred on the earth) defines the point J_{2000}. The Veis reference is also used in which the Oz axis is the axis from the centre of the earth to the north pole and the Ox axis is the projection at the date considered on the equatorial plane of the intersection of the equatorial plane and the plane of the ecliptic at 00.00 hours on 1 January 1950. This projection defines the pseudovernal point γ_{50} (so-called since it is not within the plane of the ecliptic). The Veis reference has the advantage of permitting simple transformation, by a single rotation, to the terrestrial reference in which the earth stations are located [ZAR-87].

7.1.4.6 *Location of the orbit in its plane*

The orientation of the orbit in its plane is defined by the *argument of the perigee* ω. This is the angle, taken positively from 0 to 360° in the direction of motion of the satellite, between the direction of the ascending node and the direction of the perigee (Figure 7.4).

7.1.4.7 *Conclusion*

A knowledge of the five parameters (a, e, i, Ω and ω) completely defines the trajectory of the satellite in space. The motion of the satellite in this trajectory can be defined by one of the anomalies (v, E or M).

The *nodal angular elongation* u can also be used to define the position of the satellite in its orbit. This is the angle taken positively in the direction of motion from 0 to 360° between the direction of the ascending node and the direction of the satellite: $u = \omega + v$ (Figure 7.4). This parameter is useful in the case of a circular orbit where the perigee is unknown.

7.1.5 The earth's orbit

7.1.5.1 *The earth*

In the Keplerian hypotheses, the earth is assumed to be a spherical and homogeneous body. The real earth differs from this primarily by a flattening at the poles. The terrestrial surface is equivalent, to a first approximation, to that of an ellipsoid of revolution about the line of the poles whose parameters depend on the model chosen [HUS-80]. The International Astronomical Union has recommended a value of 6378.144 km (mean equatorial radius R_E) for the semi-major axis since 1976, and for the oblateness $A = (a - b)/a$, the value 1/298.257 (b is the semi-minor axis).

7.1.5.2 *Motion of the earth about the sun*

The earth rotates around the sun (Figure 7.5) with a period of approximately 365.25 days following an ellipse of eccentricity 0.01673 and semi-major axis 149 597 870 km. This defines the astronomical unit of distance (AU). Around 2 January, the earth is nearest to the sun (the perihelion) while around 5 July it is at its aphelion (around 152 100 000 km).

The plane of the orbit is called the plane of the ecliptic. The plane of the ecliptic makes an angle of 23.44° (the obliquity of the ecliptic which decreases around 47″ per century) with the mean equatorial plane.

The apparent movement of the sun around the earth with respect to the equatorial plane is represented by a variation of the declination of the sun (the angle between the direction of the sun and the equatorial plane, see Section 7.1.5.4). The declination varies during the year between +23.44° (at the summer solstice) and −23.44° (at the winter solstice). The declination is zero at the equinoxes. The direction of the sun at the spring equinox defines the vernal point or the γ point on the celestial sphere (the geocentric sphere of infinite radius). The sun passes through it from the southern hemisphere to the northern hemisphere and the declination is zero becoming positive.

The relation between the declination of the sun δ and the date is obtained by considering the apparent movement of the sun about the earth in an orbit of ellipticity e equal to 0.01673, inclined at the equator with obliquity ε. Hence (Figure 7.6):

$$\sin\delta = \sin\varepsilon \sin u \tag{7.25}$$

with $\sin\varepsilon = \sin 23.44° = 0.39795$ and u, the nodal elongation of the sun, equal to the sum of the true anomaly of the sun and the argument of the perigee ω_{SUN}. The argument of the perigee of the orbit representing the apparent movement of the sun about the earth remains more or less constant through the years if the precession of the equinoxes is neglected and has a value around 280°.

The true anomaly of the sun is expressed as a function of its eccentric anomaly E_{SUN} by means of (7.20) and the eccentric anomaly as a function of the mean anomaly M_{SUN} by Kepler's equation (7.24). The mean anomaly is related to time by $M_{SUN} = n_{SUN}(t - t_0)$, with n_{SUN} the mean movement of the sun such that:

$$n_{SUN} = 2\pi/365.25 \text{ rad/day} = 360°/365.25 = 0.985\,626°/\text{day}$$

and t_0 the date of passing through the perihelion (about 2 January). Variation of declination with date is shown in Figure 7.7.

7.1.5.3 *Rotation of the earth*

The axis of rotation of the earth cuts its surface at the poles. The pole moves slightly with time (within a circle of about 20 metres diameter) with respect to the surface of the earth. The axis of rotation also moves in space. Its movement is a combination of periodic terms of

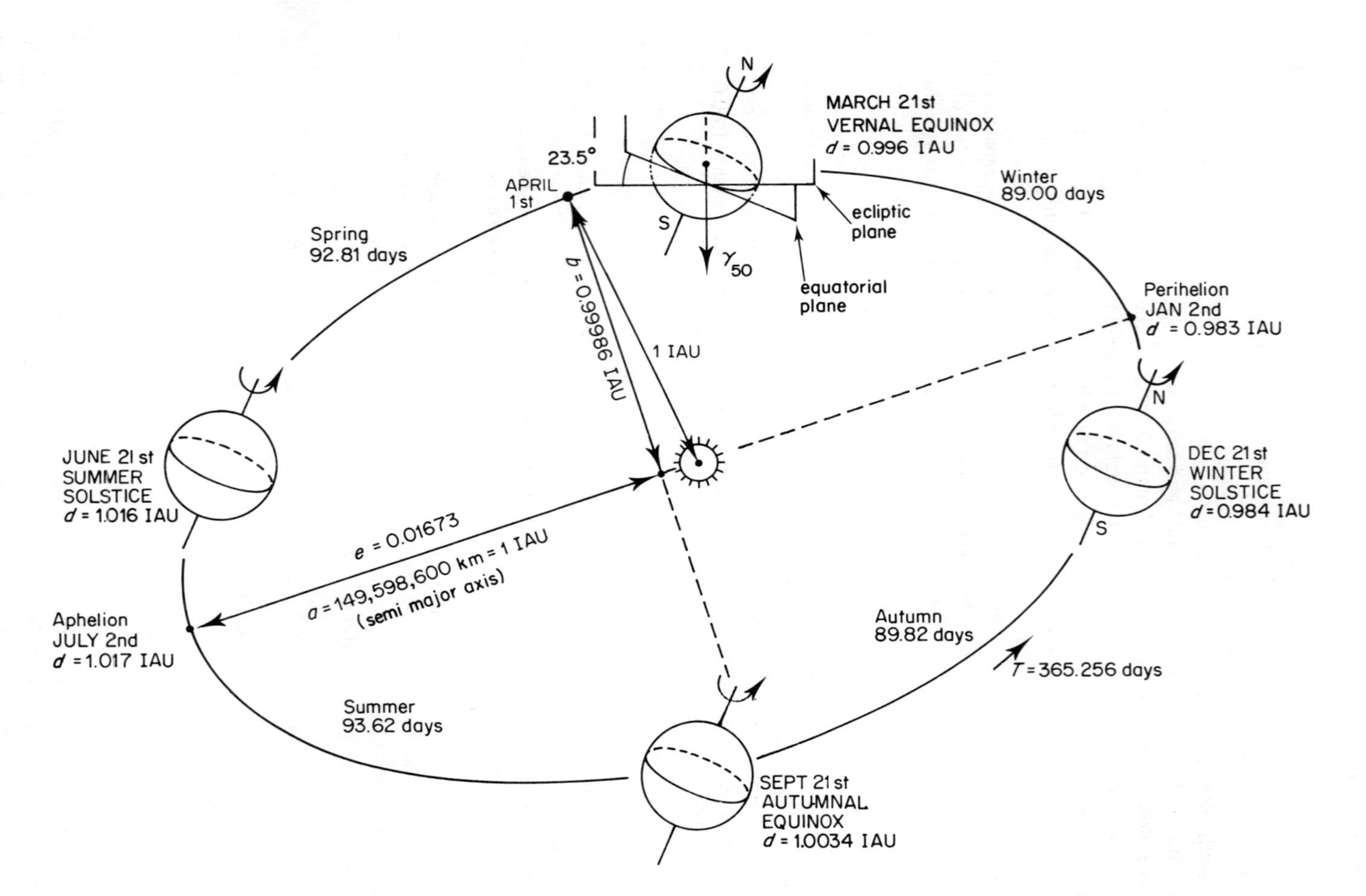

Figure 7.5 Orbit of the earth round the sun.

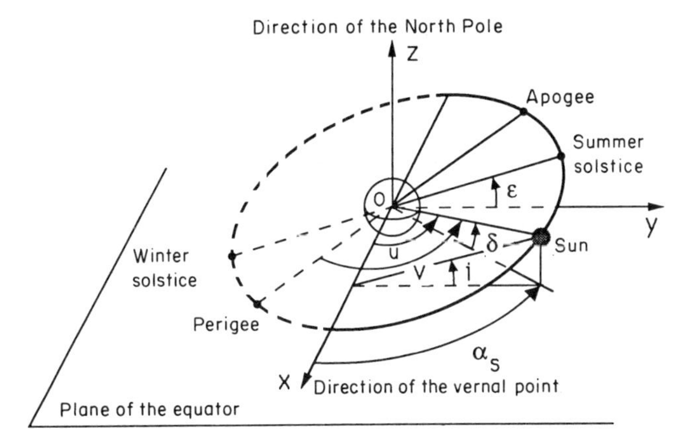

Figure 7.6 Apparent movement of the sun about the earth.

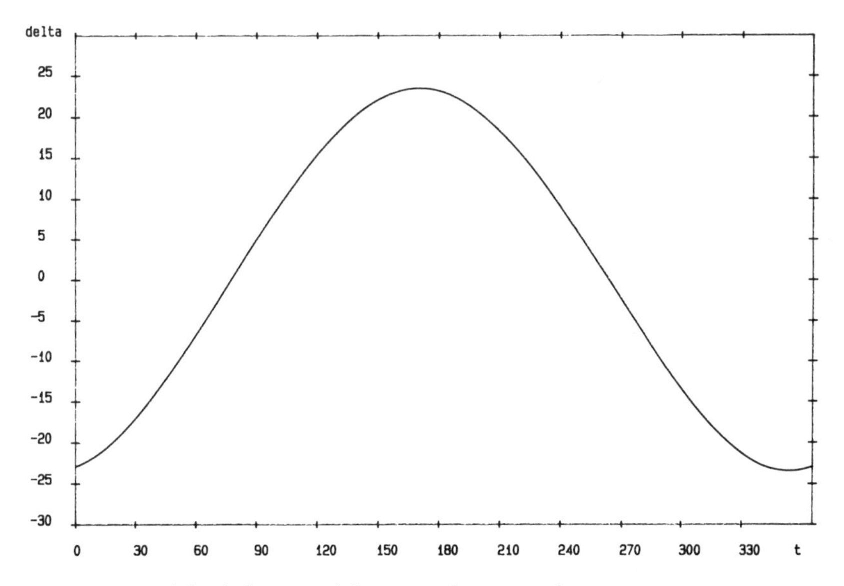

Figure 7.7 Variation of the declination of the sun in the course of a year.

limited amplitude (less than 20 seconds of arc), nutation, other non-periodic terms which have a cumulative effect and precession; precession relates to the angular momentum of the earth describing a cone, in 25 800 years, about the pole of the ecliptic (the axis normal to the plane of the ecliptic). These movements involve a variation of the plane of the equator and consequently of the point γ which thus follows the equator in the reverse direction at the rate of approximately 50″ per year. Elimination of the periodic terms permits definition of the *mean equator*, this term is also applied with the same definition to various elements (co-ordinates, planes, times etc.) which are affected by the irregularities of the rotation of the earth.

7.1.5.4 *Terrestrial, equatorial and temporal co-ordinates*

Terrestrial co-ordinates. The terrestrial co-ordinates of a location are defined by:

—The *geographical longitude* λ, the angle, in the equatorial plane, between the origin meridian and the meridian of the location, taken positively towards the east from 0° to 360°, as recommended since 1982 by the International Astronomical Union (it should be noted that this convention is not universal).
—The *geographical latitude* φ, the angle between the vertical at the location and the plane of the equator, expressed in degrees from $-90°$ (South Pole) to $+90°$ (North Pole). The meridian of the location is the intersection of the half plane passing through the line of poles containing the location and the terrestrial surface. The origin meridian of longitude is the international meridian called Greenwich.

If the earth is considered to be spherical, the vertical (the perpendicular to the local horizontal plane) of any location passes through the centre of the earth. The flattening of the earth causes the vertical of a location of latitude other than 0° or 90° to no longer pass exactly through the centre of the earth. Hence, the *geographical latitude* φ is different from the *geocentric latitude* φ' (the angle between the geocentric direction of the location and the equatorial plane). These two quantities are related by:

$$R_E^2 \tan \varphi' = b^2 \tan \varphi \tag{7.26a}$$

where b is the semi-major axis of the ellipsoid and R_E the mean equatorial radius. The distance R_C from the location to the centre of the earth is given by an approximation to the equation of the ellipse [PRI-86]

$$R_C = R_c(1 - A \sin^2 \varphi') \tag{7.26b}$$

where A is the oblateness of the earth.

Equatorial co-ordinates. The equatorial co-ordinates of a direction having its origin at the centre of the earth (the geocentric direction) are defined by:

—The *right ascension* α, taken positively in the equatorial plane from the direction of the point γ in the forward direction (that of the rotation of the earth) to the projection on the meridional plane (containing the direction and the line of poles),
—The *declination* ∂, the angle between the direction and the equatorial plane, taken positively towards the north.

Hour co-ordinates. The local hour co-ordinates of a direction are defined by:

—The *hour angle* H, taken positively in the equatorial plane in the reverse direction (towards the west) from the meridional plane passing through the location towards the meridian of the direction.
—The *declination* ∂. H is most often measured in hours (1 h = 15°, 1 min = 15′, 1 s = 15″ and conversely 1° = 4 min, 1′ = 4 s, 1″ = 0.07 s). As terrestrial rotation is direct, a fixed direction in space thus sees an hour angle which increases with time while its declination remains fixed.

The co-ordinates defined above are geocentric co-ordinates. One can also define *topocentric co-ordinates*, that is the co-ordinates of the direction of a point in space from a particular location on the surface of the earth. These co-ordinates are defined by using the plane parallel to the equator passing through the location as a reference plane. Topocentric co-ordinates differ from geocentric co-ordinates because of the *parallax* which is the angle through which the terrestrial radius of the location is seen from the point in space considered.

Ecliptic co-ordinates are the *celestial longitude* and the *celestial latitude* of a direction; the reference plane is the ecliptic instead of the equatorial plane.

Finally, the *horizontal co-ordinates* of a direction, used by astronomers, are the *azimuth*, the angle taken in the horizontal plane of the location in the reverse direction from the south towards the projection of the direction and the *height*, the angle between the direction and the horizontal plane. The *zenithal distance*, equal to the 90° complement of the height is also used. To define the pointing direction of earth station antennas, it is customary to reckon the *azimuth* from the *north* and to call the height the *elevation*.

Various expressions permit conversion from one set of co-ordinates to another; they can be found in most books on astronomy (e.g. [BDL-90]).

Sidereal time. The hour angle of the point γ is called the *local sidereal time* (LST). For a fixed direction, $H - \text{LST}$ is constant and is such that (Figure 7.8a):

$$H = \text{LST} - \alpha \qquad \text{(degrees or h min s)} \tag{7.27}$$

The *sidereal time* ST is the local sidereal time of the *international meridian*. If λ is the geographical longitude of the location (positive towards the east):

$$\text{ST} = \text{LST} - \lambda \qquad \text{(degrees or h min s)} \tag{7.28}$$

The sidereal time (of the Greenwich meridian) increases with time. Neglecting perturbations caused by variations of the fundamental planes (an error less than a hundredth of a degree) gives:

$$\text{ST} = \text{ST}_0 + \Omega_E t \tag{7.29}$$

where ST_0 is the sidereal time (of Greenwich) at 00.00 h universal time (UT) on 1 January of each year (for example 100.383° for 1990, see expression (7.33)). Ω_E is the velocity of rotation of the earth = 15.04169°/h.

Solar time. True local solar time (TT) is the hour angle of the centre of the sun. True solar time has a value 0 h when the sun passes through the meridian of the location. Mean solar time (MT) is the solar time corrected for the periodic variations ΔE associated with the irregularities of the movement of the earth. Hence:

$$\text{MT} = \text{TT} - \Delta E \tag{7.30}$$

where ΔE is called the time equation. The time equation takes account of the relative movement of the sun in an elliptical orbit with respect to the earth and the effect of the obliquity of the ecliptic. An approximate expression giving the time equation ΔE is [NOU-83]:

$$\Delta E = 460 \sin n_{\text{SUN}} t - 592 \sin 2\,(\omega_{\text{SUN}} + n_{\text{SUN}} t) \qquad \text{(s)} \tag{7.31}$$

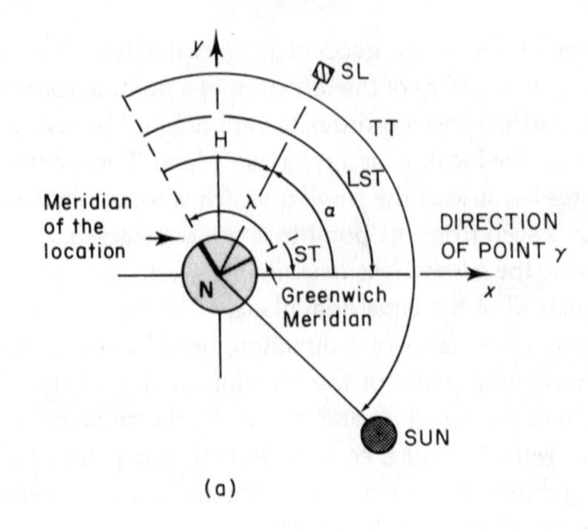

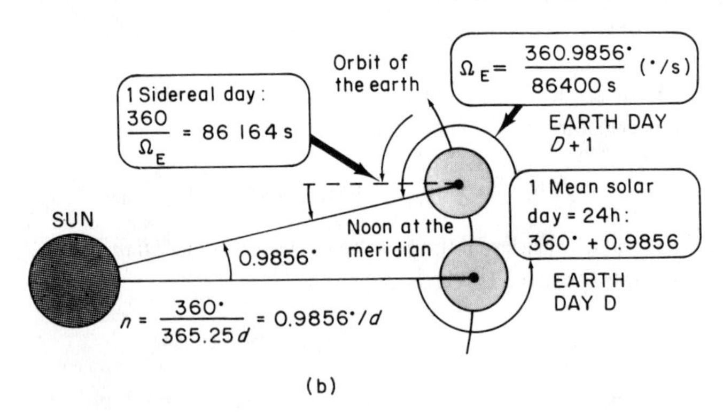

Figure 7.8 Space and time reference. (a) Definition of angles. (b) Sidereal day and mean solar day.

where t is the time in days from passing through the perihelion (around 2 January). The maximum value of ΔE is 4/15 h or 16 min.

7.1.5.5 *Time references*

Sidereal day. The various solar and sidereal times are, in spite of their name, angles. Successive returns of a fixed star, or the point γ, to the meridian of a location define a time scale in *true sidereal days*. After elimination of the periodic terms, the *mean sidereal day* is obtained. This mean sidereal day defines the period T_E of rotation of the earth and has a value of 23 h 56 min 4.1 s or 86 164.1 s.

Solar day. Successive returns of the sun to the meridian of a location provide a time scale in *true solar days* and, by elimination of the periodic terms, in *mean solar days* of duration 24 h or 86 400 s.

The sidereal day and the solar day differ because of the rotation of the earth around the sun which has a mean value of 0.9856° per day (Figure 7.8b). A time interval measured in sidereal time must be multiplied by 86 164.1/86 400 or 0.9972696 to obtain a measurement in mean time. Conversely, a time interval measured in mean time must be multiplied by 86 400/86 164.1 or 1.002 7379 to obtain a measurement in sidereal time.

Civil time and universal time. *Civil time* is mean solar time increased by 12h (the civil day starts 12 hours later than the mean solar day). To define a time which is independent of location, the civil time at Greenwich, or *universal time* (UT) (incorrectly called Greenwich Mean Time (GMT)) is used.

Universal time (UT) is related to mean sidereal time by the mean sidereal time equation which takes different forms according to the chosen reference. Newcomb's equation (used until the introduction of reference J_{2000}) gives:

$$\mathrm{ST} = \mathrm{UT} + 6\,\mathrm{h}\ 38\,\mathrm{min}\ 45.836\,\mathrm{s} + 8640\,184.542 \times T\,\mathrm{s} + 0.0929 \times T^2\,\mathrm{s} \qquad (7.32\mathrm{a})$$

where T is the number of Julian centuries (36 525 days) between the UT date and 31 December 1899 at midday UT (Julian day 2415 020).

The form compatible with reference J_{2000} is [ZAR-87]:

$$\mathrm{ST} = \mathrm{UT} + 6\,\mathrm{h}\ 41\,\mathrm{min}\ 50.548\,41\,\mathrm{s} + 864\,0184.812\,866 \times T\,\mathrm{s} + 0.093\,104 \times T^2\,\mathrm{s} - 6.2 \times 10^{-6} \times T^3\,\mathrm{s} \qquad (7.32\mathrm{b})$$

where T is the number of Julian centuries (36 525 days) between the UT date and 1 January 2000 at midday UT (Julian day 2451 545).

The Julian calendar starts at midday on 1 January 4713 BC and constitutes a system of numerical representation. Hence the day which starts at midday on 1 January 1990 is numbered 24 478 963. The sidereal time (at Greenwich) ST_0 on 1 January at 00.00h UT of each year is obtained from the previous expressions.

Example calculation. For 1 January 1990 with Newcomb's formula:

On 1 January 2000 at 00.00h, $T = 36\,524.5/36\,525$ Julian centuries,
On 1 January 1990 at 00.00h, $T = [36\,524.5 - (10 \times 365 + 2)]/36\,525 = 0.9$ century.

From which $\mathrm{ST}_0 = 6\,\mathrm{h}\ 38\,\mathrm{min}\ 45.836\,\mathrm{s} + 8640\,184.542\,\mathrm{s} \times 0.9 + 0.0929\,\mathrm{s} \times 0.9^2$
$= 6.6922$ hours (modulo 24h) $= 100.383°$.

The results for subsequent years are presented in Table 7.2.

Legal time and official time. Each country, in accordance with its region of longitude, uses a time, *legal time*, which is derived from universal time by correction by a whole number of hours according to regulations defined by law. Finally, economic considerations lead to correction of legal time according to the season (summer time) which gives the *official time*.

Table 7.2 Sidereal time ST_0 on 1 January at 00.00h UT

Year:	1991	1992	1993	...	1995	...	2000
ST_0:	100.144°	99.906°	100.653°		100.175°		99.967°

7.1.6 Earth–satellite geometry

7.1.6.1 *The satellite track*

The satellite track on the surface of the earth is the locus of the point of intersection of the earth centre–satellite vector with the surface of the earth. The track takes account of the movement of the surface of the earth with respect to the actual displacement (as a function of the true anomaly) of the earth centre–satellite vector.

The equation of the track of an orbit of fixed inclination and ellipticity is obtained from the following procedure.

From Figure 7.9, the co-ordinates (λ_{SL}, φ) of the satellite SL on the earth initially assumed to be fixed are related by the following equation (the longitude is taken with respect to a reference meridian):

$$\tan \varphi = \tan i \sin (\lambda_{SL} - \lambda_N) \tag{7.33}$$

where φ is the latitude of the satellite, λ_{SL} is the longitude with respect to a reference meridian (fixed earth), λ_N is the longitude of the node with respect to the reference meridian, and i is the inclination of the orbit.

The arc N–SL is the track of the satellite (on the fixed earth). The arc N–SL subtends an angle u (the nodal angular elongation) such that:

$$\sin \varphi = \sin i \sin u \tag{7.34a}$$

and such that:

$$\tan (\lambda_{SL} - \lambda_N) = \tan u \cos i \tag{7.34b}$$

The point on the track of highest latitude is called the vertex, the longitude of this differs from that of the node by $\pi/2$: $\lambda_v = \lambda_N + 90°$.

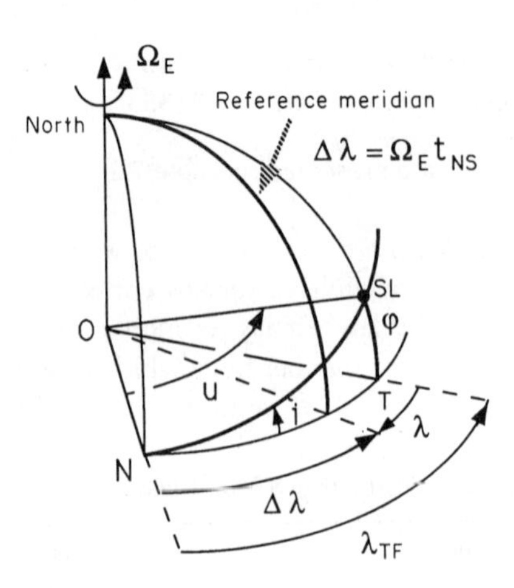

Figure 7.9 Definition of the track of a satellite.

It is necessary to take account of the rotation of the reference meridian during the satellite trajectory. Let Δt be the time elapsed since the passage of the satellite through the reference meridian (with a fixed earth) until the present time:

$$\Delta t = t_S - t_0$$

where t_S is the time elapsed since the passage of the satellite through the perigee, t_0 is the time of passage through the origin position of the reference meridian.

The time of passage through the perigee t_p is chosen as the time origin ($t_p = 0$). As a consequence of the rotation of the earth, the displacement $\Delta\lambda$ of the reference meridian towards the east during time Δt has a value $\Delta\lambda = \Omega_E \Delta t$, where Ω_E is the angular velocity of the earth:

$$\Delta\lambda = \Omega_E \Delta t = \Omega_E (t_S - t_0) = M(\Omega_E/n) - M_0(\Omega_E/n) \qquad (7.35)$$

where n is the mean movement of the satellite, M is the mean anomaly of the satellite and M_0 is the mean anomaly of the position of the origin of the meridian.

The relative longitude λ of the satellite with respect to the rotating reference meridian can thus be written:

$$\lambda = \lambda_{SL} - \Delta\lambda \qquad (7.36)$$

Longitude of the track relative to the meridian of the ascending node. If the origin position of the reference meridian is the ascending node, the longitude λ_N of the node is zero. Equations (7.32) and (7.34) become:

$$\tan\varphi = \tan i \sin\lambda_{SL}$$
$$\tan\lambda_{SL} = \tan u \cos i$$

Furthermore, M_0 is equal to M_N the mean anomaly of the ascending node. The relative longitude with respect to the satellite meridian on passing through the ascending node can thus be expressed:

$$\lambda = \lambda_{SL} - \Delta\lambda = \arcsin[(\tan\varphi)/(\tan i)] - M(\Omega_E/n) + M_N(\Omega_E/n) \qquad (7.37a)$$

or:

$$\lambda = \lambda_{SL} - \Delta\lambda = \arctan[(\tan u)/(\cos i)] - M(\Omega_E/n) + M_N(\Omega_E/n) \qquad (7.37b)$$

M and M_N, the mean anomalies of the satellite and the ascending node, are calculated from the eccentric anomalies by using Kepler's equation (7.24).

The latitude of the satellite can be eliminated from equation (7.37a) by using equation (7.33). Also, $u = \omega + v$ (modulo 2π). This gives:

$$\lambda = \arcsin\{\sin(\omega + v)\cos i[1 - \sin^2 i \sin^2(\omega + v)]^{-1/2}\}$$
$$- [(\Omega_E/n)(E - e\sin E) + (\Omega_E/n)(E_N - e\sin E_N)] \qquad (7.38a)$$

or:

$$\lambda = \arctan[\tan(\omega + v)(\cos i)] - [(\Omega_E/n)(E - e\sin E) + (\Omega_E/n)(E_N - e\sin E_N)] \quad (7.38b)$$

where E is the eccentric anomaly of the satellite and E_N is that of the ascending node.

Latitude of the satellite. The latitude φ is not modified by the rotation of the earth and therefore does not depend on the choice of the origin position of the reference meridian. It can be written:

$$\varphi = \arc\sin[\sin i \sin(\omega + v)] \tag{7.39}$$

λ and φ can be expressed as a function of only one of the parameters E or v by using one of equations (7.21).

For certain orbits, the mean movement n of the satellite is chosen to be equal to a multiple of the rank m of the angular velocity Ω_E of the earth. In the absence of perturbations, the track is thus unique (that is the satellite passes through the same points again after m revolutions) and fixed with respect to the earth. In practice, it is necessary to take account of the precession of the plane of the orbit (the drift of the right ascension of the ascending node under the effect of the asymmetry of the terrestrial potential, see Section 7.2.1).

7.1.6.2 *Satellite distance*

Distance of the satellite from a point on the earth. The co-ordinates of the satellite in Figure 7.10 are φ for the latitude (the centre angle is TOA, T being the sub-satellite point) and λ for the longitude with respect to a reference meridian. Those of the point P considered are l for the latitude (centre angle POB) with ψ for the longitude with respect to the same reference meridian. For clarity of the figure, only the difference in longitude $L = \psi - \lambda$ between the point P and the satellite (centre angle AOB) is represented. The centre angle BOT has a value ζ and the centre angle POT (in the plane of the earth centre, the satellite SL and point P) has a value ϕ. Let R be the distance from the satellite to point P, r the distance from the satellite to the earth centre and R_E the radius of the earth.

Consider the triangle OPS (S is used instead of SL for simplicity). This gives: $E^2 = R_E^2 + r^2 - 2R_E r \cos\phi$, hence:

$$R = \sqrt{(R_E^2 + r^2 - 2R_E r \cos\phi)} \tag{7.40}$$

It remains to evaluate $\cos\phi$.

In the spherical triangle TPB, the cosine law gives:

$$\cos\phi = \cos\zeta \cos l + \sin\zeta \sin l \cos \mathrm{PBT}$$

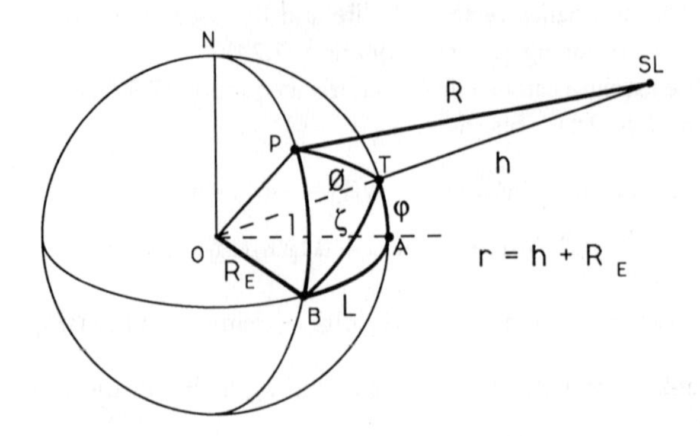

Figure 7.10 Earth–satellite geometry.

The sine rule in the triangle TAB gives:

$$(\sin \text{TAB})/(\sin \zeta) = (\sin \text{TBA})/(\sin \varphi), \text{ with TAB} = \pi/2 \text{ and TBA} = (\pi/2) - \text{PBT}$$

Hence: $\sin \zeta \cos \text{PBT} = \sin \varphi$

Furthermore, in the triangle TAB: $\cos \zeta = \cos L \cos \varphi$ from which:

$$\cos \phi = \cos L \cos \varphi \cos l + \sin \varphi \sin l \tag{7.41}$$

The proposed equations assume that the earth is spherical of radius R_E (equal mean equatorial radius). For a more precise calculation, it would be convenient to define the actual radius from the reference ellipsoid (see Section 7.1.5.1) and to use the geocentric latitude φ' of the point which is obtained from the geographic latitude φ by equation (7.26).

Satellite altitude. The altitude h of the satellite corresponds to its distance from the sub-satellite point (the distance ST in Figure 7.10). This gives:

$$h = r - R_E \tag{7.42}$$

7.1.6.3 *Satellite location—elevation and azimuth*

Two angles are necessary to locate the satellite from the point P on the surface of the earth. It is customary to use the elevation and azimuth angles.

Elevation angle. The elevation angle is the angle between the horizon at the point considered and the satellite, measured in the plane containing the point considered, the satellite and the centre of the earth. This is the angle E in Figure 7.11 where the triangle OPS of Figure 7.10 is represented. It follows, by considering the right angle OP'S (formed by extending the segment OP) and noting that the angle PSP' is equal to E:

$$\cos E = (r/R) \sin \phi \qquad \text{hence } E = \arccos[(r/R)\sin \phi] \tag{7.43a}$$

The distance r is that of the satellite from the centre of the earth, R is that of the satellite from the point P which is calculated using (7.40) and $\sin \phi$ is obtained from (7.41) using: $\sin \phi = \sqrt{(1 - \cos^2 \phi)}$.

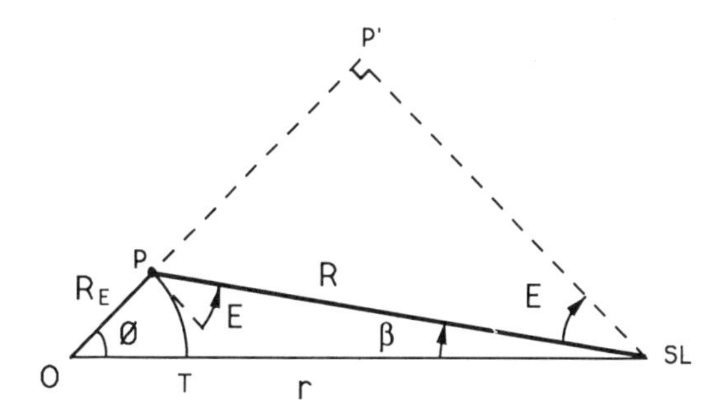

Figure 7.11 Elevation and nadir angles.

The radius of the earth R_E can be introduced with:

$$\tan E = [\cos \phi - (R_E/r)]/\sin \phi \quad (7.43b)$$

Another form is possible:

$$\sin E = [\cos \phi - (R_E/r)]/(R/r) \quad (7.43c)$$

with, from (7.40): $(R/r) = \sqrt{[1 + (R_E/r)^2 - 2(R_E/r)\cos \phi]}$

Azimuth angle. The azimuth angle A is the angle measured in the horizontal plane of the location between the direction of geographic north and the intersection of the plane containing the satellite and the centre of the earth (the plane OPS). This angle varies between 0 and 360 degrees as a function of the relative positions of the satellite and the point considered. It is the angle NPT in the spherical triangle of the same name in Figure 7.10. This gives:

$$(\sin \mathrm{NPT})/[\sin(90 - \varphi)] = (\sin \mathrm{PNT})/(\sin \phi)$$

In the triangle NBA, $(\sin \mathrm{BNA})/(\sin L) = (\sin \mathrm{BAN})/(\sin \mathrm{AON}) = 1$.

From which, since the angle BNA is equal to the angle PNT:

$$\sin \mathrm{NPT} = (\sin L \cos \varphi)/\sin \phi$$

The calculation gives an angle less than $\pi/2$, although in the case of the figure the azimuth is greater than $\pi/2$. This arises from the symmetry properties of the sine function. The result of the calculation is taken as an intermediate parameter, called a, in the determination of the azimuth ($a < \pi/2$). Thus:

$$\sin a = (\sin L \cos \varphi)/\sin \phi \quad (7.44)$$

hence:

$$a = \arcsin[(\sin L \cos \varphi)/\sin \phi] \qquad (\text{with: } \phi > 0,\ L > 0)$$

The true azimuth A is obtained from a in accordance with the position of the sub-satellite point T with respect to the point P. The various cases are summarised in Table 7.3.

7.1.6.4 *Nadir angle*

In Figure 7.11, the angle at the satellite SL between the direction of the centre of the earth O

Table 7.3 Determination of the azimuth A

Position of the sub-satellite point T with respect to point P	Relation between A and a
South-east	$A = 180° - a$
North-east	$A = a$
South-west	$A = 180° + a$
North-west	$A = 360° - a$

and the direction of the point P is called the nadir angle β. In the triangle OPS:

$$\sin\beta = (\sin\phi)R_E/R \qquad \text{hence: } \beta = \arc\sin[R_E(\sin\phi)/R] \qquad (7.45a)$$

or, if the elevation angle E is taken into consideration:

$$\phi + \beta + E = \pi/2 \qquad (7.45b)$$

which, with $\sin\beta = [\sin(\pi - \beta - \phi)]/r$, leads to:

$$\sin\beta = (\cos E)R_E/r \qquad \text{hence: } \beta = \arc\sin[R_E(\cos E)/r] \qquad (7.45c)$$

7.1.6.5 *Coverage at a given elevation angle*

The coverage zone can be specified as the region of the earth where the satellite is seen with a minimum elevation angle E. The contour of the coverage zone is defined by a set of ground locations determined by their geographical co-ordinates and hence known by their relative longitude L and latitude l values. The relationship between L and l is as follows:

$$L = \arc\cos[(\cos\phi - \sin\varphi\sin l)/\cos\varphi\cos l] \qquad (7.46)$$

where $\phi = \pi/2 - E - \arc\sin[R_E(\cos E)/r]$.

The longitudinal extent of the coverage zone with respect to the sub-satellite point is obtained by putting $l = \varphi$ in the above expression. The latitudinal extent is equal to ϕ.

7.1.6.6 *Propagation time—the Doppler effect*

Propagation time. The trajectory of radio waves on a link between an earth station and a satellite at distance R requires a propagation time τ equal to:

$$\tau = R/c \qquad \text{(s)} \qquad (7.47)$$

where c is the velocity of light (3×10^8 m/s).

Variation of relative distance—the Doppler effect. When the satellite moves with respect to the earth, the relative distance R from the satellite to a point on the surface of the earth varies. An apparent instantaneous satellite velocity $dR/dt = V_d$ can be defined in accordance with the point considered ($V_d = V\cos\xi$ where ξ is the angle between the direction of the point considered and the velocity V of the satellite). This approach or escape velocity of the satellite will cause, at the receiver, an apparent increase or decrease respectively of the frequency of the radio wave transmitted on the link (the Doppler effect). The phenomenon occurs, of course, on both the uplink and the downlink. The shift Δf_d in the frequency f of the wave on the link can be written:

$$\Delta f_d = V_r f/c = V\cos\xi\,(f/c) \qquad \text{(Hz)} \qquad (7.48)$$

where:

$c = 3 \times 10^8$ = velocity of light (m/s)
f = frequency of the transmitted wave (Hz)
V_r = relative radial velocity of the satellite (m/s)

The geometry of the system changes with the movement of the satellite with respect to the point considered, the apparent velocity of the satellite varies with time and will thus involve a variation of the Doppler shift.

An important parameter which will affect the performance of the automatic frequency control of the receiver system is the rate of variation of frequency $d(\Delta f_d)/dt$:

$$d(\Delta f_d)/dt = d/dt(V_r)\,f/c \qquad (\text{Hz/s}) \qquad (7.49)$$

Detailed calculations are given in [VIL-91]. On an equatorial circular orbit, the maximum value of Doppler shift (when the satellite appears or disappears at the horizon) can be estimated by [CCIR-Rep 214]:

$$\Delta f_d \cong \pm 1.54 \times 10^{-6} f\, m \qquad (\text{Hz})$$

where m is the number of revolutions per day of the satellite with respect to a fixed point on the earth (the period T of the orbit is equal to $24/(m+1)$ hours). For $m = 0$, the period is 24h, the satellite remains fixed with respect to the earth (a geostationary satellite, see Section 7.2.5) and the Doppler shift is theoretically zero. For $m = 3$, the period T has a value of 6h (for an altitude around 11 000 km) and the Doppler shift is of the order of 18 kHz at 6 GHz.

For an elliptical orbit, assuming that the variation of distance R is the same as the variation of radial distance r (the altitude of the satellite is large with respect to the radius of the earth and the station is on the satellite track), the velocity V_r is the radial velocity of the satellite which can be written:

$$dr/dt = (dr/\theta)(d\theta/dt),$$

with $d\theta/dt = H/r^2$ from equation (7.6) and $dr/d\theta$ which is determined from expression (7.10). This gives (with $v = \theta$)

$$V_r = dr/dt = e\sqrt{\mu}/\sqrt{[a(1-e^2)]}\,\sin v \qquad (\text{m/s}) \qquad (7.50)$$

where: e is the eccentricity, μ is GM, a is the semi-major axis and v is the true anomaly of the satellite. The radial velocity is maximum for $v = 90°$.

Apart from the problems posed by tracking variations of incident signal frequency at the receiver, variations of relative distance lead to problems of synchronisation between the signals arriving from different directions (see Section 3.4.5.4). Distance variations also cause variations of propagation time on the link [CCIR-Rep 383].

7.1.7 Eclipses of the sun

An eclipse of the sun occurs for the satellite when it passes into the conical shadow region of the earth or moon. The occurrence and duration of these eclipses depend on the characteristics of the satellite orbit. The consequences of the eclipse on the satellite are of two types. On the one hand, the electric power supply system of the satellite which almost always includes solar cells to convert solar energy into electrical energy by the photovoltaic effect must make use of an alternative energy source. On the other hand, as the satellite is no longer heated by the sun, the thermal equilibrium conditions of the satellite will be greatly modified and the temperature tends to decrease rapidly.

7.1.7.1 *Eclipses of the sun by the moon*

The orbit of the moon around the earth, with a semi-major axis of 384 400 km and a period of 27 days, has an inclination of 5.14° with respect to the ecliptic. The right ascension of the ascending node on the ecliptic is also affected by a precession in the reverse direction of period 18.6 years. The relative movement of an artificial earth satellite and the natural satellite is thus complex and determination of the dates at which the artificial satellite is aligned with the sun-moon direction cannot be formulated for the general case. Examples will be given for the orbit of geostationary satellites (Section 7.2.5.6).

Eclipses by the moon are infrequent, most often of short duration and most often do not totally obscure the solar disc. They do not generally condition the satellite design and operation unless they precede or follow an eclipse of the sun by the earth which prolongs the total time during which the satellite is in the dark.

7.1.7.2 *Eclipses of the sun by the earth*

The sun's rays are assumed to be parallel and this corresponds to a point sun at infinity. The relationship between the declination δ of the sun and the latitude l of the satellite for there to be an eclipse is as follows:

$$-\delta - \arcsin(R_E/r) < \text{latitude of the satellite} < -\delta + \arcsin(R_E/r)$$

The centre of the eclipse corresponds to a value of the nodal angular elongation u of the satellite (equal to the sum of the argument of the perigee ω and the true anomaly v of the satellite) which fulfils:

$$\text{right ascension of the sun } \alpha_{SUN} + \pi = \Omega + \arctan(\tan u \cos i)$$

where Ω is the right ascension of the ascending node of the satellite orbit.

The duration of the eclipse varies as a function of the distance r and the inclination i of the satellite orbit with respect to the declination of the sun. The longest durations will be observed when the declination of the sun is equal to the inclination of the orbit.

7.1.8 Sun–satellite conjunction

Sun–satellite conjunction occurs when the sun is aligned with the satellite as seen from an earth station. This situation leads to a very great increase of the station antenna noise temperature. This occurs for a station situated on the track of the satellite when the following two conditions are satisfied:

—The latitude of the satellite is equal to the declination of the sun,
—The hour angle (or longitude) of the satellite is equal to the hour angle (or longitude) of the sun.

The conditions for occurrence and duration of conjunction with the sun are discussed in Section 7.2 for various particular cases of orbit. The effect of sun–satellite conjunction is examined in Chapter 8.

7.2 USEFUL ORBITS FOR SPACE COMMUNICATION

In principle, the plane of the orbit can have any orientation and the orbit can have any form. The orbital parameters are determined by the initial conditions as the satellite is injected into orbit. With the Keplerian assumptions, these orbital parameters, and hence the form and orientation of the orbit in space remain constant with time. It will be seen in the following sections that, under the effect of various perturbations, the orbital parameters change with time. Hence, if it is required to maintain the satellite in a particular orbit, orbit control operations are necessary. The cost of these operations can be minimised by choosing particular values for certain orbit parameters in accordance with the constraints imposed by the telecommunications mission.

Orbits which have particularly useful properties for telecommunication purposes are either elliptical orbits, generally with a large inclination, or circular orbits in the plane of the equator. The period of these orbits is in general related to (equal or a sub-multiple of) the period of rotation of the earth about its axis.

7.2.1 Elliptical orbits with non-zero inclination

In an elliptical orbit, the velocity of the satellite is not constant. This velocity, given by equation (7.18), is maximum at the perigee and minimum at the apogee. Hence, for a given period, the satellite remains in the vicinity of the apogee for a longer time than the vicinity of the perigee and this effect increases as the eccentricity of the orbit increases. The satellite in thus visible to stations situated under the apogee for a large part of the orbital period and this permits communication links of long duration to be established.

To establish repetitive satellite communication links, it is useful for the satellite to return systematically to an apogee above the same region. The period of orbits of this kind is thus a sub-multiple of the time taken by the earth of perform one rotation with respect to the line of nodes of the orbit. On the basis of the Keplerian hypotheses, the line of nodes is fixed in space and this duration is equal to a sidereal day. In practice, it is necessary to take account of the rotation of the line of nodes (the drift of the right ascension of the ascending node) due to the effect of perturbations (see Section 7.3.2.3). The period of the orbit must thus be a sub-multiple of the time T_{EN} taken by the earth to turn through an angle equal to $(360° - \Delta\lambda)$, where $\Delta\lambda$ is the drift of the ascending node during time T_{EN}. This drift depends on the inclination, the eccentricity and the semi-major axis of the orbit.

With an orbit of non-zero inclination, the satellite passes over regions situated on each side of the equator and possibly the polar regions if the inclination of the orbit is close to 90°. By orientating the apsidal line (the line from the perigee to the apogee) in the vicinity of the perpendicular to the line of nodes (the argument of the perigee ω is close to 90° or 270°), the satellite at the apogee systematically returns above the regions of a given hemisphere. It is thus possible to establish links with stations located at high latitudes.

The apogee of the orbit is permanently situated above the same hemisphere if there is no rotation of the orbit in its plane, that is if the drift of the argument of the perigee is zero. This is the case with the Keplerian hypotheses. In reality, various perturbations cause the orbital parameters to vary. By choosing an inclination of 63.45°, the drift of the argument of the perigee becomes zero (see Section 7.3.2.3).

Although the satellite remains for several hours in the vicinity of the apogee, it does move with respect to the earth and, after a time dependent on the position of the station, the satellite disappears over the horizon as seen from the earth station. To establish permanent links, it is thus necessary to provide several suitably phased satellites in similar orbits which are spaced around the earth (with different right ascensions of the ascending node and, for example, regularly distributed between 0 and 2π) in such a way that the satellite moving away from the apogee is replaced by another satellite in the same region of the sky as seen from the stations. In this way the problems of satellite acquisition and tracking by the stations are simplified. The problem of switching the links from one satellite to another remains; the link frequencies of the various satellites can be different in order to avoid interference.

Different types of orbit can be envisaged. In the following sections, orbits of the MOLNYA (12 h) and TUNDRA (24 h) types together with various systems under investigation are discussed [BOU-90].

7.2.1.1 'MOLNYA' orbits

These orbits take their name from the communication system installed by the Soviet Union whose territories are situated in the northern hemisphere at high latitudes (see Figure 1.4). The period T of the orbit is equal to $T_{EN}/2$ or about 12 hours. The characteristics of an example orbit of this type are given in Table 7.4.

The equation of the track is determined as indicated in Section 7.1.4 by considering that the angular velocity of the earth Ω_E is approximately equal to $n/2$, where n is the mean movement of the satellite. The relative longitude with respect to the satellite meridian as it passes through the ascending node is thus expressed from (7.37) by:

$$\lambda = \lambda_{SL} - \Delta\lambda = \text{arc sin}\,[(\tan\varphi)/(\tan i)] - (M/2) + (M_N/2) \tag{7.51a}$$

or:

$$\lambda = \lambda_{SL} - \Delta\lambda = \text{arc tan}\,[(\tan u)(\cos i)] - (M/2) + (M_N/2) \tag{7.51b}$$

The latitude of the track is given by:

$$\varphi = \text{arc sin}\,[\sin i \sin u] \tag{7.39}$$

M and M_N, the mean anomalies of the satellite and the ascending node are calculated from the eccentric anomalies by using Kepler's equation (7.24). The nodal angular elongation u is equal

Table 7.4 MOLNYA orbits

Period (T)	12 h
(half sidereal day	11 h 58 min 2 s)
Semi-major axis (a)	26 556 km
Inclination (i)	63.4°
Eccentricity (e)	0.6 to 0.75
Perigee altitude h_p	$a(1-e)-R_e$
(e.g.: $e = 0.71$)	(1250 km)
Apogee altitude h_a	$a(1+e)-R_e$
(e.g.: $e = 0.71$)	(39 105 km)

to $\omega + v$, where the true anomaly v is deduced from the eccentric anomaly E by means of equation (7.20).

The track of the satellite on the surface of the earth is illustrated in Figure 7.12 for a perigee argument equal to 270°. The satellite at the apogee passes successively on each orbit above two points separated by 180° in longitude. The apogee is situated above regions of latitude 63° (the latitude of the vertex is equal to the value of the inclination and the apogee coincides with the vertex of the track when the argument of the perigee is equal to 270°). The large ellipticity of the orbit results in a transit time for the part of the orbit situated in the northern hemisphere greater than that in the southern hemisphere. As the argument of the perigee is 270°, the true anomaly v on crossing the equatorial plane has a value $v = 90°$. The eccentric anomaly E_N of the corresponding ascending node thus has a value of around 42° (calculated from equations (7.20)). Knowing the mean movement of the satellite $n = 2\pi/T$, equation (7.22) gives a transit time t_N from the perigee to the ascending node equal to 27 minutes. Hence the satellite remains for $2t_N$, or around 1h, in the southern hemisphere and for $T - 2t_N$, of the order of 11h, in the northern hemisphere. The satellite thus remains for several hours in the vicinity of the apogee and hence visible from the regions situated beneath it.

The value of inclination which makes the drift of the argument of the perigee (and thus that of the apogee) equal to zero is 63.45° (see Section 7.3.2.3). A value different from this leads to a drift which is non-zero but remains small for values of inclination which do not deviate too greatly from the nominal value. By way of example, for an inclination $i = 65°$, that is a variation of 1.55°, the drift of the argument of the perigee has a value of around 6.5° per annum.

When the argument of the perigee is different from 270°, the latitude of the satellite at the apogee is no longer the maximum latitude of the track. The variation of satellite velocity in the orbit is no longer symmetrical with respect to the point of maximum latitude and the track of the satellite on the ground loses its symmetry with respect to the meridian passing through this point; the track is inclined as illustrated in Figure 1.5. In this figure, the parts of the track for which the satellite travels towards the east (with respect to the surface of the earth)

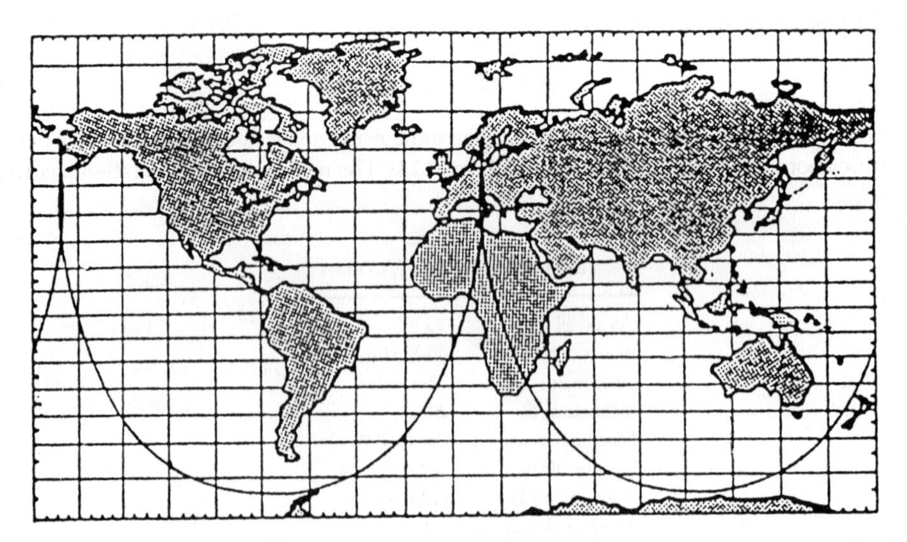

Figure 7.12 Example of the track of a MOLNYA orbit ($\omega = 270°$). (After [ASH-88]. Reproduced by permission of the Institution of Electrical Engineers.)

correspond to an angular velocity of the satellite, projected on to the equatorial plane, greater than that of the earth. After crossing the equatorial plane, the velocity decreases, passes through a value equal to that of the earth at the point with a vertical tangent and is then less than that of the earth on the part of the track oriented towards the west. In this part, it decreases up to the apogee then increases again up to the point with a vertical tangent where it again becomes greater than that of the earth. The apogee is situated between the points with a vertical tangent.

7.2.1.2 'TUNDRA' orbits

The period T of the orbit is equal to T_E, that is around 24h. The characteristics of an example orbit of this type are given in Table 7.5. An example of the track of the satellite on the surface of the earth is given in Figure 7.13 for an argument of the perigee equal to 270°.

Table 7.5 TUNDRA orbits

Period (T)	24 h
(equal to sidereal day	23 h 56 min 4 s)
Semi-major axis (a)	42 164 km
Inclination (i)	63.4°
Eccentricity (e)	0.25 to 0.4
Perigee altitude h_p	$a(1-e)-R_e$
(e.g.: $e=0.25$)	(25 231 km)
Apogee altitude h_a	$a(1+e)-R_e$
(e.g.: $e=0.25$)	(46 340 km)

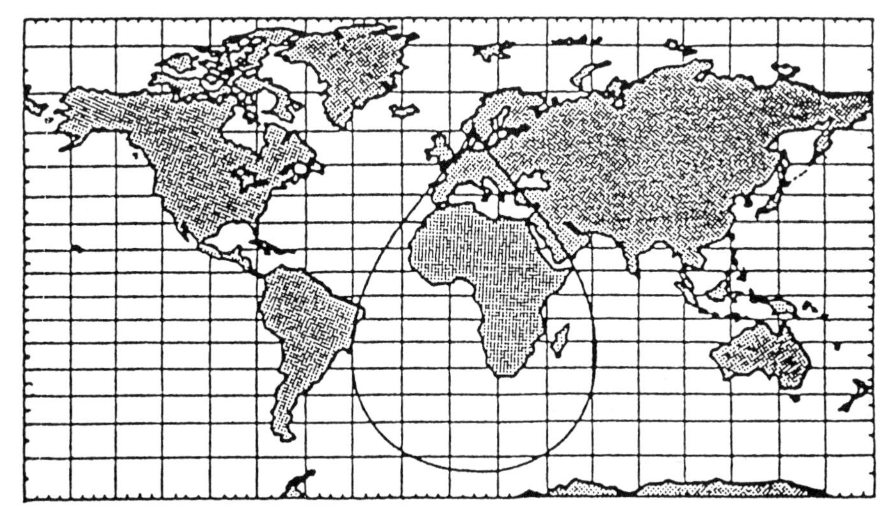

Figure 7.13 Example of the track of a TUNDRA orbit ($\omega = 270°$). (After [ASH-88]. Reproduced by permission of the Institution of Electrical Engineers.)

The equation of the track is determined by considering that the angular velocity of the earth Ω_e is very little different from the mean movement n of the satellite. The vertex is chosen as the origin of the reference meridian, since regions of high latitude are those for which the orbit is of interest. With respect to the vertex, the longitude λ_N of the node has a value $\lambda_N = -\pi/2$. Equations (7.33) and (7.34) become:

$$\tan \varphi = \tan i \cos \lambda_S$$

$$\cot \lambda_S = -\tan u \cos i$$

Also:

$$\Delta\lambda = M(\Omega_E/n) - M_0(\Omega_E/n) \cong M - M_v$$

where M_v is the mean anomaly of the vertex.

The relative longitude with respect to the satellite meridian on passing through the vertex can thus be expressed:

$$\lambda = \lambda_S - \Delta\lambda = \arccos[(\tan\varphi)/(\tan i)] - M + M_v \tag{7.52a}$$

or:

$$\lambda = \arccos\left\{[\sin(\omega + v)](\cos i)/\sqrt{[1 - \sin^2 i \sin^2(\omega + v)]}\right\} - (E - e\sin E) + (E_v - e\sin E_v) \tag{7.52b}$$

or again:

$$\lambda = -\operatorname{arc\,cotan}[(\tan u)(\cos i)] - (E - e\sin E) + (E_v - e\sin E_v) \tag{7.52c}$$

where E is the eccentric anomaly of the satellite and E_v that of the vertex.

The latitude of the track is given by:

$$\varphi = \arcsin[\sin i \sin(\omega + v)] \tag{7.39}$$

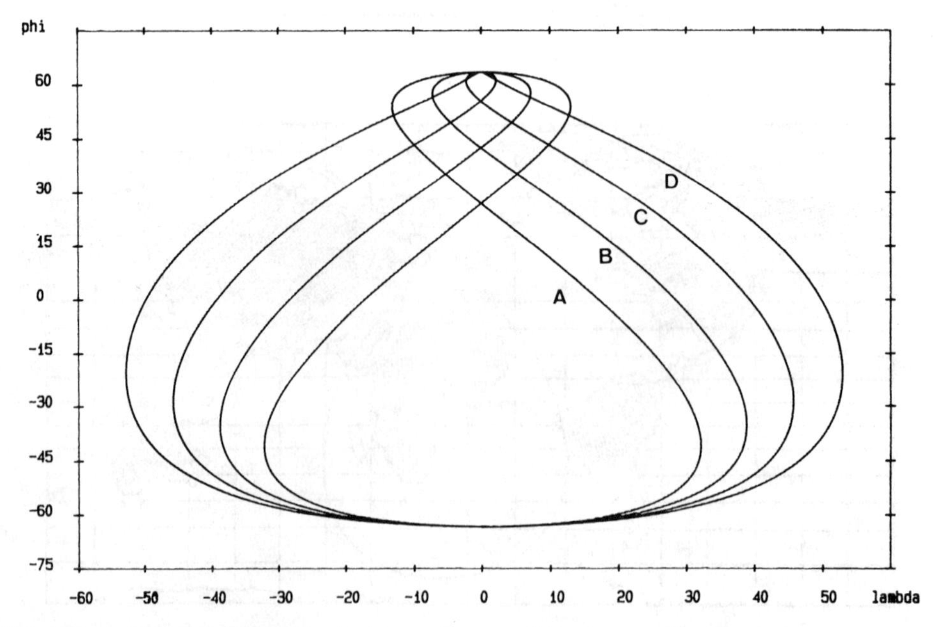

Figure 7.14 Tracks of a TUNDRA orbit ($i = 63.4°$, $\omega = 270°$) for various values of eccentricity ($e = 0.15$ (A), 0.25 (B), 0.35 (C) and 0.45 (D)).

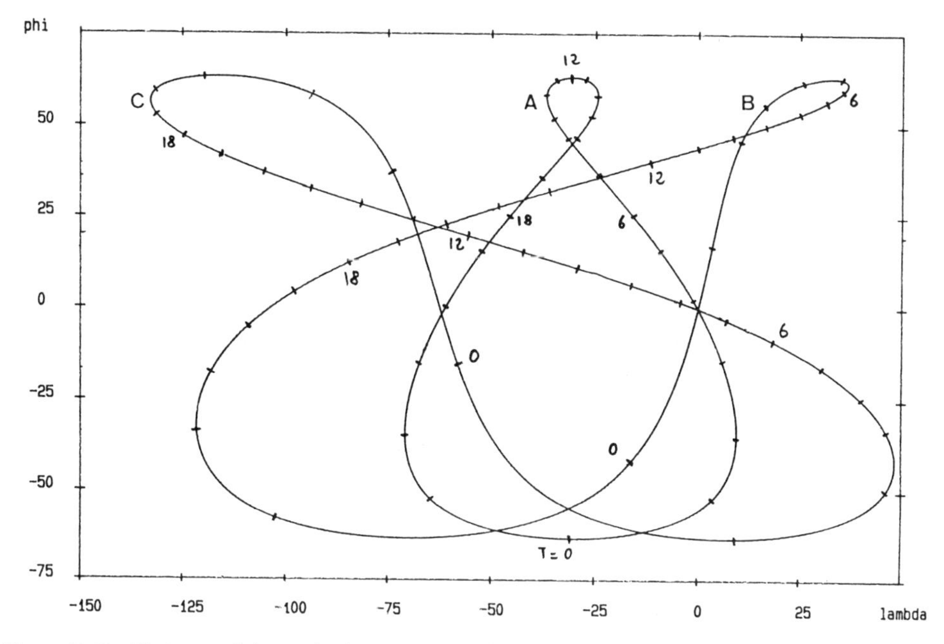

Figure 7.15 Variation of the track of a TUNDRA orbit ($i = 63.4°$) as a function of e and ω. (A) $e = 0.25$, $\omega = 270°$. (B) $e = 0.6$, $\omega = 315°$. (C) $e = 0.6$, $\omega = 202.5°$.

For $\omega = 270°$ and $i = 63.4°$, the variations of λ and φ are given in Figure 7.14 for various eccentricities. According to the value of the eccentricity, the loop above the northern hemisphere is accentuated to a greater or lesser extent. For an eccentricity equal to 0, the track has the form of a figure of eight with loops of the same size and symmetrical with respect to the equator (see Section 7.2.3). When the eccentricity increases, the upper loop decreases and the crossover point of the track is displaced towards the north. This loop disappears for a value of eccentricity of the order of 0.37. The transit time of the loop represents a substantial part of the period of the orbit and varies with eccentricity.

The position of the loop can be displaced towards the west or east with respect to the point of maximum latitude by changing the value of the argument of the perigee ω and the eccentricity. Some examples are presented in Figure 7.15 where the various tracks are represented for different values of the argument of the perigee and the eccentricity.

7.2.1.3 *Visibility of the satellite*

The elevation angle and the time of visibility are two important parameters to be considered in the choice of orbit type.

The ideal would be to have the satellite permanently at the zenith of the earth stations. For an operational system, variations of pointing angle with respect to the zenith are permissible either because the stations are equipped with a tracking system or because the antennas used have a large beamwidth. Limitations in the minimum values of elevation angle are due to the increase of antenna noise temperature (see Section 2.5.4) and to problems of blocking of the beam by obstacles (particularly for mobiles). Specification of a permissible range of variation of pointing angle with respect to the zenith for a satellite on a given orbit and at a particular point on the orbit results in the definition of a geographical region within which the satellite is visible with an elevation angle greater than a fixed minimum value. This region becomes more extensive when the satellite is further from the earth. But, within these regions, all stations do not see the satellite (which moves with respect to the earth) for the same duration. Stations situated on the track of the satellite, and particularly in the vicinity of the apogee, see the satellite for longer than the others. Regions are thus parameterised in accordance with the time of visibility.

Orbits whose tracks contain loops are particularly useful since, for regions situated under the loop, the satellite moves, during entry to the loop and during the time within the loop, in the same region of sky as seen at a large elevation angle from the earth station.

Continuous visibility. To ensure continuous coverage of these regions, a system with several satellites is required such that, for any station in the region, when the tracked satellite disappears below the minimum elevation angle, it is replaced by another satellite which is visible at an elevation angle greater than this fixed value. The orbits of these satellites are generally similar as far as the form (a, e) and inclination (i) parameters are concerned, but the values of right ascension differ since the orbits of these satellites must be in different planes to take account of the rotation of the earth with respect to the plane of the orbits. This is explained in Figure 7.16; the station situated on the meridian M(1) acquires satellite S1 at A which is thus considered to be active. It follows it during its trajectory on the orbital arc AB of length 2ρ.

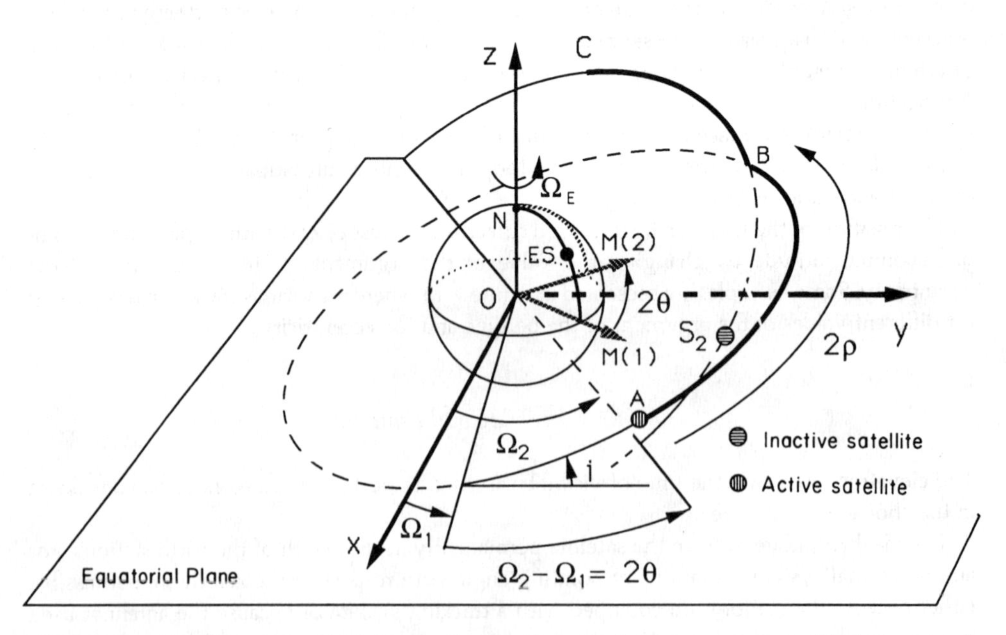

Figure 7.16 Satellites in orbits of different right ascension which are successively visible from an earth station at an elevation angle greater than a given value.

At B, the elevation angle becomes less than the required minimum and the satellite becomes inactive. The meridian M has turned during this interval and is now in position 2. By an appropriate phasing of satellite S2 on its orbit, it enters the arc CD at this time where the station can acquire it with an elevation angle greater than the required minimum (satellite 2 is thus active). For this it is necessary that the right ascension Ω_2 of the orbit of S2 is offset with respect to Ω_1 by an amount 2θ through which the meridian of the station has turned during the trajectory of the satellite from A to B (the orbital planes are assumed to be fixed in space).

The number of satellites necessary to cover a given geographical region continuously depends on the fixed minimum elevation angle and the characteristics of the orbit.

Orbits of the MOLNYA type. With an orbit of the MOLNYA type (period = 12h), a visibility duration of more than eight hours is possible with large elevation angles in regions situated under the apogees (Figure 7.17). A system with three satellites on orbits of right ascension differing by 120° thus permits continuous visibility to be ensured in these regions. Figure 7.18 shows the orbits of these satellites seen by a fixed observer. A system of this type called T-SAT has been studied by a group of British Universities and Polytechnics (London, Loughborough, Surrey, Manchester, Bradford and Portsmouth) and the Rutherford Appleton Laboratory [AGH-88], [NOR-88].

Orbits of the TUNDRA type. With an orbit of the TUNDRA type (period = 24h), a visibility duration of more than 12h is possible with high elevation angles; hence two satellites on orbits with right accension differing by 180° are sufficient. This is the principle used for the Sycomores system proposed by the CNES [ROU-88]. The form of the track is given by the curve of Figure 7.15. Figure 7.19 shows the region within which the active satellite (of the two in the

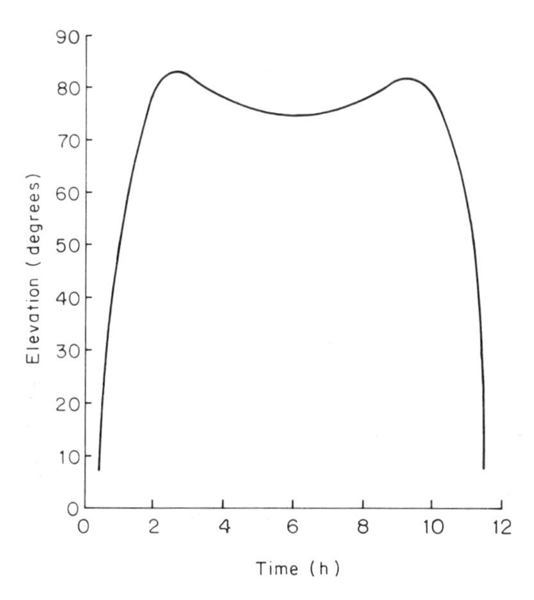

Figure 7.17 Example of the duration of visibility at the apogee in relation to the elevation angle at which the satellite is seen for a MOLNYA orbit.

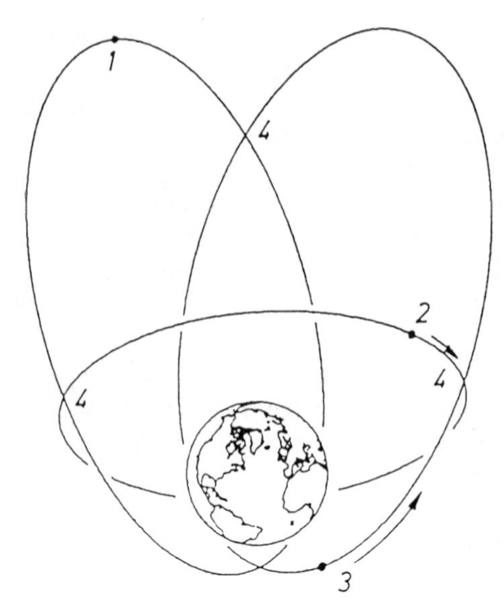

Figure 7.18 Three MOLNYA orbits whose right ascensions differ from 120° viewed from a fixed point in space. (Reproduced by permission of P. Dondl: see [DON-84].)

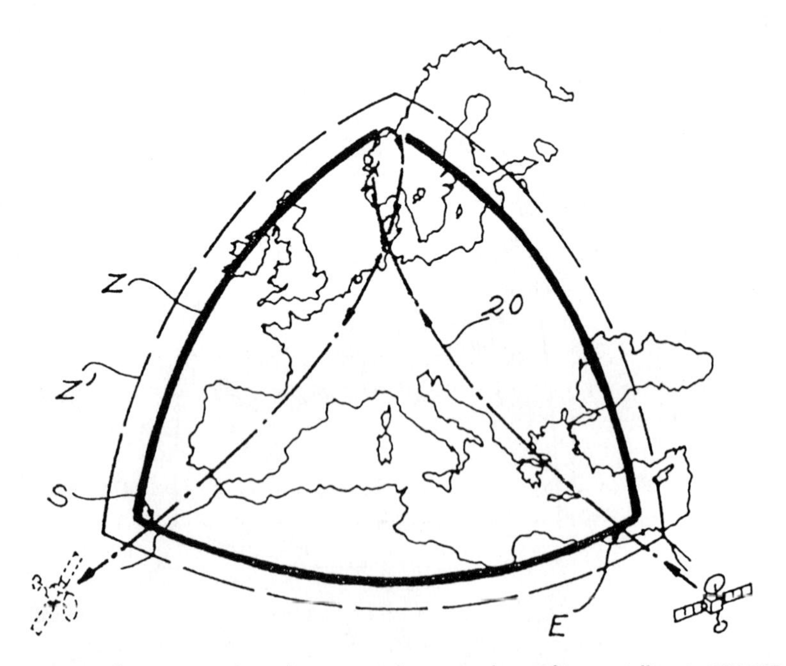

Figure 7.19 Zone of coverage with an elevation angle greater than 55°(two satellites in TUNDRA orbits). (After [ROU-88]. Reproduced by permission of the Institution of Electrical Engineers.)

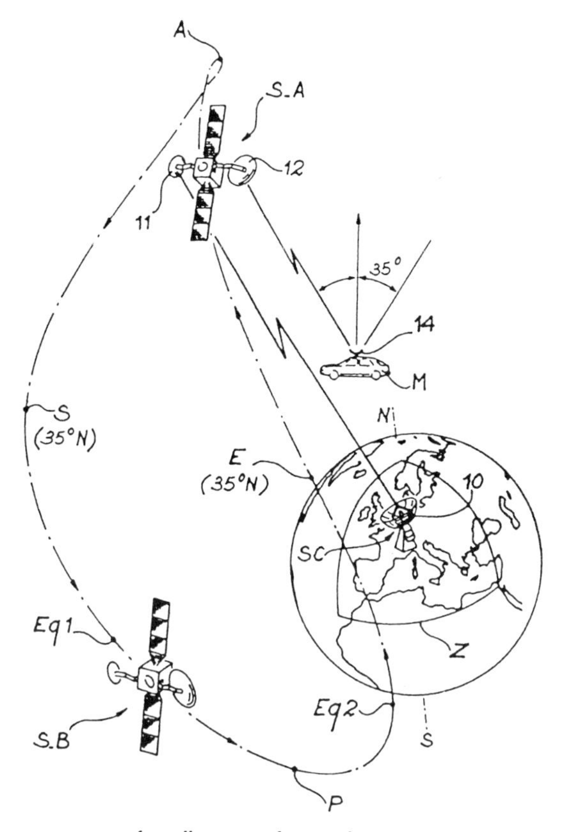

Figure 7.20 Apparent trajectory of satellites seen by an observer constrained to the rotating earth.

system) is seen with an elevation angle greater than 55°. Typical parameters of the orbit are as follows: $a = 42\,164$ km, $e = 0.35$, $i = 63.4°$, $\Omega = 270°$ (and 90°).

Figure 7.20 illustrates the apparent trajectory of the satellites seen by a distant observer restrained by the rotating earth. The two satellites succeed each other in the useful part of the trajectory on each side of the apogee.

'LOOPUS' orbits. In a system with several satellites, one of the problems encountered by the earth stations is that of repointing the antenna during the changeover from one satellite to the other. With orbits whose track contains a loop, it is possible to use only the loop as the useful part of the track of the trajectory; the satellite leaving the loop is replaced by another which enters it. Switching from one satellite to the other is thus performed at the crossover point of the track; at this instant the two satellites are seen from the earth station in exactly the same direction. It is not, therefore, necessary to repoint the antenna. This principle, called LOOPUS, is described in reference [DON-84]. To achieve continuous coverage of the region situated under the loop, the transit time of the loop must be a sub-multiple of the period of the orbit and the number of satellites necessary is equal to the rank of the sub-multiple. The coverage can be extended to one part of the hemisphere by increasing the number of satellites in orbits regularly spaced about the globe [GOS-90].

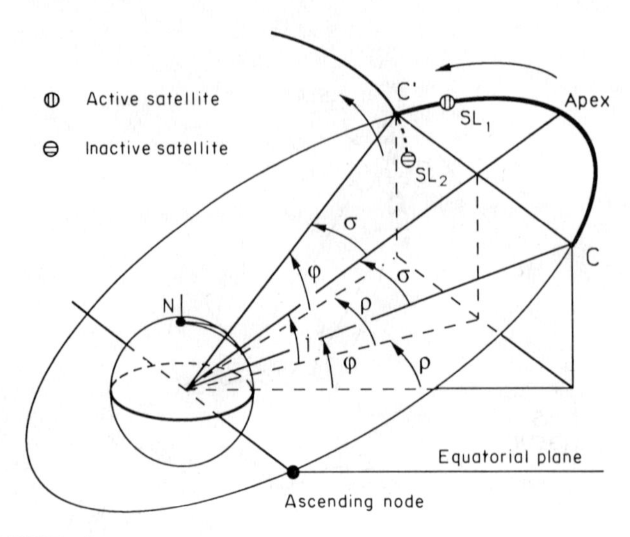

Figure 7.21 LOOPUS orbits.

To illustrate the concept, consider a system using an orbit of the MOLNYA type (12h) with the following parameters: $a = 26\,562$ km, $e = 0.72$, $i = 63.4^\circ$, $\omega = 270^\circ$. The track of this orbit on the ground contains a loop whose transit time is 8h. Three satellites thus permit continuous coverage of the region below the loop.

In Figure 7.21, the useful arc CC′ corresponds to the loop of the track. For points C and C′ to coincide on the track, it is necessary for the meridian passing through C at the initial instant to be at C′at the same time as the satellite. The earth has thus turned by an amount equal to the projection 2ρ on the equatorial plane of the variation of the true anomaly 2σ.

Hence:

$$2\rho = 360^\circ \times 24\,\text{h}/8\,\text{h} = 120^\circ$$

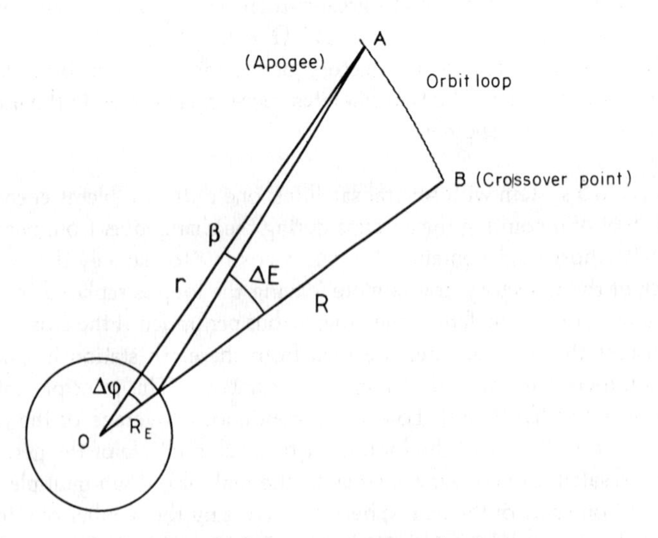

Figure 7.22 Variation of pointing angle during passage through a loop.

The variation of the anomaly σ is such that $\tan \sigma = (\arccos i) \tan \rho$, hence $\sigma = 37.8°$.

The latitude φ of the crossover point (C or C′) is given by:

$$\arcsin \varphi = (\sin \sigma)/(\sin \sigma)$$

and has a value $\varphi = 45°$.

The orbit of the succeeding satellite is such that the arc corresponding to CC′ (DD′) is reproduced from the point C′. The difference in right ascension of the nodes is thus equal to 2ρ or 120°.

Example calculation of elevation angle variation. Variation of the pointing angle of stations remains limited. Consider, for example, a station situated at the crossover point of the track. When a satellite arrives in the loop, it is at the zenith of the station. Subsequently the elevation angle decreases during the 4 h of movement of the satellite towards the north as far as the apogee. The variation ΔE of the elevation angle between the passage of the satellite through the crossover point B (at the zenith) and the apogee A of the orbit is calculated from Figure 7.22:

$$OA = a(1 + e) = r$$

$$\Delta E = \beta + \Delta\varphi$$

$$\Delta\varphi = (\text{latitude of the apogee} = i) - (\text{latitude of B} = \varphi = 45°)$$

$$(\sin \Delta\varphi)/R = (\sin \Delta E)/R_E$$

$$R = \sqrt{(R_E^2 + r^2 - 2R_E r \cos \Delta\varphi)}$$

In the system considered with $e = 0.72$, $\omega = 270°$ and $i = 63.4°$, the variation of pointing angle with respect to the zenith is of the order of 20° with a minimum elevation angle of 70° for a station situated at the crossover point of the track. The elevation angle then increases for 4 h to become equal to 90° again at the instant when the satellite leaves the loop.

7.2.1.4 *Advantages of high inclination elliptic orbits*

Large elevation angle. The main application of inclined elliptic orbits is to ensure coverage of regions at high latitude under a large elevation angle with satellites whose apparent movement with respect to the earth is small. A high elevation angle is particularly required in applications which include systems for communication with mobiles. Blocking of the beam due to occultation of the satellite by buildings and trees is minimised. Multiple trajectories caused by successive reflections by various obstacles are also reduced in comparison with systems operating with low elevation angles (geostationary satellite systems, for example). Tracking of the satellite is facilitated on account of the small apparent movement and the long visibility duration. It is even possible to use antennas whose 3 dB beamwidth is a few tens of degrees with fixed pointing towards the zenith; this permits the complexity and cost of the terminal to be reduced while retaining a high gain. Finally, the noise captured by the earth station antenna, from the ground or due to interference from other terrestrial radio systems is minimised due to the high elevation angle. This applies to all signal attenuation and noise generation effects (atmospheric gases, rain etc.) associated with the length of the oblique trajectory in the atmosphere. These advantages have led the Soviet Union to use these orbits for a long time in order to provide coverage of high latitude territories; their use is envisaged for various systems for satellite communication with mobiles. In addition to the systems already mentioned above, there is the Archimedes system proposed by the European Space Agency [ASH-88].

Sun–satellite eclipses and conjunction. In a communication system which operates satellites on elliptic orbits of high inclination, the operational part of the orbit is situated on each side of the apogee, most often coincident with the vertex of the orbit. If the inclination of the orbits is 63.4°, the maximum latitude of the satellite is 63.4° and its minimum value during the operational phase depends on the extent of the active part on each side of the vertex. This extent is reduced when the number of satellites is large. In this case the latitude of the satellite always remains high during the operational phase and the conditions for the occurrence of eclipses (Section 7.1.5.6) are not very often satisfied during this phase. This is confirmed for the MOLNYA and TUNDRA systems with three satellites. With a two satellite TUNDRA system, non-occurrence depends on the values chosen for the right ascension of the ascending node. A similar analysis can be performed in connection with solar interference.

7.2.1.5 *Disadvantages of high inclination elliptic orbits*

Traffic switching between satellites. More than one satellite in orbit is necessary in order to provide a continuous service over a given geographical region and this adversely affects the cost of the space segment. Furthermore, it is necessary periodically to switch the traffic from one satellite to another. These special procedures cause an operational load at the control centre and reduce the capacity during switching; it may be necessary to have two antennas at each earth station to point at the two satellites simultaneously and thus transfer the traffic from one satellite to the other without interruption of service.

Variation of distance. The variation of distance between the service region and the satellite during the time of activity of the latter is greater with orbits of the 'MOLNYA' type than with orbits of the 'TUNDRA' type. This distance variation has the following consequences:

—Variation of propagation time (52 ms variation for MOLNYA orbits),
—Doppler effect (14 kHz for MOLNYA orbits and 6 kHz for TUNDRA orbits in L Band (1.6 GHz) [ASH-88]),
—Variation of received signal level (4.4 dB for MOLNYA orbits) at the extremities of the up- and downlinks,
—Modification of the coverage of the satellite antennas. Figures 7.23a and 7.23b show the coverage obtained at the apogee and at the point of switching from one satellite to the other; Europe is seen through an angle of 4.9° at the apogee which changes to 8.4° at the switching point with a MOLNYA orbit (3.6° to 4.6° for a TUNDRA orbit). The antenna aperture can be optimised so that the decrease of gain at the edge of coverage at the switching point with respect to the gain at the apogee is compensated by the reduction of free space losses.

Radiation. MOLNYA orbits are characterised by a perigee at an altitude of around 1200 km which means that the satellite crosses the Van Allen belts twice per orbit (their altitude is of the order of 20 000 km, see Chapter 11); in these belts are high energy radiations which degrade the semiconductor components (such as solar cells and transistors) used in the satellite. TUNDRA orbits have the advantage of reducing the duration of crossing these bands.

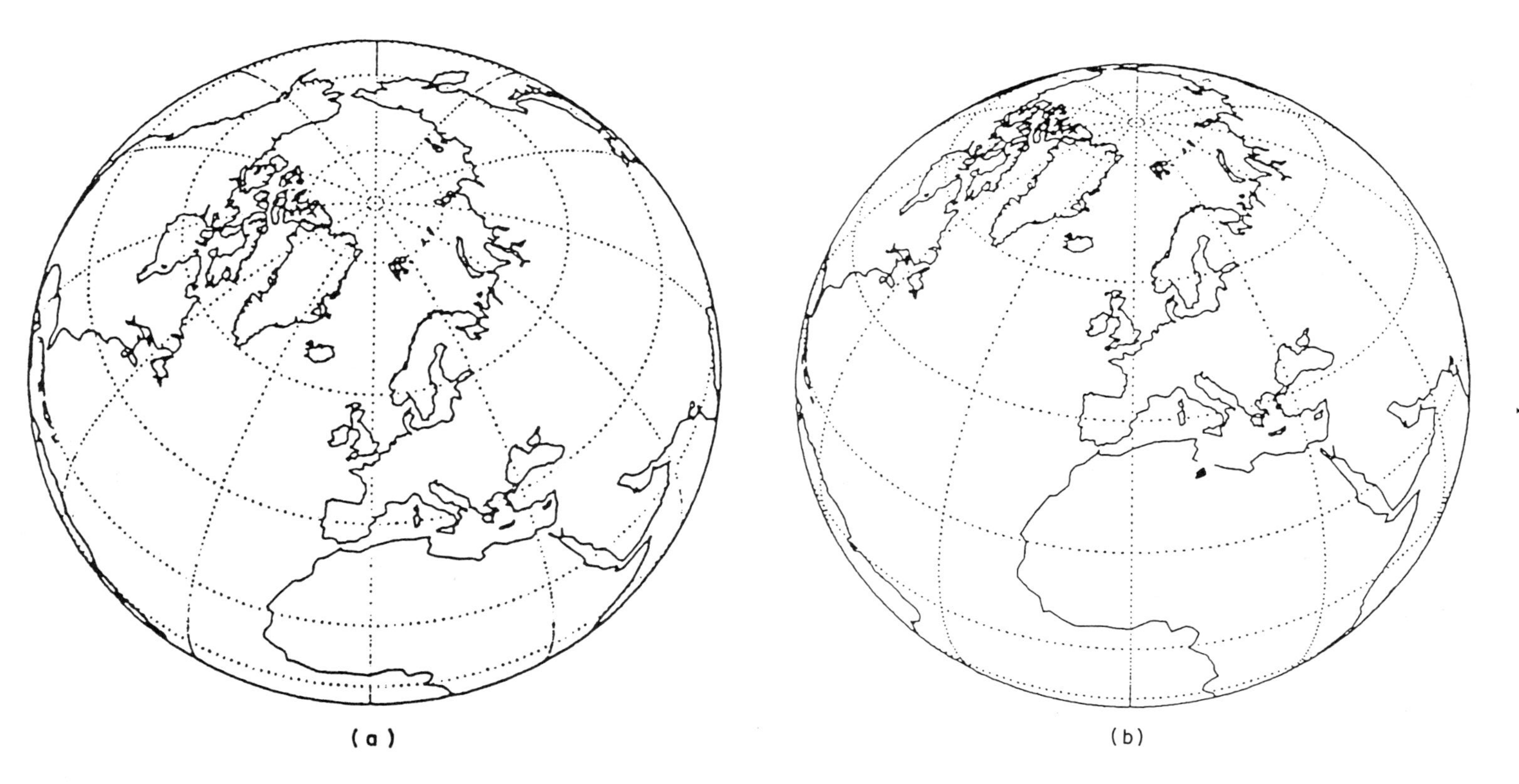

Figure 7.23 Variation of coverage (MOLNYA orbit) between the start of the activity phase and the apogee. (Reproduced by permission of P. Dondl: see [DON-84].)

Perturbations of orbit. For elliptic orbits of low altitude at the perigee, the satellite is strongly subjected to the effects of the asymmetry of the terrestrial potential and this leads to perturbations of the orbit which must be controlled.

7.2.2 Geosynchronous elliptic orbits with zero inclination

The inclination is equal to zero. The period of the orbit is equal to one sidereal day (there is no longer a drift of the ascending node). The mean movement n of the satellite is equal to Ω_E, the angular velocity of the earth. The track of the satellite remains in the equatorial plane and becomes a periodic oscillation (period T_E) about the point of longitude λ_p representing the satellite at the perigee.

The longitude of the track with respect to the perigee can be written:

$$L = \lambda - \lambda_p = v - \Omega t = v - M = \text{arc cos}\,[(\cos E - e)\,(1 - e\cos E)^{-1}] - (E - e\sin E) \tag{7.53}$$

The maximum displacement L_{max} is obtained for $d(\lambda - \lambda_p)/dE = 0$, which leads to [BIE-66]:

$$\cos E_m = [1 \pm (1 - e^2)^{1/4}]e^{-1} \tag{7.54}$$

Figure 7.24 gives the amplitude of the oscillation (longitude L_{max}) and the time t necessary to reach this point as a function of the eccentricity e.

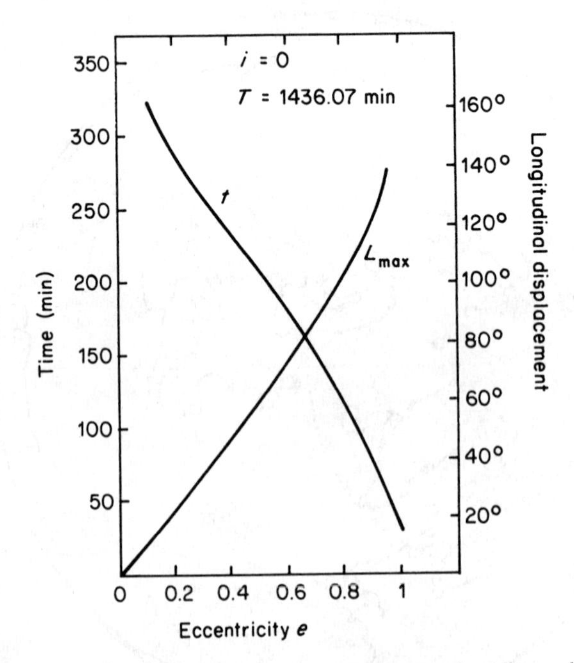

Figure 7.24 Amplitude (L_{max}) of the track of a synchronous equatorial satellite and the time t after which this amplitude is achieved as a function of the eccentricity of the orbit.

For an eccentricity less than 0.4:

$$L_{\max} \cong 2e \text{ (rad)} = 144\, e \qquad \text{(degrees)} \tag{7.55}$$

If the eccentricity is small (10^{-3}), the maximum is reached after 6 h ($T_e/4$). This time decreases for large values of eccentricity.

7.2.3 Circular geosynchronous orbits with non-zero inclination

The eccentricity is equal to zero. The period of the orbit is very little different from a sidereal day (the difference comes from the effect of the drift of the ascending node). The mean movement n of the satellite is thus very little different from Ω_E, the angular velocity of the earth. The nodal angular elongation u has a value $u = nt_{NS} \cong \Omega_E t_{NS}$. Movement of the satellite in its orbit is at constant angular velocity. On the other hand, the projection of this movement on the equatorial plane is not at constant velocity. There is thus an apparent movement of the satellite with respect to the reference meridian on the surface of the earth (that of the satellite on passing through the nodes).

The projection of the satellite orbit on the equatorial plane is illustrated in Figure 7.25 by the dotted curve. The projection of point A (the position of a fictitious satellite rotating at the velocity of the reference meridian in the plane of the equator) perpendicularly to the line of nodes cuts this curve at point B. The co-ordinates of B, in a geocentric reference in the plane of the equator such that Ox is along the line of nodes and Oy is orthogonal to Ox, are:

$$X_B = R_E \cos \Omega_E t \quad \text{and} \quad Y_B = R_E \sin \Omega_E t \cos i.$$

Hence: $\tan(\zeta t) = y_B / x_B = \cos i \tan(\Omega_E t)$

from which: $\Omega_E t = \arctan[(1/\cos i)\tan(\zeta t)]$.

Differentiating gives: $\Omega_E = d(\Omega_E t)/dt$:

$$\Omega_E = \zeta[1 + \tan^2(\zeta t)]/\{\cos i + [\tan^2(\zeta t)]/\cos i\} \tag{7.56}$$

In the vicinity of the nodes, ζt tends to 0 from which $\Omega_E \cong (1/\cos i)\,\zeta$.

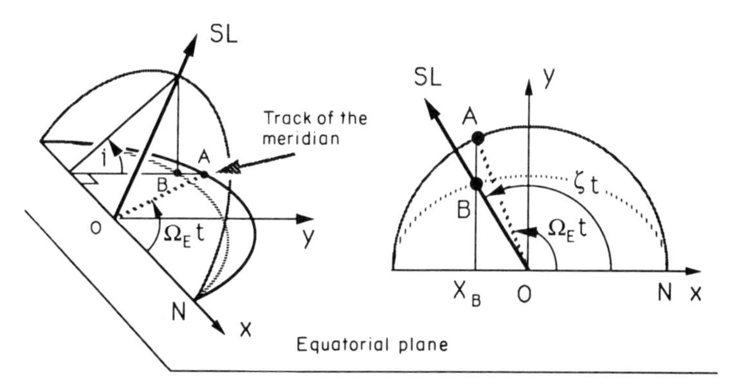

Figure 7.25 Projection of the orbit of a synchronous satellite on to a circular orbit of non-zero inclination.

Hence $\zeta \cong \Omega_E \cos i$; the angular velocity of the satellite meridian is less than that of the reference meridian and the satellite drifts towards the west.

In the vicinity of the point of maximum latitude, ζt tends to $\pi/2$ from which $\Omega_E \cong \zeta \cos i$. Hence $\zeta \cong \Omega_E/\cos i$; the angular velocity of the satellite meridian is greater than that of the reference meridian and the satellite drifts towards the east.

Taking the satellite meridian on passing through the ascending node as a reference, the relative longitude is calculated from equation (7.37):

$$\lambda = \lambda_S - \Delta\lambda = \arcsin[(\tan\varphi)/(\tan i)] - \Omega_E t_{NS} \tag{7.57a}$$

hence:

$$\lambda = \arcsin\{(\cos i \sin u)/\sqrt{(1 - \sin^2 i \sin^2 u)}\} - u \tag{7.57b}$$

and:

$$\lambda = \arctan[(\tan u)(\cos i)] - u \tag{7.57c}$$

The latitude φ has a value:

$$\varphi = \arcsin(\sin i \sin u) \tag{7.39}$$

The track of the satellite as u varies from 0 to 360° is represented in Figure 7.26 for various values of inclination i. The maximum latitude φ_m attained (the vertex) is equal to the value of inclination i of the orbit. The associated longitude λ is zero (with respect to the reference meridian).

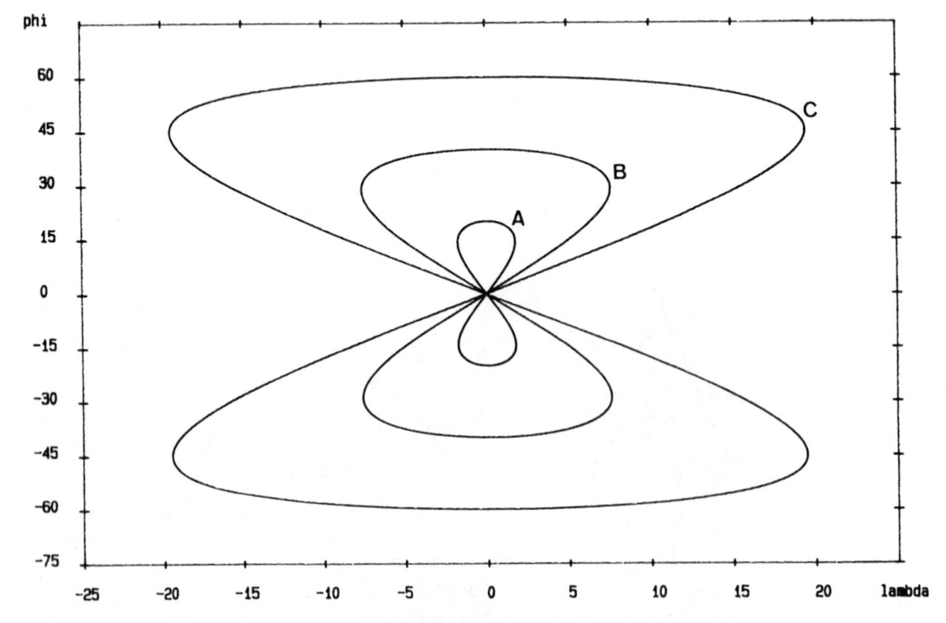

Figure 7.26 Tracks of circular synchronous orbits for various values of inclination: (A) $i = 20°$, (B) $i = 40°$ and (C) $i = 60°$.

The maximum displacement in longitude λ_{max} with respect to the reference meridian is obtained for $d\lambda/du = 0$ which leads to:

$$\tan u = 1/\sqrt{(\cos i)} \quad \text{or} \quad \sin u = 1/\sqrt{(1 + \cos i)} = 1/(\sqrt{2}) \cos (i/2)$$

from which:

$$\lambda_{max} = \arctan [\sqrt{(\cos i)}] - u = \arccos [1/\sqrt{(1 + \cos i)}] - \arcsin [1/\sqrt{(1 + \cos i)}]$$

hence:

$$\lambda_{max} = \arccos [1/(\sqrt{2}) \cos (i/2)] - \arcsin [1/(\sqrt{2}) \cos (i/2)] \qquad (7.58a)$$

or:

$$\lambda_{max} = \arcsin [(1 - \cos i)/(1 + \cos i)] = \arcsin [\sin^2(i/2)/\cos^2(i/2)] \qquad (7.58b)$$

The associated latitude φ_m at the maximum displacement λ_{max} can be written:

$$\varphi_m = \arcsin (\sin i \sin u) = \arcsin [(\sin i) 1/(\sqrt{2}) \cos (i/2)]$$

hence:

$$\varphi_m = (\sqrt{2}) \sin (i/2) \qquad (7.59)$$

For small i:

$$\lambda_{max} = i^2/4 \text{ and } \quad \varphi_m = i/\sqrt{2} \qquad \text{(rad)} \qquad (7.60)$$

7.2.4 Sub-synchronous circular orbits with zero inclination

A communication satellite must be visible from the regions concerned during the periods when it is desired to provide a communication service; this can vary from a few hours to 24 h per day. When the service is not continuous, it is desirable that the intervals during which the service is available repeat each day at the same time. A satellite following a sub-synchronous equatorial orbit can cover a given geographical region at the same local time each day. The duration of uninterrupted service which such a satellite can provide in a given region on the terrrestrial surface is a function of its altitude and the latitude of the receiver. Table 7.6 shows some typical visibility durations [CCIR-Rep 215]. Such orbits are particularly suitable for consideration for satellite broadcasting systems.

7.2.5 Geostationary satellite orbits

A particular case of the preceding section, this circular orbit ($e = 0$) in the equatorial plane ($i = 0$) is geosynchronous. The angular velocity of the satellite is the same as that of the earth ($n = \Omega_E$) and in the same direction (direct orbit). The track of the satellite is reduced to a point on the equator; the satellite remains permanently on the vertical at this point. To a terrestrial observer, the satellite appears fixed in the sky.

Table 7.6 Duration visibility for satellites in a geostationary orbit or a sub-synchronous circular equatorial orbit (non-retrogressive)

Approximate period (h)	Altude (km)	Number of transits per day above a given point	Approximate duration of visibility above the horizon on each transit (h)			
			At the equator	At $\pm 15°$ latitude	At $\pm 30°$ latitude	At $\pm 45°$ latitude
24*	35 786	Stationary	Continuous	Continuous	Continuous	Continuous
12	20 240†	1	10.1	10.0	9.9	9.3
8	13 940†	2	4.8	4.7	4.6	4.2
6	10 390†	3	3.0	2.9	2.8	2.5
3	4 190†	7	1.0	1.0	0.9	0.6

*Exact period = 23 h 56 min 4 s.
†Approximate value.

Table 7.7 Characteristics of the Keplerian orbit of geostationary satellite

Semi-major axis	$a = r$	42 164.2 km
Satellite velocity	$Vs = \sqrt{(a^3/\mu)}$	3075 m/s
Satellite altitude	R_0	35 786.1 km
Mean equatorial radius	R_E	6378.1 km
Ratio	R_0/R_E	6.614

The semi-major axis a of the orbit is such that:

$$2\pi\sqrt{(a^3/\mu)} = T_E = 1 \text{ sidereal day} = 86\,164.1 \text{ s}$$

The characteristics of the nominal Keplerian orbit are given in Table 7.7.

7.2.5.1 *Distance of the satellite from an earth station*

For a geostationary satellite, equation (7.40) can be put into the form:

$$R^2 = R_E^2 + r^2 - 2\,R_E\,r\cos\phi \quad \text{with } r = R_E + R_0,$$

hence:

$$R^2 = R_0^2 + 2R_E(R_0 + R_E)\,(1 - \cos\phi) \tag{7.61}$$

The values of R_0 and R_E give $R_E/R_0 = 0.178$ and this gives:

$$(R/R_0)^2 = 1 + 0.42(1 - \cos\phi) \tag{7.62}$$

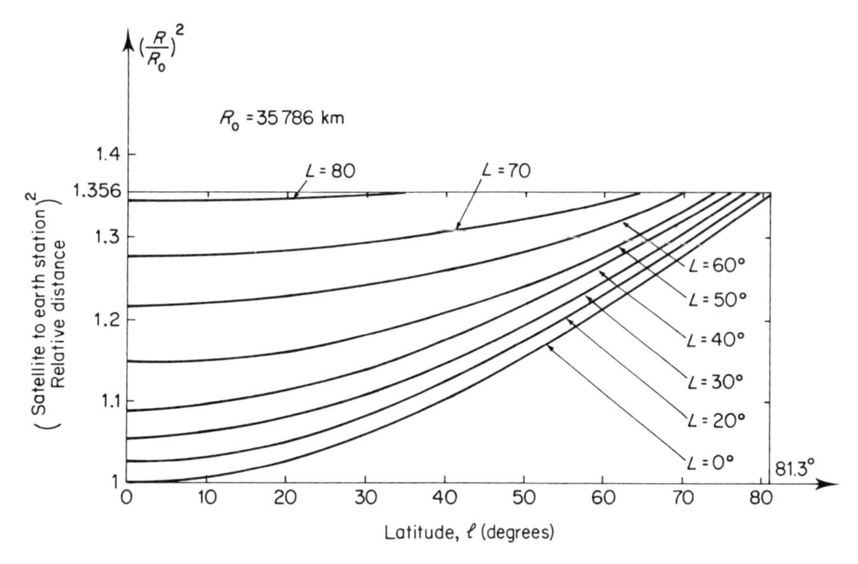

Figure 7.27 Variation of the square of the ratio of the station–satellite distance R to its nominal altitude R_0 as a function of the satellite–station latitude l and relative longitude L.

As for $\cos\phi$, knowing that $\varphi = 0$, equation (7.41) becomes:

$$\cos\phi = \cos L \cos l \tag{7.63}$$

where: l is the latitude of the station and L is the relative longitude of the satellite with respect to the station.

The variation of $(R/R_0)^2$ as a function of l is given, for various values of L, by the curves of Figure 7.27. The maximum value of $(R/R_0)^2$ is 1.356. When R^2 is replaced by R_0^2 the maximum error is 1.3 dB.

7.2.5.2 *Elevation and azimuth angle*

From an earth station whose position is defined by its latitude l and its relative longitude L with respect to the satellite, the *elevation angle* E at which the satellite is seen is obtained from equations (7.43) with $r = R_E + R_0$, hence for example:

$$E = \arctan\left\{[\cos\phi - (R_E/(R_E + R_0))]/\sqrt{(1 - \cos^2\phi)}\right\}$$

where the angle ϕ is given by (7.63).

Figure 8.14 indicates the value of elevation angle E as a function of the earth station position relative to that of the satellite. The *azimuth angle* A is obtained from the intermediate parameter a defined from equation (7.44) with $\varphi = 0$, hence:

$$a = \arcsin[\sin L/\sin\phi] \qquad (\text{with } \phi > 0,\ L > 0) \tag{7.64}$$

The true azimuth A is obtained from a as a function of the position of the station with respect to the satellite by using Table 7.8 and Figure 8.14.

Table 7.8 Determination of the azimuth A

Hemisphere of station	Position of satellite with respect to station	Relation between A and a
Northern	East	$A = 180° - a$
Northern	West	$A = 180° + a$
Southern	East	$A = a$
Southern	West	$A = 360° - a$

7.2.5.3 *Nadir angle; maximum coverage*

From equation (7.45), the nadir angle β is equal to arc sin $[R_E(\cos E)/r]$. The maximum geographical coverage is given by the portion of the earth included in a cone which is tangential to the surface of the earth and has the satellite at its vertex. The limiting elevation angle is thus zero. The vertex angle of the cone, the angle at which the earth is seen from the geostationary satellite is:

$$2\beta_{max} = \text{arc sin}\,[R_E/(R_0 + R_E)] = 17.4° \qquad (7.65)$$

The maximum latitude l_{max}, or the maximum deviation in longitude L_{max}, with respect to the satellite corresponds to the value of ϕ_{max} given by equation (7.45b) with $\beta_{max} = 8.7°$, that is $\phi_{max} = l_{max} = L_{max} = 81.3°$.

7.2.5.4 *Propagation time*

The distance between two ground stations via the satellite varies between:

$$2R_{max}\ (L = 0°, l = 81.3°) = 83\,357.60\ \text{km and } 2R_0 = 71\,572.2\ \text{km}.$$

The propagation time is greater than 0.238 s and can reach 0.278 s. The influence of the propagation time is discussed in Section 3.5.

7.2.5.5 *Eclipse of the sun by the earth*

Knowledge of the duration and periodicity of eclipses is important in the case of satellites which use solar cells as a source of energy. Furthermore, the eclipse causes a thermal shock which must be allowed for in the design of the satellite.

The duration of eclipses: The movement of the earth around the sun is represented in Figure 7.5. Figure 7.28 shows the apparent movement of the sun with respect to the equatorial plane. The satellite follows an orbit perpendicular to the plane of the figure. At the solstices, the satellite is always illuminated; but, in the vicinity of the equinoxes, it is in the earth's shadow. Considering, as a first approximation, that the sun is a point at infinity this shadow is a cylinder

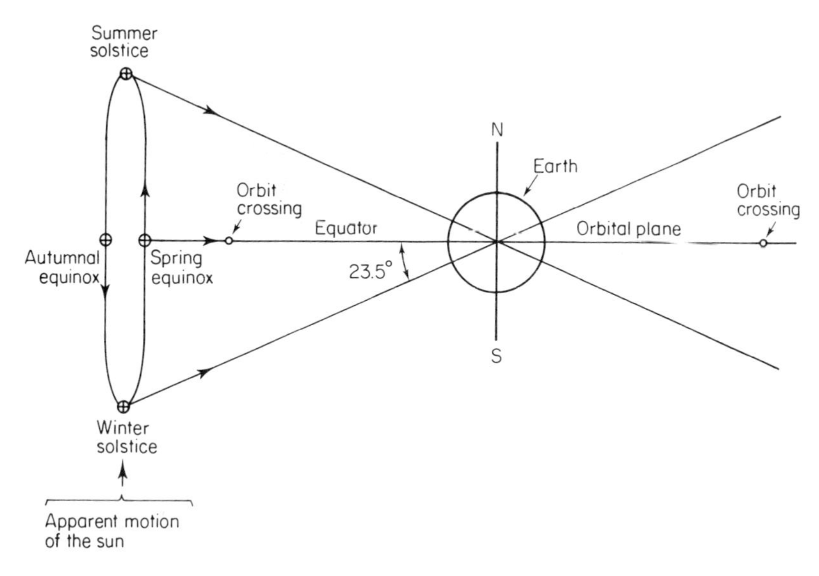

Figure 7.28 Apparent movement of the sun with respect to the orbit of geostationary satellites.

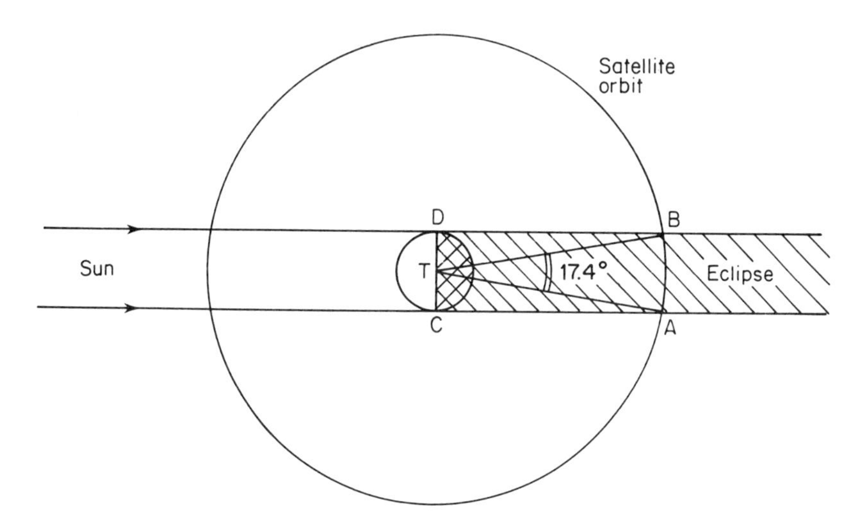

Figure 7.29 Eclipses at the equinoxes (figure in the plane of the equator).

which is tangential to the earth. On the day of the equinox, the eclipse has a maximum duration d_{max} determined from Figure 7.29 such that:

$$d_{max} = (17.4°/360°) \times (23\,\text{h} \times 60\,\text{min} + 56\,\text{min}) = 69.6\,\text{min}.$$

In reality, the sun has an apparent diameter of 0.5° as seen from the earth and there is a cone of shadow where the eclipse is total and a region of penumbra where the eclipse is partial (see Figure 10.37).

The penumbra has a width equal to the apparent diameter of the sun, that is 0.5°. In its orbit,

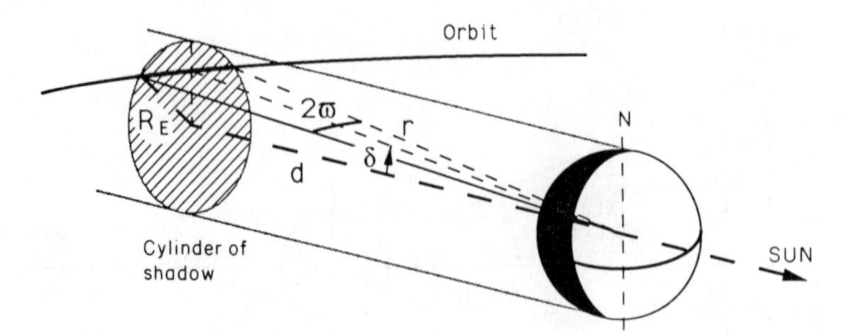

Figure 7.30 Geometry of the shadow region and the orbit out of equinox.

the satellite moves 1° in 4 min; also the total duration of the eclipse is equal to 71.5 min of which 2 min are penumbra at the start and finish.

To evaluate the duration of the eclipse other than at the equinox, consider Figure 7.30 where the arc 2ϖ is the arc of the orbit contained within the cylindrical shadow of radius R_E and δ_{SUN} is the declination of the sun. This gives:

$$\cos\varpi \cos\delta_{SUN} = d/r \quad \text{and} \quad d^2 + R_E^2 = r^2.$$

From which:

$$\cos\varpi = \sqrt{[1 - (R_E/r)^2]}/\cos\delta_{SUN} = 0.9885/\cos\delta_{SUN} \tag{7.66}$$

First and last day of the eclipse. The first day of the eclipse before the spring equinox corresponds to the relative position of the sun such that the cone of the earth's shadow is tangent to the satellite orbit. Figure 7.31 illustrates the situation before the autumn equinox as the declination of the sun decreases. The value of ϖ is thus zero ($\cos\varpi = 1$) and the declination of the sun δ_0 is such that $\cos\delta_0 = \sqrt{[1 - (R_E/r)^2]}$ or $\sin\delta_0 = R_E/r$ and hence $\delta_0 = \arcsin R_E/r = 8.7°$ (the triangle BDT is the same as the corresponding triangle of Figure 7.29).

The last day would correspond to a figure which is symmetrical with respect to the equatorial plane. A similar situation occurs again at the spring equinox. When the declination of the sun is greater than the absolute value of δ_0, the shadow does not intercept the orbit and there is no eclipse.

The first and last days of the eclipse seasons are obtained by determining the dates at which the declination of the sun has a value $\delta_0 = \pm 8.7°$, that is $\sin\delta_0 = \pm R_E/r = \pm 0.151\,28$. From equation (7.25), the values of nodal angular elongation of the sun are $u = \arcsin[\pm 0.15128/\sin\varpi]$, that is $u = \pm 22.34°$ and $u = \pm 22.34° + 180°$.

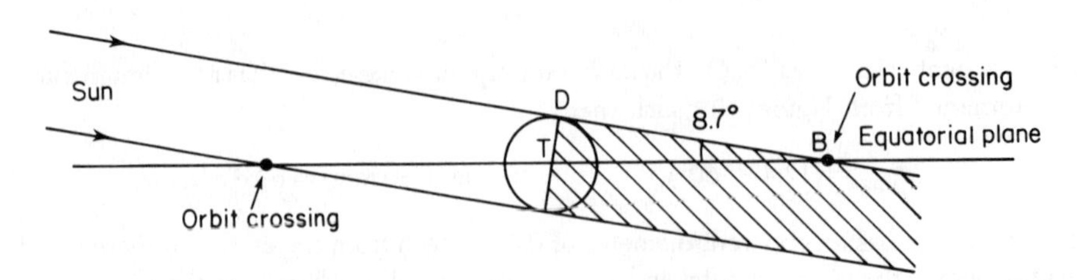

Figure 7.31 First day of the eclipse before the autumn equinox (the plane of the equator is perpendicular to the plane of the figure).

The true anomalies of the sun before and after the vernal and autumn equinoxes are deduced from $v = u + \varpi_{SUN}$. Equations (7.20) and (7.24) enable the values of the eccentric anomalies to be calculated and hence those of the associated mean anomalies: $M_1 = 54.17°$ and $M_2 = 98.57°$, $M_3 = 237.34°$ and $M_4 = 282.31°$.

The related dates of passing through the perigee are given by $t = M/n_{SUN}$ where $n_{SUN} = 2\pi/365.25$ is the mean movement of the sun in radians per day. This gives: $t_1 = 54$ d 23 h, $t_2 = 99$ d 23.5 h, $t_3 = 240$ d 19 h, $t_4 = 286$ d 10 h. Since passage through the perigee occurs between 2 and 3 January, the dates of the start and end of the periods of eclipse are 26 February and 12 April for the spring equinox and 31 August and 16 October for the autumn equinox.

Furthermore, the dates of the spring and autumn equinoxes with respect to passing through the perigee are 77 d 8 h and 263 d 18.5 h respectively; that is 21 March and 23 September. The eclipse season thus extends over about 22 days before and after the spring equinox and over around 23 days before and after the autumn equinox.

The daily duration of the eclipse is calculated from equation (7.66) by knowing the value of the declination δ and the required date. The total duration of the eclipse is $8\,\varpi$ minutes, ϖ being expressed in degrees. Figure 7.32 gives the daily duration of the eclipse assuming a cylindrical shadow.

Time of the eclipse. At half the daily duration, the satellite crosses the plane orthogonal to the equatorial plane formed by the sun and the axis of the earth. It is thus midnight true solar time at the longitude of the satellite.

The eclipse starts at $24\,\varpi/360$ (h) before true solar midnight and ends at $24\,\varpi/360$ (h) after midnight. The mean solar time from which legal time is defined is obtained by adding the time equation ΔE given by equation (7.31). The time equation varies from + 12 min to 0 during the spring equinox eclipse season and from 0 to − 15 min during the autumn equinox eclipse season.

Example calculation. Time of the start of the eclipse at the autumn equinox:

At the equinox, the total duration of the eclipse is 71.5 min.

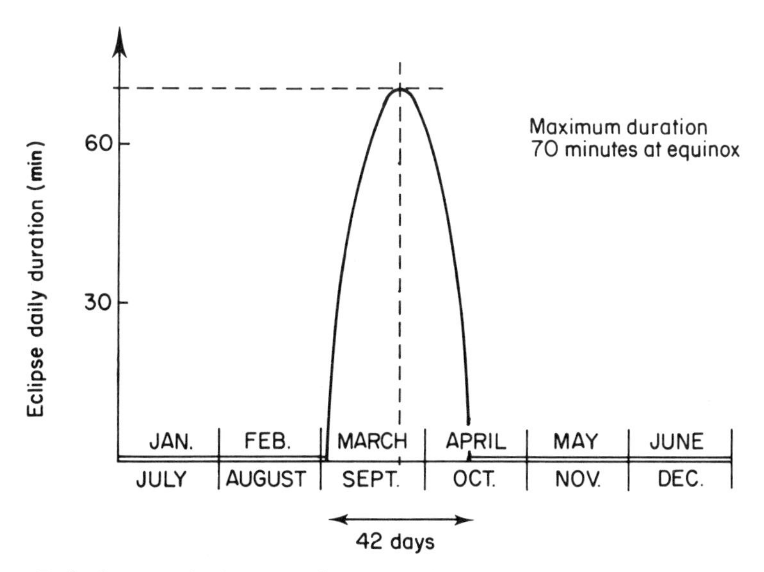

Figure 7.32 Daily duration of eclipses as a function of the date.

The true solar time of the start of the eclipse (for the meridian of the satellite) is:

$$\mathrm{TL} = 12\,\mathrm{h} - (71.5/2) = 11\,\mathrm{h}\,24.25\,\mathrm{min}\ (\text{true solar time is } 23\,\mathrm{h}\,24.25\,\mathrm{min})$$

The time equation on 23 September gives:

$$\Delta E = 460 \sin n_{SUN}t - 592 \sin 2(\omega_{SUN} + n_{SUN}t) = -453\,\mathrm{s} = -7.5\,\mathrm{min}$$

with:

$$t = 263\,\mathrm{d}\ 18.5\,\mathrm{h},\ n_{SUN} = 360/362.5 = 0.985626^\circ/\mathrm{day},\ \omega_{SUN} = 280^\circ.$$

The mean solar time of the start of the eclipse is thus:

$$\mathrm{TM} = \mathrm{TL} + \Delta E = 11\,\mathrm{h}\,(24.25 - 7.5)\,\mathrm{min} = 11\,\mathrm{h}\,16.75\,\mathrm{min}$$

For a satellite at longitude λ, the universal time UT will have a value:

$$\mathrm{UT} = 11\,\mathrm{h}\,16.75\,\mathrm{min} - 12\,\mathrm{h} + \lambda/15.$$

With, for example, $\lambda = 19^\circ$W, the universal time has a value 0 h 33 min at the start of the eclipse. The legal time of the service region differs with respect to UT by a integer number of hours; hence, for France (summer time), for example:

$$\text{Time of the start of the eclipse} = \mathrm{UT} + 2 = 02\,\mathrm{h}\,33\,\mathrm{min}.$$

Operation during an eclipse. If the satellite uses solar energy as a source of power and if the satellite must provide a continuous service, it is necessary to provide an energy storage which permits normal operation at the equinoxes for about 70 min.

One solution is to use the back-up satellite, if one exists. This is satisfactory if the two satellites are sufficiently far apart in longitude so that one is always illuminated when the other is in shadow. The separation of the two satellites must be greater than 17.4°. There are, however, two disadvantages:

- —The change of satellite involves a reorientation of the antennas on the ground and hence an interruption of service unless two antennas, or one antenna with electronic pointing, are provided.
- —The coverage is that which is common to the two satellites.

For certain types of satellite, satellites for direct television broadcasting for example, it is conceivable that a service would not be provided since eclipses always occur at night. They occur later at night when the satellite is further to the west of the region to be covered; a shift of 15° of the longitude of the satellite towards the west with respect to the longitude of the service region corresponds to an eclipse occurring at 01 h 00 true solar time of the service region, that is around 02 h 00 or 03 h 00 in civil time.

7.2.5.6 *Eclipses of the sun by the moon*

In addition to eclipses due to the earth, the solar disc as seen by a geostationary satellite can be partially or totally obscured by the moon. Compared with those due to the earth, eclipses due to the moon are of irregular occurrence and extent [SIO-81]. The number of eclipses per year due to the moon for a given orbital position varies from zero to four with a mean of two. Eclipses can occur twice in a period of 24 h. The duration of eclipses varies from several minutes to more than two hours with a mean of around 40 min (CCIR Rep 802). Figure 7.33 and Table 7.9 show an example of the occurrences, durations and extents of eclipses of the sun by the moon for a satellite at 31° W.

The satellite is liable to suffer from problems associated with excessive discharge of the batteries and a large fall in the temperature of certain parts if an eclipse of the sun due to the moon occurs just before or after an eclipse of the sun by the earth.

7.2.5.7 *Conjunction of the sun and the satellite*

Conjunction of the sun and the satellite occurs when the axis of the antenna beam from a ground station pointing towards the satellite passes through the sun. This implies that the declination of the sun is equal to the angle which the radiating axis of the antenna makes with the equatorial plane. Whatever the position of the station on the surface of the earth, this angle has a maximum value of 8.7° (see Section 7.2.5.3). Sun–satellite conjunction thus occurs at times of the year which are close to the equinoxes as follows:

—Before the spring equinox and after the autumn equinox for a station in the northern hemisphere,
—After the spring equinox and before the autumn equinox for a station in the southern hemisphere.

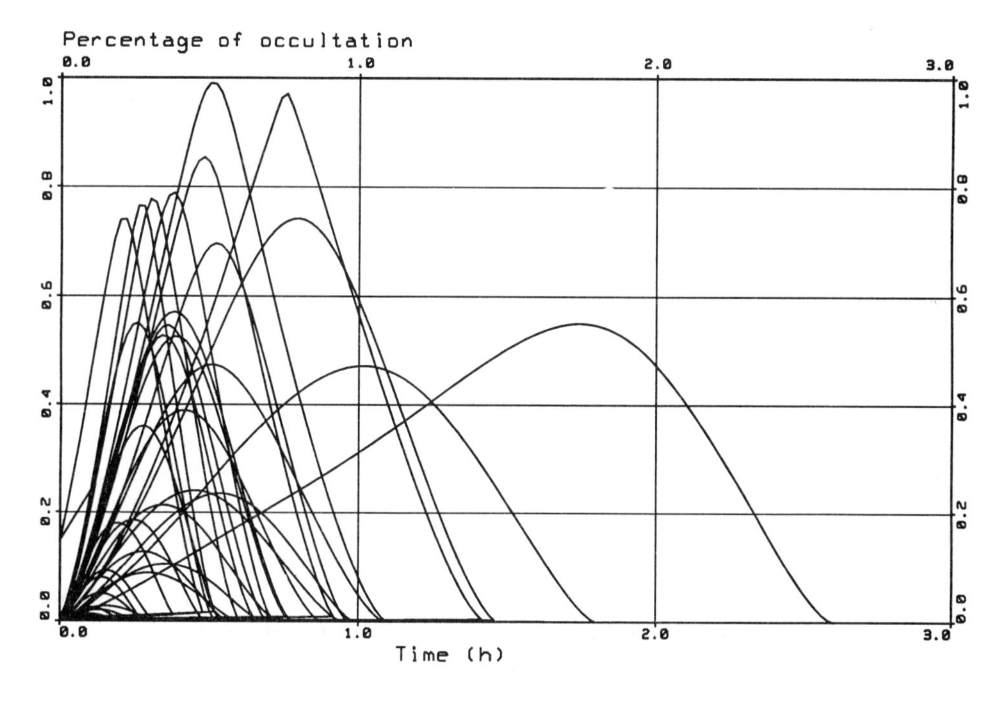

Figure 7.33 Eclipses of the sun by the moon. Percentage of sun occulation as a function of time for a geostationary satellite at longitude 31 °W from January 1992 to December 2010.

Table 7.9 Occurrence, duration, and depth of solar eclipse due to the Moon for a geostationary satellite at longitude of 31 °W from January 1992 to December 2010 (only eclipses displayed in Figure 7.33 with depth larger than 40% are listed)

Date		Time in	Time max depth	Time out	Depth
29 July	1992	8.93	9.93	10.73	0.47
21 May	1993	22.81	23.11	23.41	0.88
13 Dec	1993	16.04	16.54	17.12	0.47
12 Oct	1996	13.59	13.97	14.35	0.53
7 Apr	1997	15.63	16.13	16.73	0.99
7 Apr	1997	20.79	21.31	21.75	0.70
17 Jan	1999	12.59	13.05	13.51	0.85
11 Aug	1999	4.07	4.33	4.59	0.55
2 Jul	2000	9.61	10.41	11.07	0.74
24 Jan	2001	22.34	22.70	23.04	0.55
14 Dec	2001	9.21	9.97	10.63	0.97
25 Oct	2003	14.22	14.60	15.00	0.57
25 Oct	2003	21.73	22.11	22.45	0.79
19 Apr	2004	22.54	22.88	23.20	0.53
10 Mar	2005	18.08	19.82	20.68	0.55
9 Apr	2005	0.54	0.80	1.06	0.77
23 Aug	2006	23.87	0.22	0.48	0.74
7 Mar	2008	11.85	12.35	12.79	0.73
1 Aug	2008	3.77	4.01	4.23	0.41
26 Jan	2009	16.64	17.16	17.74	0.52
23 Jun	2009	0.17	0.41	0.63	0.62
15 Jan	2010	16.63	17.75	17.73	0.66
12 Jun	2010	4.23	4.49	4.77	0.73

Date of conjunctions. Considering a nadir angle β of 8.7° and an infinitely narrow beamwidth, the dates of sun–satellite conjunction for a station at the limit of visibility in the northern hemisphere are those corresponding to a declination of the sun equal to 8.7°. These dates have already been determined in Section 7.2.5.5 for the start of the eclipse season before the spring equinox and the end of the season after the autumn equinox; they are around 26 February and 16 October.

If the nadir angle is less than 8.7°, the dates approach the respective equinoxes. The date and time of maximum conjunction can be calculated from the angles of elevation E (obtained using (7.43a)) and azimuth A (7.64) under which the satellite is seen from the station. For greater accuracy, the flattening of the earth is taken into account by replacing the equatorial radius R_E by R_C, the distance from the station to the centre of the ellipsoid (see (7.26b)), in the calculation of the station– satellite distance R from (7.40) and by replacing the geographic latitude l by the geocentric latitude l' (see (7.26a)) in equation (7.63) for calculation of the angle ϕ [PRI-86].

The hour co-ordinates δ, declination and H, the hour angle of the satellite are obtained from the elevation and azimuth by using the formulae for conversion from horizontal co-ordinates to hour co-ordinates [BDL-90], modified to take account of the non-astronomical definition of the azimuth angle.

This gives:

$$\sin \delta = \sin l' \sin E + \cos l' \cos E \cos A \tag{7.67a}$$

and

$$\cos H = (\cos l' \sin E - \sin l' \sin \delta \cos A)/\cos \delta \tag{7.67b}$$

Conjunction with the sun will occur on the day when the declination of the sun δ_{SUN} is equal to the declination δ of the satellite. The declination of the sun is related to its right ascension α_{SUN} by $\delta_{SUN} = \tan \epsilon \sin \alpha_{SUN}$ (see Figure 7.6). Two values of right ascension of the sun correspond to a given declination. One value in the vicinity of the spring equinox (before the equinox for a station in the northern hemisphere) corresponds to:

$$\alpha_{SUN} = \arcsin [\tan \delta_{SUN}/\tan \epsilon]$$

The other value in the vicinity of the autumn equinox is such that:

$$\alpha_{SUN} = 180^\circ - \arcsin [\tan \delta_{SUN}/\tan \epsilon]$$

The relation between the date and the corresponding value of right ascension of the sun is given by (7.27) and (7.30) given that Greenwich civil time, or universal time UT, is the mean solar time increased by 12 h. Mean solar time differs from the true solar time of the time equation (7.30). True solar time is the hour angle of the sun which is related by (7.27) to sidereal time ST through the intermediary of the right ascension of the sun α_{SUN}. Finally, this gives for date t at 0 h UT (α_{SUN} and ST in hours):

$$\alpha_{SUN} = \mathrm{ST} + 12 - \Delta E \tag{7.68}$$

The sidereal time ST for the date JD at 0 h is obtained using (7.32). By simplifying equation (7.32b), for example, one can write:

$$\mathrm{ST} = (1/3600)\,[24\,110.6\,\mathrm{s} + 8640\,184.812\,866 \times T\,\mathrm{s}]$$

where T is the number of Julian centuries between the date JD at 0 h and 1 January 2000 at 12 h (adding or subtracting multiples of 24 h so that $0 < \mathrm{ST} < 24$ h).

To obtain JD from α_{SUN}, the calculation is more difficult since the time equation ΔE depends on the date. It is necessary to proceed by iterations starting from an initial date obtained by consulting astronomical tables or by determining the approximate date when the sun has the declination δ_{SUN} in Figure 7.7.

Time of conjunction. To obtain the time of conjunction with the sun on day JD, the hour angle H of the satellite must be set equal to the hour angle of the sun at the earth station. The local sidereal time LST is obtained by adding the value of right ascension of the sun on day JD:

$$\mathrm{LST} = H + \alpha_{SUN} \tag{7.27}$$

and the sidereal time on subtracting the longitude east of the station:

$$\mathrm{ST} = \mathrm{LST} - \lambda \tag{7.28}$$

The time SU of the conjunction measured in sidereal units of time is then obtained by subtracting the sidereal time at 0 h on day JD:

$$\mathrm{SU} = \mathrm{ST} - \mathrm{ST}(\mathrm{JD\ at\ 0\,h})$$

It is necessary to convert sidereal units of time SU into universal time UT (see Section 7.1.5.5):

$$\mathrm{UT} = \mathrm{SU} \times 0.997\,2696$$

The universal time UT (of Greenwich) of the maximum conjunction with the sun is thus obtained. The local time at the station will taken account of the corresponding time zone and, if necessary, the date (summer or winter time).

Number of days of interference. Conjunction is defined by alignment of earth station, satellite and sun. If the antenna beam is assumed to be infinitely narrow, conjunction corresponds to the only situation where the earth station suffers interference. If this beam has an equivalent aperture θ_i, interference will occur on several successive days around the initial date defined by the value of nadir angle β such that the declination of the sun remains between $\beta - \theta_i/2$ and $\beta + \theta_i/2$. The declination of the sun around the equinoxes varies approximately 0.4° per day. The number N_i of consecutive days of satellite–sun interference thus has a value:

$$N_i = 2.5\ \theta_i \text{ days} \tag{7.69}$$

where θ_i is the equivalent aperture of the antenna beam in degrees. By way of example, if $\theta_i = 2°$, the interference occurs for five consecutive days, that is two days before and two days after the nominal date.

Duration of the interference. The duration of interference with the sun is determined by noticing that the apparent daily movement of the sun around the earth has a value of 0.25° per minute. The duration Δt_i of the interference is thus:

$$\Delta t_i = 4\theta_i \text{ min} \tag{7.70}$$

Taking $\theta_i = 2°$, the duration of the interference is equal to 8 min.

During this interference the antenna temperature increases abruptly. The value of this increase and the method of determining the equivalent aperture θ_i are discussed in Chapter 8.

7.3 PERTURBATIONS OF THE ORBIT

Movement of the satellite in its orbit is determined by the forces acting on the centre of mass. With the Keplerian hypotheses, there is only the attraction of a central, spherical and homogeneous body which defines a conservative field of forces (equation 7.3). The trajectory obtained is plane, fixed in space and characterised by a set of constant orbital parameters. These orbital parameters can be obtained from the position and velocity vectors of the satellite by a geometric transformation. In the case of a perturbed orbit, the orbital parameters are no longer constant but are a function of the date for which the transformation is applied. Extrapolation of the orbit could be made by numerical integration of the equation of motion after taking account of the various perturbations [ZAR-84], [ZAR-87].

Perturbations of the orbit are the result of various forces which are exerted on the satellite other than the force of attraction of the central, spherical and homogeneous body. These forces consist mainly of:

—The contribution of the non-spherical components of terrestrial attraction.
—The attraction of the sun and the moon.

—Solar radiation pressure.
—Aerodynamic drag.
—Motor thrust.

The first two contributions are gravitational forces from perturbing potentials. In contrast, the other forces do not depend on the mass of the satellite and are not conservative; they are due to exchanges of the amount of movement at the surface of the satellite and depend on the aspect and geometry of the satellite and therefore provide the possibility of control.

7.3.1 The nature of the perturbations

7.3.1.1 Asymmetry of the terrestrial potential

The earth is not a spherical homogeneous body. The terrestrial potential at a point in space depends not only on the distance r to the centre of mass but also on the latitude and longitude of the point concerned and the time. This is due to the irregularities of the rotation of the earth and the mass distribution (caused by oceanic and terrestrial tides, that is movement of the surface of the oceans and the earth's crust under the effect of lunar attraction and internal geophysical phenomena). With the choice of a reference tied to the earth's crust, a simplified expansion of the static part (using mean coefficients) of the terrestrial potential is as follows:

$$U = \mu/r\left[1 - \sum_{n=2}^{\infty} (R_E/r)^n J_n P_n(\cos\varphi) + \sum_{n=2}^{\infty}\sum_{q=1}^{\infty} (R_E/r)^n J_{nq} P_{nq}(\cos\varphi)(\cos q(\lambda - \lambda_{nq}))\right] \quad (7.71)$$

where:

$\mu = 3.986 \times 10^{14}\,\mathrm{m^3 s^{-2}}$, the gravitational constant of the earth,
r = distance of the point considered with respect to the centre of the earth,
$R_E = 6378.14$ km, the mean terrestrial equatorial radius,
φ, λ = the latitude and longitude of the point considered,
J_n = zonal harmonics,
J_{nq} = tesseral harmonics,
P_n = Legendre polynomial of order n

$$P_n(x) = [1/(2^n n!)]\,d^n/dx^n[(x^2 - 1)^n]$$

P_{nq} = the associated Legendre function

$$P_{nq}(x) = (1 - x^2)^{q/2} d^q/dx^q P_n(x)$$

The J_n and J_{nq} terms are constants which are characteristic of the distribution of the mass of the earth. The J_n terms are the zonal harmonic functions of longitude. The J_2 term, due to the flattening of the earth (about 20 km) dominates all the other terms. The J_{nq} terms are the tesseral harmonic functions ($n \neq q$) of the latitude and longitude, or the sectorial ($n = q$) functions of longitude. The dominant term J_{22} is characteristic of the ellipticity of the equator (a difference of 150 m between the semi-minor and semi-major axes).

The values of the coefficients are given by various models (with many formulations for the expansion of the potential) such as those developed by the Goddard Space Flight Center and the GRIM models of the Groupe de Recherche en Geodésie Spatiale and the German Geodatische Forschung Institute [BAL-84]. Some numerical values of the coefficients are as follows (GEM 4):

$$J_2 = 1.0827 \times 10^{-3}; \quad J_{22} = 1.803 \times 10^{-6}; \quad \lambda_{22} = -14.91^\circ$$

The order of magnitude of the coefficients J_n and J_{nq} for $n > 2$ is given by $10^{-5}/n^2$ [KAU-66].

The perturbing potential is given by:

$$U_p = U - \mu/r \tag{7.72}$$

For a geostationary satellite, the ratio R_E/r is small and the latitude is close to 0 ($\cos\varphi = 1$). To a first approximation, by limiting the expansion to order 2, this gives:

$$U \approx \mu/r[1 + (R_E/r)^2 \{J_2 + 3J_{22} \cos 2(\lambda - \lambda_{22})\}] \tag{7.73}$$

with $\lambda_{22} = 14.91^\circ$, that is 15° longitude west.

7.3.1.2 *Attraction of the moon and the sun*

The moon and the sun each create a gravitational potential whose expression is of the form:

$$U_p = \mu_p\{1/\Delta - [(\mathbf{r}_p \cdot \mathbf{r})/|\mathbf{r}_p|^3]\} \qquad \text{with } \Delta^2 = |\mathbf{r}_p - \mathbf{r}| \tag{7.74}$$

where $\mathbf{r}$ is the vector from the centre of the earth to the satellite, $\mathbf{r}_p$ is the vector from the centre of the earth to the perturbing body, and $\mu_p = GM_p$ (M_p = mass of the perturbing body) is the attraction constant of the perturbing body (the moon or sun). For the moon, $\mu_p = 4.8999 \times 10^{12}\,\mathrm{m^3/s^2}$, for the sun $\mu_p = 1.345 \times 10^{20}\,\mathrm{m^3/s^2}$.

7.3.1.3 *Solar radiation pressure*

A surface element dS with normal $\mathbf{n}$ oriented in the direction of the sun and making an angle θ with the unit vector $\mathbf{u}$ directed towards the latter is subjected to the following pressure:

$$\mathbf{dF}/dS = -(W/c)[(1+\rho)(\cos\theta)^2\mathbf{n} + (1-\rho)(\cos\theta)\mathbf{n} \wedge (\mathbf{u} \wedge \mathbf{n})] \tag{7.75}$$

where ρ is the reflectivity of the surface (the ratio of the reflected and incident fluxes), W is the solar flux (power received per unit surface area) and c is the velocity of light ($W/c = 4.51 \times 10^{-6}\,\mathrm{N/m^2}$ at 1 IAU).

If the element dS is totally reflecting ($\rho = 1$), the pressure is normal to the surface:

$$\mathbf{dF}/dS = (2W/c)(\cos\theta)^2\mathbf{n}$$

If the element dS is totally absorbent ($\rho = 0$), the radiation pressure divides into a normal component $(dF/dS)_N = (W/c)(\cos\theta)^2$ and a tangential component $(dF/dS)_T = -(W/c)(\cos\theta)^2 \sin\theta$.

A satellite of apparent surface S_a in the direction of the sun and reflectivity ρ equal to 0.5 (a typical value) is subjected to a perturbing force:

$$F_p = -1.5(W/c)S_a \qquad \text{(N)} \tag{7.76}$$

If the satellite is of mass m, the acceleration due to radiation pressure is:

$$\Gamma = 6.77 \times 10^{-6} S_a/m \qquad (\text{m/s}^2)$$

The solar panels constitute practically the whole of the apparent surface of the satellite. With communication satellites of low power (1 kW), the solar panels are not extensive and the ratio S_a/m is of the order of $2 \times 10^{-2}\,\text{m}^2/\text{kg}$. This is the case, for example, of Intelsat V for which $S_a = 18\,\text{m}^2$ and $m = 1000\,\text{kg}$ from which $S_a/m = 1.8 \times 10^{-2}\,\text{m}^2/\text{kg}$. With these satellites, the acceleration due to radiation pressure is of the order of $10^{-7}\,\text{m/s}^2$ and its effect is limited.

For satellites of high electrical power on which very extensive solar panels are mounted (a surface of $100\,\text{m}^2$ for a mass of 1000 kg, for example), the ratio S_a/m is of the order of 10^{-1}; the acceleration due to radiation pressure must then be taken into account in calculating perturbations.

The main effect of solar radiation pressure is to modify the eccentricity of the orbit which evolves with a period of 1 year (Section 7.3.3.5).

For satellites in a low orbit, it is also necessary to take account of the radiation pressure of the solar flux reradiated from the surface of the earth (albedo) whose effect can be significant (20%) with respect to that of the direct solar flux.

7.3.1.4 *Aerodynamic drag*

In spite of the low value of atmospheric density encountered at the altitudes of satellites, their high velocity means that perturbations due to aerodynamic drag are very significant at low altitude (200–400 km) and are negligible only above about 3000 km. The aerodynamic force is exerted on the satellite in the opposite direction to its velocity and is of the form:

$$F_{AD} = -0.5\rho_A C_D A_e V^2 \tag{7.77}$$

where ρ_A is the density of the atmosphere, C_D is the coefficient of aerodynamic drag, A_e is the equivalent surface area of the satellite perpendicular to the velocity and V is the velocity of the satellite with respect to the atmosphere.

The density of the atmosphere depends on the altitude (the variation is exponential), the latitude, the time, solar activity etc. Various models have been developed (e.g. [JAC-77]). The coefficient of aerodynamic drag is a function of the form and nature of the surface. The velocity with respect to the atmosphere differs from the velocity of the satellite in an inertial reference since the atmosphere has some velocity as a consequence of dragging by terrestrial rotation and the phenomena of wind [ESC-84], [ZAR-84].

If the satellite is of mass m, the acceleration due to atmospheric drag is:

$$\Gamma_{AD} = -0.5\rho_A C_D V^2 A_e/m \qquad (\text{m/s}^2) \tag{7.78}$$

The main effect of atmospheric friction is a decrease of the semi-major axis of the orbit due to a reduction of the energy of the orbit. A circular orbit remains as such, but its altitude reduces whereas the velocity of the satellite increases. For an elliptical orbit, the braking occurs principally at the perigee. The altitude of the apogee decreases, the altitude of the perigee remains almost constant, the eccentricity decreases and the orbit tends to become circular. By way of example, for an elliptical orbit with perigee altitude 200 km and apogee altitude 36 000 km (a transfer orbit, see Chapter 11), the reduction of altitude of the apogee is around 5 km on each orbit.

7.3.2 The effect of perturbations; orbit perturbation

7.3.2.1 *Osculatory parameters*

The actual movement of the satellite is obtained from the fact that the satellite is in equilibrium between the inertial force $m\mathrm{d}^2r/\mathrm{d}t^2$ and the various forces which are exerted on it. The latter include:

—The force of attraction due to the potential of a spherical and homogeneous earth.
—Forces due to the various perturbing potentials.
—Non-conservative perturbing forces.

Hence:

$$m\mathrm{d}^2\mathbf{r}/\mathrm{d}t^2 = m\mu(\mathbf{r}/r^3) + m\mathrm{d}/\mathrm{d}\mathbf{r}[U_\mathrm{p}] + \mathbf{f}_\mathrm{p} \qquad (\mathbf{F},\ \mathbf{r}\text{ and }\mathbf{f}_\mathrm{p}\text{ are vectors}) \tag{7.79}$$

It is thus possible to determine the position and velocity of the satellite at each instant by integration in a geocentric reference frame. Using a geometric transformation defined by the Keplerian hypotheses, it is possible, on a given date, to obtain the six orbital parameters which are characteristic of the movement of a satellite. Unlike the Keplerian orbit where the parameters are constants, these parameters are functions of time for a perturbed orbit.

These parameters, determined for the current date t, are called *osculatory parameters.* The osculatory elements for the date t are the orbital elements of Keplerian movement which would describe the satellite if the perturbations were cancelled from the date t. The trajectory defined in this way is called the osculatory ellipse. This nomenclature is inappropriate as the curvature of the osculatory ellipse, although tangential to the trajectory on date t, is *not* the actual curvature.

The use of orbital parameters and their variations enables the progression of the trajectory (for example a rotation of the line of apsides) to be appreciated more easily than variations of distance and velocity of the satellite with respect to the expressions for Keplerian movement.

7.3.2.2 *Variation of the orbital parameters*

The variation of the orbital parameters ($\mathrm{d}a/\mathrm{d}t$, $\mathrm{d}e/\mathrm{d}t$, $\mathrm{d}i/\mathrm{d}t$, $\mathrm{d}\Omega/\mathrm{d}t$, $\mathrm{d}\omega/\mathrm{d}t$ and $\mathrm{d}M/\mathrm{d}t$) is obtained from the components of the perturbing accelerations in an orthogonal reference centred on the satellite by means of Gauss' equations.

If the perturbing accelerating field is due to a potential (only forces of gravitational origin are present), the system of differential equations can be put in a particular form as a function of the partial derivatives of the perturbing potentials with respect to the orbital parameters; these are Lagrange's equations [BAL-80], [ZAR-84], [ZAR, 87]. Integration of Lagrange's equations [KAU-66] produces the orbital parameters as the sum of a mean parameter, periodic terms with short and long periods (with respect to the period of the orbit) and, for some, a secular term (that is an increasing function of time).

7.3.2.3 *Secular variations*

Perturbations of the terrestrial potential cause secular variations which affect the ω (argument of the perigee), Ω (right ascension) and M (mean anomaly) parameters. These secular terms are a function of the even zonal harmonics, particularly J_2. This gives:

$$d\omega/dt = (3/4)n_0 AJ_2[5(\cos^2 i) - 1] \quad (7.80a)$$

$$d\Omega/dt = -(3/2)n_0 AJ_2 \cos i \quad (7.80b)$$

$$dM/dt = n_0[1 + 3/4A(1-e)^{1/2}J_2(3(\cos^2 i) - 1)] \quad (7.80c)$$

where:

$A = R_E^2/a^2(1-e^2)^2$

R_E = radius of the earth,

e, a = ecentricity and semi-major axis of the satellite orbit,

i = inclination of the orbit,

n_0 = mean movement of the satellite $= 2\pi/T = \sqrt{(\mu/a^3)}$.

For example, for an elliptical orbit of inclination 7° whose perigee altitude is 200 km and apogee altitude is 36 000 km (a transfer orbit, see Chapter 11), the derivative of the argument of the perigee $d\omega/dt$ has a value of 0.817°/day. For a circular orbit of altitude 290 km and inclination 28° (the parking orbit of STS, see Chapter 11), the derivative of the right ascension of the ascending node (the nodal regression) has a value $d\Omega/dt = 7.5°$/day. This nodal regression is zero for polar orbits ($i = 90°$).

Selection of the value of certain orbital parameters enables the derivative of another parameter to be fixed at a particular value. Hence to make the derivative of the argument of the perigee $d\omega/dt$ zero, it is acceptable to choose the value of inclination which makes the term $5(\cos^2 i) - 1$ equal to zero; that is $i = 63.4°$. There will no longer be a rotation of the perigee–apogee line in the plane of the orbit and the apogee will remain permanently above the same hemisphere. This is the reason which led to the choice of 63.4° for the inclination of the MOLNYA and TUNDRA orbits (see Section 7.2).

By choosing a pair of particular values for a and i, it is possible to obtain an orbit for which the right ascension of the ascending node varies each day by a quantity equal to the mean variation of the right ascension of the sun, that is $d\Omega/dt = 0.9856°$/day. For a circular orbit, the condition can be written $6.527a^{-7/2}\cos i = 0.985$.

The angle between the line of nodes of the orbits and the mean direction of the sun obtained in this way remains constant throughout the year. The conditions for illumination are thus always identical from one orbit to another with fluctuations due to the variation of the declination of the sun and the time equation. The satellite is said to be sun-synchronous. If the orbit is also phased, that is the period is a sub-multiple of the sidereal day or an integral number of days, the satellite passes over the same points again with a period equal to the number of days concerned. These orbits are thus particularly well suited to earth observation missions. For example, the orbit of the Spot satellite has an altitude of 820 km ($a = 7200.5$ km), an inclination of 98.7° and a period of 101.3 min; it returns over the same point after 26 days.

7.3.3 Perturbations of the orbit of geostationary satellites

The orbit of geostationary satellites has been defined in Section 7.2.5 as a direct circular orbit ($e = 0$) in the plane of the equator ($i = 0$) whose period of revolution is equal to the period of rotation of the earth ($T = 86\,164.1$ s); this gives rise to a semi-major axis a_k, calculated using the Keplerian hypotheses, of 42 164.2 km.

If a satellite is placed in an orbit defined in this way, it is observed that, due to the effect of perturbations, the parameters of the orbit do not remain constant as Kepler's equations would predict. The apparent movement of the satellite with respect to a rotating reference related to the earth is as follows:

—Displacement in the east–west plane with respect to the nominal position defined by the longitude of the satellite station (predicted to be fixed since the velocity of rotation of the satellite has been made the same as that of the earth); a modification of the radial distance is associated with this displacement.
—Displacement in the north–south plane with respect to the equatorial plane.

Examination of the orbit parameters after several weeks shows that the values of the semi-major axis, the eccentricity and the inclination of the orbit are no longer equal to the initial values. The satellite is no longer perfectly geostationary.

7.3.3.1 *Modified orbit parameters*

Conventional parameters are not well suited to characterisation of the orbit of a quasi-geostationary satellite; when the inclination tends to zero, the position of the ascending node becomes indeterminate. The same applies to the position of the perigee when the eccentricity tends to zero. It is thus logical to characterise the simultaneous progression of i and Ω and the simultaneous progression of e and $(\Omega + \omega)$.

This is obtained by introducing:

—The inclination vector i with components:

$$i_x = i \cos \Omega$$

$$i_y = i \sin \Omega$$

—The eccentricity vector $\mathbf{e}$ with components:

$$e_x = e \cos (\omega + \Omega)$$

$$e_y = e \sin (\omega + \Omega).$$

The inclination vector is represented in Figure 7.34a by a vector along the line of nodes directed towards the ascending node and of modulus equal to the inclination. The eccentricity

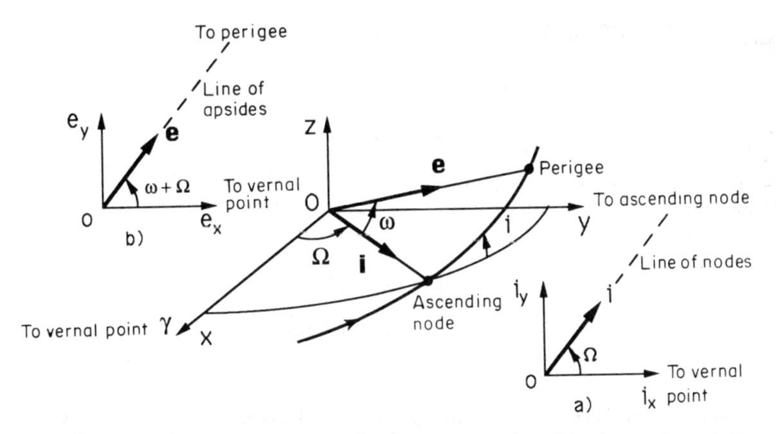

Figure 7.34 Inclination and eccentricity vectors for characterising the orbit of a quasi-geostationary satellite.

vector is represented in Figure 7.34b by a vector along the line of apsides directed towards the perigee and of modulus equal to the eccentricity.

The angle $(\Omega + \omega)$ is the sum of two angles in planes which are, in principle, different but close since the inclination remains small. Other definitions of the inclination vector can be used such as, for example, that of a vector along the axis of the angular momentum of the orbit of modulus equal to the inclination [SOO-83].

Furthermore, rather than using the mean anomaly, the position of the satellite in the orbit is characterised by the mean longitude λ_m or the true longitude λ_v which are given by:

$$\lambda_m = \omega + \Omega + M - \mathrm{ST} \tag{7.81a}$$

$$\lambda_v = \omega + \Omega + v - \mathrm{ST} \tag{7.81b}$$

where M and v are the mean and true anomalies respectively, and ST is the sidereal time of the Greenwich meridian.

As a consequence of the rotation of the earth, the sidereal time of the Greenwich meridian increases by 15.041 69° per hour (see equation 7.29). As the eccentricity is small, the relation between the mean anomaly and the true anomaly is given by:

$$\lambda_v = \lambda_m + 2e(\sin M) \tag{7.82}$$

The true longitude of the satellite thus oscillates about the mean longitude with an amplitude of $2e$ in the course of the day (see Section 7.2.2).

In conclusion, the orbital parameters adapted to the quasi-geostationary satellite are:

$$a, e_x, e_y, i_x, i_y \quad \text{and} \quad \lambda_m.$$

7.3.3.2 *Semi-major axis of a geosynchronous circular orbit*

The semi-major axis of the perturbed geosynchronous circular orbit is different from the semi-major axis a_K calculated using the Keplerian hypotheses. Lagrange's equations lead to an expression for the drift of the mean longitude which is:

$$d\lambda_m/dt = -(2/na)(dU_p/da) + n - \Omega_E \tag{7.83}$$

where Ω_E is the velocity of rotation of the earth.

For the satellite to be geostationary, it is necessary for the drift of the mean longitude $d\lambda_m/dt$ to be zero. The value of the semi-major axis a_S corresponding to a geosynchronous orbit is then obtained from:

$$a_S = a_K + 2J_2 a_K (R_E/a_K)^2 + \cdots$$
$$= a_K + 2.09\ \text{km} = 42\,166.3\ \text{km}$$

where a_K is the semi-major axis of the Keplerian orbit.

Cancellation of a long-term drift due to lunar-solar attraction leads to a further modification of the semi-major axis which has a final value:

$$a_S = 42\,165.8\ \text{km}$$

Expression (7.83) enables the derivative of the mean longitude $d\lambda_m/dt$ to be related to the variation Δa of the semi-major axis with respect to its value a_S corresponding to the synchronous orbit. This gives [ALB-83]

$$d\lambda_m/dt = -(3/2)(n_s/a_s)(\Delta a) = k_\lambda(a - a_s) \qquad \text{with } k_\lambda = -0.0128°/\text{day km} \tag{7.84}$$

7.3.3.3 *Progression of the longitude of the satellite*

For a quasi-geostationary satellite, the terrestrial perturbing potential is approximated by (7.73):

$$U_p = \mu/r[(R_E/r)^2\{J_2 + 3J_{22}\cos 2(\lambda - \lambda_{22})\}]$$

This potential creates a tangential acceleration Γ_T such that:

$$\Gamma_T = -(1/r)dU_p/d\lambda = (\mu/r^2)(R_E/r)^2 6J_{22}\sin 2(\lambda - \lambda_{22}) \tag{7.85}$$

This acceleration causes a variation of the velocity V_{SL} of the satellite in the orbit:

$$\Gamma_T = dV_{SL}/dt = d/dt[r\omega_{SL}] = (dr/dt)\omega_{SL} + r(d\omega_{SL}/dt) \tag{7.86}$$

where ω_{SL} is the angular velocity of the satellite.

As the orbit is quasi-circular, it is permissible to consider $r \approx a_S$ the semi-major axis of the geosynchronous circular orbit and $\omega_{SL} \approx n_S = \sqrt{(\mu/a_S{}^3)}$, which leads to:

$$d\omega_{SL}/\omega_{SL} \approx -(3/2)(dr/r) \qquad \text{for } r = a_S \tag{7.87}$$

Combining equations (7.86) and (7.87), this gives:

$$dr/dt\ (\text{for } r = a_S) \approx -(2/\omega_{SL})\Gamma_T \tag{7.88}$$

The longitudinal acceleration experienced by the satellite can thus be written:

$$d^2\lambda/dt^2 = d\omega_{SL}/dt \approx (3/a_S)\Gamma_T \approx D\sin 2(\lambda - \lambda_{22}) \tag{7.89}$$

where $D = 18n_S^2(R_E/a_S)^2 J_{22} = 3 \times 10^{-5}$rad/(day)$^2 = 4 \times 10^{-15}$ rad/s^2.

The longitudinal acceleration thus depends on the longitude of the satellite station. This acceleration varies sinusoidally with respect to longitude $\lambda_{22} = -14.91°$ and is zero for $\lambda = \lambda_{22} + k\pi/2$; in this way four equilibrium points are defined. Two of these points are points of stable equilibrium (that is, if the satellite is displaced from the equilibrium position, it tends to return to it); the other two are points of unstable equilibrium.

By putting $\Lambda = \lambda - \lambda_{22} \pm 90°$, the longitude of the satellite with respect to the closest point of stable equilibrium, the movement of the satellite about the point of equilibrium is governed by the equation:

$$d^2\Lambda/dt^2 = -D \sin 2(\Lambda)$$

The longitude drift $d\Lambda/dt$ as a function of longitude with respect to the point of stable equilibrium is thus of the form:

$$(d\Lambda/dt)^2 - D \cos 2\Lambda = \text{constant} \tag{7.90}$$

Curves showing the variation of drift $d\Lambda/dt$ as a function of longitude Λ about a stable equilibrium point are shown in Figure 7.35. The figures in parentheses give the period of the oscillatory movement of the satellite with respect to the point of stable equilibrium; it is at least two years. It can also be observed that, for excessive values of initial drift at the point of stable equilibrium, the natural acceleration will not cancel this drift before the satellite arrives in the vicinity of the adjoining point of unstable equilibrium. The satellite thus overshoots the point of unstable equilibrium and is attracted to the next stable equilibrium point where the same process is repeated. The drift is never cancelled, the satellite thus rotates perpetually with respect to the earth.

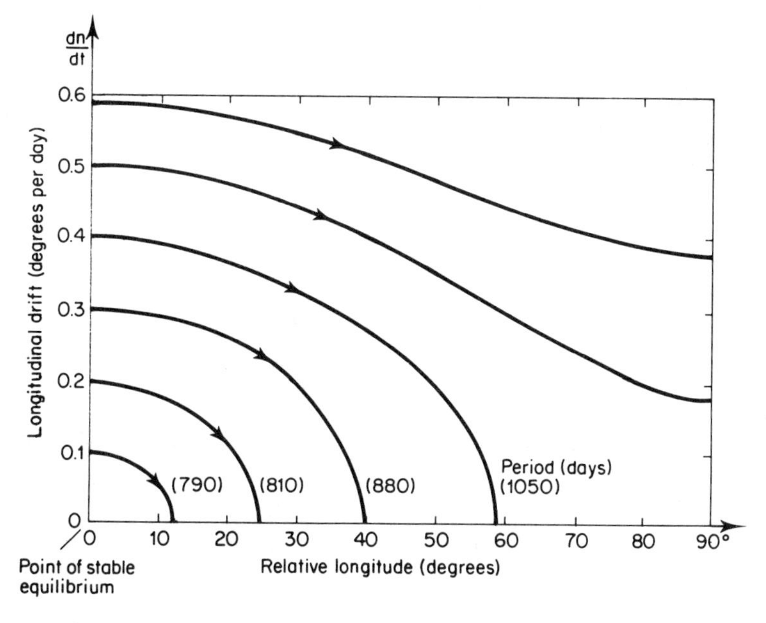

Figure 7.35 Evolution of the longitude drift as a function of the longitude with respect to a point of stable equilibrium.

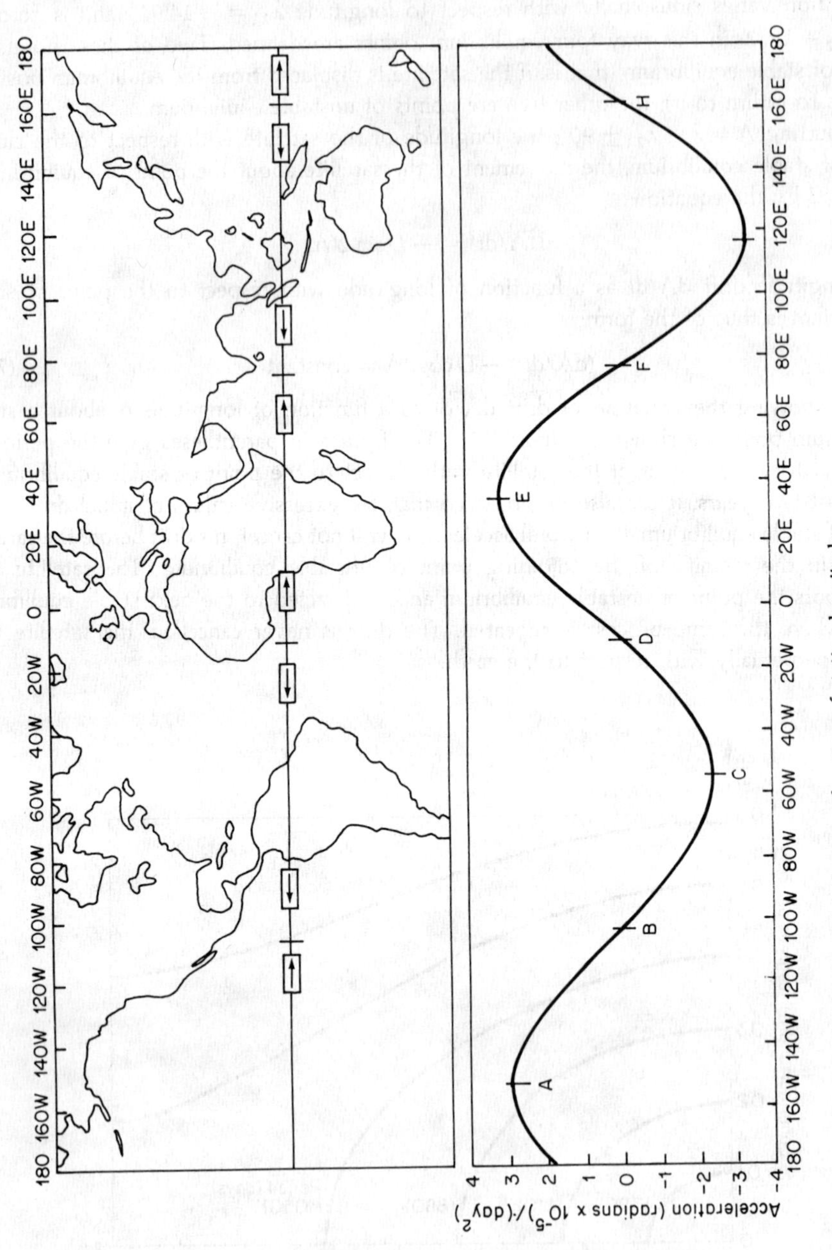

Figure 7.36 Longitudinal acceleration as a function of station longitude.

The results presented above are obtained by neglecting terms of order greater than 2 in the expansion of the perturbing potential. Figure 7.36 shows the actual longitudinal acceleration as a function of the longitude of the satellite station (CCIR-Rep 843). The longitudes of the positions of stable equilibrium are approximately 102° longitude west and 76° longitude east; those of the two positions of unstable equilibrium are 11° of longitude west and 164° of longitude east [SOO-83].

7.3.3.4 *Progression of the inclination*

The effect of the attraction of the moon and the sun can be seen in Figure 7.37 where the Oy axis is in the equatorial plane perpendicular to the direction of the vernal point (the Ox axis is not shown and is towards the front of the figure) and the Oz axis is the polar axis (see Figure 7.6). At the summer solstice, the sun (in the plane of the ecliptic) is above the equatorial plane. The plane of the moon's orbit makes an angle of 5.14° with the plane of the ecliptic. The track of the moon's orbit in the figure is within the region defined by two lines making an angle of $\pm 5.14°$ with the ecliptic; the track is a function of the value of the right ascension of the lunar orbit, which varies through 360° in 18.6 years. On the date considered, the sun and moon are assumed to be on the right of the figure (which corresponds to a new moon on the earth).

When the satellite is on the right of the figure, it is more strongly attracted by the sun and the moon than when it is on the left since the distance is less. The earth–satellite system behaves as if there were a perturbing force δF acting in one direction on half of the orbit and in the opposite direction on the other half as indicated in the figure. The same result is obtained when the moon is in the left part of the figure (full moon) after a lunar half period (27 days). The direction and magnitude of the perturbing force remains the same (except for variation of the earth–sun distance) if the sun is below the plane of the ecliptic on the left part of the figure (winter solstice) for both positions of the moon. When the moon or the sun is in the equatorial plane, the component of the perturbing force normal to the plane of the orbit caused by the body concerned is zero.

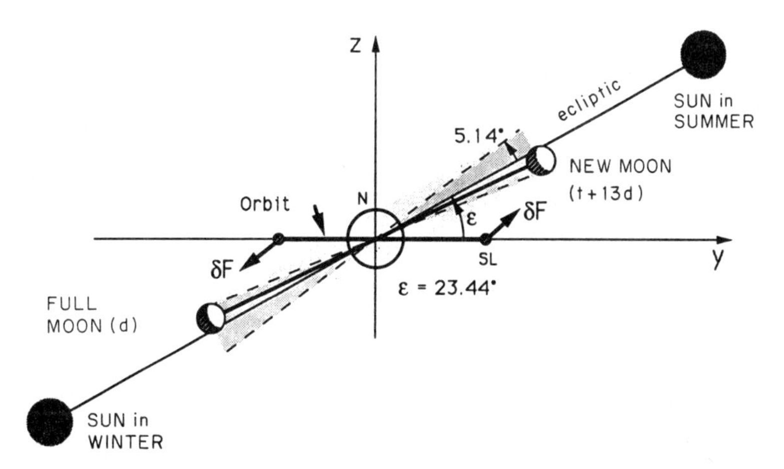

Figure 7.37 Attraction of the moon and the sun on the orbit of geostationary satellites.

The component of the perturbing force in the plane of the orbit affects the semi-major axis and eccentricity of the orbit. The long-term effect is cancelled by adjusting the value of the semi-major axis a_S to 42 165.8 km (see Section 7.3.3.2). The component of the force perpendicular to the plane of the orbit affects the inclination vector of the orbit. The effect of the sun is maximum at the solstices and zero at the equinoxes; this leads to a mean drift of the inclination vector of 0.27° per year. The component of the perturbation normal to the plane of the orbit due to the moon is maximum twice per lunar period and passes through zero between. The effect leads to a mean drift of the inclination vector between 0.48°/year and 0.68°/year as a function of the value of right ascension of the ascending node of the lunar orbit within its period of 18.6 years.

The combined effects of lunar-solar attraction on the inclination vector of the orbit of a quasi-geostationary satellite show the following principal effects:

—an oscillation of period 13.66 days and amplitude 0.0035°,
—an oscillation of period 182.65 days and amplitude 0.023°,
—a long-term variation

The components of the long-term variation are given by:

$$di_x/dt = H = (-3.6 \sin \Omega_M) \times 10^{-4} \text{ degrees/day}$$

$$di_y/dt = K = (23.4 + 2.7 \sin \Omega_M) \times 10^{-4} \text{ degrees/day}$$

where Ω_M is the right ascension of the ascending node of the lunar orbit for the period concerned and is given by:

$$\Omega_M = 12.111 - 0.052\,954\, T \qquad (T = \text{days since } 1/1/1950)$$

Figure 7.38 shows the secular progression of the inclination vector on a given date. The direction of drift Ω_D and the value of the derivative $\Delta i/\Delta t$ between the initial date t_0 and date t are such that:

$$\cos \Omega_D = H/\sqrt{(H^2 + K^2)} \qquad \Delta i/\Delta t = \sqrt{(H^2 + K^2)} \qquad \text{(7.91a and b)}$$

As a function of the epoch within the 18.6-year period of Ω_M:

—Ω_D varies from 81.8 degrees/year to 98.9 degrees/year,
—$\Delta i/\Delta t$ varies from 0.75 degrees/year to 0.95 degrees/year.

The zonal terms of terrestrial potential also affect the progression of the inclination vector and cause a recession of the right ascension of node Ω equal to 4.9 degrees/year.

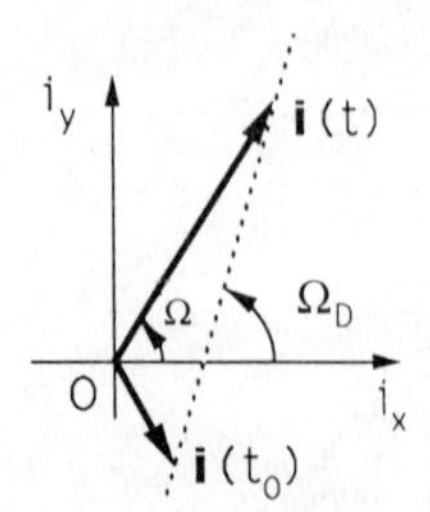

Figure 7.38 Secular evolution of the inclination vector.

The various contributions cause the extremity of the inclination vector on average to describe a circle in 54 years about a point with co-ordinates $\mathbf{i}_x = -7.4°$ and $\mathbf{i}_y = 0°$. This point constitutes a point of stable equilibrium for the long-term drift of the plane of the orbit corresponding to an inclination i equal to 7.4° and a right ascension Ω equal to 0° [KAM-78].

7.3.3.5 Progression of the eccentricity

Solar radiation pressure creates a force which acts in the direction of the velocity of the satellite on one half of the orbit and in the opposite direction on the other half. In this way, a circular orbit tends to become elliptical (Figure 7.39). The apsidal line of the orbit is perpendicular to the direction of the sun.

The ellipticity of the orbit does not increase constantly. With the movement of the earth about the sun, since the apsidal line remains perpendicular to the direction of the sun, the ellipse deforms continuously and the eccentricity may remain limited to a maximum value.

The progression of the eccentricity and the argument of the perigee is represented by that of the eccentricity vector which is obtained from Lagrange's equations involving e, ω and Ω to a first order in e and i by considering that the perturbing acceleration derives from a pseudo-potential. Assuming that the apparent orbit of the sun is circular and equatorial and neglecting short period terms (one day period), calculation leads to:

$$\mathrm{d}\mathbf{e}_x/\mathrm{d}t = -(3/2)(C/n_s a_S)\sin\alpha_S$$

$$\mathrm{d}\mathbf{e}_y/\mathrm{d}t = (3/2)(C/n_S a_S)\cos\alpha_S$$

where:

α_S = right ascension of the sun,
$C = \rho(S_a/m)(W/c)$,

with

S_a = apparent surface area of the satellite in the direction of the sun (m^2),
m = mass of the satellite (kg),
ρ = coefficient of reflectivity ≈ 1.5,
$W/c = 4.51 \times 10^{-6}\,\mathrm{N/m^2}$.

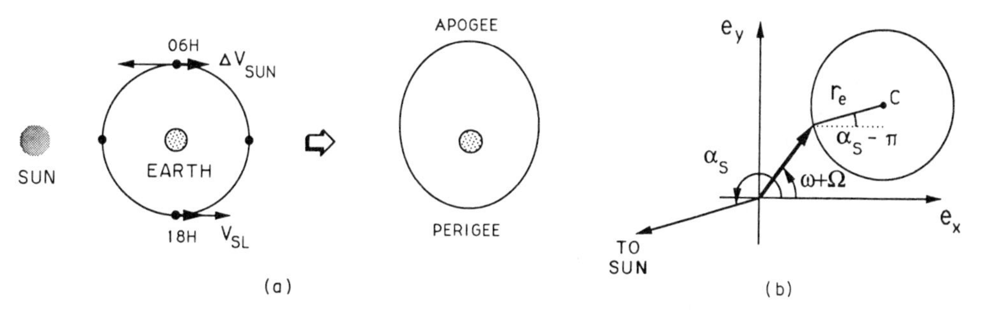

Figure 7.39 Effect of solar radiation pressure on the eccentricity of the orbit. (a) Deformation of the orbit. (b) Circle of natural eccentricity.

Assuming C to be constant, the extremity of the eccentricity vector thus describes, in one year, a circle of radius r_e whose centre has co-ordinates (Figure 7.39b):

$$C_X = e_x(t_0) - r_e \cos \alpha_S(t_0)$$

$$C_Y = e_Y(t_0) - r_e \cos \alpha_S(t_0)$$

This circle is called the circle of natural eccentricity. Its radius has a value:

$$r_e = (3/2)C/n_S a_S \Omega_{SUN}$$

with:
Ω_{SUN} = mean angular velocity of the sun = $d\alpha_S/dt = 0.9856°/\text{day}$,
Hence: $r_e = 1.105 \times 10^{-2} \rho(S_a/m)$, with $\rho \approx 1.5$, S_a in m^2 and m in kg.

The radius of the circle of natural eccentricity is of the order of 5×10^{-4} for a satellite of 2 kW and 1000 kg in orbit. The eccentricity vector is such that the vector from the centre of the circle to the extremity of the eccentricity vector is directed towards the sun.

Finally, lunar-solar attraction also causes a perturbation with a period of about one month and of amplitude of the order of 3.5×10^{-5} which is superimposed on the progression of the eccentricity vector under the effect of radiation pressure.

7.3.4 Orbit corrections: station keeping of geostationary satellites

As a consequence of perturbations, the orbit parameters of geostationary satellites differ from the nominal parameters. The orbit is characterised by an inclination i, an eccentricity e and a longitude drift $d\lambda/dt$ which are small but not zero. The effect of these parameters on the position of the satellite will first be analysed in order to determine the station-keeping requirements. Correction procedures will then be presented.

7.3.4.1 *Position and velocity of the satellite*

In a geocentric rotating reference, the spherical coordinates of the satellite are the radius r, the declination or latitude φ and the latitude l. Since the eccentricity e and the inclination i are small, these co-ordinates are related to the orbital parameters by:

$$r = a_s + \Delta a - a_s e \cos v = a_s + \Delta a - a_s e \cos (\alpha_{SL} - (\omega + \Omega))$$

$$= a_s + \Delta a + a_s(e_X \cos \alpha_{SL} + e_y \sin \alpha_{SL}) \tag{7.92a}$$

with Δa, the difference between the actual half axis and the synchronous half axis, equal to—(2/3) (a_S/n_S) $d\lambda_m/dt$ (cf. equation (7.84)) and α_{SL}, the right ascension of the satellite, equal to $v + \omega + \Omega$ since i is small.

$$\lambda = \lambda_m + 2e \sin M = \lambda_m + 2e \sin [\alpha_{SL} - (\omega + \Omega)] \tag{7.92b}$$

$$\lambda = \lambda_m + 2e_X \sin \alpha_{SL} - 2e_Y \cos \alpha_{SL}$$

$$\varphi = \text{arc} \sin [\sin (\omega + v) \sin i]$$

$$= i_x \sin \alpha_{SL} - e_y \cos \alpha_{SL} \tag{7.92c}$$

The velocity of the satellite can be resolved into the components V_N perpendicular to the orbital plane towards the north, V_R in the earth–satellite direction and V_T perpendicular to the radius vector in the plane of the orbit in the direction of the velocity.

This gives, with Ω_E the velocity of rotation of the earth:

$$V_r = V_s(e_X \sin \alpha_{SL} - e_Y \cos \alpha_{SL}) \tag{7.93a}$$

$$V_T = a_S[(d\lambda_m/dt)/3 + a_S\Omega_E] + V_S(e_X \cos \alpha_{SL} + e_Y \sin \alpha_{SL}) \tag{7.93b}$$

$$V_N = V_S(i_X \cos \alpha_{SL} + i_Y \sin \alpha_{SL}) \tag{7.93c}$$

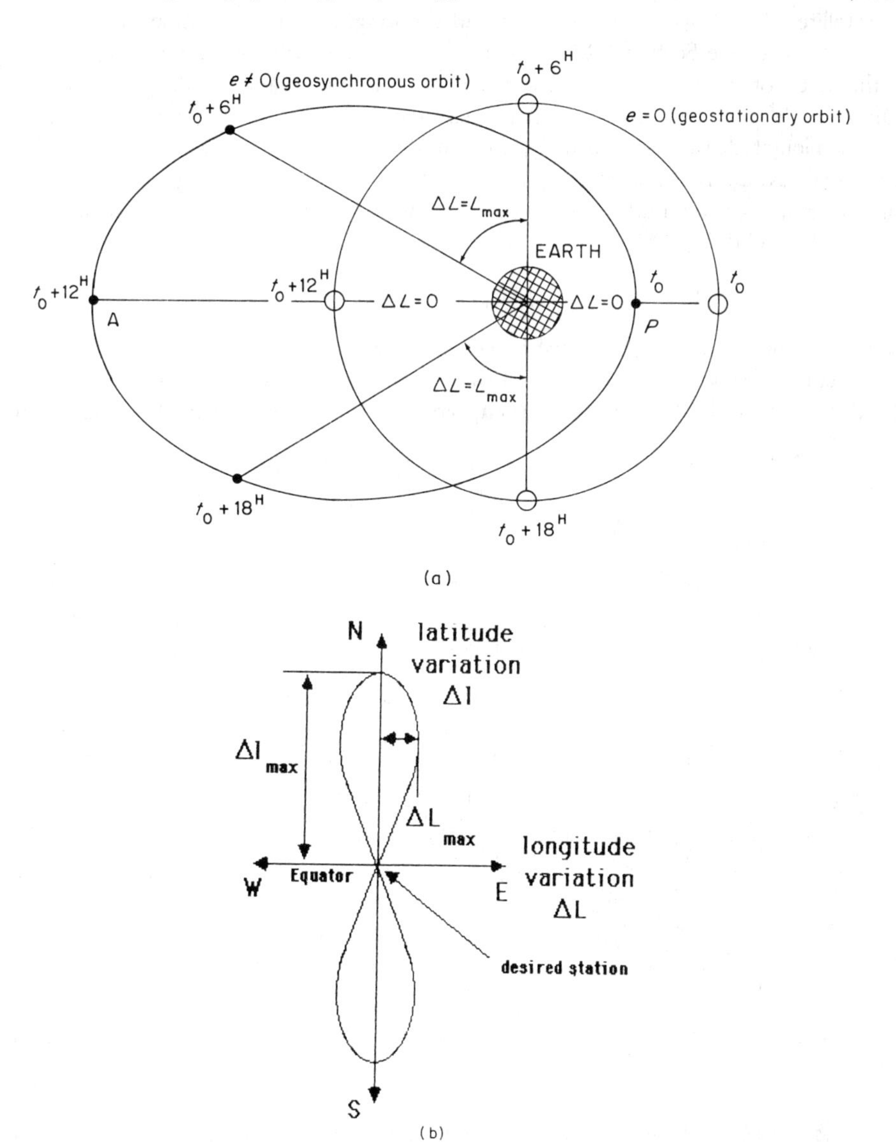

Figure 7.40 The effect of eccentricity (a) and residual inclination (b).

7.3.4.2 *Effect of residual eccentricity and inclination*

The residual eccentricity leads to an oscillation of the longitude of the satellite about the mean longitude of its station. This is illustrated in Figure 7.40a where the successive positions of two satellites are represented; one is on a circular orbit of period one sidereal day and the other is on an elliptical orbit of the same period. The difference in longitude $\Delta\lambda$ is obtained from equation (7.82) and has a value $\Delta\lambda = \lambda_v - \lambda_m = 2e \sin M$. The maximum difference in longitude $\Delta\lambda_{max}$ thus has a value $2e$ radians, that is $2\pi e/180 = 114e$ degrees (see Section 7.2.2).

The residual inclination causes an apparent daily movement, as shown in Figure 7.40b, of the satellite with respect to the equator and the longitude of the station in the form of a 'figure of eight' (see Section 7.2.3). The amplitude of the latitude variation $\Delta\varphi_{max}$ is equal to the value of the inclination i. The maximum longitude variation $\Delta\lambda_{max}$ has a value of $4.36 \times 10^{-3} i^2$ (degrees) and the related latitude φ_m is 0.707 (see relation (7.59)). This maximum longitude shift is reached at the end of the time t such that $(2\pi t/T) = 1/[\sqrt{2}\cos(i/2)]$. For i small, $\Delta\lambda_{max}$ is negligible (e.g. $i = 1°$, $\Delta\lambda_{max} = 4.36 \times 10^{-3}$ degrees). The maximum daily variation of longitude due to eccentricity and inclination of orbit is illustrated in Figure 7.41 (CCIR-Rep 563).

7.3.4.3 *The station-keeping window*

To fulfil its mission, the satellite must remain stationary with respect to the earth and occupy a well defined position on the equator. However, the combined effect of oscillations of period 24 h due to the inclination and eccentricity and the long-term drift of the mean longitude

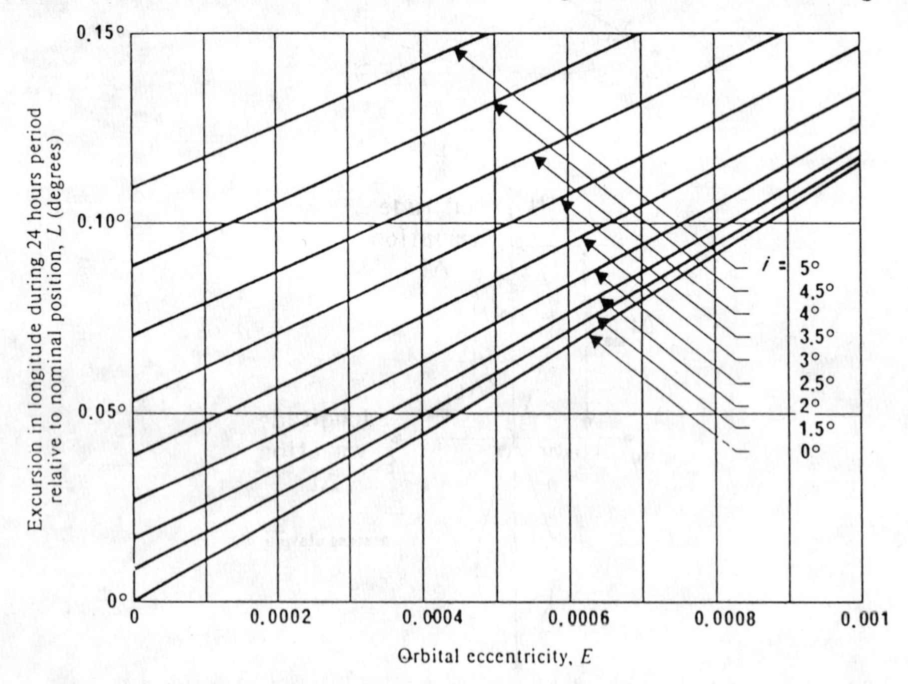

Figure 7.41 Daily variation of longitude due to the effect of residual eccentricity e and inclination i (the peak-to-peak variation is equal to twice the value given in the figure). (From CCIR-Rep 556-4. Reproduced by permission of the ITU.)

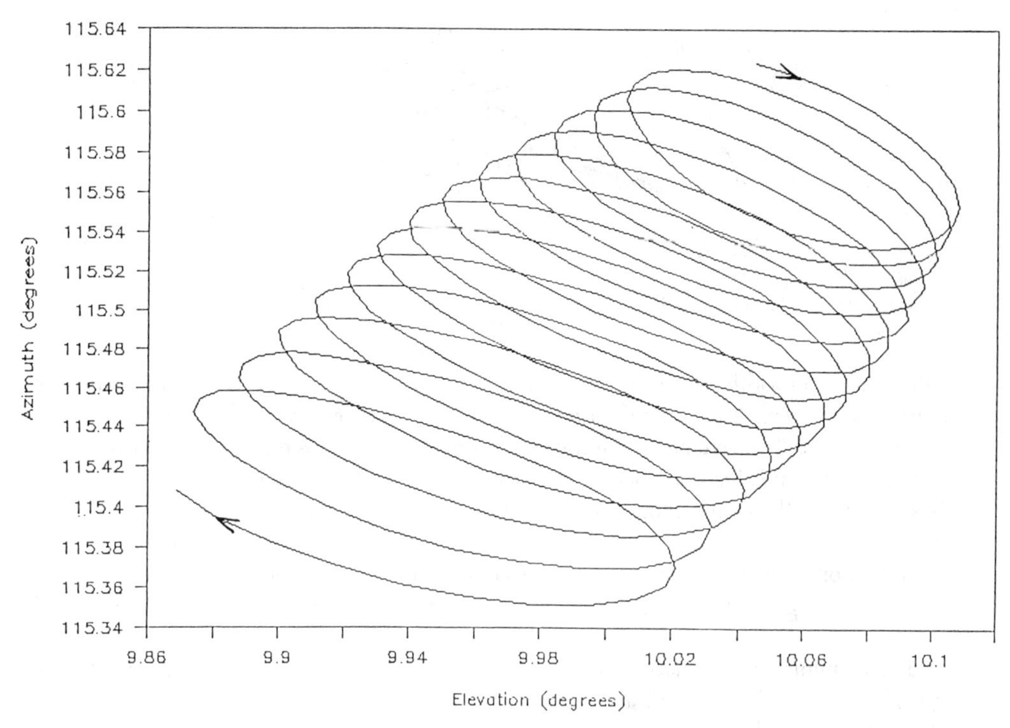

Figure 7.42 Apparent movement of the satellite due to the combined effect of residual eccentricity and inclination.

leads to an apparent movement of the satellite with respect to its nominal position. Figure 7.42 shows the relative movement of the satellite with respect to its nominal position for an orbit of semi-major axis 42 164.57 km, eccentricity 2×10^{-4} and inclination 0.058. As it is in practice impossible to maintain the satellite absolutely immobile with respect to the earth, a station-keeping 'window' is defined.

The station-keeping window represents the maximum permitted values of the excursions of the satellite in longitude and latitude. It can be represented as a pyramidal solid angle, whose vertex is at the centre of the earth, within which the satellite must remain at all times. The station-keeping window is defined by the two half-angles at the vertex, one within the plane of the equator (E–W width), and the other in the plane of the satellite meridian (N–S width). The maximum value of the residual eccentricity determines the overshoot of the radial distance. Figure 7.43 shows the volume available for relative displacement of the satellite with respect to its original central position for a window with a typical specification of $\pm 0.05°$ in longitude and latitude.

The objective of station keeping is to control the progression of the orbital parameters under the effect of perturbations by applying periodic orbit corrections in the most economic manner so that the satellite remains within the window.

The dimensions of this 'window' are fixed by the mission. They are determined by the following considerations:

—As the dimensions become smaller, the ground station antenna pointing and tracking systems become simpler.

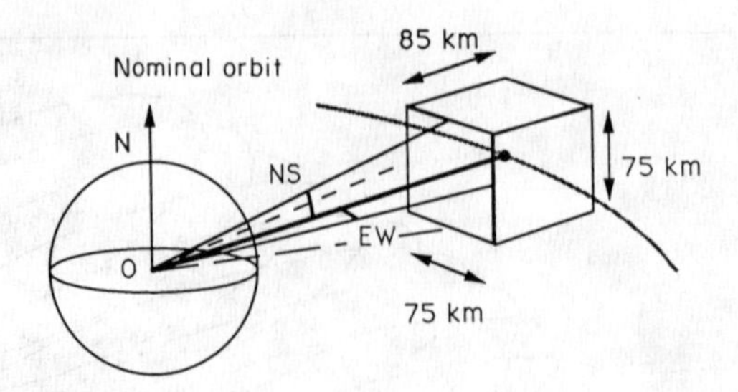

Figure 7.43 Station-keeping 'window'.

—When the beamwidth of the ground station antennas is large or when the station is mounted on a vehicle (an aircraft, a boat or a lorry) which contains a pointing system to take account of its movement, 'windows' of fairly large dimensions are acceptable.
—Geostationary satellites equipped with narrow beam antennas pointing towards specific sites on the earth require more and more precise station keeping as the beams become narrower. This precision also permits the use of ground station antennas with fixed pointing.
—The adoption of a strict station keeping tolerance for satellites permits better utilisation of the orbit of geostationary satellites and the radio-frequency spectrum [CCIR-Rec 484].

The radiocommunication regulations, revised by CAMR 79, impose a station-keeping accuracy of $\pm 0.1°$ in longitude for fixed and broadcast service satellites. A tolerance of $\pm 0.5°$ in longitude is permitted for satellites which do not use the frequency bands allocated to fixed or broadcast satellite services.

7.3.4.4 *The effect of orbit corrections*

Orbit corrections are achieved by applying 'velocity increments' ΔV to the satellite at a point in the orbit. These velocity increments are the result of forces acting in particular directions on the centre of mass of the satellite for sufficiently short periods (compared with the period of the orbit) for them to be considered as impulses. The impulse applied can be radial, tangential or normal to the orbit in accordance with the definitions given above for the velocity at a point in the orbit defined by r, λ and φ. The impulse does not change the values of r, λ and φ instantaneously but modifies the component of velocity concerned by a quantity ΔV. The effect of this velocity increment on the orbit parameters is determined from equations (7.93a, b and c) [DON-84]. It can be shown that a normal impulse modifies the inclination, a radial impulse modifies the longitude and the eccentricity, and a tangential impulse modifies the drift and the eccentricity.

Actuators are, therefore, mounted on the satellite and are capable of producing forces perpendicular to the orbit to control the inclination and tangential forces (parallel to the velocity). There is no need to generate radial thrusts since a modification of the longitude is obtained from a drift created by the tangential impulses which also permit the eccentricity to be controlled at lower cost.

The actuators thus permit independent control of movements out of the plane of the orbit ('north–south' station keeping by control of the inclination) and movements in the plane of the orbit ('east–west' station keeping by control of the drift and, if required, the eccentricity).

There could, however, be coupling due to inaccuracy of the satellite attitude control and bias of the actuator mountings; this could, for example, cause a thrust which should be oriented perpendicularly to the orbit not to be perfectly so. A component acting in the plane of the orbit is thus generated. The most commonly used actuators generate thrusts by burning chemicals, called propellants. The quantity of propellant consumed is proportional to the velocity increment provided (see Section 10.3.2).

7.3.4.5 *North–south station keeping*

North–south station keeping is achieved by thrusts acting perpendicularly to the plane of the orbit thereby modifying its inclination. Only the long-term drift of the inclination vector is corrected since the amplitude of periodic perturbations (2×10^{-2} degree) remains less than the normal size (0.1°) of the window.

The optimum procedure is to induce a modification of the inclination vector in the opposite direction to that of the drift $\mathbf{\Omega}_D$ defined by equation (7.91). This conditions the value of right ascension of the point of the orbit where the manoeuvre is performed (Figure 7.44a). In this figure, the exterior circle represents the north–south width of the window. The maximum permitted value of inclination is represented by the interior circle which differs from the previous one by a margin calculated to take account of measurement and orbit restoration inaccuracies.

Modification of the i_x and i_y components of the inclination vector as a function of the value of the normal velocity increment ΔV_N at a point of right ascension α_{SL} is given by:

$$\Delta i_x = \Delta V_N \cos \alpha_{SL} / V_S \tag{7.94}$$

$$\Delta i_y = \Delta V_N \sin \alpha_{SL} / V_S$$

where V_S is the velocity of the satellite in the orbit and is equal to 3075 m/s. The relation

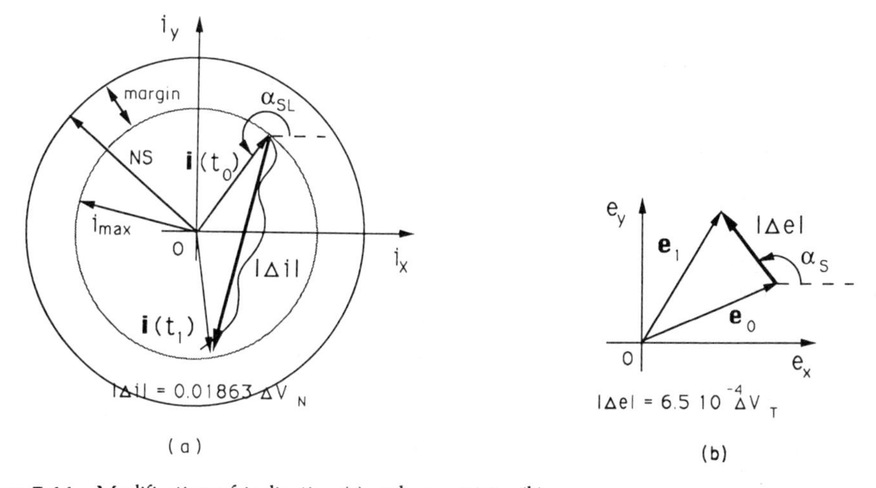

Figure 7.44 Modification of inclination (a) and eccentricity (b).

between the modulus of the inclination correction $|\Delta i|$ and the modulus of the normal velocity increment V_N is thus:

$$|\Delta i| = 0.018\,63\,|\Delta V_N| \tag{7.95}$$

where

$$|\Delta V_N| = 53.7\,|\Delta i| \qquad (\Delta V\text{: m/s, } \Delta i\text{: degrees})$$

It is thus feasible to calculate the velocity increments necessary to compensate the corresponding inclination drift each year; the drift is calculated from equation (7.91). The results are presented in Table 7.10.

In the case where only the long-term drift is corrected, the cost of north–south control is independent of the number of manoeuvres performed.

Various strategies are possible:

—One strategy is to allow the inclination vector to drift up to the maximum permitted value then to apply a correction in the opposite direction to that of the drift such that the inclination vector returns to the opposite position of the permitted region. This strategy minimises the number of manoeuvres, but requires large velocity increments which may lead to problems of coupling with east–west station keeping.

—It may be useful, from the point of view of operational load at the satellite control centre, to coordinate north–south manoeuvres with those of east–west station keeping whose period of recurrence is different and shorter. In this case, corrections are made before the inclination reaches the limiting value [SOO-85].

—With the above two strategies, for the corrections to be optimum the values of right ascension of the points where they are performed are imposed by the position of the drift vector at the end of the cycle. Depending on the techniques used for attitude control during the correction, it can be that the manoeuvres are not permitted in certain sections of the orbit (the earth–sun–satellite geometry may affect the accuracy of attitude

Table 7.10 Annual inclination drift and the velocity increment required to correct it

Year	Inclination drift (°/year)	Annual ΔV (m/s)	Year	Inclination drift (°/year)	Annual ΔV (m/s)
1990	0.909	48.8	2001	0.843	45.3
1991	0.884	47.5	2002	0.874	46.9
1992	0.854	45.8	2003	0.901	48.4
1993	0.821	44.1	2004	0.920	49.4
1994	0.788	42.3	2005	0.931	50.0
1995	0.762	40.9	2006	0.935	50.2
1996	0.745	40.0	2007	0.931	50.0
1997	0.743	39.9	2008	0.918	49.3
1998	0.755	40.5	2009	0.898	48.2
1999	0.779	41.8	2010	0.871	46.8
2000	0.811	43.5	2011	0.840	45.1

measurement, for example). The corrections cannot, therefore, be made in an optimum manner during certain periods of the year when the value of right ascension falls in the prohibited regions. The strategy thus consists of positioning the extremity of the inclination vector, just before entry into the critical period, so that the greatest area of the permitted region is available to the inclination vector drift [BEL-83].

It should be noted that the right ascension of the point where the correction is performed does not necessarily correspond to a node of the orbit. This position is obligatory only if it is wished to make the inclination zero and this is generally not the case for the strategies described above. However, the general direction of the drift tends to lead the inclination vector to the y axis (Ω close to 90°) in the reference frame of Figure 7.4. Control of the drift is, therefore, performed globally by compensating rotation of the plane of the orbit by means of thrusts directed towards the south with Ω close to 90° and/or thursts directed towards the north with Ω close to 270°. The time of day depends on the season. In summer the thrust to the south is performed towards midday and in winter towards midnight; in spring and autumn it is performed in the evening and morning respectively. The thrust towards the north is performed with an offset of 12 h.

Finally, in order to reduce the cost of station keeping, it is possible not to provide inclination control by allowing a larger value for the maximum permitted inclination, for example 3°. At the start of the satellite's life, an inclination equal to the maximum permitted value and a right ascension of the ascending node of the orbit are imposed such that the initial inclination vector is parallel to and opposed to the mean direction of the natural drift for a period corresponding to the lifetime of the satellite [ULI-85]. The inclination decreases for about half of the lifetime, passes through zero and then increases until it reaches the maximum value which determines the end of the operatonal life of the satellite. As the mean annual drift is of the order of 0.85° the lifetime could, for example, be chosen as around seven years.

The principal consequences of a non-zero inclination are a north–south oscillation of the satellite (see Section 7.2.3) as seen from earth stations and a displacement of the coverage of the satellite antennas. This displacement can be compensated by using a steerable antenna or acting on the satellite attitude control ('Comsat manoeuvre') [ATI-90].

A similar strategy can be used when a satellite is put into orbit before the operational requirements become effective (as a consequence of an excessively early launch slot reservation). At launch the satellite is put into a parking orbit whose inclination is chosen to be such that, as a consequence of natural drift, the inclination is zero on the date when the satellite is put into service.

7.3.4.6 *East–west station keeping*

East–west station keeping is provided by thrusts acting tangentially to the orbit. It is divided into control of the drift (maintenance of mean longitude) and, if necessary, control of the eccentricity. Maintenance of longitude consists of compensating the longitudinal drift due to the ellipticity of the equator and hence the value depends on the orbital position of the satellite. Control of eccentricity consists of maintaining the modulus of the eccentricity less than the maximum permitted eccentricity.

An isolated tangential impulse modifies both the semi-major axis, and hence the drift, and the eccentricity of the orbit. The modification of the semi-major axis Δa as a function of the value of the tangential velocity increment ΔV_T has a value:

$$\Delta a = -(2/\Omega_E)\Delta V_T \quad \text{(m)} \tag{7.96}$$

where Ω_E is the velocity of rotation of the earth.

The modification $\Delta d = \Delta d\lambda_m/dt$ of the drift as a function of Δa or ΔV_T has a value:

$$\Delta d = -(3\Omega_E/2a_S)\Delta a \tag{7.97a}$$

where a_S is the semi-major axis of the orbit.

$$\Delta d = -(3\Omega_E/V_S)\Delta V_T \quad \text{(degrees/day)}$$
$$= -(3/a_S)\Delta V_T \tag{7.97b}$$

The modification of the e_X and e_Y components of the eccentricity vector as a function of the value of the tangential velocity increment ΔV_T at a point of right ascension α_{SL} is given by (Figure 7.44b):

$$\Delta e_X = 2\Delta V_T(\cos\alpha_{SL}/V_S)$$
$$\Delta e_Y = 2\Delta V_T(\sin\alpha_{SL}/V_S) \tag{7.98a}$$

where V_S is the velocity of the satellite in the orbit equal to 3075 m/s. The relation between the modulus of the inclination correction $|\Delta e|$ and the modulus of the normal velocity increment V_T is then:

$$|\Delta e| = 2\Delta_T/V_S = 6.5 \times 10^{-4}|\Delta V_T|$$

$$|\Delta V_T| = |1538.5|\Delta e| \quad (\Delta V\text{:m/s}) \tag{7.98b}$$

7.3.4.6.1 Modification of the satellite drift

A velocity increment $\Delta V = 1$ m/s leads to a modification of the semi-major axis $\Delta a = 27.4$ km, a drift increment $\Delta d = -0.352°$/day and a modification of the eccentricity $|\Delta e| = 0.65 \times 10^{-3}$.

The mean longitude is not instantaneously modified by the applied velocity increment; it changes progressively as a consequence of the combined effect of drift and eccentricity. Figure 7.45 illustrates the modification of the orbit and the variation of longitude as a function

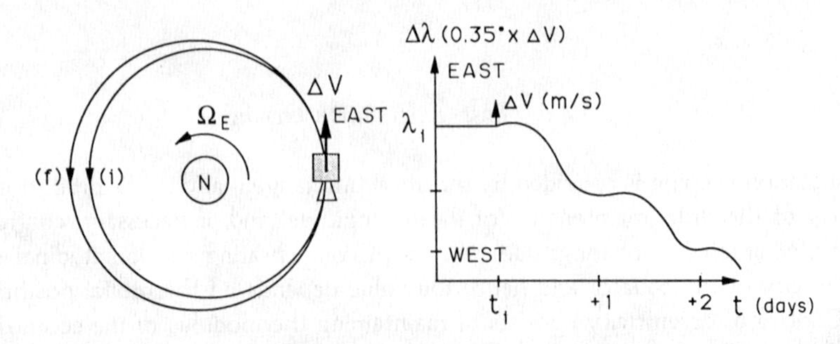

Figure 7.45 The effect of an impulse tangential to the orbit.

of time for an initially geostationary satellite to which a thrust and hence a velocity increment ΔV has been applied directed towards the east. Following the impulse, the longitude increases slightly towards the east then decreases continuously towards the west [SOO–83]. The new final orbit (f) is an elliptical orbit with an apogee altitude above the initial orbit, and the perigee being located where the impulse was applied.

Modification of the value of the drift alone, without changing the eccentricity, or the inverse, can be achieved by applying two thrusts in opposite directions separated by half a sidereal day, that is at two points of right ascension α_{SL} and $\alpha_{SL} + \pi$. This gives:

$$\Delta d = -(3\Omega_E/V_S)(\Delta V_{T1} + \Delta V_{T2})$$

$$\Delta e = (2/V_S)(\Delta V_{T1} - \Delta V_{T2})(\cos\alpha_{SL} + \sin\alpha_{SL})$$

The strategies used to control the natural drift depend on the size of the radius of the circle of natural eccentricity with respect to the maximum permitted eccentricity.

7.3.4.6.2 Control of the natural drift

If the radius of the circle of natural eccentricity is less than the permitted eccentricity, only the progression of the drift is controlled. To obtain an eccentricity of the orbit which is always less than the limiting value, it is necessary to locate the centre of the circle of eccentricity at the origin of the reference frame and to orient the eccentricity vector in the direction of the sun. This means that a non-zero eccentricity equal to the radius of eccentricity must be imposed on the orbit on injection of the satellite into orbit; the perigee of this orbit must be in the direction of the sun. The orientation of the orbit follows the direction of the sun with the rotation of the earth about the sun and the eccentricity remains constant and hence less than the limiting value.

The strategy used to control the drift alone depends on the position of the satellite with respect to the stable points. Movement in the equatorial plane due to the asymmetry of the terrestrial potential obeys equation (7.90) recalled below:

$$(\mathrm{d}\Lambda/\mathrm{d}t)^2 - D\cos 2\Lambda = \text{constant}$$

where $\Lambda = \lambda - \lambda_{22} \pm 90^\circ$ is the longitude of the satellite measured with respect to the closest stable equilibrium point and $D = 18n_S^2(R_E/a_S)^2 J_{22} = 4 \times 10^{-15}\,\text{rad/s}^2$.

The satellite must be maintained at the nominal longitude Λ_N, measured with respect to the position of the nearest stable equilibrium point, while tolerating a small maximum deviation $\epsilon/2$ on each side of Λ_N. This deviation $\pm\epsilon/2$ is determined from the dimensions in longitude of the station-keeping window by deducting a margin which is intended to absorb orbit restoration errors, manoeuvre inaccuracies and short period oscillations.

If Δ is the longitude measured from Λ_N ($\Delta = \Lambda - \Lambda_N$), this gives:

$$\cos 2\Lambda_N = \cos 2(\Lambda_N + \Delta) = \cos 2\Lambda_N - 2\Delta\sin 2\Lambda_N$$

and equation (7.90) can be written:

$$(\mathrm{d}\Delta/\mathrm{d}t)^2 - 2D\Delta\sin 2\Lambda_N = \text{constant} \tag{7.99}$$

The curve representing the drift $\mathrm{d}\Delta/\mathrm{d}t$ as a function of Δ is a parabola defined by the nominal longitude Λ_N and the initial conditions. If the satellite is initially at the point I (nominal

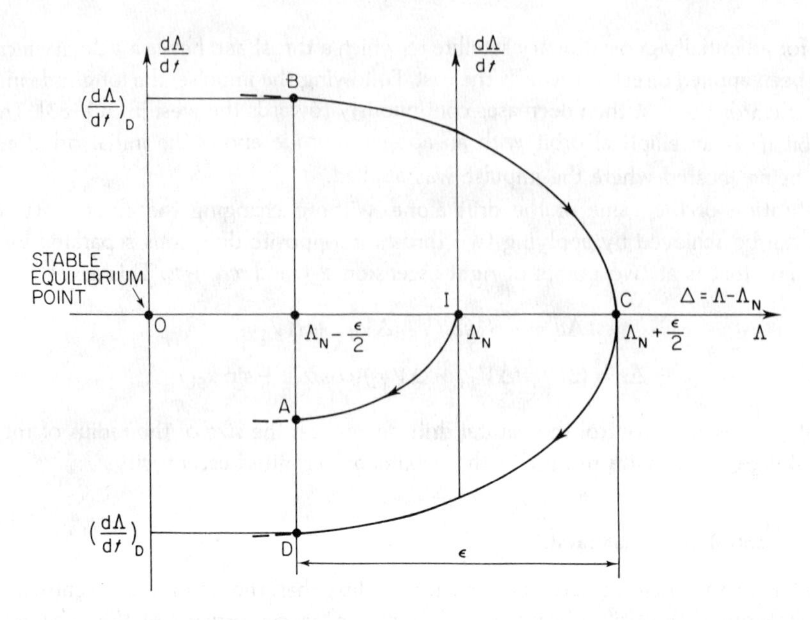

Figure 7.46 Strategy for maintaining longitude away from a point of equilibrium.

position) with a zero drift (Figure 7.46), it moves towards the point of stable equilibrium by following the parabola of equation (7.99) which has its vertex at I.

Strategy 'of the inclined plane'. When the longitude reaches the value $\Lambda = \Lambda_N - \epsilon/2$ (point A, the west limit), a velocity increment is applied to the satellite to cause it to describe, at the point representing the satellite, a parabola which has its vertex at $\Lambda = \Lambda_N + \epsilon/2$ (point C), the eastern limit of movement.

From C, the satellite moves nearer to the point of stable equilibrium to reach point D ($\Lambda = \Lambda_N - \epsilon/2$, the western limit). By applying a suitable velocity impulse ΔV_T at this point, the drift $(d\Lambda/dt)_D$ changes sign and the representative point in Figure 7.46 changes from D to B. Then the cycle starts again.

East–west station keeping thus consists of causing the satellite to describe the cuve BCD of Figure 7.46 at the point representing the satellite.

The velocity increment to be applied. In equation (7.99), the constant is calculated from a location at point C ($\Delta = \epsilon/2$, $d\Lambda/dt = 0$). Hence:

$$\text{Constant} = 2D(\epsilon/2)\sin 2\Lambda_N$$

Equation (7.99) can be written:

$$(\Delta d)^2 = (d\Lambda/dt)^2 = -2D(\Delta - \epsilon/2)\sin 2\Lambda_N \qquad (7.100)$$

At the point D ($\Delta = -\epsilon/2$):

$$(d\Lambda/dt)_D^2 = 2D\epsilon \sin 2\Lambda_N$$

from which:

$$(d\Lambda/dt)_D = -\sqrt{(2D\epsilon \sin 2\Lambda_N)} \quad (7.101)$$

To move from point D to point B it is necessary to impose a variation of drift $d\Delta/dt$ equal to $-2(d\Delta/dt)_D$. Equation (7.97b) enables the corresponding velocity increment Δv_T to be calculated:

$$\Delta V_T = -(V_S/3\Omega_E)\Delta d \qquad (\text{m/s; rad/s})$$

hence for $\Delta d = -2(d\Lambda/dt)_D$

$$|\Delta v_T| = (V_S/3\Omega_E)2\sqrt{(2D\epsilon \sin 2\Lambda_N)}$$

Hence:

$$|\Delta v_T| = 2.5\sqrt{(\epsilon \sin 2\Lambda_N)} \qquad (\text{m/s; rad}) \quad (7.102)$$

As the satellite always drifts in the same direction, this velocity increment is always applied in the same direction (that of the natural drift).

Correction periods. The period of application of the velocity impulses is equal to the duration T of the parabolic path. This time is calculated by integrating equation (7.100) from $\epsilon/2$ to $-\epsilon/2$:

$$T = [(2\sqrt{2})/\sqrt{D}]\sqrt{(\epsilon/\sin 2\Lambda_N)} = 516\sqrt{(\epsilon/\sin 2\Lambda_N)} \quad \text{days} \quad (7.103)$$

where ϵ is in radians.

The velocity impulse and the cycle duration depend on the position of the satellite with respect to the stable equilibrium point and the longitudinal dimension of the window.

Annual velocity impulse. The velocity impulse to be applied per year is $\Delta V_T = \Delta v_T\ (365/T)$, hence:

$$\Delta V_T = 1.77 \sin 2\Lambda_N \qquad (\text{m/s}) \quad (7.104)$$

It depends on the longitude of the point about which the satellite is maintained and not on the total window width (if only the long-term drift is corrected and not the short-term variations). At maximum it is of the order of 2 m/s.

The preceding calculations are based on the equations of Section 7.3.3.3 which were obtained by neglecting, in particular, terms of order greater than 2 in the expansion of the terrestrial potential. When the actual acceleration to which the satellite is subjected is considered (Figure 7.36), the required annual velocity increment is given by Figure 7.47.

The 'return to centre' strategy. Equation (7.104) and Figure 7.47 indicate that the annual velocity increment depends on the longitude of the position in the nominal orbit of the satellite. In the vicinity of a point of equilibrium, this velocity increment is very small. In practice, it is not the drift due to asymmetry of the terrestrial potential which must be compensated, but the east–west component of the velocity increments induced by corrections perpendicular to the equatorial plane (north–south control). The strategy adopted in this case consists of locating the satellite at the centre of the 'window' to perform the north–south correction in order to provide the maximum margin in longitude. A correction in longitude is then

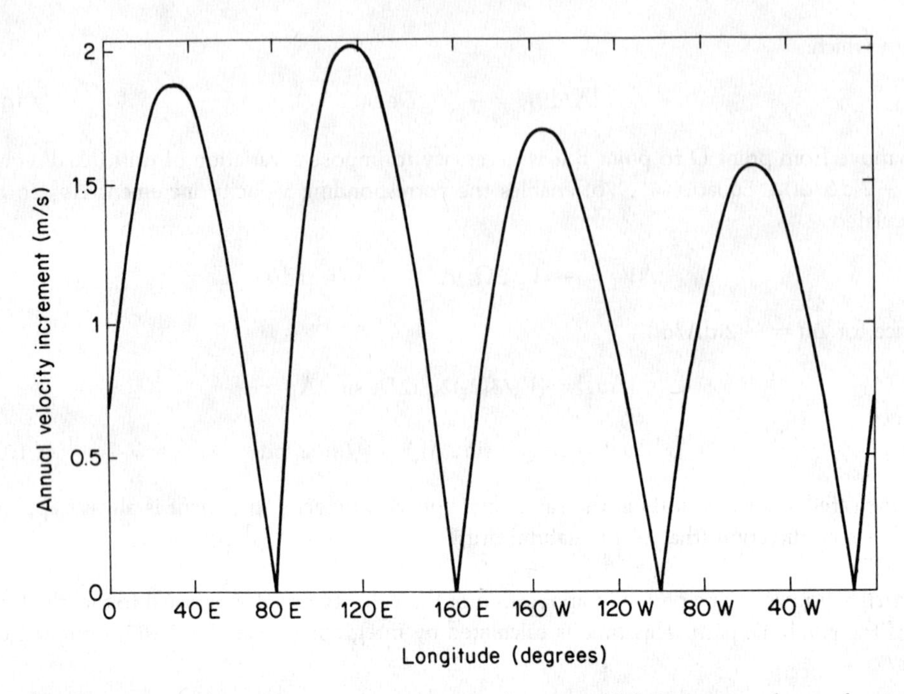

Figure 7.47 Annual velocity increment required to control east–west drift as a function of station longitude.

performed before the satellite reaches the limit of the window in order to return it to the centre of the window. A second correction in longitude cancels the drift caused by the first.

Examples of the strategies. Figure 7.48a illustrates the variations of longitude of a satellite with the 'inclined plane' strategy. The nominal position is at longitude 49° east (Indian Ocean and the value of $\Lambda_N = |\lambda - \lambda_{22} - 90°| = 26°$. The natural drift of the longitude is directed towards the east. To guarantee a window of $\pm 0.5°$ ($\epsilon = 1° = 0.017$ rad) the correction period must be 75 days and the annual velocity increment is 1.5 m/s.

Velocity increments are applied when the satellite approaches the eastern edge of the window in such a way as to impose a drift which becomes zero before it reaches the western edge of the window [LEG-80].

Figure 7.48b shows the return to centre strategy. The longitude of the satellite is 11.5° west, that is near a point of unstable equilibrium. The effect of a north–south correction on the initial longitude drift can be observed. An east–west correction creates a drift which returns the satellite to the centre of the window. A second east–west correction cancels this drift.

7.3.4.6.3 Control of eccentricity

The preceding strategies are applicable when the radius of eccentricity r_e of the orbit is sufficiently small for the eccentricity of the orbit not to be controlled. When this is not the case, that is for satellites with large solar panels, the natural eccentricity ($e_n = r_e$) of the orbit

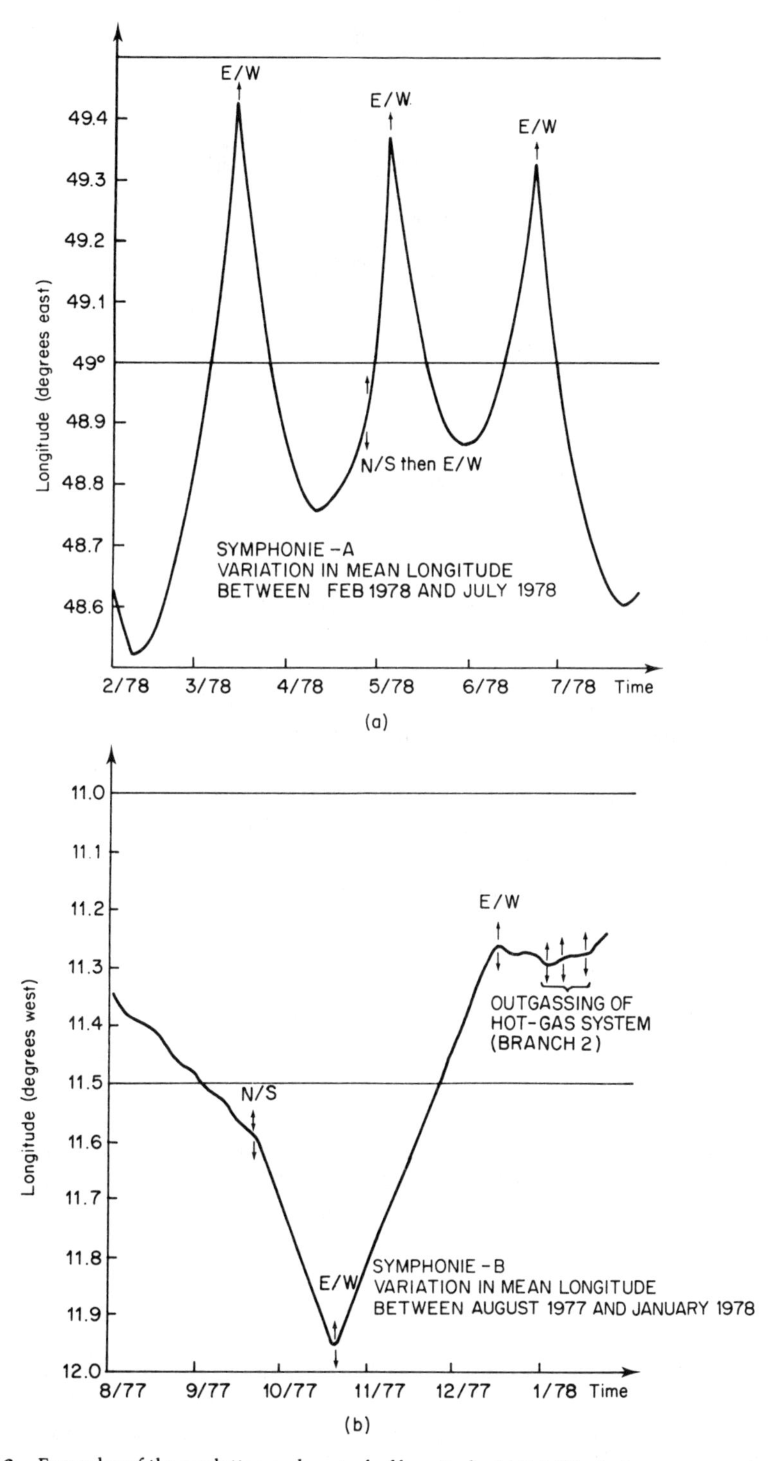

Figure 7.48 Examples of the evolution and control of longitude. (a) Satellite far from a point of equilibrium. (b) Satellite close to a point of unstable equilibrium.

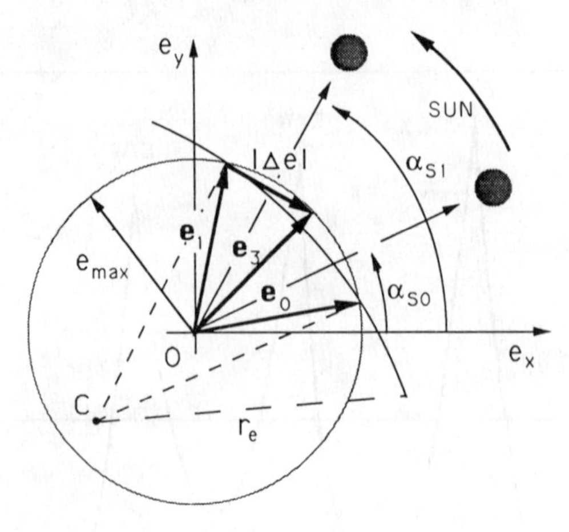

Figure 7.49 Strategy for controlling eccentricity.

is such that the induced movement in longitude ($\Delta\lambda = 2e_n \sin M$) occupies an excessively large part of the window.

It is then necessary to prevent the eccentricity from exceeding a value e_{max} such that $4e_{max}$ is the part of the window allocated to oscillation of longitude under the effect of the eccentricity. This interval is determined from the dimension in longitude of the station-keeping window; from this is deducted a band which permits variation of the mean longitude under the effect of long-term drift during the expected time between two east–west manoeuvres, a margin to take account of inaccuracies of manoeuvring and orbit restoration and the east–west effects induced by north–south control.

The strategy consists of locating the eccentricity vector ($\mathbf{e}_0$) on the boundary circle which defines the region of maximum permitted variation of eccentricity (centre 0, radius e_{max}) at the start; the right ascension of the sun is $\alpha_{SUN\,0}$ (Figure 7.49). The eccentricity varies on a circle of radius r_e and centre C such that the vector Ce remains parallel to the direction of the sun. When the eccentricity vector again reaches the circle of constraint ($\mathbf{e}_1$), the right ascension of the sun is $\alpha_{SUN\,1}$. A velocity increment is then applied which puts the extremity of the eccentricity vector on the circle of constraint in the same position ($\mathbf{e}_2$) with respect to the sun as at the start of the cycle.

The station-keeping strategy cycle is thus determined by the duration of these corrections and the cost is high when the maximum permissible eccentricity is small. It is possible to combine the eccentricity control manoeuvres with those of drift control [BRO-90].

7.3.4.7 *The operational aspect—correction cycles*

The control strategies aim to provide satellite station keeping while minimising the amount of propellant consumed to generate the required velocity increments. It is also necessary to take account of constraints imposed by operational aspects (such as obligations to personnel) and security. Hence repetitive correction cycles with a period which is a multiple of a week are well suited.

A typical example is that of a 14-day cycle which includes north–south and east–west corrections. The organisation of the cycle is as follows:

—Inclination correction at the start of the cycle.
—Measurement and restoration of the orbit (about two days of measurements are necessary to determine the orbit with sufficient accuracy to calculate the corrections to be performed).
—Eccentricity or drift corrections.
—Measurement and restoration of the orbit, verification of the result of the corrections.
—Natural progression of the orbit.
—Measurement and restoration of the orbit at the end of the cycle in preparation for the following cycle.

7.3.4.8 *The overall cost of station keeping*

As long as the amplitude of periodic perturbations is compatible with the size of the station-keeping window, the effect of these perturbations is not corrected and does not affect the station-keeping budget.

By correcting only long-term drifts, the budget is of the order of:

—43–48 m/s per year for north–south control (inclination correction),
—1 to 5 m/s per year for east–west control (longitude drift and eccentricity corrections).

The actual total cost depends on:

—the date of the start of station keeping,
—the longitude of the station,
—the S_a/m ratio of the satellite,
—the dimensions of the window.

7.3.4.9 *Termination of station keeping at the end of life*

Satellite station keeping is possible by means of propellants (indispensable for the operation of thrusters) which are stored in reservoirs. When the propellants are consumed, station keeping is no longer provided and the satellite drifts under the effect of the various perturbations. In particular, it adopts an oscillatory movement in longitude about the point of stable equilibrium (see Section 7.3.3.3) which causes it to sweep a portion of space close to the orbit of geostationary statellites. Although small, the associated probability of collision is not zero (around 10^{-6} per year [HEC-81]).

Consequently a special procedure is adopted which aims to remove satellites from the geostationary orbit at the end of their lifetime. Manoeuvres are performed using a small quantity of propellant which is reserved for this purpose before complete exhaustion of the reservoirs. These manoeuvres place the satellite in an orbit of higher altitude (about 150 km; that is, a ΔV requirement of 5.4 m/s) than that of geostationary satellites (an orbit of lower altitude is undesirable because of the possible danger of collision during operations to install geostationary satellites in orbit). This operation requires a quantity of propellant of the order of 2.3 kg for a satellite of 1000 kg [CCIR-Rep 1004]. This quantity of propellant represents approximately six weeks of normal station keeping and hence potential availability of the satellite.

The major difficulty lies in estimating the quantity of propellant remaining in the reservoirs at a given instant. This estimate is made from the pressure variation of the pressurising gas in the propellant reservoirs and by integration of the operating time of the thrusters during the lifetime of the satellite. The error is large and can reach the quantity of propellant required to provide satellite station keeping for six months.

7.3.4.10 *Measurement and orbit restoration*

Determination of the position of a geostationary satellite depends on two types of measurement [SAI-87]:

—distance measurement,
—angular measurement.

Distance measurement. Measurement of satellite distance depends on measurement of the propagation time of an electromagnetic wave between the ground and the satellite. The distance d between a transmitter and a receiver is deduced from a measurement of the phase shift $\Delta\Phi$ between the transmitted and received waves:

$$\Delta\Phi = 2\pi f(d/c) \qquad \text{(rad)}$$

In practice, the phase shift between a sinusoidal signal of frequency f which modulates the command carrier and the same signal after retransmission by the satellite in the form of modulation of the telemetry carrier is measured (see Section 10.5).

Angular measurement. Several procedures are possible:

—By measuring the antenna pointing angles; the directional properties of the antenna receiving the telemetry signal from the satellite are used. The antenna direction is controlled in such a way that the satellite is on the axis of the principal lobe of the antenna. The accuracy of measurement is of the order of 0.005 to 1 degree according to the mechanical characteristics of the antenna.
—By interferometry; two stations A and B separated by a distance L, called the base, receive a signal from the satellite, the telemetry carrier of frequency f, for example (Figure 7.50). The trajectory difference Δd between the satellite and each of the stations A and B gives rise to a propagation time difference $\Delta t = \Delta d/c$ and is measured by a phase shift $\Delta\Phi = 2\pi f \times \Delta t$ between the received signals:

$$\Delta\Phi = (2\pi L \cos E)/\lambda$$

The value of the elevation angle E is deduced from this expression and the satellite is situated on a cone of axis AB and half angle E at the vertex. Combination of two bases enables the satellite to be located on the common generating line of the two cones. The accuracy of the measurement is of the order of 0.01°. It is not sufficient to determine the final orbit. This type of measurement is, however, useful during the launch into orbit phase.

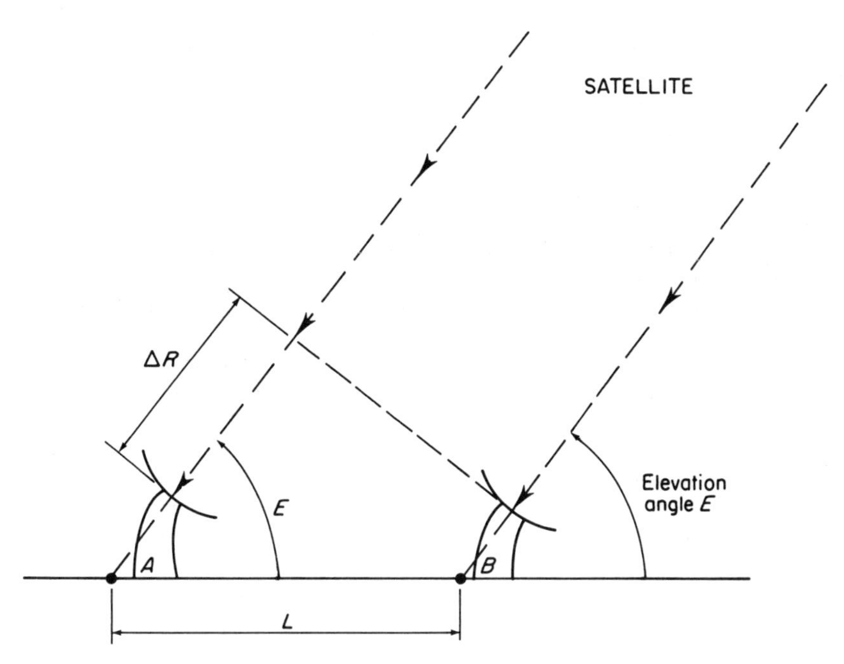

Figure 7.50 Angle measurement by interferometry.

Restoration of the orbit parameters is made by means of a series of distance and/or angle measurements. Various methods are possible in accordance with the number of operational stations. For geostationary satellites it is current practice to use only one measurement station by combining a series of distance and angle measurements made at different times.

The orders of magnitude of accuracy of restoration of orbit parameters are as follows:

—Semi-major axis 60 m
—Eccentricity 10^{-5}
—Inclination 3×10^{-3} degree
—Longitude 2×10^{-3} degree

7.4 CONCLUSION

The aspects of orbit geometry presented in this chapter are fundamental to comprehension of the design and modes of operation of artificial satellites. In the context of satellite communication systems, they determine the launching and orbit control procedures, the design of the platform (including attitude control, thermal control, the electric power supply system, the propulsion system, etc.) and the characteristics of the radio-frequency links (such as space losses, propagation times, antenna pointing, satellite–sun conjunction, etc.).

REFERENCES

[AGH-88] H. Aghvami *et al.* (1988) Land mobile satellites using the highly elliptic orbits—the UK T-SAT mobile payload, *IEE Conference on Satellite Mobile Communications, Brighton*, Sept., pp. 147–153.

[ALB-83] F. Alby (1983) Les Perturbations de l'Orbite Geostationnaire, *Le Mouvement du Véhicule Spatial en Orbite (Cours de Technologie Spatiale du CNES)*, Cepadues Editions, Toulouse.

[ASH-88] C.J. Ashston (1988) Archimedes-land mobile communications from highly inclined satellite orbits, *IEE Conference on Satellite Mobile Communications, Brighton*, Sept., pp. 133–137.

[ATI-90] A. Atia, S. Day, L. Westerlund (1990) Communications satellite operation in inclined orbit 'the comsat maneuver', *13th International Communication Satellite Systems. Conference, Los Angeles*, March, pp. 452–455.

[BAL-80] G. Balmino (1980) Le mouvement elliptique perturbé, *Le Mouvement du Véhicule Spatial en Orbite (Cours de Technologie Spatiale)*, CNES, Toulouse.

[BAL-84] G. Balmino (1984) Modéles de potentiel, *Mathématiques Spatiales (Cours de Technologie Spatiale du CNES)*, Cepadues Editions, Toulouse.

[BEL-83] B. Belon (1983) Stratégies de Maintien à Poste des Satellites Géostationnaires, Le *Mouvement du Véhicule Spatial en Orbite (Cours de Technologie Spatiale du CNES)*, Cepadues Editions, Toulouse.

[BIE-66] P. Bielkowicz (1966) Ground tracks of earth-period satellites. *AIAA Journal*, **4**, No. 12, Dec., pp. 2190–2195.

[BDL-90] Bureau des Longitudes (1990) *Ephémérides Astronomiques*, Masson, Paris.

[BOU-90] M. Bousquet, G. Maral (1990) Orbital aspects and useful relations from earth satellite geometry in the frame of future mobile satellite systems, *13th International Communication Satellite Systems Conference, Los Angeles*, pp. 783–789, March.

[BRO-90] J. Brock *et al* (1990) Simultaneous eccentricity and longitude control for Intelsat VII, *13th International Communication Satellite Systems Conference, Loss Angeles*, pp. 466–472, March.

[CCIR-REP 214] Report 214-4 (1990) *The Effects of Doppler Frequency Shifts and Switching Discontinuities in the Fixed-Satellite Service*, Vol. IV–Part 1, Dusseldorf.

[CCIR-REP 215] Report 215-6 (1990) *Systems for the Broadcasting-Satellite Service*, Vol. X & XI–Part 2, Dusseldorf.

[CCIR-REC 484] Recommendation 484-2 (1990) *Station-Keeping in Longitude of Geostationary Satellites Using Frequency Bands Allocated to Fixed-Satellite Service*, Vol. IV–Part 1, Dusseldorf.

[CCIR-REP 383] Report 383-4 (1990) *The Effects of Transmission Delay in the Fixed-Satellite Service*, Vol. IV–Part 1, Dusseldorf.

[CCIR-REP 453] Report 453-4 (1990) *Technical Factors Influencing the Efficiency of Use of the Geostationary-Satellite Orbit by Radiocommunication Satellites Sharing the Same Frequency Bands*, Vol. IV–Part 1, Dusseldorf.

[CCIR-REP 556] Report 556-3 (1990) *Factors Affecting Station-Keeping of Geostationary-Satellites of the Fixed Satellite Service*, Vol. IV–Part 1, Dusseldorf.

[CCIR-REP 808] Report 808-2 (1990) *Broadcasting-Satellite Service*, Vol. X & XI—Part 2, Dusseldorf.

[CCIR-REP 1004] Report 1004 (1990) *Physical Interference in the Geostationary-satellites Orbit*, Vol. IV—Part 1, Dusseldorf.

[DON-84] P. Dondl (1984) Loopus opens a new dimension in satellite communications, *Int. Journal of Satellite Communications*, **2**, No. 4, pp. 241–250.

[DUR-87] E.J. Durwen (1987) Determination of sun interference periods for geostationary satellite communications links, *Space Communications and Broadcasting*, **5** No. 3, July, pp. 183–195.

[ESC-84] P. Escudier (1984) Rentrée naturelle des satellite, *Mathématiques Spatiales (Cours de Technologie Spatiale du CNES)*, Cepadues Editions, Toulouse.

[GOS-90] W. Goschel, J. Nauck, P. Horn (1990) LOOPUS-Mobile-D—A new mobile communication satellite system, *13th International Communication Satellite Systems Conference, Los Angeles*, pp. 886–899, March.

[HEC-81] M. Hechler, J.C. Van Der Ha (1981) Probability of collisions in the geostationary ring, *Journal of Spacecraft*, **18**, No. 4, July–August.

[HUS-80] J.C. Husson (1980) Compléments sur les Repères de l'Espace et du Temps, *Le Mouvement du Véhicule Spatial en Orbite (Cours de Technologie Spatiale)*, CNES, Toulouse.

[JAÇ-77] L. Jacchia *et al.* (1977) Thermospheric temperature, density and composition: new models, *Smithsonian Astrophysical Observatory Special Report No. 375*, New York.

KAM-78] A. Kamel (1978) Synchronous satellite ephemeris due to earth's triaxiality and luni-solar effects, *AIAA Astrodynamics Conference, Palo Alto*, Aug.

[KAU-66] W. Kaula (1966) *Theory of Satellite Geodesy*, Blaisdell, Waltham.

[KAW-86] S. Kawase, E.M. Soop (1986) Ground antenna pointing performance for geostationary orbit determination, *ESA Journal*, **10**, No. 1, pp. 71–84.

[LEG-80] P. Legendre (1980) Maintien à poste des satellite géostationnaires; Stratégies des corrections d'orbite, *Le Mouvement du Véhicule Spatial en Orbite* (*Cours de Technologie Spatiale*), CNES, Toulouse.

[MAR-79] J. Marec (1979) *Optimal Space Trajectories*, Elsevier, Amsterdam.

[NOR-88] J.R. Norbury *et al.* (1988) Land mobile satellite service provision from the Molniya orbit, *IEE Conference on Satellite Mobile Communications, Brighton*, Sept. pp. 143–146.

[NOU-83] F. Nouel (1983) Les Repères de l'Espace et du Temps, *Le Mouvement du Véhicule Spatial en Orbite* (*Cours de Technologie Spatiale du CNES*), Cépadues Editions, Toulouse.

[POC-87] J.J. Pocha (1987) *Mission Design for Geostationary Satellites*, Reidel.

[PRI-86] W.L. Pritchard, J.A. Sciulli (1986) *Satellite Communications System Engineering*, Prentice-Hall.

[ROU-88] D. Rouffet, J.F. Dulck, R. Larregola, G. Mariet (1988) SYCOMORES: a new concept for land mobile satellite communications, *IEE Conference on Satellite Mobile Communications, Brighton*, Sept., pp. 138–142.

[SAI-87] J. Saint-Etienne (1987) Localisation spatiale, Cours de l'Option toulousaine de Télécom 'Télécommunications et Systèmes Aérospatiaux', *Note CNES*, March.

[SIO-81] C.A. Siocos (1981) Broadcasting satellite power blackouts from solar eclipses due to the Moon, *IEEE Trans. on Broadcasting*, **BC-27**, No. 2, pp. 25–28.

[SOO-83] E.M. Soop (1983) Introduction to geostationary orbits, *ESA SP-1053*, ESTEC, Noordwijk.

[SOO-85] E.M. Soop (1985) Geostationary orbit inclination strategy, *ESA Journal*, **9**, pp. 65–74.

[SOO-87] E.M. Soop (1987) Orbit control of geostationary spacecraft from dedicated control centres, *ESA Bulletin*, No. 52, Nov., pp. 42–46.

[STU-88] J.R. Stuart *et al.* (1988) Mobile satellite communications from highly inclined elliptical orbits, *AIAA 12th International Communication Satellite Systems Conference, Arlington*, March 13–17, pp. 537–541.

[ULI-85] C. Ulivieri, A. Agneni (1985) An investigation on passively controlled Geosynchronous orbits, *Mécanique Spatiale pour les satellites Géostationnaires* (*Cours de Technologie Spatiale du CNES*), Cepadues Editions, Toulouse.

[VIL-91] E. Vilar, J. Austin (1991) Analysis and correction techniques of Doppler shift for non-geosynchronous communication Satellites, *International Journal of Satellite Communications*, **9**, No. 2, pp. 122–136, March.

[ZAR-84] O. Zarrouati (1984) Perturbations d'orbites; Extrapolation d'orbites perturbées, *Mathématiques Spatiales* (*Cours de Technologie Spatiale du CNES*), Cepadues Editions, Toulouse.

[ZAR-87] O. Zarrouati (1987) *Trajectories Spatiales*, Cepadues Editions, Toulouse.

8 EARTH STATIONS

This chapter is devoted to the organisation of earth stations. In particular it treats the various aspects and sub-systems which determine the performance of the satellite communication system. Consequently, the characteristics of the antenna, transmitting and receiving sub-systems and communication common equipment are examined. On the other hand, equipment whose specification is not directly associated with satellite communication, such as that for interfacing with the terrestrial network, switching, multiplexing and supplying power, is considered only in its functional aspect.

8.1 STATION ORGANISATION

The general organisation of a satellite communication system is shown in Figure 8.1. The system consists principally of an antenna sub-system, with an associated tracking system, a transmitting section and a receiving section. It also includes equipment to interface with the terrestrial network together with various monitoring and electricity supply installations. This

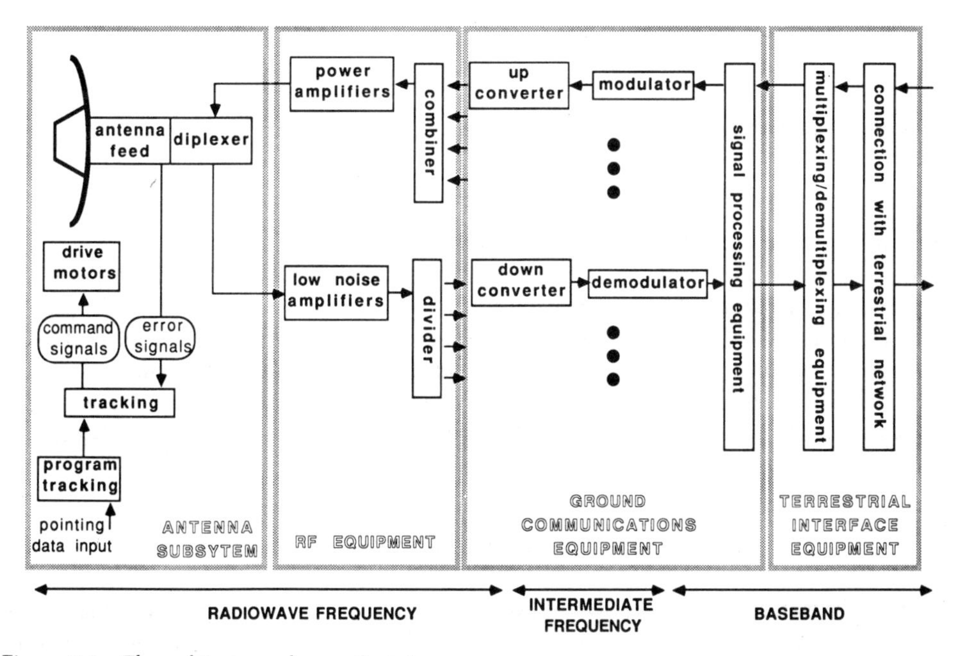

Figure 8.1 The architecture of an earth station.

organisation is not, in principle, fundamentally different from that of other telecommunication stations such as those for terrestrial microwave links. The principal special feature arises from the tracking system which can, in certain cases, be particularly simple.

The antenna is generally common to transmission and reception for reasons of cost and bulk. Separation of transmission directions is achieved by means of a diplexer. Antennas are often capable of transmitting and receiving on orthogonal polarisations (circular or planar) in order to permit re-use of frequencies (see Section 2.1.4).

The tracking system keeps the antenna pointing in the direction of the satellite in spite of the relative movement of the satellite and the station. Even in the case of a geostationary satellite, orbital perturbations cause apparent displacements of the satellite which are, however, limited to the 'station-keeping window' (see Section 7.3.4). Furthermore the station can be installed on a mobile vehicle whose location and direction vary with time.

The performance required of the tracking system varies in accordance with the characteristics of the antenna beam and the satellite orbit. For small antennas, the tracking system can be eliminated (fixed mounting) and this enables costs to be reduced.

The size and complexity of stations depend on the service to be provided and the power radiated by the satellite. The simplest stations permit reception only and are equipped with a parabolic antenna which may have a diameter of less than one metre [GOR-80]. The largest are first built Intelsat Standard A stations with antennas of 32 metres diameter.

8.2 RADIO-FREQUENCY CHARACTERISTICS

The characteristics which determine the radio-frequency performance of earth stations occur in the link budget expressions for the uplink and the downlink which have been discussed in Chapter 2:

$$(C/N_0)_U = (P_T G_T)_{ES}(1/L_U)(G/T)_{SL}(1/k) \qquad \text{(Hz)} \tag{8.1}$$

where $(P_T G_T)_{ES}$ is the effective isotropic radiated power of the earth station,

$$(C/N_0)_D = (P_T G_T)_{SL}(1/L_D)(G/T)_{ES}(1/k) \qquad \text{(Hz)} \tag{8.2}$$

where $(G/T)_{ES}$ is the figure of merit of the earth station.

The expressions are established for a particular link as characterised by the frequency of its carrier, the polarisation of the wave, the type of modulation and the bandwidth occupied. An earth station usually transmits and receives several carriers for which the characteristics of the stations, particularly the effective isotropic radiated power, can be different.

It should also be noted that the transmitting and receiving gains which arise in these expressions are related to each other since the same antenna is used for transmitting and receiving.

8.2.1 Effective isotropic radiated power

The effective isotropic radiated power (EIRP) is the product $(P_T G_T)_{ES}$ of the available power of the carrier considered at the antenna input and the transmission gain of the antenna in the direction of the satellite at the frequency considered.

8.2.1.1 *Available power P_T*

The carrier power P_T available at the antenna input is a function of the power $(P_{HPA})_{ES}$ of the power amplifier, the connection losses $(L_{FTX})_{ES}$ between the amplifier output and the antenna interface and the power reduction $(L_{MC})_{ES}$ caused by multicarrier operation:

$$P_T = (P_{HPA})_{ES}(1/L_{FTX})_{ES}(1/L_{MC})_{ES} \qquad \text{(W)} \tag{8.3}$$

In most applications, a particular earth station transmits more than one carrier to the satellite concerned. This situation exists when access to the satellite is in frequency division multiple access (FDMA) with an arrangement of the 'one-carrier-per-link' type (see Section 4.4.1.1) or with a multibeam satellite when destination selection is by transponder hopping (see Section 5.2). The configuration of the earth station amplifying system depends on the way in which the carriers to be transmitted are coupled (see Section 8.4.2). $(P_{HPA})(1/L_{MC})$ is the available power on the considered carrier at the output of the power amplifying system referred to for simplicity as the transmitting amplifier power (P_{TX}) in Chapter 2. The power reduction L_{MC} is then either the amplifier output back-off OBO (for coupling before amplification) or the loss L_C of the coupling device (for coupling after amplification). The power P_{HPA} of the amplifier is the output power at saturation with one carrier.

8.2.1.2 *Transmission gain G_T*

The maximum gain G_{Tmax} for an antenna of given diameter D is defined for a carrier frequency f_T by (cf. expression (2.2)):

$$G_{Tmax} = \eta_T(\pi D f_T/c)^2 \tag{8.4a}$$

where η_T is the transmission efficiency of the antenna and c is the velocity of light. As the antenna never points perfectly at the satellite, the actual transmission gain G_T differs from the maximum gain G_{Tmax} by a factor L_T which is a function of the pointing angle error:

$$G_T = (G_{Tmax}/L_T)_{ES} \tag{8.4b}$$

The gain reduction L_T has a value (cf. expression (2.4)):

$$L_T = 10^{1.2(\alpha T/\theta_{3dB})^2}$$

where α_T is the pointing error angle whose value depends on the type of tracking system. The influence of the type of tracking is examined in Section 8.3.7.6.

8.2.1.3 *Limitation of the effective isotropic radiated power*

In order to limit interference between satellite systems, the CCIR specifies a limit to the value of effective isotropic radiated power in the direction of geostationary satellites other than that to which the antenna is pointed [CCIR-Rep 524], [CCIR-Rep 1001].

Hence, for example, in Band C (6 GHz) for directions having an angle ϕ greater than 2.5° with respect to the principal axis of the antenna and deviating less than 3° from the geo-stationary satellite orbit, the effective isotropic radiated power in a bandwidth of 4 kHz must not exceed:

$$\mathrm{EIRP}_0 - 25\log\phi \quad \text{(dBW)} \quad \text{for } 2.5° \leqslant \phi \leqslant 25°$$

$$\mathrm{EIRP}_0 - 35 \quad \text{(dBW)} \quad \text{for } 25° \leqslant \phi \leqslant 180°$$

where the value of EIRP_0 is taken to be between 32 and 38.5 dBW. An expression for the limit in Ku band (12 GHz) is in preparation.

8.2.2 Figure of merit of the station

The figure of merit $(G/T)_{ES}$ is defined at the station receiver input as the ratio of the equivalent receiver gain G to the system noise temperature T of the earth station.

8.2.2.1 *Equivalent receiver gain G*

The equivalent receiver gain G is determined from the actual receiver gain G_R by allowing for the losses L_{FRX} suffered by the signal in the connection between the antenna interface and the receiver. The real receiver gain G_R differs from the maximum gain $G_{R\max}$ by a factor L_R which is a function of the pointing angle error α_R. The maximum gain $G_{R\max}$ for an antenna of given diameter D is defined at a carrier frequency f_R by $G_{R\max} = \eta_R(\pi D f_R/c)^2$, where η_R is the efficiency of the receiving antenna:

$$G = (G_R/L_{FRX})_{ES} = (G_{R\max}/L_R)_{ES}/(1/L_{FRX})_{ES} \tag{8.5}$$

The values of the pointing error angles α_R and α_T are identical. On the other hand, the corresponding gain losses L_R and L_T are not the same for receiving and transmitting since, for a given value of $\alpha_R = \alpha_T$, L_R and L_T depend on the receiving and transmitting frequencies f_R and f_T, which are different, and on the type of tracking. The influence of the type of tracking on the gain is examined in Section 8.3.7.6.

8.2.2.2 *System noise temperature T*

The concept of system noise temperature T has been explained in Section 2.5.3. The noise temperature is given by:

$$T = (T_A/L_{LRX})_{ES} + T_F(1 - 1/L_{FRX})_{ES} + T_R \quad \text{(K)} \tag{8.6}$$

The system noise temperature T is a function of the antenna noise temperature T_A, the connection losses L_{FRX} between the antenna interface and the receiver input, the thermo-dynamic temperature T_F of this connection and the effective noise temperature T_R of the receiver. Recall that the antenna noise temperature T_A of an earth station depends on the meteorological conditions, attenuation due to hydrometeorites causes an increase in the noise temperature of the sky (see Section 2.5.4). The antenna noise temperature also depends on the elevation angle.

The figure of merit G/T of the station is thus defined for a minimum elevation angle and clear sky conditions. Taking account of rain conditions in the link budget leads to a reduction $\Delta(G/T)$ of the earth station figure of merit.

8.2.3 Standards defined by international organisations

The international organisations for international satellite communication have defined various standards for earth stations operating in connection with the satellites they operate. These standards specify numerous parameters, particularly the figure of merit G/T, for different services and applications.

8.2.3.1 INTELSAT standards [RAN-89]

The characteristics of earth stations used in the INTELSAT system are grouped in the form of modules (IESS: INTELSAT Earth Station Standards) in accordance with the following categories:

—Series 100: Introduction.
—Series 200: Classes of authorised stations (antenna performance, G/T, side lobe level etc.).
—Series 300: Access, modulation and coding, carrier EIRP.
—Series 400: Additional specifications such as the characteristics of the satellites, geographical advantage, intermodulation levels, service circuits.
—Series 500: Digital circuit multiplication equipment (DCME).

The much lower performance of the first communication satellites (INTELSAT I—Early Bird—in 1965) imposed large dimensions on earth stations. Standard A INTELSAT stations are thus characterised by a figure of merit G/T of 40.7 dB K^{-1} (the antenna of 32 m diameter is equipped with a monopulse tracking system and a receiving amplifier with a noise temperature less than 30 K using a maser followed by a parametric amplifier). These stations are provided with transmitting amplifiers of several kilowatts.

These characteristics lead to stations of very high cost (of the order of £10 M); however this forms only a small part of the total cost in comparison with the cost of the space segment. More than 250 stations of this type are operated in the world.

In the mid 1970s, a new standard of earth station appeared, Standard B, which was intended to facilitate the creation of low traffic links. With a G/T ratio of 31.7 dB K^{-1} (9 dB less than Standard A), the Standard B station permits the use of an antenna of around 11 m diameter with a simplified tracking system (step by step). This type of station, operated in SCPC/PSK at 64 kbit/s or in FDM/CFM, makes routes of the order of tens of channels profitable which is not the case with Standard A stations.

The availability of frequency bands at 14/11 GHz, with the INTELSAT V generation (in 1981) has led to the specification of a third standard earth station, Standard C, characterised by a clear sky ratio G/T of 39 dB K^{-1}. Its antenna diameter, of the order of 15 m, makes it the equivalent at 14/11 GHz of a Standard A station. It is an expensive type of station (more than £5 M) principally suited to heavy traffic routes.

From 1983, enhancement of the characteristics of the INTELSAT satellites (higher EIRP and greater re-use of frequencies), extensive use of digital techniques (particularly error correcting codes) and a proliferation of new services have led to the introduction of new earth station standards in association with more and more varied access and modulation systems:

—Standard D for telephone and low density links (the VISTA service) realised in SCPC mode using frequency modulation with companding or digital transmission with a transmission rate of 16 kbit/s. Standard D includes two sub-standards, D1 and D2; VISTA networks are mainly envisaged as star networks with a central station of D2 type and peripheral stations of D1 type.

—Standards E and F for IBS services in Ku and C band respectively. Each of these standards is divided into three sub-standards: E1, E2, E3, F1, F2, and F3 which correpond to increasing values of G/T.

—Standards Z and G for transponder reservation services for national and international use respectively. Standards Z and G are not part of the IESS and contain only general specifications intended to avoid interference with transmissions of other standards; items include the antenna radiation pattern, polarisation purity, off-axis EIRP density etc.

Following the increase of satellite EIRP, the specification of G/T for new Standard A and C stations has been reduced from 40.7 to 35 dB K^{-1} and from 39 to 37 dB K^{-1} respectively since 1986. The specification of the associated antenna side lobes has been modified.

Table 8.1(a) shows the characteristics of the various INTELSAT standards. Table 8.1(b) provides more detailed information on the associated IDR service to Standard E. Table 8.1(c) provides several examples of Standard Z earth stations operating at 6/4 GHz [KEL-84].

Access and modulation modes. Analogue transmission is operated in multiple access using frequency division (FDMA) with the following carrier types:

—FDM/MF (IESS 301). The capacity extends from 12 channels (in 1.25 MHz) to 972 channels (in 36 MHz). Speech concentration enables the capacity to be increased by around 50% per carrier without a significant power increase.

—FDM/CFM (IESS 302). The use of channel-by-channel companders enables the capacity of carriers which support from 24 channels (in 1.25 MHz) to 792 channels (in 25 MHz) to be increased.

—SCPC/CFM (IESS 305). This type of transmission is used on very low density routes within the VISTA service with Standard D1 and D2 stations. It enables a single channel carrier in an allocated band of 30 kHz to be routed under quality conditions inferior to those of a conventional FDM/FM service.

—TV/MF (IESS 306). The carriers use an allocated bandwidth of 17.5, 20 or 30 MHz.

Digital links provided by the INTELSAT system (TDMA or FDMA) use the following types of carrier:

—SCPC/PSK (IESS 303). The first INTELSAT digital system (since 1975), this mode uses QPSK carriers at 64 kbit/s (56 kbit/s useful rate); transmission is generally controlled by the activity on the channel.

—SPADE (IESS 304). An SCPC system with a demand assignment arrangement for application to routes with very low traffic (see Section 4.7.3).

—TDMA (IESS 307). Commissioned at the end of 1985, this system provides a capacity of 120 Mbit/s with a BCH error correcting code of coding ratio 7/8. It is applied to packets destined for revised Standard A stations or stations situated at the boundary of the coverage area. A digital speech concentrator (DSI: digital speech interpolation) enables the transmission capacity per transponder to be doubled. Designed initially to replace FDM/FM/FDMA, this system is of importance principally for routes with heavy traffic and high connectivity. For routes of low connectivity (of the point-to-point type) or low traffic (which does not justify over-dimensioning the earth station for a capacity of 120 Mbit/s), FDMA in digital form is rapidly replacing FDM/FM under the name of the IDR (intermediate data rate) system.

—IDR (IESS 308). This system, whose commissioning started in 1986, uses digital FDMA carriers with intermediate data rates between the two previous types (64 Kbit/s and 120 Mbit/s). Taking account of existing digital hierarchies, the most common data rates are 1.544 Mbit/s, 2.048 Mbit/s, 6.312 Mbit/s and 8.448 Mbit/s. The bits arising from the use of a convolutional error correcting code, with a coding ratio of 3/4 and Viterbi decoding at the receiver, must be added to these useful information rates. This coding enables the application of IDR carriers to be extended from Standards A and C to Standards E2, E3 and F3.

—IBS (International Business Service) (IESS 309). Defined in 1983 in response to requirements for International Business Services, this systems, initially of the closed network type, has developed to cover open networks (that is those which can be interconnected). In both cases, it consists of an FDMA system with characteristics similar to those of the IDR system.

IBS closed network: Information rate 64 kbit/s to 8.448 Mbit/s, additional capacity of 10% for service channels and possible coding, convolutional coding with coding ratio 3/4, Viterbi decoding.

IBS open network: information rate 64 kbit/s to 1.920 Mbit/s, additional capacity of 32/30 for service channels and possible encryption, convolutional coding of coding ratio 1/2, Viterbi decoding, CCITT terrestrial interfaces, compatibility with the EUTELSAT SMS standard.

Table 8.1 Characteristics of the INTELSAT standards

(*a*) *General characteristics*

Standard	Frequency (GHz)	G/T (dB K^{-1})	EIRP (dB W)	Diameter (m)	Service
A	6/4	⩾40.7	70 to 90	30	TV or FDM/FM/FDMA or TDM/PSK/TDMA
A Revised		⩾35		16	(IESS 201)
B	6/4	⩾31.7	60 to 85	11 to 14	TV or SCPC/QPSK or FDM/FM/FDMA (IESS 202)
C	14/11	⩾39	72 to 87	14 to 18	TV or FDM/FM/FDMA or TDM/PSK/TDMA
C Revised		⩾35		11 to 13	or (IESS 203)
D1	6/4	⩾22.7	53 to 57	5	SCPC/FM
D2		⩾31.7		11	VISTA (IESS 204).
E	14/11				IBS, IDR (IES 205).
F1	6/4	⩾22.7	63 to 91	4.5 to 5	IBS in TDM/QPSK/FDMA
F2		⩾27	60 to 87	7.5 to 8	IDR in FDM/CFM with F3
F3		⩾29	59 to 86	9 to 10	(IESS 206)
Z	6/4 14/11				Leased channels

(*b*) *Standard E (IBS/IDR Services, 14/11 GHz)*

	$E-1$	$E-2$	$E-3$
Antenna diameter (m)	3.5	5.5	8 to 10
EIRP (dBW)	57–86	55–83	49–77
G/T (dB K^{-1})	25	29	34
Equivalent number of channels at 64 kbit/s	400	700	1000
Equivalent number of carriers of 1.5 Mbit/s capacity	16	28	42
Highest information rate per station (Mbit/s) (72 MHz channels)	2048	4096	6144
System performance during 99% of the time (BER)	10^{-6}	10^{-6}	10^{-6}
Clear sky link bit error rate	10^{-8}	10^{-8}	10^{-8}

Whether open or closed, IBS networks are distinguished from IDR by the fact that they do not benefit from authorisation for connection to the public switched telephone network. In Ku band, the service quality provided by IBS services is inferior to that of IDR (10^{-6} during 99% of the time instead of 10^{-7} during 80% of the time and 10^{-3} during 99.93 of the time).

Users wishing to benefit from a higher quality service can, since the end of 1986, make use of the 'super IBS' service for which the quality is made equivalent to that of IDR by increasing the EIRP level by around 2.5 dB.

Table 8.1 *(continued)*

(c) Standard Z (Leased channels, 6/4 GHz)

	'Large'	'Small'	TV/Radio (receive only)
Antenna size (m)	11.0–13.0	6.0–8.0	4.5–5.0
Transmission gain (dB)	54.5–56.0	49.3–51.7	
Reception gain (dB)	51.5–53.0	46.3–48.7	44.0
LNA temperature (K)	45–80	100	100
G/T at 10° elevation angle (dB/K)	31.7–33.0	24.5–26.9	22.0
Tracking	Automatic	Manual	Manual
Polarisation (with frequency re-use)	Circular	Circular	Circular
Ellipticity ratio (Tx/Rx)	1.06/1.09	1.06/1.09	–/1.4
Number of channels	More than 12	2–12	1 Video + audio 1 radio
HPA type	TWT or klystron	TWT	—
HPA power	1–3 kW	50–400 W	—
Carrier types	FDM/(C)FM SCPC/DM TV/FM	SCPC/CFM SCPC/DM	Video/audio-FM Radio SCPC/CFM
Power per SCPC channel			
to a small station	<1 W	1 W	—
to a large station	1 W	10 W	—

Development of the INTELSAT earth segment. Developments in progress in connection with INTELSAT earth stations involve the following areas:

—Channel multiplication (IESS 501). The use of digital circuit multiplication equipment (DCME) permits multiplication of the number of channels routed by satellite by more than four with respect to non-concentrated terrestrial channels at 64 kbit/s (see Section 8.6.4),

—Introduction of carriers at 140 Mbit/s using modems of advanced technology such as 8 state phase modulation with Viterbi decoding of coding ratio 7/9.

—Development of VSAT networks within the INTELNET service.

—Digitisation of the VISTA service.

8.2.3.2 EUTELSAT standards

Earth stations of the EUTELSAT European satellite communication system are of several types:

—Stations for transmission of high capacity telephone trunks and television, similar to INTELSAT Standard C stations (14/11 GHz, 14 to 18 m antenna, EIRP between 72 and 87 dBW, clear sky G/T greater than 39 dB K^{-1}).

—Stations used within the Satellite Multi Services (SMS) system which conform to the specifications of two EUTELSAT standards: Standard 1 with antennas of the order of 5 m and a clear sky G/T of 30.4 dB K^{-1} and Standard 2 with antennas of 3.7 m and a G/T of 27.4 dB K^{-1}.
—Fixed stations with diameters between 8 and 12 m and transportables of the order of 4 m used for various applications such as the establishment of an emergency transmission network in disaster areas with transportable stations.
—Antennas for television reception (RCVO: Receive Only) of diameter between 2 and 6 m.

8.2.3.3 *INMARSAT standards*

The international organisation for mobile maritime telecommunication services INMARSAT proposes different standards for mobile stations which operate in Band L [DOR-79]. The band used in the mobile–satellite direction (transmission) is between 1626.5 and 1646.5 MHz and the band used in the satellite–mobile direction (reception) is between 1530 and 1545 MHz. The polarisation used is right circular for the uplink and left circular for the downlink. The frequencies used for satellite–coast station links (feeder links) belong to the Fixed Satellite Service (FSS) and are operated at C band.

Different standards have been defined for stations on ships:

—Standard A, characterised by a clear sky G/T greater than -4 dB K^{-1} for an elevation angle greater than 10°, EIRP of 37 dB W (transmitter power 500 W), parabolic antenna of diameter of the order of 90 cm mounted on a stabilised platform and provided with a tracking device. The station permits transmission of several telephone channels in SCPC/FM mode and data (Telex 50 bauds in SCPC/BPSK/TDMA).
—Standard B, characterised by a clear sky G/T greater than -4 dB K^{-1} (using a parabolic antenna of diameter of the order of 90 cm) for transmission of telephony with data rate compression to 16 kbit/s in SCPC/BPSK and low rate data (1.2 to 16 kbit/s in SCPC/BPSK/TDMA).
—Standard C, characterised by a clear sky G/T greater than -23 dB K^{-1} and an EIRP greater than 12 dB W. The antenna is omnidirectional and should provide the required EIRP and G/T values for elevation angles as low as $-15°$. The station permits transmission of very low rate messages (600 bit/s in SCPC/BPSK). The 1200 symbol/s transmission uses convolutional error correcting coding of ratio 1/2 with interleaving and automatic repetition request [HIG-88], [TEL-88], [MEU-89]. Standard C provides the possibility of exchanging messages between a boat and a terrestrial subscriber and between boats, transmitting distress signals within a worldwide distress and maritime security system, broadcasting messages and so on.

Aeronautical stations, characterised by a G/T greater than -12 dB K^{-1} and an EIRP greater than 29.5 dB W permit transmission of data from 300 to 600 bit/s and telephony at 7.2 kbit/s.

For terrestrial mobiles, stations of Standard C type can be used for data transmission from 300 to 600 bit/s. Standard M, under development, would permit telephone transmission.

8.2.3.4 *Other systems*

On the fringe of the INMARSAT system, experiments for communication with aircraft should be noted [BRO-78], for example in Japan using the ETS-V satellites [HAS-89], [YAS-89],

[KAD-89] and at the European Space Agency, in order to test the Prodat system [ROG-89]. Numerous projects involving communication systems with portable earth terminals are also in progress [EST-89].

8.3 THE ANTENNA SUBSYSTEM

The characteristics required for an earth station antenna are as follows [CCIR-Rep 390]:

—High directivity along the axis of the antenna, in principle directed towards the nominal satellite position (for useful signals).
—Low directivity in other directions which can correspond to satellites other than that with which it is required to establish a link (for undesirable signals).
—Antenna efficiency as high as possible for both frequency bands (uplinks and downlinks) on which the antenna operates.
—High isolation between orthogonal polarisations.
—The lowest possible antenna noise temperature.
—Continuous pointing in the direction of the satellite with the required accuracy.
—Limitation, as far as possible, of the effect of local meteorological conditions (such as wind, temperature, etc.) on the overall performance.

8.3.1 Radiation characteristics (major lobe)

The antennas used, usually parabolic reflectors, are of the type with a radiating aperture. The performance and properties of radiating apertures are treated in numerous works [COM-88], [JON-84], [RUD-82], [MIY-82, Ch. 5]. The characteristic parameters of an antenna have been recalled in Section 2.1. For an earth station antenna the important parameters which characterise the radiation of the major lobe are the gain, the angular beamwidth and the polarisation isolation.

The antenna gain arises directly in the expressions for the effective isotropic radiated power (EIRP) and the figure of merit (G/T) of the station. The antenna beamwidth determines the type of tracking system used in accordance with the particular characteristics of the satellite orbit. The value of polarisation isolation determines the capacity of an antenna to operate in a system where frequency re-use by orthogonal polarisation is used [GOR-88]. Assuming that the carrier powers of orthogonal polarisations are the same, the interference introduced by the antenna from one carrier to the other is equal to the polarisation isolation which must, therefore, be greater than a specified value. By way of example, INTELSAT advocates, for certain standards and applications, a value less than 1.06 for the axial ratio (AR) in the direction of a satellite with new antennas. This corresponds to a carrier power to interference power ratio $(C/N)_I$ greater than 30.7 dB.

8.3.2 Slide-lobe radiation

Most of the power is radiated (or acquired) in the major lobe. However, a non-negligible amount of power is dispersed by the side lobes. The side lobes of an earth station antenna determine the level of interference with other satellites in orbit. This is particularly important

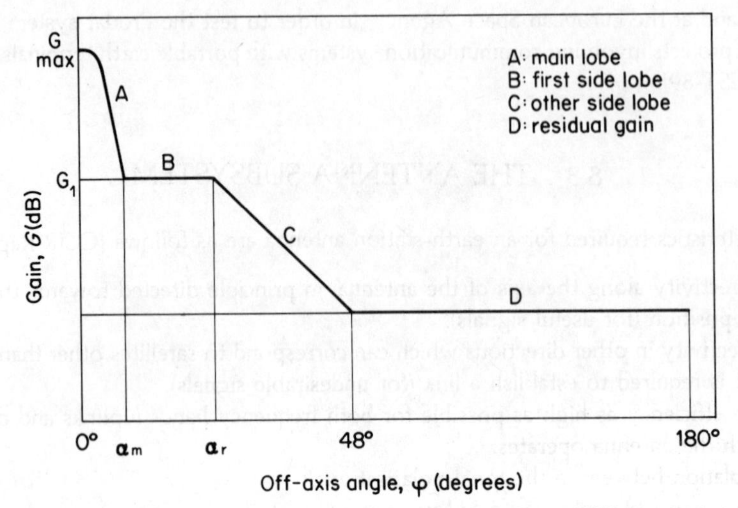

Figure 8.2 Reference radiation pattern of an earth station antenna. (From CCIR-Rep 391–6. Reproduced by permission of the ITU.)

for the side lobes which illuminate an area of several degrees of the orbit of geostationary satellites. To limit these effects, a reference diagram has been proposed in the Radiocommunication Regulations issued by the International Telecommunications Union. This proposal is presented in Figure 8.2 and is defined by the following expressions [CCIR-Rep 391]:

—For an antenna of diameter D greater than 100 λ where λ is the wavelength:

$$G(\alpha) = G_{max} - 2.5\,10^{-3}(\alpha D/\lambda)^2 \quad \text{for } 0 < \alpha < \alpha_m \quad \text{(dB)}$$

$$G(\alpha) = G_1 = 2 + 15\log(D/\lambda) \quad \text{for } \alpha_m \leqslant \alpha < \alpha_r \quad \text{(dB)}$$

where G_1 is the gain in the first lobe

$$G(\alpha) = 32 - 25\log\alpha \quad \text{for } \alpha_r \leqslant \alpha < 48° \quad \text{(dB)}$$

$$G(\alpha) = -10 \quad \text{for } 48° \leqslant \alpha < 180° \quad \text{(dB)}$$

$$\alpha_m = (20\lambda/D)\sqrt{(G_{max} - G_1)} \quad \text{(degrees)}$$

$$\alpha_r = 15.85\,(D/\lambda)^{-0.6} \quad \text{(degrees)} \tag{8.7a}$$

—For an antenna of diameter D less than 100 λ:

$$G(\alpha) = G_{max} - 2.5\,10^{-3}(\alpha D/\lambda)^2 \quad \text{for } 0 < \alpha < \alpha_m \quad \text{(dB)}$$

$$G(\alpha) = G_1 = 2 + 15\log(D/\lambda) \quad \text{for } \alpha_m \leqslant \alpha < 100\,\lambda/D \quad \text{(dB)}$$

$$G(\alpha) = 52 - 10\log(D/\lambda) - 25\log\alpha \quad \text{for } 100\,\lambda/D \leqslant \alpha < 48° \quad \text{(dB)}$$

$$G(\alpha) = 10 - 10\log(D/\lambda) \quad \text{for } 48° \leqslant \alpha < 180° \quad \text{(dB)} \tag{8.7b}$$

In order to further limit the interference between adjacent satellites, the CCIR [CCIR-Rec 580] encourages antenna manufacturers to produce antennas such that the gain G of at least

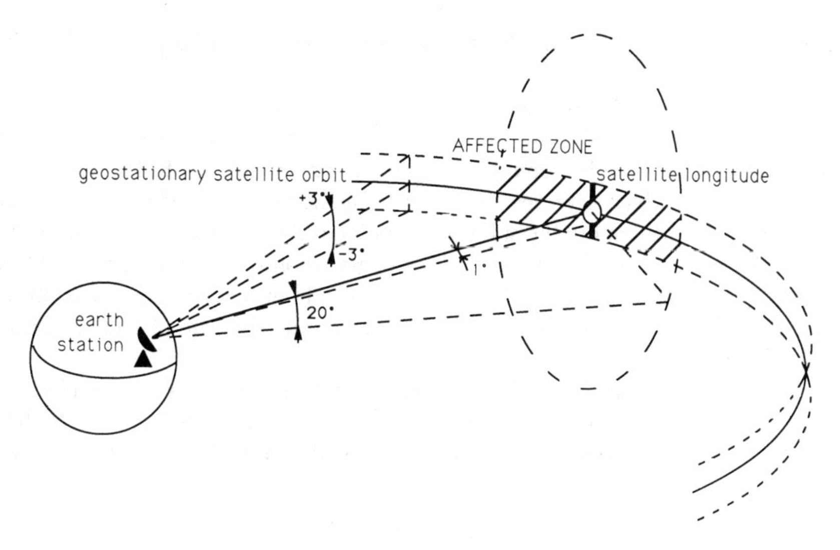

Figure 8.3. Zone around the geostationary satellite orbit to which the design objective for earth station antennas applies.

90% of the side-lobe peaks does not exceed:

$$G = 29 - 25 \log \varphi \qquad \text{(dB)} \tag{8.8a}$$

where φ is the angle of the direction considered with respect to the axis of the diagram for $1^\circ \leqslant \varphi \leqslant 20^\circ$ and for regions of space which make an angle of 3° with respect to the geostationary orbit (Figure 8.3). This recommendation is valid for antennas having a D/λ greater than 100 installed after 1991.

For antennas such that D/λ is less than 100 (and greater than 35), the gain G of at least 90% of the secondary lobe peaks must not exceed:

$$G = 49 - 10 \log(D/\lambda) - 25 \log \varphi \tag{8.8b}$$

The use of offset mountings with two reflectors seems well suited to obtaining both good radio-frequency characteristics (gain and polarisation isolation) for the major lobe and low side-lobe levels [ANS-89], [CCIR-Rep 998]. Reduction of side-lobe level can also be obtained by combining radiation patterns generated by auxiliary sources [ARN-83], [BRU-89], [HAR-89], [SCH-89].

8.3.3 Antenna noise temperature

The concept of antenna noise temperature has been presented in Section 2.5.4. For an earth station, the noise acquired by the antenna originates from the sky and surrounding ground radiation (Figure 2.15) [CCIR-Rep 868]. It depends on the frequency, the elevation angle and the atmospheric conditions (clear sky or rain). The type of antenna mounting also has an influence on the contribution of radiation from the ground and this will be discussed in Section 8.3.4. Before that an important phenomenon characterised by an increase of antenna temperature during conjunction of the sun and the satellite will be presented [VUO-83], [RAU-85], [MOH-88].

Conjunction of the sun and the satellite corresponds to a situation where the earth station, the satellite and the sun are aligned. In practice, the increase of antenna noise temperature also occurs in the vicinity of the conditions of precise conjunction due to the non-zero width of the antenna beam which captures the noise from the sun even when it is not exactly behind the satellite. Furthermore the sun is not a point source. Hence, an increase of antenna noise temperature causes significant performance degradation which may lead to link unavailability when the direction of the centre of the sun is within a solid angle of width θ_i (see Section 8.3.3.4); that is once per day for several days around the date of precise conjunction (geometric alignment).

The geometric considerations relating to the occurrence conditions of this phenomenon are discussed in [LUN-70], [LOE-83], [GAR-84], [DUR-87] and in Section 7.2.5.7 where expressions are given which permit the number of days and the daily duration to be calculated as a function of θ_i.

8.3.3.1 *Noise temperature of the sun*

The increase of antenna noise temperature during conjunction depends on the noise temperature of the sun in the band of frequencies concerned. The brightness temperature of a point on the surface of the sun varies as a function of the wavelength, its position within the solar disc and solar activity. Various models have been developed to estimate the mean brightness temperature of the sun as a function of the wavelength. An approximate expression for the mean brightness temperature of the sun excluding periods of solar activity is proposed in reference [RAU-85] for operation in C band:

$$T_{SUN} = (1.96\ 10^5/f)[1 + (\sin 2\pi\{[\log6(f - 0.1)]/2.3\})/2.3] \quad (K) \qquad (8.9)$$

where f is the frequency in GHz.

Another approximate expression [VUO-83] leads to values in K band which are close to those given by the models of Van de Hulst and Allen [SHI-68]:

$$T_{SUN} = 120\,000 f^{-0.75} \quad (K) \qquad (8.10)$$

where f is the frequency in GHz.

The above expressions give mean values of the brightness temperature on the solar surface which is assumed to be quiet. In reality, variations from one point to another are large and greater at low frequencies; at 4 GHz, the temperature varies from 25 000 K to 70 000 K. At 12 GHz, the temperature of the centre of the sun (the cold point) is around 12 000 K and the mean temperature over the whole disc is of the order of 16 000 to 19 000 K. These brightness temperature variations of the sun as a function of wavelength for the radio-frequency domain are illustrated by the curves of Figure 8.4 [BOI-83].

In periods of intense solar activity, a considerable increase in the brightness temperature can be observed, particularly at low frequencies—more than 50% during 1% of the time in C band [CCIR-Rep 390]. At 12 GHz, the brightness temperature can reach and exceed 28 000 K.

The apparent diameter of the sun itself also depends on the wavelength of the radiation and varies inversely with frequency, particularly below 10 GHz. In Ku band, it is slightly greater than the apparent diameter of the solar disc in the visible region, that is around 0.5°.

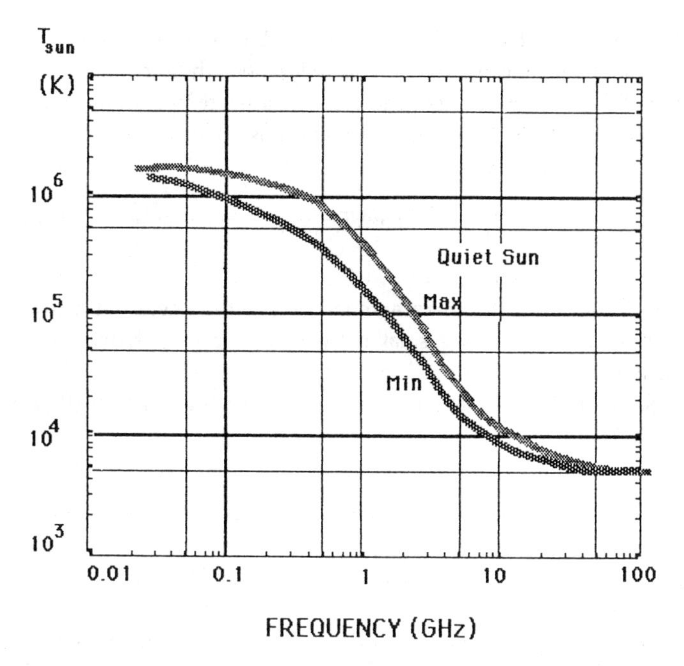

Figure 8.4 Sun temperature as a function of frequency (quiet sun).

8.3.3.2 Increase of the noise temperature during conjunction

The increase of antenna noise temperature is obtained by integrating the product of the brightness temperature $T_{SUN}(\theta, \varphi)$ and the antenna gain $G(\theta, \varphi)$ (with the functions defined in spherical co-ordinates θ,φ) over the solid angle through which the sun is viewed [HO-61]:

$$\Delta Ta = (1/4\pi) \iint_{\text{solar disc}} T_{SUN}(\theta, \varphi) G(\theta, \varphi) \sin\theta \, d\varphi \qquad \text{(K)} \qquad (8.11)$$

An approximate estimate of the increase in noise temperature is obtained by considering that the antenna radiation pattern is concentrated within a beam of equivalent width θ_e and considering the mean brightness temperature of the sun. The increase in the antenna noise temperature is then proportional to the ratio of the solid angle through which the sun is viewed to the solid angle which corresponds to the equivalent width θ_e of the antenna beam if this width is greater than the apparent diameter of the sun. Otherwise, the increase in antenna noise temperature is equal to the noise temperature of the sun:

$$\Delta T_A = T_{SUN}(0.5/\theta_e)^2 \qquad \text{if } \theta_e > 0.5^\circ \qquad \text{(K)}$$

$$\Delta T_{A\,max,unpol} = T_{SUN} \qquad \text{if } \theta_e < 0.5^\circ \qquad \text{(K)} \qquad (8.12)$$

The apparent diameter of the sun is taken to be 0.5°. The equivalent width θ_e of the beam can be taken to be θ_{3dB}.

Electromagnetic waves originating from the sun have a random polarisation. The source of an earth station antenna operating with frequency re-use by orthogonal polarisation is equipped with a polariser which enables waves arriving with the proper polarisation at the input of the corresponding receiver to be isolated. Under these conditions, the noise power acquired from the sun, and in turn the increase of noise temperature of the corresponding antenna, are reduced by half. So, when a polariser is used, the increase of antenna noise temperature has the following value:

$$\Delta T_{A\,max} = 0.5\,\Delta T_{A\,max,\,unpol} \quad \text{(K)} \qquad (8.13)$$

with $\Delta T_{A\,max,\,unpol}$ given by expression (8.12).

Expressions (8.12) show that, for an antenna of small diameter (at a given wavelength), the increase in noise temperature is smaller than for a large antenna. For example, in Ku band ($f = 12\,\text{GHz}$) an antenna of 1.2 m with a polariser ($\theta_{3dB} = 1.5°$) has an antenna temperature increase of 890 K while an antenna of 5 m with a polariser ($\theta_{3dB} = 0.35°$) suffers an increase of 8000 K.

8.3.3.3 *Permissible increase of antenna noise temperature with solar conjunction*

The link budget established under normal operating conditions (clear sky) often retains a margin M_1 with respect to the ratio C/N_0 (the ratio of carrier power-to-noise power spectral density) which is necessary in order to obtain the nominal required service quality (for a given percentage of the time). The quality objectives with rain generally provide for a permissible degradation for smaller percentages of the time. This value of degradation acts as an additional margin M_2 on the ratio C/N_0 with respect to that necessary under normal operating conditions. These margins $M_1 + M_2$ (in dB), when referred to the figure of merit G/T of the earth station, permit an increase $\Delta T_{A\,all}$ (all = allowable) to be accepted in the antenna noise temperature of the earth station.

For the margin $M = M_1 + M_2$, the permissible increase $\Delta T_{A\,all}$ in the antenna noise temperature is determined from expression (8.6):

$$\Delta T_{A\,all} = T(10^{0.1M} - 1)L_{FRX} \quad \text{(K)} \qquad (8.14)$$

where: M is the available margin expressed in dB, T is the clear sky system noise temperature, and L_{FRX} is the connection losses between the antenna interface and the receiver input.

Conversely, for an increase ΔT_A in the antenna temperature, the degradation $\Delta(C/N)$ or $\Delta(C/N_0)$ of the carrier to noise or carrier to noise spectral density ratio is equal to the relative increase $\Delta T/T$ of the system noise temperature and is given by:

$$\Delta(C/N) = \Delta(C/N_0) = \Delta T/T = 10\log\left[TL_{FRX}/(TL_{FRX} + \Delta T_A)\right] \qquad (8.15)$$

where T is the clear sky system noise temperature and L_{FRX} represents the connection losses between the antenna interface and the receiver input.

8.3.3.4 *Angular diameter of the region of interference*

The region of interference is defined as the region of the sky such that, when the centre of the sun is within it, the increase in antenna noise temperature ΔT_A caused by the sun exceeds the acceptable limit $\Delta T_{A\,all}$.

The variation of antenna temperature when the apparent movement of the sun crosses the antenna beam will be examined (Figure 8.5) (this figure assumes that maximum solar interference is obtained at antenna boresight, i.e. the day when solar conjunction occurs):

—As long as the sun is far from the antenna axis, the antenna temperature is equal to its nominal 'clear sky' value (the service quality is greater than the nominal value as a consequence of the margin M_1);
—As the sun approaches the beam, the antenna noise temperature, and hence the system noise temperature, increase. At first the available margin M_1 compensates for the increase in system temperature and the nominal quality objective remains satisfied;
—When the increase in system temperature exceeds the available margin M_1, the quality falls below the nominal objective. As long as the quality remains greater than the value corresponding to degraded operation and, with the reservation that the cumulative total of the corresponding durations is less than the specified percentage of time for the degraded mode, operation of the system remains assured. The additional increase of the permitted system temperature corresponds to the margin M_2;
—When the total increase in the antenna temperature is such that the quality objective in degraded mode is no longer satisfied ($\Delta T_A = \Delta T_{A\,all}$, a function of $M_1 + M_2$), interruption of the service occurs. The position of the centre of the sun with respect to the axis of the antenna beam thus defines the angular half diameter of the interference region (Figure 8.5);
—During traversal of the interference region, the increase in noise temperature is greater than the maximum permissible $\Delta T_{A\,all}$ and the service is interrupted;
—The service is re-established when the quality again becomes greater than the objective in degraded mode as the sun leaves the interference region.

The angular diameter θ_i of the interference region thus depends on the diameter of the antenna beam and the ratio of the increase in permissible antenna noise temperature to the

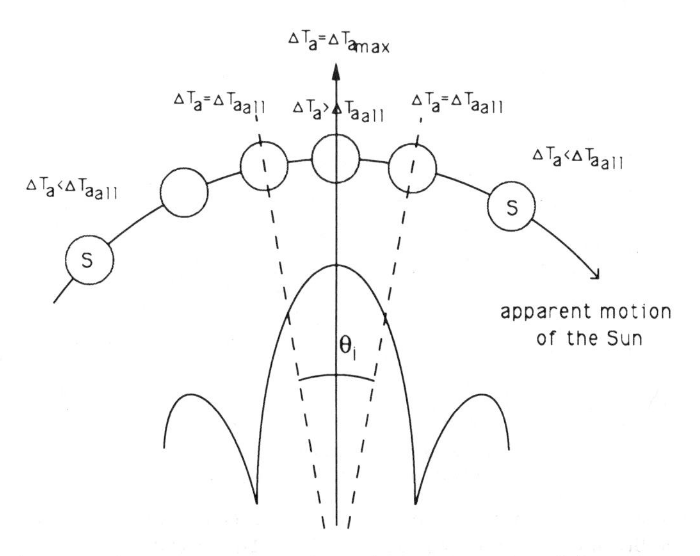

Figure 8.5 Antenna noise temperature variation with apparent motion of the sun.

maximum increase in the antenna noise temperature during conjunction. Figure 8.6 gives the value of the angular diameter θ_i as a function of this ratio and the value of the ratio D/λ for the antenna [CCIR-Rep 390].

Example. Consider the receiver system characterised by:

—System temperature $T = 400\,K$
—Total margin $M = M_1 + M_2 = 5\,dB$
—Connection losses $L_{FRX} = 0.5\,dB$

The permissible increase $\Delta T_{A\,all}$ in the antenna temperature calculated from expression (8.14) has a value of 970 K.

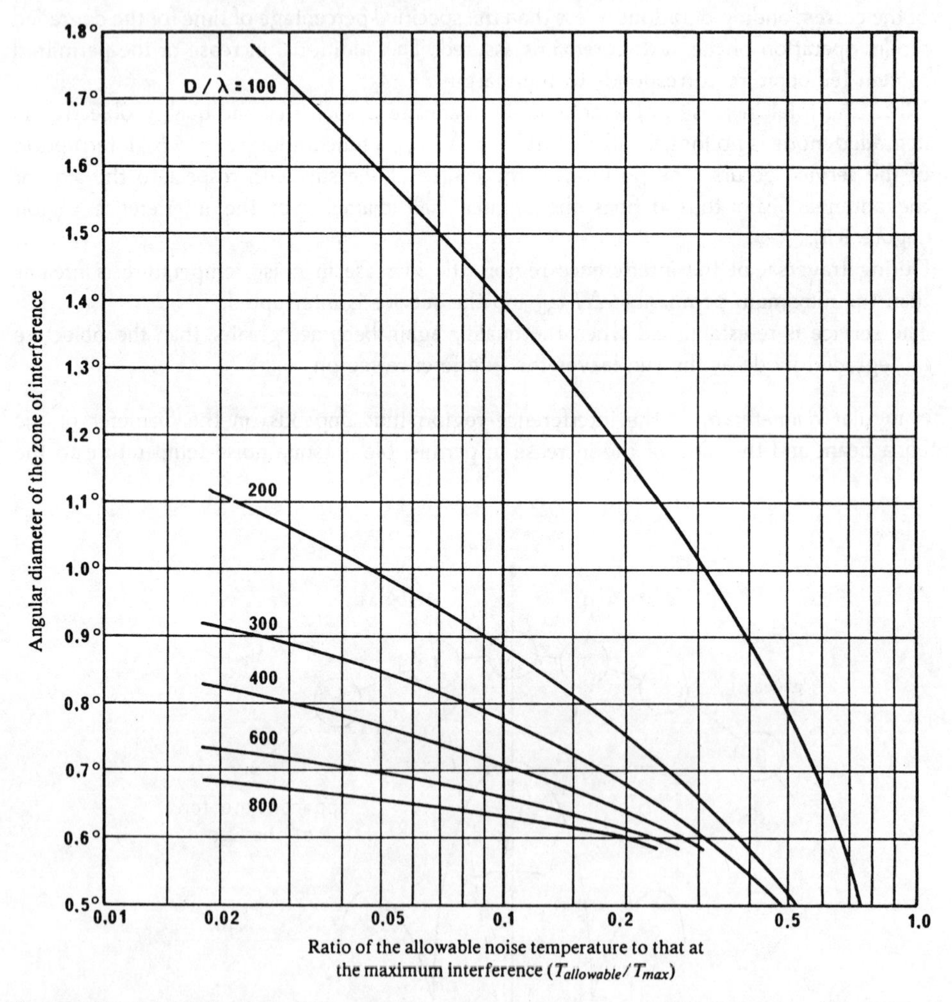

Figure 8.6 Angular diameter of the zone of interference as a function of the ratio of the allowable noise temperature (ΔT_{all}) to that at the maximum interference (ΔT_{max}). (From [CCIR-Rep. 390–6]. Reproduced by permission of the ITU.)

A station equipped with an antenna of 1.2 m diameter with a polariser operating at 12 GHz whose antenna temperature increase ΔT_{Amax} is 890 K when in sun–satellite conjunction (see the example of Section 8.3.3.2) thus continues to function (by satisfying the quality objective in degraded mode) in spite of the solar interference.

On the other hand, a station equipped with an antenna of greater diameter suffers an interruption of service. The ratio between the increase in permissible antenna noise temperature and the maximum increase of antenna noise temperature during conjunction has a value of $970/8000 = 0.12$. Consider an antenna of 5 m diameter ($D/\lambda = 200$), from Figure 8.6, the angular diameter θ_i of the interference region is 0.85°.

The duration of service interruption for a geostationary satellite under these conditions will be calculated. The apparent movement of the sun is $360°/24 \times 60 = 0.25°/\text{min}$ (see Section 7.2.5.7), if the aiming direction of the antenna remains fixed, the maximum duration T_i of service interruption is:

$$T_i = (\theta_i/0.25)\text{min} = 4\,\theta_i \qquad \text{(min)} \tag{8.16}$$

θ_i, the angular diameter of the interference region, is expressed in degrees. For the example considered, the duration T_i has a value of $4 \times 0.85°$ or 3.4 min.

When the available margin is small, it can be assumed that interference which leads to an interruption of service occurs as soon as the solar disc penetrates the major lobe of the antenna. The antenna beamwidth to be considered is thus of the order of $2\theta_{3dB}$. The angular diameter θ_i of the interference region can thus be expressed approximately as:

$$\theta_i = 2\theta_{3dB} + 0.5° \tag{8.17}$$

8.3.4 Types of antenna

The antennas considered belong to various classes:

—the horn antenna,
—the parabolic antenna,
—the phased array antenna.

The horn antenna allows a high figure of merit to be achieved but it is expensive and cumbersome even if its bulk can be reduced with a folded horn. This type of antenna was used at the start of space communication for experimental links with the Telstar satellite (Pleumeur Bodou 1). This technology is no longer in use.

Phased array antennas have an advantage when the beam is in constant movement as is the case for stations mounted on mobiles; however the technology remains relatively difficult and costly which limits the use of this type of antenna.

The most used antennas are those with parabolic reflectors. The three principal mountings are:

—symmetrical mounting,
—offset mounting,
—Cassegrain mounting.

8.3.4.1 *Antenna with symmetrical parabolic reflector*

Figure 8.7 illustrates the mounting of an antenna with a parabolic reflector which has symmetry of rotation with respect to the principal axis on which the primary feed is placed at the focus. The main weakness of this mounting is that the feed supports and the feed itself have a masking effect on the radiating aperture (aperture blocking). This blocking leads to a reduction of antenna efficiency and an increase in the level of the side lobes due to diffraction by the obstacles.

Furthermore, the primary feed faces towards the earth and that part of the radiation pattern of the primary feed which does not intercept the reflector (spillover) easily captures the radiation emitted by the ground and this makes a relatively large contribution to the antenna noise temperature (several tens to around 100 K).

Spillover is attenuated if the amplitude of the primary feed radiation is reduced from the centre towards the perimeter. To obtain a low noise temperature, a directional primary feed and a long focal length are necessary. The antenna is thus cumbersome and badly suited to the installation of microwave circuits immediately behind the feed; the bulk of such circuits would have a substantial masking effect.

8.3.4.2 *Offset mounting*

Offset reflector mounting enables microwave circuits to be located immediately behind the primary feed without masking effects. It does not involve, as the name might suggest, offsetting the feed with respect to the focus but the use of that part of the parabola situated on one side of the vertex for the reflector profile (Figure 8.8). Presently used for antennas of small diameter (1 to 4 m), this mounting is little used for large antennas for which the Cassegrain mounting is preferred. With offset mounting, the spillover remains orientated towards the ground and the antenna temperature remains high.

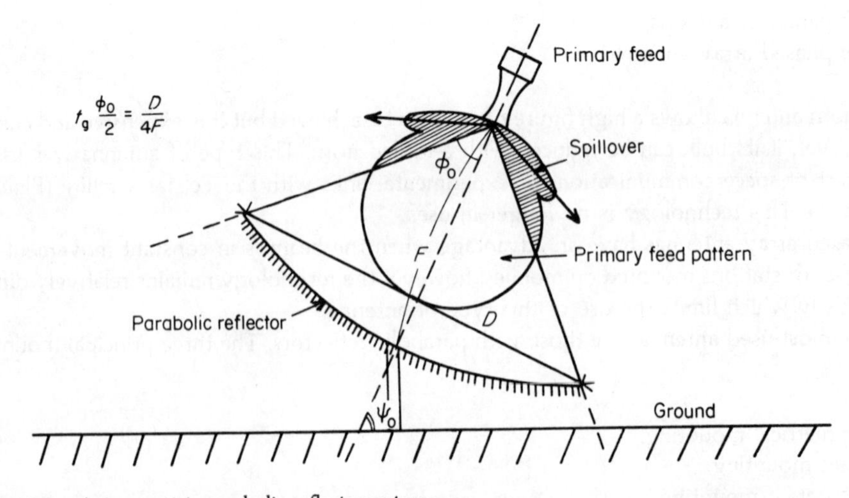

Figure 8.7 Axisymmetric parabolic reflector antenna.

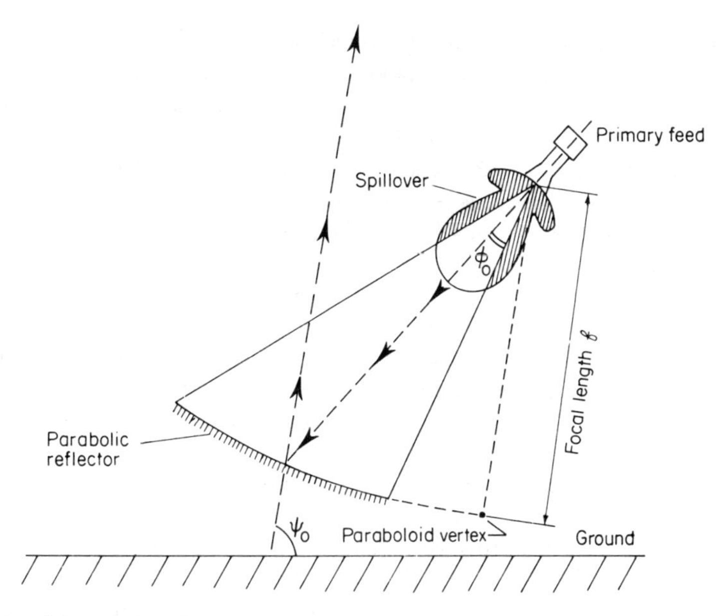

Figure 8.8 Offset-fed parabolic reflector antenna.

8.3.4.3 Cassegrain mounting

With Cassegrain mounting (Figure 8.9), the phase centre of the primary feed is situated at the first focus S of an auxiliary hyperbolic reflector. The other focus R of the auxiliary reflector coincides with the focus of the main parabolic reflector. If D is the aperture diameter of the parabolic reflector and f_d its focal length, the solid apex angle Φ_0 under which the reflector is viewed is given by the half angle:

$$\tan(\Phi_0/2) = D/4f_d \tag{8.18}$$

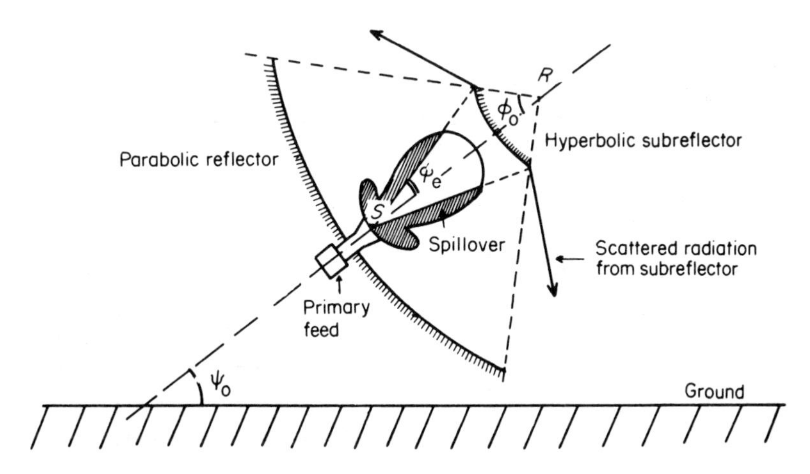

Figure 8.9 Dual reflector Cassegrain antenna.

The performance of a Cassegrain antenna is evaluated by using the concept of an equivalent paraboloid [COM-88]. An equivalent parabolic reflector antenna is defined as an antenna which has a single reflector of identical diameter to that of the main reflector of the Cassegrain antenna and a focal length equal to that of the Cassegrain assembly. The equivalent paraboloid is thus of diameter D, focal length f_e and characterised by the apex angle $2\Phi_e$ of the auxiliary reflector as viewed from the focus S (Figure 8.10).

The Cassegrain antenna is thus less cumbersome although it retains the advantage of antennas with a long focal length:

—The antenna noise temperature is low firstly since the greatest part of the spillover is no longer directed towards the ground but towards the sky and secondly because the spillover is reduced; the high value of equivalent focal length permits the use of directional primary feeds and the low values of the actual focal lengths f_d and f_s of the parabolic and hyperbolic reflectors attenuate the residual spillover.
—The gain of the antenna is little affected by surface imperfections.

Another advantage is that microwave circuits can easily be located immediately behind the primary feed which is located behind the main reflector. The effect of losses in the link are thus limited. However, for antennas of large diameter (e.g. 30 m), this equipment is situated at a significant height above the ground. To facilitate maintenance, it is possible to install it at ground level in a building under the antenna by using a system of microwave mirrors to guide the radio waves from the primary feed at ground level to the focus S of the reflector (Figure 8.11). This arrangement (beam focused waveguide) enables the high losses inherent in a coaxial cable or waveguide to be avoided while permitting rotation of the antenna about

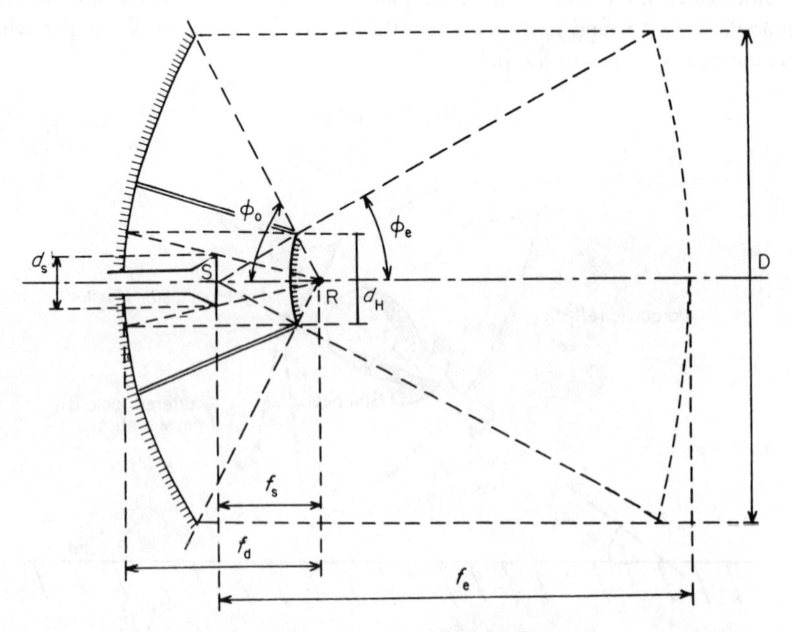

Figure 8.10 Single reflector equivalent focal length of a dual reflector Cassegrain antenna.

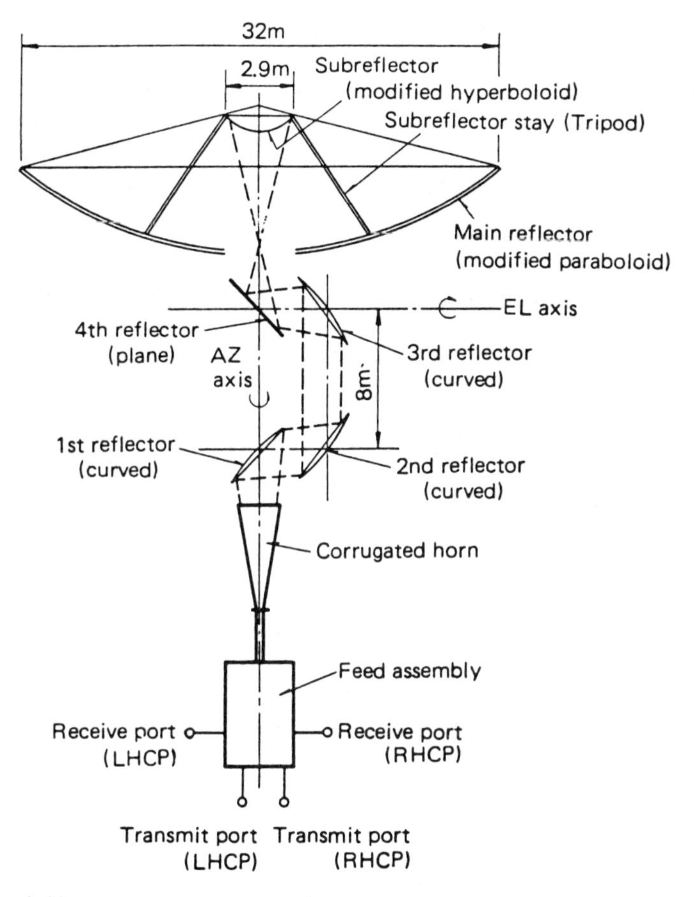

Figure 8.11 Guided-beam system.

two orthogonal axes and allowing feed and radio-frequency equipment to remain fixed. Because of the high cost, and the reduction in antenna diameter (the new Intelsat Standard A for example) which makes installation of radio-frequency equipment at the focus of the antenna easier, this periscope mounting is less and less used.

A disadvantage of the Cassegrain mounting is the masking effect of the auxiliary reflector. The perturbation caused by the auxiliary mirror leads to a slight reduction of gain and the 3 dB beamwidth, a noticeable increase in the level of the first side lobe and a modification of the level or a broadening of subsequent side lobes. These effects are negligible for a small d_H/D ratio (d_H is the diameter of the auxiliary reflector). For medium sized antennas they can be minimised by selecting dimensions such that:

$$f_s/f_d = d_S/d_H$$

$$d_s = (2f_d\lambda/\eta_s)^{1/2} \qquad \text{(m)} \tag{8.19}$$

where the notation is that of Figure 8.10, η_s is the efficiency of the primary feed.

The masking effect of the auxiliary reflector can be overcome by choosing an offset Cassegrain mounting.

8.3.4.4 *Multibeam antennas*

A small antenna suitable for the reception of several satellites grouped in one portion of the geostationary orbit has been developed by COMSAT Laboratories [KRE-80]. This multiple-beam torus antenna (MBTA) is equivalent, for each beam, to an antenna of 9.8 m aperture and has a gain of the order of 50 dB at 4–6 GHz with a noise temperature of the order of 30 K for an elevation angle of 20°.

8.3.5 Pointing angles of an earth station antenna

The elevation and azimuth angles which specify the direction of a satellite from a point on the earth's surface have been defined in Chapter 7 as a function of the relative co-ordinates of the satellite and the point concerned.

This section is devoted to one application of the determination of these angles for the case of a geostationary satellite and a practical presentation in the form of a nomogram. The polarisation angle, which completes the characterisation of radio-frequency antenna pointing when plane polarised waves are used, is also defined.

8.3.5.1 *Elevation and azimuth angles*

The orientation of the axis of an antenna pointed towards a satellite is defined by two angles—the azimuth A and elevation angle E. These two angles are specified as a function

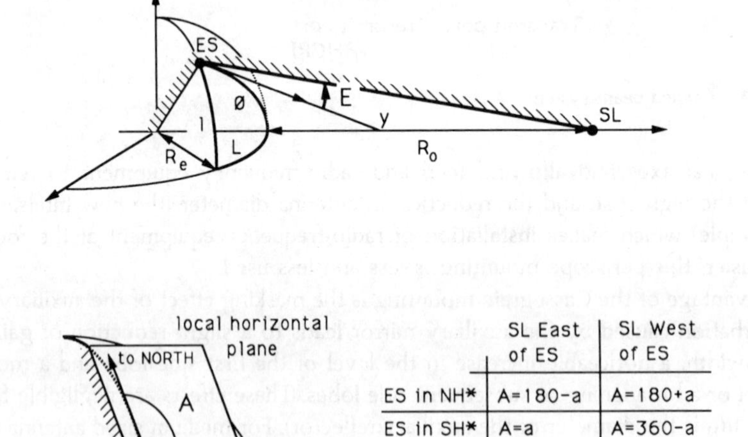

Figure 8.12 Azimuth and elevation angles.

of the latitude l and the relative longitude L of the station (L is the absolute value of the difference between the longitude of the satellite and that of the earth station).

The azimuth angle is the angle about a vertical axis through which the antenna must be turned, in the opposite direction to the trigonometric direction, from the geographical north to bring the axis of the antenna into the vertical plane which contains the direction of the satellite. This plane passes through the centre of the earth, the station and the satellite (Figure 8.12). The azimuth angle A has a value between 0° and 360°. Its value is obtained from Figure 8.13 by means of an intermediate parameter a determined from the family of curves and used to deduce the angle A using the table inserted into the figure. The curves result from the following relation which could be used for greater accuracy:

$$a = \arctan(\tan L/\sin l) \tag{8.20}$$

The elevation angle E is the angle through which the antenna must be turned in the vertical plane containing the satellite to bring the boresight of the antenna from the horizontal to the direction of the satellite (Figure 8.12). The elevation angle E is obtained from the

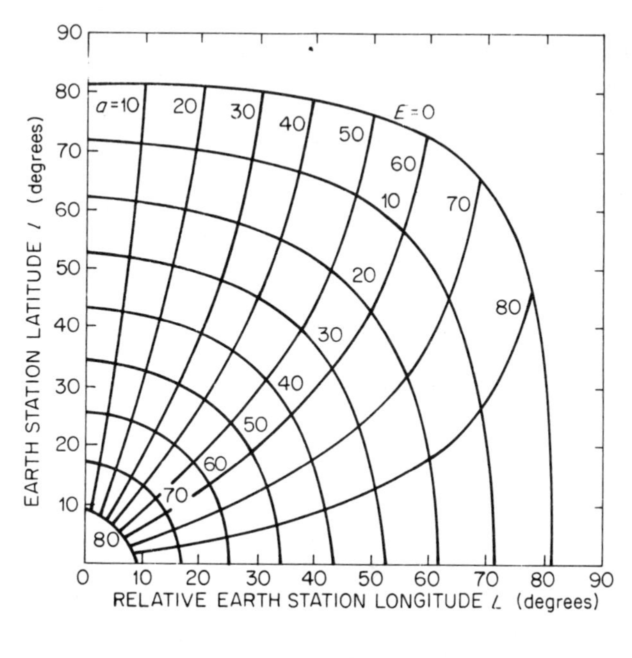

	SL EAST OF ES	SL WEST OF ES
NORTH HEMISPHERE	$A=180-a$	$A=180+a$
SOUTH HEMISPHERE	$A=a$	$A=360-a$

Figure 8.13 Azimuth and elevation angles as a function of the earth station latitude and satellite relative longitude.

corresponding family of curves of Figure 8.13 which follow from the relation:

$$E = \arctan[(\cos\phi - R_E/(R_E + R_0))/(1 - \cos^2\phi)^{1/2}] \quad (8.21)$$

where:

$\cos\phi = \cos l \cos L$,
R_E = radius of the earth = 6378 km,
R_0 = altitude of the satellite = 35 786 km.

8.3.5.2 *Polarisation angle*

The polarisation angle ψ is the angle which defines the orientation of the plane of polarisation of a linearly polarised wave transmitted (or received) by the satellite with respect to the plane defining the polarisation of the wave received (or transmitted) by the antenna when the antenna boresight is pointed towards the satellite. Assuming that the plane of polarisation of the wave from the satellite is perpendicular to the plane of the orbit, that is to the equatorial plane for a nominal orbit, and the plane defining the antenna polarisation contains the vertical location at its origin, the polarisation angle is the angle through which the plane defining the antenna polarisation must be turned about the direction of the satellite in order to match the polarisation of the wave from the satellite. The angle is determined from the curves of Figure 8.14 which are established by means of the relation:

$$\cos\psi = \frac{\sin l\{1 - [R_E/(R_E + R_0)]\cos L \cos l\}}{[(\cos^2 l \sin^2 L + \sin^2 l)\{[R_E/(R_E + R_0)]^2 \cos^2 l + 1 - 2[R_E/(R_E + R_0)]\cos L \cos l\}]^{1/2}} \quad (8.22a)$$

A more convenient relation can also be derived:

$$\tan\psi = \sin L/\tan l \quad (8.22b)$$

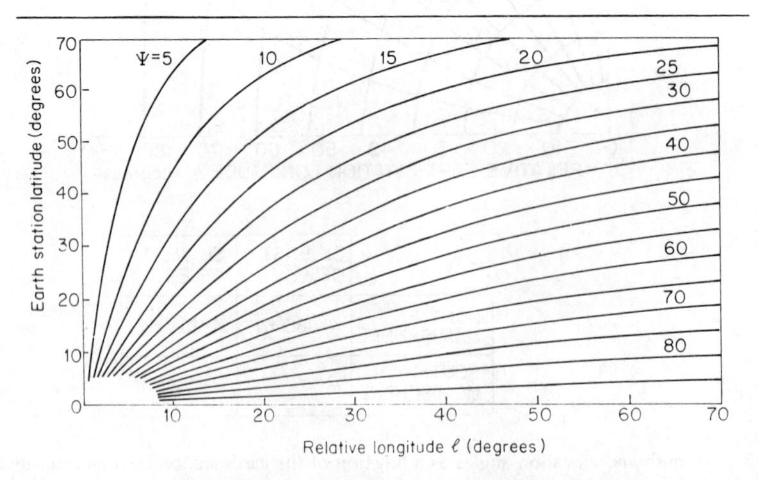

Figure 8.14 Polarisation angles as a function of earth station latitude and satellite relative longitude.

For an observer situated behind the antenna, the rotation is made in the reverse trigonometric direction for a station to the east of the satellite meridian and in the direct direction for a station situated to the west of the satellite meridian.

8.3.6 Mountings to permit antenna pointing

For fixed stations which operate with a specific geostationary satellite, the range of angles through which the antenna is likely to be pointed is small. Its magnitude must, however, be sufficient to permit repointing to a standby satellite in case of breakdown of the first. In the more general case, it is desirable to provide equipment capable of providing beam pointing in any direction in order to be able to establish links with different geostationary satellites or non-geostationary satellites.

Movement of the antenna results from movement about two axes—a primary axis which is fixed with respect to the earth and a secondary axis which rotates about the first.

8.3.6.1 Azimuth-elevation mounting

Azimuth-elevation mounting corresponds to a vertical fixed primary axis and a horizontal secondary axis (Figure 8.15) constrained to rotate about the vertical axis. Rotation of the antenna support about the vertical axis enables the azimuth angle A to be adjusted and rotation of the antenna about the associated horizontal axis of the support then permits the elevation angle E to be adjusted. This is the mounting most commonly used for antennas of steerable earth stations.

With the previous mounting, the secondary axis may not be in the horizontal plane and hence may be at an angle other than 90° with the primary axis; this is non-orthogonal 'azimuth-elevation' mounting. This mounting is useful for Cassegrain antennas since the volume within which the antenna operates for different pointing angles is reduced with respect to conventional 'azimuth-elevation' mounting. On the other hand, coupling between the rotation about the axes is introduced and angular displacements about the axes no longer correspond to the azimuth and elevation angles previously defined.

'Azimuth-elevation' mounting has the disadvantage of leading to high angular velocities when tracking a satellite passing through the vicinity of the zenith. The elevation angle then reaches 90° which generally corresponds to a mechanical stop to prevent overtravel of the antenna about the secondary axis. To track the satellite, the antenna must thus perform a rapid rotation of 180° about the primary axis.

This constraint can be avoided by giving the pointing system an additional degree of freedom. This permits the introduction of a bias on the secondary axis support with respect to the vertical (Figure 8.16). This function may be realised, for example, by a relative rotation of two half cylinders initially with the same principal axis coincident with the primary axis and whose contact faces are not orthogonal to the axis. Once introduced, rotation of the antenna about the secondary axis for a given elevation pointing angle is thus equal to the elevation angle less the bias. Hence for pointing at the zenith ($E = 90°$), the maximum travel is not reached. This mounting is used for example for stations mounted on mobiles.

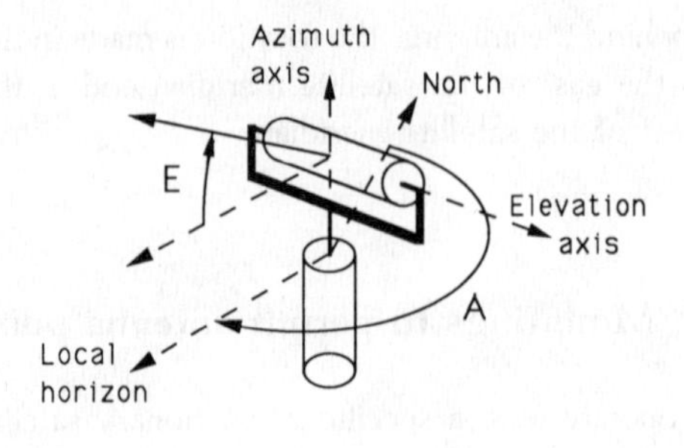
Azimuth
axis
North
E
Elevation
axis
A
Local
horizon

(a)

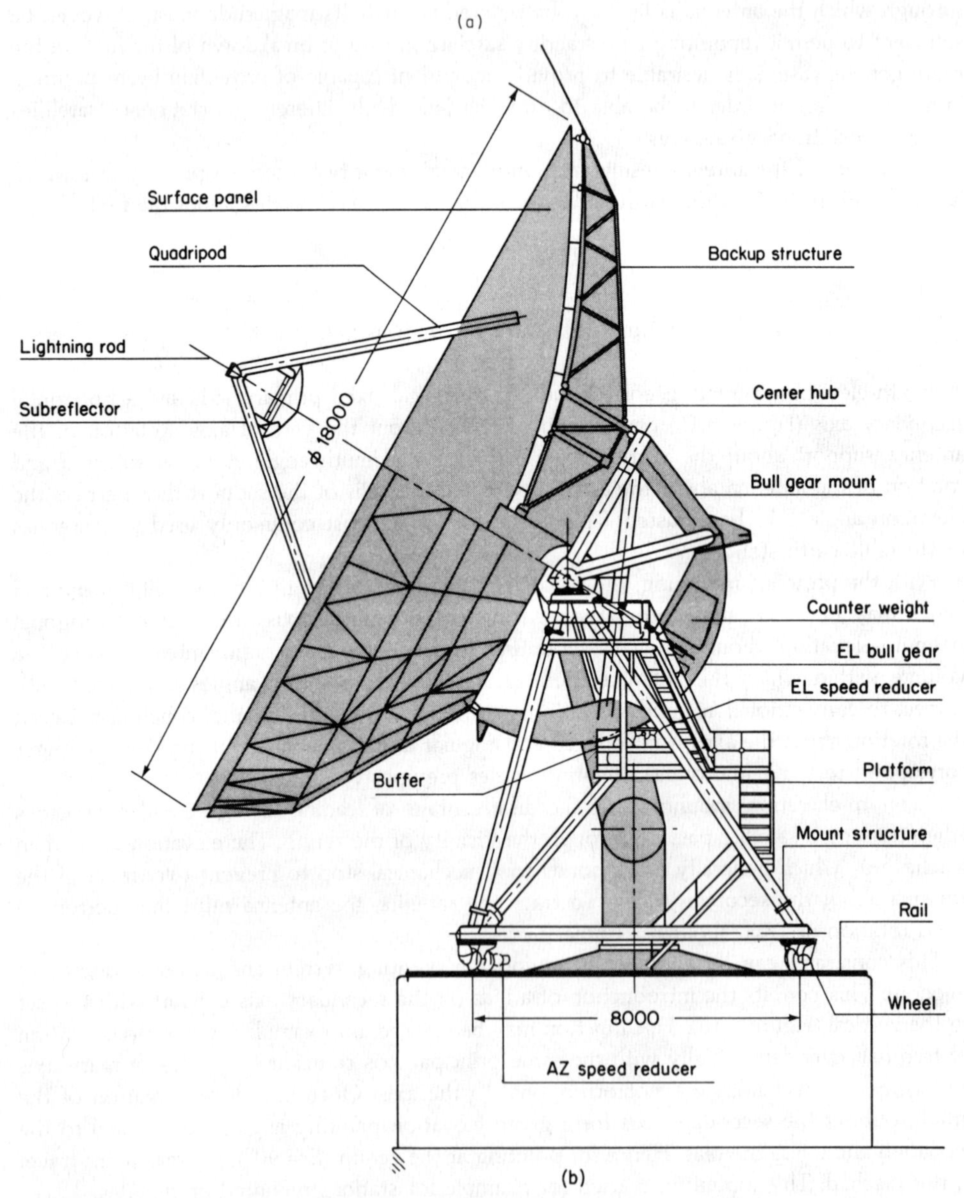
Surface panel
Quadripod
Backup structure
Lightning rod
Subreflector
ϕ 18000
Center hub
Bull gear mount
Counter weight
EL bull gear
EL speed reducer
Buffer
Platform
Mount structure
Rail
Wheel
8000
AZ speed reducer

(b)

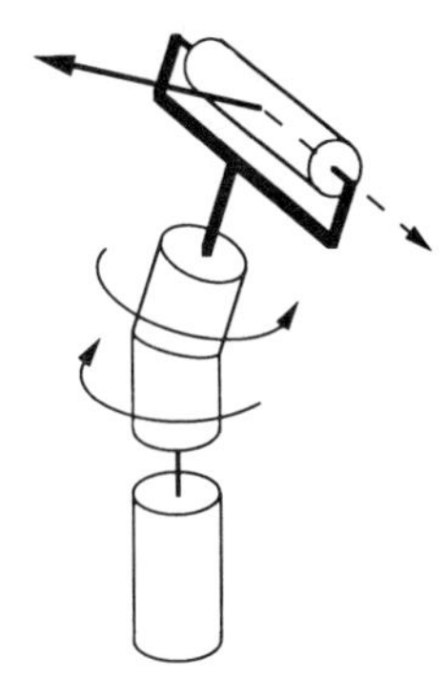

Figure 8.16 Modified azimuth-elevation mount; relative rotation of 180° of the two parts of the secondary axis support introduces an offset with respect to the vertical of the upper part of the primary axis.

8.3.6.2 X–Y mounting

X–Y mounting has a fixed horizontal primary axis and a dependent secondary axis which rotates about the primary axis and is orthogonal to it (Figure 8.17). This mounting does not have the disadvantage of the azimuth-elevation mounting when the satellite passes through the zenith (a high speed of rotation about the primary axis). X–Y mounting is thus useful for satellites in low orbits rather than for geostationary satellites and stations mounted on mobiles.

For a station in the northern hemisphere, the pointing angles X (rotation about the primary axis from the local horizontal) and Y (rotation about the secondary axis from the plane perpendicular to the primary axis) are given as a function of the latitude l and the relative longitude L of the station by:

$$X = \arctan[(\tan E)/\sin A_R] \tag{8.23a}$$

$$Y = \arcsin[-\cos A_R \cos E] \tag{8.23b}$$

with A_R, the azimuth of the satellite relative to the primary axis of the mounting (X axis), such that $A_R = A - A_X$ and where:

—E is the elevation angle of the satellite obtained from equation (8.21),
—A is the azimuth of the satellite obtained from equation (8.20) and the table of Figure 8.13),
—A_X is the orientation of the X axis with respect to the north.

The angles and their projection on the horizontal plane are taken as positive in the inverse trigonometric direction.

Figure 8.15 Azimuth-elevation mount. (a) Axes of rotation; antenna pointing in the direction of the satellite is obtained by rotation through an angle equal to the azimuth A about the vertical primary axis (from the direction of the north to the direction contained in the plane of the satellite, the station and the centre of the earth), then by rotation through an angle E (the elevation angle) about the horizontal secondary axis. (b) An example of implementation with a Standard C antenna.

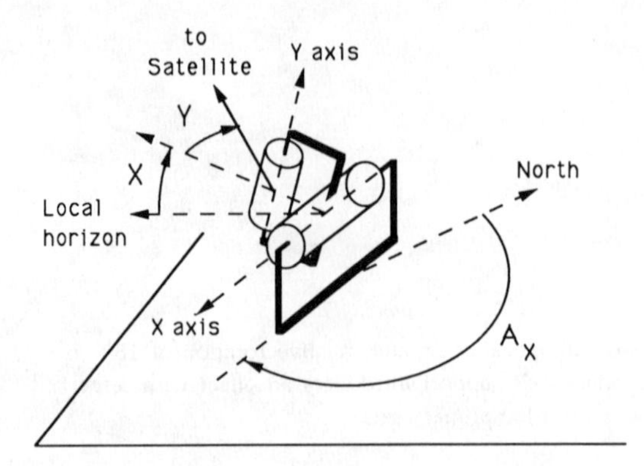

Figure 8.17 X–Y mount; antenna pointing in the direction of the satellite S is obtained by rotation through an angle X (from the direction of the zenith to the direction S_1 contained in the plane of the satellite, the station and the earth centre) about the horizontal primary axis orientated towards the south, then by rotation through an angle Y about the associated secondary axis of the part which rotates about the primary axis.

8.3.6.3 *Equatorial mounting*

Equatorial or polar mounting corresponds to a primary axis (the 'hour axis') parallel to the axis of rotation of the earth and a secondary axis (the 'declination axis') perpendicular to the former (Figure 8.18). This mounting is used for telescopes since it permits tracking of the apparent movement of stars by rotation only about the hour axis which thus compensates for the rotation of the earth about the polar axis.

This mounting is useful for links with geostationary satellites since it is possible to point the antenna at several satellites successively by rotation about the hour axis. However, the

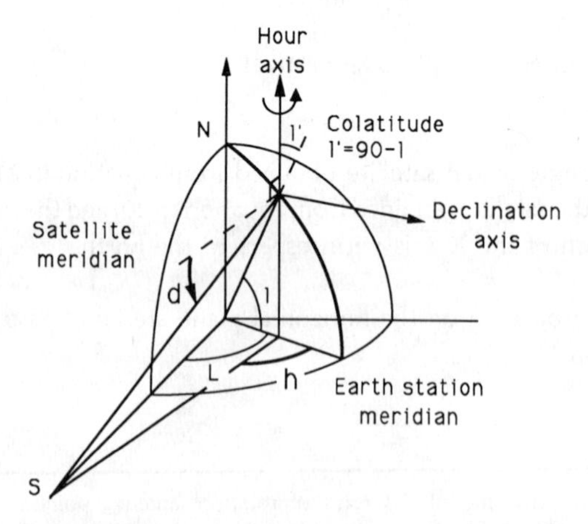

Figure 8.18 Polar mount: definition of the hour angle (the angle between the projection of the station in the equatorial plane from the station meridian and the satellite) and declination (the angle between the parallel to the equatorial plane and the satellite in the plane perpendicular to the equator which contains the station and the satellite).

fact that the satellites are not at infinity necessitates, in principle, slight adjustment of orientation about the declination axis.

The expressions for the hour angle h (the rotation about the hour axis from due south to the plane containing the satellite) and the declination d (the rotation in the plane of the hour axis containing the satellite from the perpendicular to the hour axis to the direction of the satellite) as functions of the latitude l and the relative longitude L of the station are:

$$h = \tan^{-1}[\sin L/(\cos L - 0.15126 \cos l)] \quad (8.24a)$$

$$d = \tan^{-1}[-0.151\,26 \sin l \sin h/\sin L] \quad (8.24b)$$

The hour angle is positive towards the east if the relative longitude L is defined as the algebraic difference between the longitude of the satellite (degrees east) and the longitude of the station (degrees east). The value 0.151 26 corresponds to the ratio $R_E/(R_0 + R_E)$ for the nominal values of the terrestrial radius ($R_E = 6378$ km) and the nominal altitude of the geostationary satellite.

For $L = 0$, the hour angle is zero and expression (8.24b) is not defined. Direct determination of the declination angle leads to:

$$d_{L=0} = \tan^{-1}[-l/(6.610\,78 - \cos l)] \quad (8.24c)$$

where the coefficient 6.610 78 corresponds to the nominal value of the ratio $(R_0 + R_E)/R_E$.

The curves of Figure 8.19 provide the values of the hour and declination angles. In practice the orientation of the hour axis is obtained by inclining the axis with respect to the vertical in the plane of the local meridian by a value equal to 90° minus the latitude of the station (the co-latitude).

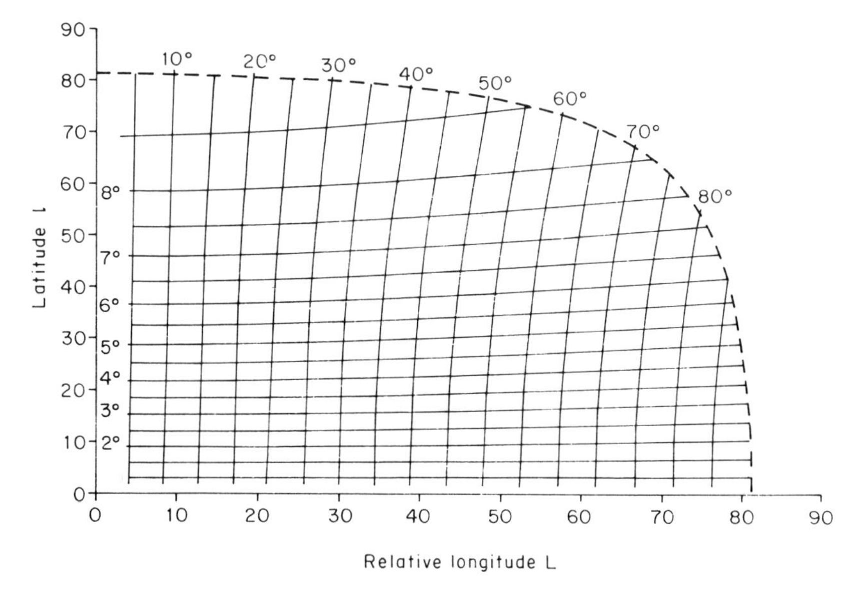

Figure 8.19 Hour angle h and declination d as a function of the latitude l of the station and its relative longitude L with respect to the satellite.

For antennas of small diameter having a sufficiently large main lobe, it is possible to point at several satellites by rotation only about the hour axis. This approach is often adopted for semi-professional antennas intended for the reception of television signals transmitted by various satellites.

As the declination remains fixed, a pointing error occurs which depends on the latitude of the station and the angular separation between the satellites. Figure 8.20 gives the declination error as a function of the latitude l of the station and the relative longitude L of the satellite with respect to the station (the nominal value of the declination is that of a satellite situated at the longitude of the station, that is $d_{L=0}$). It should be noted that it is at medium latitudes (40°) that this error is greatest.

A reduced pointing error is obtained by fixing the declination angle at a value which results from a compromise between the values corresponding to the various satellites considered. For example, for a station of latitude 40°, a sweep of 50° of the geostationary arc on both sides of the station meridian leads to a declination error of 0.3° at the limit of the excursion if the declination is correct for a direction in the meridian plane of the station towards a satellite at the same longitude. By adjusting the declination at a value less than $0.3°/2 = 0.15°$ to that corresponding to correct pointing if the satellite were in the plane of the meridian, that is $d_{L=0}(l = 40°) - 0.15°$, the maximum pointing error is equal to 0.15° when at the extreme ends of the sweep and in the plane of the meridian, and less than 0.15° when operating elsewhere.

The compromise can be further improved by introducing a bias on the orientation of the hour axis. The hour axis is no longer parallel to the polar axis but is inclined towards the

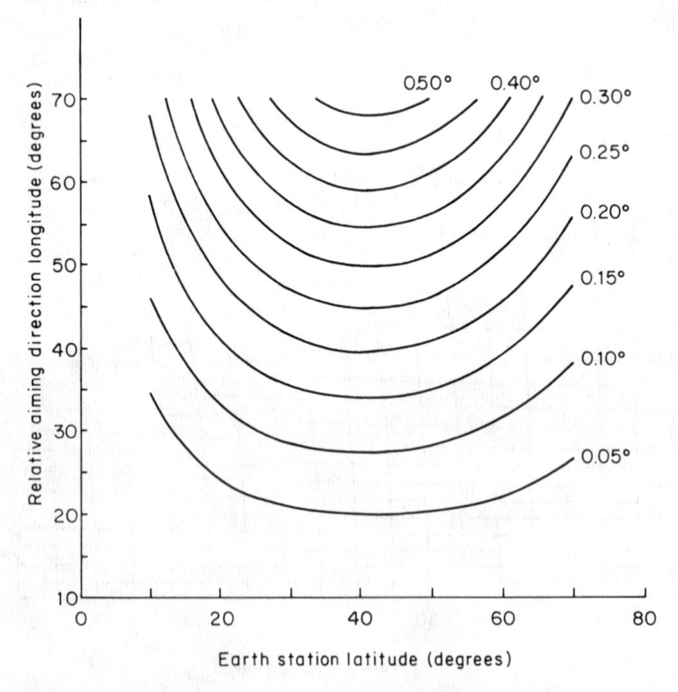

Figure 8.20 Pointing error with respect to the orbit of geostationary satellites as a function of the earth station latitude and the relative longitude of the aiming direction.

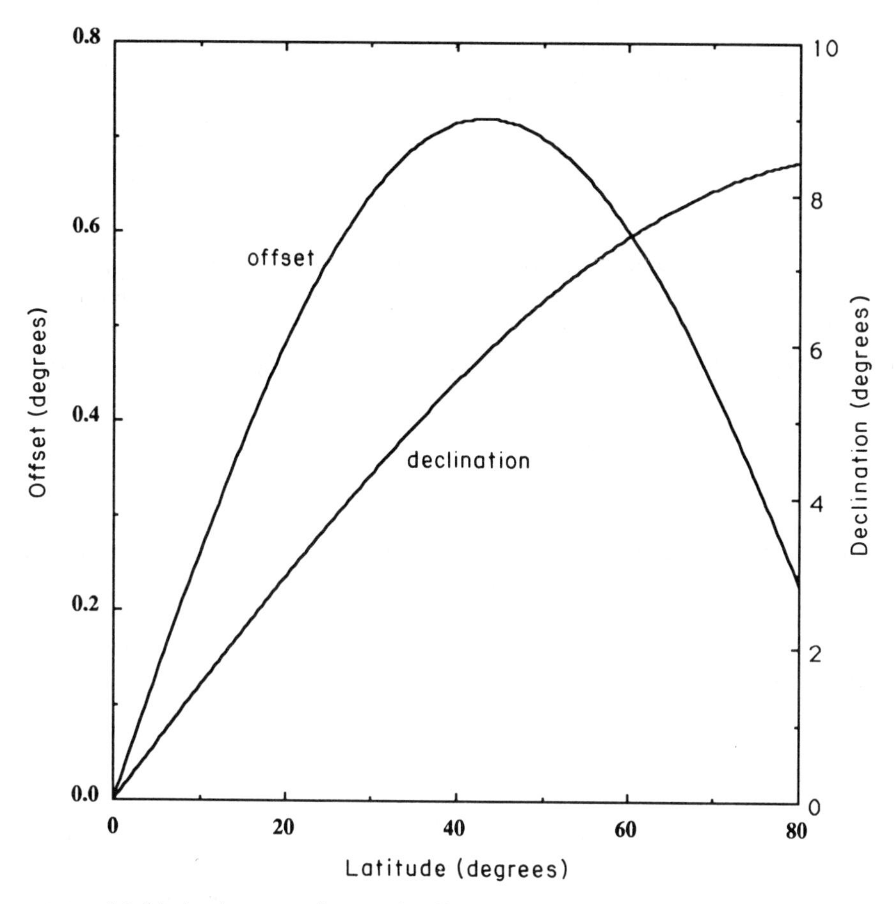

Figure 8.21 Modified polar mount: hour angle offset and declination angle as a function of the latitude of the earth station.

exterior in the plane of the meridian in such a way as to cancel the declination error when operating in the plane of the meridian while minimising it over the swept portion of the orbit (modified polar mounting). The bias to be introduced depends on the latitude of the station and the required magnitude of the sweep in longitude. Figure 8.21 gives the value of bias and declination adjustment (rotation with respect to the perpendicular to the mounting hour axis) as a function of the station latitude for an orbit sweep of $\pm 45^{\circ}$ on each side of the meridian. The resultant pointing error is less than 0.1°.

These values are obtained by taking account of the oblateness of the earth which leads to the replacement of the latitude l in equations (8.24) by l' (the geocentric latitude) and $R_E/(R_0 + R_E)$ by $R_E(1 - A \sin^2 l')/(R_0 + R_E)$, where A is the oblateness coefficient equal to 1/298.257 or 3.352×10^{-3} (see Section 7.3).

8.3.6.4 *Tripod mounting*

Tripod mounting is an approach suited to geostationary satellites. The antenna is fixed to the support by means of three legs of which two are of variable length. According to the mounting used, pointing in elevation and azimuth may or may not be independent. Mounting is simple but the magnitude of pointing variation is limited (for example 10° about a mean direction).

8.3.7 Tracking

Tracking consists of maintaining the axis of the antenna beam in the direction of the satellite in spite of movement of the satellite or station. Several types of tracking are possible and are characterised by their tracking error (pointing angle error). Choice of the type of tracking depends on the antenna beamwidth and the magnitude of apparent movement of the satellite.

8.3.7.1 *The influence of antenna characteristics*

The angular width of the beam directly affects selection of the type of tracking. It should be noted that, at the frequencies used, the 3 dB angular beamwidth can be small. By way of example, Figure 8.22 gives the 3 dB angular width for different frequencies as a function of antenna diameter.

The loss of gain $\Delta G(\alpha)$ for a depointing α with respect to the direction of maximum gain is given by (see equation (2.4)):

$$\Delta G(\alpha) = -12(\alpha/\theta_{3\text{dB}})^2 \quad \text{(dB)} \tag{8.25}$$

Depointing is associated with relative movement of the satellite and the direction of maximum gain of the antenna. Decisions relating to antenna installation and tracking procedure depend on the beamwidth in relation to the magnitude of apparent movement of the satellite; the determining criterion is the variation of antenna gain with depointing.

Another antenna characteristic which is associated with its diameter and directly affects the performance of orientating devices is its mass. For small antennas, the mass of the parabolic reflector ranges from a few tens to several hundreds of kilograms. For large antennas it is several tonnes; the moving part of the 32.5 metre diameter antenna of Pleumeur Bodou IV, for example, weighs 185 tonnes. Meteorological conditions (wind speed) and the mass of the antenna itself cause deformations of it which vary with elevation angle.

8.3.7.2 *Apparent movement of the satellite*

The apparent movement of the satellite has been examined in Chapter 7 as a function of the type of orbit.

For satellites on inclined elliptical orbits, movement causes a variation of elevation angle whose magnitude about the zenith varies according to the type of orbit, the number of satellites in the system and the location of the service area.

For geostationary satellites the apparent movement is contained within the station-keeping window whose dimensions characterise the accuracy of station keeping ($\pm 0.1^\circ$ north–south and east–west, for example). The actual movement within the window is a combination of north–south movement with a period of 24 h due to residual orbit inclination (in the form of a figure of eight for a large inclination), east–west movement of the same period due to

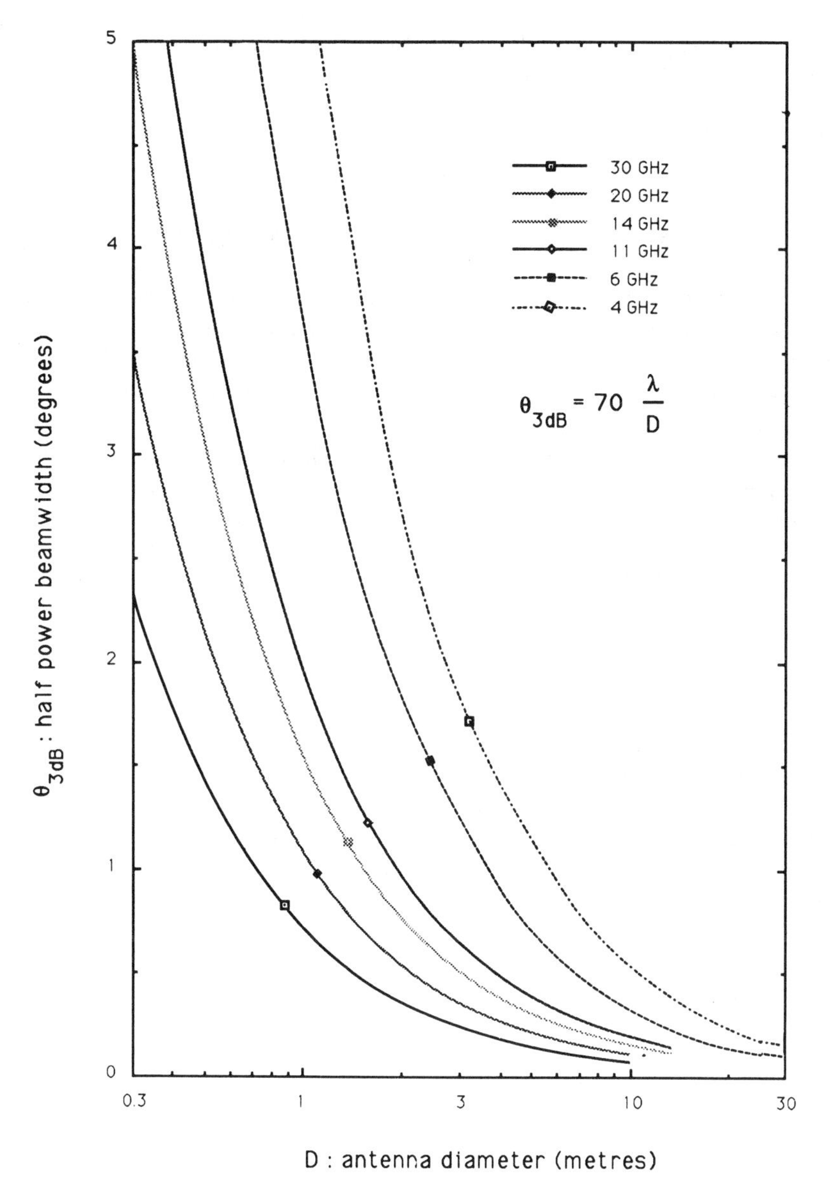

Figure 8.22 Half-power beamwidth θ_{3dB} versus antenna diameter D.

eccentricity and a drift towards the east or west whose size and direction depend on the longitude of the satellite position. The apparent speed of movement of the satellite does not exceed 2°/h; for an orbit inclination of 5°, the maximum angular velocity is 7×10^{-4} degree/s (2.5°/h).

At a given time, the satellite can be anywhere within the window. Only the satellite control station knows the position of the satellite as a function of time with some uncertainty. Distributed orbital data tables permit the times when the satellite will be close to the centre of the window to be predicted (in some cases this permits the maximum depointing of the earth station antenna to be minimised).

8.3.7.3 *Fixed antenna without tracking*

Tracking is not necessary when the antenna beamwidth is large in comparison with the station-keeping window of a geostationary satellite or for the case of a system of satellites on inclined elliptical orbits when the antenna beamwidth greatly exceeds the solid angle which contains the apparent movement of the active orbiting satellite.

The usable part of the beam can be defined at -0.1, -0.5, -1 or $-n$ dB in accordance with the acceptable loss of gain; the choice results from optimisation of the characteristics of the links between the satellite and the stations.

In the case of pointing towards a geostationary satellite, the maximum depointing angle which determines the dimensions of the system can be minimised, for a given window size and a given θ_{3dB} (or λ/D ratio), by performing initial pointing when the satellite is closest to the centre of the window. Coarse orientation of the antenna is achieved from pointing angles determined with the help of abacs or the expressions given in Section 8.3.5. Fine pointing is then obtained by searching for the maximum beacon signal level from the satellite concerned by displacing the pointing direction on each side of the axis which is assumed to correspond to maximum gain (locating the angular values corresponding to a given decrease of level and pointing in the corresponding direction of the centre point). As a result of small variations of gain in the vicinity of the electromagnetic axis, the *initial pointing error* (IPE) is of the order of 0.1 to 0.2 θ_{3dB}.

Considering that initial pointing is achieved when the satellite is assumed to be at the centre of the station-keeping window of half-width SKW, and designating the uncertainty in angular position of the satellite θ_{SPU}, the maximum value of depointing angle α_{MAX} is determined with the help of Figure 8.23:

$$\alpha_{MAX} = \text{SKW}\sqrt{2} + \theta_{SPU} + \alpha_{IPE} \tag{8.26}$$

where α_{IPE} of the form $b\,\theta_{3dB}$ is the initial pointing error.

α_{MAX} is thus of the form $a + b\,\theta_{3dB}$ (a = constant = SKW $\sqrt{2} + \theta_{SPU}$).

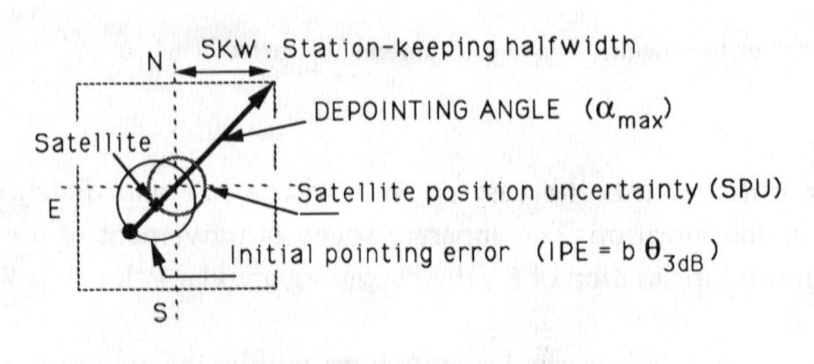

Figure 8.23 Maximum depointing angle with a fixed mount antenna.

8.3.7.4 *Programmed tracking*

With programmed tracking, antenna pointing is achieved by providing the antenna orientation control system with the corresponding values of azimuth and elevation angles at each instant. These azimuth and elevation angles are calculated in advance for successive instants, taking account of the predicted apparent movement of the satellite, and the values are stored in a memory. Pointing is then performed in open loop without determination of the pointing error between the actual direction of the satellite and the aiming direction at each instant.

The pointing error thus depends on the accuracy of knowledge of the apparent movement of the satellite from which the various pointing angles are calculated and the accuracy with which pointing in a given direction is achieved by the antenna. (Errors arise from inaccuracies in local references and the coding of the antenna orientation, feedback control errors and so on.)

Programmed tracking is mainly used for orbiting satellites with earth station antennas of large λ/D ratio which thus have a relatively large major lobe which does not require very high pointing accuracy. If high pointing accuracy is required (a small λ/D ratio) with an orbiting satellite, programmed tracking serves to preposition the antenna in an area of the sky where the satellite will appear in such a way as to ensure acquisition by a feedback tracking system operating on the satellite beacon.

This type of tracking is little used for geostationary satellites since if the value of λ/D is large, a fixed mounting is usually sufficient and if λ/D is small, it is preferable to use an automatic beacon tracking device.

8.3.7.5 *Computed tracking*

This system is a variant of the previous case and is well suited to tracking geostationary satellites for antennas having an intermediate value of λ/D which does not justify the use of automatic beacon tracking.

With computed tracking, a computer incorporated in the pointing system evaluates the antenna orientation control parameters. The computer uses the orbit parameters (inclination, semi-major axis, eccentricity, right ascension of the ascendant node, argument of the perigee, discrepancy) and, if necessary, a model of their progression. The data in memory are, if necessary, refreshed periodically (after a few days). The system can also extrapolate the progression of the orbit parameters from daily satellite displacements which are stored in memory.

8.3.7.6 *Automatic closed loop tracking*

With antennas having a small value of λ/D, and hence a small angular antenna beamwidth with respect to the apparent magnitude of satellite movement, precise tracking of the satellite is obtained by continuously aligning the antenna direction to that of a beacon located on the satellite.

The accuracy depends on the method used to determine the direction of arrival of the beacon signal, the deviation between the direction of arrival of the beacon signal and the actual direction of the satellite (caused by propagation aberrations) and the precision of the feedback control system.

In addition to an accuracy which can be very high (the tracking error can be less than 0.005° with a monopulse system), an advantage of this procedure is its autonomy since tracking information does not come from the ground. Moreover, it is the only conceivable system for mobile stations whose antenna movement cannot be known a priori (if the itinerary of the mobile is known, programmed tracking could conceivably be used).

Two techniques are used for beacon tracking—tracking by sequential amplitude detection and monopulse tracking [HAW-88].

8.3.7.6.1 Sequential amplitude detection

Sequential amplitude detection tracking systems make use of variations in received signal level as a consequence of controlled displacement of the antenna pointing axis. The level variations generated in this way enable the direction of maximum gain, which corresponds to the highest received signal level, to be determined. The main source of error arises from the fact that the system is incapable of distinguishing a level variation due to antenna depointing from a level variation caused by a change of electromagnetic wave propagation conditions. Various procedures are used—conical scanning, step-by-step tracking and electronic depointing.

Tracking by conical scanning. The antenna beam rotates continuously about an axis which makes a given angle (small compared with $\theta_{3dB}/2$) with respect to the axis of maximum gain. When the direction of the satellite differs from the direction of the axis of rotation, the received level is modulated at the rate of rotation of the antenna as a function of the angular deviation between the two directions. The tracking receiver correlates this modulation with the antenna rotation to generate orientation control signals. The modulation of the received signal becomes zero when the direction of the satellite coincides with the axis of rotation. The tracking error is between $0.2\,\theta_{3dB}$ and $0.05\,\theta_{3dB}$.

This technique, which has been used for a long time, particularly for small antennas has been progressively abandoned in favour of step-by-step tracking which enables a tracking accuracy of the same order of magnitude to be obtained with less mechanical complexity. Furthermore the modulation of the uplink signal at the rate of antenna rotation can be aggravating.

Step-by-step tracking. Antenna pointing is achieved by searching for the maximum received beacon signal. This proceeds by successive displacements (steps) of the antenna about each of the axes of rotation (the method is also known as step-track or hill-climbing). The direction of the subsequent displacement is determined by comparison of the received signal level before and after the step. If the signal increases, the displacement is made in the same direction. If the signal decreases, the direction of displacement is reversed. The procedure is performed alternately on each of the two axes of rotation of the antenna [TOM-70].

There are several limits to the accuracy of tracking as follows:

—The uncertainty in the direction of the maximum may be greater than the step size which must therefore be chosen to be sufficiently small (of the order of $\theta_{3dB}/10$) [RIC-86].

—The gain of the antenna (and hence the level of the received signal) about the direction of maximum gain varies slowly with depointing angle (the lobe has a flat top). Determination of the direction of maximum gain is thus less precise than determination of the pronounced null of the gain characteristic of monopulse systems (see below). The accuracy with which

the direction of maximum gain is determined is fundamentally a function of the θ_{3dB} beamwidth and hence λ/D.
—The system has a limited dynamic response and one must wait for the reflector displacement before detecting each variation of the received signal.
—Finally, as with all systems where depointing information is obtained from variations of received signal level, step-by-step tracking is affected by spurious amplitude modulation of the signal level. Furthermore, it is necessary to have a sufficient C/N (carrier power-to-noise power) ratio at the input of the tracking receiver (typically 30 dB).

Taking account of these limitations, the tracking error is between 0.05 θ_{3dB} and 0.15 θ_{3dB}.

For antennas used with geostationary satellites, the system can be used either in continuous tracking mode or in 'point and rest' mode where, after pointing the antenna in the direction of the satellite, the positions of the axes of rotation are clamped until pointing is performed again. This repointing is activated either periodically at regular intervals or on detection of a reduction of received signal level. This point and rest mode reduces the operations and stresses suffered by the pointing servos and the orientation motors which permits an increased lifetime of the hardware and reduced maintenance.

The performance of a step-by-step tracking system can be improved by reducing the sensitivity to amplitude fluctuations of the received signal by filtering the control signals (smoothed step-track). One solution can consist of combining computed tracking with a step-by-step system. An estimate of the pointing direction is thus obtained from a simplified model of the apparent movement of the satellite. The step-by-step tracking system then improves the pointing with respect to the direction obtained. The direction of the satellite is then known and this enables the model of apparent movement to be updated [EDW-83].

As the system has an estimate of the satellite direction at each instant, possible errors caused by fluctuations of signal amplitude can be detected and this permits incorrect commands to be cancelled. Furthermore, step-by-step tracking need not be continuously activated, its role is to update the movement model periodically and this updating enables good accuracy of computed tracking to be obtained.

Tracking by electronic deviation. This recent technique is comparable with step-by-step tracking. The difference lies in the technique used for successive displacement of the beam in the four cardinal directions, since this is realised electronically. Depointing by a given angle is obtained by varying the impedance of four microwave devices coupled to the source waveguide; these devices are located symmetrically on each side of the waveguide in two perpendicular planes [WAT-86]

Successive deviation of the beam in four directions enables the magnitude and direction of depointing to be evaluated if the received signal does not arrive along the principal axis of the antenna. The signal is actually received with different levels in accordance with the direction of deviation of the beam. An error signal is derived from a combination of successive signals and this enables the antenna orientation to be controlled in such a way as to reduce depointing.

The system thus uses a tracking receiver with a single channel as for a step-by-step system. Furthermore, determination of pointing error is achieved without mechanical displacement of the antenna and stresses on the antenna orientation mechanism are reduced. Finally, the tracking accuracy obtained is greater as a consequence of the virtual simultaneity (on the timescale of the apparent movement of the satellite) of level measurements in each of the directions of deviation. This results in a rapid dynamic response. The tracking error obtained can fall to $0.01\,\theta_{3dB}$ [DAN-85].

8.3.7.6.2 The monopulse technique

Excitation of an antenna pattern which is specifically intended for tracking and contains a zero on the axis permits the antenna to be orientated in such a way as to cancel the received signal. The orientation control signals are generated by comparison, in a 'monopulse' tracking receiver, of a reference signal and the error angle measurement signals. The reference signal is the beacon signal (the 'sum' channel Σ) extracted from the system which generates the depointing signals in the antenna (see Figure 8.1). The error angle measurement signals result from depointing as measured in two orthogonal planes (the 'difference' channels Δ at the output of the depointing signal generating system in the antenna). In this way error signals (Δ/Σ) are available which are independent of the received signal level.

The error angle measurement signals are provided either by comparison of the waves received from four sources located around the electromagnetic axis of the antenna (amplitude or phase monopulse) or by detection of the higher order modes generated by depointing of the antenna in the waveguide coupled to the primary source (mode extraction).

Multiple source monopulse. Each source in a multiple source monopulse system has a radiation pattern which is slightly shifted with respect to the principal axis of the antenna (Figure 8.24). In each of the two orthogonal planes, the difference between the two signals received from the two sources becomes zero for a wave which arrives parallel to the principal axis. This difference is, to a first approximation (in the vicinity of correct pointing), proportional to the depointing angle. Insensitivity to variations in incident signal level is ensured by normalising the difference signals with respect to the sum signal. Tracking accuracy can reach 0.01° for an antenna having a 3 dB beamwidth of 2° (an accuracy of $5 \times 10^{-3}\theta_{3dB}$)[AND-63]. This type of system, which was used on the first large earth stations (Pleumeur-Bodou 2 for example), has been abandoned in favour of mode extraction systems which are simpler to implement.

Mode extraction monopulse. The mode extraction system (Figure 8.25) uses the special propagation properties of modes of order greater than the fundamental TE_{11} in a waveguide

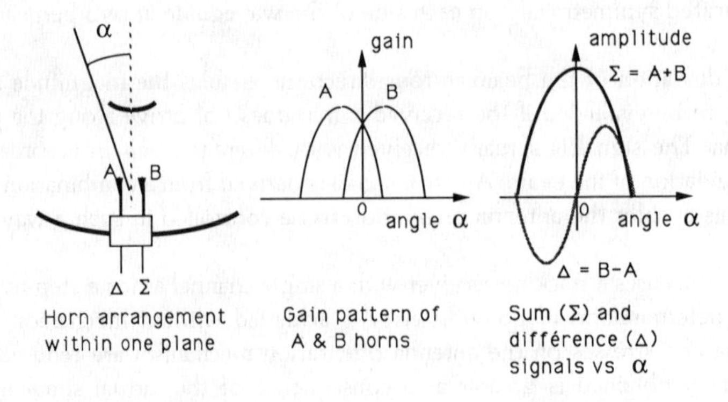

Figure 8.24 Multihorn monopulse tracking system.

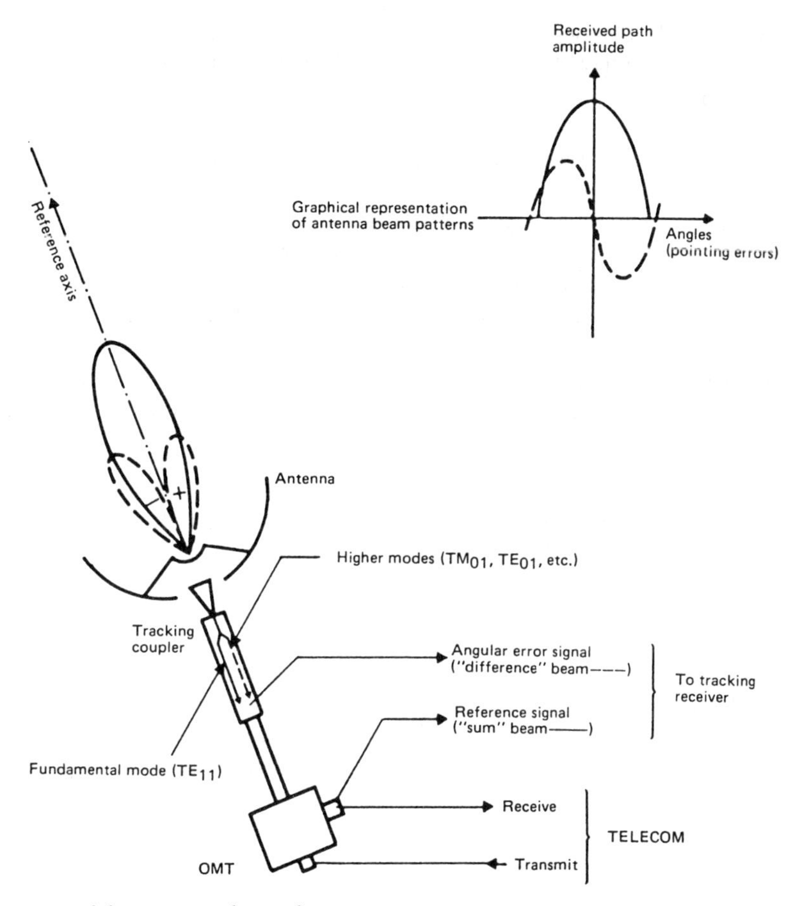

Figure 8.25 Multihorn monopulse tracking system. (From CCIR-88. Reproduced by permission of the ITU.)

(TE (Transverse Electric) signifies a wave whose magnetic field component H along the direction of propagation is zero). For a circularly or linearly polarised wave, only the TE_{11} mode propagates if the incident wave arrives along the axis of the guide; if the wave arrives with an angular deviation with respect to the axis of the guide, TM_{01} and TE_{21} modes are generated and the amplitude of the TE_{11} mode decreases slightly. The TM_{01} and TE_{21} modes are odd functions of depointing and are orthogonal. With a linearly polarised wave, use of the TM_{01} and TE_{21} modes permits depointing to be determined in the two reference planes. With a circularly polarised wave, depointing is determined in each plane by the phase shift between one of the TM_{01} or TE_{21} modes (only one of the two is used) and the principal TE_{11} mode.

Monopulse tracking systems are characterised by excellent performance in terms of tracking accuracy and speed of response. Tracking accuracy is between $0.02\,\theta_{3dB}$ and $0.05\,\theta_{3dB}$. In contrast, these systems are costly since they require coherent tracking receivers with several channels and sources which are difficult to fabricate. They are principally used in stations of the former Intelsat Standard A type (30 m diameter in C band). For antennas of smaller diameter (for example the new Standard A of 16 m diameter), step-by-step systems are economically more attractive and the performance degradation is scarcely noticeable.

8.3.7.7 *The influence of tracking type on antenna gain*

Table 8.2 summarises the various types of tracking and the accuracies obtained. According to the type of tracking used, the maximum depointing angle has a particular value which may be independent of λ/D (for the case of programmed or computed tracking), the sum of a constant term and a term which is a function of λ/D (for fixed mounting) or a function of λ/D (for most cases of automatic beacon tracking).

The loss in gain with respect to the maximum gain can be evaluated for the various cases from equation (8.25) as a function of the maximum value α_{MAX} of the depointing angle:

$$\Delta G = -12(\alpha_{MAX}/\theta_{3dB})^2 \qquad \text{(dB)}$$

The minimum gain of the antenna for the maximum depointing condition α_{MAX} as a function of λ/D and the efficiency η is thus of the form:

$$G_{MIN} = \eta(\pi D/\lambda)^2 10^{-1.2[\alpha_{MAX} D/70\lambda]^2} \qquad (8.27)$$

Fixed mounting. For a fixed mounting used with a geostationary satellite, the maximum value α_{MAX} of the depointing angle is given by equation (8.26) and is of the form $a + b\,\theta_{3dB}$ (a is the sum of the semi-diagonal $SKW\sqrt{2}$ of the station-keeping window and the uncertainty θ_{SPU} in the satellite position with respect to the centre of the window; $b\,\theta_{3dB}$ is the initial pointing error IPE).

The loss of gain is thus:

$$\Delta G = -12(b + a/\theta_{3dB})^2 \qquad \text{(dB)}$$

This loss of gain is not the same on the up- and downlinks since the 3 dB beamwidths (θ_{3dB}) of the antenna beams are different.

Table 8.2 Tracking system performance

Tracking type	Accuracy	Gain loss
None (fixed mounting)	Initial pointing error: IPE = 0.1–0.2 θ_{3dB}	A function of the station-keeping window
Programmed or calculated	Typical: 0.01°	A function of D/λ
Conical scanning	0.05–0.2 θ_{3dB} (typical: 0.01°)	$\Delta G = 0.03$–0.5 dB
Step-by-step	0.05–0.15 θ_{3dB} (typical: 0.01°)	$\Delta G = 0.03$–0.3 dB
Electronic deviation	0.01–0.05 θ_{3dB} (typical: 0.005°)	$\Delta G = 0.001$–0.03 dB
Monopulse	0.02–0.05 θ_{3dB} (typical: 0.005°)	$\Delta G = 0.005$–0.03 dB

The expression for the gain G_{MIN} of the antenna for conditions of maximum depointing α_{MAX} as a function of λ/D, efficiency η and the parameters a and b defined above can be put in the form:

$$G_{MIN} = \eta(\pi D/\lambda)^2 10^{-1.2[b + (aD/70\lambda)]^2} \tag{8.28}$$

With fixed mounting and a satellite whose apparent movement is relatively large (as for inclined elliptical orbits), the minimum gain of the antenna is given by expression (8.27) after determination of the maximum depointing angle between the fixed antenna aiming direction and the greatest distance corresponding to the entry into (or departure from) service of the active satellite.

Programmed tracking. For programmed or computed tracking, the loss of gain depends on the beamwidth θ_{3dB}, that is λ/D and the maximum value α_{MAX} of the depointing angle:

$$\Delta G = -12(\alpha_{MAX}/\theta_{3dB})^2 \qquad \text{(dB)}$$

The minimum gain of the antenna for maximum depointing conditions α_{MAX} is given by equation (8.27).

Automatic tracking. Finally, with automatic beacon tracking, the tracking error is very often a function of the 3 dB beamwidth. The depointing angle α_{MAX} is thus of the form $\alpha_{MAX} = c\theta_{3dB}$ and the corresponding loss of gain is given by:

$$\Delta G = -12(c)^2 \qquad \text{(dB)}$$

The loss of gain is thus constant and independent of frequency and antenna efficiency. It is therefore the same on the up- and downlinks. Hence, for a step-by-step tracking system where the tracking error is between 0.05 and 0.15 θ_{3dB} ($0.05 \leqslant c \leqslant 0.15$), for example, the loss of gain is between 0.03 and 0.3 dB.

The expression for the antenna gain G_{MIN} for a tracking error of the form $c\theta_{3dB}$ as a function of λ/D and efficiency η is thus:

$$G_{MIN} = \eta(\pi D/\lambda)^2 10^{-1.2[c]^2} \tag{8.29}$$

Conclusion. To summarise the influence of depointing error on gain, it must be emphasised that an increase in antenna diameter does not necessarily lead to an increase in antenna gain in the direction of the satellite. In fact, for a station operating with a fixed depointing angle α which is not directly proportional to λ/D (the case of programmed or computed tracking or fixed mounting), an increase in λ/D leads to an increase in on-axis gain. On the other hand, depending on the value of a, an increase in λ/D does not always lead to an increase in gain in a direction which makes an angle α with the direction of maximum gain (Figure 8.26). This is due to the reduction of angular beamwidth as λ/D increases.

As far as tracking of geostationary satellites is concerned, typical accuracies of orbit control ($\pm 0.05°$) make the use of a tracking system unnecessary (i.e. fixed mounting) up to diameters around 4 m for antennas operating in Ku band (14/11 GHz). Between 4 and 6 m, the choice between fixed mounting and a system of computed or step-by-step tracking is determined

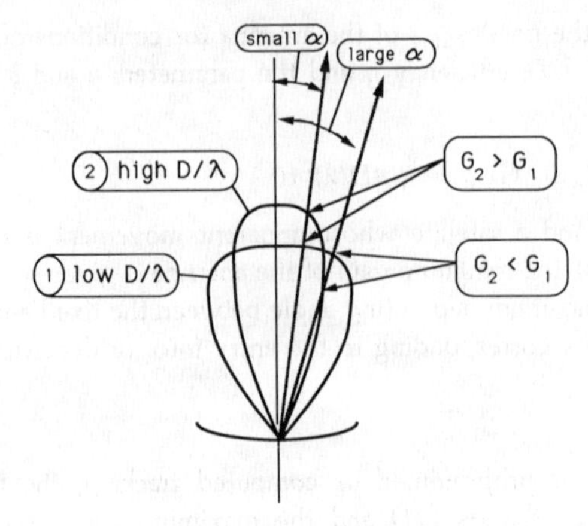

Figure 8.26 Gain variations versus D/λ as a function of the depointing angle α

by a more detailed analysis. Above 6 m diameter, the use of a tracking system is virtually obligatory. The use of step-by-step tracking systems, possibly associated with computed tracking, is tending to become universal for both medium and large stations. They are more economic and are well suited to the operating conditions [KEO-88].

8.3.7.8 *Antennas mounted on mobiles*

With a directional antenna, automatic tracking can only be by feedback control of the antenna axis to the direction of a beacon mounted on the satellite.

The difficulties of acquiring and maintaining locking of the servo loop may also require the use of an inertially stabilised platform, particularly for antennas mounted on-board ships. Movement of the antenna with respect to the vessel is determined from information provided by the tracking receiver and the inertial platform.

Orientation of the beam can also be achieved with electronically controlled antennas. It may be worth controlling antenna orientation in only one axis with the other retaining a fixed direction. One approach of this type has been used for testing links with airborne mobiles using an array antenna which is electronically pointed in azimuth and mounted on the fuselage of an aircraft [NAK-89].

Finally, particularly for terrestrial mobiles, the use of fixed zenith pointing antennas which have a sufficiently large 3 dB beamwidth (to the detriment of the gain) can be considered; this applies particularly to the case of systems using satellites in inclined elliptical orbits where the elevation angle under which the active satellite is viewed remains high (e.g. greater than 45° or 60°, see Section 7.3).

For links with geostationary satellites where the elevation angles are smaller, an omnidirectional antenna avoids the complexity and cost of a tracking system and also occupies less space. This approach has been used in the Prodat system proposed by the European Space Agency [ROG-89].

8.4 THE RADIO-FREQUENCY SUB-SYSTEM

The radio-frequency sub-system contains the following:

—On the receiving side; low noise amplifying equipment and equipment for routing the received carriers to the demodulating channels.
—On the transmitting side: equipment for coupling the transmitted carriers and power amplifiers.

In each direction, frequency converters form the interface with the telecommunication sub-system which operates at intermediate frequency.

8.4.1 Receiving equipment

The earth station figure of merit G/T is determined by the value of system noise temperature T which is given by expression (8.6):

$$T = (T_A/L_{LRX}) + T_F(1 - 1/L_{FRX}) + T_R$$

where T_A is the antenna temperature, L_{FRX} is the connection loss between the antenna interface and the receiver input, T_F is the physical temperature of this connection and T_R is the equivalent noise temperature referred to the receiver input.

Antenna temperature has already been mentioned in Section 8.3.3. At a given antenna temperature, the system noise temperature T is reduced by minimising the connection loss between the antenna interface and the receiver input and limiting the equivalent noise temperature referred to the receiver input.

Connection loss is most satisfactorily reduced by locating the first stage of the receiver as close as possible to the antenna feed. The equivalent noise temperature T_R referred to the receiver input is of the form (cf. expression (2.21)):

$$T_R = T_{LNA} + (L_1 - 1)T_F/G_{LNA} + T_{MX}L_1/G_{LNA} + (L_2 - 1)T_FL_1/G_{LNA}G_{MX} + T_{IF}L_2L_1/G_{LNA}G_{MX} + \cdots \quad (8.30)$$

where the effective input noise temperatures of the various stages are included together with connection losses between these stages (Figure 8.27).

Equation (8.30) shows that it is necessary to use receiving equipment whose first stage has low noise and sufficiently high gain to mask the noise introduced by the following stages. In accordance with case (a) or (b) of Figure 8.27 (see Section 8.4.1.4) the loss L_1 from the LNA output, located close to the antenna feeder, to the frequency conversion equipment, often located at some distance from the LNA, includes either only connection loss ($L_1 = L_F$), or connection loss (L_F) associated with the influence of the power splitter ($L_1 = L_FL_{PS}n$, where L_{PS} is the insertion loss and n the power division ratio).

For small stations, frequency conversion and low noise amplification can be combined in equipment which is mounted behind the source (L_1 is close to zero); but the attenuation L_2

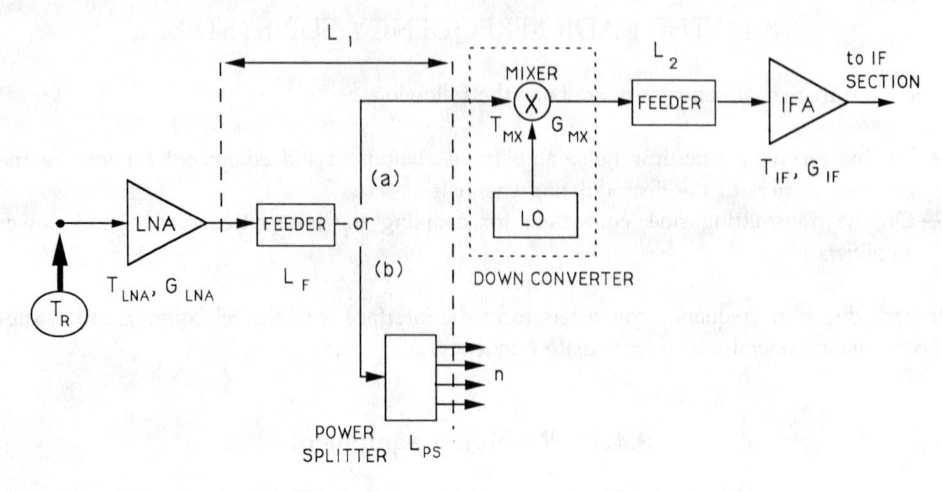

Figure 8.27 Receiver front end block diagram. (a) Conversion *en bloc.* (b) Carrier by carrier conversion.

of the link between the converter and the following stages can then make a non-negligible contribution.

The first Intelsat Standard A stations used cryogenically cooled maser amplification systems. The very high operating cost has caused these systems to be abandoned in favour of parametric amplifiers. The use of transistor amplifiers then became important first in C band and subsequently in Ku band.

8.4.1.1 *The parametric amplifier*

Figure 8.28 shows the structure of a parametric amplifier. A parametric amplifier is a reflection amplifier, that is the amplified signal is obtained by reflection of the incident signal from the active element. Direction and separation of the signals is achieved by means of a circulator. The active element is a varactor (variable capacitance diode) biased in such a way as to present a negative impedance to the incident signal.

The varactor is coupled by three ports which are tuned to the frequency F_s of the signal to be received, the pump (oscillator) frequency F_p and the image frequency F_i. When $F_p > F_s$ and $F_i = F_p - F_s$, the Manley Rowe equations show that amplification occurs at frequency F_s; the power delivered to the external circuit is $P_s = -P_p(F_s/F_p)$, where P_s and $-P_p$ are the powers *given out* by the varactor at frequencies F_s and F_p. There is thus a transfer of power from the pump to the signal. The amplified signal can leave only by the port tuned to F_s, that is the incident path. The circulator, which initially routed the signal delivered by the antenna from port 1 to port 2, routes the reflected and amplified signal from port 2 to port 3.

The advantage of the varactor is that it permits amplification by means of a reactance which is theoretically devoid of noisy resistive elements. The equivalent noise temperature referred to the input T_E is given by:

$$T_E = T_F[(F_i/Q^2F_s) + F_s/F_i] \qquad (K) \qquad (8.31)$$

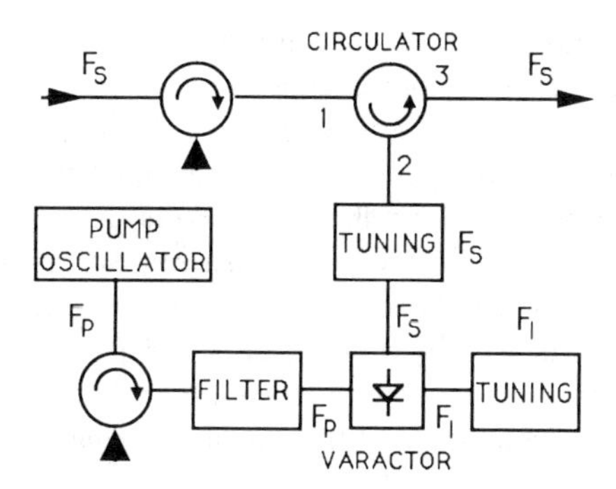

Figure 8.28 Parametric amplifier block diagram.

where T_F is the physical temperature of the diode and Q is the quality factor of the circuit. It is advantageous to cool the varactor in order to obtain a low noise temperature; for example, for $T_F = 20\,\text{K}$, $Q = 10$, $F_i = 28\,\text{GHz}$ and $F_s = 4\,\text{GHz}$, the temperature T_E is 5 K. Under the same conditions, if $T_F = 300\,\text{K}$, $T_E = 67\,\text{K}$.

If B_s and B_i are the bandwidths of the circuits tuned to the signal and image frequencies respectively, the power amplifier gain G and its bandwidth B are approximately related by:

$$B\sqrt{G} = 2/[(1/B_s) + (1/B_s)] \qquad \text{(Hz)} \tag{8.32}$$

8.4.1.2 *Transistor amplifiers*

Transistor amplifiers operate in transmission; that is the signal at the input of a four port network appears amplified at its output. The difference between the output and input powers is provided by the amplifier power supply circuit. An active element (the transistor) is necessary in order to obtain a gain greater than unity but this element also generates noise. As a consequence of their junction structure, bipolar transistors cause (shot) noise other than thermal noise and can provide mediocre performance at high frequencies.

On the other hand, the noise due to field effect transistors is mainly of thermal origin and can be reduced by selecting the type of semiconductor used and the geometric characteristics of the transistor. The performance in terms of noise factor is continuously improving due to the use of gallium arsenide (GaAs) and submicron lithography. Finally, the appearance of high mobility electron transistors (HMET) has enabled the noise temperature of receiving equipment to be further reduced, particularly at high frequencies (20 GHz) [TOK-84], [SHI-85], [IWA-85].

8.4.1.3 *Cooled amplifiers*

Whether the amplifier is parametric or transistor, cooling of the active element enables the effective noise temperature of the equipment to be reduced. By using liquid helium, it is possible to cool the equipment to a few tens of degrees Kelvin. Under these conditions, the receivers of the first Standard A stations had noise temperatures of the order of 15 K at 4 GHz. The disadvantages of cryogenic cooling systems are their high installation and operating costs and maintenance problems. Cryogenic cooling is no longer used in earth stations in C and Ku band since the performance of equipment at ambient temperature or with Peltier cooling is adequate. However, it remains useful for obtaining low values of noise temperature (less than 80 K) with field effect transistors operating in Ka band (20 GHz) [OHM-87]. Cooling of field effect amplifiers from ambient temperature (300 K) to 20 K enables the noise temperature to be reduced by a factor of 3 to 5 [WEI-80].

Peltier thermoelectric devices enable the temperature of the active element to be reduced to around −50°C; the noise temperature of the amplifier is thus reduced in comparison with operation at ambient temperature.

Table 8.3 gives performance orders of magnitude (for noise temperature and gain) of low noise amplifiers for various frequency bands when cooled and when operating at ambient temperature. The gain depends on the number of elementary stages used and is between 10 and 25 dB per stage according to the technology used and the bandwidth required. The useful bandwidth of amplifiers varies between a few hundred MHz and 2 GHz and is greatly affected by the performance in terms of noise temperature, gain stability and standing wave ratio.

For a particular application, it is important to check that variations of noise temperature and gain over the operating band remain within acceptable limits. This also applies to the standing wave ratio (SWR) which indicates the matching conditions of the amplifier input impedance to that of the link to the antenna. The value of SWR can be reduced by using an isolator at the low noise amplifier input to the detriment of noise temperature.

8.4.1.4 *Distribution of carriers and frequency conversion*

Once low noise amplification has been performed, the carriers received in the frequency band used on the link are converted to an intermediate frequency where the operations of filtering and signal processing are simpler (see Section 8.5). The conversion may be realised either *en bloc* on the whole of the frequency band used in the receiver (case (a) of Figure 8.27) or carrier by carrier (case (b) of Figure 8.27).

Conversion *en bloc* of the frequency band is used in equipment intended for the reception of single channel carriers (SCPC). Distribution of carriers to different demodulators is performed at intermediate frequency (typically 140 MHz) and selection of a particular carrier frequency (of narrow bandwidth, typically 30 kHz) is achieved by alignment of the demodulator. Conversion *en bloc* of the frequency band is also usual for small antennas intended for television signal reception or data transmission. In this case the frequency converter is usually integrated with the low noise amplifier; the combination is mounted on the feeder which is located at the focus of the antenna. The converter output frequency is of the order of one gigahertz (900 MHz to 1700 MHz) and this permits reduction of losses in the co-axial cable between the converter and the remainder of the equipment which may be distant from the antenna.

Carrier-by-carrier conversion involves the use of frequency conversion equipment which permits the carrier concerned to be selected and converted to the intermediate frequency. This intermediate frequency is the same regardless of the frequency of the received

Table 8.3 Low noise amplifier performance

Technology	Frequency	Noise temperature	Bandwidth	Gain	No. of stages
Cryogenic	4 GHz	13 K	0.5 GHz	30 dB	2
Parametric	20 GHz	< 100 K	0.5 GHz	30 dB	2
Cooled	4 GHz	35 K	0.5 GHz	30 dB	2
Parametric (Peltier)	12 GHz	85 K	0.5 GHz	30 dB	2
	20 GHz	150 K	0.5 GHz	30 dB	2
Ambient	4 GHz	55 K	0.5 GHz	30 dB	2
Parametric	12 GHz	150 K	0.5 GHz	30 dB	2
	20 GHz	200 K	0.5 GHz	30 dB	2
FET Cryogenic	20 GHz	75 K	0.5 GHz	45 dB	4
FET Cooled	4 GHz	40 K	0.6 GHz	60 dB	5
(Peltier)	12 GHz	120 K	0.75 GHz	60 dB	7
	12 GHz	160 K	2 GHz	60 dB	7
	20 GHz	180 K	1 GHz	45 dB	3
FET Ambient	4 GHz	70 K	0.6 GHz	60 dB	5
	12 GHz	130 K	0.75 GHz	60 dB	7
	12 GHz	180 K	2 GHz	60 dB	7
	20 GHz	350 K	1 GHz	22 dB	2
HMET	20 GHz	300 K	1.6 GHz	16 dB	1

radio-frequency carrier; tuning is performed at the converter by controlling the local oscillator frequency. This permits standardisation of intermediate frequency equipment and hence cost reduction and simplified maintenance. Only the bandwidth of this equipment must be matched to that of the particular received carrier. Common values of intermediate frequency are 70 MHz and 140 MHz.

When an earth station must demodulate several carriers simultaneously, it is necessary to distribute the power at the low noise amplifier (LNA) output among the various converter channels. This is performed by a power splitter using passive devices (hybrid couplers or power dividers). The power splitter insertion loss L_{PS} adds to the connection loss L_F from the LNA output to the power splitter input. The power split among the n converter channels translates into an attenuation by a factor n. Hence the total loss L_I from the LNA output to any converter input is $L_I = L_F L_{PS} n$.

8.4.2 Transmission equipment

The power per carrier P_T provided by the transmission equipment determines the value of the equivalent isotropic radiated power (EIRP) which is a characteristic of the earth station for the link considered. The available carrier power P_T at the antenna input depends on the power P_{HPA} of the power amplifier, the connection losses L_{FTX} between the output of the amplifier and the antenna interface and the power loss L_{MC} entailed in multiple carrier operation as specified by expression (8.3):

$$P_T = (P_{HPA})(1/L_{FTX})(1/L_{MC}) \qquad (W)$$

The power amplifier characteristics vary according to the technology used which may be travelling wave tube, klystron or transistor. The magnitude and nature of the power loss L_{MC} entailed in multicarrier operation depends on the type of coupling which may be performed before or after power amplification.

8.4.2.1 *Power amplifiers*

The power amplifier sub-system uses a tube or transistor power stage which may be associated with a preamplifier and a lineariser. This sub-system also includes protection and control equipment and possibly a cooling system. Table 8.4 presents the main characteristics of these amplifiers.

8.4.2.1.1 Tube amplifiers

The amplifier tubes used in earth stations are klystrons or travelling wave tubes. The general organisation of these devices is similar; they consist of an electron gun, a system for focusing the electrons which enables an extended cylindrical beam to be obtained, a device which enables the kinetic energy of the electrons to be converted into electromagnetic energy and a collector of the electrons in the beam.

In a klystron, the conversion device consists of a series of cavities, which are microwave resonant circuits and are traversed by the electron beam. The low level electromagnetic wave which excites the first cavity causes modulation of the velocity of the electrons which cross it. This modulation creates an induced wave in the second cavity which in turn increases the modulation of the electron beam. The process repeats itself and is amplified in the

Table 8.4 Power amplifier characteristics

Technology	Frequency (GHz)	Power (kW)	Efficiency	Bandwidth (MHz)	Gain (dB)
Klystron	6	1–5	40	45	40
	14	0.5–3	25	85	40
	18	1.5	25	60	40
	30	0.5	20	100	40
Travelling wave tube (TWT)	6	0.1–3	30	600	45
	14	0.1–2.5	40	500	50
	18	0.5	40	800	50
	30	0.05–0.15	40	2000	50
FET	6	5–50	30	600	50
	14	1–10	20	500	50

following cavities. A radio-frequency wave at high level is thus produced at the output of the last cavity. The powers obtained range from a few hundreds of watts (around 800 W) to several kilowatts (5 kW). The bandwidth of the klystron is limited by the presence of resonant cavities in the amplification process. It is of the order of 40 to 80 MHz in C band (6 GHz) and 80 to 100 MHz in Ku band (14 GHz).

In the travelling wave tube, the energy transfer device is arranged around a helix which surrounds the electron beam and along which the electromagnetic wave propagates (see Figure 9.14). The helix effectively slows the wave so that the axial component of the electromagnetic wave velocity (equal to the product of the velocity of light and the ratio of the helix step length to the length of one turn) is approximately equal to that of the electrons. Consequently, a continuous mechanism of energy transfer occurs along the helix. The electromagnetic wave gains the kinetic energy given up by the electrons. The power obtained ranges from several tens of watts (e.g. 35 W) to several kilowatts (e.g. 3 kW). The bandwidth of the travelling wave tube is large—about 600 MHz in C band at 6 GHz and about 3 GHz in Ka band at 30 GHz.

Tube amplifiers enable high powers to be produced and are therefore widely used in earth stations. The choice between travelling wave tubes and klystrons depends on the required bandwidth; for equal powers, the cost advantage is with the klystron. Available powers depend only slightly on frequency; tubes are available at 17 GHz (for links with broadcast satellites) and 30 GHz.

Tube amplifiers require a suitable power supply to deliver the various voltages (up to 10 kV) required on the electrodes. These voltages must be adequately regulated (to a relative value of 10^{-3}) in order to avoid fluctuations in the radio-frequency output of the tube. For high powers, it is necessary to provide forced air (up to around 3 kW) or circulating liquid cooling arrangements. In spite of a high power gain (40 to 50 dB), the power required at the input of the tube usually requires the use of a pre-amplifier. This pre-amplifier uses a low power travelling wave tube (from a few watts to a few tens of watts) or several transistor stages.

8.4.2.1.2 Transistor amplifiers

Semiconductor amplifiers provide powers of several tens of watts in C band (6 GHz) and several watts in Ku band (14 GHz). These amplifiers usually use gallium arsenide (GaAs) field effect transistors. In spite of the low powers available (which are continuously increasing with progress in technology), transistor amplifiers are increasingly used because of their low cost, linearity and wide bandwidth.

8.4.2.1.3 Power amplifier characteristics

Non-linearity. Power amplifiers are non-linear. As shown in Figures 2.28, 4.76 and 9.2, as the carrier power applied to the input of an amplifier is increased, there is a region of quasi-linear operation at low level after which the high level output power no longer increases in proportion to the input power. The maximum power obtained at the output corresponds to saturation (unless a limit on power dissipation prevents the saturation point being reached which is the case for solid state amplifiers in particular. The maximum output power at saturation in single carrier operation $(P_{01})_{SAT}$ is the amplifier power given in the manufacturer's catalogues (P_{HPA}). When operating with several carriers, intermodulation products appear at frequencies corresponding to linear combinations of the input carrier frequencies (see Chapter 4, Section 4.4.3). When the carriers are modulated, the intermodulation products which fall within the useful bandwidth of the amplifier behave as noise which is quantified for the bandwidth of each carrier by the value of intermodulation power spectral density $(N_0)_{IM}$.

To limit intermodulation noise when several carriers are amplified simultaneously to a value compatible with the link budget specification (see Section 4.4.4), it is advisable to operate the amplifier below the saturation region. The output back-off (OBO), defined as the ratio of the output power delivered on one of the N carriers (P_{ON}) to the saturation power, determines the position of the operating point (see Section 4.4.5). The power delivered at the amplifier output for the carrier concerned is thus equal to:

$$P_{O}^{N} = P_{HPA} \times \text{OBO} \qquad \text{(W)} \tag{8.33}$$

The value of back-off depends on the minimum allowed value of the earth station's contribution to the carrier power to intermodulation power spectral density ratio $(C/N_0)_{IM}$ of the link, the number of carriers and the input/output characteristic of the amplifier.

Comment. The total back-off is sometimes defined as the ratio of the total power available on all N carriers to the saturation power in single carrier operation; when the carriers are of equal level, the power per carrier can be obtained by dividing the product of the amplifier saturation output power and the total output back-off by N. Additional specifications (such as the point of compression to 1 dB, the third order intercept point and the conversion and AM/PM transfer coefficients) relating to the characteristics of amplifier non-linearities are given in Chapter 9.

Gain variations. The specified gain of a power amplifier is susceptible to variation as a function of various parameters. It is important to specify the stability of the gain, that is the magnitude of permitted variations as a function of the various parameters. Hence the following are specified for a particular application:

—The stability of the gain as a function of time (e.g. $\leqslant 0.4$ dB/24 h) at constant input level.

—The magnitude of gain variations as a function of frequency within the bandwidth for a given power level (e.g. $\leqslant 4$ dB in 500 MHz).

—The maximum rate of change of gain fluctuations as a function of frequency in a specified portion of the band (e.g. $\leqslant 0.05$ dB per MHz).

Standing Wave Ratio (SWR) and propagation time. The maximum standing wave ratio is specified at the input and output of the amplifier and also for the load driven by it. The group propagation time within the frequency band of a power amplifier varies as a function of frequency. Meeting the specification can require the installation of propagation time equalisers in the transmission channel, usually in the stages operating at intermediate frequency.

Spurious modulation, noise and radiation. In order to limit the effects of satellite amplifier non-linearity, it is necessary to limit spurious amplitude modulation of the carrier at the amplifier output. For example, the Intelsat specifications limit the relative level of components to $-20(1+\log f)$ dB in any interval of 4 kHz between 4 kHz and 500 kHz, measured at the carrier frequency, and to -74 dB above 500 kHz. Spurious amplitude modulation is mainly due to fluctuations in the power supply.

The maximum level of harmonics (frequencies which are a multiple of the carrier), noise and other spurious signals are also specified (e.g. -65 dBW in any band of 4 kHz).

8.4.2.2 *Linearisers*

The use of linearisers is becoming more common in order to limit the effects of amplifier non-linearity. Combined with the pre-amplifier, or located before it, most linearisers produce amplitude and phase distortion of the signal in order to compensate for the specific characteristics of the power amplifier (Figure 8.29). For a given level of intermodulation noise, the lineariser permits a reduction of back-off (in absolute value); that is the amplifier is operated closer to saturation and the rate of phase change is limited. The reduction of back-off provides a considerably greater available carrier power for an amplifier of given saturation power and hence cost, power consumption and bulk.

8.4.2.3 *Carrier pre-coupling*

As stated in the introduction to this chapter, an earth station very often transmits several carriers (at different frequencies) to the satellite concerned. As the antenna interface generally has only a single input (for a given polarisation), it is necessary to multiplex these carriers, which have been modulated separately, by frequency division in order to combine them on the same physical connection.

Carrier coupling can be performed at low power level (for example by using hybrid couplers) before power amplification (Figure 8.30). Power amplification then operates in a multicarrier regime and must be operated with an output back-off in order to limit intermodulation noise power. The power loss L_{MC} caused by multicarrier operation in this case has a value:

$$(L_{MC})_{ES} = -(OBO)_{ES} \qquad \text{(dB)}$$

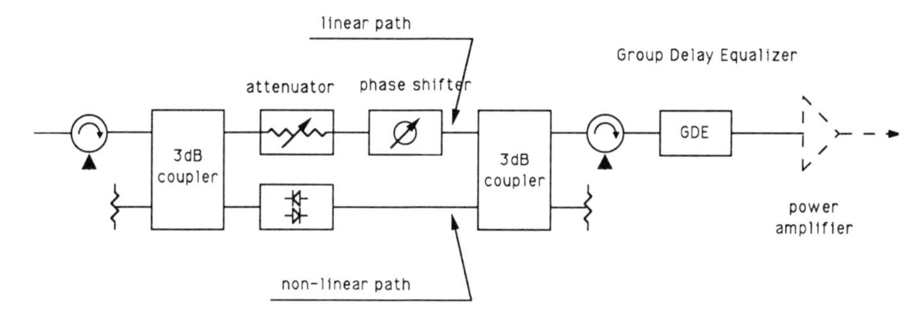

Figure 8.29 Non-linear predistoration type lineariser.

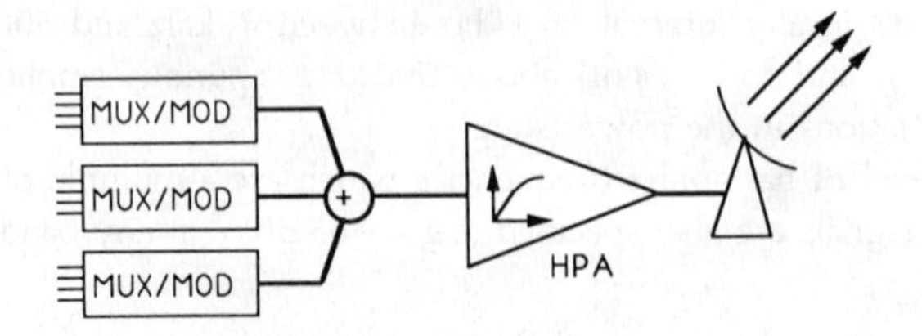

Figure 30 Carrier coupling prior to power amplification.

where OBO is the output back-off defined as the ratio of the available power of the carrier concerned (one of N) to the output power at saturation in single carrier operation.

The advantages of pre-coupling lie in the simplicity of coupling and the flexibility to adapt to changes in the number and bandwidth of carriers. The number of amplifiers is also minimised. On the other hand, this mode of coupling introduces a source of intermodulation noise in the earth segment which affects the overall link budget. Limitation of intermodulation noise to an acceptable value requires the use of an amplifier with sufficient back-off and this leads to the use of a device with a saturation power much greater than the required power. Furthermore, the amplifier must have a sufficient bandwidth to amplify the different carriers (this can prohibit the use of a klystron which would otherwise be more economic).

8.4.2.4 Carrier post-coupling

Coupling can also be performed after separate amplification of each carrier (Figure 8.31). It is then necessary to have as many amplifiers as there are carriers (plus any replicated equipment). Each amplifier amplifies only one carrier; the amplifiers can therefore operate at saturation.

However, the coupling device introduces losses L_C which can be identified with the reduction in power L_{MC} caused by multicarrier operation in expression (8.3)

$$(L_{MC})_{EC} = L_C \qquad \text{(dB)}$$

Since each carrier is amplified separately, the required bandwidth is limited. Furthermore, operation to saturation enables low power, and hence lower cost, amplifiers to be used. It is advisable, however, to perform the carrier coupling operation at high level with a minimum of loss.

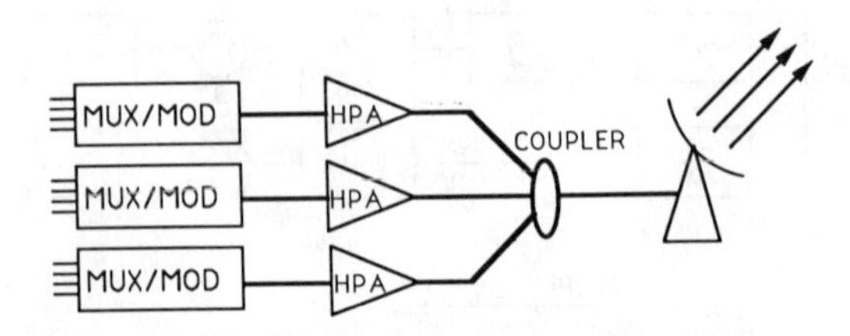

Figure 8.31 Post-amplification carrier coupling.

Two types of device permit coupling to be realised:

—Hybrid couplers (aperiodic coupling).
—Band-pass filter multiplexers.

Aperiodic coupling. To couple the amplifier outputs while observing the impedance conditions, it is possible to use conventional hybrid couplers. This approach permits considerable flexibility of use since coupling is wide band but this is accompanied by large losses. In fact a hybrid coupler permits two signals to be combined but the power of each signal is shared between the two outputs (in equal parts for a 3 dB coupler). As a single output is used, the power on the unused output is dissipated in the matched load (half of the total power for a 3 dB coupler). This loss occurs again when coupling the sum of the first two signals with the third and so on.

Coupling by multiplexer. The use of multiplexers involving band-pass filters tuned to each carrier enables the losses to be minimised to the detriment of system flexibility. Two techniques are used to combine the signals:

—Circulators route the signals after band-pass filtering and reflection on to the filter outputs in accordance with a principle similar to that used for satellite output multiplexers (OMUX) (see Section 9.2.3.2).

—The use of hybrid couplers as illustrated in Figure 8.32. The principle of operation is as follows. The signal of amplitude A corresponding to carrier 1 is divided into two components of amplitude $A/\sqrt{2}$ and phase shifted by 90° by hybrid coupler 1. These two components pass through the band-pass filters tuned to the frequency and band of carrier 1. The two components present at the input of coupler 2 are summed in phase after division by $\sqrt{2}$ at the antenna port and in phase opposition on the other port (port B) due to their phase shift of 90°. The signal of amplitude B corresponding to carrier 2 is divided into two components of amplitude $B/\sqrt{2}$ and phase shifted by 90° by hybrid coupler 2. These two components are reflected on to the outputs of the band-pass filters tuned to the frequency and band of carrier 1. The reflected components appear at the ports of coupler 2 and are also summed in phase after division by $\sqrt{2}$ on the antenna port.

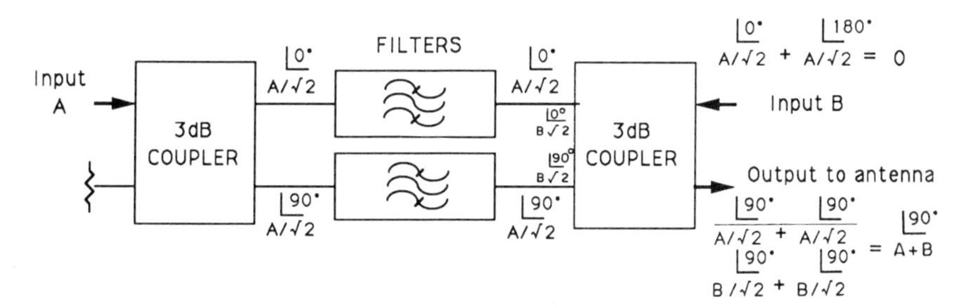

Figure 8.32 Carrier coupling with filter combiner.

The disadvantages of coupling by multiplexer are the loss of flexibility associated with the need to have band-pass filters which are perfectly matched to the characteristics of the carriers to be combined. On the other hand, the losses are small, of the order of a decibel.

8.4.2.5 *Mixed coupling (pre- and post-coupling of carriers)*

A combination of the two types of coupling is often used. Each amplifier amplifies a restricted number of carriers and the outputs of these amplifiers are coupled by one of the techniques (or a combination of the two) described in the previous section. This permits, for example, all carriers that are routed by a given transponder to be coupled into the same amplifier. Then, at the earth station, each amplifier can be associated with a particular satellite transponder.

8.4.3 Redundancy

To satisfy the objectives of reliability and specified availability, it is often necessary to replicate the radio-frequency equipment of an earth station (see Chapter 13).

As far as the input stages are concerned, the use of a redundant receiver is normal except for small stations which usually do not incorporate redundancy. Since the operation of the station is monitored, it is rare to have more than one stand-by system since maintenance of the station is guaranteed.

For the output stages, the redundancy arrangement depends on the type of coupling. With pre-coupling of the carriers, the power amplifier (single except in the case of mixed coupling) is usually replicated. With post-coupling of the carriers, it is not useful to replicate each carrier amplifier and the use of back-up equipment shared among several active units is usual.

8.5 COMMUNICATION SUB-SYSTEMS

The communication sub-system on the transmission side consists of equipment for converting baseband signals to radio-frequency carriers for amplification; conversely, on the reception side, it converts the carriers at the output of the low noise amplifier to baseband signals.

The baseband signal may be either analogue or digital. In the analogue case, it can be a telephone channel in the case of a single channel per carrier system (SCPC) transmission system, a multiplex of telephone channels, a television signal or a radio broadcast. In the digital case, it is usually in the form of a bit stream which corresponds to one or a multiplex of telephone channels or data packets.

The functions to be realised on the receiving side are as follows:

—conversion of the carrier frequency (RF) to an intermediate frequency (IF),
—filtering and equalisation of group propagation delay,
—carrier demodulation.

In the case of transmission using time division multiple access (TDMA) it is also necessary to re-establish a continuous digital stream from the packets of the received frame.

On the transmission side, if time division multiple access is used it is necessary to group the bits of the baseband signal into packets which are inserted in the proper time slots provided in the frame. Finally, as for analogue signals, the following operations are performed:

—modulation of a carrier at an intermediate frequency,
—filtering and equalisation of group propagation delay,
—conversion of the modulated carriers to radio frequency.

8.5.1 Frequency translation

The function of the frequency translation sub-system is to select a particular in-band carrier at the output of the low noise amplifier and translate the spectrum of this carrier to the chosen intermediate frequency. An intermediate frequency of the same conventional value for every channel permits the use of standardised equipment. The choice of intermediate frequency is determined by the following considerations: on the one hand the value must be greater than the spectral width occupied by the modulated carrier but on the other hand, this value must be sufficiently low to permit selective band-pass filtering of the modulated carrier. The selectivity Δf of a filter represented by its quality factor Q is defined by the ratio f/Q, where f is the central frequency of the filter. Assuming a quality factor of 500, if it is required to isolate a signal occupying a bandwidth of 1 MHz, the operating frequency of the filter will be 500 MHz. Common values of intermediate frequency are 70 MHz and 140 MHz.

Frequency translation can be of the single or dual conversion type. The organisation of the two types of frequency translation system for reception is described in the following sections. System architectures on the transmission side are similar.

8.5.1.1 *Single frequency conversion*

The frequency translation system consists of a band-pass filter centred on the radio-frequency carrier to be received and a mixer which also receives the signal from a local oscillator (Figure 8.33). The input filter serves to eliminate the image frequency which is a characteristic of

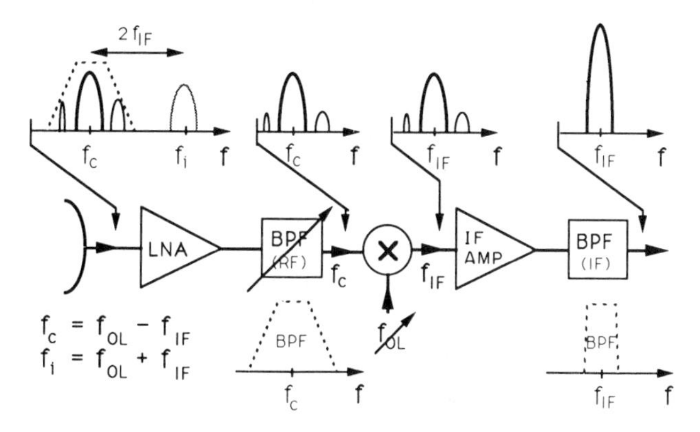

Figure 8.33 Single conversion down-converter.

the frequency conversion process. If f_{LO} is the frequency of the local oscillator, two carriers of frequency $f_{LO}+f_{IF}$ and $f_{LO}-f_{IF}$ are translated to the intermediate frequency f_{IF}. Only one of these two frequencies corresponds to the frequency f_C of the carrier to be received. The other frequency $f_I=f_C+2f_{IF}$, which is called the image frequency, may correspond to another carrier and must be eliminated. The separation between the required carrier and its image frequency is equal to $2f_{IF}$. For a low value of intermediate frequency (e.g. 70 MHz), it is necessary to provide a radio-frequency filter with high selectivity tuned to the frequency of the carrier to be received; for example, if $f_{IF}=70$ MHz and $f_C=4$ GHz, the selectivity must be of the order of 50. At 12 GHz, it is of the order of 200.

Selection of the received carrier is achieved by changing the frequency of the local oscillator and the centre frequency of the image frequency rejection filter. Realisation of a tunable and easily controllable radio-frequency filter is difficult and the double frequency changing structure is often preferred.

8.5.1.2 *Dual frequency conversion*

To provide frequency agility without tuning the input filter to the carrier to be received, it is necessary to keep the image frequency outside the band of frequencies within which the carrier to be received can occur (Figure 8.34).

The frequency obtained after translation must be as high as the receiving bandwidth is wide. For example, for the band $f_1=3.625$ to $f_2=4.2$ GHz (width $f_2-f_1=575$ MHz), detecting the translated frequency as $f_{IF1}=1400$ MHz permits the use of an input band-pass filter with a fixed bandwidth of 575 MHz; this ensures sufficient rejection of the image frequency which, in the worst case (when the carrier to be received is at the lower edge of the band i.e. $f_c=f_1$) is at $f_i=f_c+2f_{IF1}=f_1+2f_{IF1}=3625+2\times1400=6425$ MHz, that is 2225 MHz greater than f_2 and hence outside the passband.

The desired intermediate frequency f_{IF2} (e.g. 70 MHz) is obtained by a second frequency translation which is performed after band-pass filtering centred on the value of the first

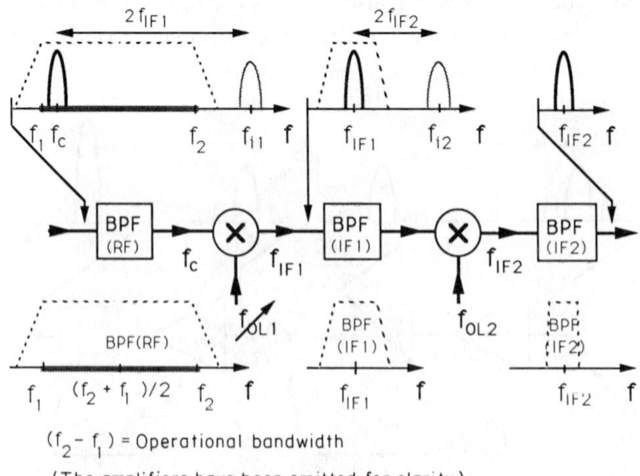

Figure 8.34 Dual conversion down-converter.

intermediate frequency f_{IF1}. This filter has a sufficient bandwidth (e.g. 40 MHz) to allow any type of modulated carrier to pass. It eliminates the image frequency in the second translation since the image frequency $f_{i2} = 1540$ MHz is situated at 140 MHz above the first intermediate frequency $f_{IF1} = 1400$ MHz if the second local oscillator generates a frequency of $f_{LO2} = 1470$ MHz (if $f_{IF2} = 70$ MHz).

Selection of the received carrier frequency f_C is accomplished by setting the frequency of the first local oscillator to $f_C + 1400$ MHz. A (radio-frequency) frequency synthesiser is often used (with frequency variation in steps, for example 125 kHz). The frequency of the second oscillator remains fixed as does the central frequency of the various band-pass filters.

Some configurations use a fixed radio-frequency source for the first oscillator. Tuning is then achieved by adjusting the frequency of the second oscillator (a synthesiser) which operates at a low frequency and is thus easier to design. On the other hand, the first intermediate frequency is not of a fixed value and it is necessary for the associated band-pass filter to be tunable.

8.5.1.3 *Translation of the whole of the useful band*

The organisation of the translation system described above assumes that the carriers are separated before the translation equipment and there is only one carrier involved per IF channel. It is also possible to translate the whole of the received frequency band, and hence all the carriers, to the intermediate frequency band at the same time. This architecture is used particularly for SCPC systems.

By way of example, Figure 8.35 shows the block diagram of a sub-system for transmission and reception translation with a double change of frequency; modulated carriers in the 52–88 MHz band are translated into the 5.850–6.425 GHz band with a double frequency conversion using a second intermediate frequency of 825 MHz. On the receiving side, received carriers in the 3.625–4.200 GHz band are translated to the 52–88 MHz band (with a second intermediate frequency of 1400 MHz). The two translations use a single frequency synthesiser.

8.5.1.4 *Characteristics of frequency translation sub-systems*

Apart from the capacity for frequency agility discussed above, which determines the range of acceptable frequencies for input and output signals, the specified characteristics of a frequency translation sub-system are as follows:

—frequency stability of the local oscillators (long term and phase noise),
—maximum level of spurious frequency components,
—long-term gain stability in the frequency band,
—linearity (the level of intermodulation products or intercept point).

8.5.2 Amplification, filtering and equalisation

The functions of amplification, filtering and group propagation delay equalisation are realised at intermediate frequency. These operations are facilitated by retaining a fixed intermediate frequency regardless of the radio-frequency carrier concerned.

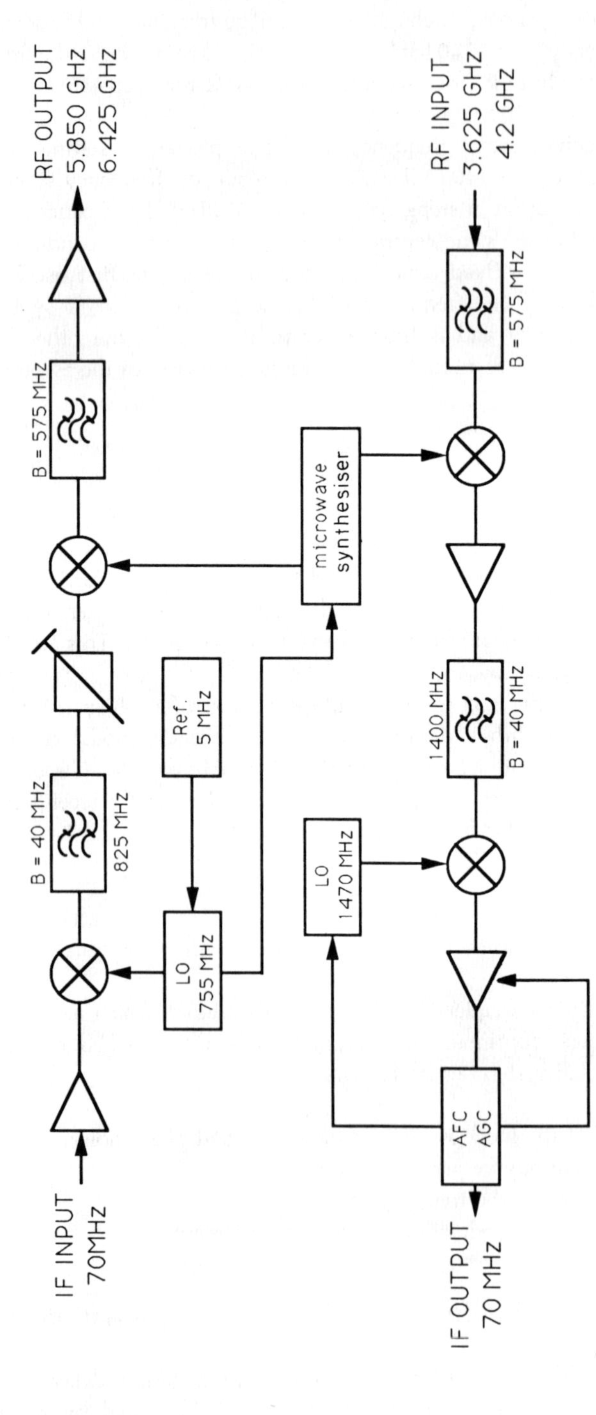

Figure 8.35 Architecture of a transmit and receive frequency converter.

On reception, the intermediate frequency amplifier includes automatic gain control so that a constant level is provided at the input of the demodulation sub-system. On the transmission side, gain control enables the level (or the back-off) at the input of the radio-frequency amplifier to be adjusted.

Band-pass filtering at intermediate frequency defines the spectrum of the modulated carrier and limits the noise bandwidth. The characteristics of this filter depend on the modulation characteristics of the carrier concerned. These filters are usually designed with transfer functions of the Butterworth or Chebyshev type using capacitors and inductors. The filter elements in the transmission and reception channels, the power amplification stages and the satellite transponder all introduce group delay variations as a function of frequency. These variations are corrected within the useful bandwidth by means of group propagation delay equalisers. These equalisers are integrated into the band-pass filter or realised separately by means of LC cells of the bridged T type. By way of example, Figure 8.36 and Table 8.5 give the INTELSAT specifications for filtering and group delay equalisation as a function of the bandwidth occupied by the modulated carrier.

Table 8.5 Characteristics of the INTELSAT filtering specification

Amplitude specification

Carrier bandwidth (MHz)	A (MHz)	B (MHz)	C (MHz)	D (MHz)	a (dB)	b (dB)	c (dB)	d (dB)	e (dB)
1.25	0.9	1.13	1.15	4.0	0.7	1.15	3.0	25	0.0
2.5	1.8	2.25	2.75	8.0	0.7	1.5	2.5	25	0.0
5.0	3.6	4.50	5.25	13.0	0.5	2.0	3.0	25	0.0
10.0	7.2	9.00	10.25	19.0	0.3	2.5	5.0	25	0.1
20.0	14.4	18.00	20.50	28.0	0.3	2.5	7.5	25	0.1
36.0	28.8	36.00	45.25	60.0	0.6	2.5	10.0	25	0.3
Video	12.6	15.75	18.00	26.5	0.3	2.5	6.5	25	0.1
Video	24.0	30.00	—	—	0.5	2.5	—	—	0.3

Group decay specification

Carrier bandwidth (MHz)	A (MHz)	H (MHz)	f (ns)	g (ns)	h (ns)
1.25	0.9	1.13	24	24	30
2.5	1.8	2.1	16	16	20
5.0	3.6	4.1	12	12	20
10.0	7.2	8.3	9	9	18
20.0	14.4	16.6	4	5	15
36.0	28.8	33.1	3	5	15
Video	12.6	14.2	6	6	15
Video	24.0	30.0	5	5	15

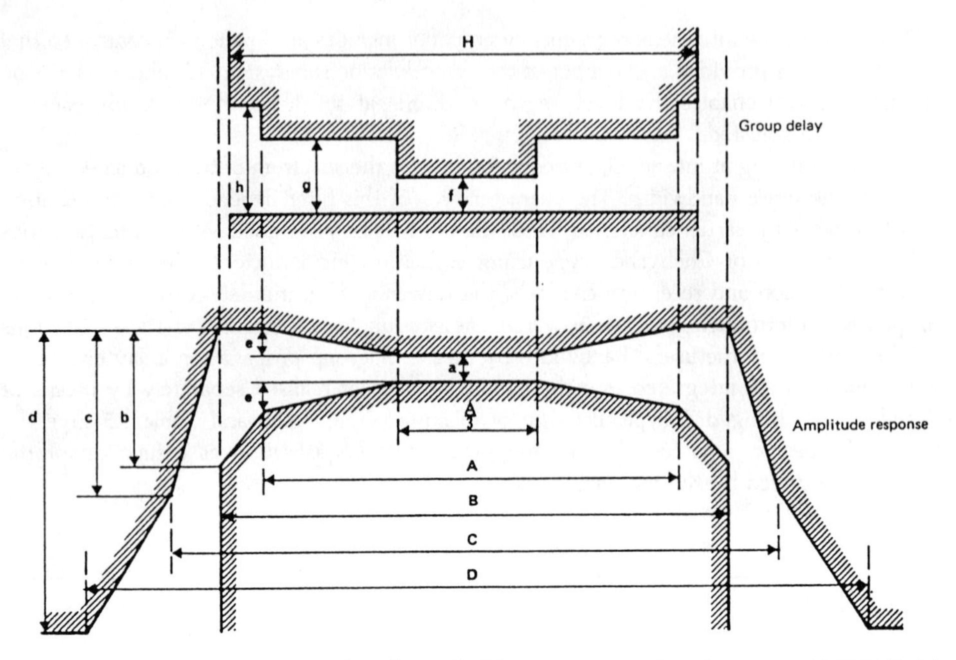

Figure 8.36 Amplitude and group delay limits for filtering at intermediate frequency (INTELSAT). The values of the parameters shown in the figure are given in Table 8.5 as a function of the spectral occupation of the modulated carrier.

8.5.3 Modulation and demodulation

The operations of modulation (on the transmission side) and demodulation (on the receiving side) are realised at intermediate frequency. Modulation and demodulation sub-systems are realised in accordance with the multiple access mode (frequency division (FDMA) or time division (TDMA)), the type of baseband signal (multiplex or single channel), the type of modulation of the carrier by the baseband signal (analogue frequency modulation or digital phase modulation).

8.5.3.1 *Analogue transmission*

With analogue transmission, frequency modulation is the most used. The characteristic parameters of frequency modulation have been defined in Chapter 3. The baseband signal may be an isolated telephone channel (SCPC), a frequency division multiplex of telephone channels or a television signal. The modulation and demodulation sub-systems used must have good linearity and constancy of group propagation delay over a frequency band which can be as wide as that of the satellite channel for a high capacity transmission in which the modulated carrier occupies all of the transponder.

On the transmission side, the modulators used are based on an LC oscillator where the capacitor is a diode whose capacitance varies as a function of the applied voltage (the modulating signal). Automatic frequency control (AFC) causes the mean value of the intermediate frequency to follow the specified value.

On reception, the frequency demodulator produces a voltage proportional to the difference between the instantaneous frequency and the reference intermediate frequency. Conventional demodulator structures using a discriminator are often used. However, these techniques suffer from a high demodulation threshold (10 to 12 dB); that is, it is necessary for the modulated carrier-to-noise power ratio (C/N) at the demodulator input to be greater than the threshold in order to ensure correct operation (the signal-to-noise ratio at the demodulator output is proportional to that at the input). Below the threshold, the signal-to-noise ratio at the output is no longer proportional to that at the input and degrades rapidly. Demodulators with an improved threshold permit operation with a lower modulated carrier-to-noise power ratio (C/N), of the order of 6 to 9 dB. The principle of improved threshold demodulators is to limit the noise bandwidth to the instantaneous spectrum of the carrier. Several techniques are possible such as negative feedback of frequency, controlled filters and phase-locked loops [BEE-85].

8.5.3.2 *Digital transmission*

With digital transmission, phase modulation with two or four states (BPSK or QPSK) is used. The principle of these modulation types has been presented in Section 3.7.7. To obtain a modulated carrier with two phase states, the intermediate frequency signal from a quartz crystal oscillator is multiplied by an NRZ format bit stream. The multiplier is realised using a double balanced mixer. Band-pass filtering limits the spectrum of the modulated carrier. Four state phase modulation is obtained by combining two signals modulated with two phase states in quadrature [FEH-81].

The architecture of the demodulator is similar to that of the modulator. With coherent demodulation the modulated carrier is multiplied by an unmodulated carrier which is generated locally. Carrier recovery is achieved by passing the (modulated) received carrier through a non-linear circuit which produces components at frequencies in the spectrum which are multiples of that of the carrier, then by filtering one of these components and frequency dividing.

Frequency division introduces a phase ambiguity which must be resolved for correct detection of the signal [LIN-71], [BIC-86]. Filtering of the spectral component is performed either by a phase-locked loop or a passive filter. The latter approach can be preferable in the case of transmission using time division multiple access (TDMA) where the equipment operates in bursts. Passive filtering enables lower carrier acquisition times to be achieved.

8.5.4 Additional functions

A number of additional functions are realised before modulation (on the transmission side) and after demodulation (on the receiving side) in accordance with the type of signal and the nature of the transmission.

8.5.4.1 *Analogue transmission*

With frequency modulation, pre- and de-emphasis of the baseband signal permits the quality of the link to be improved (see Section 3.6.1). For multiplexed transmission, an energy dispersion device (see Section 3.6.11) and special devices to permit monitoring of transmission quality are also provided.

On transmission, the pre-emphasis filter increases the amplitude of the high frequency components of the signal to be transmitted. On reception, the de-emphasis filter, with a transfer function inverse to that used on transmission, attenuates the high frequency noise components and re-establishes the initial amplitude/frequency distribution of the signal.

Addition at baseband of a low frequency triangular signal to the composite signal which constitutes the multiplex of telephone channels causes spreading of the carrier spectrum. This signal is activated when the effective level of multiplexing becomes too low for the density specification to be satisfied.

Also in the case of a multiplexed transmission, addition of a 60 kHz pilot tone on the transmission side permits level monitoring on reception. Furthermore, a band-stop filter at the transmitter has the purpose of forming a window above the highest frequency of the multiplex which is empty of all signals. This permits measurement at the receiver of the noise contributed by the link (out of band noise (OBN)).

8.5.4.2 *Digital transmission*

As with analogue transmission, energy dispersion avoids the appearance of discrete frequencies in the spectrum of the modulated signal. The digital form of the signal also allows incorporation of error correcting coding if necessary.

Energy dispersion is realised by scrambling the bit stream to be transmitted before modulation (see Section 3.7.10). Scrambling is performed by modulo 2 addition of the bit stream and a pseudo-random sequence which is generated by a set of shift registers with appropriate feedback. On reception, the scrambled sequence recovered by the demodulator (containing erroneous bits) is combined with the same pseudo-random sequence which is generated locally and appropriately synchronised.

Error correcting coding involves the introduction of redundant bits into the information bit stream (see Section 3.7.9). Two types of coding are used—block coding and convolutional coding. With block coding, the coder associates r bits of redundancy with each block of n information bits; each block is coded independently of the others. The code bits are generated by linear combination of the information bits of the corresponding block. Cyclic codes, particularly the codes of Reed–Solomon and BCH (Bose, Chaudhari and Hocquenghem) for which every code word is a multiple of a generating polynomial, are the most used. For a convolutional code, $(n + r)$ bits are generated by the coder from the $(N - 1)$ preceding packets of n bits of information; the product $N(n + r)$ defines the constraint length of the code. The coder consists of shift registers and adders of the 'exclusive or' type.

Various possibilities are available for decoding block and convolutional codes. With block cyclic codes, one of the conventional methods uses the calculation and processing of syndromes resulting from division of the received block by the generating polynomial; this is zero if the transmission is error free. For convolutional codes, the best performance is obtained with the Viterbi decoding algorithm.

8.5.5 Time division multiple access terminals

When time division multiple access is used, the TDMA terminal is situated between the terrestrial network and the radio-frequency equipment of the earth station. In the direction of transmission, the terminal receives the baseband signals to be transmitted in continuous mode and supplies packets of information (stored in buffer memory) to the radio-frequency equipment which transmits these packets in bursts at the instants which correspond to their assignment of bursts in the frame. In the receiving direction, the terminal accepts the modulated carrier in bursts and reconstitutes a continuous bit stream. The common TDMA terminal equipment (CTTE) is divided into two operational sub-systems—the intermediate frequency sub-system (IFSS) and the common logic equipment (CLE).

8.5.5.1 *The intermediate frequency sub-system*

This sub-system provides the following functions:

—phase modulation, usually with four states, of the carrier transmitted to the satellite by the digital data in packets,
—demodulation, usually coherent, of the bursts received from the satellite and recovery of the binary data,
—transponder hopping by directing the packets from the modulator to different converters (on the transmitting side) and by multiplexing into the demodulator of bursts received from different converters (on the receiving side).

The functions of modulation and demodulation have been described above. Operation in burst mode involves a number of particular difficulties as follows.

On reception, the bursts received in succession are from different earth stations and thus have different phases and amplitudes. The use of a passive filter (with automatic control of the centre frequency of the filter which corrects slow drifts of the input frequency) resolves the problem of rapid phase recovery at the start of each burst. Automatic phase control compensates, on demodulation, for the rapid variation of frequency between consecutive bursts. Automatic gain control with a rapid response time (less than a microsecond) permits burst-to-burst amplitude variations to be compensated.

On transmission, when bursts are not being transmitted, the intermediate frequency channel output level should be sufficiently low to avoid interference (a typical rejection ratio is greater than 60 dB).

The transponder hopping technique involves selection of a particular satellite channel through which the burst concerned passes according to its destination (see Section 5.2). This selection is made by modifying the frequency of the transmitted carrier; each channel corresponds to a different frequency. The station is thus provided with n channels at intermediate frequency corresponding to n different converters which can produce n different radio-frequency carriers. On transmission, the frequency switching sub-system routes the bursts from the modulator to the n intermediate frequency channels. On reception, it provides multiplexing of the received bursts from the n intermediate frequency channels in order to send them to the demodulator.

8.5.5.2 *Common logic equipment*

An example block diagram of a common logic equipment sub-system is shown in Figure 8.37 [BAR-85].

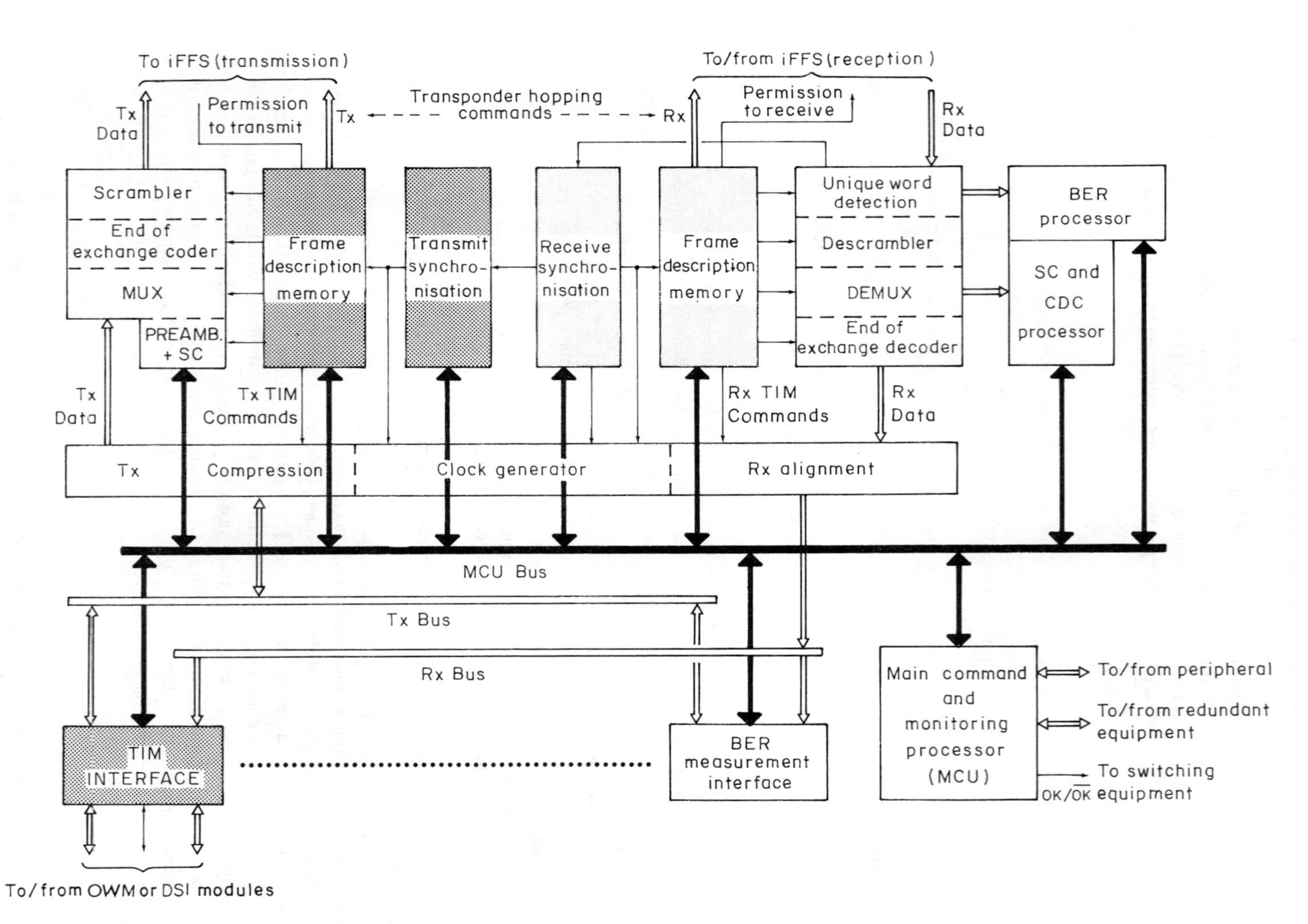

Figure 8.37 Block diagram of a common logic equipment (CLE) sub-system of a TDMA terminal.

The sub-system consists of various modules as follows:

—The receiver controller performs alignment of the data at symbol level using signals from the TDMA terminal clock, recognition of unique words and resolution of ambiguity in data demodulation, synchronisation of the buffers containing packetised data to the received burst time plan, descrambling and error correction decoding, demultiplexing of service channels (SC) and control delay channels (CDC).
—The timebase generates the terminal clock. This module includes a voltage controlled quartz crystal oscillator which synchronises itself to the reference packet clock. Other time references are obtained by division of the main clock.
—The transmission controller provides functions similar to those of the receiver controller: synchronisation of the buffers containing packetised data to the transmit burst time plan, generation of the preamble (header), data multiplexing, application of error correcting coding and scrambling.
—The main processor provides the following functions: acquisition and synchronisation of the network, management and processing of burst time plans, operator dialogue and possibly: automatic test and diagnostics, management of redundancy and assistance with maintenance.
—The auxiliary processor extracts the messages from the service channels (SC) and the control delay channel (CDC) which have been demultiplexed on reception and routes them to the main processor for execution. It also participates in the acquisition and synchronisation procedures.

8.6 THE NETWORK INTERFACE SUB-SYSTEM

This sub-system is the interface between baseband signals produced by, or destined for, the communication common equipment and baseband signals in the terrestrial network format. The main functions are multiplexing (and demultiplexing) of telephone channels (which may include digital speech concentration and channel multiplication), suppression (or cancellation) of echoes and various functions particular to single channel transmission (SCPC).

8.6.1 Multiplexing and demultiplexing

Even if the telephone channels on the terrestrial network are already multiplexed, it is almost always necessary to rearrange the distribution of telephone channels as the multiplexing standards used on the terrestrial network and on satellite links are slightly different. Furthermore, the telephone channels arriving at the earth station on the terrestrial link do not all have the same destination. Telephone channels having the same destination are grouped into a single multiplex which modulates a carrier and is transmitted to this particular destination. Similarly, on reception only one section of the telephone channels present in a received multiplex relate to the terrestrial network connected to the earth station concerned. These channels (or groups or supergroups of channels) are separated from the others and combined with those from other multiplexes received on the various multidestination carriers to form a terrestrial standard multiplex destined for the switching exchange connected to the earth station concerned.

8.6.1.1 *Frequency division multiplexing*

With analogue transmission, the arrangement of multiplexed telephone channels according to CCITT recommendations G. 322 and G. 423 is as follows. A primary CCITT group consists of 12 telephone channels translated in frequency with an interval of 4 kHz between 60 and 108 kHz. A CCITT supergroup is obtained by translating five primary groups into the 312 to 552 kHz band.

For transmission by satellite (INTELSAT), the arrangement of channels is such that the first group of 12 channels (group A) is translated into the 12 to 60 kHz band. A 24-channel multiplex (12 to 108 kHz) is obtained by adding the fifth group of a translated and inverted supergroup to group A. A 36-channel multiplex (12 to 156 kHz) is obtained by adding the fifth and fourth groups of a translated and inverted supergroup to group A. This continues in steps of 12 channels up to 72 channels (12 to 300 kHz). A multiplex of 96 channels associates group A, the first supergroup and two groups of the following supergroup (12 to 408 kHz). For a greater number of channels, the arrangements associate group A with an increasing number of supergroups (see Table 3.1). The frequency band between 0 and 12 kHz is used for transmission of telephone and telex service channels.

8.6.1.2 *Time division multiplexing*

Digital transmission multiplexing standards have been presented in Section 3.7.2. There are two types of hierarchy and these have a base of either 24 telephone channels (1.544 Mbit/s) or 30 telephone channels (CEPT, 2.048 Mbit/s). These hierarchies are used in terrestrial networks and on satellite links.

The multiplexing equipment in the earth stations combines bit streams from various origins which have the same destination. Problems arise due to lack of synchronisation of the earth station clock and the bit stream clocks at diverse origins. This lack of synchronism is due to instability and drift of the oscillators and variations of propagation times on the links.

When the bit strems are not synchronous, it is necessary to use buffer memories and possibly to add stuffing bits in order to obtain exactly the same bit rates before multiplexing. The operation is facilitated when the bit streams are synchronous (synchronised clocks) or plesiochronous (clocks which are not synchronised, but whose precision is better than 10^{-10} or 10^{-11}). In the latter case, the technique of frame slipping permits periodic readjustment of the bit streams (CCITT, Recommendation G. 811) and there is no need for stuffing bits (see Section 3.7.4).

8.6.2 Digital speech interpolation (DSI)

Digital speech interpolation exploits the silences in a telephone channel to insert bits representing the active speech on another channel into these silences. In this way a number m of telephone channels from the terrestrial network can be carried in a multiplex with a capacity of n digital telephone channels where $m > n$ [CAM-76], [KEP-89].

Figure 8.38 illustrates this principle. Speech detectors are necessary to identify the silences in the terrestrial network channels. The DSI equipment assigns one channel of the satellite link (the bearer channel) to each active terrestrial channel and a connection network performs

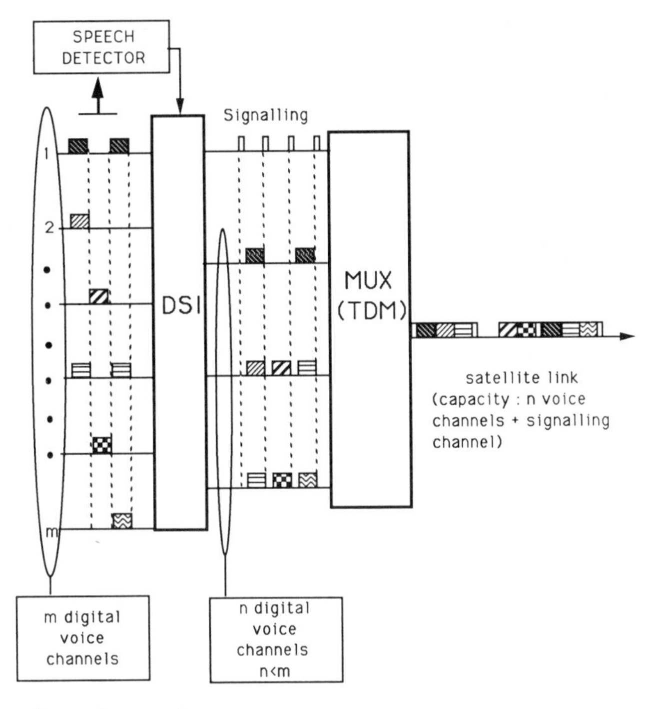

Figure 8.38 Digital speech interpolation (DSI).

the corresponding branching operations. The assignment is arbitrary and can change from one talk spurt to another on the same terrestrial channel. The assignment information is transmitted on a signalling channel.

On reception, dedicated equipment establishes the connections between the bearer channels and the terrestrial channels, in accordance with the signalling messages received, in such a way as to route the bits to their proper destination.

The performance of a digital speech interpolation system is measured by the gain m/n. This gain is greatest when the number of channels to be concentrated is large. If the number of channels exceeds 60, the gain obtained can reach 2.5. Limitation of the gain arises from the following:

—Degradation of quality associated with clipping when a burst of speech cannot be routed because at that instant all the bearer channels of the satellite link are occupied.
—Degradation associated with temporary overloading of the signalling channel.
—A large number of terrestrial telephone channels which have an activity rate greater than that of speech (when used for data transmission, for example).

The performance of DSI equipment can be improved by using a 'bit stealing' technique; the least significant bit in the quantised speech sample on seven channels of the terrestrial network is not transmitted. For these seven channels, quantisation is performed to seven bits instead of eight; this momentarily increases the quantisation noise, but the quality degradation on the seven channels is less than that which would be observed on the clipped channel.

8.6.3 Digital circuit multiplication equipment (DCME)

Digital circuit multiplication equipment (DCME) permits an improvement on digital speech interpolation in the commercial exploitation of satellite telephone channels [INTELSAT Specification IESS 501], [ABO-89], [BAR-89], [CAM-89], [FOR-89], [YAT-89].

The system combines two techniques for multiplying the number of telephone channels which can be transmitted on the same satellite channel; these are digital speech interpolation (DSI) and adaptive differential pulse code modulation (ADPCM).

In comparison with the technique of digital speech interpolation described in the previous section, a further factor of 2 is obtained by means of adaptive differential coding (CCITT Rec. G. 721/G. 723). Four bits are used to code samples of the voice signal instead of eight bits. Hence a given number of bearer channels on the satellite link can convey a greater number of terrestrial channels (of the order of five times more).

Telephone channels carrying data (voice band data (VBD)) are processed separately. They are identified in the speech concentration equipment by detecting the presence of a 2100 Hz pilot tone which serves to inhibit echo suppression (CCITT Rec. G. 164) and by analysis of the amplitude and frequency spectrum. A specially optimised form of differential adaptive coding is then applied. Coding depends on the equipment and may be 5 bits per sample (transmission rate 40 kbit/s, CCITT Rec. G. 722) or even 4 bits (transmission rate 32 kbit/s) for bit rates in the telephone channel up to 9.6 kbit/s (V29 modem). Digital channels at 64 kbit/s (CCITT No. 6 and 7 signalling) can also be transmitted in limited numbers. These channels are transmitted directly without speech concentration or adaptive coding (clear channels).

8.6.3.1 DCME architecture

An example of the architecture of circuit multiplication equipment is shown in Figure 8.39. This architecture includes the following.

Input data link interface (DLI). This interface equipment receives signals at 1.544 Mbit/s or 2.048 Mbit/s and transforms them into NRZ format bits at 2.048 Mbit/s; it also provides clock recovery and frame synchronisation in plesiochronous mode.

Time slot interchange (TSI). This equipment rearranges the bits when signals at 1.544 Mbit/s are present at the input of the data interface. When converting 1.544 Mbit/s to 2.048 Mbit/s, the DLI introduces stuffing bits and only 24 bits out of 31 correspond to information bits. The TSI then groups ten streams at 2.048 MHz (initially at 1.544 Mbit/s) in order to obtain eight streams at 2.048 Mbit/s which now contain only information bits and the control bits which are also generated by the equipment.

Digital speech interpolation (DSI): The digital speech interpolation equipment consists of speech detectors associated with a noise level monitor, possibly a delay line to anticipate speech detection, a 2100 Hz pilot tone detector and a device to distinguish between speech and data signals.

The equipment considered permits combination of typically 150 terrestrial channels into 62 bearer channels which are arranged in three bit streams at 2.048 Mbit/s. However, at a given point in the traffic, the number of bursts of conversation can exceed 62 and the equipment can produce up to 96 simultaneous samples.

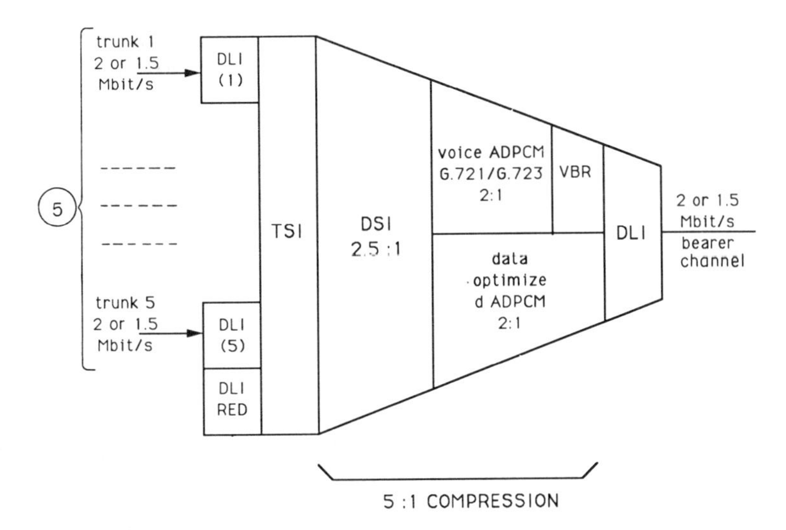

Figure 8.39 Block diagram of digital circuit multiplication equipment (DCME).

Adaptive pulse code modulation (ADPCM). These differential adaptive encoders use an appropriate coding algorithm which meets CCITT Recommendations G. 721 and G. 723. The encoders regroup the three bit streams at 2.048 MHz which correspond to the 62 bearer channels into a bit stream at 2.048 Mbit/s. Under normal conditions, PCM speech signal samples of μ or A law are encoded in four bits.

When the DSI equipment delivers more than 62 simultaneous samples, additional encoders are activated in order to create temporary bearer channels. All the encoders must then share the output bit rate which cannot exceed the maximum value of 2.048 Mbit/s. The encoders of certain channels (other than those carrying data) selected randomly from one sample to the next then operate on three bits instead of four in accordance with a variable bit rate (VBR) procedure.

Data signals transmitted within the bandwidth of a telephone channel (up to 9.6 kbit/s) are processed in specially optimised encoders and coded as speech signals at 32 kbit/s.

Output data link interface. This equipment realises the interface between the output of the ADPCM encoders at 2.048 Mbit/s and a standard 2.048 Mbit/s (or 1.544 Mbit/s) PCM link.

8.6.3.2 *Overall multiplication factor*

In the case where the terrestrial circuits feeding the equipment are distributed over a wide geographical area and, in particular, include different time zones, traffic peaks are spread over a period of time and the probability of having a large number of channels active at the same time is low. It is thus possible to increase the number of terrestrial channels connected to the equipment, for example up to 240 channels, which are transmitted in the available 62 bearer channels. The gain of the speech concentration system thus becomes four by taking advantage of the dispersion of the periods of activity among the 240 channels; the overall multiplication factor under these conditions may reach a value of 8.

Nominal operation of a DCME system uses additional equipment at both ends of a point-to-point satellite link. Modification to the equipment for transmissions of the point-to-multipoint type are possible. There are two types of operation—'multidestination operation' and 'multi-clique operation'.

8.6.3.3 Multidestination operation

In multidestination mode, the DCME installed in the earth station is capable of the following:

—On the receiving side, extraction of the telephone channels relating to the earth station from bearer channels of the incoming 2.048 Mbit/s streams from other stations.
—On the transmission side, concentration of terrestrial channels on to bearer channels by indicating the destination of each sample so that the receiving equipment at each destination can identify the samples which relate to it and reconstruct the corresponding telephone channels.

Streams at 2.048 Mbit/s conveying the bearer channels are either routed directly on continuous intermediate data rate (IDR) carriers or within bursts of a time division multiple access (TDMA) frame. Telephone channels on the terrestrial link to the switching centre are not concentrated and the bit rate on this link is thus greater than the rate transmitted on satellite links. For example, a terrestrial link at 8.448 Mbit/s will be necessary to carry traffic exchanged on a 2.048 Mbit/s satellite link with four other locations.

8.6.3.4 Multi-clique operation

In multi-clique operation, the samples transmitted on the bearer channels are arranged in several groups (sample groups) which are combined into the frame forming the bit stream at 2.048 Mbit/s. Each group is associated with a particular destination and contains talk spurt assignment information from the DSI process. The INTELSAT system uses two groups of samples per frame.

This approach enables the channel multiplication equipment (DCME) to be located in the switching centre associated with the earth station. Hence the terrestrial link between the switching centre and the earth station also benefits from the concentration gain which may, however, be smaller due to the reduced number of channels per group.

The following operations are performed at the earth station:

—On the transmission side, the bearer channel stream (which may have been multiplexed with other similar streams at the switching centre) directly modulates the multidestination carrier (IDR carrier) or is transmitted in a sub-burst of the TDMA frame.
—On the receiving side, the bit stream received from a particular station contains several

groups (two in the case of INTELSAT) of which only one is destinated for the station concerned. Only the corresponding bits are accepted and multiplexed with those from another station to form the bit stream which is routed to the switching centre. The operation is simple to realise since the destination of samples is known from their position in the frame. This operation is realised by dedicated equipment (clique sorting facility (CSF)) in the case of IDR transmissions or by terrestrial network interface equipment (digital non-interpolated (DNI) and direct digital interface (DDI)) in the case of TDMA operation. It is not necessary to deconcentrate bearer channels at station level and this permits advantage to be taken of the concentration gain on the terrestrial link.

8.6.4 Echo suppression and cancellation

A link between two earth stations is characterised by a long propagation delay particularly if a geostationary satellite link is involved for which the propagation delay can reach 270 ms (see Section 7.2.5.4). Echoes are generated as a consequence of the poor impedance matching of the end user's two-wire line at the interconnection of a four-wire link and a two-wire link in the switching centre (see Section 3.3.5 and Figure 3.8).

In order to avoid the occurrence of an echo, that is retransmission towards the first user

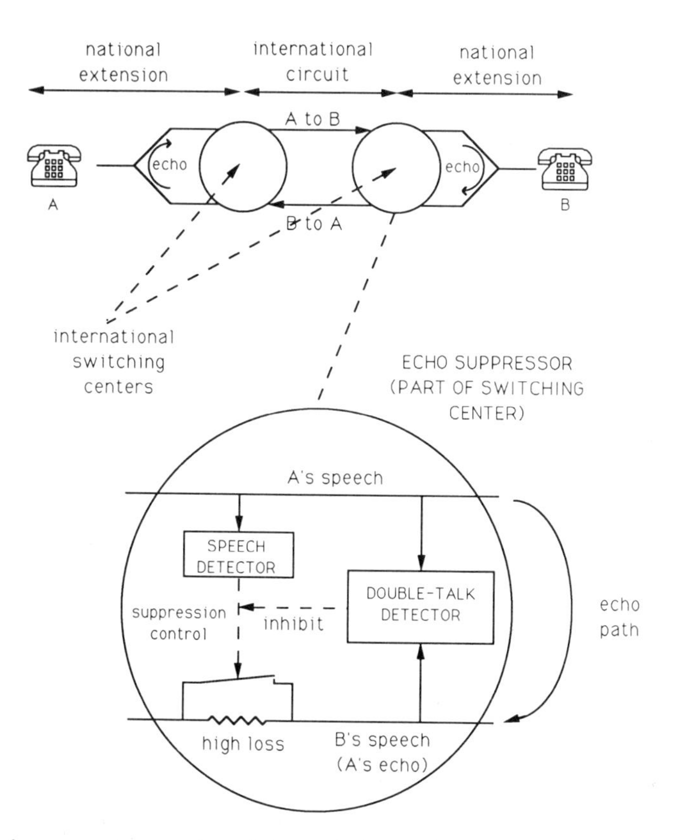

Figure 8.40 Echo suppression on a long-distance telephone link.

of his voice signal on the return channel, an echo control device is installed as shown in Figure 8.40. The simplest procedure consists of opening the return channel when the outward channel is active. In this case, the second user cannot make himself heard to the first user as long as the latter continues to speak. A compromise is obtained by inserting an impedance in the return channel. The value is chosen so as to attenuate the echo, but not the direct signal, sufficiently so that the first user can hear the second when the latter indicates a desire to take over the conversation. This technique is known as echo suppression [SUY-76].

A more sophisticated technique is increasingly being used. This technique consists of generating a replica of the echo and injecting it in phase opposition to the real echo on the return channel; this leads to cancellation of the echo [DEM-77]. The echo cancelling technique requires a large calculation and signal processing facility for each channel. It has been made possible only by progress in large-scale integration which has provided substantial computing power in a small volume.

8.6.5 Equipment specific to SCPC transmission

With SCPC, dedicated equipment may provide activation of the carrier by speech and companding of the signal.

The use of a speech detector permits activation of the transmitted carriers only when speech is present on the channel concerned. This permits a reduction of the number of carriers passing through the satellite transponder at a given instant and hence a higher EIRP per carrier and less intermodulation noise.

A syllabic compressor (CCITT Rec. G.162) on each transmission channel and an associated expander on the receiving side permit a subjective improvement of transmission quality (see Section 3.6). The compressor introduces a gain which varies as a function of the signal amplitude. Above a reference level, the gain decreases as the amplitude increases; below the reference level the direction of gain variation is reversed. The response of the expander to the amplitude of the received signal is the inverse of that of the compressor.

8.7 MONITORING AND CONTROL; AUXILIARY EQUIPMENT

Monitoring of correct operation and control of the earth station are the purpose of a dedicated sub-system. Several specifications are given in this section concerning the earth station electrical power supply which is included in the auxiliary equipment.

8.7.1 Monitoring alarms and control (MAC)

The monitoring, alarm and control equipment of the earth station has the following purposes:

- To provide the operators with the necessary information for monitoring and controlling the station (this includes measured parameters, equipment in service, switch positions etc.) and managing the traffic.
- To initiate alarms in case of incorrect operation or an incident affecting the main station equipment or the link performance and permit identification of the equipment which is involved.

—To permit control of the station equipment; this includes bringing equipment into service, adjustment of parameters, switching of redundant equipment and so on.

Monitoring and control functions can be provided locally, in a centralised manner, or under the control of a computer. Locally, the functions are provided on the equipment itself by means of warning lights, indicators and control push-buttons. With centralised control, the various functions are combined at one control centre. All the monitored parameters are available at this centre and are presented to the operator by means of various display devices (such as screens, indicators and warning lamps). Control of a variety of equipment is possible from the console. This control and monitoring centre is situated at some distance from the equipment.

The next step is to transfer all the monitoring operations to a computer which records the parameters, selects the most important parameters for display on a standard screen, detects abnormal situations, prepares special commands and executes them either automatically without human intervention (such as bringing replicated equipment into operation) or after approval by the operator.

With centralised or computer-aided management, it is possible to have a station without permanent staff; monitoring and control information can be routed to a distant common network control centre by means of dedicated terrestrial lines or service channels on the satellite links.

8.7.2 Electrical power

Electrical energy is necessary for operation of the earth station equipment. This energy is obtained in most cases from the general energy distribution network. According to the specified criteria of availability, it is often necessary to take precautions against interruption of this source of energy. Three types of energy are generally available to an earth station as follows:

—Uninterruptible: this feeds all equipment which must operate without interruption such as radio-frequency communication equipment, emergency lighting etc.
—Stand-by: this feeds devices which tolerate supply interruptions which can last for several minutes (such as antenna servos etc.).
—Without stand-by: this supplies non-critical circuits which can tolerate interruptions of several hours (such as air conditioning, antenna de-icing etc.). It is generally possible, in the case of prolonged power failures, to be able to supply, on demand, some circuits which are normally without stand-by from the stand-by supply.

Energy without stand-by is provided by the general distribution network (the mains). This also applies to stand-by energy when the mains is available. The uninterruptible supply is realised using batteries which continuously supply the equipment concerned either directly as DC or by way of a converter if an alternating current is necessary. The general energy distribution network provides float charging of the batteries by means of rectifiers. In case of power failure an electrical generator is started up automatically. This generator feeds the equipment connected to the stand-by energy circuits. It also replaces the mains to ensure that the batteries remain charged; the rectifier supply circuits are automatically disconnected from the mains when it fails. At the end of the failure, the generator stops and the circuits using stand-by energy are switched to the mains.

8.8 CONCLUSION

From the start of the satellite communication era, earth stations have developed continuously although the general architecture of the stations has remained unchanged.

This development has been evidenced by a reduction in the size of earth stations. The diameter of antennas, initially more than 30 m, can now in some cases be less than one metre. This is due to the increase of equivalent isotropic radiated power (EIRP) of communication satellites in association with the use of high performance transmission techniques. This reduction is also evident in the size of the equipment used in the stations and has been made possible by the use of digital techniques and large-scale integration of components.

Use of these technologies has also enabled the processing capacity and complexity of equipment to be greatly increased. This has resulted in an increase in performance. In this way the use of sophisticated transmission techniques such as time division multiple access, spread spectrum transmission, error correcting coding and so on has been made possible. Much greater ease of operation and maintenance has resulted from the use of these technologies for equipment design. For example, at the frequency translation stage, programmable frequency synthesisers permit rapid carrier frequency selection and high stability of the displayed frequency. Monitoring under computer control ensures continuous checking of the operation of diverse systems and rapid detection of faulty equipment and even its replacement by replicated equipment.

Simultaneously, the appearance of new systems has permitted better exploitation of the particular characteristics of satellites such as broadcasting capacities and the possibility of access to widespread users without additional cost. These systems open up the possibility of numerous telecommunication services in areas as varied as business communication, rural telecommunication, video data distribution, data broadcasting, interactive transfers and communication with mobiles.

Many of these systems make use of small earth stations which are installed on the user's premises and provide direct telephone links (rural communication), data links (very small aperture terminals (VSAT) on private networks) and video reception. For communication with mobiles, it is of course necessary that the earth stations can be installed without too many constraints on the motion of the mobile concerned.

For these systems to be developed, the cost of earth stations must be reduced as much as possible. The existence of small earth stations of moderate cost opens the door to the general use of satellite communication.

REFERENCES

[ABO-89] A. Aboaf, R. Bailly (1989) Advanced techniques for circuits multiplication: Celtic 3G, *ICDSC 8th International Conference on Digital Satellite Communications, Pointe à Pître*, pp. 813–822.

[AND-63] J.V. Anders et al. (1963) The precision tracker, *Bell System Technical Journal*, pp.1309–1356.

[ANS-89] H. Ansorge (1989) 2.4 m offset dual reflector for a transportable 20/30 GHz earth station, *1st European Conference on Satellite Communications*, Munich, pp. 165–176, Nov.

[ARN-83] J. Arnbak, M. Herben, R. Van Spaendonk (1983) Improved orbit utilization using auxiliary feeds in existing earth terminals, *Space Communication and Broadcasting*, **1**, pp. 405–416.

[BAR-72] C.S. Barham (1972) Review of design and performance of microwave multiplexers, *Marconi Review*, **35**, No. 184, pp. 1–23.

[BAR-85] C. Bareyt, J. Salomon, C. Sergent, D. Vautier (1985) Accès multiple a repartition dans le temps a 120 Mbit/s: un nouveau terminal pour les systèmes de télécommunications par satellite, *Commutation et Transmission*, No. 2, pp. 5–22.

[BAR-89] C. Bareyt, A. Karas, J. Salomon (1989) Earth station configuration for digital circuit multiplication equipment, *ICDSC 8th International Conference on Digital Satellite Communications, Pointe à Pître*, pp. 543–549.

[BEE-85] Beech Moor et al. (1985) Threshold extension techniques, *IBA Report 130/84*, July.

[BIC-86] J.C. Bic, D. Duponteil, J.C. Imbeaux (1986) *Eléments de Communications Numériques* (2 tomes), Dunod.

[BOI-83] L. Boithias (1983) *Propagation des ondes radio dans environment terrestre*, Dunod.

[BRE-74] C. Bremenson, J. Jaubert (1974) Réseau linéariseur pour tube à ondes progressives, *Revue technique THOMSON-CSF*, **6**, No. 2, pp. 529–548.

[BRI-90] R. Briskman, D. Keyser (1990) Mobile RDSS terminal performance, *AIAA 13th International Communication Satellite Systems Conference, Los Angeles*, pp. 682–683, March.

[BRO-78] D.L. Brown, G.E. Swan (1978) A study of aerosat payload configurations, *Int. Conf. Maritime and Aeronautical Satellite Communication and Navigation*, pp. 109–115.

[BRO-90] J. Broughton *et al.* (1990) An aircraft earth station for general aviation, *International Mobile Satellite Conference IMSC '90, Ottawa*, pp. 156, June.

[BRU-89] A. Brunner, H. Thiere, New symmetrical low sidelobe antennas for German earth stations, *1st European Conference on Satellite Communications, Munich*, pp. 155–164, Nov.

[BUC-89] J.D. Buchs, J. Czech, M. Wassermann (1989) Comsat technology, *Space*, **2**, No. 3.

[CAH-84] D. Cahana, J.R. Potukuchi, R.G. Marshaler, Linearized transponder technology for satellite communications, *Comsat Technical Review*, **15**, No. 2A, pp. 277–339.

[CAM-76] S.J. Campanella (1976) Digital speech interpolation, *Comsat Technical Review*, **6**, No. 1, pp. 127–158, Spring.

[CAM-89] J.S. Campanella, J.H. Rieser (1989) Operating aspects of new LRE/DSI DCM, *ICDSC 8th International Conference on Digital Satellite Communications, Pointe à Pître*, pp. 803–810.

[CCIR-REC 524] REC. 524 (1986) Maximum permissible levels of off-axis EIRP density from earth stations in the fixed satellite service transmitting in the 6 GHz frequency band, CCIR Vol. IV–1, Dubrovnik, pp. 242–2.

[COM-88] P.F. Combes (1988) *Transmission en espace libre et sur les lignes*, Dunod.

[DAN-85] R. Dang, B.K. Watson, Davis (1985) Electronic tracking systems for satellite ground stations, *15th European Microwave Conference, Paris*, pp. 681–687.

[DEM-77] N. Demytko, K. English (1977) Echo cancellation on time variant circuits, *Proceedings of the IEEE*, **55**, No. 3, pp. 444–453.

[DOR-79] C. Dorian (1979) The marisat system, *Trends in Communication Satellite*, Pergamon Press.

[DUR-87] E.J. Durwen (1987) *Determination of sun interference periods for geostationary satellite communication links*, pp. 183–195, Elsevier Science.

[EDW-83] D.J. Edwards, P.M. Terrell (1983) The smoothed step-track antenna controller, *International Journal of Satellite Communications*, **1**, pp. 133–139.

[EST-89] P. Estabrook et al. (1989) A 20/30 GHz personal access satellite system design, *ICC 89*, pp. 7.4.1–7.4.7.

[FEH-81] K. Feher (1981) *Digital Communications—Microwave Applications*, Prentice-Hall.

[FEH-83] K. Feher (1983) *Digital Communications—Earth Station Engineering*, Prentice-Hall.

[FOR-89] G. Forcina, W.S. Oei, T. Oishi, J. Phiel (1989) Intelsat digital circuit multiplication equipment, *ICDSC 8th International Conference on Digital Satellite Communications, Pointe à Pître*, pp. 795–803.

[GAR-84]H. Garcia (1984) Geometric aspects of solar disruption in satellite communications, *IEEE Transactions on Broadcasting*, **BC-30**, No. 2, pp. 44–49, June.

[GHO-88] A. Ghorbani, N.J. McEwan (1988) Propagation theory in adaptative cancellation of cross-polarization, *International Journal of Satellite Communications*, **6**, No. 1, pp. 25–28.

[GIL-86] A.S. Gilmour Jr (1986) *Microwave Tubes*, Artech House.

[GOR-80] W.D. Gorton (1980) Transportable earth stations for multiple applications, *Wescom*, pp.9.3.1–9.3.7.

[HAR-89] E. Hartinger, W. Rebhan (1989) Transportable elliptical 11/14 GHz ground station antenna with improved sidelobe in the satellite orbit plane, *1st European Conference on Satellite Communications, Munich*, pp.145–154, Nov.

[HAS-89] Y. Hase et al. (1989) ETS-V/EMSS Experiments on aeronautical communications, *IEEE International Conference on Communications, Boston*, pp. 7.1.1–7.16, June.

[HAW-88] G.J. Hawkins et al. (1988) Tracking systems for satellite communications, *IEE Proceedings*, **135**, No. 5, pp. 393–407.

[HAW-90] G.J. Hawkins et al. (1990) Standard-M mobile satellite terminal employing electronic beam squint tracking, *International Mobile Satellite Conference IMSC'90, Ottawa*, pp. 535–539, June.

[HIG-88] T. Higuchi, T. Shinohara (1988) Experiment of Inmarsat standard-C system, *4th International Conference on Satellite Systems for Mobile Communication and Navigation*, pp. 47–51, October.

[HO-61] H.C. Ho (1961) On the determination of the disk temperature and the flux density of a radio source using high gain antennas, *IRE Transactions on Antennas and Propagation*, pp. 500–510.

[IWA-85] M. Iwakumi et al. (1985) A20 GHz Peltier-cooled low-noise HEMT amplifier, *IEEE MTT-S International Microwave Symposium, St Louis*, June.

[JOH-84] R.C. Johnson, H. Jasik (1984) *Antenna Engineering Handbook*, McGraw-Hill.

[KAD-89] N. Kadowaki (1989) ETS-V/EMSS experiments on message communications with hand-held terminal, *ICC 89*, pp. 7.31–7.35.

[KEL-84] T.M. Kelley (1984) Leased services on the Intelsat system: domestic service and international television, *International Journal of Satellite Communications*, **2**, No. 1, pp. 29–40.

[KEO-88] K. Keough (1988) Techniques and capabilities of the new generation antenna control unit, *AIAA 12th International Communication Satellite Systems Conference, Arlington*, Paper 88–0792, March.

[KEP-89] W.R. Kepley, A. Kwan (1989) DSI development for 16 kbit/s voice systems, *ICDSC 8th International Conference on Digital Satellite Communications, Pointe à Pître*, pp. 551–559.

[KRE-80] R.W. Kreutel, J.B. Potts (1980) The multiple-beam Torus earth stations antennas, *International Conference on Communications ICC 80, Seattle*, pp. 25.4.1–25.4.3, June.

[KUD-87] C.M. Kudsia (1987) High power combiners for satellite earth terminals, *Canadian Satellite User Conference, Ottawa*, pp. 214–220, May 25–28.

[KUN-65] M.R. Kundu (1965) *Solar Radio Astronomy*, Wiley, New York.

[LAY-90] N. Lay et al. (1990) Description and performance of a digital mobile satellite terminal, *International Mobile Satellite Conference IMSC '90, Ottawa*, pp. 272–278, June.

[LHO-89] J.Y. L'Honnen (1989) A new global approach to earth station power amplifiers, *1st European Conference on Satellite Communications, Munich*, pp. 635–644, Nov.

[LIN-71] W.C. Lindsey, M.K. Simon (1971) Data-aided carrier tracking loops, *IEEE Transaction on Communication*, **COM 19** No. 2, pp. 157–168.

[LIN-89] K.T. Lin, L.J. Yang (1989) A sun interference prediction program, *Comsat Technical Review*, **19**, No. 2, pp. 311–332, Fall.

[LOE-83] J. Loeffler (1983) Planning for solar outages, *Satellite Communications*, pp. 38–40, April.

[LUN-70] C.W. Lundgren (1970) A satellite system for avoiding serial sun-transit outages and eclipses, *Bell Technical Journal*, pp. 1943–1957, October.

[MAR-90] M. Maritan, M. Borgford (1990) A high gain antenna system for airborne satellite communication applications, *International Mobile Satellite Conference IMSC '90, Ottawa*, pp. 150–155, June.

[MAT-83] M. Mathieu (1983) *Télécommunication par faisceau hertzien*, Dunod.

[MEU-89] F. Meuleman, A. Glavieux (1989) Mobile satellite digital communications: Prodat and standard C, *ICDSC 8th International Conference on Digital Satellite Communications, Pointe à Pître*, pp. 499–506, April.

[MIL-90] R. Milne (1990) An adaptative array antenna for mobile satellite communications, *Internatinal Mobile Satellite Conference IMSC '90, Ottawa*, pp. 529–534, June.

[MIY-82] K. Miya (1982) *Satellite Communications Technology*, KDD Tokyo.

[MOH-88] F. Mohamadi, D. Lyon, P. Murrell (1988) Effects of solar transit on Ku-band Vsat systems, *International Journal of Satellite Communications*, **6**, pp. 65–71.

[NAK-89] H. Nakamura et al. (1989) Field trial to aeronautical satellite communication system, *ICDSC 8th International Conference on Digital Satellite Communications, Pointe à Pître*, pp. 483–491.

[OHM-87] G. Ohm, M. Alberty (1987) Cryogenically cooled 20 GHz FET amplifier, *Space Communication and Broadcasting*, **5**, pp. 197–205.

[OHM-90] S. Ohmori *et al.* (1990) A phased array tracking antenna for vehicles, *International Mobile Satellite Conference IMSC '90, Ottawa*, pp. 519–522, June.

[OTA-88] S. Otani *et al.* (1988) NEXTART advanced VSAT terminal for satellite communications systems, *AIAA 12th International Communication Satellite Systems Conference, Arlington*, Paper 88-0818, March.

[RAN-89] J.F. Rancy (1989) Le sytème Intelsat, Notes de cours de l'option "Télécommunications et Systémes Aérospatiaux", *Télécom Paris*, Site de Toulouse.

[RAU-85] D.B. Rauthan, V.K. Garg (1985) Geostationary satellite signal degradation due to sun interference, *Journal of Aero. Soc. India*, **37** No. 2, pp. 137–143.

[REI-90] M. O'Reilly *et al.* (1990) LRB-2 Earth station for the ACTS program, *AIAA 13th International Communication Satellite Systems Conference, Los Angeles*, pp. 514–521, March.

[RIC-86] M. Richaria (1986) Design considerations for an earth station step-track system, *Space Communications and Broadcasting*, **4**, pp. 215–228.

[ROG-89] R. Rogard, A. Jongejans, C. Loisy (1989) Mobile communications by satellite: results of field trails conducted in Europe with the PRODAT system, *ICDSC 8th International Conference on Digital Communications, Pointe à Pître*, pp. 713–720.

[ROS-90] P. Rossiter *et al.* (1990) L-Band briefcase terminal network operation, *International Mobile Satellite Conference IMSC '90, Ottawa*, pp. 279–284, June.

[RUD-82] A.W. Rudge (1982) *The Handbook of Antenna Design*, Peter Peregrinus.

[SAA-88] T. Saam (1988) Microsat, the economic benefits of VSATs, *AIAA 12th International Communication Satellite Systems Conference, Arlington*, Paper 88–0819, March.

[SCH-89] B. Schlobohm F. Arndt (1989) Small-earth station antenna synthesized by a direct PO method, *1st European Conference on Satellite Communications, Munich*, pp. 133–144, Nov.

[SET-88] Setam. I. Ayukawa (1988) A study on the transmitting power control for earth stations, *AIAA 12th International Communication Satellite Systems Conference, Arlington*, Paper 88–0791, March.

[SHI-68] F. Shimbukuro, J.M. Tracey (1968) Brightness temperature of quiet sun at centimeter and millimeter wavelengths, *The Astrophysical Journal*, **6**, pp. 777–782, June.

[SHI-71] O. Shimbo (1971) Effects of intermodulation AM-PM conversion and additive noise in multicarrier TWT systems, *Proceedings of the IEEE*, **59**, 230–238.

[SHI-85] K. Shibata, B. Abe, H. Kawasaki (1985) Broadband HEMT and Gaas FET amplifiers for 18–26-5 GHz, *IEEE MTT-s International Microwave Symposium, St Louis*, June.

[SUT-90] C. Sutherland (1990) A satellite data terminal for land mobile use, *International Mobile Satellite Conference IMSC '90, Ottawa*, pp. 2261–266, June.

[SUY-76] Suyderhoud *et al.* (1976) Echo control in telephone communications, *National Telecom Conf.*, **1**, pp 8.1.1–8.1.5.

[SWA-81] G. Swarvas H. Suyderhoud (1981) Enhancement of FDM-FM satellite capacity by use of compandors, *COMSAT Technical Review*, **11**, No. 1, Spring.

[TEL-88] N. Teller, K. Phillips (1988) The standard communication system, *4th International Conference on Satellite Systems for Mobile Communication and Navigation, Brighton*, pp. 43–46.

[TOK-84] Y. Tokumitsu, M. Niori (1984) A 20-GHz low-noise HEMT amplifier for satellite communications, *Space Communication and Broadcasting*, **2**, pp. 71–76.

[TOM-70] N. Tom (1970) Autotracking of communication satellite by the steptrack technique, *IEE Conference Proceedings on Earth Station Technology*, pp. 121–126.

[VUO-83a] X.T. Vuong, R.J. Forsey (1983) C/N- Degradation due to sun transit in an operational communication satellite system, *Satellite Communication Conference SCC-83, Ottawa*, pp. 11.3.1–11.3.4.

[VUO-83b] X.T. Vuong, R.J. Forsey (1983) Prediction of sun transit outages in an operational communication satellite system, *IEEE Transaction Broadcasting, BC-29*, **4**, pp. 134–139.

[WAT-86] Watson, M. Hart (1986) A primary-feed for electronic tracking with circularly-polarised beacons, *Proceedings of Military Microwaves, Brighton*, pp. 261–266, June.

[WEI-80] S. Weinreb (1980) Low-noise cooled gasfet amplifiers, *IEEE Transactions on Microwave Theory and Techniques*, **MTT-28**, October.

[WHE-84] J. Whelehan (1984) Low-noise amplifiers for satellite communications, *Microwave Journal.*

[WOO-88] D. Woodring (1988) Multi-band transportable terminal, *AIAA 12th International Communication Satellite Systems Conference, Arlington*, Paper 88–0793, March.

[YAS-89] Y. Yasuda *et al.* Field experiment on digital maritime and aeronautic satellite Communication systems using ETS-V, *IEEE International Conference on Communications, ICC-89*, pp. 7.2.1–7.27, June.

[YAT-89] Y. Yatsuzuka (1989) A design of 64kbps DCME with variable rate coding and packet discarding, *ICDSC 8th International Conference on Digital Satellite Communications, Pointe à Pître*, pp. 547–551.

9 THE COMMUNICATION PAYLOAD

The communication payload provides a radio relay for links between earth stations. This chapter is devoted to a description of the payload with the emphasis on design principles, characteristic parameters and the technologies used for the equipment.

Prior to the 1990s, the payload could be considered to consist of two distinct parts with well defined interfaces—the repeater and the antennas. This dichotomy became less clear with the appearance of active antennas which closely associate the radiating elements and the amplifiers. For clarity of presentation, active antennas will be considered in the part of the chapter devoted to antennas after amplifier technology has been discussed in the sections pertaining to the repeater.

In this work, the word repeater designates the electronic equipment which performs a range of functions on the signals from the receiving antenna before sending them to the transmitting antenna. The repeater usually supports several channels which share the same frequency band.The architecture differs for conventional repeaters, multibeam satellite repeaters and regenerative repeaters. Each of these architectures is the subject of a particular sub-section. The functions and characteristic parameters of the payload are presented first.

9.1 MISSION AND CHARACTERISTICS OF THE PAYLOAD

9.1.1 Functions of the payload

The main functions of the communications payload of a satellite are as follows:

—To capture the radio signals transmitted, in a given frequency band and with a given polarisation, by the earth stations of the network concerned. These stations are situated within a particular region on the surface of the earth and are seen from the satellite within an angle of a few degrees; this angle determines the angular width of the antenna beam,
—To capture as few undesirable signals as possible; these do not conform to the specifications stated above in that they are from a different region or do not have the specified values of frequency or polarisation,
—To amplify the received signals while limiting noise and distortion as much as possible (the level of the received signal is of the order of a few tens of picowatts),
—To change the frequency of the carriers received on the uplinks into those required for the carriers transmitted on the downlinks (for example from 14 GHz to 11 GHz),

—To provide the power required in a given frequency band at the interface with the transmitting antenna (the power to be provided ranges from tens to hundreds of watts),
—To transmit radio signals in a given frequency band and with a given polarisation (which are characteristic of the downlink antenna beam) to the destination which is a particular region on the surface of the earth.

These functions are to be realised regardless of the architecture of the transponder. For a multibeam satellite repeater, it will also be necessary to route the carriers received on each link to their destination regions [BEA-88]. The regenerative repeater must also provide demodulation and remodulation of the carriers in addition to other specific functions (see Chapter 6 and Section 9.4).

The band of frequencies on which the repeater must operate (as defined by the bandwidth allocated by the Radiocommunication Regulations) is large—from 500 MHz to 2 GHz according to the frequency band. To facilitate power amplification, this band is usually divided into a number of sub-bands (or channels also called transponders) with which separate amplification chains are associated. The bandwidth of these channels is several tens of MHz.

9.1.2 Characterisation of the payload

The characteristic parameters of a communication satellite payload are as follows:

—The transmitting and receiving frequency bands and polarisations for the various channels.
—The transmission and reception coverages.
—The equivalent isotropic radiated power (EIRP) or the power flux density achieved in a given region.
—The power flux density required at the receiving antenna in order to produce the performance specified above at the transponder output (this can depend on the channel or group of channels concerned).
—The figure of merit (G/T) of the receiving system in a given region.
—The non-linear characteristics.
—The reliability after N years for a specified number (or percentage) of channels in good working order.

Antenna coverage is specified in terms of the radio-frequency characteristics to be obtained at a set of reference points on the surface of the earth. The beamwidth is obtained by taking account of the various sources of antenna beam pointing error and is specified by the permissible gain reduction (often taken as 3 dB) at the boundary of the coverage area (see Sections 9.5 and 9.6). The receiving and transmitting coverages are generally different.

The equivalent isotropic radiated power (EIRP) or power flux density produced in a given region is generally specified at the boundary of the coverage area under particular operating conditions for one repeater channel. It usually involves operation of the amplifier at saturation. It should be noted that there is a statutory limit to the power flux density which may be produced on the surface of the earth by communication satellites [CCIR-Rep. 558].

The minimum power flux density at the satellite receiving antenna is defined for a specified coverage and under particular operating conditions for the amplifier of the repeater channel (usually to obtain saturation of the amplifier).

The figure of merit (G/T) of the receiving system is also defined for a specified coverage (for example by a minimum value at the coverage boundary).

Characterisation of non-linearities includes, for example, the level of third order intermodulation products for operation with a given number of carriers of the same amplitude and particular values of output back-off (see Section 9.2.1).

Reliability is the subject of Chapter 13.

9.1.3 The relationship between the radio-frequency characteristics

The principal parameters which characterise the payload from the point of view of the link budget are the equivalent isotropic radiated power (EIRP) for the downlink and the figure of merit (G/T) for the uplink. Although they characterise different links, these parameters are not independent. By considering, for simplicity, the case of a station-to-station link through the payload in the absence of interference and in single access, the carrier power-to noise power spectral density ratio $(C/N_0)_T$ for the complete link can be written (see equation (2.50)):

$$(C/N_0)_T^{-1} = (C/N_0)_U^{-1} + (C/N_0)_D^{-1}$$

In this expression, $(C/N_0)_U$ is the carrier-to-noise density ratio on the uplink and this is propotional to the figure of merit G/T of the satellite. $(C/N_0)_D$ is the carrier-to-noise density ratio which characterises the downlink alone and is proportional to the equivalent isotropic radiated power (EIRP) of the channel.

For a fixed performance objective which determines the value of $(C/N_0)_T$, the ratios $(C/N_0)_U$ and $(C/N_0)_D$, and hence the values of the figure of merit G/T and the radiated power (EIRP), are combined by an expression of the following type:

$$C = A(G/T)^{-1} + B(\text{EIRP})^{-1} \tag{9.1}$$

where A, B and C are constants for a given configuration. This relationship is illustrated in Figure 9.1. As for earth stations there is, therefore, the possibility of a compromise in the choice of parameter values. By assuming that the gain of the receiving and transmitting antennas is fixed for a given coverage, the compromise between the power P_{TX} delivered by the output amplifier and the system noise temperature T is finally established.

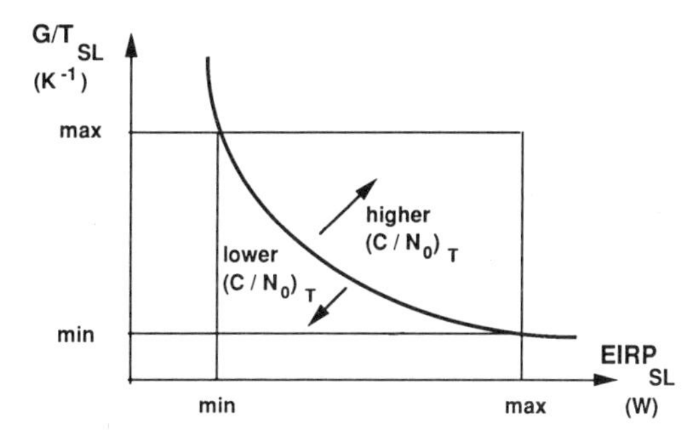

Figure 9.1 Satellite G/T versus EIRP for a given overall performance objective $(C/N)_T$.

For a specified performance objective, it is thus possible to compensate for an increase in system noise temperature by an increase in the power of the channel output amplifier (an exchange of power for noise temperature) in accordance with the constraints of power and noise factor limitation. This remains valid (with a more complex formulation) for a link with interference and intermodulation noise.

9.2 CONVENTIONAL TRANSPONDERS

This part of the chapter is devoted to a presentation of the architecture and technologies of conventional repeater equipment in a single beam network, that is where all the sections are situated in the same coverage region. There is, therefore, only one antenna access for each transmission direction (one output on the receiving side and one input on the transmitting side). Since the architecture of the repeater is largely determined by the problems arising from equipment non-linearities, characterisation of these non-linearities is presented first.

9.2.1 Characterisation of non-linearities

The payload equipment demonstrates non-linear characteristics. The behaviour of the equipment thus depends on the signal level applied at the input. This applies particularly to equipment which uses active components such as travelling wave tubes (TWT), klystrons and transistors. However, it also applies under certain conditions, notably at high power levels, to passive equipment such as filters and antennas; in this case one refers to passive intermodulation (PIM) products [AUG-84], [HOE-86], [TAN-90].

9.2.1.1 *The non-linear characteristics of an amplifier*

One of the main functions of the repeater is amplification of the signal power level. The input–output characteristic of an ideal amplifier is merely a coefficient of proportionality. In practice, the output voltage, particularly at high level, does not vary in proportion to the amplitude of the input signal. Various models of this phenomenon can be devised; one of the simplest is to consider the instantaneous amplitude S_o of the output signal as a polynomial function of the instantaneous amplitude S_i of the input signal:

$$S_o = aS_i + bS_i^3 + cS_i^5 + \cdots \tag{9.2}$$

where a, b, c etc. are constants. These constants, and the order of the polynomial (only odd powers are necessary if only odd intermodulation products are considered, see Section 9.2.1.3), are selected to represent the actual characteristic of the amplifier as closely as possible.

Non-linear phenomena also affect the phase of the output signal and this will depend on the amplitude of the input signal. These phenomena are not taken into account in polynomial modelling. The relative phase variation $\Delta\phi$ as a function of input power P_i can be modelled independently. For example [BER-70]:

$$\Delta\phi = a[1 - \exp(-bP_i)] + cP_i \tag{9.3}$$

where a, b and c are constants chosen to approximate the actual characteristic.

To include the non-linear amplitude and phase effects, which occur simultaneously, it is convenient to characterise the non-linear element by a complex function. This complex function produces the amplitude $g(S)$ and phase $f(S)$ characteristic functions of the amplitude S of the input signal. It is conventional to approximate this function by a series [FUE-73]. This then gives:

$$g(S)\exp[jf(S)] = \sum_{P=1}^{P} b_p J_1(\alpha p S) \tag{9.4}$$

where $J_1(x)$ is the first order Bessel function of the first kind, b_p is the series of complex coefficients, and α is a real coefficient.

The coefficients and the order P of the series are chosen to represent the actual characteristics of the non-linear element as closely as possible. The input will thus be considered to be a narrow band signal of the form [DUP-89]:

$$S_i(t) = \mathrm{Re}\{S(t)\exp[j(2\pi f_0 t + \phi(t))]\} \tag{9.5}$$

where $S(t)$ and $\phi(t)$ are the instantaneous amplitude and phase of the input signal. This formulation enables the case of a modulated carrier to be considered by introducing the appropriate expressions for $S(t)$ and/or $\phi(t)$; the multicarrier case can be taken into account by summing the signals.

The aim of the following sections is to define various parameters which are currently used. Polynomial modelling, although imperfect, will be used since it permits most of the principal phenomena to be easily illustrated.

9.2.1.2 *Power transfer characteristic in single carrier operation*

If an unmodulated carrier, whose instantaneous amplitude is expressed in the form $S_i(t) = A\sin\omega_1 t$, is applied at the input of a device, expansion of the instantaneous amplitude of the output signal S_0 using equation (9.2) produces a sum of terms; one has angular frequency ω_1 and the others consist of harmonics with frequencies which are multiples of ω_1. In the applications considered, the bandwidth of the equipment is less than the nominal frequency. The harmonics are thus eliminated by filtering. The output power of the equipment for the carrier is obtained by taking half of the square of the amplitude of the resulting signal of angular frequency ω_1. Hence, with polynomial modelling:

$$P_o^1 = (1/2)(aA + 3bA^3/4 + 15cA^5/24 + \cdots)^2 \tag{9.6}$$

where P_o^1 designates the output power (subscript o = output) in single carrier operation (subscript 1).

Introducing the input signal power $P_i^1 = A^2/2$ (the powers are defined across standard 1 ohm loads) gives:

$$P_o^1 = P_i^1[a + (3b/2)P_i^1 + (15c/6)(P_i^1)^2 + \cdots]^2 \tag{9.7}$$

This relation constitutes the power transfer characteristic which thus represents the output power P_o^1 at the carrier frequency as a function of the carrier input power P_i^1.

The curve representing this characteristic has a maximum for a particular value $(P_i^1)_{sat}$ of carrier power applied at the input [BAU-85]. This maximum corresponds to the *saturation output power (in single carrier operation)* $(P_o^1)_{sat}$. The *saturation power* in single carrier operation is the value used to characterise an amplifier (particularly amplifiers with travelling wave tubes or klystrons) in the manufacturer's data sheet.

The saturation output power $(P_o^1)_{sat}$ is related to the corresponding input power $(P_i^1)_{sat}$ by:

$$(P_o^1)_{sat} = G_{sat}(P_o^1)_{sat} \tag{9.8}$$

where G_{sat} is the *saturation power gain of the device.*

Normalised characteristics; input and output back-off. A particular operating point (*Q*) of the amplifier is characterised by the pair of input and output powers $(P_i^1, P_o^1)_Q$. It is convenient to normalise these quantities with respect to the saturation output power $(P_o^1)_{sat}$ and the input power $(P_i^1)_{sat}$ required to obtain saturation respectively.

The normalised characteristic thus relates the magnitude $Y = P_o^1/(P_o^1)_{sat}$ and the magnitude $X = P_i^1/(P_i^1)_{sat}$. The upper curve of Figure 9.2 shows such a characteristic which

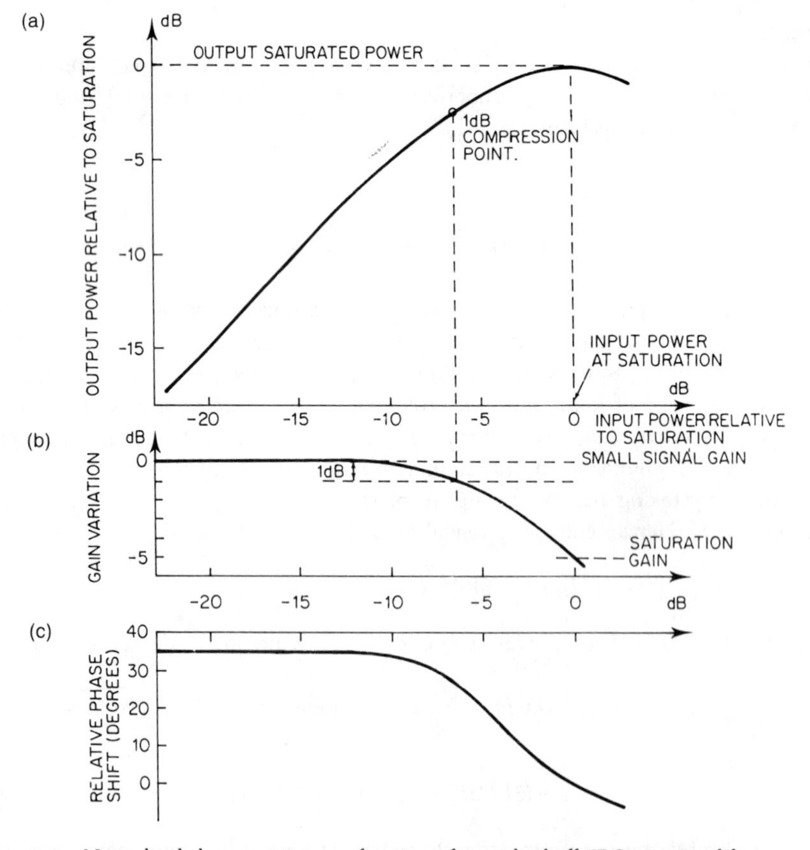

Figure 9.2 Normalised characteristics as a function of input backoff (IBO): (a) Amplifier power transfer in single carrier operation. (b) Power gain. (c) Relative phase shift between input and output.

relates the Y and X values in decibels for a typical travelling wave tube. For a particular operating point defined by $(P_i^I, P_o^I)_Q$, $(X)_Q$ and $(Y)_Q$ represent the *input back-off* (IBO) and the *output back-off* (OBO) respectively (see Section 2.9.1.2).

A simplified model of the normalised characteristic can be obtained by considering equation (9.7) restricted to fifth order for simplicity. Differentiating with respect to the input power and setting this derivative to zero for $P_i^I = (P_i^I)_{sat}$ to give the saturation power gives:

$$(P_o^I) = (G_{sat}/4)(P_i^I)_{sat}[3 - (P_i^I)/(P_i^I)_{sat}]^2 \qquad \text{(W)} \tag{9.9a}$$

Normalisation in this simplified case leads to:

$$Y = (X/4)(3 - X)^2 \tag{9.9b}$$

AM/AM conversion coefficient. For X small, equation (9.9), with values expressed in decibels, reduces to $(Y)_{dB} = (X)_{dB} +$ constant. The slope of the characteristic (in dB) in the diagram is thus equal to 1, that is for 1 dB variation of the input power, the output power also varies by 1 dB (in the linear region). This property can be visualised on the upper curve of Figure 9.2.

The slope of the characteristic is called the *AM/AM conversion coefficient* and is expressed in dB per dB. This conversion coefficient, therefore, has a value of unity when the absolute value of backoff is large (for example an output backoff less than − 15 dB with a travelling wave tube). The AM/AM conversion coefficient decreases as input power increases up to saturation where it becomes zero.

Power gain. The ratio of output power P_o to input power P_i is the *power gain.* It is constant in the linear part of the characteristic, which corresponds to a low power level, where it is called the *small signal power gain* G_{ss}. The gain then decreases with the approach of saturation as illustrated by the intermediate curve of Figure 9.2. At saturation, it takes the particular value G_{sat}, *the saturation power gain of the device.*

Point of compression to 1 dB. The output power obtained when the actual characteristic deviates by 1 dB from an extension of the linear part defines the *point of compression to 1 dB.* This point corresponds to a reduction of 1 dB in power gain.

This parameter is used to define the part of the characteristic which can be considered to be linear. In order to obtain quasi-linear operation for an amplifier module, a signal level greater than a value defined with respect to the point of compression to 1 dB (for example 10 dB) must be prohibited. The point of compression to 1 dB is often used in the manufacturer's technical data sheet to characterise the power of a transistor amplifier (with these amplifiers it can be difficult to reach the saturation power without component damage).

AM/PM conversion factor K_p. The effect of non-linearity also appears in the phase of the signal. The device concerned introduces phase shift between the input and the output. The relative variation of phase shift with respect to that corresponding to saturation as a function of input signal level is illustrated by the lower curve of Figure 9.2. The slope of this characteristic, called the *amplitude modulation to phase modulation (AM/PM) conversion factor* K_p is given by:

$$K_p = \Delta\phi/\Delta P_i^I \qquad (°/\text{dB}) \tag{9.10}$$

The conversion factor, expressed in degree/dB, is maximum for a value of input power less than the saturation value by a few dB.

9.2.1.3 *Power transfer characteristic in multicarrier operation*

The signal applied at the input of the device is now considered as the sum of sinusoidal signals. It is, therefore, put in the form:

$$S_i = A \sin \omega_1 t + B \sin \omega_2 t + C \sin \omega_3 t + \cdots$$

Expansion of the instantaneous amplitude of the output signal S_o using equation (9.2) causes the appearance of terms at the input angular frequencies (ω_1, ω_2, etc.) and at frequencies corresponding to linear combinations of these frequencies or *intermodulation products.* These intermodulation products have been defined in Section 4.4.3. Only odd intermodulation products occur in the vicinity of the input frequencies. On the other hand, the amplitude of these intermodulation products decreases with their order. The most troublesome are third order products of the form $2f_i - f_j$ and $f_i + f_j - f_k$.

In the case of n unmodulated carriers of equal amplitude A_i, a general expression for the amplitude A_{on} of each of the n components of the output signal at the same frequencies as those of the n input signal components is given by [PRI-86]:

$$A_{on} = aA_i[1 + (3b/2a)(n - 1/2)A_i^2 + (15c/4a)(n^2 - 3n/2 + 2/3)A_i^4 + \cdots] \quad (9.11)$$

Under the same conditions, the amplitude A_{IMn} of the intermodulation products of the form $2f_i - f_j$ is given by:

$$A_{IMn} = (3b/4)A_i^3\{1 + (2c/6b)A_i^2[(25/2) + 15(n - 2)] + \cdots\} \quad (9.12)$$

and the amplitude A'_{IMn} of the products of the form $f_i + f_j - f_k$ is given by:

$$A'_{IMn} = (3b/2)A_i^3\{1 + (10c/2b)A_i^2[(3/2) + (n - 3)] + \cdots\} \quad (9.13)$$

9.2.1.4 *Characterisation of non-linearities using two unmodulated carriers of equal amplitude*

To characterise a device, it is customary to consider two input signal components of equal amplitude ($A = B$). The power P_o^2 of one of the output components at frequency ω_1 or ω_2 can then be expressed as a function of the input power ($P_i^2 = A^2/2$) of one of the two components. The curve representing output power variation for one of the two carriers as a function of the power of one of the two input components can then be plotted after normalising the output and input magnitudes with respect to $(P_o^1)_{sat}$ and $(P_i^1)_{sat}$ respectively (Figure 9.3).

In comparison with the curve representing single carrier operation, saturation corresponds to an output power less than $(P_o^1)_{sat}$. The maximum power which the device can deliver is effectively shared between the two carriers and the various intermodulation products. For a travelling wave tube, the difference between the maximum power in single carrier operation and the maximum power of one of the two carriers is of the order of 4 to 5 dB (see Figure 9.3).

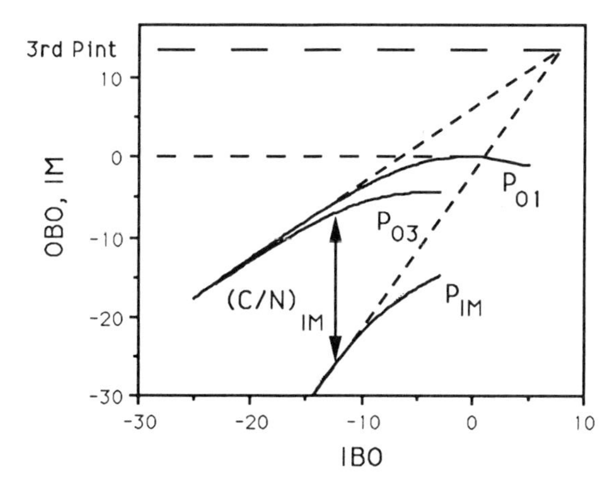

Figure 9.3 Normalised power transfer characteristic with two equal amplitude carriers.

The effect can be illustrated using the polynomical model. By considering two input carriers of equal amplitude and angular frequencies ω_1 and ω_2, equation (9.2) leads to:

$$P_o^2 = P_i^1[a + (9b/2)P_i^2 + (15c/6)(P_i^2)^2 + \cdots]^2 \qquad \text{(W)} \tag{9.14}$$

where P_o^2 is the power of one of the output components at frequency ω_1 or ω_2 and $P_i^2 = A^2/2$ is the power of one of the input components.

The coefficients of this relation can be expressed as a function of the characteristic parameters for single carrier operation. By restricting to third order, as in the previous section, this gives:

$$(P_o^2) = (9G_{sat}/4)(P_i^2)[1 - (P_i^2)/(P_i^1)_{sat}]^2 \qquad \text{(W)} \tag{9.15a}$$

Normalising the output and input magnitudes with respect to $(P_o^1)_{sat}$ and $(P_i^1)_{sat}$ respectively, the following simplification is obtained:

$$Y' = X'(9/4)(1 - X')^2 \tag{9.15b}$$

where $Y' = P_o^2/(P_i^1)_{sat}$ and $P_i^2/(P_i^1)_{sat}$ are the normalised magnitudes. Saturation is reached in this simplified case for $X' = 1/3$. The normalised output power is then $Y' = 1/3$.

Input and output back-off. For a particular operating point defined by $(P_i^2, P_o^2)_Q$, X' and Y' represent the *input back-off (IBO)* and the *output back-off (OBO)* respectively. Notice that the back-offs are defined in multicarrier operation with respect to the saturation powers in single carrier operation. In the example used for illustration, the input and output back-offs corresponding to saturation in two carrier operation are -5 dB ($X' = Y' = 1/3$).

The total (input or output) back-off may also be defined. It is the ratio of the sum of the powers (input or output) of the various carriers to the saturation power in single carrier operation. In the above example, the total input and output back-off corresponding to saturation with two carriers would be -2 dB ($X'_T = Y'_T = 2 \times 1/3$).

Third order intermodulation. The power P_{IM}^2 of one of the third order intermodulation products can also be plotted on the normalised diagram with respect to $(P_o^1)_{sat}$ and $(P_i^1)_{sat}$ (Figure 9.3).

It can be seen that the linear part of this curve (for small values of X') has a slope of three with a decibel scale. This can be verified from the simplified model. The power of the terms of angular frequency $2\omega_1 - \omega_2$ and $2\omega_2 - \omega_1$ in the expansion obtained from equation (9.2) is given by:

$$P^2_{IM} = (P^2_i)^3[(3b/2) + (25c/2)(P^2_i) + \cdots]^2 \qquad \text{(W)} \tag{9.16}$$

where P^2_{IM} is the output power of one of the third order intermodulation products for operation with two carriers of equal amplitude. The coefficients can be expressed as a function of the characteristic parameters of single carrier operation. By restricting to third order, as in the previous section, this gives:

$$(P^2_{IM}) = (P^2_i)^3 G_{sat}/[(P^1_i)_{sat}]^2 \qquad \text{(W)} \tag{9.17a}$$

Normalising the output and input magnitudes with respect to $(P^1_o)_{sat}$ and $(P^1_i)_{sat}$, this simplified case gives:

$$IM = (1/4)(X')^3 \tag{9.17b}$$

where $IM = P^2_{IM}/(P^1_i)_{sat}$ and $X' = P^2_i/(P^1_i)_{sat}$ are the normalised magnitudes. Equation (9.17), with the values expressed in decibels, can be put in the form: $(IM)_{dB} = 3(X')_{dB} + \text{constant}$. The slope of the characteristic (in dB) in the diagram is thus equal to 3, that is for 1 dB variation of carrier power at the input, the power of one of the intermodulation products varies by 3dB (in the linear region).

On the other hand, equations (9.17) do not show a reduction of the slope of the intermodulation product characteristic as the normalised input power increases. Third order modelling of the characteristic is not sufficiently representative in the vicinity of saturation. By considering the fifth order model, for example, an expression of the following form would be obtained:

$$IM = p(X')^3(q + rX')^2 \tag{9.18}$$

This shows a saturation effect when X' becomes large (p, q and r are constants which are functions of the characteristic parameters of single carrier representation).

Relative level of third order intermodulation products. The ratio $(C/N)_{IM}$ of the output power of one of the two carriers to the power of one of the third order intermodulation products for different values of input backoff characterises the *relative level of the third order intermodulation products.* A table of values is often provided in the technical data sheet of an amplifier to characterise the effect of the nonlinearity.

Third order intercept point. Another parameter is currently used to characterise the effect of nonlinearity, particularly with transistors. The straight lines obtained by extending the linear parts of the characteristics of the useful signal (one of the two carriers) and one of the third order intermodulation products meet at a point called the *third order intercept point.*

The ordinate ($P_{int,3rd}$) of this point (expressed as the absolute value of the power) enables the linearity of several devices to be compared; the higher the value at the intercept, the more linear the device (for a given power). The value of the third order intercept is greater by about 10 dB than the value of the point of compression to 1 dB.

The third order intercept also enables the ratio $(C/N)_{IM}$ to be determined for a given output power on one of the two carriers. By considering the linear part of equations (9.15) and (9.17), expressed in decibels, the following is obtained:

$$(P_o^2)_{dB} = 10\log(9G_{sat}/4) + (P_i^2)_{dB} = K_1 + (P_i^2)_{dB}$$

and

$$(P_{IM}^2)_{dB} = 10\log\{G_{sat}/[(P_i^1)_{sat}]^2\} + 3(P_i^2)_{dB} = K_2 + 3(P_i^2)_{dB}$$

The value of the intercept is such that:

$$P_{int,3rd} = K_1 + (P_i^2)_{dB} = K_2 + 3(P_i^2)_{dB} \qquad \text{(dB)}$$

hence

$$P_{int,3rd} = (3K_1 - K_2)/2 \qquad \text{(dB)}$$

Furthermore:

$$(P_o^2)_{dB} - P_{int,3rd} = [(K_2 - K_1)/2] + (P_i^2)_{dB}$$

The difference (in decibels) between the output power on one of the two carriers and the power of one of the two third order intermodulation products is then:

$$(P_o^2)_{dB} - P_{IM} = [K_1 + (P_i^2)_{dB}] - [K_2 + 3(P_i^2)_{dB}] = K_1 - K_2 - 2(P_i^2)_{dB}$$

from which:

$$(C/N)_{IM} = (P_o^2)_{dB} - (P_{IM})_{dB} = 2[(P_{int,3rd})_{dB} - (P_o^2)_{dB}] \qquad \text{(dB)} \qquad (9.19)$$

This relation is obtained by considering the linear parts of equations (9.15) and (9.17) and is thus valid well below saturation (below the point of compression).

It should be noted that the values of $(C/N)_{IM}$ involved here are used to characterise the device (the amplifier) and correspond to operation of the amplifier in a specific mode (two unmodulated carriers of equal amplitude).

To determine the value of $(C/N)_{IM}$ to be used in the link budget equation (equation (4.6)) for a particular application, the corresponding situation in terms of the number of carriers, the relative level of these carriers, the modulation used and so on should be considered either by experimental characterisation or by simulation based on modelling using Bessel functions, for example (see Section 9.2.1.1).

The AM/PM conversion characteristic of the tube also causes generation of intermodulation products which are additional to those caused by amplitude non-linearity [FUE-73]. A degradation of several decibels of the $(C/N)_{IM}$ ratio, as determined by considering only amplitude non-linearity, can thus be introduced particularly for large values of backoff.

Transfer coefficient K_t. In multicarrier operation, non-linear phase effects also cause transfer of the amplitude modulation of one carrier into phase modulation of other carriers. In the case of operation with two carriers, *the amplitude modulation to phase modulation transfer coefficient K_t from one carrier to another* is defined as the slope of the relative variation (with respect to the phase at saturation) of the phase of one output carrier (whose input amplitude is held constant) when the input amplitude of the other is varied. This phenomenon gives rise to intelligible crosstalk and this source of distortion is reduced if the factor K_t of the device concerned is small.

9.2.1.5 *The capture effect*

Consider a non-linear device in multicarrier operation for which the power of one of the input carriers is less than that of the other carriers and differs from them by a quantity $\Delta(P_i)$. At the output, the difference $\Delta(P_o)$ between the power of the other carriers and that of the carrier considered is increased. This phenomenon is called the *capture effect*. It can occur with two carriers and the simplified modelling used above by comparing the amplitudes of the output components when the amplitudes of the two input components are different. The input signal is of the form:

$$S_i = A \sin \omega_1 t + B \sin \omega_2 t$$

From equation (9.2), determination of the components $(A_o^2)_{\omega 1}$ and $(B_o^2)_{\omega 2}$ of the output signal at angular frequencies ω_1 and ω_2 gives:

$$(A_o^2)_{\omega 1} = A[a + (3b/4)A^2 + (3b/2)B^2]$$

hence

$$(P_o^2)_{\omega 1} = (P_i^2)_{\omega 1}\{1 + (3b/a)[(P_i^2)_{\omega 1}/2 + (P_i^2)_{\omega 2}]\}^2 \qquad \text{(W)}$$

and

$$(B_o^2)_{\omega 2} = B[a + (3b/4)B^2 + (3b/2)A^2]$$

hence

$$(P_o^2)_{\omega 2} = (P_i^2)_{\omega 2}\{1 + (3b/a)[(P_i^2)_{\omega 2}/2 + (P_i^2)_{\omega 1}]\}^2 \qquad \text{(W)}$$

Normalising with respect to $(P_i^1)_{sat}$ and $(P_i^1)_{sat}$:

$$(Y')_{\omega 1} = (X')_{\omega 1}\{1 - (1/3)[(X')_{\omega 1} + 2(X')_{\omega 2}]\}^2$$

where $(X')_{\omega 1}$ and $(Y')_{\omega 1}$ are the input and output powers at angular frequency ω_1 normalised with respect to the single carrier saturation power and:

$$(Y')_{\omega 2} = (X')_{\omega 2}\{1 - (1/3)[(X')_{\omega 2} + 2(X')_{\omega 1}]\}^2$$

where $(X')_{\omega 2}$ and $(Y')_{\omega 2}$ are the input and output powers at angular frequency ω_2 normalised with respect to the single carrier saturation power.

The ratio ΔP_i of the input signal power at frequency ω_1 to that at frequency ω_2 is given by:

$$\Delta P_i = (P_i^2)_{\omega 1}/(P_i^2)_{\omega 2} = (X')_{\omega 1}/(X')_{\omega 2} = (A/B)^2$$

The ratio ΔP_o of the output powers is given by:

$$\Delta(P_o) = \Delta(P_i)\{[1 - ((X')_{\omega 1} + 2(X')_{\omega 2})/3]/[1 - ((X')_{\omega 1} + 2(X')_{\omega 2})/3]\}^2 \qquad (9.20)$$

Since the normalised magnitudes $(X')_{\omega 1}$ and $(X')_{\omega 2}$ are by definition less than unity, the quantity in square brackets is always greater than unity if $(X')_{\omega 1}$ is greater than $(X')_{\omega 2}$. Since the coefficient of $\Delta(P_i)$ is greater than unity, the ratio of the output powers $\Delta(P_o)$ is thus greater than the ratio of the input powers $\Delta(P_i)$.

The capture effect Δ is defined by:

$$\Delta = \Delta P_o/\Delta P_i \quad \text{or} \quad (\Delta)_{dB} = (\Delta P_o)_{dB} - (\Delta P_i)_{dB} \qquad (9.21)$$

Equation (9.20) shows that when the input backoffs are much less than unity (a large absolute value of backoff), the capture effect disappears ($(\Delta)_{dB} = 0$ dB). The capture effect is greatest when the difference between the input signal powers is large and the absolute value of backoff is small. For example, with $(\Delta P_i)_{dB} = 10$ dB, the capture effect Δ will be 0.25 dB for a total input backoff of -15 dB and 5 dB at saturation (zero total backoff). With $(\Delta P_i)_{dB} = 2$ dB, the capture effect Δ is only 1.5 dB at saturation.

9.2.2 Repeater architecture

The organisation of the repeater is determined by the mission specification and technological constraints. A large power gain, together with a low equivalent input noise temperature and a high output power over a wide frequency band, is to be provided. Frequency conversion of the signals must also be performed.

9.2.2.1 *Low noise amplification and frequency conversion*

Frequency conversion between the uplink and the downlink enables decoupling between the input and output of the repeater to be ensured. Re-injection of signals radiated by the output into the transponder input can thus be avoided by filtering.

Frequency conversion can be envisaged as the first operation performed on the signals from the antenna (a front-end mixer). However, except in special cases, this arrangement does not enable the required system noise temperature specification to be satisfied because of the high noise figures of mixers. Furthermore, it is preferable to share the power gain between two amplifier units operating with different input and output frequencies. This enables the danger of instability, which is inherent in a very high gain amplifier where all the stages operate at the same frequency, to be limited.

The repeater thus consists firstly of a low noise amplifier which provides the required value of effective input noise temperature at the uplink frequency. A high gain (20 to 40 dB) minimises the contribution to noise of the mixer which follows the amplifier.

A mixer associated with a local oscillator then provides frequency conversion (Figure 9.4a). The position of the mixer in the chain is determined by the level of the signal to be converted since this must remain sufficiently low to minimise non-linear effects.

9.2.2.2 *Single and double frequency conversion*

After frequency conversion, and taking account of the gain of the low noise amplifier and the conversion losses of the mixer, a particular value of gain remains to be provided in order to obtain the total required power gain. Depending on the frequency bands concerned, technological considerations can make it difficult to obtain high power gain at the downlink frequency. An architecture with a double change of frequency, using an intermediate frequency of lower value than the downlink frequency, is then used (Figure 9.4b). The uplink signals are first converted to the intermediate frequency (several GHz) where amplification is performed. Another stage of frequency conversion then increases the frequency to that of the downlink.

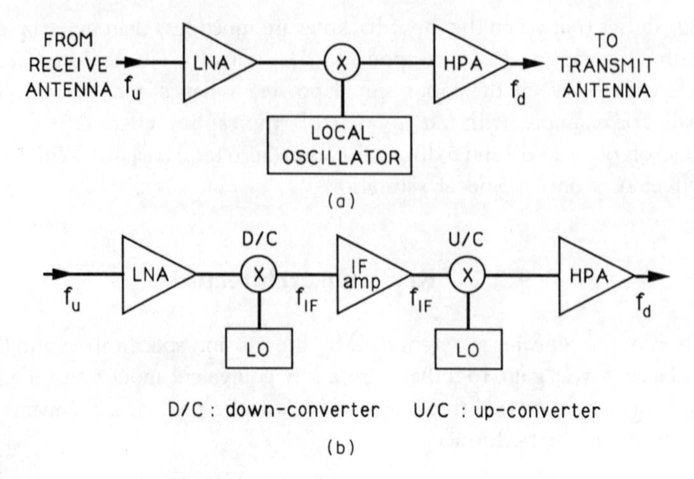

Figure 9.4 Repeater architecture. (a) Single frequency conversion. (b) Dual frequency conversion.

This double frequency conversion architecture was used for the first satellites operating in Ku band (14/12 GHz). It remains appropriate for proposed satellites operating in Ka band (30/20 GHz) and above.

The INTELSAT satellites have a payload in Ku band with double frequency conversion and an intermediate frequency of 4 GHz. This architecture is convenient for interconnection of the Ku band payload and the payload operating in C band (6/4 GHz).

9.2.2.3 *Amplification after frequency conversion*

The signal is further amplified after conversion. The signal level increases as it progresses through the amplifying stages of the repeater. The operating point on the transfer characteristic of each stage moves progressively towards the non-linear region (Figure 9.5a). The level of

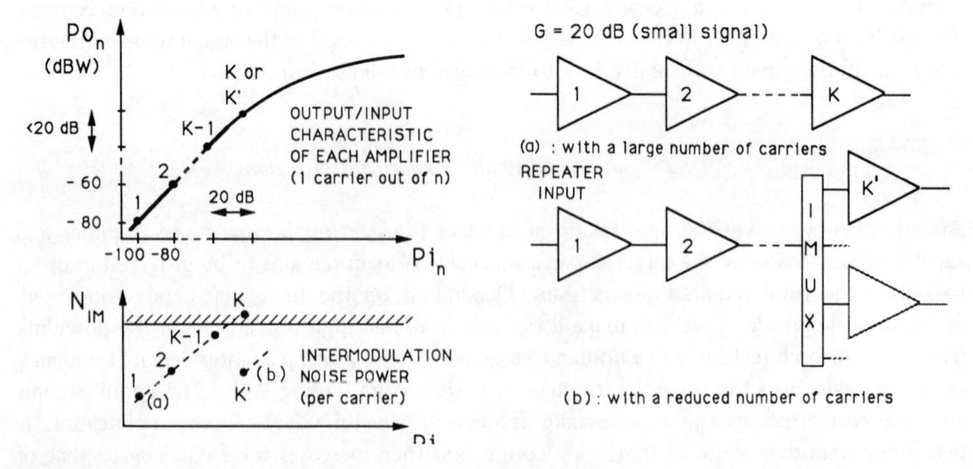

Figure 9.5 Reduction of intermodulation noise by channelisation of the frequency band.

intermodulation noise is negligible for the input stages, which operate at very low level, and subsequently increases. According to the technology used, as characterised by its intercept or compression point, the specified maximum intermodulation noise level may be exceeded when the signal has passed through a certain number of stages. In this case amplification cannot be continued under the same conditions; it is necessary either to choose a technology with a higher intercept or to operate under particular conditions which permit intermodulation noise to be reduced at a given power level. Since the performance of available active components is rapidly reaching a limit, intermodulation noise must be reduced.

9.2.2.4 *Channelisation of the repeater*

Intermodulation noise. The input stages of the repeater operate on the whole of the system frequency band, several hundreds of MHz. Several tens of carriers share this band and will thus give rise to a large number of intermodulation products as these signals pass through a non-linear device. A reduction in the number of intermodulation products, and hence the level of intermodulation noise, is obtained by limiting the number of carriers passing through the same amplifier.

Hence when the level of intermodulation noise tends to become excessive for wide band amplification, the system frequency band is then divided into several sub-bands which are amplified separately (Figure 9.5b).

Channelisation. The purpose of channelisation of the repeater is to create channels (sub-bands) of reduced width. Since the number of carriers in each sub-band is less, the intermodulation noise generated by the sub-band amplification stage is much less than the intermodulation noise which would have been generated if the amplifying stage had operated on the total system bandwidth. Figures 9.5a and b illustrate the comparison qualitatively.

Amplification of the signal continues within the channel until the required power level is obtained. The amplifier power is shared among the various carriers which occupy the channel. The maximum power available with existing equipment developed for space applications is limited. Without channelisation this maximum power must be shared among all the carriers occupying the system bandwidth. With channelisation, a limited number of carriers shares this maximum power. The usable power per carrier is thus greater.

The advantages of channelisation are thus twofold:

—To permit power amplification with a limited increase of intermodulation noise due to the reduced number of carriers per amplifier.
—To increase the total power transmitted by the repeater since the maximum power available with the technology used can be delivered, if necessary, in each channel.

As the band is shared in parallel channels; distortion occurs when part of the energy of a signal feeds into the channels adjacent to the nominal one within which the spectrum of the signal should be contained. These adjacent channel interference (ACI) effects are minimised by a guard band of sufficient width between channels and by the use of filters which restrict the channel widths as closely as possible to those of ideal band-pass filters.

Channel separation is obtained by means of a set of bandpass filters called input (de)multiplexers (IMUX). The bandwidths of the channels range from a few tens of MHz to around a hundred MHz (e.g. 36, 40, 72 and 120 MHz). The various sub-bands are recombined in the output multiplexer (OMUX) after amplification in each channel. The word 'transponder' is sometimes used instead of 'channel' to designate the equipment which operates within a given sub-band.

Adjacent and alternate channels. The output multiplexer may or may not be of the adjacent channel type. With non-adjacent channels, the various channels which are combined by a given multiplexer are separated by a wide guard band, equal, for example, to the width of one channel. Realisation of the multiplexer is thus facilitated to the detriment of efficient use of the radio-frequency band. To use the bandwidth profitably, it is divided into alternate even and odd channels by means of separate IMUXs at the beginning to the channelised section (see Figure 9.9). The channels of each group (even and odd) are then recombined by a different OMUX for each group. The OMUX outputs are in this case connected either to two different transmitting antennas (e.g. Telecom 1) or to the two inputs of a dual mode antenna through a hybrid coupler [KUD-92].

The adjacent channel multiplexer allows recombination of adjacent channels. Obtaining a good performance (i.e. narrow guard bandwidth, low insertion loss and high isolation between channels) imposes severe constraints on the characteristics of the bandpass filters used and leads to substantial complexity in the design and optimisation of the output multiplexer [KUD-80], [KUD-88], [CAM-90a].

9.2.2.5 *Channel amplification*

Amplification within the channel uses a preamplifier which provides the power required to drive the output stage. This preamplifier, called the channel or driver amplifier, is generally associated with a variable gain device which can be adjusted by telecommand. This permits compensation for variations of power amplifier gain during the lifetime of the satellite.

The power amplifier provides the power delivered to the OMUX inputs at the output of each channel.

9.2.2.6 *Input and output filtering*

At the repeater input, a band-pass filter limits the noise bandwidth and provides a high rejection at the downlink frequencies. At the output, a band-pass filter eliminates the harmonics generated by the non-linear elements and provides additional repeater output–input isolation. These filters must introduce the lowest possible insertion losses for useful signals. High input filter insertion losses cause degradation of the repeater figure of merit (G/T) and output filter losses cause a reduction of the isotropic radiated power (EIRP).

9.2.2.7 *General organisation*

The complete architecture of a repeater with a single frequency conversion in accordance with the considerations discussed above is presented in Figure 9.6. The equipment which operates over the entire system bandwidth constitutes the receiver. It is followed by the input multiplexer which defines the start of the channelised section, the channel amplifiers and the output multiplexer.

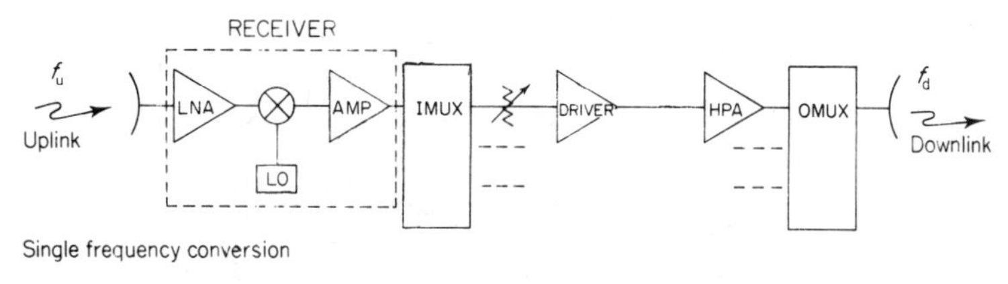

Figure 9.6. Repeater block diagram. Single frequency conversion.

When a double frequency conversion architecture is required, uplink frequency conversion can be performed either in the receiver on the entire system bandwidth with a single mixer, or in the channelised part which then requires as many mixers as channels. In the first case, the whole of the channelised part operates at the downlink frequency and the mixer operates at low level over the whole band. With the second configuration, demultiplexing and part of the channel amplification is performed at intermediate frequency. However, a large number of mixers, which also operate at high level, is required.

9.2.2.8 Redundancy

The architecture presented in Figure 9.6 does not include any replicated equipment. In order to guarantee the required reliability at the end of life, this architecture is modified to limit single failure points as far as possible; these are elements whose failure causes loss of the mission (see Chapter 13).

The input and output multiplexers are not replicated since these are passive elements whose failure rate is very low and replication is difficult. The receiver is generally duplicated with an identical stand-by unit; this is 2/1 redundancy—two installed units for one active one. A switch operated by telecommand routes the signal from the antenna to the receiver in use; the receiver outputs are connected to the IMUX by means of a hybrid coupler which provides passive routing of the signals. Various redundancy strategies (e.g. 4/2) can be used when the satellite contains several payloads.

The channel amplifying equipment is replicated (Figure 9.7). In conventional arrangements a given small number of IMUX output channels is shared among a larger number of amplification chains. For example, with 3/2 redundancy, a switch with two inputs and three outputs routes

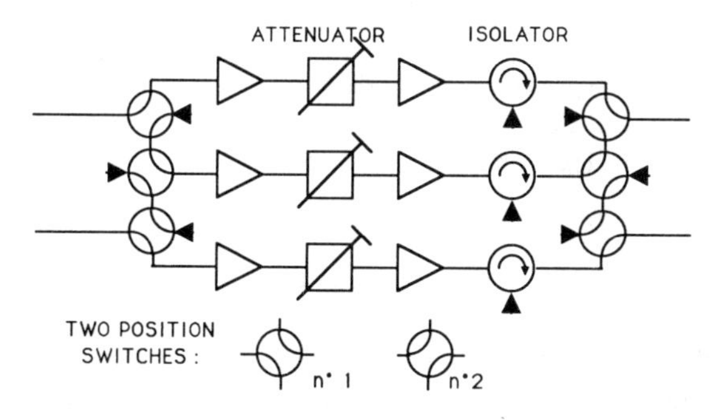

Figure 9.7 2/3 conventional channel redundancy.

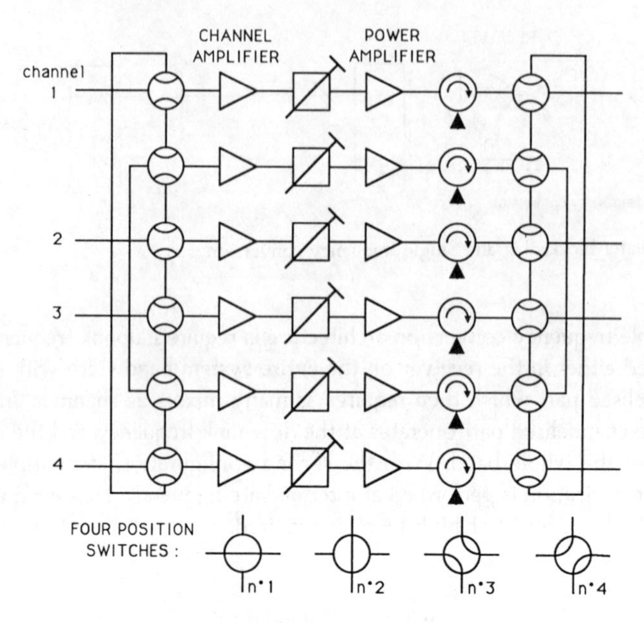

Figure 9.8 4/6 ring redundancy.

the signals available on two outputs of the IMUX to two out of the three installed chains. A switch with three inputs and two outputs routes the signals from the two active chains to two of the OMUX inputs. In case of failure of one of the active amplifiers, the stand-by unit replaces it, but a further failure involves loss of a channel. Therefore in order to increase the reliability of the channelised part, a more complex system called a redundancy ring is often used on modern payloads. With this arrangement, all the channels share a larger number of installed chains; for example 12/8 redundancy where 8 channels share 12 installed chains. Each channel at the output of the IMUX can be directed to the input of several chains by means of a set of interconnected multi-position switches (Figure 9.8). A large number of possible ways of substituting for failed equipment is thus available; in this way a high reliability at the end of life can be obtained [JES-86].

9.2.3 Equipment characteristics

The performance of the repeater equipment determines that of the payload. The following sections will review the principal units with an emphasis on performance, the influence of this performance at system level and the technologies used.

9.2.3.1 *The receiver*

The receiver consists of an amplifier at the uplink frequency, a frequency conversion stage and amplification after conversion. These elements are generally assembled in the same housing using a modular design.

Table 9.1 Low noise amplifier (LNA) characteristics

Noise figure	6 GHz	14 GHz	30 GHz	Gain
Parametric amplifier		2 dB	3 dB	15 dB
Field effect transistor	2 dB	3 dB	5 dB	20 dB
High electron mobility transistor	1.8 dB	2.5 dB		30 dB

The latest technology employs hybrid circuits using unencapsulated chips (active components). At intermediate frequencies, these hybrid circuits are formed by the deposition of several successive layers to realise passive components (resistors and capacitors) and interconnections on an alumina substrate (thick film technology). The bare chips are then bonded onto the substrate. At radio frequencies, the process used is similar except for a greater required precision of the tracks; the conductors (microstrip circuits) are produced by photolithography of the substrate (alumina, sapphire etc.) which is covered with a layer of metallic conductor (gold) obtained by deposition under vacuum (thin film technology). For low frequency functions (oscillators, power supplies), one technology used consists of mounting the components on the surface of a polyamide printed circuit.

These technologies, which are conventional in terrestrial applications, have been approved for space applications and conform to the specifications of the PSS 01-606 procedure of the European Space Agency for example (see Chapter 13). Their use permits the mass and volume of modern receivers to be reduced (From 2 to 1 kg) [CER-88]. The typical bulk of a receiver is 30 × 20 × 10 cm and its power consumption is between 5 and 15 W.

The input amplifier. The amplifier at the uplink frequency is the main element which determines the figure of merit G/T of the transponder. This amplifier must thus have a low noise temperature and a high gain in order to limit the contribution of the noise temperature of subsequent stages. The first satellites used tunnel diode amplifiers. Subsequently parametric amplifiers were used; their principle of operation is based on the reflection of the signal onto a negative resistance (see Section 8.4.1.1).

Field effect amplifiers, whose technology is constantly advancing, are now used. Typical values of noise figures obtained in the various frequency bands are given in Table 9.1. In C and Ku bands, field effect amplifiers are dominant due to the use of GaAs and high mobility electron (HMET) technologies [HAN-86]. At high frequencies (30 Ghz), the parametric amplifier remains of use in order to obtain a low noise figure. Several stages are cascaded in order to obtain a gain of the order of 30 dB before frequency conversion.

The frequency conversion stage. The conversion stage consists of a mixer, a local oscillator and filters. The frequency of the local oscillator is the difference between the centre frequency of the uplink band and the centre frequency of the downlink band (for single frequency conversion architecture and assuming a continuous frequency band). In C band it is of the order of 2.2 GHz. In Ku band it is, for example, 1.5, 2.58 or 3.8 GHz according to the frequency band used on the downlink (10.95–11.2 GHz, 11.54–11.7 GHz or 12.5–12.75 GHz) for an uplink in the 14–14.5 GHz band.

The principal characteristic parameters are:

—The conversion loss, i e. the ratio of the input power (at the frequency of the uplink) to the output power (at the frequency after conversion) and the noise figure (typical values are 5 to 10 dB).

—The stability of the frequency generated by the local oscillator. The stability is expressed in the long term over the lifetime of the satellite (the typical value of relative frequency variation must be less than ± 1 to $\pm 5 \times 10^{-6}$) and in the short term ($< 10^{-6}$) under the conditions of the specified temperature range.

—The amplitude of undesirable signals—residual input and output signals at the oscillator frequency and its harmonics (typically < -60 dBm), spurious output signals at frequencies close to that of the main signal (typically < -70 dBc in a bandwidth of 4 kHz situated more than 10 kHz from the signal).

Traditionally the mixer is of the double balanced type and uses Schottky diodes. Modern technologies use diodes with beam leaded connections associated with coplanar slotted lines.

The local oscillator uses a frequency reference obtained from a quartz oscillator (which may be temperature stabilised or provided with a trimming circuit to permit adjustment of the frequency by telecommand).

The local oscillator frequency required is obtained either in a conventional manner by multiplication of the reference frequency (using a succession of amplifying and frequency multiplication stages) or by direct generation using a frequency synthesiser based on a phase-locked loop (PLL). The frequency is produced by a voltage controlled oscillator locked to the quartz reference frequency.

Amplification after frequency conversion. Amplification after conversion provides gain which complements that before the channelised part. This multi-stage amplifier may contain a device (for instance, a PIN diode attenuator) for gain control by telecommand. High linearity is the main characteristic required for these stages. They operate over a wide bandwidth, and hence with a large number of carrier signals whose levels may be sufficient to cause non-linear effects. Typically, the level of third order intermodulation products must remain less than that of the carrier by more than 40 dB (with characterisation based on two carriers of equal amplitude at the input).

The overall receiver gain is of the order of 60 to 70 dB. This gain must be constant with frequency over the useful bandwidth in order to avoid distortion associated with non-linearity of the transponder output stages (amplitude-phase transfer, see Sections 9.2.1.2 to 9.2.1.4). Typically, the ripple should not exceed 0.5 dB over a range of 500 MHz. To obtain such a small ripple, matching between the various stages must be meticulous in order to minimise the standing wave ratio. This minimisation is facilitated by inserting an isolator (a circulator with a matched load) between each stage; this dissipates waves reflected at the interface.

9.2.3.2 *Input and output multiplexers*

These devices define the input and output of the channelized part. The term multiplexer in this case designates a passive device which is used to combine signals at different frequencies from different sources onto a single output or to route signals from a single source to different outputs according to the frequency of the signal. The multiplexers are configured as interconnected high selectivity band-pass filters. The various architectures and performance obtained (such as channel spacing, insertion loss and isolation between channels) depend on the interconnection technique used [BAR-72].

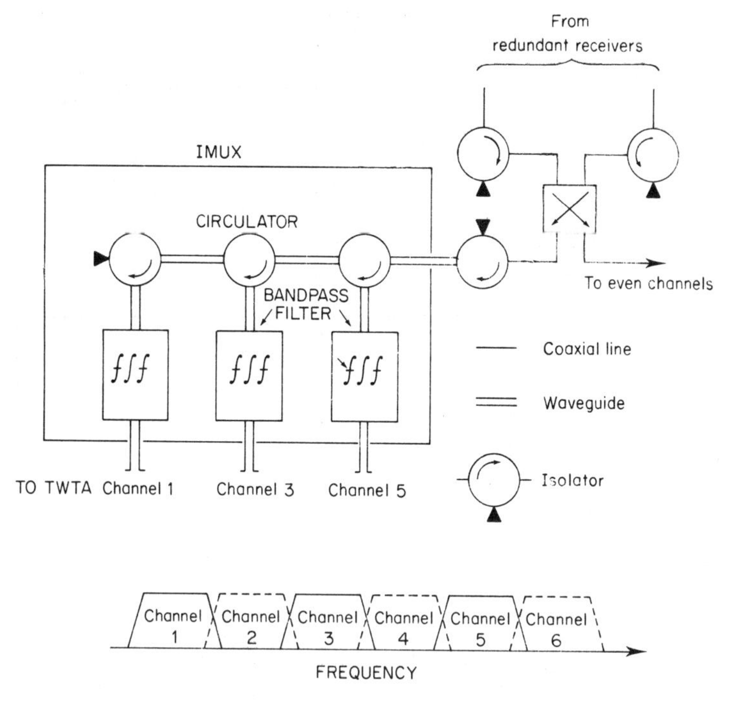

Figure 9.9 Arrangement of an input multiplexer (IMUX).

Input multiplexer. The input multiplexer (IMUX) divides the total system bandwidth into different sub-bands. The band-pass filters used define the bandwidth of the various channels. A typical configuration involves a battery of band-pass filters fed through circulators. Figure 9.9 shows an example where the channels are organised in two groups of even and odd channels. The IMUX is then divided into two parts which share the power available at the receiver output by means of a hybrid. As shown in the figure, this hybrid also permits the signals to be delivered to the IMUX to be provided from the redundant receiver thereby avoiding selection by means of a switch.

The losses in the multiplexer depend on the number of times the signal concerned passes through a circulator and the number of reflections at the band-pass filter inputs (the loss per element is of the order of 0.1 dB). The losses thus differ from one channel to the other and are maximum for the channel furthest from the IMUX input. Division of the IMUX into several parts, each supporting a limited number of channels, enables the difference in losses between channels to be reduced; the losses themselves are not critical since they are compensated by the channel amplification.

The output multiplexer. The output multiplexer (OMUX) recombines the channels after power amplification. Unlike the IMUX, losses in the OMUX are critical since they lead directly to a reduction of the radiated power. Instead of using circulators, which are bulky and introduce losses, output coupling of the band-pass filters is achieved by mounting the filters on a common waveguide (a manifold) of which one end is short-circuited (Figure 9.10). The output

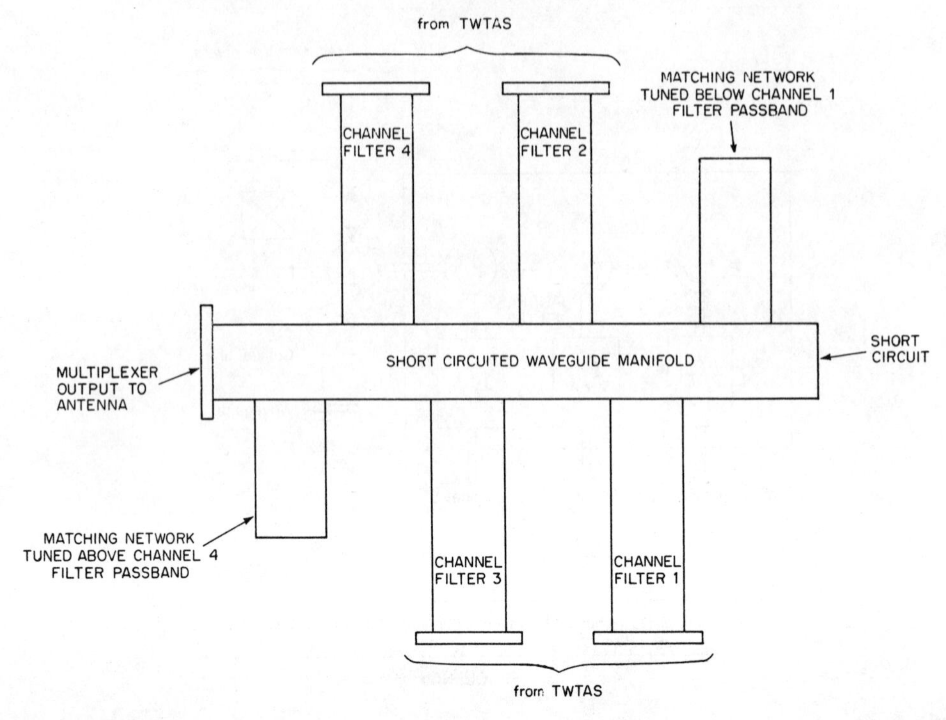

Figure 9.10 Output multiplexer (OMUX) using a common waveguide (manifold).

of each filter, coupled to the common waveguide through an iris, must constitute a short-circuit for out-of-band signals which originate from other channels. The characteristics of each filter thus influence operation of the whole system due to interactions.

Design and optimisation of the OMUX are difficult, particularly with a narrow guard band between each channel which is characteristic of the adjacent channel multiplexers used on current satellites. Previously, organisation of the channelised part into even and odd channels left a guard band between each channel with a width equal to that of one channel for each group; this involved less severe constraints on the specification of the output multiplexers associated with each group of channels.

For certain applications (for example in the case of a back-up satellite common to several systems using channels of different frequencies), multiplexers with tunable channels may be proposed. The frequency of each channel is then changed by telecommand by adjusting the resonant frequency of the band-pass filters using a tuning device operated by an electric motor [ROS-88].

Band-pass filters. The characteristics of the band-pass filters used are defined as a function of frequency by the amplitude and group delay specifications (Figure 9.11). The amplitude specifications indicate:

—The amplitude and maximum slope of the ripples of the amplitude of the transfer function within the passband.
—The rate of decrease of the amplitude at the limit of the passband.
—The minimum value of attenuation outside the passband.

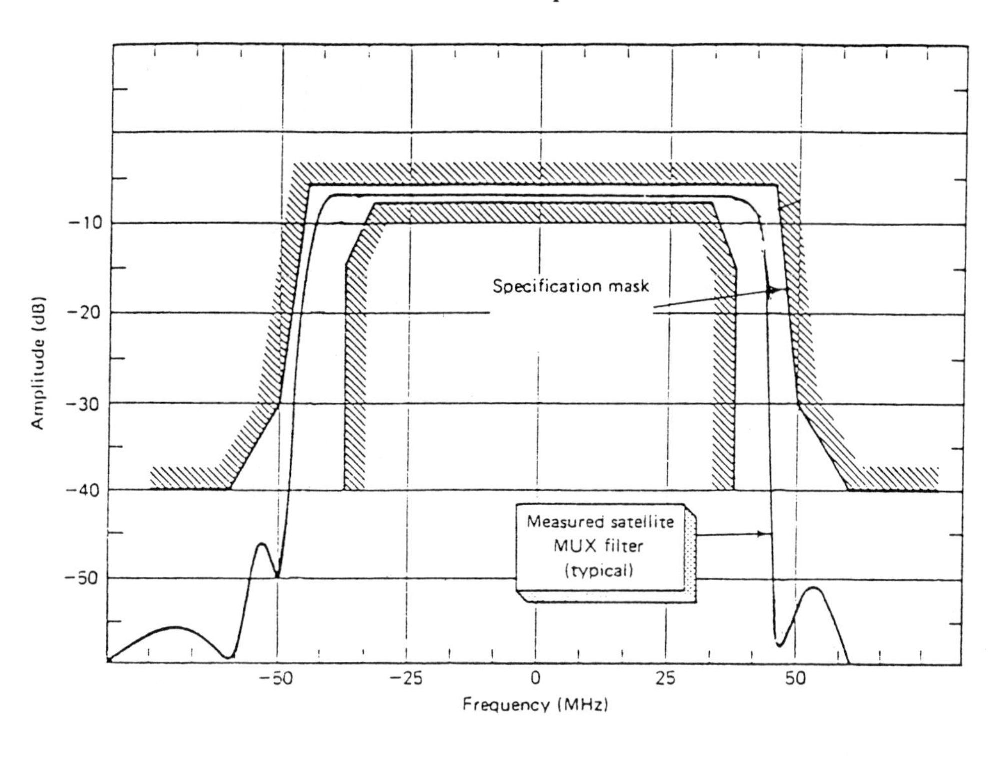

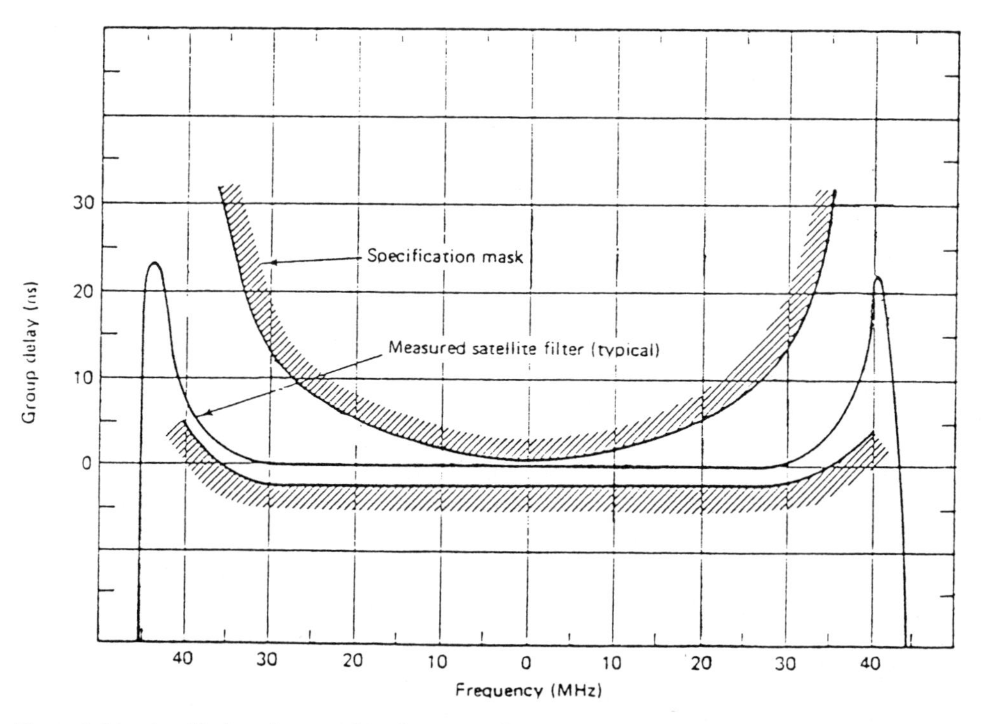

Figure 9.11 Amplitude and group delay limits as a function of frequency.

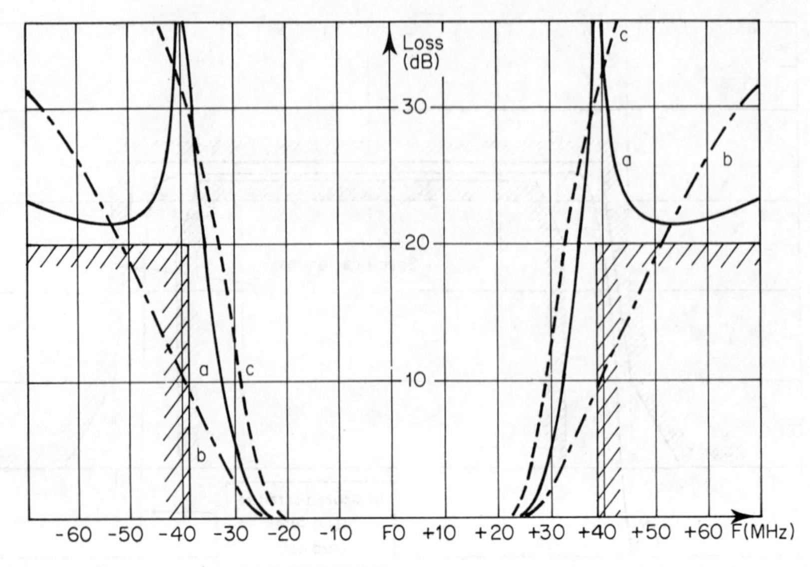

Figure 9.12 Examples of filter responses: (a) Four pole elliptic. (b) Four pole Chebyshev. (c) Six pole Chebyshev.

Gain ripples within the band are particularly critical at the input multiplexer which is located before the channel power amplifiers. The ripples give rise to spurious modulation of the signal amplitude. Due to AM/PM conversion effects in the power amplifiers, this amplitude modulation causes spurious phase modulation of the signals. This disturbs the operation of the frequency or phase demodulators of the earth station receivers and hence causes a degradation of the quality of the link.

A high slope at the extremity of the passband allows narrow guard bands between channels and thus permits maximum utilisation of the frequency bands employed. High out-of-band attenuation is necessary to avoid interference between channels.

The group delay specification defines the maximum permissible variation of group delay within the passband. Variation of group delay causes phase shift between the spectral components of a wideband signal and hence distortion.

The transfer functions are of the Chebyshev or elliptic type with several poles (four to eight) [ACC-86]. Examples of amplitude responses are given in Figure 9.12. Group delay equalisers associated with the filter elements enable the required characteristics to be obtained.

Waveguide cavity filters permit the high Q factors imposed by the amplitude specifications to be obtained [HAU-88]. Although single mode filters were used for early realisations, bi-mode techniques, where two resonant modes are excited in the same cavity have become dominant. This technique permits reduction by a factor of two in the number (and hence mass and volume) of cavities required for a transfer function with a given number of poles. Tri-mode cavities and even quadri-mode cavities have also been developed.

Transverse electric modes (TE_{11n}) are usually used in cavities. Coupling of resonant modes between adjacent cavities is realised by irises. Coupling of modes of different kinds (TE and TM (transverse magnetic)) is also foreseen and offers new possibilities for the realisation of multimode filters [ROS-89] [KUD-92].

In order to limit the drift of the centre frequency of the filter (typically less than 2.5×10^{-4} over the lifetime), it is important to avoid dimensional variations of the cavity as a consequence

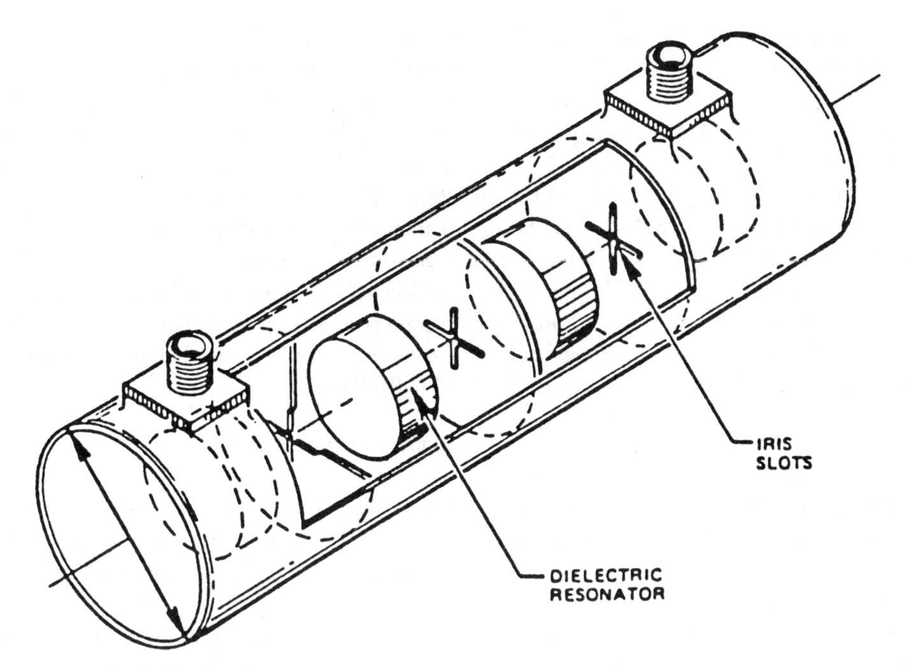

Figure 9.13 Example of realisation of a bimode cavity filter with resonator [CAM-90a]. (© AIAA. Used with permission.)

of ageing and thermal expansion. The material used must, therefore, have high mechanical stability and a low coefficient of expansion. It must also be light and a good conductor.

Aluminium, in spite of its high coefficient of expansion ($22 \times 10^{-6}/°C$) which requires precise temperature control to avoid thermal deformation, is used since its good conductivity and low density (2.7) permit cavities to be produced. Realisation of resin impregnated carbon fibre cavities seems promising due to the low coefficient of expansion ($-1.6 \times 10^{-6}/°C$), high rigidity and a low density (1.6). The C band payload multiplexers of the Telecom 1 satellite were produced with this material [FRA-82]. The complexity of the fabrication process, however, is limiting development of this technology. Invar, an alloy of 36% steel and 64% nickel (coefficient of expansion $1.6 \times 10^{-6}/°C$) has a high density (8.05) but its rigidity permits the use of cavities with thin wall manufacturing. These properties cause the material to be widely used. A coating of silver within the cavity ensures good conductivity and a good surface state which are required to obtain a high Q factor.

The size of the cavity is directly determined by the propagating wavelength within the medium which constitutes the interior. Traditionally, the interior of the cavity is empty and the cavities are large, particularly at low frequencies (C band). Hence the input multiplexers of the INTELSAT VI satellite contain 50 band-pass filters whose dimensions are 7 × 10 × 3 cm and whose mass totals 28 kg. The adjacent channel output multiplexers are provided by filters with three bimode cavities mounted on a common waveguide. The total mass of the 10 output multiplexers is 26 kg [THO-84].

Use of a material of high permittivity within the cavity (a resonator) and concentration of the field lines in a reduced volume permit fabrication of smaller cavities [BER-89]. Figure 9.13 shows an example of filter realisation using bimode cavities coupled together by an iris [CAM-90a]. Further reduction in the volume and mass of the multiplexer can be expected using superconducting microwave devices enabling construction of planar thin film microstrip type filters. The mass of a 44 channel input multiplexer could be as small as 3.3 kg (about 5 kg with the cryocooler) compared to 18.7 kg for the INTELSAT VII 44 channel IMUX using dielectric resonator technology. This related reduction in volume is more than 90%.

Another technique for realising multiplexers is the surface acoustic wave (SAW) technology. This technology can provide extremely sharp filters and inherent linear phase in the passband in a very small size and mass. It has the disadvantage that its preferred frequency band of operation is from 50 to 500 MHz and the passband has large group delay ripples ($\simeq 25$ ns) of short frequency intervals. The band of operation would require double conversion for C- or Ku-band satellite systems and probably a single conversion for mobile L-band systems. SAW-based IF processors, including banks of SAW channel filters, are being designed and qualified for the INMARSAT 3 satellites [KOV-91], [KUD-91].

9.2.3.3 *The channel amplifier*

At the output of the wideband amplifiers, in view of their non-linear characteristics and the permissible level of intermodulation products (see Section 9.2.2.3), there is a maximum signal level which can be obtained after frequency conversion. The effect of power splitters and losses in the multiplexer then determine the available signal level at the output of the IMUX. This level is generally insufficient to drive the channel output stage on account of the power which it delivers and its power gain.

The channel amplifier provides the required power gain, conventionally of the order of 20 to 50 dB. Good linearity is required in spite of the reduced number of carriers in the channel in order to avoid an excessive contribution to the system intermodulation noise. The amplifier is realised using bipolar or, more frequently, field effect transistors. The use of hybrid technology with bare chips and monolithic microwave integrated circuits (MMIC) permits compact and light realisation.

The amplifier is often associated with an attenuator which enables the gain to be adjusted over a range from 0 to several dB in steps of tenths of dB. This attenuator, which is controllable via the payload TTC links, is typically realised with PIN diodes whose bias is adjusted in order to vary the conductivity.

It can also be associated with a lineariser which compensates the non-linear amplitude and phase characteristics of the output stage; several techniques can be used for linearisation [BRE-74]. The predistortion technique which involves passing the signal through a circuit with a transfer function opposite to that of the device to be linearised seems to be the most appropriate (see Figure 8.31) [CZE-84].

9.2.3.4 *The output stage*

The output stage provides the power at the output of each channel and this determines the value of equivalent isotropic radiated power (EIRP) of the channel concerned. The nominal power is defined by the single carrier saturation power of the amplifier used. The operating point is adjusted in accordance with the nature of the signals transmitted in the channel in such a way as to obtain intermodulation noise which is compatible with the specification.

The chosen operating point, as defined by the corresponding input backoff IBO (or output backoff OBO), results from a compromise between the available output power and the level of intermodulation noise:

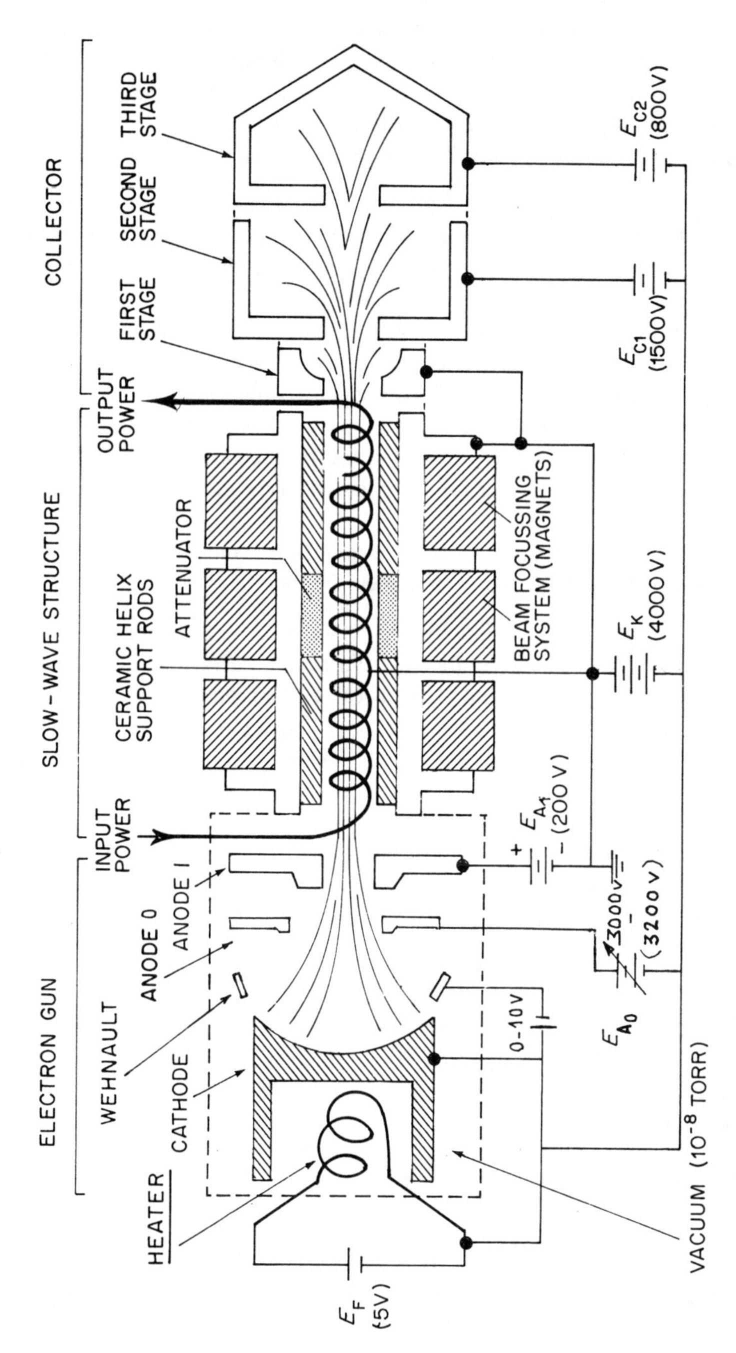

Figure 9.14 The arrangement of a travelling wave tube.

—A small (absolute) backoff (operation close to saturation) has the benefit of high power but intermodulation noise is high since the device operates in a highly non-linear region. It should be noted that, in multicarrier operation, zero input backoff corresponds to an operating point on the transfer characteristic beyond saturation for one of the carriers and hence the maximum power per carrier is obtained for a non-zero input back-off (see, for example, the curve for one of the two carriers in Figure 9.3).
—A large (absolute) backoff limits intermodulation noise, but the available output power is reduced.

The procedure used to determine the backoff generally consists of optimising the backoff for which the carrier power-to-noise power spectral density ratio $(C/N_0)_T$ of the total link (station-to-station) is maximised (see Section 4.4.4).

A particularly important parameter of the output stage is the efficiency. The efficiency is defined as the ratio of radio-frequency output power to electric power consumed. The difference is dissipated in the form of heat. A high value of efficiency thus leads to a reduction of electricity consumption and hence the size and mass of the satellite electrical system; the performance required of the thermal control system (as specified in terms of the heat extraction capacity) is also reduced. The efficiency of an amplifier is generally maximum close to saturation.

Two types of power amplifier are used on satellites—travelling wave tube amplifiers (TWTA) and transistor solid state power amplifiers (SSPA).

Travelling wave tube amplifiers. Travelling wave tubes operate by interaction between an electron beam and the radio wave. Figure 9.14 illustrates the organisation of a travelling wave tube. The electron beam, generated by a cathode raised to a high temperature, is focused and accelerated by a pair of anodes. The wave propagates along a helix; the electron beam, whose focus is maintained by concentrically located magnets, flows within the helix. The axial velocity of the wave is artificially reduced by the helix to a value close to the velocity of the electrons. The interaction leads to a slowing of the electrons which give up their kinetic energy. A collector receives the electrons at the output of the helix. Division of the collector into several stages at different potentials permits better matching to the dispersion of the residual energy of the electrons and hence an increase in the efficiency of the tube [DAY-88], [PEL-88].

Typical values of the characteristics of tubes used are:

—Power at saturation: from 8 to 50 W (100 to 250 W for applications in direct broadcast television satellites).
—Efficiency: 40 to 50%.
—Gain at saturation: around 55 dB.
—$(C/N)_{IM}$ at saturation: 10 to 12 dB.
—AM/PM conversion coefficient K_p: around 4.5°/dB.

An electric power supply (electric power conditioner (EPC)) generates the various voltages (up to 4000 V) required for operation of the tube [HUB-88]. The efficiency is of the order of 80% which leads to a global efficiency of 40%. The total mass is around 2.2 kg (tube, 0.7 kg; power supply, 1.5 kg).

At high frequencies (Ka band, 20 GHz), propagation conditions can cause large variations (from 5 to 25 dB) of link attenuation. Travelling wave tube amplifiers operating at EHF (extra high frequency) can be designed to deliver different saturation powers (e.g. 5, 17 and 50 W)

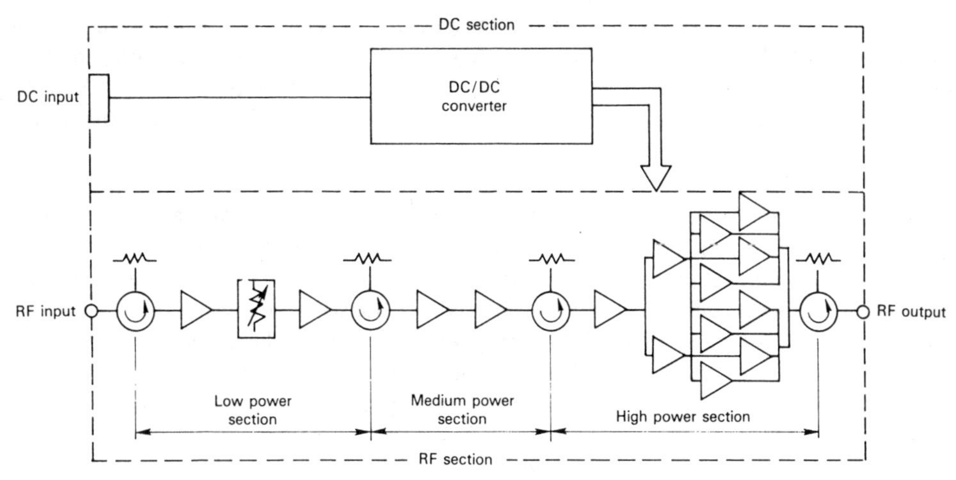

Figure 9.15 Block diagram of a transistor power amplifier.

in order to be able to match the channel amplifier power to the propagation conditions by telecommand.

Solid state amplifier. These amplifiers use field effect transistors. The available power and operating frequency are continuously increasing as technology progresses. The powers required are obtained by connecting transistors in parallel in the output stages (Figure 9.15). Transistor amplifiers have been used operationally in C band since the beginning of the 1980s with powers of the order of 10 W to complement tube amplifiers; they are now appearing in Ku band with powers around 30 W [FRE-84], [LAP-86], [ONO-88].

The main characteristics of transistor amplifiers are:

—Power: 10 W at 4 GHz, 30 W at 12 GHz (depending on the number of transistors in parallel).
—Efficiency: 20 to 35%.
—Gain at saturation: 50 dB (depending on the number of stages).
—$(C/N)_{IM}$ at saturation: 14 to 18 dB.
—AM/PM conversion coefficient K_p: around 2°/dB.

Table 9.2 Performance comparison of the SSPA versus the TWTA

	TWTA	SSPA
Operating frequency range	3.7–4.2 GHz	3.7–4.2 GHz
Saturated power output	10–15 W	10–15 W
Gain at saturation	58 dB	58 dB
Third order intermodulation product relative level $(C/N)^*_{IM}$	11 dB	15 dB
AM/PM conversion coefficient K_p	4.5°/dB	2°/dB
DC to RF efficiency including EPC[†]	40–45%	30–35%
Mass including EPC	2.2 kg	0.9 kg
Failure in 10^9 h	> 2000	< 500

*Close to saturation
[†]EPC: Electric power conditioning.

The power supply associated with the transistor amplifier generates the required supply and bias voltages (which range from one volt to a few tens of volts). Temperature compensation is necessary to avoid thermal drifts. The efficiency is of the order of 85 to 90% and this leads to an overall efficiency of 20 to 35% according to the frequency band. The total mass varies from 1 to 2 kg according to the power.

Comparison between tube and transistor amplifier. Table 9.2 shows typical values of various parameters for a travelling wave tube amplifier and a transistor amplifier. It can be seen that the transistor amplifier is more linear than the tube amplifier. For the same installed power, the available carrier power will thus be greater with a transistor amplifier since the backoff required will be less for the same level of intermodulation noise. This greater linearity leads to an increase in the capacity of the transponder [FRE-84]. The transistor amplifier (more precisely the amplifier and its power supply) is also lighter than the tube amplifier. On the other hand, the efficiency of a transistor amplifier is less than that of a tube amplifier; the power consumption and thermal dissipation are thus greater.

Finally, one characteristic which has not been mentioned above is reliability. The reliability of transistor amplifiers seems to be greater than that of tube amplifiers, although substantial experience of operation in space is not yet available. The use of redundancy with a smaller total number of units can thus be envisaged and this leads to a further mass reduction.

9.3 TRANSPONDERS FOR MULTIBEAM SATELLITES

A multibeam satellite features several antenna beams which provide coverage of different service regions. As received on board the satellite, the signals appear at the outputs of one or more receiving antennas. The signals at the repeater output must be fed to the various transmitting antennas. Two basic configurations are possible:

- —The receiver–transmitter combinations constitute independent networks.
- —The stations within different coverage regions are associated within the same network and links must be established between any pair of stations situated in different regions.

In the first case, the satellite contains as many independent repeaters as there are receiver–transmitters. These repeaters operate in different frequency bands (e.g. 6/4 and 14/12 GHz), possibly with two orthogonal polarisations for two receiver–transmitters in the same frequency band. This is the situation with the Telecom 1 satellite, for example.

The second configuration corresponds particularly to the concept of the multibeam satellite discussed in Chapter 5. Interconnectivity between the different beams must be established. This is achieved by transponder hopping for conventional (non-regenerative) satellites; the regenerative satellite case is considered in Section 9.4. This is the situation with the Eutelsat satellites, for example.

Combinations of these two configurations are also encountered. For example, the Intelsat satellites have payloads at 6/4 and 14/12 GHz with the possibility of interconnection at intermediate frequency by means of microwave switches at 4 GHz (which is the intermediate frequency of the 14/12 GHz payload).

In the following part of this section, a system containing M receiving beams (uplinks) in the same or different frequency band and N transmitting beams (downlinks) with the possibility of interconnecting any pair of beams will be considered.

9.3.1 Fixed interconnection

The possible interconnections between beams are established once and for all during manufacture. The receiver coverage is often common to all regions, possibly using two orthogonal polarisations ($M = 1$ or 2 according to whether one or two polarisations are used).

The satellite contains as many active receivers as there are uplink beams. At the receiver outputs, the input multiplexers (IMUX) divide the frequency band into different channels; the number of channels is, a priori, a multiple of the number N of transmitting regions if the traffic between regions is balanced and the channel widths are the same.

Unlike the single beam satellite where all channels are grouped on transmission with a unique destination region, the multibeam repeater contains as many output multiplexers as transmitting beams; each of these multiplexers combines the channels which are allocated to the beam concerned. Selection of the destination region is achieved by choosing the uplink carrier frequency so that, after frequency conversion, it is within the band of one of the channels allocated to the region concerned (transponder hopping, see Section 5.2).

9.3.2 Reconfigurable (semi-fixed) interconnection

Unlike the arrangement of the previous section, association of channels with transmitting antenna inputs is not explicit. Using switches controlled by telecommand (mechanically actuated switches in the waveguide), it is possible to reconfigure the payload by changing the branching between the channel output and the inputs of the multiplexers associated with the inputs to the transmitting antennas. This enables the capacity of a beam (that is the width or number of channels allocated to the beam) to be adapted in accordance with changing traffic demand in the service regions during the lifetime of the satellite. Of course, the number of possible configurations is limited and is predefined during manufacture.

These facilities are available on the INTELSAT satellites for example. They are further exploited on the EUTELSAT II satellites by permitting the frequency band used for the downlink beams to be selected. This is possible due to three conversion units which convert the uplink frequencies (14–14.5 GHz) into three separate frequency bands which exploit the various segments available for downlinks in Ku band in Region 1 (10.95–11.2 GHz, 11.45–11.7 GHz and 12.5–12.75 GHz). The two orthogonal polarisations are used on each uplink and downlink frequency band. The organisation of the transponder is represented in Figure 9.16. Three transmitting beams are generated, two with the 'west' antenna and one with the 'east' antenna. The 'east' antenna is also used as a receiving antenna. The coverage of each of the three transmitted beams can be switched independently between narrow beam coverage at high gain on the central part of Western Europe and a wider European coverage (see Figure 9.40). The channels are arranged in three groups corresponding to the downlink frequency sub-bands for each polarisation and fed to the inputs of the transmitting antennas. It is possible to modify the branching of certain channels (3, 4, 5, 11, 12 and 13) by controlling the switches located between the multiplexers and the channel amplifiers. A high degree of flexibility of channel management according to the type of signal to be transmitted (such as

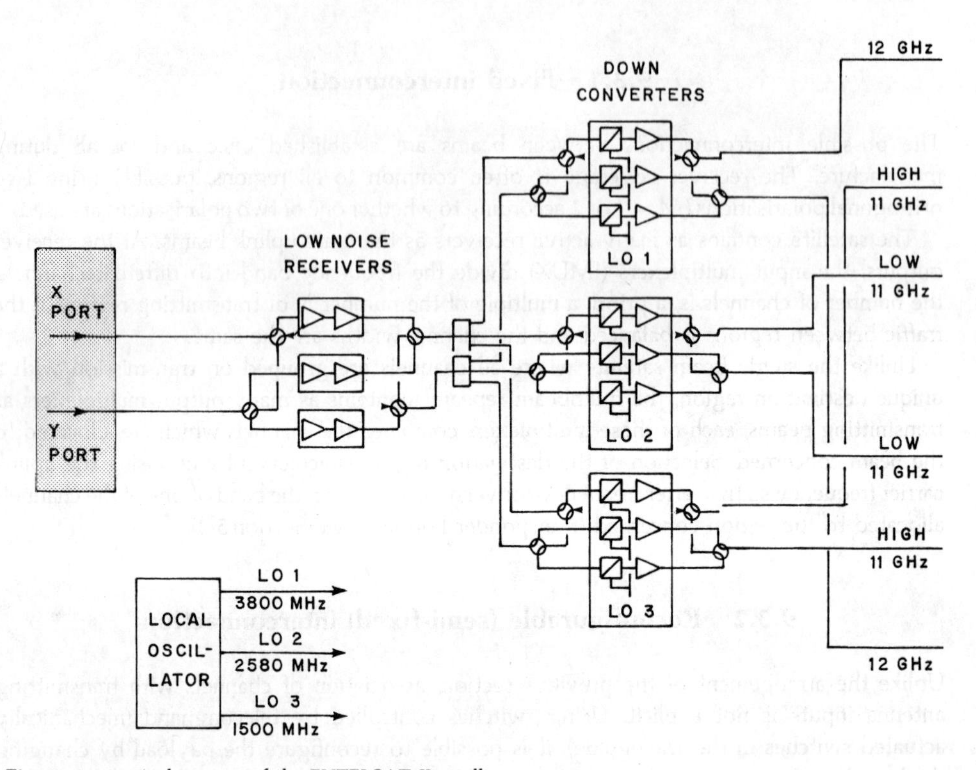

Figure 9.16 Architecture of the EUTELSAT II satellite repeater.

telephony or television) is thus obtained. The frequency chart given in Figure 9.17 illustrates frequency re-use by orthogonal polarisation and the effect of switching on the channels concerned.

9.3.3 On-board switching

Multibeam satellites may require rapid reconfiguration (within a few hundred nanoseconds) of the intercommunications between beams. They must, therefore, be equipped with a fast switching device. A switching matrix is used to interconnect the receivers associated with uplink beams and the downlink beam transmitters sequentially. Fast switching implies the use of:

—Switches using active elements.
—An on-board device to control the switching sequence (Distribution Control Unit (DCU)).

9.3.3.1 Active switching elements

The first switching matrices to be developed used PIN diodes as the switching elements. This is the case for the 6 × 6 (actually 10 × 6 due to redundancy) switching matrix used on Intelsat

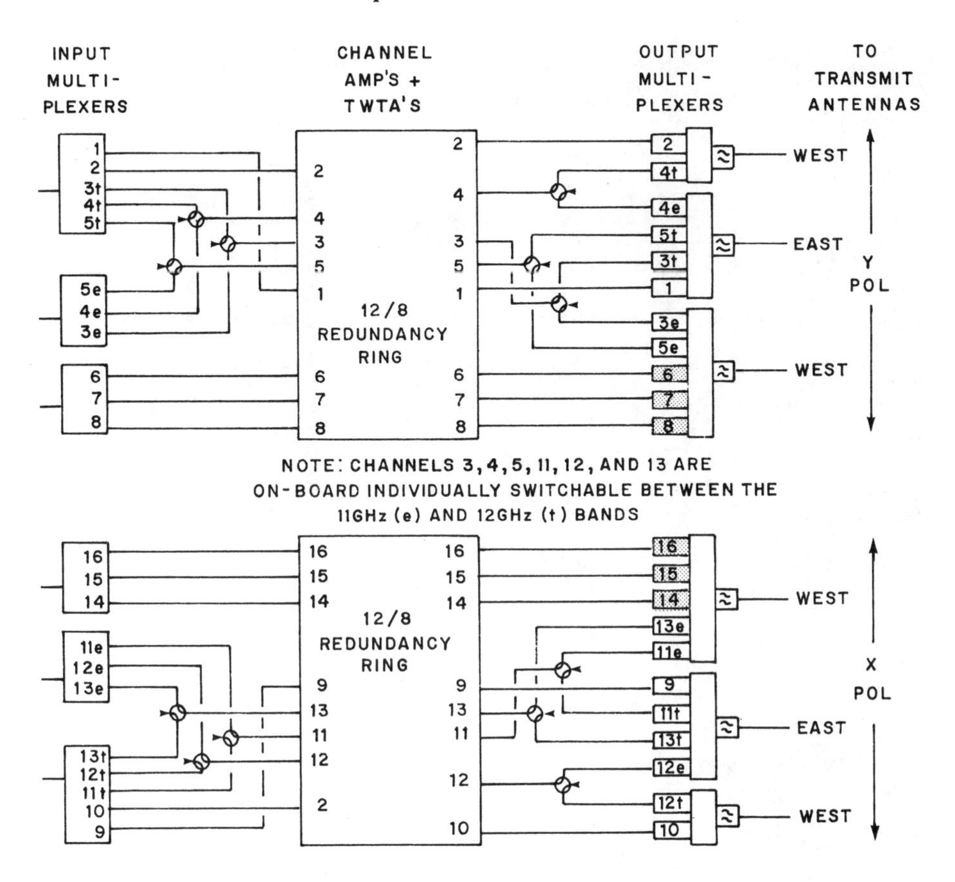

VI [ASS-81]. PIN diodes have now been replaced by field effect transistors which provide better isolation (60 dB), a shorter switching time (less than 0.1 ns instead of 10 to 100 ns) and gain (of the order of 15 dB with two stages in cascade); this enables the losses inherent in the architecture to be partially compensated. Developed systems have used single gate transistors [OPP-87] and dual gate transistors, when available, whose two electrodes facilitate matching, polarisation and control [BET-87].

9.3.3.2 *Optical switching*

The use of optical switching elements has also been investivated [KRE-80]. With this approach, the radio-frequency carriers modulate optical sources (lasers). The light signals are dynamically switched with optical switches and are then detected to regenerate the radio-frequency carriers.

Various approaches are possible for realisation of the switching elements; these include opto-electronic switching using an intermediate detector [HAR-80], [KIE-81], modification of the coupling between two adjacent optical waveguides (delta-beta switch) and transmission or total internal reflection (TIR) where two optical waveguides cross by variation of the dielectric constant [TSA-78]. A combination of the last two techniques has permitted the development of an optical switch which provides good performance in terms of isolation (58 dB) and switching time (several nanoseconds) and can be used to realise switching matrices

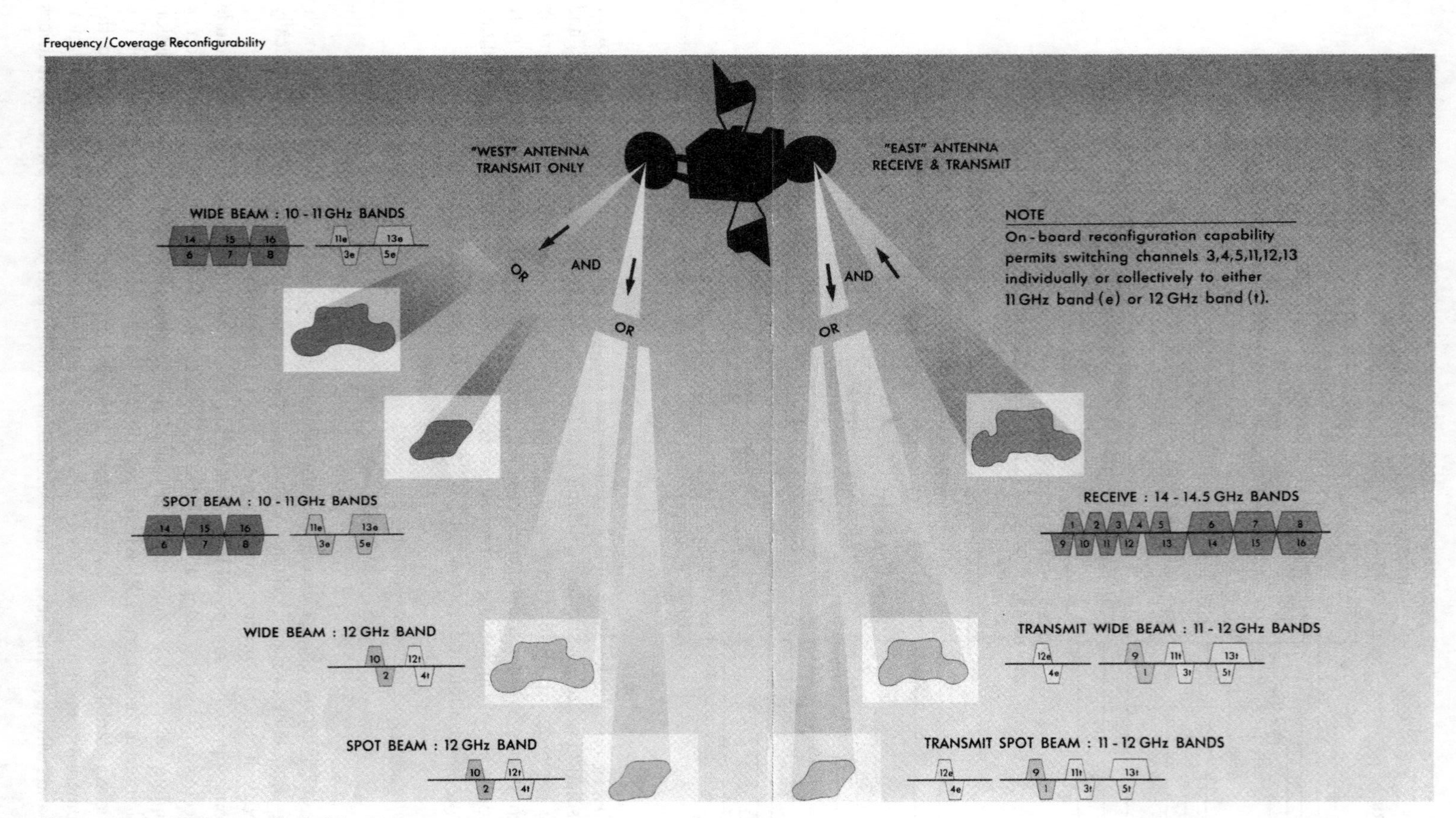

Figure 9.17 Frequency plan of the EUTELSAT II satellite.

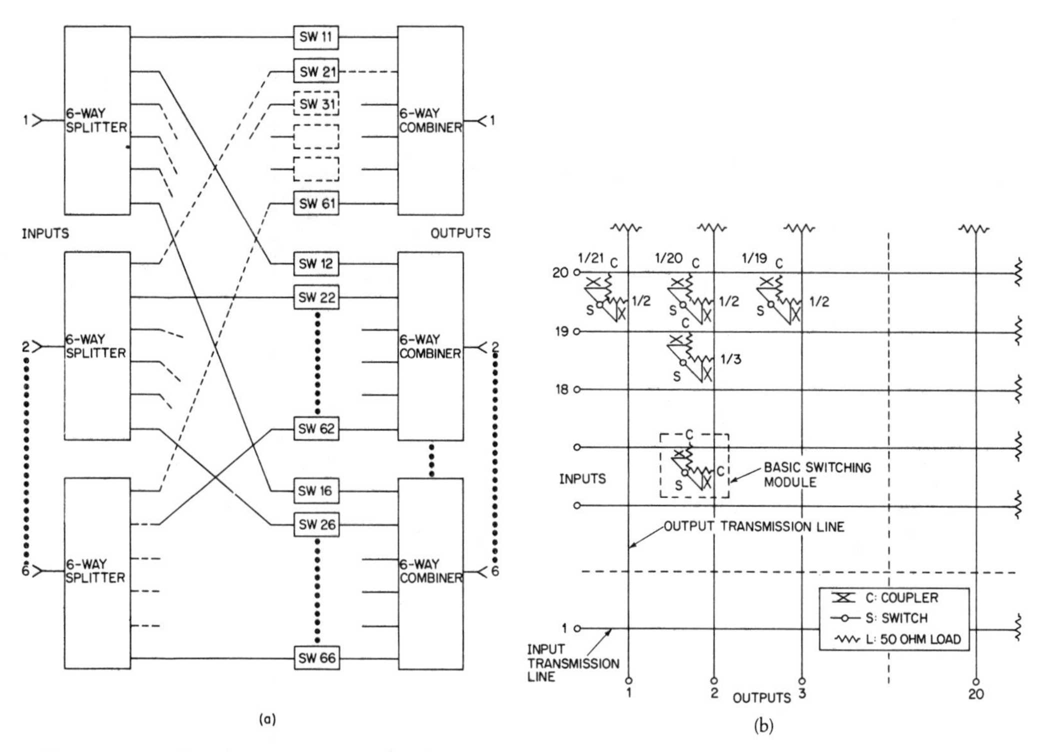

Figure 9.18 Switch matrixes: (a) with splitters and combiners, (b) cross-bar.

of different architectures [PET-87]. In spite of the high insertion losses (60 dB), the advantage of an optical switching matrix lies in the associated reduction of size and mass (e.g. for Intelsat VI, each of the two 10×6 microwave matrices measures $30 \times 40 \times 2.5\,\text{cm}^3$ and weighs 7.3 kg; a 10×6 matrix realised using optical technology would measure $9 \times 1 \times 0.5\,\text{cm}^3$ and would weigh 1 kg).

9.3.3.3 Switching matrix architectures

Of the various conceivable architectures for an $N \times N$ (or $M \times N$) matrix, only two are capable of distributing the information present on one of the uplink beams onto several downlinks (broadcast mode). These are the architectures with power splitters and combiners on each input and output (power divider–combiner architecture) and with cascaded directional couplers (coupler cross-bar architecture). These architectures are presented in Figure 9.18.

The first architecture uses $N(M)$ power dividers to separate the input into N channels and N power combiners with $N(M)$ inputs at the output. The various divider outputs are connected to one input of each of the combiners by way of an on–off switch. The connection between the input of the divider concerned and the output of the combiner may be made or not according to the closed or open state of the switch. The matrix has a cubic form which makes access to the interior elements difficult and favours coupling between channels (Figure 9.18a).

Table 9.3 Performance of several switch matrix implementations

	Comsat[a]	Thomson[b]	General Electric[c]	Ford[c]	NTT[d]
Array size	8 × 8	8 × 8	20 × 20	20 × 20	4 × 4
Type of architecture	Divider combiner	Divider combiner	Coupler crossbar	Coupler crossbar	Divider combiner
Switching device	Pin diode	Pin diode	Dual gate FET	Dual gate FET	Dual gate FET
Bandwidth	3.5–6.5 GHz	3.7–4.2 GHz	6–7 GHz	3.5–6 GHz	1.8 GHz ± 140 MHz
Switching time	⩽60 ns	⩽50 ns	25 ns	15 ns	⩽100 ns
Insertion loss	⩽23 dB	⩽28 dB	18 dB	20.7 dB	⩽17 dB
Ins. loss variation (any 500 MHz)	⩽1 dB	⩽1 dB	1 dB	1 dB	—
Path to path insert. loss scatter	⩽1.7 dB	⩽1.5 dB	—	—	⩽3 dB
Path isolation	⩾50 dB	⩾50 dB	⩾50 dB	⩾45 dB	⩾53 dB
Intermodulation $(C/N)_{IM}$	⩾45 dB	⩾45 dB	—	—	—
Group delay variation	⩽0.5 ns	⩽1 ns	—	⩽0.5 ns	—
Switch matrix mass	2.95 kg	2.3 kg	11 kg	8.7 kg	—
Switch matrix size (cm)	15 × 16 × 11	10.5 × 12 × 12	47 × 48 × 17	47 × 48 × 8.5	27 × 20 × 20 × 15
Power consumption	<7.5 W	8.5 W	33 W	5.7 W	1.7 W

[a] [ASS-82]
[b] [ROZ-76]
[c] SPI-83]
[d] [KAT-83]

The matrix architecture (cross-bar) uses $N(M)$ input lines and N output lines with interconnecting elements located at the intersections (Figure 9.18b). These elements consist of two directional couplers with an on–off switch between them. The resultant matrix has the advantage of being planar and well suited to the use of microwave integrated circuit technologies.

There are two possible approaches to realising the couplers for the switching elements—identical and distributed coupling coefficients. In the first approach, all the directional couplers of one row or one column have the same coupling coefficient. Realisation of couplers is simplified, but the power varies from one crossover point to the other.

The other approach enables minimum insertion loss to be obtained by operating all switches at the same power level. In order to distribute the power equally among the N crossover points on the same input line, the couplers have different coupling coefficients equal to $1/(N+1)$, $1/N, \ldots, 1/2$ for the 1st, 2nd, ..., Nth crossover points. It is thus necessary to realise couplers with different coupling coefficients of a precise value. Furthermore the performance is affected by impedance variations between switches.

Correct operation of the matrix in spite of the failure of one or more switching elements is obtained by producing matrices which contain more rows and/or columns than necessary (more than the number of uplink and downlink beams). Redundancy is thus provided and switches located at the input (and/or output) are used to route signals from the inputs to the active lines and/or from the active columns to the outputs). A fault detecting device can be integrated into the switching matrix [BET-87].

Table 9.3 compares the performance of different realisations of switching matrices.

9.3.3.4 *Frequency domain switching*

Another multiple access switching technique is possible with multi-beam satellites; this is frequency division multiple access with on-board switching (satellite switched frequency division multiple access (SS/FDMA)).

In the context of a non-regenerative repeater, a battery of band-pass filters is used as a switch and routing is performed according to the frequency of the uplink. The difference compared with the transponder hopping techniques used with current satellites lies in the bandwidth and number of filters used; interconnections between the beams of current satellites are realised on the basis of the width of one satellite channel (typically 36 to 72 MHz) and the number of channels is limited by considerations of bulk and mass. An SS/FDMA system could contain a large number of filters with a bandwidth matched to the capacity of the beams concerned. The use of variable bandwidth filters or selection of those whose bandwidth is matched from a group of filters permits on-demand assignment of the resource (the frequency band) as a function of the traffic to be transmitted from one beam to the other [SAN-87]. Surface acoustic wave (SAW) and magnetostatic surface wave (MSW) technologies permit realisation of narrow bandwidth filters of low mass and bulk. The bandwidth can be made variable by combining band-pass filters of adjacent widths [AND-88], [KOV-91].

9.4 REGENERATIVE TRANSPONDERS

Figure 6.1 shows the essential functions of a regenerative transponder. Fundamentally, a regenerative transponder performs carrier demodulation and remodulation of the resulting baseband signals before transmission at the downlink frequency. In addition, the availability of baseband signals allows specific processing, as examined in Chapter 6, to be considered.

Two applications are evident at the present time:

—Satellite systems for fixed stations with link switching at baseband using time division multiple access (Baseband SS-TDMA); the advantage of the regenerative satellite lies in the possibility of interconnecting networks with different data rates (see Section 6.3.2.2).
—Satellite systems for mobiles with time division multiplexing on the downlinks and single channel uplinks which access the satellite using frequency division multiple access (SCPC-FDMA/TDMA). Using regenerative satellites it is possible to reduce the required EIRP for mobile stations and to operate the satellite repeater in the vicinity of saturation (see Section 6.4.1).

9.4.1 Examples of satellites with on-board regeneration

The ITALSAT (national satellites developed by Italy) and ACTS (the NASA Advanced Communications Technology Satellite) satellites are two examples of satellites with on-board regeneration.

9.4.1.1 *The ITALSAT satellite*

The ITALSAT satellite contains three payloads of which one is regenerative with baseband switching which interconnects six narrow beams at 30/20 GHz. The block diagram of the

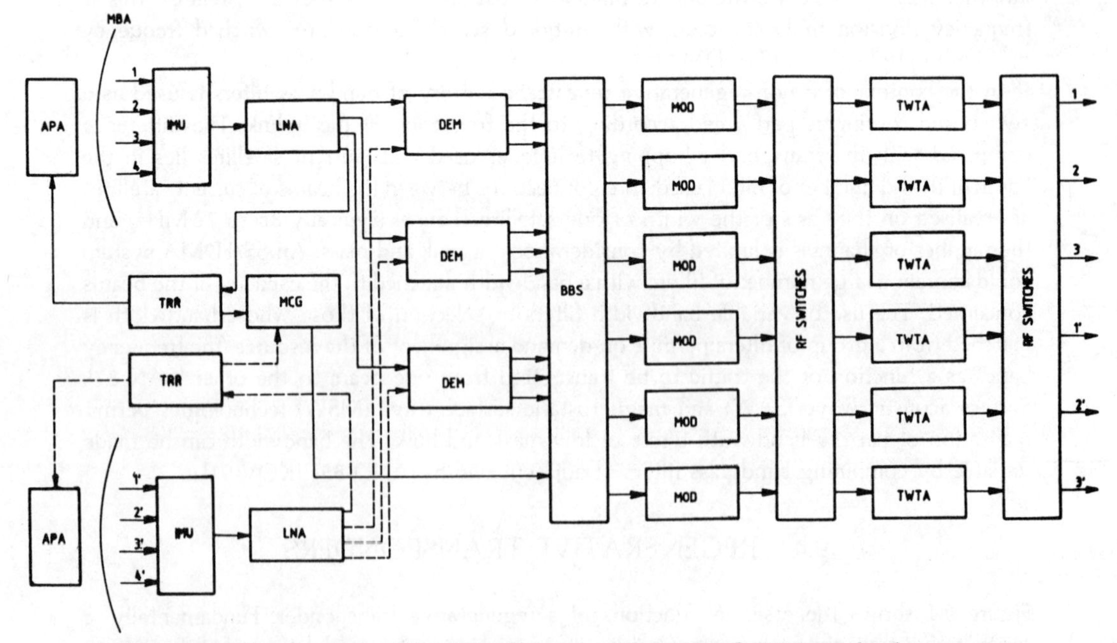

Figure 9.19 Architecture of the ITALSAT satellite payload [MOR-88]. (© AIAA. Used with permission.)

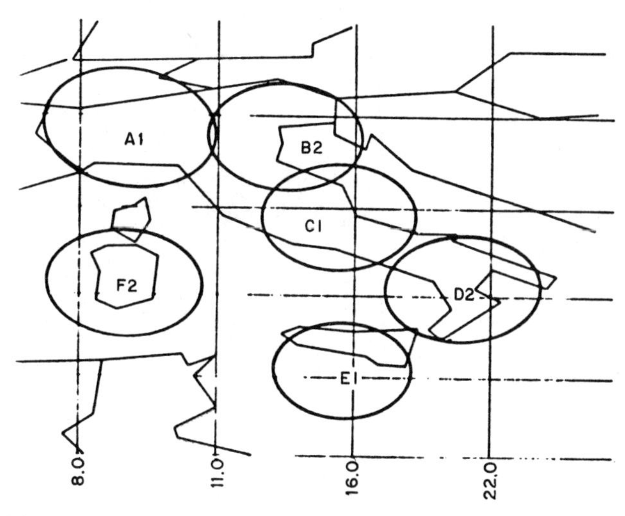

Figure 9.20 Multibeam coverage of the ITALSAT satellite [MOR-88]. (© AIAA. Used with permission.)

regenerative payload is given in Figure 9.19 and contains six 110 MHz bandwidth channels configured in groups of three with two deployable antennas which are used for both transmission and reception [SAG-87], [MOR-88a]. Each antenna generates three narrow beams to provide the coverage illustrated in Figure 9.20.

The carriers received at 30 GHz with a data rate of 147 Mbit/s from the three beams are frequency multiplexed, amplified and then shifted in frequency (LNA). After amplification at the intermediate frequency of 12 GHz, the carriers are demodulated using coherent demodulation (DEM). Baseband switching connects the six demodulators to six modulators.

The functions of the baseband switch matrix (BBS) are as follows:

—Synchronisation of the demodulated bursts with the on-board clock.
—Routing of bursts to the appropriate demodulator.
—Generation of reference bursts to synchronise the earth station network.

The baseband signals at the output of the matrix modulate six carriers at 20 GHz with four phase states and a data rate of 147 Mbit/s (MOD). These carriers are then amplified by six of the eight TWT amplifiers before being routed to the diplexers of the sources which illuminate the two reflectors. These reflectors can be oriented and steered by a pointing system (APA) which uses a terrestrial radio-frequency beacon and a four source error angle measurement system (TRR) in order to guarantee the required pointing accuracy (of the order of 0.03°).

9.4.1.2 *The ACTS satellite*

The ACTS satellite contains a payload at 30/20 GHz which uses two multibeam antennas, one for reception and one for transmission; each generates three fixed beams and two beams

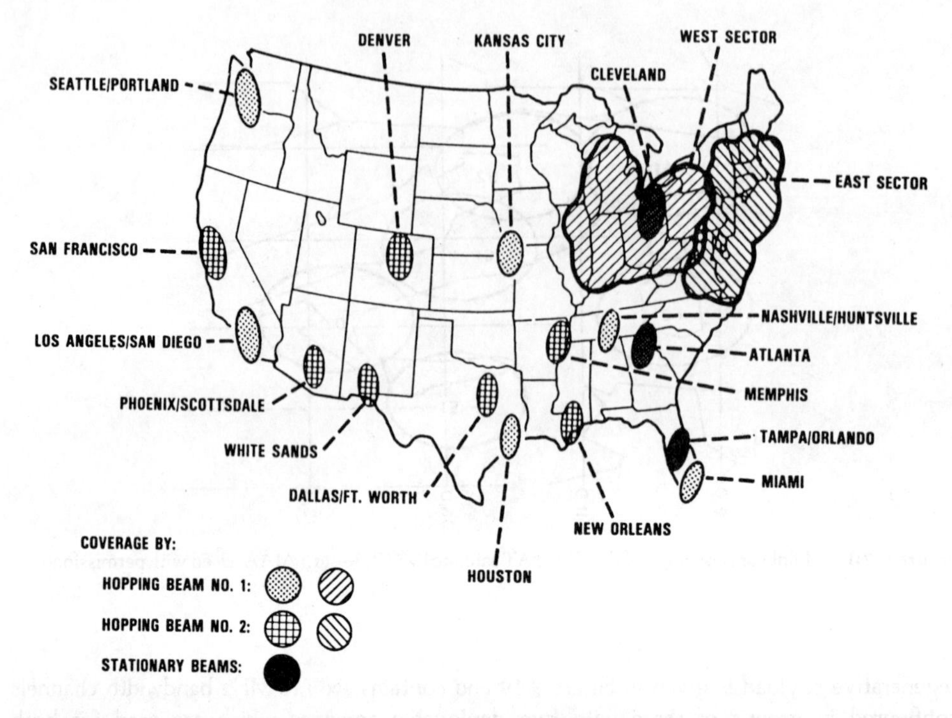

Figure 9.21 Coverage of the ACTS satellite antennas [NAD-88]. (© AIAA. Used with permission.)

with electronic pointing and an aperture of 0.3°. The various coverages obtained are presented in Figure 9.21 [NAD-88]. In addition, a 1° beam with mechanical pointing can be oriented to any point on the earth which is visible from the satellite.

The received carriers are routed to the switching device with a frequency change to 3 GHz by four receiving and amplifying units (Figure 9.22). The switching device consists of an intermediate frequency matrix and a baseband processor; one or the other is used according to the mode of operation of the payload.

In the first mode of operation, the IF switching matrix interconnects the uplinks and downlinks at 220 Mbit/s using time division multiple access of the three fixed beams.

The baseband processor is used with the two scanning beams in a mode of operation which uses baseband switching and temporary data storage in buffer memories (see Section 6.3.2). Each beam accepts either one link at 110 Mbit/s or two frequency division multiplexed links at 27.5 Mbit/s. The links use time division multiple access (TDMA) with a 1 ms frame. Three demodulators (one at 100 Mbit/s and two at 27.5 Mbit/s) recover the bursts which are organised in words of 64 bits. The words are stored in memory before routing by the baseband switching matrix to the output memories from which they are extracted for multiplexing before remodulation. The maximum capacity per beam is 110.592 Mbit/s which corresponds to 1728 words of 64 bits per frame. In order to compensate for the large attenuation in Ka band during rain, the data rates of the earth stations concerned are divided by four and an error correcting code of coding ratio 1/2 is also activated (see Section 3.7.9). The demodulators then operate at one half of the nominal data rate (55 Mbit/s and 13.75 Mbit/s); the bit rate transmitted in the bursts concerned is halved. A maximum likelihood

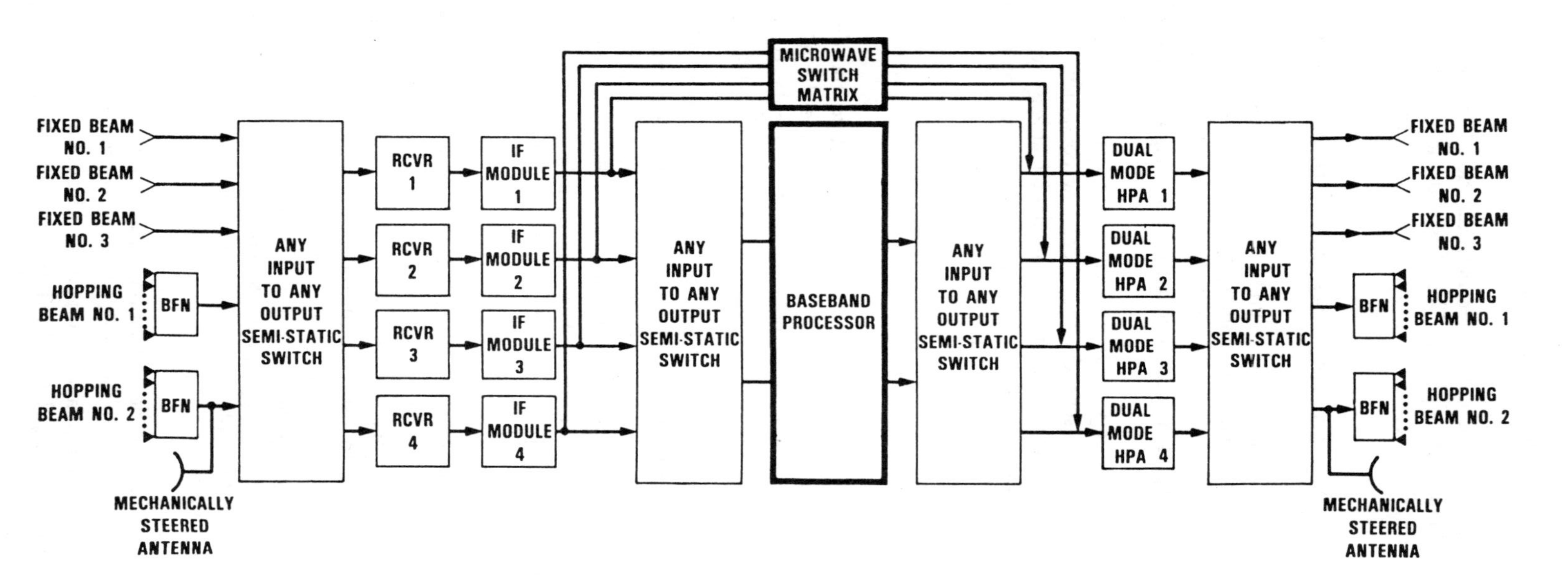

Figure 9.22 Architecture of the ACTS satellite payload [NAD-88]. (© AIAA. Used with permission.)

decoder decodes the convolutional code after demodulation. An encoder performs the complementary function before remodulation. Decoders and encoders are activated only for the bursts concerned in the frame; control is provided by the network control station.

9.4.2 Equipment for regenerative transponders

A regenerative transponder contains various units of which some, such as the low noise amplifiers, mixers, intermediate frequency amplifiers, power amplifiers and radio-frequency filters are similar to those of a conventional transponder. These units are described in Section 9.2. Modern technologies are envisaged for this equipment [MAH-87], [ASS-88], [ACC-89], [EVA-89].

The equipment specific to a regenerative satellite consists of the demodulating and remodulating equipment and the baseband signal processing equipment.

The signals carried by a regenerative transponder are digital (at least at baseband, since the noise which is added to radio-frequency links gives the received carriers an analogue character). The specific equipment is thus designed to process digital signals.

9.4.2.1 Demodulators

Demodulation can be either coherent or differential according to the digital modulation anticipated for the uplink. In particular, for applications which use the uplink in frequency division multiple access mode (FDMA), possibly with a single channel per carrier (SCPC/FDMA), it is useful to demodulate the various carriers simultaneously.

Coherent demodulation. Four state phase modulation (QPSK) in association with a time division access mode provides good performance. Conventional QPSK demodulator structures are well suited. For carrier recovery, conventional structures use phase-locked loops whose acquisition times are too long for operation in burst mode (the phenomenon of 'hang-up'). Specific architectures such as that using a phase cancellation method may be considered [KUR-81]. It is also necessary to resolve the phase ambiguity of the recovered carrier. This can be achieved by using unique word detection. Finally, the demodulator contains circuits for digital clock recovery and digital signal restoration.

Signal filtering at intermediate frequency before demodulation must be optimised in accordance with the link characteristics; this applies particularly to transmission filtering at the earth station to limit degradation due to inter-symbol interference. This degradation leads to an increase of the bit error rate with respect to the theoretical value. Filters of the raised cosine type are currently used.

The ACTS satellite uses a particular form of frequency modulation of index 1/2 (serial minimum shift keying (SMSK)). An appropriate coherent demodulator has been developed [STI-82]. The performance and complexity of the demodulators are more or less equivalent to those of a QPSK demodulator.

Differential demodulation. Coherent demodulation requires a carrier recovery arrangement whose complexity involves an increase of mass and power consumption and poses reliability problems. Modulation using differential coding permits differential demodulation and avoids carrier recovery (see Section 3.7.8.). This simplification of the demodulator architecture is paid for with an increase of the order of 2.5 dB in the power required on the uplink. Differential demodulation is based on comparison of the symbol duration between the received waveform and this same waveform delayed by the symbol duration [ACC-82], [ACC-83]. It is, therefore, necessary to provide a delay line with a stable performance, particularly with respect to temperature. Various technologies have been developed and include filters on silica substrates [LEE-78a], [CHI-81a], [CHI-81b], microstrip lines on dielectric substrates, waveguide filters [OHM-81] etc.

Multicarrier demodulation. One of the advantages of a regenerative satellite is the ability to use frequency division multiple access on the uplinks while having time division multiple access on the downlinks. This enables the transmitting power of earth stations to be reduced and the maximum benefit to be gained from the power of a transponder which operates close to saturation (see Section 6.4.1). A large number of carriers is thus present at the input of the satellite transponder and these must be demodulated.

One approach is to use a bank of band-pass filters centred on the various carriers and to follow each filter with a demodulator. This leads to a high mass and power consumption when a large number of carriers is concerned.

When the various uplink carriers have the same data rate and are equally distributed in frequency, block demodulation of all the carriers may be considered. The approach and complexity depend strongly on the existence, or otherwise, of synchronisation of the symbol clock of the digital signals carried on the various carriers [IZU-84], [PER-87].

Several techniques are available for realising a multicarrier demodulator (MCD) [ANA-89]. One uses baseband processing of the signal. After changing the frequency of the carriers to the vicinity of the baseband, time samples of the composite signal are taken and analysed by a digital signal processing algorithm. This processing can be performed on all carriers by combining multiphase networks and using the Fast Fourier Transform (FFT) [BEL-74], [BEL-86], [CAM-89] or on each carrier after demultiplexing using an array of digital filters or a tree partition with successive divisions of the spectrum by two [GOC-88], [ALB-89].

The processing can be performed at intermediate frequency by using the 'Chirp' Fourier Transform by means of surface acoustic wave filters (SAW) mounted in transmultiplexers [ANA-85], [HOD-88], [BAK-89], [KOV-91]. The filter output signal is a time domain representation of the short-term spectrum of the frequency multiplexed input signal. The use of optical techniques in association with surface acoustic wave devices permits demultiplexing, and in some circumstances demodulation, of several carriers within the same circuit [ANA-86], [ANA-88].

Instead of permanently assigning a carrier at each station for transmission (or otherwise) of information, it is also possible to share a set of links at different frequencies with the same data rate between stations. Time division multiple access is then used and stations select a frequency for each burst for which a time interval in the frame is available (multi-frequency time division multiple access (MF-TDMA) or multi-carrier TDMA (MC-TDMA). In order to limit the complexity of the transmitting station, the digital data rate must remain low (e.g. 2 Mbit/s) as must the frame duration (e.g. 0.5 ms). The number of symbols per frame is thus limited (to 500 with the values selected). To retain adequate efficiency, it is thus necessary to limit the maximum size of each burst header.

In this context, operation of a multicarrier demodulator on board the satellite is not simple. The bursts from different stations have different frequencies and clock rates and reduction or even suppression of the preamble at the head of each burst does not permit frequency and clock rate recovery from one burst to the other. It is thus necessary to ensure synchronisation of the earth station clocks, for example by controlling these clocks to a reference on board the satellite [IZU-84], [ANA-87], [COL-87].

The various techniques presented above remain appropriate to the realisation of a multi-carrier demodulator in this context. The carrier recovery circuit in particular must be optimised when the preamble is suppressed. The circuit can exploit the coherence of bursts between successive frames or, for example, use a non-linear estimation method [VIT-83]. In other cases (INTELSAT IBS and IDR) the preamble has been retained and modified structures developed [RHO-89].

9.4.2.2 *Modulators*

The structure of a modulator is simpler than that of a demodulator. Usually four state phase modulation is used. Modulation can be realised either at intermediate frequency before changing frequency to the downlink value [STI-82] or directly at the downlink frequency [KOG-77], [OHM-82], [ANA-84], [ALB-89]. The demodulator contains a local oscillator whose frequency must be stabilised, particularly with respect to temperature variations. The active components used are PIN diodes or field effect transistors.

9.4.2.3 *The baseband processor*

The baseband switching device routes the packets from a particular uplink beam to the appropriate downlink beam. Different architectures are possible for performing this function. Among these, three stage structures of the time space time (TST) type and single stage T type structures are evident. Baseband switching implies, a priori, organisation of the data in the form of a frame. Clock realignment circuits using buffer memories may be necessary [INU-83].

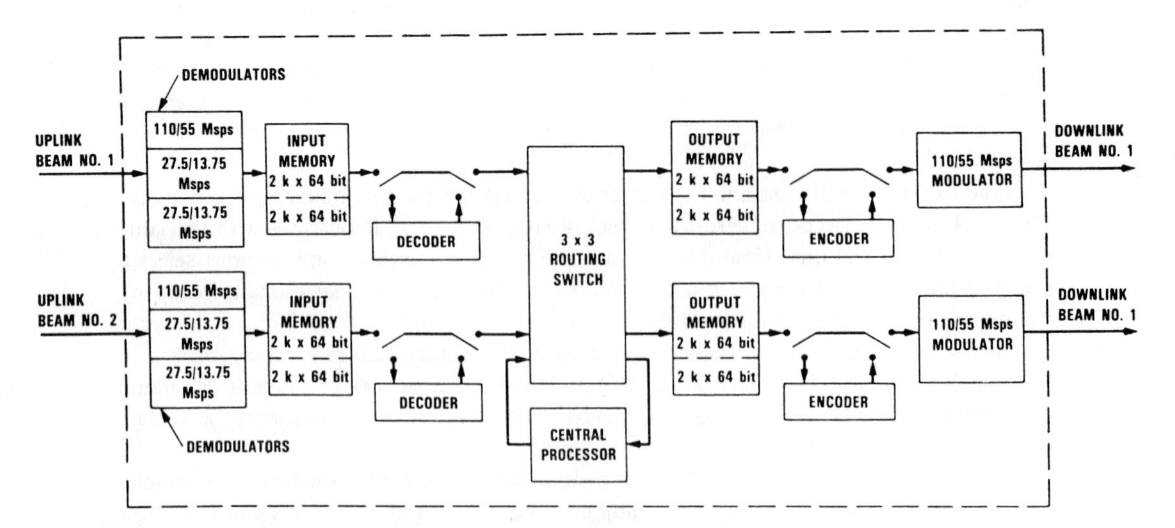

Figure 9.23 Architecture of TST type switching [NAD-88]. (© AIAA. Used with permission.)

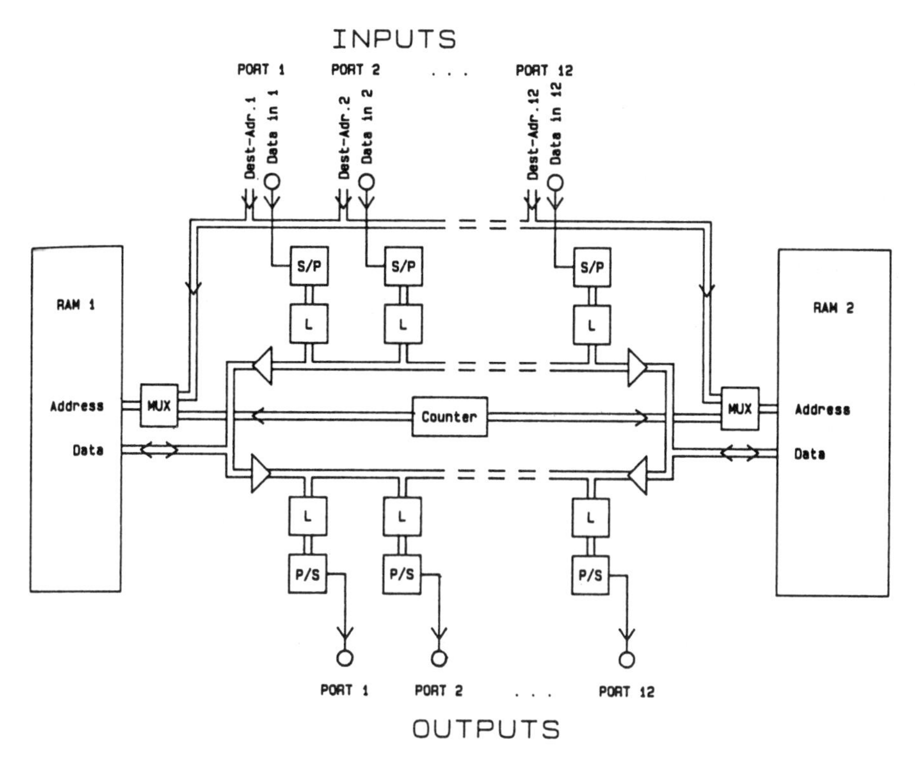

Figure 9.24 Block diagram of a single stage (baseband) switch [BER-87]. L = latch, S/P = serial/parallel converter, P/S = parallel/serial converter. (Reproduced by permission of Elsevier Science Publishers B.V.)

In a TST structure, the bursts from the uplinks are stored in buffer memories for the duration of one frame. These bursts are then extracted from the memories and physically routed by a switching network to the output buffer memories associated with the downlinks [SAB-81], [PEN-84], [MOA-86]. Figure 9.23 is an example of such an architecture [NAD-88].

The TST structure suffers from complexity due to the need to find a path through three stages for all bursts to be routed and the increased number of components required to realise a non-blocking switching network (where it is always possible to find a path) in comparison with a blocking network.

These disadvantages can be avoided by means of a single stage network where routing can be provided in a simple manner [EVA-87]. Synchronous binary words arriving at the various inputs are first converted into parallel form and then transferred successively, via a time multiplexed bus, to various sections of memory. These words are represented by the number of the time interval which corresponds to a combination of the number of the input on which the word was present and the number of the word in the frame. The process is repeated until all the words in the frame on each input are stored in memory. Words are written into memory at successive addresses in accordance with the destination coordinates of the word as defined by the number of the output and the number of the word in the frame. Hence the word to be transmitted on the first output in the first time interval is stored at address 0, the word to be transmitted on the second output in the first time interval is stored at address 1 and so on. Reading and transfer of words on to the bus is thus realised simply under the control of a counter. The write address for each word is provided, on arrival of the word at the corresponding input, in a synchronous manner by a control device. Figure 9.24 illustrates one realisation of such a switching device [BER-87].

A single stage structure also suffers from certain failings; in particular it does not permit multidestination broadcasting of bursts. Other structures (such as a modified T type stage and an S type stage with a buffer or time multiplexing) are possible in accordance with the particular context [KUM-89].

The presence of the baseband signal offers additional possibilities such as changing the digital data rate [MOA-82], [INU-83] and using error correcting coding to combat attenuation due to rain. Convolutional decoders have been developed for this purpose [CLA-82].

9.4.3 Equipment technology

The digital form of the signal permits the use of digital technologies (i.e. integrated circuits). The high processing power required together with the constraints on mass, bulk and power consumption require large or very large scale integration of the components (LSI or VLSI) with full custom or semicustom (ASIC or gate array) realisation. Hybrid circuits and surface mount components are also highly suitable. In comparison with the technologies available for equipment used on the ground, it is necessary to take account of the constraints due to the specific environment in the design of on-board satellite equipment.

9.4.3.1 *The specific environment*

This environment (see Chapter 12) is characterised by:

—High radiation dosage during the lifetime of the equipment (typically 10^5 rad in geostationary orbit for a lifetime of seven to ten years). This radiation contains, in particular, heavy ions which are capable of causing disruptions (single event upsets (SEU) or bit-flips [GUP-86]) and can involve latch-up and soft errors in the logic state of gates,
—The absence of convection which causes problems in evacuating the heat dissipated by components in a reduced volume. The use of heatsinks (strips of copper bonded to components) enables the heat generated to be routed to the equipment casing and the structure of the satellite.
—The requirement of a long lifetime without the possibility of intervention for maintenance demands high reliability.

9.4.3.2 *Required properties*

The technologies used must have the following properties—radiation resistant, high noise immunity, high speed, low power consumption and high integration density. Some of these properties are incompatible; a compromise must be adopted according to the application concerned.

Semiconductor technologies. Silicon (Si) and gallium arsenide (GaAs) are the semiconductor materials used to produce logic circuits. Bipolar and MOS technologies are used with silicon. In bipolar technology, emitter coupled logic (ECL) enables high speeds to be achieved. In MOS technology, CMOS is characterised primarily by low power consumption. CMOS on a sapphire substrate (CMOS/SOS) is well suited to the production of radiation hardened circuits of high reliability.

Radiation resistance. The components used must be able to withstand radiation of the order of 5×10^3 rad, taking account of a 4 to 5 mm aluminium shield, for missions of 10 years in geostationary orbit.

CMOS on silicon provides good resistance to cumulative doses but is sensitive to latch-up phenomena caused by heavy ions. Hardening can be provided by an epitaxial layer and a guard ring. CMOS on a sapphire substrate sustains the passage of heavy ions well but its resistance to cumulative doses is limited.

Bipolar and GaAs sustain cumulative doses well and are little affected by the latch-up phenomenon. Passage of heavy ions, in contrast, leads to transient logic errors or soft errors.

Speed and power consumption. GaAs permits the highst speeds to be obtained (delays of 0.1 ns per gate) followed by bipolar ECL and CMOS (there is a ratio of 10 between GaAs and SOS/CMOS).

CMOS, on the other hand, provides the lowest power consumption although this consumption increases linearly with switching frequency (a CMOS gate consumes current only during a change of state). The consumption of fast CMOS logic remains less (10%) than that of bipolar transistor logic which is of the same order as that of GaAs.

Integration density. In connection with integration density, the greatest number of gates per chip is obtained with CMOS on silicon technology (CMOS/BULK) (20 000 gates per chip), followed by bipolar ECL (9000 gates) and CMOS/SOS (8000 gates). GaAs allows only 1000 gates per chip [MOR-88b].

In conclusion, CMOS components dominate for realisation of circuits with a high integration density and limited operating speed; these will be associated with circuits using GaAs components when high speeds are required.

Bipolar technology (ECL for high speed and TTL for medium speeds) is also widely used for realisation of ASIC circuits [MAR-87]. Special precautions at logic level, such as coding and refreshing of data in memory, permit the effects of heavy ions (single event upsets) to be combated.

9.5 ANTENNA COVERAGE

A satellite communication mission specifies the coverage performance of a geographical region in terms of minimum radio-frequency objectives (EIRP or flux at the ground, G/T or flux at the satellite). This performance must be realised at a range of locations specified by their geographical coordinates. These locations are the coverage reference points. Several concepts must be considered:

—Geographical coverage; the contour joining the reference points as they are seen from the nominal position of the satellite.

—Geometric coverage; a contour including the apparent displacements of the geographical coverage due to the effects of movement of the satellite.
—Radio-frequency footprint; the contour representing a given value of radio-frequency performance.

In this part of the chapter only geometrical aspects of coverage will be considered. Radio-frequency aspects (radio-frequency footprints) are presented in Section 9.6.

9.5.1 Geographical coverage

The geographical region of the earth is seen from the satellite within a solid angle which depends on the relative positions of the satellite and the region concerned.

The coverage reference points are identified in an axis system which has the satellite as the origin (Figure 9.25). The axis system is centred on the satellite and contains an axis oriented in the satellite–earth centre direction (the z axis), an axis perpendicular to the first in a plane parallel to the equatorial plane and oriented towards the east (the x axis) and an axis normal to the other two such that the axis system is direct (the y axis is oriented towards the north for a satellite in the equatorial plane). A point on the surface of the earth with geographical coordinates λ (longitude with respect to the Greenwich meridian) and φ (latitude) has coordinates x_p, y_p and z_p in the satellite reference. These coordinates are calculated as a function of the altitude h of the satellite and the coordinates (latitude λ_{SL} and longitude φ_{SL}) of the sub-satellite point. In the case of the geostationary satellite, the relations can be obtained from the relative longitude L of the satellite and the point concerned and the latitude l of this point.

$$x_p = R_0 + R_E(1 - \cos l \cos L) \tag{9.23}$$

$$y_p = R_E \cos l \sin L$$

$$z_p = R_E \sin l$$

9.5.1.1 *True view angles*

The directions of the reference points as seen from the satellite are defined by angles. Among the various possible pairs of angles (see Section 9.5.3), it is convenient to choose those which

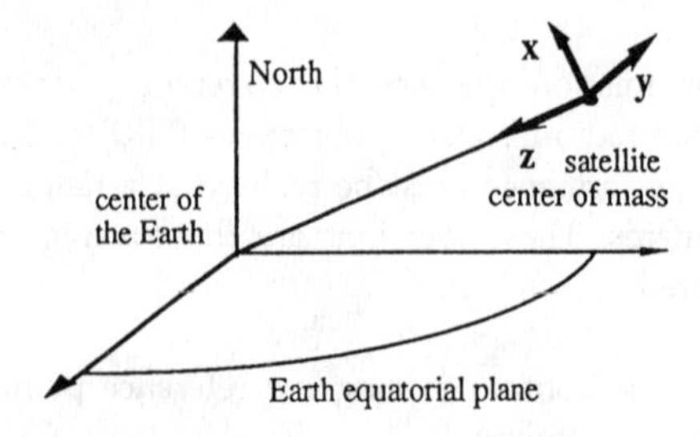

Figure 9.25 Reference co-ordinate system for coverage definition.

correspond to the angles under which the direction of a point is seen from the satellite ('true view' angles). The two angles defining a point of the coverage consist of:

—The angle θ between the direction of the centre of the earth and the direction of the point concerned.
—The angle φ between two planes, one defined by the direction of the centre of the earth and the x axis and the other defined by the direction of the centre of the earth and the point concerned.

The true view angles are obtained from the coordinates x_p, y_p and z_p of the point concerned in the satellite reference by the following relations:

$$\theta = \arctan (x_p/y_p)$$

$$\varphi = \arctan [z_p/(x_p^2 + y_p^2)^{1/2}] \tag{9.24}$$

In the case of a geostationary satellite, the true view angles can be obtained directly from the relative longitude L of the satellite and the point concerned and the latitude of this point (see Section 9.5.4).

9.5.1.2 *Representation of the coverage*

Representation of coverage on a map poses the problem of converting from three-dimensional space to a plane.

One representation consists of using a reference plane tangential to the surface of the earth

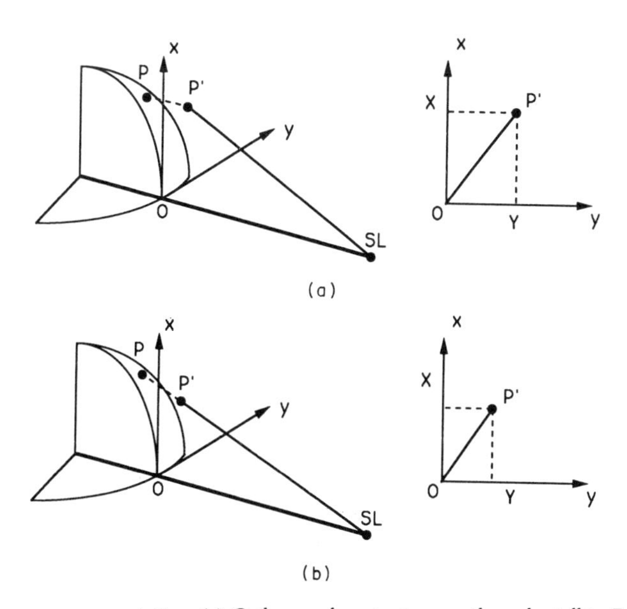

Figure 9.26 Coverage representation. (a) Orthogonal projection on the subsatellite Earth tangent plane, (b) oblique projection on the subsatellite Earth tangent plane.

at the point under the satellite and performing a projection of the points on the surface of the earth on to this plane (Figure 9.26a). The result bears little resemblance to the surface of the earth seen from the satellite. In particular, with a sub-satellite point on the equator (e.g. a geostationary satellite), the poles are apparent although in reality they are not visible.

A more realistic representation is obtained by using the same plane tangential to the surface of the earth at the sub-satellite point and hence perpendicular to the direction to the satellite–earth centre direction, but defining the coverage as the track on this plane of the lines joining the satellite and the various points concerned (Figure 9.26b). The map obtained in this way is a faithful representation of the view from the satellite only for points situated in the vicinity of the tangential point. However, the oblique projections limit the error at the extremity of the part of the earth visible from the satellite. Hence, with a geostationary satellite, the poles are not visible.

The X and Y coordinates of a point on the surface of the earth are defined in an axis system centred on the sub-satellite point; the $0x$ axis is oriented towards the north and the $0y$ axis towards the east (this reference is currently adopted to represent antenna radiation patterns, the Z axis is then the direction of propagation and the set X, Y and Z form a direct reference). The X and Y coordinates are obtained from the relative geographical coordinates L (the relative longitude between the sub-satellite point and the meridian of the point concerned) and l, the latitude of the point, using the following relations for the particular case of a geostationary satellite:

$$X = K \sin l \tag{9.25}$$

$$Y = K \cos l \sin L$$

with $K = 1/(6.62 - \cos l \cos L)$ and inversely by:

$$l = \arc\sin\{\cos[\arc\tan(Y/X)] \sin\kappa\}$$

$$L = \arc\tan\{\sin[\arc\tan(Y/X)] \tan\kappa\} \tag{9.26}$$

with: $\kappa = \arc\sin(6.62 \sin\nu) - \nu$ and $\nu = \arc\tan\{Y/\sin[\arc\tan(Y/X)\}$

Finally, the true view angles defined in Section 9.5.1.1 can be represented on a plane. For example, a point P on the coverage can be represented by inserting the quantities θ_x and θ_y into a system of rectangular axes defined by:

$$\theta_x = \theta \cos\varphi$$

$$\theta_y = \theta \sin\varphi \tag{9.27}$$

The axis system used is centred on the satellite–earth centre direction; the $0x$ axis is in the north direction and the $0y$ axis in the east direction. The angle θ is thus represented by the distance $0P$ and the angle φ by the angle between $0x$ and $0P$ (Figure 9.27).

The representation obtained well expresses any apparent displacement of the point in the reference frame; for example, a rotation about the axis of the earth (a variation of φ with θ constant) is represented by the arc of a circle centred on 0. On the other hand, the relative positions of points are not respected. Consider a direction represented by the angles (θ, φ). This direction is defined by a point P on a sphere centred on the reference frame. Consider also a point P' corresponding to a direction $(\theta', 0)$. The angular distance χ

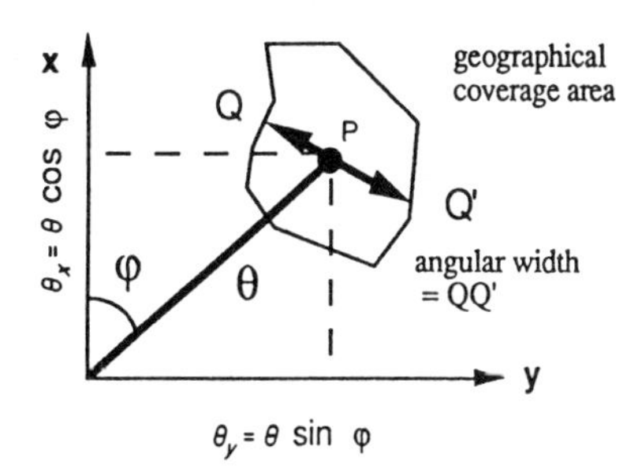

Figure 9.27 'True view' representation.

between P and P' seen from the centre of the sphere (the relations are in the spherical triangle defined by the path of the z axis, the point P and the point P') is such that:

$$\cos \chi = \cos \theta \cos \theta' + \sin \theta \sin\theta' \cos \varphi$$

hence:

$$\chi = \arccos [\cos \theta \cos \theta' + \sin \theta \sin \theta' \cos \varphi]$$

In the representation chosen, the angular distance between P and P' is obtained from the plane triangle OPP' by:

$$PP' = [\theta^2 + \theta'^2 - 2\theta\theta' \cos \varphi]^{1/2}$$

The relative error ϵ in this representation is thus given by:

$$\epsilon = (PP' - \chi)/\chi$$

For a geostationary satellite the nadir angle α is at most equal to 8.7°; by considering two points at the limit of visibility, one on the meridian of the satellite ($\theta' = 8.7°, \gamma = 0$) and the other on the equator ($\theta = 8.7°, \gamma = 90°$), the angular difference as determined using this representation is 12.304° while the actual angle at the centre is 12.280°. The error incurred is less than 0.024° and the relative error is less than 2×10^{-3}. It is therefore permissible, at least for a geostationary satellite, to assume that relative positions are respected. Geometrical transformation (such as translation) can, therefore, be performed using this representation.

9.5.2 Geometric coverage

Geometric coverage is defined by assigning a zone of uncertainty, which represents the apparent movement of the point concerned in the reference frame associated with the antenna, to each point of the geographical coverage as seen from the satellite. This zone is centred on the point concerned and has a radius equal to the angular pointing error α_p. The geometrical coverage covers all the points defined by the geographical coverage by including the zones of uncertainty.

Geometric coverage includes the combined effects of depointing of the boresight due to satellite movement and deformation of the geometric coverage due to displacement of the satellite with respect to the geographical region.

When the antenna is provided with a pointing control mechanism, the pointing error of the principal axis of the beam is, in principle, eliminated. However, due to the effect of apparent movement of the satellite with respect to the geographical region to be served, the view angles of the geographical region vary. This also occurs when coverage of a given geographical region must be obtained from two different orbital positions. This is the case for a system with two geostationary satellites of identical design, or having a common stand-by satellite.

9.5.3 Global coverage

Global coverage of the earth's surface as seen from the satellite is required when it is wished to establish links between stations which may be situated anywhere on the earth's surface.

9.5.3.1 *Maximum geographical coverage*

Geographical coverage is limited by the terrestrial curve along which the cone having the satellite as its vertex is tangent to the earth. This cone has a vertex angle 2β equal to 17.4° for a geostationary satellite.

9.5.3.2 *Coverage at minimum elevation angle*

Stations located on the limiting curve of the geographical coverage would have their antennas pointing horizontally. Severe performance degradation of the links would result as a consequence of increased antenna noise temperature and propagation attenuation due to the greater length of the path in the atmosphere. The limit of geographical coverage is often defined by the curve along which the direction of the satellite makes an angle E with the horizontal (minimum elevation angle).

The resulting geographical coverage for different elevation angles is represented on the planisphere of Figure 9.28 for the case of a geostationary satellite [CCIR-Rep 206]. The coverage generally used corresponds to an elevation angle E of 5°.

Figure 9.29 gives the elevation angle E of the earth station antenna and the angle θ which the station–satellite direction makes with the satellite–earth centre direction as a function of the angle ϕ between the direction of the station and the satellite as seen from the centre of the earth. The angle ϕ is obtained as a function of the latitude l of the station and the relative longitude L by means of equation (7.63):

$$\cos\phi = \cos l \cos L$$

The angular width of the geographical region seen from a geostationary satellite for a minimum elevation angle E_{min} is fixed by the value 2θ determined for $E = E_{min}$ in the equation:

$$2\theta = 2\text{arc sin}\,[(R_E \cos E)/(R_0 + R_E)]$$

$$= 2\text{arc sin}\,(0.15 \cos E_{min}) \qquad \text{(degree)} \qquad (9.28)$$

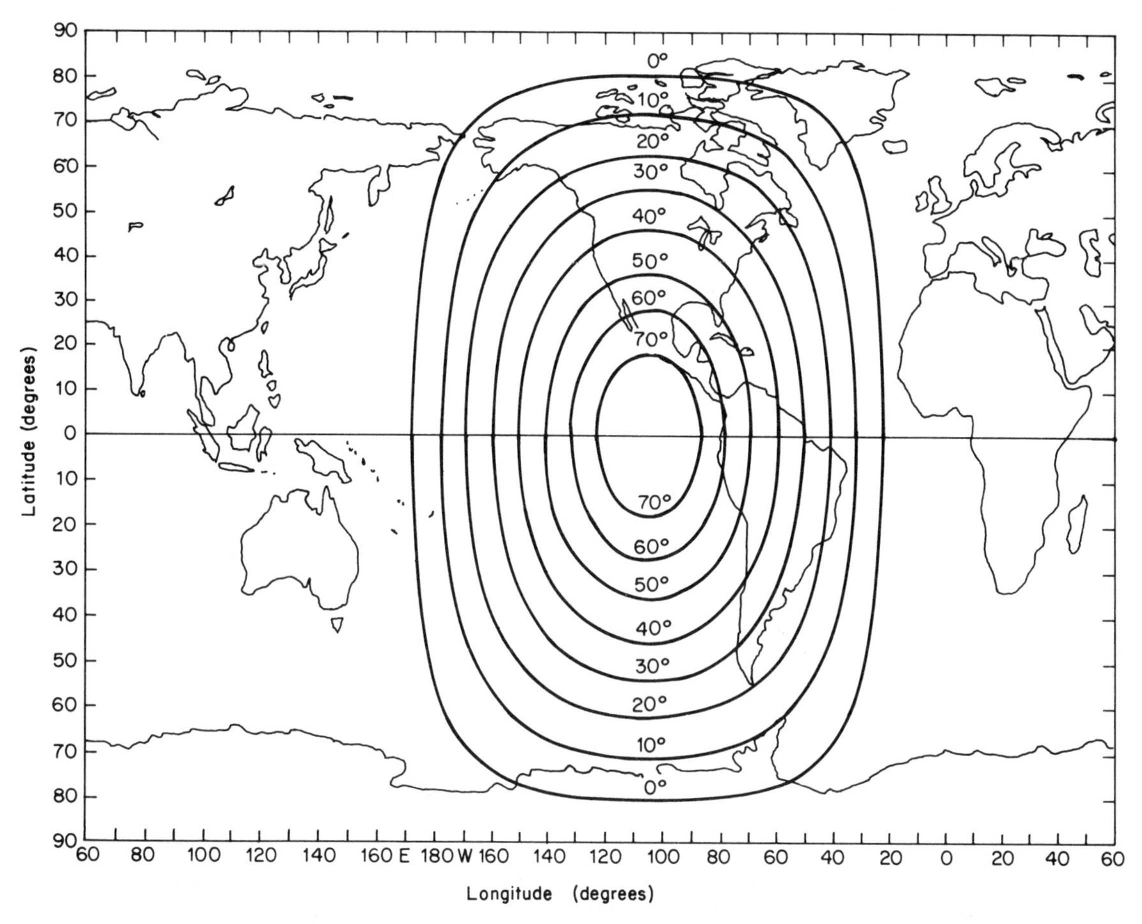

Figure 9.28 Global coverage for a given elevation angle.

The maximum latitude of the limit of the geographical zone is given by:

$$l_{\text{max}} = 90^\circ - (\theta + E_{\text{min}}) \qquad \text{(degree)} \tag{9.29}$$

The antenna illuminating the geographic zone is assumed to have its boresight directed towards the centre of the earth. If the pointing error α_p of the boresight is taken into account, the angular width of the geometric zone is then equal to $2\theta + 2\alpha_p$. If, for example, $\alpha_p = 1$ degree, the angular width of the global geometric coverage for a minimum elevation angle $E_{\text{min}} = 10^\circ$ will be equal to $2 \arcsin(0.15 \cos 10^\circ) + 2\alpha_p = 19^\circ$.

9.5.3.3 *The INTELSAT network*

The INTELSAT network provides almost worldwide coverage with the aid of three groups of satellites located above the Atlantic, Pacific and Indian Oceans. Only the polar regions and the centre of the United States are not covered since the system, is intended for intercontinental

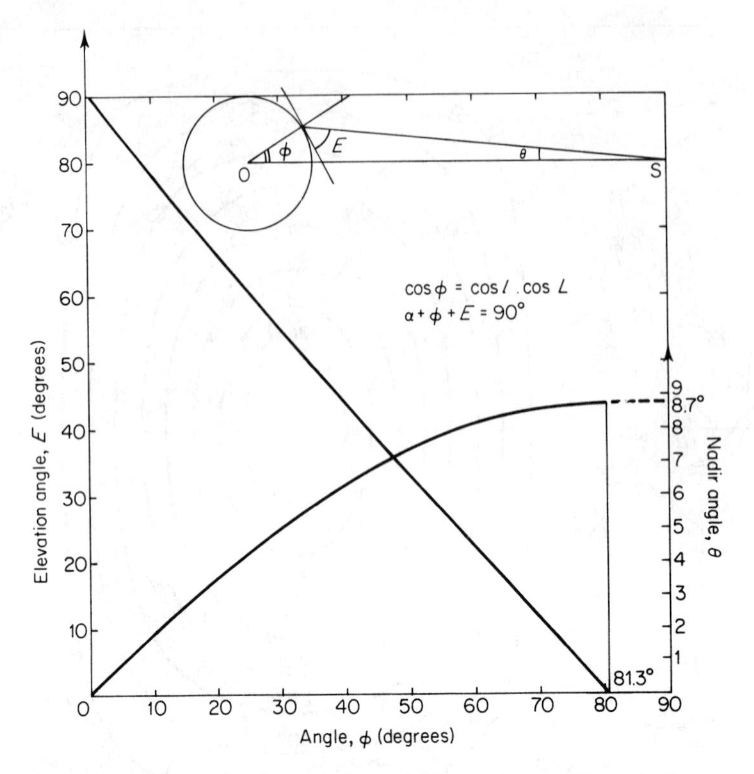

Figure 9.29 Elevation angle E and nadir angle θ as a function of geocentric angle Φ.

links. However, a double hop is not recommended for commercial telephony due to the excessive propagation delay. The maximum range is thus limited under these conditions to around 17 000 kilometres.

When propagation time is not a disadvantage, the range can be increased by using several hops by means of relay satellites. It is also possible to use intersatellite links (see Chapter 5).

9.5.4 Reduced or spot coverage

When the coverage is not global, it must relate to a particular region of the earth as seen from the satellite. The antenna boresight does not pass through the centre of the earth but through a reference point on the surface of the earth which a priori defines the centre of coverage.

Antenna pointing is characterised by the nominal direction (in the absence of pointing error) of the boresight as defined by two angles in a reference frame associated with the satellite. This pointing direction can be defined by the true view angles in the reference frame specified in Section 9.5.1.1.

The geometry of the system for a geostationary satellite is illustrated in Figure 9.30. Various angles enable the pointing direction represented by the relative coordinates (latitude l and relative longitude L) to be defined from the aiming point on the surface of the earth. R_0 is the nominal altitude of the geostationary satellite and is equal to 35 786 km and

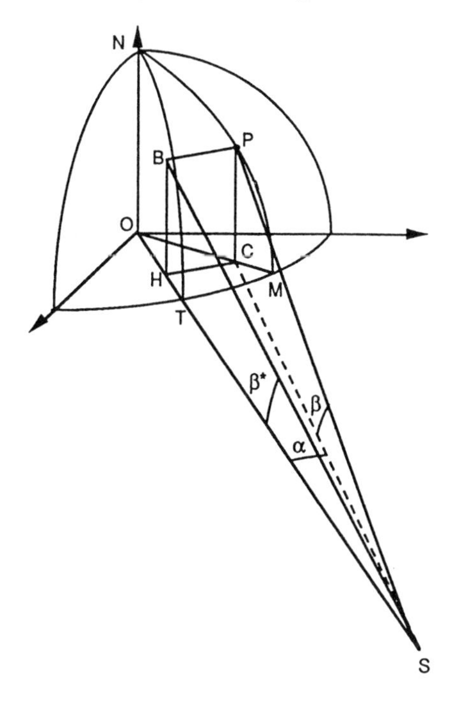

Figure 9.30 Geostationary satellite–Earth station geometry.

$R_E = 6378$ km is the mean equatorial radius. The angle α between the axis of the earth and the projection of the aiming axis on the orbital plane has a value given by:

$$\alpha = \arctan \frac{R_E \cos l \sin L}{R_0 + R_E(1 - \cos l \cos L)} \tag{9.30}$$

The angle β between the projection of the aiming axis on the orbital plane and the boresight is given by:

$$\beta = \arctan \frac{R_E \sin l \cos \alpha}{R_0 + R_E(1 - \cos l \cos L)} \tag{9.31}$$

The angle β^* between the axis of the earth and the projection of the boresight on the plane passing through the poles and containing the satellite is given by:

$$\beta^* = \arctan \frac{R_E \sin l}{R_0 + R_E(1 - \cos l \cos L)} \tag{9.32}$$

The angle α^* between the projection of the boresight on the plane which contains the satellite and passes through the poles and the boresight is given by:

$$\alpha^* = \arctan \frac{R_E \cos l \sin L \cos \beta^*}{R_0 + R_E(1 - \cos l \cos L)} \tag{9.33}$$

The true view angles remain to be defined. The angle θ between the axis of the earth and the boresight is obtained from the preceding expressions by:

$$\cos\theta = \cos\alpha\cos\beta \tag{9.34a}$$

or by:

$$\cos\theta = \frac{R_0 + R_E(1 - \cos l \cos L)}{R} \tag{9.34b}$$

where R is the distance from the satellite to the point concerned and is obtained using:

$$R = \sqrt{R_0^2 + 2R_E(R_E + R_0)(1 - \cos l \cos L)}$$

Finally, the angle φ between the meridional plane of the satellite and the plane defined by the direction of the centre of the earth and the boresight is obtained from:

$$\varphi = \arctan\frac{\sin L}{\tan l} \tag{9.35a}$$

if the nominal location the antenna is aiming at is in the northern hemisphere, or

$$\varphi = \pi + \arctan\frac{\sin L}{\tan l} \tag{9.35b}$$

if the nominal location the antenna is aiming at is in the southern hemisphere.

9.5.5 Evaluation of antenna pointing errror

Antenna pointing has been defined for a nominal orbital position of the satellite and by assuming a reference frame aligned with the local orbital reference. Attitude movements of the satellite with respect to the local orbital reference, and of the satellite with respect to the nominal orbital position, lead to depointing of the aiming angle. Depointing is defined as the angle through which the initially defined aiming direction must be turned in order to make it again point towards the reference point on the earth's surface. This angle is defined in different planes in accordance with the displacement concerned. In order to combine the different contributions, the depointing angle is defined by its projections on to two reference planes which are the $z0x$ and $z0y$ planes of the reference frame represented in Figure 9.25.

9.5.5.1 *Depointing due to attitude movement*

Attitude movement of the satellite is represented by rotation about the local orbital reference axes. The nominal attitude corresponds to alignment of the mechanical axes of the satellite with the reference (see Section 10.2). Displacement with a given amplitude (a control error or an intentional bias) about each of the axes causes displacement of the antenna boresight which is assumed to be rigidly mounted with respect to the body of the satellite. Depointing for each axis is expressed by the angular displacement projected on to the two reference planes.

Depointing due to roll. Depointing in roll is equivalent to rotation of the boresight about the roll axis; this describes a cone of vertex half angle ϵ_R. The depointing components are given by:

$$\Delta\Psi_{R,x} = \arctan[\tan(\beta^* + \epsilon_r)\cos\alpha] - \beta \tag{9.36a}$$

$$\Delta\Psi_{R,y} = \arctan\left[\frac{\cos^2\beta^*}{\cos(\beta^* + \epsilon_r)}\tan\alpha\right] - \alpha^* \tag{9.36b}$$

A simplified formulation taking account of the actual values of the angles for a geostationary satellite leads to:

$$\Delta\Psi_{R,x} = \epsilon_r \cos\alpha^* \tag{9.37a}$$

$$\Delta\Psi_{R,y} = 0 \tag{9.37b}$$

Depointing due to pitch. Depointing in pitch is equivalent to rotation of the boresight about the pitch axis; this describes a cone of vertex half angle ϵ_P. The depointing components are given by:

$$\Delta\Psi_{P,x} = \arctan\left[\frac{\cos^2\alpha}{\cos(\alpha + \epsilon_P)}\tan\beta^*\right] - \beta \tag{9.38a}$$

$$\Delta\Psi_{P,y} = \arctan[\tan(\alpha + \epsilon_P)\cos\beta^*] - \alpha^* \tag{9.38b}$$

A simplified formulation taking account of the actual values of the angles for a geostationary satellite leads to:

$$\Delta\Psi_{P,x} = 0 \tag{9.39a}$$

$$\Delta\Psi_{P,y} = \epsilon_P \cos\beta \tag{9.39b}$$

Depointing due to yaw. Depointing in yaw is equivalent to rotation of the aiming axis about the yaw axis; this describes a cone of vertex half angle ϵ_Y. The depointing components are given by:

$$\Delta\Psi_x = -\arctan\left[\frac{\cos(\varphi + \epsilon_y)\tan\beta}{\cos\varphi}\right] + \beta \tag{9.40a}$$

$$\Delta\Psi_y = -\arctan\left[\frac{\sin(\varphi + \epsilon_y)\tan\alpha^*}{\sin\varphi}\right] - \alpha^* \tag{9.40b}$$

A simplified formulation taking account of the actual values of the angles for a geostationary satellite leads to:

$$\Delta\Psi_x = \frac{R_{Er}}{R}\sin\phi\sin\varphi\cos\beta\epsilon_y \tag{9.41a}$$

$$\Delta\Psi_y = \frac{R_E}{R}\sin\phi\cos\varphi\cos\alpha^*\epsilon_y \tag{9.41b}$$

9.5.5.2 *Depointing due to displacement in orbit*

Displacement of the centre of mass of the satellite will modify the direction of the reference point in the satellite reference frame and hence introduce depointing. For any orbit, it is necessary to calculate the direction of the reference point for each point of the orbit.

For an ideal geostationary satellite, the direction of the reference point is fixed. In reality, the satellite is not perfectly geostationary and moves with respect to its nominal position within the station-keeping window (see Section 7.3). This movement results from non-zero eccentricity and inclination.

North–south and east–west movements each make a contribution to depointing. The inclination, although determining the north–south amplitude of the window, also makes a contribution which is not taken into account in the window. This also applies to the eccentricity. Depointing for each of the effects is expressed by the angular displacements projected on to the two reference planes.

Depointing due to north–south movement. Depointing due to north–south movement in the window is equivalent to rotation of the reference frame about one axis in the equatorial plane perpendicular to the satellite–earth centre direction. The depointing components as a function of the angular half-height NS of the window are given by:

$$\Delta\Psi_{\mathrm{NS},x} = \arctan \frac{2\cos\alpha^* \sin\beta^* \sin\dfrac{\mathrm{NS}}{2}\cos\left(l' + \beta^* - \dfrac{\mathrm{NS}}{2}\right)}{\sin l' - 2\sin\beta^* \sin\dfrac{\mathrm{NS}}{2}\sin\left(l' + \beta^* - \dfrac{\mathrm{NS}}{2}\right)} \quad (9.42a)$$

$$\Delta\Psi_{\mathrm{NS},y} = \arctan\left[\frac{\sin L}{\tan l\left[\dfrac{1}{\sin\beta^*} - 2\sin\dfrac{\mathrm{NS}}{2}\sin\left(l' - \dfrac{\mathrm{NS}}{2} + \beta^*\right)\right]\Big/\sin l'}\right] - \alpha^* \quad (9.42b)$$

with $l' = \arctan(\tan l/\cos L)$.

A simplified formulation taking account of the actual values of the angles for a geostationary satellite leads to:

$$\Delta\Psi_{\mathrm{NS},x} = \frac{R_{\mathrm{E}}}{R}\cos[\arcsin(\cos l \sin L)]\cos\left[\arctan\left(\frac{\tan l}{\sin L}\right) + \beta^*\right]\mathrm{NS} \quad (9.43a)$$

$$\Delta\Psi_{\mathrm{NS},y} = \frac{R_{\mathrm{E}}}{R}\cos[\arcsin(\cos l \sin L)]\sin\left[\arctan\left(\frac{\tan l}{\cos L}\right) + \beta^*\right]\sin\alpha^*\mathrm{NS} \quad (9.43b)$$

Depointing due to east–west movements. Depointing due to east–west movement in the window is equivalent to rotation of the reference frame about the polar axis. The depointing components as a function of the angular half-width EW of the window are given by:

$$\Delta\Psi_{\mathrm{EW},x} = \arctan\left[\frac{\tan l}{\dfrac{\sin L}{\sin l} - 2\sin\dfrac{\mathrm{EW}}{2}\sin\left(L + \alpha - \dfrac{\mathrm{EW}}{2}\right)}\right] - \beta \quad (9.44a)$$

$$\Delta\Psi_{\mathrm{EW},y} = \arctan \frac{2\sin\alpha\cos\beta\sin\dfrac{\mathrm{EW}}{2}\cos\left(L+\alpha-\dfrac{\mathrm{EW}}{2}\right)}{\sin L - 2\sin\alpha\sin\dfrac{\mathrm{EW}}{2}\sin\left(L+\alpha-\dfrac{\mathrm{EW}}{2}\right)} \tag{9.44b}$$

A simplified formulation taking account of the actual values of the angles for a geostationary satellite leads to:

$$\Delta\Psi_{\mathrm{EW},x} = \frac{R_\mathrm{E}}{R}\cos l \sin(L+\alpha)\sin\beta\ \mathrm{EW} \tag{9.45a}$$

$$\Delta\Psi_{\mathrm{EW},y} = \frac{R_\mathrm{E}}{R}\cos l \cos(L+\alpha)\mathrm{EW} \tag{9.45b}$$

Depointing due to eccentricity. Eccentricity leads to an east–west movement in the window which has already been taken into account. This motion is the result of the variations of the altitude of the satellite. The depointing components induced by radial displacement as a function of the eccentricity e are given by:

$$\Delta\Psi_{\mathrm{e},x} = \arctan\left[\frac{R_0 + R_\mathrm{E}(1-\cos l\cos L)}{R_0 + R_\mathrm{E}(1-\cos l\cos L) - (R_0+R_\mathrm{E})e}\tan\beta^*\cos\alpha\right] - \beta \tag{9.46a}$$

$$\Delta\Psi_{\mathrm{e},y} = \arctan\left[\frac{R_0 + R_\mathrm{E}(1-\cos l\cos L)}{R_0 + R_\mathrm{E}(1-\cos l\cos L) - (R_0+R_\mathrm{E})e}\tan\alpha\cos\beta^*\right] - \alpha^* \tag{9.46b}$$

A simplified formulation taking account of the actual values of the angles for a geostationary satellite leads to:

$$\Delta\Psi_{\mathrm{e},x} = \frac{R_0 + R_\mathrm{E}}{R}\tan\beta^*\cos\beta\frac{180}{\pi}e \tag{9.47a}$$

$$\Delta\Psi_{\mathrm{e},y} = \frac{R_0 + R_\mathrm{E}}{R}\tan\alpha\cos\alpha^*\ \frac{180}{\pi}e \tag{9.47b}$$

The influence of the inclination. The influence of the inclination on the geometry of the satellite–earth system manifests itself differently according to the position of the satellite in the orbit. This is illustrated in Figure 9.31 for the case of a quasi-geostationary satellite whose orbit has an inclination i.

The inclination is equivalent to a north–south movement in the window for which the depointing effect has already been determined. At the node of the orbit (at the maximum 90° north–south displacement) the inclination is also equivalent to an apparent rotation of the polar axis about the satellite–earth centre direction.

This rotation causes:

—Depointing equivalent to that caused by a movement about the yaw axis. The depointing

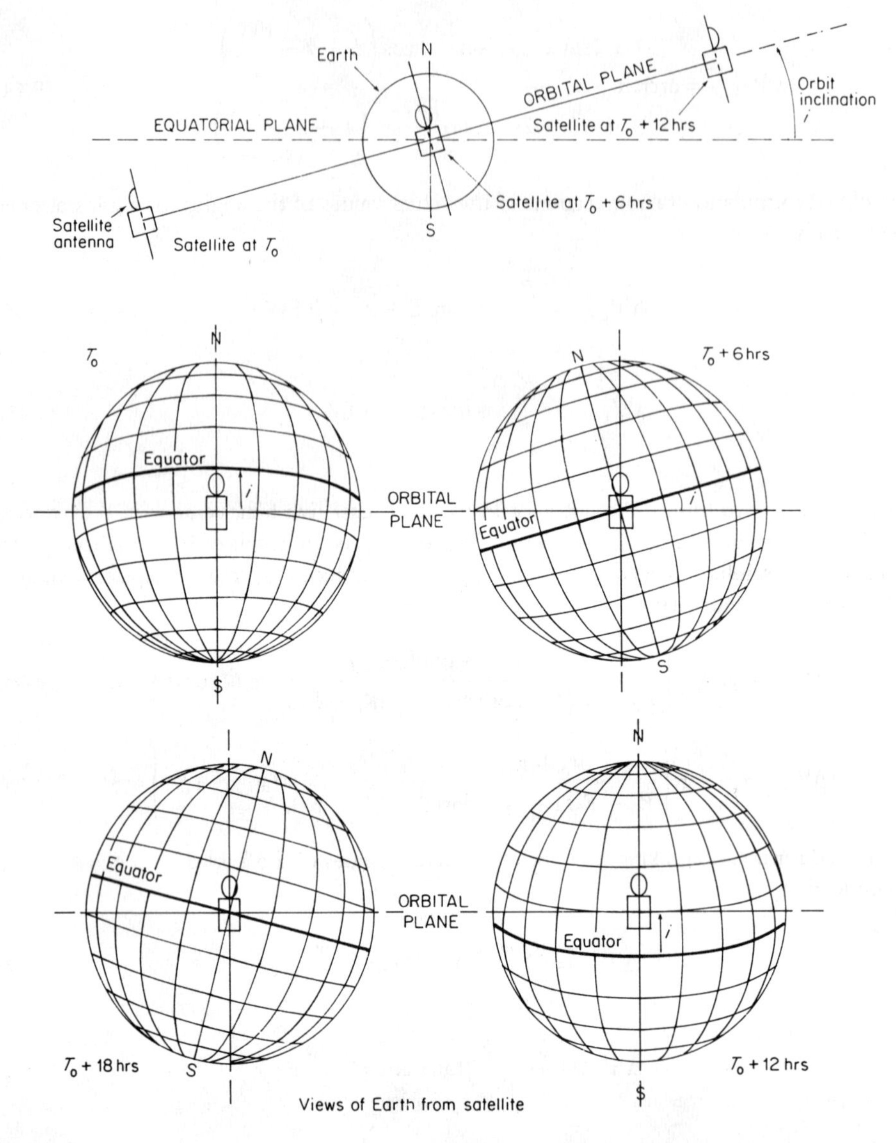

Figure 9.31 Apparent movement of the Earth seen from a quasi-geostationary satellite subject to residual inclination.

components as a function of the inclination, taking into account the usually small value of orbit inclination, are given by:

$$\Delta\Psi_{i,x} = \frac{R_E}{R} \sin\phi \sin\varphi \cos\beta i \tag{9.48a}$$

$$\Delta\Psi_{i,y} = \frac{R_E}{R} \sin\phi \cos\varphi \cos\alpha^* i \tag{9.48b}$$

—An apparent rotation of the antenna beam with respect to an earth station about the line which joins the satellite and the station concerned. The maximum value of the angle of rotation is equal to the inclination i for a station situated on the equator at the longitude of the satellite. In the case where the links are established using waves with orthogonal polarisation, it adds a component of the orthogonally polarised wave to the wave with nominal polarisation. The maximum relative interference power level in the case of plane polarisation is $(C/N)_I = \sin^2 i$, that is 25.6 dB for $i = 3°$. Such a level makes re-use of the same band of frequencies with two orthogonal plane polarisations difficult. Re-use of frequencies by crossed plane polarisation, which enables the capacity of a link in a given band of frequencies to be doubled, assumes that the satellite is maintained on an orbit with sufficiently small inclination.

9.5.5.3 *Depointing associated with the antenna*

The direction of the actual electromagnetic axis of the antenna can deviate from the pointing direction as determined in the reference frame which defines the orientation of the antenna axis. There is, first, an uncertainty in the direction of the actual electromagnetic axis with respect to the principal axis of the antenna (the reference axis of the reflector, for example) which defines the orientation. This uncertainty results from inaccurate measurement of radiation patterns.

The antenna mounting on the satellite can cause an uncertainty in the orientation of the antenna reference with respect to the reference associated with the mechanical axes of the satellite. Depending on the procedures used, this uncertainty may or may not be included in the uncertainty discussed above.

Finally, deformation of the antenna reflector involves depointing of the boresight with respect to the nominal direction. For an antenna mounted on a satellite, this deformation is caused mainly by temperature differences related to the antenna illumination conditions by the sun.

The depointing direction which results from these various sources of uncertainty is not known a priori.

9.5.5.4 *Resulting overall depointing*

Pointing errors are essentially random variables with varying degrees of correlation. A definition of pointing error must take into consideration the probability of occurrence of the event which has led to this value being obtained. A random variable is usually specified by its mean value and its standard deviation σ; the 3σ value corresponds to a probability of 99.73% that the value is not exceeded considering Gaussian variables.

When the values of the various components have been determined, the overall depointing must be evaluated by combining them. Depointing due to displacement of the satellite has been decomposed into perpendicular axes oriented along the north–south and east–west axes. Depointing due to uncertainty in the direction of the electromagnetic axis may be oriented in any direction and is, a priori, independent of depointing due to displacement of the satellite.

The above calculations provide the individual depointing components along the x and y axes. It is required to estimate the overall contribution to depointing along each of these axes. The overall depointing angle is the resultant of these two orthogonal contributions. Two problems are to be addressed:

—What is the appropriate combination of the individual components along each x and y axis?
—How is the overall depointing angle constructed from its two x and y components?

These problems have been addressed in a paper [BEN-86] where the error components considered are identified at subsystem level. These errors are then grouped in several classes according to their temporal characteristics—constant, long-tem varying, diurnally varying and short-term varying. In the present case the individual components are identified at system level only and a different approach has to be followed [BOU-90], [BOU-91].

Combination of individual components along each axis. In evaluating the overall contribution, attention must be paid to the fact that some components are mutually exclusive. For instance, a non-zero inclination generates a yaw-like depointing when the satellite passes the nodes of the orbit and turns into a depointing induced by the NS latitudinal displacement when the satellite is 90° away from the orbit nodes (the vertex of the orbit). These two depointing components are mutually exclusive as they do not occur at the same time.

The simplest approach is to add all non-exclusive individual depointing components along each axis. However, as some of these components are random in nature, this is a pessimistic approach as the probability of occurrence of such a combination is very low. The components which can be considered to be random are those which arise from satellite attitude displacement, initial boresight misalignment, deformation due to mechanical and thermal constraints and so on.

In practice, the magnitudes of satellite displacements are small enough for the approximations indicated above to apply and proportionality between a depointing component and satellite displacement can be assumed. As a result of this proportionality, the σ value of the motion translates into the σ value of the depointing component.

The overall depointing component along any axis is a combination of deterministic and random variables. The worst case value of the deterministic component is the algebraic sum of the maximum values of the non-exclusive deterministic individual components. Assuming that the individual random components are independent, the σ value of the overall random depointing component along any axis is obtained as the square root of the sum of the σ^2 values of all the individual random components. Along each axis, the value of the overall component $\Delta\Psi x$ and $\Delta\Psi y$ is obtained by adding the 3σ value to the deterministic one. The value obtained corresponds to the worst case of the deterministic component and a probability of 99.73% that the random one is not exceeded.

Combination of the overall x and y components. The simplest approach is to consider the square root of the sum of the squares of the overall x and y components:

$$\Delta\Psi = [(\Delta\Psi x)^2 + (\Delta\Psi y)^2]^{1/2}$$

If the component values $\Delta\Psi x$ and $\Delta\Psi y$ as defined above, which correspond to a probability of 99.73% of not being exceeded, are used, the probability that the overall depointing will not exceed this value is greater. This approach may be too pessimistic.

Estimating a depointing value with a given probability of not being exceeded is not a simple matter; if the two components are independent and have a Gaussian distribution with zero mean and the same variance σ^2, it is known that the distribution of the depointing angle Ψ is a Rayleigh distribution. The value of depointing which corresponds to a given value of not being exceeded can be obtained from the Rayleigh cumulative distribution function:

$$F(\Psi) = 1 - \exp(-\Psi^2/\sigma^2) \tag{9.49}$$

For instance, a depointing angle of 3.44σ, where σ is the standard deviation of each of the two components, is not exceeded with a probability of 99.73%. If the two components have different means and a common variance, the distribution of the depointing angle is a Rice distribution.

In practice, the two components have different means and variances and there is no general expression for the cumulative distribution function. An upper bound for the depointing angle could be the sum of the squares of the means plus the 3σ values of the overall x and y components, in the sense that this value would not be exceeded with a probability at most equal to 99.73%. A lower bound is the mean plus the 3σ value of the largest component, as this value corresponds to a probability of not being exceeded of 99.73% for this axis only.

Example. An example calculation will now be given in order to illustrate the above derivations. This example concerns a geostationary satellite with the following parameters:

Satellite attitude control accuracy (3σ values):

$$\epsilon_R = 0.05^\circ, \quad \epsilon_P = 0.03^\circ, \quad \epsilon_Y = 0.5^\circ$$

Orbit inclination: $i = 0.07^\circ$
Orbit eccentricity: $e = 5 \times 10^{-4}$

The station-keeping window is NS = EW = $\pm 0.1^\circ$ and is large enough to accommodate the daily orbital displacement with some margin (maximum latitudinal displacement due to inclination = $\pm 0.07^\circ$, maximum longitudinal displacement due to eccentricity = $\pm 0.06^\circ$). In the nominal location the antenna is aiming at $l = 45^\circ$ and L (relative longitude with respect to the satellite) = 60°.

The values given in Table 9.4 are computed from the equations given in Sections 9.5.4 and 9.5.5:

—The set of angles (α, β), (α, β^*) and (θ, φ) any of which can be used to define the nominal direction in which the antenna should point,
—The individual components of depointing along the x and y axes.

As discussed above, the standard deviation of depointing due to attitude control is obtained as the square root of the sum of the σ^2 values of the individual components. The corresponding 3σ values are:

$$\Delta\Psi_{AC,x} = 0.069^\circ \qquad \Delta\Psi_{AC,y} = 0.0629^\circ$$

The effect of inclination is considered only when the satellite is passing the node of the orbit:

$$\Delta\Psi_{i,x} = 0.0067^\circ \qquad \Delta\Psi_{i,y} = 0.078^\circ$$

Indeed, the components at that point are larger than those induced at the vertex of the orbit considering the NS half-width of the window.

Table 9.4 Satellite antenna pointing angles and depointing components under the conditions of the example of Section 9.5.4.4

$\alpha = 5.5895°$	$\alpha = 5.5895°$	$\theta = 8.5°$
$\beta = 6.4169°$	$\beta^* = 6.4474°$	$\varphi = 40.9°$
$\alpha^* = 5.5543°$		$\phi = 69.3°$
$\Delta\Psi_{R,x} = 0.0498°$		$\Delta\Psi_{R,y} = 0.0005°$
$\Delta\Psi_{P,x} = 0.0003°$		$\Delta\Psi_{P,y} = 0.0298°$
$\Delta\Psi_{Y,x} = 0.0483°$		$\Delta\Psi_{Y,y} = 0.0554°$
$\Delta\Psi_{NS,x} = 0.0043°$		$\Delta\Psi_{NS,y} = 0.0011°$
$\Delta\Psi_{i,x} = 0.0067°$		$\Delta\Psi_{i,y} = 0.078°$
$\Delta\Psi_{EW,x} = 0.0011°$		$\Delta\Psi_{EW,y} = 0.0046°$
$\Delta\Psi_{e,x} = 0.0034°$		$\Delta\Psi_{e,y} = 0.0029°$

The depointing originating from the longitudinal displacement EW within the window is:

$$\Delta\Psi_{EW,x} = 0.0011° \qquad \Delta\Psi_{EW,y} = 0.0046°$$

As stated earlier, the effects of eccentricity are twofold—a longitudinal displacement and a radial displacement. The longitudinal displacement is contained in the width of the window and the depointing is determined by the radial displacement. Hence only the influence of radial displacement is considered:

$$\Delta\Psi_{e,x} = 0.0034° \qquad \Delta\Psi_{e,y} = 0.0029°$$

The overall x and y components of depointing are then:

$$\Delta\Psi_x = 0.086° \qquad \Delta\Psi_y = 0.0782°$$

The overall depointing due to attitude and orbit control is now obtained by combining the overall x and y components. A pessimistic value is:

$$\Delta\Psi = [(\Delta\Psi_x)^2 + (\Delta\Psi_y)^2]^{1/2} = 0.115°$$

An optimistic approach would be to consider the depointing to be equal to the value of the larger overall component:

$$\Delta\Psi = \Delta\Psi_x = 0.086°$$

If the overall x and y components are considered as independent zero mean random variables and assumed to represent the 3σ values for the depointing components, then the standard deviation of each overall component is $\sigma \cong \Delta\Psi_x/3 \cong \Delta\Psi_y/3 = 0.027°$.

Hence the depointing corresponding to a probability of 99.73% of not being exceeded is 3.44σ, i.e. $\Delta\Psi = 0.095°$.

9.5.5.5 *Antenna provided with a pointing control system*

In order to limit the pointing error, the antenna can be equipped with a system to control the antenna pointing direction to a beacon located on the ground. An error angle measuring device located on the satellite antenna determines the deviation of the antenna pointing direction with respect to the direction of the beacon on the ground. The error signals generated are used to control the antenna pointing mechanism (APM).

The principles of operation of error angle detectors have been described in Section 8.3.7.6. A system with several sources easily integrates into the network of radiating elements of a multisource antenna (e.g. TDF-1). A mode extraction system is more practical when the antenna contains only a single source (e.g. TV-SAT). According to the performance of the error angle detector and the dynamics of the pointing system, the pointing accuracy of an antenna provided with a control system is between 0.1° and 0.03° [BIN-84], [PER-85].

9.5.6 Conclusion

As two thirds of the surface of the globe is submerged, global coverage is not well suited to service by stations mounted on the ground. A reduction of coverage area leads to an increase of the gain of the antenna which provides this coverage. Furthermore, limitation of coverage to the region to be served makes frequency re-use using space diversity possible; two antenna beams sufficiently separated can use the same frequency with reduced mutual interference.

For international links it is thus preferable to limit the coverage to continents (coverage of a hemisphere), regions (regional or zonal coverage), or even to zones of small extent dispersed at several points on the globe (multiple point coverage). In the context of a national network, the coverage will be limited to the country's territory (national coverage). These zones must of course be within the coverage represented in Figure 9.28 in the case of a geostationary satellite.

For example, the INTELSAT VII satellite, in addition to global coverage, provides reduced coverage (hemispheres and zones) for links at frequencies of 6/4 GHz and point coverage for zones of small extent between which a large traffic flows (14/11 GHz). This coverage is illustrated in Figure 9.32 [MAD-90], [THO-90].

9.6 ANTENNA CHARACTERISTICS

9.6.1 Antenna functions and characteristics

The main functions of satellite antennas are as follows:

- —To collect the radio waves transmitted, in a given frequency band and with a given polarisation, by the ground stations situated within a particular region on the surface of the earth.
- —To capture as few undesirable signals as possible; these do not correspond to the specifications stated above (they may be from a different region or not have the specified values of frequency or polarisation).
- —To transmit radio waves, in a given frequency band and with a given polarization, to a particular region on the surface of the earth.
- —To transmit minimum power outside the specified region.

The link budget between the satellite and the ground depends on the equivalent isotropic radiated power (EIRP). For an available transmission power P_T, the EIRP increases with the

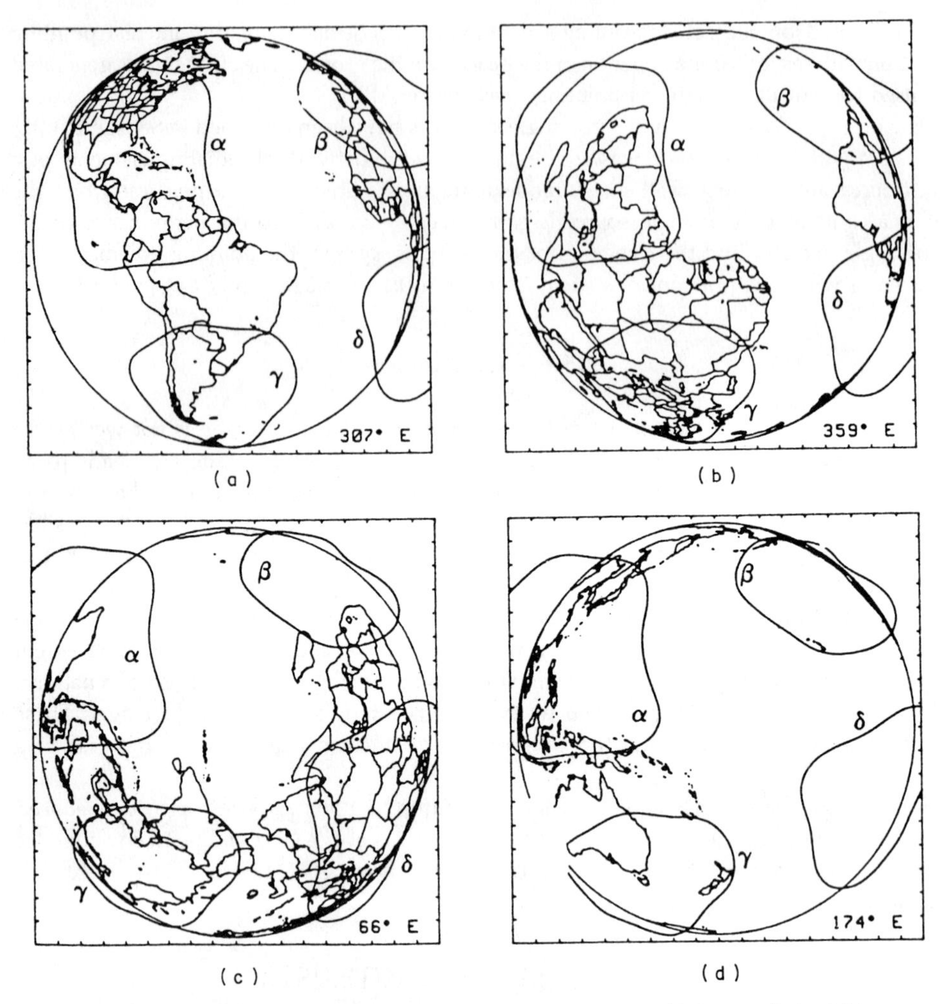

Figure 9.32 INTELSAT VII C-Band zone coverage [NEY-90]. (a) At 307° E Longitude (upright spacecraft attitude). (b) At 359° E Longitude (inverted spacecraft attitude). (c) At 66° E Longitude (inverted spacecraft attitude). (d) At 174° E Longitude (upright spacecraft attitude). (© AIAA. Used with permission.)

gain G_T of the transmitting antenna. Similarly on the uplink, a high G/T for the satellite requires a high value of receiving antenna gain.

A high value of antenna gain is obtained with a directional antenna. The required directivity depends on the mission to be performed—global coverage of the earth, zone or spot coverage. Obtaining a high directivity in association with conformity of the beam to the geographical region to be covered permits frequency re-use by space diversity and hence better use of the spectrum due to economy of frequency bands.

This re-use of frequencies requires antennas with reduced side lobes in order to limit interference. The CCIR provides a reference mask for the antenna radiation pattern which is presented in Figure 9.33 [CCIR-Rep 558]. A standard radiation pattern mask (such as the one given in Figure 8.3) requires a well-defined centre axis for the beam as the gain variations are defined versus the off-axis angle α. This is not the situation when shaped beams are considered (see Section 9.6.6). The mask proposed by the CCIR defines the required gain

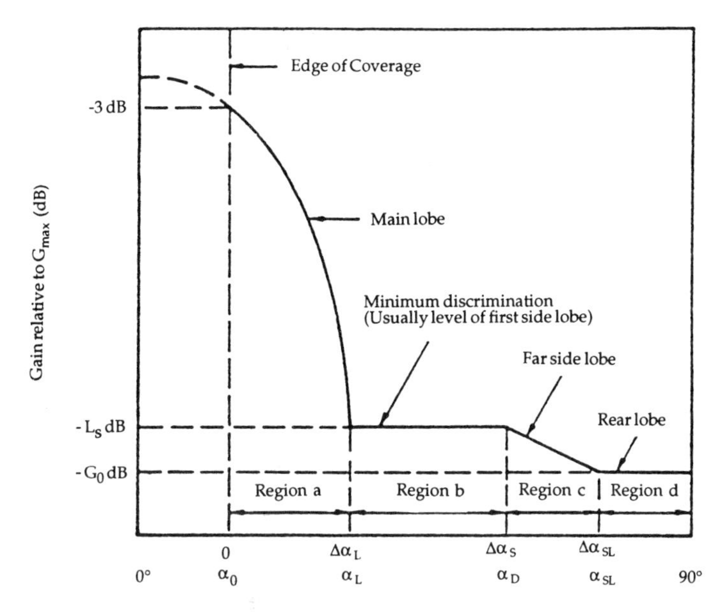

Figure 9.33 Reference limits for the antennas of a fixed service satellite. (From CCIR-Rep558-4. Reproduced by permission of the ITU.)

decrease as a function of the angular distance from the edge of the coverage.

Region A corresponds to the part of the main lobe outside the coverage, the typical gain variation being expressed as:

$$G(\alpha) = G_{max} - 3(\alpha/\alpha_0)^2 \qquad \text{(dB)}$$

Region B is such that the discrimination is large enough to allow satellites operating at the same orbital location to provide coverage. The minimum discrimination $-L_s$ could be about -20 to -30 dB.

Region C incorporates the far side lobes. Gain within region C is equal to $-G_0 = 0$ dB. The proposed mask is still under investigation and valuable information is provided in the CCIR report. Frequency re-use is also achieved by using orthogonal polarisation. A high value of polarisation isolation is thus necessary to limit interference.

In summary, the important characteristics of the antenna sub-system are:

—conformity of the beam to the region to be covered,
—an antenna radiation pattern with reduced side lobes,
—high isolation between orthogonal polarisations,
—accurate beam pointing.

Coverage and minimum beamwidth are closely associated with the satellite attitude and orbit stabilisation procedure (CCIR-Rep 453). Narrow beams and strict pointing specifications can require the use of an active antenna pointing system.

9.6.2 The radio-frequency footprint

Having determined the geometrical coverage (the set of reference points for which a radio-frequency performance objective must be satisfied when depointing is taken into account), it is necessary to define the antenna beam which enables this objective to be achieved.

This definition depends on the nature of the specified objective. Usually an objective of minimum EIRP for transmission coverage and minimum G/T for reception coverage are specified. The beam which maximises the gain at the specified points at the edge of coverage is then sought. In this case it should be noted that, even if the antenna gain is the same for the points specified at the edge of coverage, the power received by the stations located at these points differs from one to the other. The distance to the satellite and the elevation angle vary with the station considered; these lead to variations of free space loss and atmospheric attenuation respectively.

Hence, if optimisation of power flux over a given geometrical coverage is the specified objective, it is necessary to weight each coverage reference point with a coefficient which represents the relative attenuation variations. Definition of the beam is more complex and its angular width can differ significantly from that obtained from the geometrical coverage.

Various types of antenna beam are used to illuminate the geometrical coverage region:

—a beam of circular cross-section,
—a beam of elliptical cross-section,
—a contoured beam,
—multiple beams.

The form of the antenna beam does not always perfectly encompass the geometrical coverage. The beam is characterised in different planes by its N dB beamwidth which is defined by the solid angle at the edge of which the gain has fallen by N dB with respect to the maximum gain (the boresight) within the coverage. Its representation on the map showing the geometrical coverage gives the radio-frequency coverage or footprint of the beam (curves of equal gain).

The form of footprint obtained depends on the chosen representation. Hence, a beam of circular cross-section appears as an ellipse on every representation corresponding to a projection on to a plane except, of course, when the axis of the beam is perpendicular to the plane. In particular, with representation on a plane tangential to the earth at the sub-satellite point, an antenna beam of circular cross-section is correctly represented if the boresight coincides with the direction of the earth centre. The representation is erroneous (an elliptical trace) if the aiming axis is different. For a geostationary satellite, the maximum angle between the boresight and the plane of projection is 8.7° and the error is small, of the order of 1%.

On the other hand, the representation defined from the true view angles faithfully follows the form of the antenna beam independently of the boresight direction and the altitude of the satellite.

9.6.3 Circular beam

The cross-section of the beam is circular. It is the same as the radiating aperture of the antenna which is usually reflecting.

9.6.3.1 *Radio-frequency coverage to 3 dB*

The angular 3 dB beamwidth is taken as the angle θ_{3dB} under which the geometrical coverage is seen from the satellite. The on-axis antenna gain in this case is equal to:

$$G_{max} = 48360\,\eta/\theta_{3dB}^2 \tag{9.50}$$

and the gain at the coverage boundary is thus:

$$G(\theta_{3dB}/2) = G_{max}/2 = 24180\,\eta/\theta_{3dB}^2$$

where η is the efficiency of the antenna and θ_{3dB} is expressed in degrees.

To calculate the numerical coefficients of these expressions, the coefficient k which relates the 3 dB beamwidth to the ratio $\lambda/D\,(\theta_{3dB} = k\lambda/D)$ is taken to be 70 (for a relecting antenna, k varies between 57 for uniform illumination and 80 when the main lobe of the source is entirely intercepted by the reflector). The illuminating efficiency η_i of the reflector is thus maximum and of the order of 0.75. Assuming that the antenna efficiency η coincides with the illumination efficiency η_i (no ohmic losses), the maximum gain at the edge of a beam of angular width θ_{3dB} corresponding to a relative fall of gain of 3 dB thus has a value:

$$G(\theta_{3dB}/2)_{dB} = 42.5 - 20\log\theta_{3dB} \qquad \text{(dB)} \tag{9.51}$$

Variations of antenna gain and hence EIRP and G/T in the coverage region are thus limited to 3 dB.

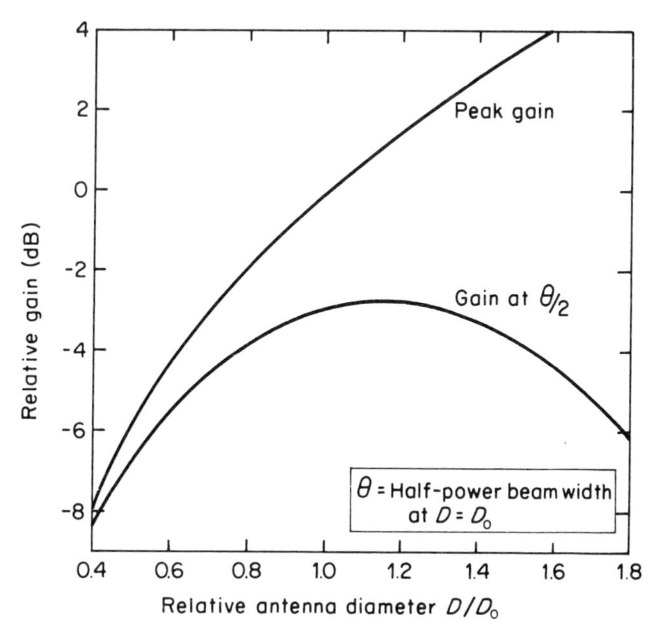

Figure 9.34 Variation of on axis gain and gain in the $\theta/2$ direction as a function of D/D_0 ($\theta = \theta_{3dB}$ for $D = D_0$).

9.6.3.2 *Radio-frequency coverage providing maximum gain at the edge of geometrical coverage*

The curves of Figure 9.34 show how the gain on the axis and in the $\theta/2$ direction vary as a function of the ratio D/D_0, where D is the diameter of the antenna and D_0 is the particular value corresponding to a beam of 3 dB angular width such that $\theta_{3dB} = \theta$ [HAT-69]. It appears that for D between D_0 and 1.3 D_0, the gain in the $\theta/2$ direction remains greater than its initial value and passes through a maximum. The on-axis gain itself increases by more than 2 dB.

An antenna exists, therefore, for which the combination of on-axis gain and the fall in gain (N dB) in the $\theta/2$ direction with respect to the on-axis gain is associated with maximum gain in the $\theta/2$ direction. Hence, the gain at the coverage boundary (minimum gain over the region) is maximised by choosing to provide a geometrical coverage (defined by an angular width θ seen from the satellite) width a beam of N dB width equal to the angular width θ of the region.

The optimum value of the relative fall of gain (N dB) which maximises the gain at the boundary of a geometrical coverage of angular width $\theta = \theta_{NdB}$ can be determined as a function of the 3 dB beamwidth θ_{3dB}. Assuming that the variation of gain about the axis is parabolic (see equation (2.3)):

$$\Delta G = -12(\alpha/\theta_{3dB})^2, \quad \text{hence for } \alpha = (\theta_{NdB}/2) \text{ and } \Delta G = N \text{ (dB):}$$

$$N = -12[(\theta_{NdB}/2)/\theta_{3dB}]^2 = -3(\theta_{NdB}/\theta_{3dB})^2$$

Hence:

$$\theta_{NdB} = \theta_{3dB}\sqrt{(N/3)} = k(\lambda/D)\sqrt{(N/3)} \tag{9.52}$$

The gain on the axis of the beam as a function of N and θ_N is given by:

$$G_{max} = \eta(\pi D/\lambda)^2 = \eta[(k\pi)^2/3][N/(\theta_{NdB})^2] \tag{9.53}$$

The gain $G(\theta_{NdB}/2)$ at the edge of a beam of angular width $\theta = \theta_{NdB}$ corresponding to a fall of gain of N dB thus has a value:

$$G(\theta_{NdB}/2)_{dB} = (G_{max})_{dB} - N = 10\log[\eta(k\pi)^2/3] + 10\log[N/\theta_{NdB})^2] - N$$

Hence:

$$G(\theta_{NdB}/2)_{dB} = 10\log[\eta(k\pi)^2/3] + 10\log N - 20\log\theta_{NdB} - N$$

To obtain the maximum gain at the edge of a beam of fixed angular width $\theta = \theta_{NdB}$, it is necessary to maximise $10\log[\eta(k\pi)^2/3]$ and $10\log N - N$. Maximisation of $10\log N - N$ leads to:

$$[10/(N\ln 10)] - 1 = 0, \qquad \text{hence } N = 10/\ln 10 = 4.34$$

The gain is thus maximised at the edge of a beam when the angular width corresponds to a fall of gain $N = 4.3$ dB.

It remains to maximise $10\log[\eta(k\pi)^2/3]$ in order to obtain the highest possible gain. When a reflecting antenna is used, the efficiency η of the antenna depends in particular on the efficiency of illumination η_i of the reflector; this in turn determines the factor k which defines the 3 dB angular width (k and η_i vary in opposite directions).

The value currently used for k is 70 and this corresponds to a source radiation pattern which illuminates the edge of the reflector with a relative level with respect to the centre of around -12 dB; the efficiency of illumination η_i is thus of the order of 0.75. On the other hand, the factor k is maximum and approximately equal to 80 for a source radiation pattern whose main lobe is entirely intercepted by the reflector. The efficiency of illumination η_i is no greater than 0.6 under these conditions. Nevertheless, the latter scheme maximises the product $\eta_i k^2$, to of the order of 3800, and consequently maximises the quantity $10 \log[\eta(k\pi)^2/3]$.

Assuming that the antenna efficiency η coincides with the efficiency of illumination η_i (no ohmic losses), the maximum gain at the edge of a beam of angular width θ corresponding to a relative fall of gain $N = 4.3$ dB thus has a value:

$$G(\theta/2)_{\text{dB}} = 10 \log[3800\pi^2/3] + 10 \log 4.3 - 4.3 - 20 \log\theta_N$$

Hence:

$$G(\theta_{4.3\text{dB}}/2)_{\text{dB}} = 43 - 20 \log\theta_{4.3\text{dB}} \quad \text{(dB)} \tag{9.54}$$

The gain on the axis is 4.3 dB greater.

Under these conditions, the diameter of the reflector obtained from equation (9.53) is given by:

$$D = \lambda(k/\theta_N)\sqrt{(N/3)} = 95(\lambda/\theta)$$

where θ corresponds to the angular width at 4.3 dB. Recall that, in the case where θ corresponds to the angular width at -3 dB and the illumination efficiency is maximum with a value of K of the order of 70, the diameter is given by the conventional relation $D = 70(\lambda/\theta)$.

Maximisation of the gain at the edge of a beam of specified angular width θ, obtained by choosing a relative gain level at the edge of -4.3 dB with respect to the on-axis gain and an illumination efficiency of 0.6, leads to an increase of the reflector diameter of around 35%. This increase of diameter entails an increase of mass of the order of 50%. The gain at the edge is then around 0.5 dB greater than when the relative gain at the edge is -3 dB with respect to the on-axis gain. The on-axis gain itself is around 1.3 dB greater.

9.6.4 Elliptical beams

A narrow beam of elliptical cross-section provides greater flexibility for matching the coverage zone. The beam is characterised by two angular widths θ_A and θ_B, which correspond to the major axis A and the minor axis B of the ellipse, and the orientation of the ellipse with respect to the reference frame (Figure 9.35).

9.6.4.1 Radio-frequency coverage to 3 dB

Assuming that the 3 dB angular widths of the beam correspond to the angles θ_A and θ_B, the on-axis gain is:

$$G_{\max} = 48360\eta/(\theta_A\theta_B) \tag{9.55}$$

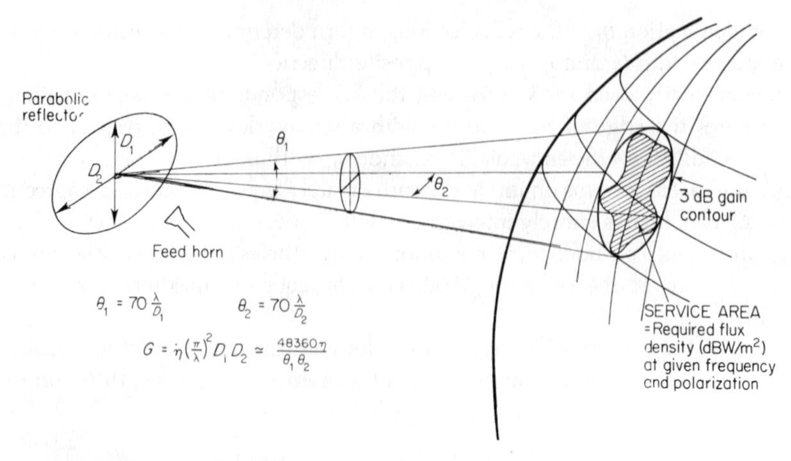

Figure 9.35 Characterisation of an elliptic antenna beam.

where η is the efficiency of the antenna and θ_A and θ_B are expressed in degrees. The angles θ_A and θ_B are related to the corresponding diameters of the radiating aperture by the conventional relation $\theta = 70(\lambda/D)$ (Figure 9.35).

9.6.4.2 *Radio-frequency footprint at 4.3 dB*

It is possible, as for the beam with a circular cross-section, to define angular widths which maximise the gain at the edge of the footprint. The angles θ_A and θ_B then correspond to the 4.3 dB angular widths. Under these conditions the on-axis gain is:

$$G_{max} = 69320\,\eta/(\theta_A\theta_B) \tag{9.55a}$$

where η is the antenna efficiency and θ_A and θ_B are expressed in degrees.

9.6.4.3 *Optimisation*

In the general case, the reference points of the geographical coverage define a region which is shifted in longitude with respect to the satellite. It is then necessary to find the parameters of an ellipse which corresponds to a coverage to 3 dB (or another value) and maximises the gain at the reference points at the edge of the coverage. The parameters of the ellipse which characterise the footprint of the beam are the major axis A, the minor axis B, the inclination (tilt) of the major axis T and the position (X_0, Y_0) of the centre of the ellipse with respect to the sub-satellite point in a true view representation (Figure 9.36). An optimisation procedure can involve choosing four extreme points on the coverage and finding the ellipses of minimum angular width which pass through these points and cover all the other points. A constraint on the minimum value of the ellipticity a/b of the beam can be introduced to take account of the feasibility of realising the antenna in hardware.

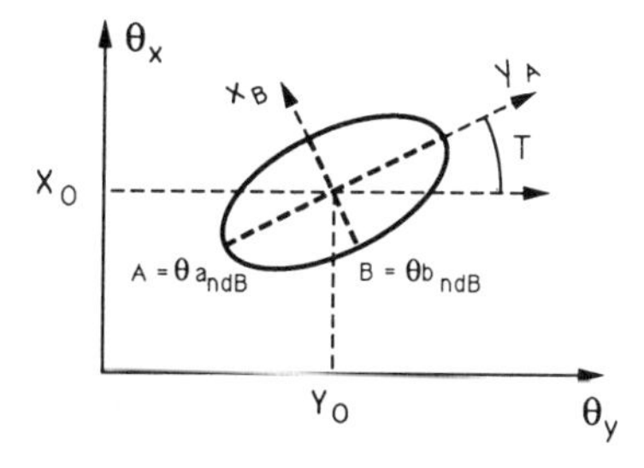

Figure 9.36 The defining parameters of elliptic coverage.

9.6.5 The influence of depointing

The loss of gain associated with depointing depends on the form in which the performance objectives are specified.

9.6.5.1 Performance specified in terms of minimum EIRP over a region

Geometric coverage takes account of the pointing error of the antenna beam; a circle of radius equal to the pointing error is placed on each reference point of the geographical coverage using true view angle representation.

Taking pointing errors into account for circular or elliptical beams leads to a broadening by twice the pointing error (Figure 9.37). Broadening the antenna beam which just covers the zone to be served by twice the depointing is equivalent to defining the geometric region

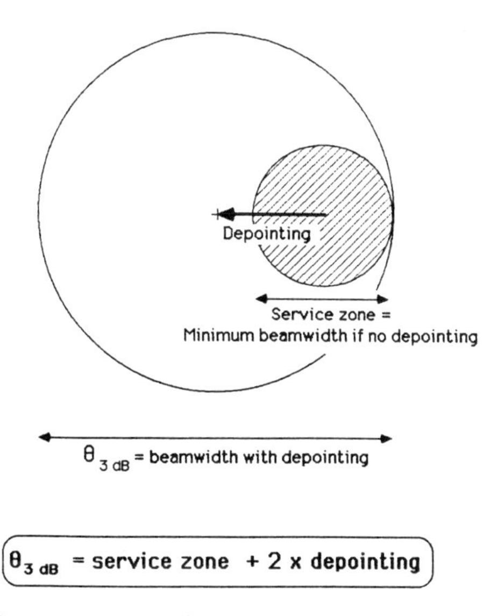

Figure 9.37 Extension of a beam on account of depointing.

by assigning a circle of uncertainty of radius equal to the depointing to all points on the boundary.

Considering an elliptical beam and assuming that a 3 dB coverage which takes account of the depointing α is used, the antenna gain on the coverage axis is (equation (9.55)):

$$G_{max} = 48\,360\,\eta/(\theta_A \theta_B)$$

If the antenna beam just covered the region to be served (not taking depointing into account), the on-axis gain would be:

$$G'_{max} = 48\,360\,\eta/[(\theta_A - 2\alpha)(\theta_B - 2\alpha)]$$

The loss of gain A due to broadening of the beam to take account of depointing is thus:

$$A = G_{max}/G'_{max} = [(\theta_A - 2\alpha)(\theta_B - 2\alpha)]/(\theta_A \theta_B) = (1 - 2\alpha/\theta_A)(1 - 2\alpha/\theta_B)$$

Hence, in decibels:

$$A_{dB} = 10 \log(1 - 2\alpha/\theta_A) + 10 \log(1 - 2\alpha/\theta_B) \qquad \text{dB}$$

The above loss of gain results from a broadening by 2α of the 3 dB beamwidth to take account of depointing. To obtain the same value of EIRP at the limit of coverage, the transmitter power must be increased by $-A$ dB.

If α is small compared with θ_A and θ_B, this gives:

$$A_{dB} = -(10/2.3)[(2\alpha/\theta_A) + (2\alpha/\theta_B)] = -8.7[(\theta_A + \theta_B)/(\theta_A \theta_B)]\alpha \qquad \text{(dB)} \qquad (9.56)$$

With a circular beam (and coverage to 3 dB), $\theta_A = \theta_B = \theta_{3dB}$. This gives:

$$A_{dB} = -17.4(\alpha/\theta_{3dB}) \qquad \text{(dB)} \qquad (9.57)$$

It should be noted that this expression differs from the expression $\Delta G = -12(\alpha/\theta_{3dB})^2$ which characterises the loss of gain due to depointing in the vicinity of the boresight of the antenna. The problem posed here is different; it is to evaluate the loss of gain due to broadening of the beam by 2α to ensure coverage of a geographical region with a minimum EIRP. Typically, if α is equal to $(\theta_{3dB}/10)$ the loss is -1.7 dB.

9.6.5.2 *Performance specified in terms of minimum EIRP at a particular point*

In this case, the service must provide a minimum power at a specified point (a single receiving station) situated nominally on the axis of the antenna beam. Depointing of the satellite antenna causes a loss of power as defined by the law of variation of gain as a function of the components α_A and α_B of the direction α with respect to the axes y_A and x_B which define the antenna coverage (see Figure 9.36):

$$G(\alpha) = G_{max} - 12[(\alpha_A/\theta_{A3dB})^2 + (\alpha_B/\theta_{B3dB})^2] \qquad \text{(dB)} \qquad (9.58)$$

Hence there is a loss of gain for depointing with components α_A and α_B which has a value:

$$A_{dB} = -12[(\alpha_A/\theta_{A3dB})^2 + (\alpha_B/\theta_{B3dB})^2] \qquad \text{(dB)}$$

With a circular beam, $\theta_{A\,3dB} = \theta_{B\,3dB} = \theta_{3dB}$. This gives:

$$A_{dB} = -12(\alpha/\theta_{3dB})^2 \qquad \text{(dB)} \tag{9.59}$$

The above loss of gain results from variation of the satellite antenna pointing direction due to the effect of displacement of the satellite. To provide the minimum power at the earth station situated nominally on the antenna axis, the transmitter power must be increased by $-A_{dB}$.

9.6.5.3 *Example*

Consider an antenna of 3 dB width equal to 1°. Depointing is estimated at 0.3°. In the first case (equation (9.58)) the loss of gain is -5.2 dB. In the second case (equation (9.59)), the loss of gain is -1.1 dB.

The example shows that the power margin to be provided is greater, for the same resulting depointing, in the case of a service which provides minimum EIRP over a region (such as direct television broadcasting or VSAT) than in the case of a service to a fixed station. If this margin is deemed to be too large in view of the value of depointing, it would be advisable to provide an antenna beam pointing control system which limits the variation of antenna coverage (e.g. TDF-1 [FRA-82]).

9.6.6 Shaped beams

The elliptical beam is a first step towards matching the antenna radiation diagram to the coverage zone. However, except for special cases, it does not permit total conformity of the beam and the region to be served to be obtained since the latter is not usually of simple geometric form. This leads to both interference outside the specified coverage and a loss of gain with respect to the maximum which is theoretically possible over a coverage of given angular area.

9.6.6.1 *Limiting value of gain over a given coverage*

The theoretical limiting value of gain G_{lim} over a coverage of complex form is obtained by considering an ideal lossless antenna whose beam conforms exactly to the solid angle Ω (in steradian) which defines the coverage (it is zero to the exterior of the coverage). This gain is by definition equal to:

$$G_{lim} = 4\pi/\Omega \tag{9.60}$$

The solid angle can be approximated by the angular area (in radian^2) under which the coverage is seen. By considering a beam of circular cross-section and angular width θ, the corresponding solid angle Ω is equal to $2\pi[1 - \cos(\theta/2)]$ and the angular area S has a value $\pi(\theta/2)^2$. In the case of global coverage of the earth by a geostationary satellite ($\theta = 17.4°$), the error is less than 2 parts in 1000.

9.6.6.2 *Determination of the angular area*

The angular area S of the coverage is defined by a set of n points which constitute a polygon and can be calculated from the true view coordinates β_x and β_y of the points defined by equations (9.26). The angular area S of the polygon is given by $S = \Sigma_n S_i$, where the area S_i is the algebraic area defined by the line joining point P_i to point P_{i+1} (Figure 9.38). This gives:

$$S = \Sigma_n\{\tfrac{1}{2}[(\theta_x)_i + (\theta_x)_{i+1}][(\theta_y)_i - (\theta_y)_{i+1}]\} \quad (\text{degree}^2) \tag{9.61}$$

The pointing error is represented by a disc of uncertainty of radius equal to the depointing α centred on each vertex of the polygon. Since the sum of the exterior angles of a polygon is equal to 2π, the increase of area ΔS to take the pointing error into account is given by:

$$\Delta S = \alpha(p + \pi\alpha) \quad (\text{degree}^2) \tag{9.62}$$

where P is the perimeter of the polygon which defines the geographical coverage and is determined by:

$$P = \Sigma_n P_i = \Sigma_n\{[(\theta_x)_{i+1} - (\beta_x)_i]^2 + [(\theta_y)_{i+1} - (\theta_y)_i]^2\}_{1/2} \quad (\text{degree}) \tag{9.63}$$

9.6.6.3 *Beam shaping techniques*

Beam shaping can be obtained using two different methods whose principles are given below. The accuracy of realisation is the subject of Section 9.6.8.

The first methods consists of modifying the power distribution within a beam generated by a single source. Shaping of the beam is achieved by modifying the mechanical form of the reflector—the form of the radiating aperture or the nature of its profile (see Section 9.6.8.2). Whatever technique is used, the shape of the beam cannot be further modified after the mechanical design of the antenna has been completed, particularly when the satellite is in orbit.

With the second technique, shaping of the beam is obtained by combining the radiation of several elementary beams. These beams are generated by an antenna with several radiating elements which are excited by coherent signals having a given amplitude and phase distribution (Figure 9.39). The array of radiating elements can be located at the focus of an antenna reflector or lens. It can also generate the antenna beam directly (an array antenna, see Section 9.6.8.4).

The latter technique enables a beam of any form to be obtained and its gain distribution

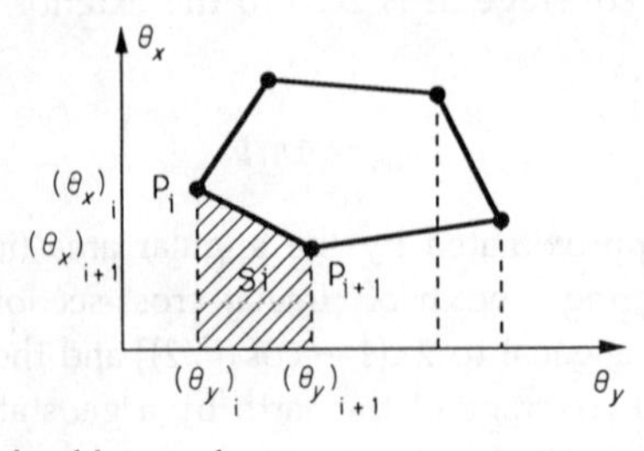

Figure 9.38 Coverage area defined by a polygon.

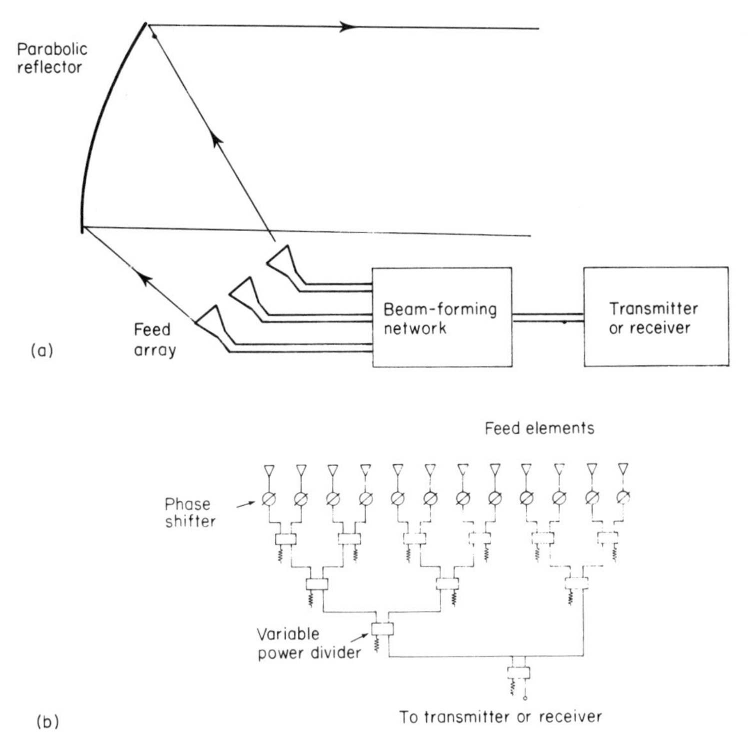

Figure 9.39 Shaped beam antenna using a phased array.

over the coverage area can be adapted on demand. Since the resultant beam is obtained by combining elementary beams of smaller angular width, it is possible to obtain a gain close to the maximum over a large part of the interior of the coverage area; a decrease occurs only at the edge of the coverage. Furthermore, the decrease outside the coverage area is rapid. So even in the case of illumination of a coverage area of simple geometric form (circular in the limit), the compound beam has a definite advantage (in terms of gain over the coverage area and reduction of interference outside it) over illumination by a single beam.

Another major advantage lies in the possibility of modifying the radio coverage of the antenna by controlling the amplitude and phase distribution of the radiating elements. This possibility can be effective even when the satellite is in orbit by configuring the beam forming network with elements which can be controlled by telecommand.

The disadvantages associated with the above technique result from the added complexity of the antenna and this is accompanied by an increase in the size of the radiating aperture (due to generation of individual beams which are narrower than the angular width of the coverage area).

9.6.6.4 Examples of shaped radio-frequency coverage

The coverages of the EUTELSAT II satellite are illustrated in Figure 9.40. Two different coverages can be obtained by switching the antenna feed circuits [DUR-88]. This also applies

Transmit coverage east antenna

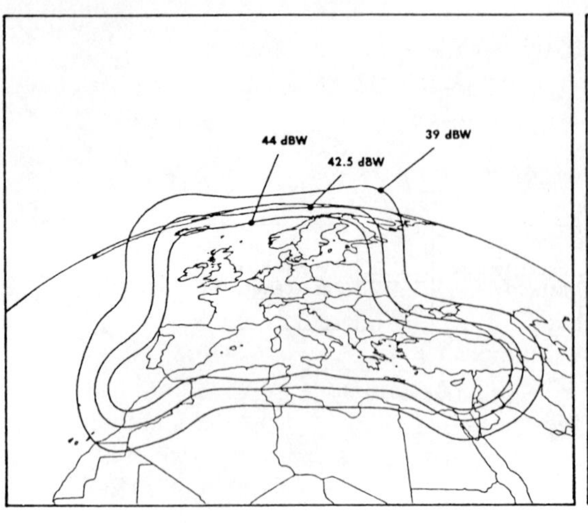

High gain transmit coverage east or west antenna

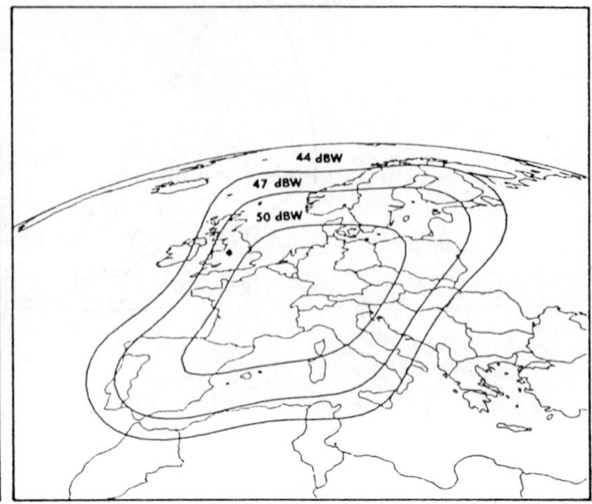

Transmit coverage west antenna

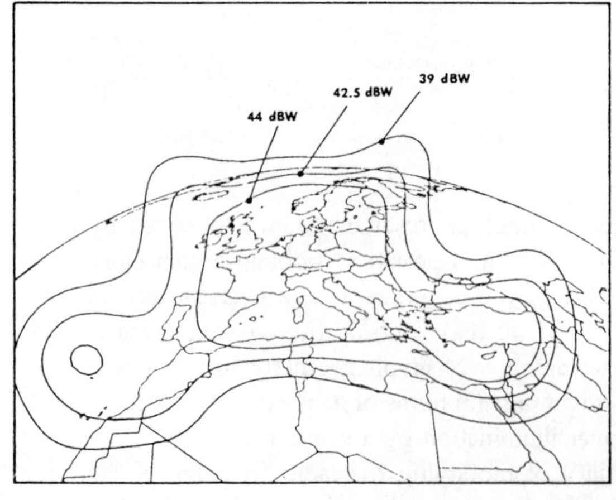

Receive coverage east antenna

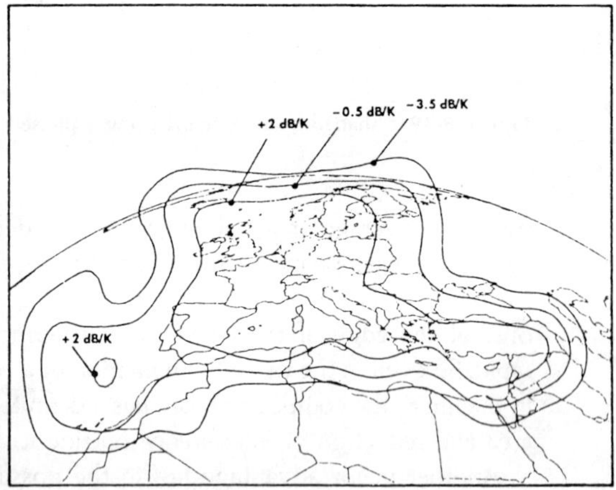

Figure 9.40 Coverage of the EUTELSAT II satellite.

to the INTELSAT VII satellites for which a choice of three types of coverage is possible according to the orbital position (see Figure 9.32). A change of coverage is obtained by selecting one of the three sets of power splitters and phase shifters which are preset on integration of the satellite [THO-83]. On the other hand, the coverage of the DSCS III satellites can be dynamically adapted by continuous (and not discrete) action of the array elements [TRA-88].

This technique can also be used to produce scanning beams whose orientation is controlled by varying the phases in the feed network of the radiating elements. In this way the ACTS satellite is provided with two scanning beams (see Section 9.4).

9.6.7 Multiple beams

Unlike the preceding coverages which use a single beam in a given frequency band with a given polarisation, multiple beam coverage implies generation of several beams which may be in different frequency bands and have different polarisations.

9.6.7.1 Separate multiple beams

This coverage consists of a set of geographical regions which are separated from each other. These regions are of simple geometric form in true view angle representation and are illuminated by beams of narrow circular cross-section. The regions could correspond to large towns between which it is required to establish high capacity links. The beams can thus share the same frequency when their angular separation is sufficient. The use of orthogonal polarisation enables the isolation between links to be increased if the angular separation is too small. Figure 9.41 illustrates this concept in the context of coverage of regions of Europe with large communication requirements.

9.6.7.2 Contiguous beams

Service with a given geometric coverage can be provided by a set of narrow contiguous beams rather than a single narrow beam or a matched beam (Figure 9.42). Since each of the beams is narrower than a beam which would cover the whole of the region to be served,

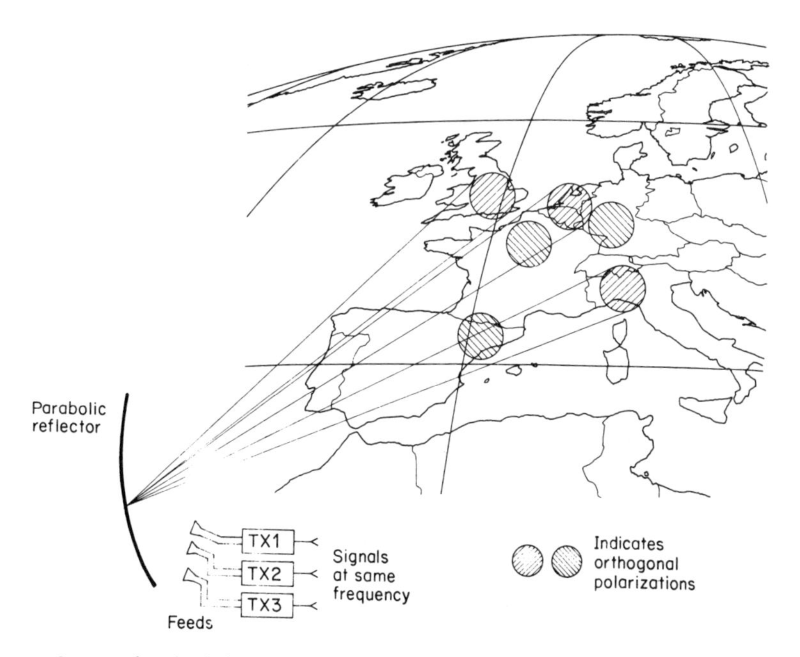

Figure 9.41 Separated multiple beams.

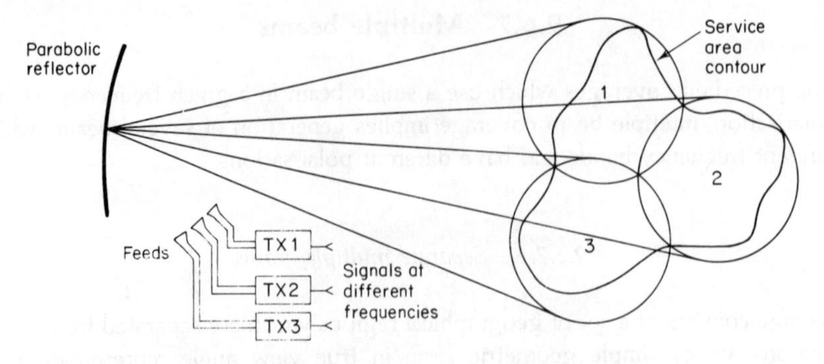

Figure 9.42 Contiguous multiple beams.

the corresponding gain is higher. It is thus conceivable to use earth stations with small diameter antennas.

Since the beams partially overlap, the frequencies used must differ from one beam to the other. The beams thus share the total available system bandwidth as determined by the radiocommunication regulations. The capacity per beam is thus limited, more severely when the number of beams is large. Another disadvantage of this concept is associated with the fact that the information transmitted differs from one beam to the other. To ensure interconnectivity, a specific procedure to ensure routing of the signals between beams must be operated. These procedures, specific to multibeam satellite networks, have been described in Sections 5.2 and 5.3.

9.6.7.3 *Beam lattice*

The concept presented above can be associated with that of frequency re-use to obtain a lattice of narrow beams which can support a greatly extended coverage. Frequency re-use enables the reduction of available bandwidth per beam to be reduced when the number of beams becomes large. A basic pattern formed from beams using a set of different frequencies is regularly repeated over the coverage (Figure 9.43).

This figure also shows the variation of angular distance between beams which re-use the same frequency as a function of the number of frequencies used. The angular distance determines the interference induced in the frequency band concerned by spatial frequency re-use (frequency sharing). Figure 9.44 shows how interference occurs on a particular beam in a three frequency lattice. The greatest interference occurs at the edge of the beam; here the level of the interfering signal is highest and the level of the useful signal is least due to the decrease of gain at the edge. The contributions of the six beams of the adjacent patterns which share the same frequency must be considered.

By using a larger number of beams (and frequencies) in the basic pattern, the angular distance between beams using the same frequency is increased; this leads to a decrease of interference in the system. On the other hand, the usable bandwidth, and hence the capacity per beam, consequently decrease. An example of European coverage by a three frequency beam lattice is given in Figure 9.45.

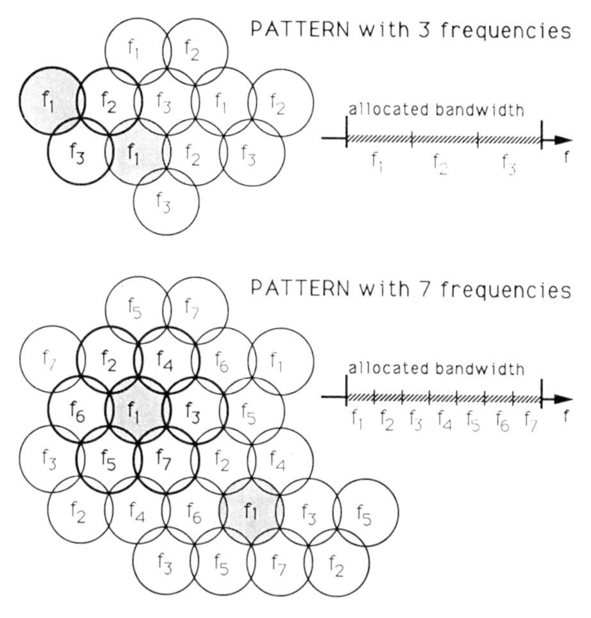

Figure 9.43 Lattice coverage with a three frequency pattern (a) and a seven frequency pattern (b).

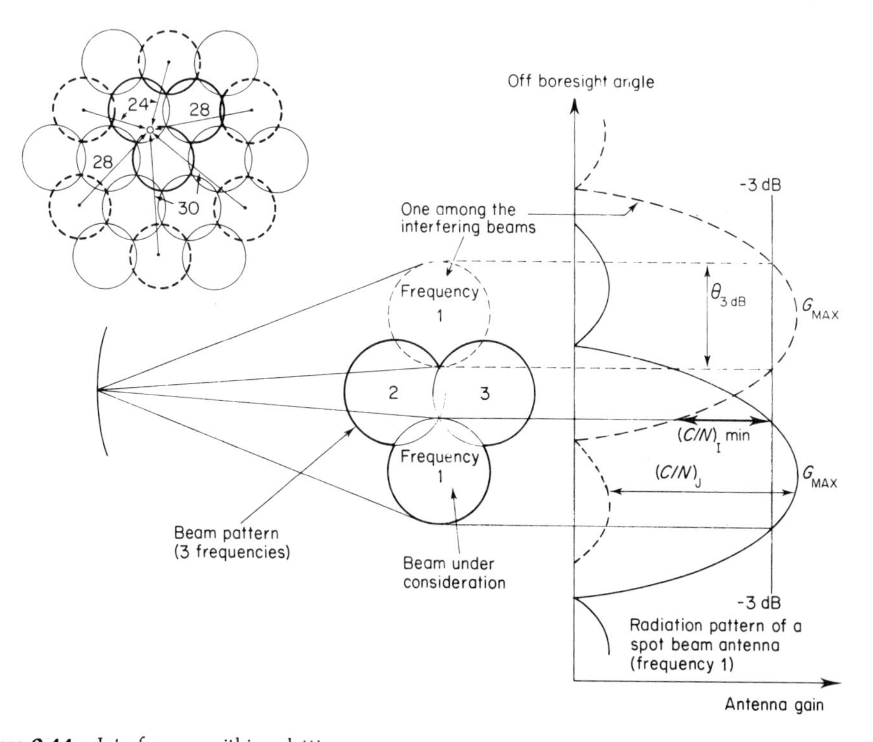

Figure 9.44 Interference within a lattice.

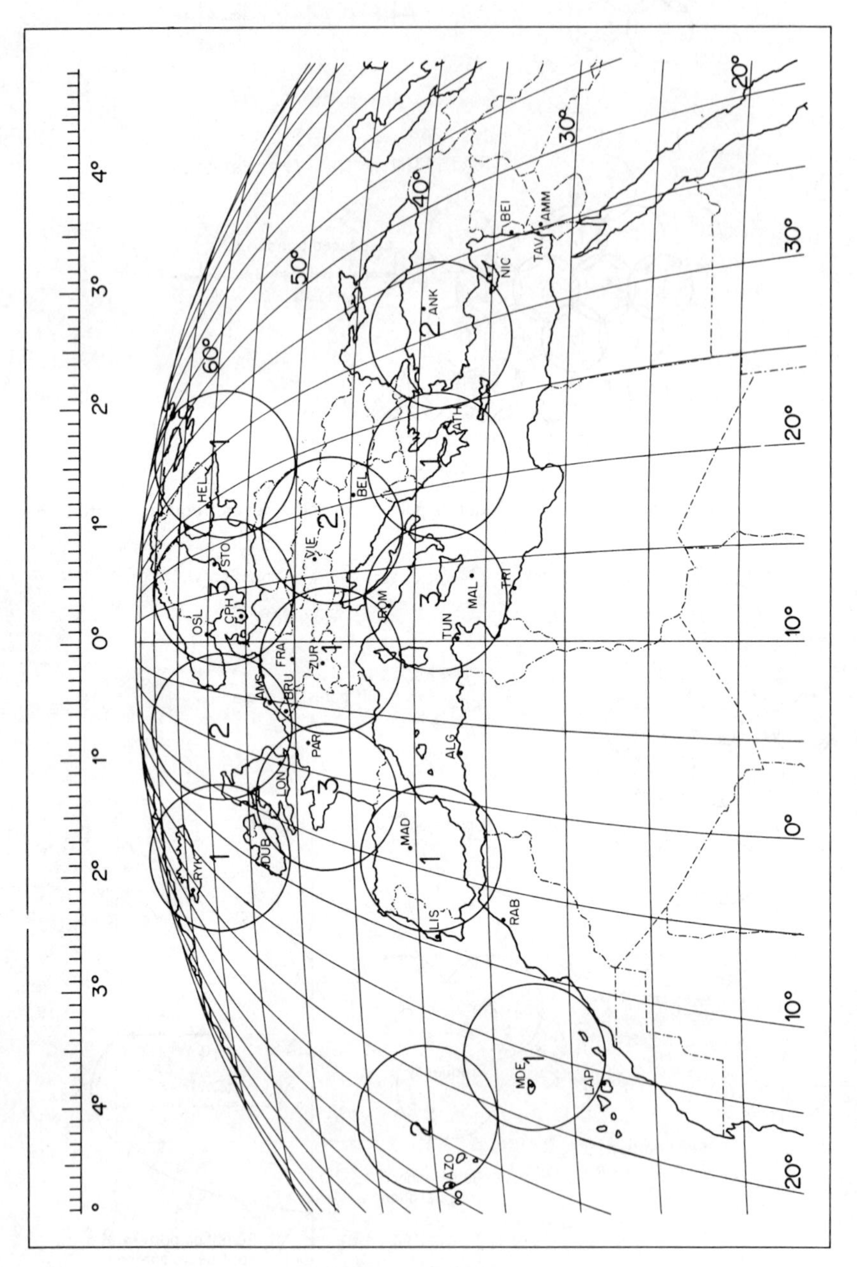

Figure 9.45 Example of European coverage with a three frequency lattice.

9.6.8 Types of antenna

The types of antenna used differ in accordance with the principle used to control the satellite attitude. A simple method of providing attitude stabilisation consists of causing the satellite to rotate about an axis perpendicular to the plane of the orbit (spin stabilisation, see Section 10.2). The antennas can be mounted directly on the rotating satellite or on a platform which maintains a constant orientation with respect to the earth. The attitude of the satellite as a whole can also be controlled in order to maintain a fixed orientation with respect to the earth (three-axis stabilisation, see Section 10.2).

In the case where the antennas are mounted on a platform which is rotating with respect to the earth, the antenna must have a toroidal radiation pattern or generate rotation of the pattern in such a way as to compensate for that of the platform.

Nowadays communication satellites have a platform which supports the payload and whose attitude is stabilised with respect to the earth. The radiating aperture of the antenna thus maintains a fixed orientation with respect to the direction of aiming and this provides flexibility in antenna realisation.

9.6.8.1 *Antennas with a toroidal radiation pattern*

For a spin stabilised satellite, the simplest antenna generates a radiation pattern of revolution about the axis of rotation. To ensure global coverage, the beamwidth of the toroidal pattern is the order of 17°. The antenna gain is only a few decibels.

A toroidal pattern can be obtained by means of a set of radiating filaments (wire antennas). This procedure was used on the first operational satellites; for Intelsat I and II, for example, the antenna gain is from 4 to 5 dB receiving and around 9 dB transmitting.

9.6.8.2 *Despun antennas*

To increase the antenna gain, it is necessary to concentrate the beam on the region to be covered and then to ensure that its orientation remains fixed with respect to the earth. The antenna beam thus turns in the opposite direction to the rotation of the satellite; the antenna is said to be 'despun'.

Mechanical despun antennas. This approach consists of rotating the antenna assembly about the axis of rotation of the satellite by means of an electric motor in such a way as to keep the antenna axis pointing towards the earth [DON-69]. The presence of bearings whose lubrication is difficult and rotating couplings between the antenna and the radio equipment are the source of reliability problems and performance degradation.

Electronic despun antennas. The electronic scanning antenna provides an elegant solution to the difficulties of mechanical origin indicated above. The antenna consists of a set of radiating elements mounted on the exterior of a cylinder. These radiating elements are fed sequentially with a phase which varies as a function of the rotation of the satellite. The failing of this type of antenna, other than the losses in the feeds to the radiating elements, lies in the amplitude and phase discontinuities which appear in the antenna radiation pattern during successive switchings. This type of antenna is used on the Meteosat satellite, for example.

9.6.8.3 *The stabilised platform*

The satellite is provided with a platform on which the antennas and the payload transponders are mounted. This platform maintains a fixed orientation with respect to the earth.

In the case of a rotation stabilised satellite, this platform consists of the upper part of the satellite which is driven in contra-rotation with respect to the lower part which itself rotates about an axis perpendicular to the plane of the orbit (a 'dual spin satellite'). This approach permits installation of high performance antennas and avoids the problems of rotating couplings between the antennas and the radio equipment. However, the problems associated with the presence of a mechanical bearing (such as lubrication and mechanical friction which disturbs the gyroscopic effect) and sliding contacts to transfer electrical energy remain. The INTELSAT VI satellite is an example of this type of architecture [THO-83].

The three axis stabilised satellite itself forms the platform on which the antennas are mounted. Greater freedom is thus provided for mounting large antennas.

Whatever the type of attitude control of the stabilised platform, if the antenna mounting is rigid, the pointing accuracy of the antennas is that of the attitude stabilisation (down to 0.05°). Greater pointing accuracy requires the use of systems which control antenna pointing using a beacon on the ground (e.g. the TDFI satellite [FRA-82]).

9.6.9 Antenna technologies

The frequency bands used by communication satellites are such that the wavelength is small compared with the mechanical size of the antenna. The antennas used are of the radiating aperture type—horn, reflecting, lens and array antennas.

9.6.9.1 *The horn antenna*

The horn antenna is one of the simplest types of directional antenna. It is well suited to, and widely used for, global coverage of the earth. A 3 dB beamwidth of 17.5° is obtained at 4 GHz from a horn whose aperture diameter is 30 cm.

A beam of smaller width would require a horn with a larger aperture and proportionally greater length thereby making installation on the satellite difficult. Furthermore, the horn antenna has poor side-lobe characteristics. These characteristics are improved by corrugation (annular discontinuities) of the interior of the horn. The length of the horn can be reduced by using an excitation system employing a microstrip antenna.

Horns are, however, currently used as a primary source in reflecting antennas.

9.6.9.2 *Reflecting antennas*

This type of antenna is the most commonly used to obtain spot beams or shaped beams. The antenna consists of a parabolic reflector illuminated by one or more radiating elements located at the focal point.

The technique of reflector realisation usually consists of bonding two carbon fibre skins impregnated with resin on each side of a core of aluminium honeycomb. This technique allows excellent results to be obtained in terms of profile realisation accuracy, dimensional stability and rigidity in spite of the mechanical and thermal constraints. Reflection losses are low, less than 0.1 dB in Ku band.

It is possible to modify the pointing direction of the beam in orbit by telecommand by providing the antenna with a control device for the mechanical orientation of the reflector. With a multisource antenna, pointing can also be achieved by modifying the phase distribution of the radiating element feed.

Two reflector mounting. A two reflector mounting in which the main reflector is illuminated by an auxiliary reflector which is itself illuminated by the radiating element or elements (a Cassegrain or Gregorian mounting according to whether the auxiliary reflector is hyperbolic or parabolic) can also be used.

Two reflector mounting, on account of the compactness of the antenna obtained, has an advantage in respect of mechanical mounting of the antenna on the satellite. In certain cases it also facilitates antenna design (e.g. for shaped beams).

Offset mounting. Symmetrical mountings suffer from blocking of the aperture by the radiating elements or the auxiliary reflector and their supports; this leads to a degradation of efficiency and an increase in the level of side lobes.

Use of a portion of the reflector which is offset with respect to the principal axis of the parabola avoids blocking the aperture of the paraboloid (an offset mount). Offset illumination can be used with a one or two reflector mounting (Figure 9.46a and b). Offset mounting also permits easier integration of the antenna onto the satellite, particularly with large reflectors which require folding of the antenna for the launch phase.

When the offset reflector is illuminated by a linearly polarised wave, an orthogonally polarised component originates in the reflector due to the asymmetry with respect to the antenna axis. Offset mounting is thus characterised by a low value (of the order of 20 to 25 dB) of antenna polarisation discrimination. When circular polarisation is used, beam squinting is also observed.

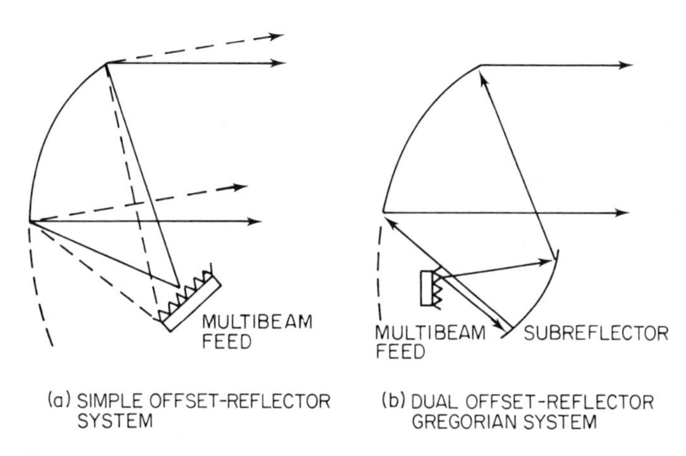

Figure 9.46 Reflecting antennas with offset mounting. (a) Simple reflector. (b) Dual reflector (Gregorian).

Beam shaping by reflector configuration. A reflecting antenna of circular form generates in principle a beam of circular cross-section. Simple beam shaping is achieved by modifying the form of the reflector aperture. Hence an elliptical reflector generates a beam of elliptical cross-section (see Figure 9.35).

Beam shaping achieved by modifying the form of the radiating aperture of the antenna is, in practice, limited to a beam of elliptical cross-section. A reflector of excessively complicated form would lead to difficulty in correctly matching the illumination pattern of the primary source since this would involve low illumination efficiency and a high level of side lobes.

It is also possible to use a reflector of circular form whose profile is parabolic in one plane and cylindrical in the other. The beam obtained in this way is no longer of circular cross-section but is approximately elliptical. The aperture remains circular and this facilitates optimisation of the illumination law.

A different technique consists of modifying the profile of the reflector which is then no longer parabolic. A circular reflector whose profile deviates from parabolic at the edges enables the relative gain at the coverage boundary to be increased; in this way gain variations within the coverage are limited.

Finally it is possible to synthesise a reflector profile which enables a beam to be obtained from the primary source pattern whose spatial power distribution corresponds to that required to ensure illumination of the coverage area. These synthesis techniques are complex and difficult to apply. Furthermore, realisation of a reflector with the required profile is particularly difficult.

Multisource antennas. By locating an array of sources at the focus of the antenna, it is possible to obtain either a shaped beam or multiple beams [ROE-83].

If the array of sources is fed from the same signal with a particular amplitude and phase law, a shaped beam is obtained. This distribution is obtained by means of a set of phase shifters, couplers and power splitters—the Beam Forming Network (BFN) [ANG-88], [CAR-89].

By way of example, an array of 22 horns located at the focus of a reflector with an aperture of 1.6 m is used to generate the antenna beams of the EUTELSAT II satellite (see Figure 9.40). The various converages (two hemispheres, four regions) of INTELSAT VII (Figure 9.32) are obtained from two arrays of 120 sources located at the foci of two reflectors of 2.44 m and 1.57 m diameter.

Independently fed sources permit generation of separates beams which are characterised by their frequency and polarisation. A multibeam coverage such as that illustrated in Figure 9.41 can thus be obtained.

This technique is also used to generate lattice coverage. The size of the source array increases with the size of the lattice and sources located at the edge of the array are situated far from the focal point; this leads to degradation of the corresponding radiation pattern. When the number of beams becomes large, a reduction in the number of sources can be obtained by sharing the sources among several beams; a beam in a given direction is obtained by an appropriate amplitude and phase distribution of a carrier from several sources.

Dual-grid antenna. To obtain an antenna radiation pattern having very high polarisation purity, one approach is to use a reflector consisting of a grid, that is an array of conductors parallel to the required linear polarisation. When the grid is illuminated by a radio wave, only the component of the electric field parallel to the grid is reflected. Current can flow only along the conductors and the field component orthogonal to the grid cannot exist. The

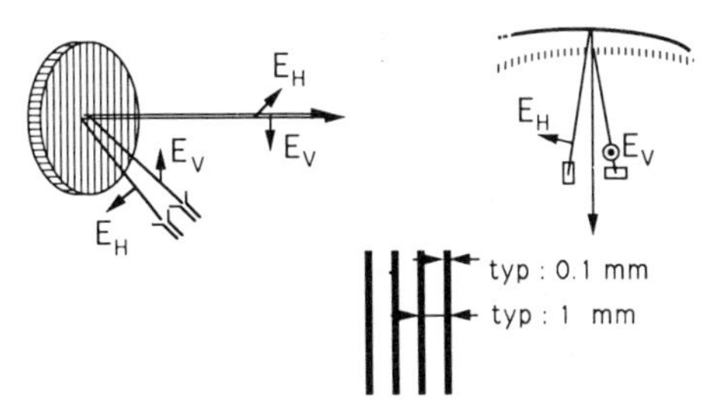

Figure 9.47 Grided reflector antenna organisation.

reflected wave is thus characterised by a very high polarisation purity. Two separate antennas whose grids are perpendicular can be used to generate two beams with linear orthogonal polarisation.

In order to reduce bulk and mass, the dual-grid antenna concept has been developed and is used on EUTELSAT II, for example [RIS-85], [DUR-88]. The antenna consists of two reflectors with offset feed mounted one behind the other with a slight bias so that their foci are located at close but not coincident points (Figure 9.47). To illuminate each reflector, it is thus easy to locate one or more sources operating with a given polarisation at the corresponding focus. The front reflector is formed from a material which is transparent to radio waves on which an array of conductors is arranged parallel to the electric field of the waves generated by the associated illuminating sources. These waves are thus reflected by the first reflector. The waves radiated by the other source array have the orthogonal polarisation. Hence they pass through the front reflector and are reflected by the rear one before again passing through the front reflector. The rear reflector can be either a grid (whose orientation is orthogonal to the first) or a conventional reflector. Even if a component with polarisation orthogonal to the nominal polarisation is generated due to offset mounting, this orthogonal component, which is parallel to the grid of the front reflector, will be blocked at the rear of the grid.

The front reflector, and the structure between the two reflectors, must be transparent to radio waves but capable of withstanding mechanical and thermal stresses. The use of Kelvar skins on each side of a honeycomb core of the same material is highly suitable.

The array of conductors can be embedded in the composite material and produced either by chemical etching of a conducting deposit or mechanically cutting a copper deposit formed on the Kevlar skin before bonding to the honeycomb [FLE-88]. For a Ku band antenna, typical dimensions are of the order 0.2 mm for the conductor width with a step width of the order of a millimetre. The polarisation isolation obtained with this type of antenna exceeds 36 dB.

Dichroic reflectors. A dichroic surface is reflecting to radio waves within a given band of frequencies and transparent outside this band. To obtain such a surface, an array of dipoles whose dimensions are characteristic of the frequency to be reflected is arranged on a substrate which is transparent to electromagnetic waves.

By realising the auxiliary reflector as an assembly of two reflectors using this technique, the antenna has two focal points which depend on the frequency of operation. This permits the same reflector to be used in two different frequency bands; the difficulty associated with mounting a matched source for each frequency band at the focus is thereby resolved. Figure 9.48 shows an example where the dichroic reflector is reflecting in Ka band and transparent

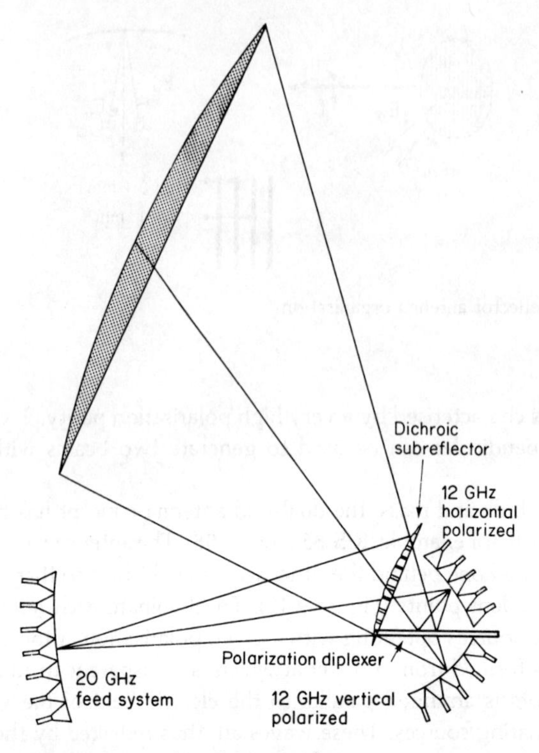

Figure 9.48 Dual frequency antenna with a dichroic surface.

in Ku band. Furthermore, a surface which is polarisation selective can be used to generate two different foci according to the polarisation of the wave in Ku band [LOP-82].

9.6.9.3 *Lens antennas*

Antennas of this kind associate one or more radiating elements with a 'lens' which focuses the radiated electromagnetic energy [SLE-88]. Lens antennas have the advantage with respect to symmetrical reflecting antennas of having the source array situated behind the radiating aperture and this eliminates blocking of the beam. This characteristic is particularly useful when a large set of sources associated with a high performance (and hence bulky) source array is required to support the creation of a large number of multiple beams or high performance beam forming, for example.

The principle of the lens is to produce a propagation delay which is maximum along the axis and reduces towards the periphery where it becomes zero. The spherical wave generated by the source is thus transformed into a plane wave. Several approaches to realisation of the lens can be envisaged:

—A homogeneous dielectric material. The lens obtained in this way has the advantage of a wide passband but has a high mass.
—An assembly of metallic waveguides (a stepped or zoned lens) whose length is arranged

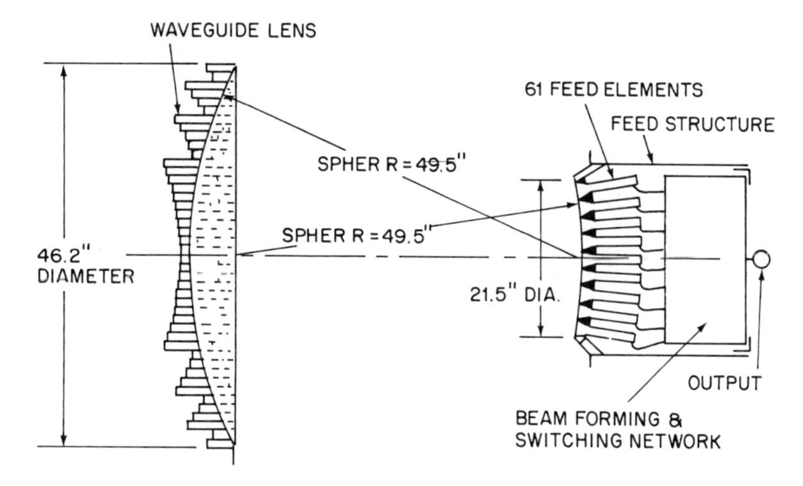

Figure 9.49 Lens antenna formed from an assembly of waveguide sections.

in such a way as to produce the required phase advance to transform the incident spherical wave into a plane wave (Figure 9.49) [SCO-76]. These lenses are light but have a relatively narrow bandwidth (of the order of 5%).

—An assembly of delay lines terminated by radiating elements [MAT-76]. The bandwidth of the 'bootlace' lens is wide and the weight is intermediate between waveguide lenses and dielectric lenses.

Lens antennas suffer from a large mass and bulk. Their application seems to be reserved at present for military satellites (DSCS III) where the capacity for dynamic reconfiguration enables an antenna radiation pattern to be obtained which has zero gain in any particular direction in order to protect against interference. The American satellite DSCS III is thus equipped with a receiving antenna which produces 61 beams of 2° and two transmitting antennas each generating 19 beams.

9.6.9.4 Array antennas

An array antenna uses a large number of radiating elements distributed over the area which constitutes the radiating aperture (Figure 9.50). The overall radiation pattern results from a combination in amplitude and phase of the waves radiated by the array of elements. In principle, operation of the array is similar to that of a source array located at the focus of a reflecting antenna. The difference lies mainly in the number of radiating elements and the surface area; these are determined by the required gain and width of the antenna beam which is radiated directly by the array. The radiating elements can be horns, dipoles, resonant cavities, printed elements, etc. [MAI-82].

Properties of array antennas. The distance between the radiating elements is typically of the order of 0.6 λ. The radiation pattern is adjusted by modifying the phase and amplitude of the supply to the radiating elements by means of controllable power dividers and phase shifters.

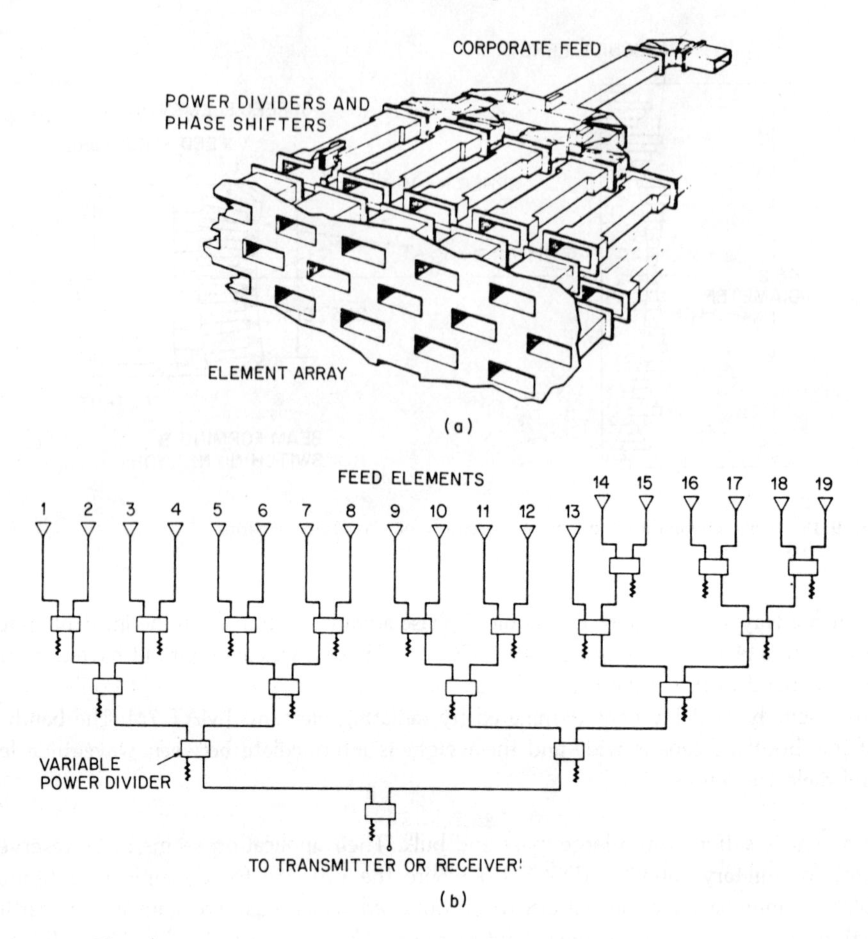

Figure 9.50 Phased array antenna.

For example, by feeding all the radiating elements in phase with the same amplitude, the beam obtained has characteristics similar to those of a beam generated by a reflecting antenna with uniform illumination; the maximum gain is proportional to $(\pi D/\lambda)^2$ and the 3 dB beamwidth is of the order of λ/D in radians, that is around 60 λ/D in degrees. By attenuating the amplitude on the periphery of the radiating aperture, the side-lobe level is reduced and the beamwidth increased. On the other hand, as a consequence of the reduction of illumination efficiency which is no longer unity, the on-axis gain decreases.

By feeding the elements with a phase which varies linearly from one element to the next from one edge of the array to the other, an inclination of the phase plane with respect to the surface of the array can be introduced and this modifies the orientation of the beam.

Feeding an array antenna. With a conventional array antenna, the antenna input power is provided by a conventional power amplifier. Of course, on account of the law of reciprocity, the antenna operates in a similar manner on reception and a low noise amplifier is connected at the output of the beam forming array.

The antenna efficiency is a function of the illumination efficiency (0.8 to 1) and is determined by the amplitude weighting at the edge of the array and the ohmic losses in the power splitters and phase shifters (from one to several dB according to complexity). The ohmic losses in the power distribution constitute a critical parameter.

A shaped beam is obtained by feeding the radiating elements with a particular amplitude and phase distribution of the power available at the antenna input. Dynamic control of the beam is obtained by using controllable power dividers and phase shifters.

9.6.9.5 *Active antennas*

An active antenna is an antenna in which the supply to the radiating elements is realised directly by an amplifier module. According to the total power to be radiated, the power available per amplifier module and the number of radiating elements, an active module can be connected to a single radiating element or a small group of radiating elements.

In the case of an antenna performing both transmitting and receiving functions, the active module also includes the low noise amplifier and transmit-receive separation which is performed at the module input by a circulator [POU-89].

An active antenna constitutes in principle a directly radiating array. It is however conceivable to combine the active array with a one or two reflector mounting. Use of reflector mounting enables a large radiating aperture to be obtained without the size of the array becoming prohibitive (and causing problems of folding etc). For example, by illuminating an auxiliary parabolic reflector, whose focal point coincides with that of the main reflector, with the near field of the array, a magnified image of the array is obtained (Figure 9.51) [SOR-88]. The choice between an array antenna with direct radiation and an array with illumination by reflector depends on considerations associated with the number of beams, the state of the art of the technology, the powers used, etc. [GIA-89].

Beam shaping. The beam shaping elements (attenuators and phase shifters) form an integral part of the active modules. They are located upstream of the power amplifying elements and

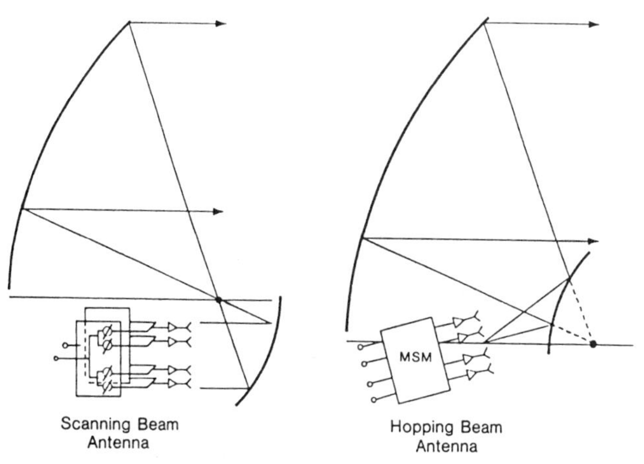

Figure 9.51 Antenna combining a phased array with a single or dual reflector mounting [SOR-88]. (© AIAA. Used with permission.)

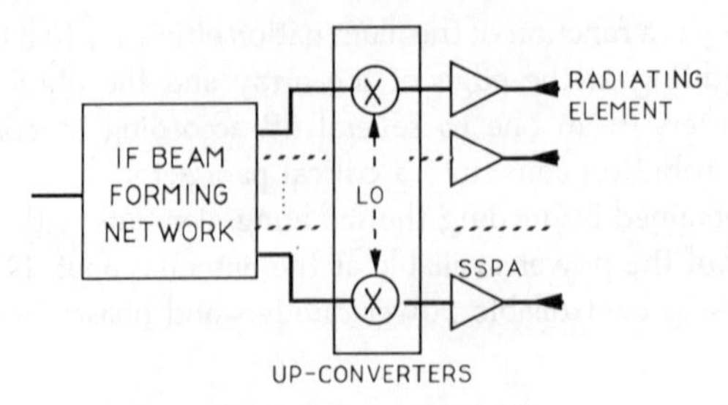

Figure 9.52 Active antenna with beam forming network operating at intermediate frequency.

downstream of the low noise amplifiers. Beam shaping can be realised either at the operating frequency of the antenna or at intermediate frequency. In the latter case, frequency conversion is performed between the beam forming elements which define the amplitude and phase distribution and the radiating elements (Figure 9.52). The latter approach facilitates realisation of the elements, particularly when the antenna operates at high frequency.

Advantages of active antennas. The advantages lie in the use of low power solid state amplifiers which enable good linearity to be obtained, the high reliability associated with realising a function by parallel connection of a large number of identical elements (failure of elements leads to progressive degradation with little noticeable effect on performance) and the possibility of reproducing components.

Active antennas also permit high EIRPs to be obtained; these are characteristic of systems for direct broadcasting or communication with mobiles. For a given antenna (as defined by the coverage), limitation of the radiated power results from technical limitations of the power of a single amplifier per channel (around 250 W with a TWT). With an active antenna, the radiated power depends on the power of the active module and the number of radiating elements.

The disadvantages of the active antenna arise mainly in connection with the losses in the modules (ohmic losses and the limited efficiency of amplifiers). Losses between the amplifier module and the radiating element are limited due to their proximity.

Technology of active antennas. Modern techniques of high frequency circuit integration, particularly those of monolithic microwave integrated circuits (MMIC), permit realisation of active modules of small bulk and low mass [SOR-86]. These modules can incorporate beam shaping elements (controllable attenuators and phase shifters) [SOR-88]. It is also possible to realise beam distribution and formation functions using optical techniques [KUN-86], [TAM-87].

The radiating elements themselves can be realised using microwave integrated circuit technologies on dielectric substrates [GAR-79], [SME-86]. An array of printed radiating elements (a patch) can be produced by etching a conducting layer which has been deposited on a dielectric substrate over a ground plane (the microstrip techniques) [MUN-74], [DER-76]. The problems of printed antennas lie in the narrowness of the passband. Furthermore, the array radiates in a direction perpendicular to the plane of the printed antenna and this poses difficulties in integrating the active modules and the supply to the radiating elements. One solution to the narrowness of the passband is to feed the printed element via electromagnetic coupling (electromagnetically coupled patch EMCP). The passband can attain 15% of the operating frequency.

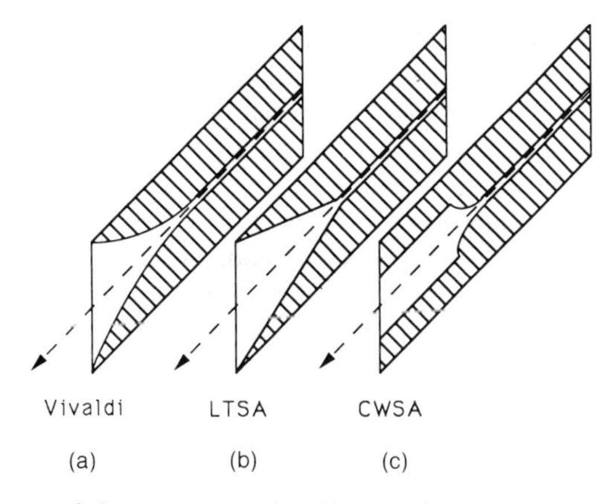

Figure 9.53 End-fire tapered slot antennas. (a) Vivaldi type, (b) LTSA type, (c) CWSA type.

Radiating elements realised from slotted lines can also be considered. A slotted line consists of two parallel conductors on the same face of a dielectric substrate deposited on a ground plane. To cause radiation, the interval between the two conductors is modified as the extremity of the substrate is approached. Different types of slot antenna on dielectric substrates (tapered slot antennas TSA) are obtained according to the profile used (Figure 9.53):

—The Vivaldi antenna, where the width of the slot varies exponentially [GIB-79], [THU-82].
—The linearly tapered slotline antenna (LTSA) where the width of the slot varies linearly [PRA-79].
—The constant width slot antenna (CWSA) which has a discontinuity in the slot width [KOL-83].

Slot antennas on dielectric substrates belong to the class of travelling wave antennas. Propagation of the wave occurs along the slot and the antenna radiates with linear polarisation in the direction of the extremity of the slot, in the plane parallel to the substrate. It is thus simple to locate various radiating elements side by side to produce an array. Furthermore, each slot can be fed from an active module integrated on to the same substrate.

9.7 CONCLUSION

Development of antenna technology has led to the realisation of antennas with shaped beams and multiple beams which can be mobile. These antennas permit frequency re-use using space and/or polarisation diversity. The search for high directivity leads to the use of radiating apertures of large dimensions. Large reflectors which can be deployed in orbit are envisaged [KEL-86]. Increased directivity is directly associated with increased pointing precision. Limited precision of attitude and orbit control and pointing uncertainty can require the use of a pointing control system.

In connection with the transponder, technological progress has permitted a reduction of mass and power consumption while permitting an increase in the complexity of the functions realised [MAH-87]. Hybrid and monolithic microwave integration technologies have been qualified for space applications and are increasingly employed.

The use of resonators permits a large reduction of the size and mass of multiplexing filters. Semiconductor power amplifiers are increasingly used and their high-frequency performan continues to improve. A particularly large effort is being devoted to an increase of efficiency in order to reduce the power consumption of both tube amplifiers, which retain some advantages, and semiconductor amplifiers.

Finally, the complexity of antennas, and other payload equipment, poses numerous problems of analysis, interference reduction and losses in addition to difficulties of realisation. The use of simulation software, to optimise the radiation patterns of antennas and the performance of microwave circuits, and computer-aided design to facilitate mechanical integration are indispensable.

REFERENCES

[ACC-82] L. Accantino, A. Angelucci (1982) A 14 GHz DCPSK direct demodulator for satellite applications, *MTT-S International Microwave Symposium, Dallas*, pp. 313–316, June.

[ACC-83] L. Accatino, A. Angelucci, F. Pattini (1983) Low level high dynamics on-board 40-PSK receiver operating at 14 GHz, *Globecom 83, San Diego*, pp. 19–23, November.

[ACC-86] L. Accatino, R.J. Cameron (1986) Microwave shaping filter for digital satellite communications, *European Microwave Conference*, pp. 822–826.

[ACC-89] L. Accatino, A. Angelucci, B. Piovano (1989) RF techniques for advanced on-board processing of digital signals, *IEEE International Conference on Communication ICC/89, Boston*, pp. 34.5–34.5.6, June.

[ALA-85a] G.B. Alaria *et al* (1985) On-board processor for a TST/SS-TDMA telecommunications system, *ESA Journal*, **9** No 1, pp. 29–37.

[ALA-85b] G.B. Alaria *et al* (1985) TST onboard processor, *GLOBECOM 85, New Orleans*, pp. 6.7.1–6.7.5, Dec.

[ALB-89a] M. Alberty *et al* (1989) 12 GHZ FET QPSK modulator with advanced technology, *ICDSC 8th International Conference on Digital Satellite Communications, Pointe à Pître*, pp. 333–337, April.

[ALB-89b] T. Alberty *et al* (1989) A digital multicarrier demodulator with synchronisation for mobile SCPC satellite communications, *ICDSC 8th International Conference on Digital Satellite Communications, Pointe à Pître*, pp. 491–498, April.

[ALB-89c] T. Alberty, M. Spinnler, G. (1989) OHM, QPSK/SQPSK modulators in advanced technology for regenerative repeaters, *1st European Conference on Satellite Communications, Munich*, pp. 399–408, Nov.

[AMB-86] A. D'Ambrosio, B. Baccetti (1986) High performance burst-mode QPSK demodulator for on board applications, *ICSDC 7th International Conference on Digital Communications, Munich*, pp. 567–572, May.

[AMB-88] A. D'Ambrosio, G. Alletto (1988) Burst-mode quaternary PSK coherent demodulators for regenerative payloads, *AIAA 12th Communication Satellite Systems Conference Arlington*, pp. 569–574, March.

[AME-88] S. Ames, P. Monte, C.F. Hoeber, F. Chethik (1988) Bandwidth and power efficient satellite TDMA demodulator and decoder, *AIAA 12th Communication Satellite Systems Conference, Arlington*, pp. 304–312, March.

[ANA-84] F. Ananasso, L. Lo Presti, M. Pent, E. Saggesse (1984) Simulation-aided design of the ITALSAT satellite regenetative transmission channel, *AIAA 10th Communication Satellite System Conference, Orlando*, Paper 84-0709, pp. 731–741, March.

[ANA-85a] F. Ananasso, E. Saggese (1985) A survey on the technology of multicarrier demodulators for FDMA/TDM user-oriented satellite systems, *GLOBECOM 85, New Orleans*, pp. 6.1.1–6.1.7, Dec.

[ANA-85b] F. Ananasso, E. Saggese (1985) On-board burst-mode modems: a simulation-aided analysis, *GLOBECOM 85, New Orleans*, pp. 6.2.1–6.2.6, Dec.

[ANA-86] M.E. Anagnostou (1986) Retransmission strategies for satellite systems, *Proceedings of the IEEE (25th Conference)*, pp. 2092–2093, Dec.

[ANA-87] F. Ananasso, S. Bellaccini (1987) Integration and testing of as SS-TDMA 120 Mbit/s regenerative repeater for advanced communication satellites, *Proceedings of the IEEE*, **134**, No 5, pp. 499–509, May.

[ANA-88] F. Ananasso, I. Bennion (1988) Integrated-optics for on-board processing in advanced communication satellite, *IEEE/GLOBECOM 88, Miami*, Nov.

[ANA-89] F. Ananasso, E. Saggese (1989) System and technology issues in on-board processing communication satellites of 1990's, *ICDSC 8th International Conference on Digital Satellite Communications, Pointe à Pitre*, pp. 609–616, April.

[AND-88] O. Andreassen, A.L. Viddal (1988) Bandwidth switchable SAW filters on quartz, *Proceedings of the IEEE Ultrasonics Symposium.*

[ANG-88] A. Angelucci, R. Buroco (1988) Optimized Synthesis of microstrip branch-line couplers, *International Microwave Symposium, New York*, pp. 411–414, May.

[APP-90] T. Appleby (1990) Current capabilities and near term improvements in space TWTAs covering L-band to Ka band, *AIAA 13th Communication Satellite Systems Conference, Los Angeles*, March.

[ARI-88] Y. Arimoto *et al* (1988) A multibeam antenna for 27/22 GHz band satellite broadcasting, *AIAA 12th International Communication Satellite Systems Conference, Arlington*, pp. 670–678, March.

[ASS-81] F. Assal, K. Betaharon, A. Zaghloul, R. Gupta (1981) Wideband microwave switch matrix for SS-TDMA systems, *ICDSC 5th International Conference on Digital Satellite Communications, Genoa*, pp. 421–427, March.

[ASS-82] F. Assal, R. Gupta, J. Apple, A. Lopatin (1982) Satellite switching center for SS/TDMA communications, *COMSAT Technical Review*, **12**, No. 1, pp. 29–68, Spring.

[ASS-88a] F. Assal, C.E. Mahle (1988) Hardware development for future commercial communications satellites, *AIAA 12th Communication Satellite System Conference, Arlington*, pp. 332–343, March.

[ASS-88b] F. Assal, A.I. Zaghloul, R. Sorbello (1988) Multiple spot-beam systems for satellite communications, AIAA 12th Communication Satellite Systems Conference, Arlington, pp. 322–331, March.

[ATT-82] S. Attwood, D. Sabourin (1982) Baseband-processed SS-TDMA communication system architecture and design concepts, *AIAA 9th Communication Satellite Systems Conference, San Diego*, pp. 188–195, March.

[AUG-84] G. August (1984) Multipactor breakdown—lessons unlearned, *AIAA 10th Communication Satellite Systems Conference, Orlando, Paper 84-0688*, March.

[BAK-89] P.M. Bakken, K. Grythe, P.M. Ronnekuev (1989) The on-board probe saw/digital signal processor, *ICDSC 8th International Conference on Digital Satellite Communications, Pointe à Pître*, pp. 617–624, April.

[BAN-88] S. Bangara, A. Patacchini (1988) EUTELSAT II: a second generation European regional satellite system, *AIAA 12th Communication Satellite Systems Conference, Arlington, Paper 88-0765*, March.

[BAR-72] C.S. Barham (1972) Review of design and performance of microwave multiplexers, *Marconi Review*, **35**, No. 184, pp. 1–23.

[BAU-85] R. Bauer, W. Steiner, W. WÜrscher (1985) Method and instrumentation for the precise measurement of satellite transponder saturation point, *International Journal for Satellite Communications*, **3**, pp. 265–270.

[BEA-88] J. Beaucher (1988) Répéteurs pour les satellites de télécommunications des années 1990, *Revue des Télécommunications*, **62**, No. 1, pp. 55–60.

[BEL-74] M. Bellanger, J.L. Daguet (1974) TDM-FDM transmultiplexers: digital polyphase and FFT, *IEEE Transactions on Communications*, **COM-22**, No. 9, Sep.

[BEL-86] S. Bellini, G. Tartara (1986) On board multicarrier digital demodulation in regenerative satellites, *AIAA 11th Communication Satellite System Conference, San Diego*, pp. 667–671, March.

[BEL-89a]L. Bella (1989) An experimental satellite system with on-board processing and low to medium rate access, *IEEE International Conference on Communication ICC/89, Boston*, pp. 34.2.1–34.2.6, June.

[BEL-89b] L. Bella, F. Fournon, E. Del Ré (1989) On-board processor oriented to low rate satellite services, *ICDSC 8th International Conference on Digital Communications, Pointe à Pître*, pp. 823–828, April.

[BEN-84] G. Benelli *et al* (1984) Performance of uplink random-access and downlink TDMA techniques for packet satellite networks, *Proceedings of the IEEE*, **72**, No. 11, pp. 1583–1593, Nov.

[BEN-86] C.A. Benet, R.D. Dewell (1986) Antenna beam pointing error budget analysis for communications satellites, *Space Communications and Broadcasting*, **4**, No. 3, pp. 205–214, September.

[BEN-85] W.R. Bennett (1985) Statistics of regenerative digital transmission. *The Bell System Technical Journal*, pp. 1501–1542.

[BER-71] A. Berman, C.E. Mahle (1970) Non linear phase shift in travelling-wave tubes as applied to multiple access communications satellite, *IEEE Transactions on Communication Technology*, **COM-18**, pp. 37–48, February.

[BER-87] W. Berner, W. Grassman (1987) A baseband switch for future space applications, *Space Communication and Broadcasting*, **5**, pp. 71–78.

[BER-89a] W. Berner, M. Piontek (1989) Switching structures for on-board processing, *1st European Conference on Satellite Communications, Munich*, pp. 217–224, Nov.

[BER-89b] G. Bertin, L. Accatino (1989) Analysis of dielectric-loaded cavities including the mounting structure, *ICEAA 89, Torino.*

[BER-90] W. Berner *et al* (1990) Switching and network control on-board advanced satellite, *AIAA 13th International Communication Satellite Systems Conference, Los Angeles*, pp. 649–654, March.

[BET-86] K. Betaharon *et al* (1986) On-board processing for communication satellite systems technologies and implementation, *ICDSC 7th International Conference on Digital Satellite Communications, Munich*, pp. 421–426, May.

[BET-87a] K. Betaharon, K. Kinuhata, P.P. Nuspl, R. Peters (1987) On-board Processing for Communication Satellites Technologies and implementations, *International Journal for Satellite Communications*, **5**, No. 2, pp. 139–146, April–June.

[BET-87b] K. Betaharon *et al* (1987) Advanced satellite switching centre with on-board diagnostics, *International Journal for Satellite Communications*, **5**, No. 2, pp. 147–154, April–June.

[BHA-86] K.B. Bhasin et al (1986) Optical techniques to feed and control GaAs MMIC modules for phased array antenna applications, *AIAA 11th Communications Satellite Systems Conference, San Diego*, pp. 506–513, March.

[BIN-84] N. Binghan, A.D. Craig, L. Flook (1984) Evolution of European telecommunication satellite pointing performance, *AIAA 10th Communication Satellite Systems Conference, Orlando*, pp. 336–343, March.

[BJO-89] G. Bjornstrom *et al* (1989) SAW technology for multicarrier demodulation in advanced payloads, *1st European Conference on Satellite Communications, Munich*, pp. 409–420, Nov.

[BOU-88] M.L. Boucherot (1988) Démodulation de porteuses multiples pour répéteurs non-transparents à bord d'un satellite, *Revue des Télécommunications*, **62**, No. 3/4, pp. 302–309.

[BOU-90] M. Bousquet, G. Maral (1990) Analysis of antenna depointing as a result of satellite attitude control and station keeping errors, *AIAA 13th International Communication Satellite Systems Conference, Los Angeles*, Paper 90-0818 (March 11–15, 1990).

[BOU-91] M. Bousquet, G. Maral (1991) Satellite antenna depointing originating from attitude control and station keeping specifications, *International Journal of Satellite Communications*, **9**, No. 2, pp. 85–92.

[BRA-88] R. Braff (1988) Ranging and processing mobile satellite, *IEEE Transactions on Aerospace Electronics*, **24**, No. 1, pp. 14–22, Jan.

[BRE-74] C. Bremenson, J. Jaubert (1974) Réseau linéariseur pour tube à ondes progressives, *Revue technique THOMSON-CSF*, **6**, No. 2, pp. 529–548.

[BUC-89] J.D. Buchs, J. Czdech, M. Wassermann (1989) Comsat Technology Advances, *Space*, No. 3, pp. 25–30, June.

[BUR-90] C. Burgio, J. Dumesnil (1990) The EUTELSAT II programme, *AIAA 13th Communication Satellite Systems Conference, Los Angeles*, pp. 355–359, March.

[CAC-84] E. Cacciamani *et al* (1984) The emergence of satellite systems for rural communications, *Proceedings of the IEEE*, **72**, No. 11, pp. 1520–1525.

[CAH-85] D. Cahana *et al* (1985) Linearised transponder technology for satellite communications, Part I (part II, see LEE-85), *Comsat Technical Review*, **15**, No. 2A, pp. 277–3083, Fall.

[CAM-88a] S.J. Campanella, B. Pontano, H. Chalmers (1988) Future switching satellites, *AIAA 12th Communication Satellite Systems Conference, Arlington*, pp. 264–273, March.

[CAM-88b] S.J. Campanella (1988) A flexible on-board demultiplexer/demodulator, *AIAA 12th Communication Satellite Systems Conference, Arlington*, pp. 299–303, March.

[CAM-90a] R.J. Cameron, W.C. Tang and C.M. Kudsia (1990) Advances in dielectric loaded filters and multiplexers for communications satellites, *AIAA 13th Communication Satellite Systems Conference, Los Angeles*, pp. 264–273, March.

[CAM-90b] J. Camanella, S. Sayegh, M. Elamin (1990) A study of on-board multicarrier digital demultiplexer for a multi-beam mobile satellite payload, *AIAA 13th Communication Satellite Sytems Conference, Los Angles*, pp. 638–648, March.

[CAR-89] P. Carle (1989) Multiport branch-waveguide couplers with arbitrary power splitting, *IEEE MTT-S Digest*, pp. 317–329.

[CCIR-Rep 558] CCIR Report 558-3 (1986) Satellite antenna patterns in the fixed satellite service, *Dubrovnik*, Vol 4—Part 1.

[CER-88] A. Cerro, D. Parise (1988) Nouveaux récepteurs embarqués hybrides, *Revue des télécommunications*, **62**, No. 1, pp. 61–66.

[CHI-81a] W. Childs *et al* (1981) A 120/Mbit/s 14 GHz regenerative receiver for spacecraft applications, *Comsat Technical Review*, **11**, No. 1.

[CHI-81b] W.H. Childs, P.A. Carlton, R. Egri, C.E. Mahle, A.E. Williams (1981) A 14 GHz regenerative receiver for spacecraft application, *5th International Conference on Digital Satellite Communications, Genoa, 1981*, pp. 453–459.

[CLA-77] J.B. Clarricoats, G.T. Poulton (1977) High efficiency microwave reflector antennas: a review, *Proceedings of the IEEE*, **65**, No. 10.

[CLA-82] R.T. Clark, R.D. McCallister (1982) Development of an LSI maximum likelihood convolution decoder for advanced forward error correction capability on the NASA 30/20 GHz program, *AIAA 9th Communication Satellite Systems Conference, San Diego*, Paper 82-0459, March.

[COA-89] F.P. Coakley, C.R. Shashidhar (1989) Concurrent fault detection and diagnosis of TST switches for on-board satellite systems, *1st European Conference on Satellite Communications, Munich*, pp. 463–464, Nov.

[COB-83] E. Coban, *et al* (1983) High speed wideband 20/20 microwave switch matrix, *International Conference on Communication ICC/83, Boston*, pp. B1.6.1–B1.6.6.

[COL-87] G. Colombo, G. Pennoni (1987) Advanced on-board processing for user oriented communication system, *International Journal for Satellite Communications*, **5**, No. 2, pp. 77–84, April–June.

[CRO-90] G. Crone *et al* (1990) Technology advances in reconfigurable contoured beam reflector antennas, *AIAA 13th Communication Satellite Systems Conference, Los Angeles*, pp. 255–263, March.

[CZE-84] J. Czech (1984) A linearized 4 GHz, wideband FET power amplifier for communications satellites, *AIAA 10th Communication Satellite Systems Conference, Orlando*, Paper 84–0766, March.

[DAY-88] J. Dayton (1988) High efficiency longlife traveling wave tubes for future communications satellites, *AIAA 12th Communication Satellite Systems Conference, Arlington*, pp. 452–456, March 1988.

[DEC-84] R. Decristoforo, A.L. McFarthing (1984) A 120 Mbit/s QPSK modem designed for the Intelsat TDMA network, *International Journal of Satellite Communications*, **3**, 1984

[DEL-86] E. Del Re, R. Fantacci (1986) Design of a demultiplexer for a regenerative satellite, *EURASIP Signal Processing* III, pp. 1095–1098.

[DEL-89] E. Del Re, R. Fantacci (1989) A simplified I-Q digital multicarrier demodulator, *IEEE International Conference on Communications ICC/89, Boston*, pp. 34.3.1–34.3.5, June.

[DER-76] A.G. Derneryd (1976) Microstrip array antenna, *6th European Microwave Conference.*

[DER-86] M.A. McDermott, R.N. Tamashiro (1986) A 20 GHz 70 watt 48 percent efficient space

communications TWT, *AIAA 11th Communications Satellites Systems Conference, San Diego*, pp. 600–605, March.
[DON-69] E. Donnelly, *et al* (1969) The design of mechanically despun antenna for Intelsat III communications satellite, *IEEE Trans. Antennas & Propagation*, **AP 17**, pp. 407–414.
[DUP-89] J. Dupraz (1989) *Signaux bruits et modulation*, Eyrolles.
[DUR-88] G. Duret, T. Guillemin (1988) Antenne multidaisceau à configuration modifiable pour les satellites Eutelsat II, *Revue des tékécommunications*, **62**, No. 1, pp. 75–80.
[EDR-90] M. Edridge *et al* (1990) Features of the Intelsat VII repeater, *AIAA 13th Communication Satellite Systems Conference, Los Angeles*, pp. 111–119, March.
[EGR-86] R.G. Egri, Karimullah, F.T. Assal (1986) A 120-M bit/s TDMA QPSK modern for on-board applications, *COSMAT Technical Review*, **17**, No. 1, pp. 23–54.
[ELA-86] EL-Amin, T. Chung, B.G. Evans (1986) Access protocol performance for onboard processing business satellite systems, *International Journal of Satellite Communications*, **4**, No. 4, pp. 203–210.
[ENG-90] K. Engel, I. Hastings and G. McDonald (1990) Recent advances in microwave switching technologies for satellite applications, *AIAA 13th Communication Satellite Systems Conference, Los Angeles*, March.
[EST-89] P. Estabrook *et al* (1989) A 20/30 GHz personal access satellite system design, *IEEE International Conference on Communications ICC 89, Boston*, pp. 7.4.1–7.4.7, June.
[EVA-84] B.G. Evans (1984) Satellite onboard processing, *Electronics & Power*, pp. 533–536.
[EVA-86a] B.G. Evans (1986) An access protocol for an on-board processing business satellite system, *ICDSC 7th International Conference on Digital Satellite Communications, Munich*, pp. 149–154, May.
[EVA-86b] B.G. Evans (1986) An on-board processing satellite payload for European land-mobile communication, *ICDSC 7th International Conference on Digital Satellite Communications, Munich*, pp. 545–550, May.
[EVA-86c] B.G. Evans *et al* (1986) Baseband switches and transmultiplexers for use in an on-board processing mobile/business, *ICDSC 7th International Conference on Digital Satellite Communications, Munich*, pp. 587–592, May.
[EVA-87] B.J. Evans (1987) *Satellite Communication Systems*, Peter Perigrinus.
[EVA-89] J,V. Evans, C.E. Mahle (1989) Research into advanced satellite technology at Comsat laboratories, *ICDSC 8th International Conference on Digital Satellite Communications, Pointee à Pître*, pp. 593–600, April.
[EVA-90] J. Evans, C. Mahle (1990) Research into advanced satellite technology at COMSAT laboratories, *AIAA 13th International Communication Satellite Systems Conference, Los Angeles*, pp. 798–806, March.
[FLE-88] A. Flechais, C. Rouyer, J.P. Clariou, G. Brazzini (1988) New trends in making antenna reflectors, *AIAA 12th Communication Satellite Systems Conference, Arlington*, pp. 88–0875, March.
[FRA-82] M. Franchini, B. Fabis, K. Schneider, P. Gault (1982) The Franco-German broadcasting satellite program, *IAF 33th Congress, Paris*, pp. IAF 82–84, Sep.
[FRE-84] M.R. Freeling, A.W. Weinrich (1984) RCA advanced Satcom: the first all-solid-state communications satellite, *AIAA 10th Communication Satellite Systems Conference, Orlando*, Paper 84–0715, March.
[FUE-73] J.C. Fuenzalida, O. Shinbo, O. W.L. Look (1973) Time domain analysis of intermodulation effects caused by nonlinear amplifiers, *COMSAT Technical Review*, **3**(1), 89–143.
[GAR-79a] F.M. Gardner (1979) *Phase-lock Loop Techniques*, Wiley.
[GAR-79b] R. Garg, K. Gupta (1979) *Microstrip Line and Slot Line*, Artech House.
[GAR-90] F. Garione (1990) ACTS hardware—a pictorial, *AIAA 13th International Communication Satellite Systems Conference, Los Angles*, pp. 490–496, March.
[GED-89] R.T. Gedney, R.J. Schertler (1989) Advanced communications technology Satellite, *IEEE International Conference on Communication, ICC 89, Boston*, pp. 52.1.1–52.1.6, June.
[GEL-88] B. Geller, P. Goettle (1988) Quasi-monolithic 4-GHz power amplifiers with 86% power-added efficiency, *1988 IEEE MTT-S International Microwave Symposium, New York*, May.

[GHO-90] S. Ghosh, C. Lee-Yow, H. Viskum (1990) An advanced multi-function Ku-band antenna for satellite applications, *IEEE AP-S International Symposium, Texas,* May 7–11.

[GIA-89] F. Giannini, G. Leuzzi, M. Ruggieri (1989) On-board non-linear circuitry for Ku-band active arrays, *International Journal on Satellite Communications,* **7**, No. 2, pp. 103–110, April.

[GIB-79] P.J. Gibson (1974) The Vivaldi aerial, *9th European Microwave Conference,* pp. 101–105, Sept.

[GIL-86] A.S. Gilmour Jr. (1986) *Microwave Tubes,* Artech House.

[GLA-89] F.-J. Glandorf, P. Kuck (1989) 30 GHz low-noise HEMT amplifier for satellite applications, *1st European Conference on Satellite Communications, Munich,* pp. 381–390, Nov.

[GOC-88] H. Gockler (1988) A modular multistage approach to digital FDM demultiplexing for mobile SCPC satellite communications, *International Journal on Satellite Communications,* **6**, No. 3, pp. 283–288.

[GOL-84] A.M. Goldman (1984) An electro-optical communications satellite transponder, *AIAA 10th International Communication Satellite Systems Conference, Orlando,* pp. 262–263, March.

[GUP-88] K.C. Gupta, A. Benalla (1988) *Microstrip Antenna Design,* Artech House.

[GRA-86] J. Graebner, W. Cashman (1986) Advanced communication technology satellite system description, *GLOBECOM 86, Houseton,* pp. 16.1.1–16.1.9. Dec.

[GRA-90] J. Graebner, W. Cashman (1990) ACTS multibeam communications package: technology for the 1990s, *AIAA 13th International Communication Satellite Systems Conference, Los Angeles,* pp. 497–507, March.

[GUP-86] R. Gupta, J. Potokuchi, F. Assal (1986) Ground-based failure diagnostic/ monitoring schemes for SS/TDMA satellite, *7th International Conference on Digital Satellite Communicaions, Munich,* pp. 393–400, May.

[HAN-86] K. Handa *et al* (1984) Trends and development of low noise amplifiers using new FET device, *AIAA 11th International Communication Satellite Systems Conference, San Diego,* pp. 308–312, March.

[HAR-80] E.H. Hara, R.I. McDonald (1980) A broadband opto-electronic microwave switch, *IEEE Transaction on Microwave Theory and Techniques,* MTT-28, pp. 662–663.

[HAS-89] Y. Hase *et al* (1989) ETS-V/EMSS experiments on aeronautical communications, *IEEE International Conference on Communications ICC/89, Boston,* pp. 7.1.1–7.1.6, June.

[HAT-69] G.W. Hatch (1969) Communications subsystem design trends for the DSC program, *IEEE Transactions on Aerospace and Electronic Systems,* **AES-5** (5), Sept.

[HAU-88] W. Hauth, R. Keller, V. Rosenberg (1988) The corrugated-waveguide band-pass filter—a new type of waveguide filter, *18th European Microwave Conference, Stockholm,* pp. 945–949, September.

[HEI-86] W. Heine (1986) System concept for a digital repeater intended for a future generation of German communication satellite, *ICDSC 7th International Satellite Systems Conference, Munich,* pp. 179–186, May.

[HEI-89] J. Heichler (1989) A fault tolerant bus for space applications, *1st European Conference on Satellite Communications, Munich,* pp. 445–462, Nov.

[HO-82] P. Ho, J. Wisniewski, J. Pelose, H. Perasso (1982) Dynamic switch matrix for the TDMA satellite switching system, *AIAA 9th International Communications Satellite Systems Conference, San Diego,* Paper 82-0458, March.

[HOD-88] D.M. Hodson (1988) A satellite FSK demodulator using an analog saw chirp Fourier transformer, *AIAA 12th Communication Satellite Systems Conference, Arligton,* pp. 561–568, March.

[HOE-86] C.F. Hoeber, D.L. Pollard, R.R. Nicholas (1986) Passive intermodulation product generation in high power communications satellites, *AIAA 11th International Communication Satellite Systems Conference, San Diego,* pp. 361–375, March.

[HOE-86] C.F. Hoeber *et al* (1986) Bandwidth and power efficient satellite TDMA demodulator, *ICDSC 7th Conference on Digital Satellite Communication, Munich,* pp. 573–578, May.

[HUB-88] K. Hubner (1988) An advanced electronic power conditioner for 12 GHz 100–150 W traveling wave tube amplifiers, *AIAA 12th Communication Satellite Systems Conference, Arlington,* pp. 457–462, March.

[INU-81] T. Inukai, S.J. Campanella (1981) Onboard clock correction for SS/TDMA and baseband processing satellites, *COMSAT Technical Review,* **11**, No. 1, pp. 77–102 Spring.

[INU-83] T. Inukai, S.J. Campanella, T. Dobyns (1983) On board baseband processing: rate conversion, *ICDSC, 6th International Conference on Digital Satellite Communications, Phoenix*, pp. XI–24 XI–31.

[INU-88] T. Inukai, D. Jupin, R. Lindstrom, D. Meadows (1988) ACTS TDMA network control architecture, *AIAA 12th Communication Satellite Systems Conference, Arlington*, pp. 225–239, March.

[ISH-86] K. Ishida, I. Oka, I. Endo (1986) Cochannel interference canceling effects of MSE processing and Viterbi decoding in onboard baseband processor, *IEEE Transactions on Communications*, **COM 14** No. 10, pp. 1049–1053.

[IZU-84] T. Izumisawa, S. Kato, T. Kohri (1984) Regenerative SCPC satellite communications systems, *AIAA 10th Communication Satellite Systems Conference, Orlando*, pp. 269–275, March.

[JAR-84] K. Jarett (1984) Operational aspects of INTELSAT VI satellite-swiched TDMA communication system, *AIAA 10th International Communication Satellite Systems Conference, Orlando*, pp. 107–111, March.

[JES-86] K. Jesche (1986) Calculation of the reliability of highly complex redundancy networks, *Space Communication and Broadcasting*, **4**, No. 2, pp. 141–148, June.

[KAD-89] N. Kadowaki *et al.* (1989) ETS-V/EMSS experiments on message communications with hand-held terminal, *EEE International Conference on Communications ICC/89, Boston*, pp. 7.3.1–7.3.5, June.

[KAS-86] Y. Kasekami *et al.* (1986) A design approach for on-board modems, *IEEE International Conference on Communications ICC/86, Toronto*, pp. 56.8.1–56.8.7, June.

[KAT-83] H. Kato, S. Okasaka, K. Kondoh (1983) Multibeam satellite communication system for Japanese domestic communications, *Satellite Communications Conference SCC/83, Ottawa*, pp. 22.2.1–22.2.4, June.

[KAZ-85] Y. Kazekami, H. Sawada (1985) Hardware implementation of on-board modem, *Globecom 85 New Orleans*, pp. 6.5.1–6.7.5, Dec.

[KEI-86] J. Keigler, L. Muhleflder (1986) Optimum antenna beam pointing for communications satellites, *AIAA 11th Communication Satellite Systems Conference, San Diego*, pp. 70–78, March.

[KEL-86] H. Kellermeier, Vorbrugg, K. Pontoppidan (1986) The MBB unfurlable mesh antenna (UMA) design and development, *AIAA 11th Communication Satellite Systems Conference, San Diego*, pp. 417–426, March.

[KEO-88] B. Keough (1988) Techniques and capabilities of the new generation antenna control unit, *AIAA 12th Communication Satellite Systems Conference, Arlington*, pp. 88–0792, March.

[KIE-81] R.A. Kiehl, D.M. Drury (1981) Performance of optically coupled microwave switching devices, *IEEE Transactions on Microwave Theory and Techniques, MTT-29*, pp. 1004–1010.

[KOB-88] Y. Kobayashi *et al* (1988) Application of front fed offset cassegrain antennas for communication satellites, *AIAA 12th International Communication Satellite Systems Conference, Arlington*, pp. 678–684, March.

[KOG-77] K. Koga, T. Muratani, A. Ogawa (1977) Onboard regenerative repeaters applied to digital satellite communications, *Proceedings of the IEEE*, **65**, No. 3, pp. 401–410.

[KOL-83] E.L. Kolberg, J. Johansson, T. Thungrent (1983) New results on tapered slot end fire antennas on dielectric substrates, *8th COIMW*.

[KOV-91] R. Kovac *et al.* (1991) SAW-Based IF processors for mobile communications satellites, *IAF Congress, Montreal*, October.

[KRE-80] R.W. Kreutel *et al.* (1980) Optical transmission technology in satellite applications, *COMSAT Tech. Review*, **10**, pp. 321.

[KRI-85a] W. Kriedte, A. Vernucci (1989) Advanced regional mobile satellite system for the nineties, *GLOBECOM 85, New orleans*, pp 38.1.1–38.1.6, Dec.

[KRI-85b] M. Kristiansen, H. Berthelot (1985) An evaluation of the time-domain communication system simulator TOPSIM III, *ESA Journal*, **9**, No. 3, pp. 361–373.

[KUB-89] S. Kubota, S. Kato, T. Shitani (1989) Compact high-speed and high-coding-gain general purpose FEC encoder/decoder-NUFEC CODEC, *IEEE International Conference on Communication ICC/89, Boston*, pp. 25.9.1–25.3.6, June.

[KUD-80] C.M. Kudsia, K.R. Ainsworth, M.V. O'Donovan (1990) Microwave filters and multiplexing networks for communications satellites in the 1980's, *AIAA 8th Communications Satellite Systems Conference, Orlando*, Paper 80–0522, March.

[KUD-85] M. Kudoh, I. Eguchi (1985) GaAs baseband switching matrix for on-board signal processing, *GLOBECOM 85, New Orleans*, pp 6.4.1–6.4.4, Dec.

[KUD-88] C. Kudsia (1988) Optimisation of satellite transponder for maximum spectral efficiency, *AIAA 12th Communication Satellite Systems Conference, Arlington*, March.

[KUD-92] C. Kudsia, R. Cameron, W.C. Tang (1992) Innovations in microwave filters and multiplexing networks for communications satellite systems, *IEEE Transactions on Microwave Theory and Techniques*, **40**, No. 6, pp. 1133–1149.

[KUM-89] T. Kumagai *et al* (1989) On-board processor for Intelsat Business Services, *ICDSC 8th International Conference on Digital Satellite Communications, Pointe à Pitre*, pp. 829–836, April.

[KUN-86] R. Kunath, K. Bhasin (1986) Optically controlled phased-array antenna concepts using GaAs MMIC, *IEEE Antenna and Propagation International Symposium, Philadelphia*, **1**, pp. 353–357.

[KUR-81] H. Kurihara *et al.* (1981) Carrier recovery circuit with low cycle skipping rate for QPSK/TDMA systems, *5th ICDSC International Conference on Digital Satellite Communications, Genoa*, pp. 319–324, March.

[LAP-86] N. Laprade, M. Caporossi, L. Dolan (1986) Ku-Band SSPA for communications satellites, *AIAA 11th Communication Satellite Systems Conference, San Diego*, pp. 321–326, March.

[LEE-78a] Y.S. Lee (1978) 14 GHz MIC 16 ns delay filter for differentially coherent QPSK regenerative repeater, *IEEE International Microwave Symposium Digest*, pp. 37–40.

[LEE-78b] Y.S. Lee, W.H. Childs (1978) Temperature compensated $BaTi_4O_9$ microstrip delay line, *IEEE International Microwave Symposium Digest*, pp. 419–421.

[LEE-85] Y.S. Lee, I. Brelian, A. Atia (1985) Linearised transponder technology for satellite communications, Part II (part I, see CAH-85), *Comsat Technical Review*, **15**, No. 2A, pp. 277–3083, Fall.

[LIE-89] M. Lieke, N. Nathrath (1989) Antennas for DBS and FSS satellites: realised and planned new concepts, *1st European conference on Satellite Communications, Munich*, pp. 537–550, Nov.

[LOC-88] A. Lockwood (1988) The adaptive multiplexer—an exciting new multiplexing technique, *AIAA 12th International Communication Satellite Systems Conference*, pp. 285–291.

[LOP-82] Lopriore, M.A. Saitto, G.K. Smith (1982) A unifying concept for future fixed satellite service payloads for Europe, *ESA Journal*, **6**, No. 4, pp. 371–396.

[LOP-90] M. Lopriore, T. Jones (1990) Ku-band payload trade-offs for ISDN services in Eurpoe, *AIAA 13th International Communication Satellite Systems Conference, Los Angeles*, pp. 264–270, March.

[LOS-88] G. Losquadro (1988) Multiple access system for a data relay satellite using a phased array antenna, *AIAA 12th International Communication Satellite Systems Conference*, pp. 88–0768, March.

[MAD-90] P. Madon D. Sachdev (1990) Intelsat VII program and the future, *AIAA 13th Communication Satellite Systems Conference, Los Angeles*, pp. 75–77, March.

[MAH-87] C. Mahle, G. Hyde, T. Inukai (1987) Satellite scenarios and technology for the 1990's, *IEEE Journal on Selected Areas in Communications*, **SAC-5**, No. 4, pp. 556–570.

[MAI-82] R.J. Mailloux (1982) Phased array theory and technology, *Proceedings of the IEEE*, **70**, No. 3, March.

[MAI-89] H. Maier *et al.* (1989) A modular on-board multiprocessor for a switching system, *1st European Conference on Satellite Communications, Munich*, pp. 477–494, Nov.

[MAN-69] J.M. Manley (1969) The generation and accumulation of timing noise in PCM systems-An experimental & theoretical study, *The Bell System Technical Journal*, **48**, pp.541–613, March.

[MAR-84] G. Maral, M. Bousquet (1984) Performance of regenerative/conventional satellite systems, *International Journal of Satellite Communications*, **2**, No. 3, pp. 199–20.

[MAR-87] F. Marconicchio, A. DI Cecca, A. Vernucci, M. Donati (1987) The ITALSAT satellite on board baseband processor, concept and technologies, *GLOBECOM 87, Tokyo*, pp. 24.5.1–24.5.7, Dec.

[MAR-89] F. Marconicchio (1987) The on-based processor for ITALSAT SS-TDMA multibeam package, *IEEE International Conference on Communication ICC/89, Boston*, pp. 34.4.1–34.4.6, June.

[MAT-76] W.W. Matthews, W.G. Scott, C.C. Har (1976) Advances in multibeam satellite antenna technology, *Record of the IEEE Eascon*, pp. 132A–132D.

[MIT-83] R. Mittra, W.A. Imbriale, E.J. Maanders (1983) *Satellite Communication Antenna Technology*, North-Holland.

[MIZ-86] T. Mizuno *et al* (1986) International problems in digital satellite communications, *ICDSC 7th International Conference on Digital Satellite Communication, Munich*, pp. 659–666, May.

[MOA-82] R. Moat, D. Sabourin, G. Stilwell, R. McCallister, M. Borota (1982) Baseband processor development for the advanced communications satellite program, *National Telesystem Conference NTC/82, Galveston*, pp. A2.4.1–A2.4.4.

[MOA-86] R. Moat (1986) ACTS baseband processor, *IEEE GLOBECOM 86*, Houston, pp. 16.4.51–16.4.6, Dec.

[MOR-88] G. Moreli, Matitti, T. (1988) The ITALSAT satellite program, *AIAA 12th Communication Satellite Systems Conference, Arlington*, pp. 112–122, March.

[MOR-88] M. Morikura, K. Enomoto, S. Kato (1988) High speed onboard digital signal processing and LSI implementation, *IEEE International Conference on Communications ICC 88, Philadephia*, pp. 16.1.1–16.1.6, June.

[MOR-90] N. Morita *et al* (1990) C-band solid state power amplifier, *AIAA 13th Communication Satellite Systems Conference, Los Angeles*, March.

[MOY-86] L.L. Moy *et al* (1986) On the effectiveness of on board processing, *AIAA 11th Communication Satellite Systems Conference, San Diego*, pp. 672–675, March.

[MUE-88] F. Muennemann *et al* (1988) Dual-mode downlink TWT for the ACTS satellite, *Microwave Systems News*, **18**, No. 5, pp. 50–61, May.

[MUN-74] R.E. Munson (1974) Conformal microstrip antenna, *IEEE Transactions on Antennas and Propagation*, **AP-22**.

[NAD-88] M. Naderi, P. Kelly (1988) NASA'S advanced communications technology satellite (ACTS): an overview of the satellite, the network, and the underlying techniques, *AIAA 12th Communication Satellite Systems Conference, Arlington*, pp. 204–224, March.

[NAK-82] M. Nakamura, Y. Watanabe, K. Kohiyama, K. Kondo, N. Ishida, K. Miyauchi (1982) Future advanced satellite communications systems with integrated transponders, *AIAA 9th Communication Satellite Systems Conference San Diego*, pp. 227–233, March.

[NEY-90] P. Neyret, L. Dest, K. Betaharon, E. Hunter, L. Templetonn (1990) The INTELSAT VII spacecraft, *AIAA 13th Communication Satellite Systems Conference, Los Angeles*, pp. 95–110, March.

[NUS-86] P. Nuspl (1986) On board processing for communications satellite systems:systems and benefits, *ICDSC 7th International Conference on Digital Satellite Communication, Munich*, pp. 137–148, May.

[OHM-81] G. Ohm (1981) Experimental 14-11 Ghz regenerative repeater for communication satellite, *ICDSC 5th International Conference on Digital Satellite Communications, Genoa*, pp. 445–451, March.

[OHM-82] G. Ohm, M. Alberty (1982) 11 GHz QPSK modulator for regenerative satellite repeater, *IEEE Transactions on Microwave Theory and Techniques*, **30**, No. 11, pp. 1921–1926.

[OHS-86] T. Ohsawa, J. Namiki (1086) Digital group demodulation system of multiple PSK carriers, *AIAA 11th Communications Satellite Systems Conference, San Diego*, pp. 313–320, March.

[OHT-86] K. Ohtani *et al.* (1986) An onboard digital demodulator for regenerative SCPC satellite communication systems, *IEEE International Conference on Communications ICC/86, Toronto*, pp. 56.6.1–56.6.6, June.

[OLM-90] D. Olmstead, R. Schertler (1990) Advanced communications technology satellite (ACTS), *AIAA 13th International Communication Satellite Systems Conference, Los Angeles*, pp. 522–528, March.

[ONO-88] T. Ono, H. Hiros, T. Imatani (1988) Ku-band solid state power amplifier, *AIAA 12th International Communication Satellite Systems Conference, Arlington*, Paper 88-0834, March.

[OPP-87] A. Oppetit, D. Thebault, J.L. Rousson (1987) Real-time configurable SS/TDMA equipment for second generation French telecommunications, *International Journal for Satellite Communications*, **5**, No. 2, pp. 133–138, April–June.

[PAP-89] A. Papiernik (1989) Les activitiés du Groupement de Recherche Microantennes du CNRS, *Onde Electrique*, **69**, No. 2, pp. 15–21, March–April.

[PAT-85] F. Pattini, P. Porzio Giusto (1985) A synchronization technique for the on-board master clock of a regenerative TDMA satellite, *IEEE International Conference on Communications ICC 85 Chicago*, pp. 32.3.1–32.3.6, June.

[PEA-90] R. Peach, A. Malarky (1990) Enhanced spectral efficiency using bandwith switchable SAW filtering for mobile satellite communications systems, *2nd International Mobile Satellite Conference, Ottawa*, June.

[PEL-87] J.N. Pelton, W.W. Wu (1987) The challenge of 21st century satellite communications, INTELSAT enters the second millennium, *IEEE Journal on Selected Arcasin Communications*, **SAC-5**, No. 4, pp. 571–59.

[PEL-88] A. Pelletier, C. Stael (1988) Ku-Band modern TWTA's for satellite transponders, *AIAA 12th International Communication Satellite Systems Conference, Arlington*, pp. 438–444, March.

[PEN-84] G. Pennoni (1984) A TST/SS-TDMA telecommunications system: from cable to switchboard in the sky, *ESA Journal*, **8**, pp. 151–162.

[PEN-85] G. Pennoni (1985) Bit and burst synchronisation in regenerative SS-TDMA systems, *IEEE International Conference on Communications ICC/85, Chicago*, pp. 32.8.1–32.8.4, June.

[PEN-86] G. Pennoni, G. Colombo (1986) Advanced on board processing for user oriented communication systems, *ICDSC 7th International Conference on Digital Satellite Communication, Munich*, pp. 171–177, May.

[PER-85] G. Perrotta (1985) Accuracy limitations of RF sensor fine pointing systems in multibeam antennas, *Space Communication and Broadcasting*, **3**, pp. 131–150.

[PER-86] G. Perrotta (1986) FDMA/TDM satellite communication systems for domestic/business services, *ICDSC 7th International Conference on Digital Satellite Communication, Munich*, pp. 152–162, May.

[PET-87] R.A. Peters (1987) SS-TDMA switch matrices using optical technology, *International Journal of Satellite Communications*, **5**, No. 2, pp. 155–162, April–June.

[POU-89] J.L. Pourailly, C. Guerin (1989) Avenir des antennes réseaux actives, *Onde Electrique*, **69**, No. 2, pp. 7–17, March–April.

[PRA-79] S.N. Prasad, S. Mahapatra (1979) A novel MIC slot-line antenna, *9th European Microwave Conference*, Sept.

[PRI-86] W.L. Pritchard, J.A. Sciulli (1986) *Satellite Communications—Systems Engineering*, Prentice Hall.

[PRO-83] J. Proakis (1983) *Digital Communications*, McGraw-Hill.

[RAA-86] A.R. Raab, C.E. Profera (1986) Wideband beam-forming networks for satellite antenna systems: a mass, schedule and cost-effective route to better RF performance, *AIAA 11th Communication Satellite Systems Conference, San Diego*, pp. 426–430, March.

[REI-82] S. Reisenfeld (1982) Onboard processing for 30/20 GHz communications satellite, *IEEE International Conference on Communications ICC'82, Philadelphia*, pp. 5E.3.1–5E.3.4.

[RHO-89] S.A. Rhodes, S.I. Sayegh (1989) TDMA/QPSK channels,, *ICDSC 8th International Conference on Digital Satellite Communications, Pointe à Pître*, pp. 845–852, April.

[RIS-85] F. Rispoli, R. Jorgensen, G. Doro, J.C. Magnan (1985) Antenna subsystem for new European satellite telecommunications Eutelsat II generation, *GLOBECOM '85, New Orieans*, pp. 14.1.1–14.1.6, Dec.

[ROE-83] A.G. Roederer (1983) Antennes embarquées à réflecteurs multisources, *JINA 83, Nice*, Nov.

[ROS-88] V. Rosenberg, P. Rosowsky, W. Rümmer, D. Wolk (1988) Tunable mainfold multiplexers a new possibility for satellite redundancy philosophy, *18th European Microwave Conference, Stockholm*, pp. 870–875, Sep.

[ROS-89] V. Rosenberg, D. Wolk (1989) New possibilities of cavity-filter design by a novel TE-TM mode iris coupling, *1st European Conference on Satellite Communications, Munich*, Dec.

[ROS-90] U. Rosenberg *et al* (1990) High performance output multiplexers for Ku-band satellites, *AIAA 13th Communication Satellite Systems Conference, Los Angeles*, pp. 747–754, March.

[ROZ-76] X. Rozec, F. Assal (1976) Microwave switch matrix for communications satellites, *IEEE International Conference on Communications ICC/76, Philadelphia.*

[ROZ-88] X. Rozec (1988) Charge utile du satellite TDF1 de diffusion directe de télévision, *Revue des télécommunications*, **62**, No. 1.

[SAB-81] D. Sabourin, R. Jirberg (1981) Baseband processor development for SS-TDMA communication systems, *International Telemetering Conference ICT 81, San Diego*, Oct.

[SAI-89] G. Saldi, E. Matiello (1989) Prototype of a digital on-board processor for satellite communications of mobile terminals, *1st European Conference on Satellite Communications, Munich*, pp. 465–476, Nov.

[SAG-86] E. Saggese, V. Speziale (1986) In orbit testing of digital regenerative satellite: the Italsat planned test procedures., *ICDSC 7th International Conference on Digital Satellite Communications, Munich*, pp. 411–420, May.

[SAG-87] E. Saggese, V. Speziale (1987) In-orbit testing of digital regenerative satellite: the ITALSAT planned test procedures, *International Journal for Satellite Communications*, **5**, No. 2, pp. 183–190, April–June.

[SAN-87] P.V. De Santis (1987) Non-regenerative satellite-switched FDMA (SS/FDMA) payload technologies, *International Journal for Satellite Communications*, **5**, No. 2, pp. 171–182, April–June.

[SAN-89] P.V. De Santiàs (1989) Reconfigurable on-board connectivity for Intelsat thin route service, *ICDSC 8th International Conference on Digital Satellite Communications, Pointe Pître*, pp. 625–632, April.

[SCH-88] W.G. Schmidt (1988) The ACTS LBR system: a technology test bed for future VSAT/TDMA networking applications, *AIAA 12th Communication Satellite Systems Conference, Arlington*, pp. 259–265, March.

[SCH-90] G. Schennum *et al* (1990) The Intelsat VII antenna farm, *AIAA 13th International Communication Satellite Systems Conference, Los Angeles*, pp. 120–125, March.

[SCO-76] W.G. Scott *et al.* (1976) Development of multiple beam lens antennas, *AIAA, 6th CSSC*, Montreal, Paper 76–250.

[SHA-83] J.T. Shaneyfelt, S. Attwood (1983) Satellite baseband processor test performance summary, *EASCON 83, Washington*, pp. 75–83.

[SHA-89a] H.M. Sharifi, M. Arozullah (1986) Multiple satellite networks with on-board processing, *IEEE International Conference on Communication ICC/86, Toronto*, pp. 30.1.1–30.1.6, June.

[SHA-89b] M.D. Shaw *et al* (1989) SAW chirp filter techology for satellite on-board processing applications, *International Journal of Satellite Communications*, **7**, pp. 263–282.

[SHI-71] O. Shimbo (1971) Effects of intermodulation AM-PM conversion and additive noise in multicarrier TWT Systems, *Proceedings of the IEEE*, No, pp. 230–238.

[SHI-89] O. Shimbo, J. Lee, G. Lo (1989) Transmission design of two 45 Mbps digital carriers in a nonlinear satellite channel, *IEEE International Conference on Communication ICC/89, Boston*, pp. 25.5.1–25.5.6, June.

[SKO-70] Skolnik (1970) *Radar Handbook*, McGraw-Hill.

[SLE-88] C.J. Seletten (1988) *Reflector and Lens Antennas*, Artech House.

[SME-86] J. Smetana, T.J. Kascak, R.E. Alexovich (1986) MMIC antenna technology developement in the 30/20 GHz band, *AIAA 11th Communications Satellite Systems Conference, San Diego*, pp. 430–444, March.

[SMO-84] A.E. Smoll, T.E. Roberts, E.W. Matthews, E.A. Lee and C.C. Han (1984) A new multiple beam satellite antenna for 30/20 GHz communications coverage of conus-Experimental Evaluation, *AIAA 10th International Communication Satellite Systems Conference, Orlando*, pp. 33–42, March.

[SOR-86] R.M. Sorbello, A.I. Zaghloul, S. Lee, S. Siddiqi B.D. Geller (1986) 20-GHz phased-array-fed antennas utilizing distributed MMIC modules, *Comsat Technical Review*, **16**, No. 2, pp. 339–372, Fall.

[SOR-88] R. Sorbello (1988) Advanced satellite antenna developments, *AIAA 12th Internatioonal Communication Satellite Systems Conference, Arlington*, pp. 652–659, March.

[SPI-83] E.W. Spisz (1983) NASA development of a satellite switched SS/TDMA IF switch matrix, *CECON'83, Cleveland*, pp. 19–27.

[STI-82] J.H. Stiulwell (1982) Serial MSK modem for the advanced Communications satellite, *National Telesystem Conference NTC'82, Galveston*, pp. A2.5.1–A2.5.5.

[TAM-87] M. Tamburrini *et al* (1987) Optical feed for a phased-array microwave antenna, *Electronics Letters*, **23**, No. 13, pp. 680–681, June.

[TAN-90] W.C. Tang, C.M. Kudsia (1990) Multipactor breakdown and passive intermodulation in microwave equipment for satellite applications, *IEEE Military Communications Conference MILCOM 90, Monterey*, Sept. 30 to Oct. 3.

[THO-84] R.E. Thomas, D.R. Caroll (1984) The baseband processor in future satellite communication systems, *EASCON 84, Washington*, pp. 151–155.

[THO-83] P.T. Thompson, J.C. Johnston (1983) INTELSAT VI: a new satellite generation for 1986–2000, *International Journal of Satellite Communications*, **1**, no. 1, pp. 3–14, January.

[THO-90] P.T. Thompson, R. Silk (1990) INTELSAT VII: another step in the development of global communications, Journal of British Interplanetary Society, **43**, pp. 353–364.

[THO-71] L. Thourel (1971) *Less antennes*, Dunod.

[THU-82] T. Thungren, E.L. Kollberg (1982) Vivaldi antenna for single beam integrated receivers, *12th European Microwave Conference*, Sep.

[TOM-89] H. Tomita, J. Namiki (1989) Preambleless demodulator for satellite communications, *IEEE International Conference on Communication ICC/89, Boston*, pp. 16.4.1–16.4.6, June.

[TON-84] R. Tong, C.M. Kudsia (1990) Enhanced performance and increased EIRP in Communications satellites using cointiguous multiplexers, *AIAA 13th Communication Satellite Systems Conference, Los Angeles*, March.

[TRA-88] J.J. Travormina (1988) Multiple-beam antennas for military satellite communications, *Microwave Systems News*, pp. 20–30, October.

[TSA-78] C.S. Tsai, B. Kim, F.R. EL-Akkari (1978) Optical channel waveguide switch and coupler using total internal reflection, *IEEE Journal of Quantum Electronics*, **QE-14**, pp. 513–517.

[VID-88] Vidal Saint-andre B (1988) Antennes: domaine technique clé des satellities, *Revue des télécommunications*, **62**, No. 1, pp. 67–74.

[VIT-83] A.J. Viterbi, A.M. Viterbi (1983) Non-linear estimator of PSK modulated carrier phase with application to burst digital transmission, *IEEE Transactions on Information Theory*, **IT-29**, No. 4, July.

[WEL-89] E. Weller *et al* (1989) On-board processing techniques for a regional system an experiment using a double-hop satellite link, *ICDSC 8th International Conference on Digital Satellite communications, Pointe à Pître*, pp. 633–640, April.

[WHI-90] N Whittaker *et al* (1990) The design of a linear L-band high power amplifier for mobile communication satellites, *International Mobile Satellite Conference, Ottawa*, pp. 384–389, June.

[WU-88] J. Wu, A. Roederer (1988) A maximin optimisation method for contoured-beam satellite antennas, *ESA Journal*, **12**, No. 2, pp. 289–298.

[YAS-89] Y. Yasuda *et al* (1989) Field experiment on digital maritime and aeronautical satellite communication systems using ETS-V, *IEEE International Conference on Communications ICC/89, Boston*, pp. 7.2.1–7.2.7, June.

[YIM-89] W.H. Yim *et al* (1989) On-board multicarrier demodulator for mobile applications using DSP implementation, *1st European Conference on Satellite Communications, Munich*, pp. 437–444, Nov.

10 ORGANISATION OF TELECOMMUNICATION PLATFORMS

The organisation of a telecommunication satellite is determined mainly by the following:

—the requirements of the telecommunications payload,
—the nature and effects of the space environment,
—the performance of launchers and the constraints which they impose.

The characteristics of the telecommunications mission take precedence in defining the payload; as far as the platform is concerned they result in requirements which are expressed in terms of electrical power to be provided, antenna pointing accuracy, thermal power to be extracted, space required for equipment mounting, the number of TTC (telemetry, tracking and command) channels and so on.

The nature and effects of the space environment affect orbit control, subsystem organisation and the choice of materials and components.

Launchers impose mechanical constraints on the structure of the satellite. Their performance limits the mass which can be injected into orbit and influences the specification of the propulsion subsystem. Furthermore, the limited enclosed volume which is available requires the solar panels and antennas to be folded.

The basic information necessary to determine the required electrical and thermal power to be dissipated by the payload has been provided in Chapter 9. The nature and effects of the environment are the subject of Chapter 12. Finally, the launch procedures and the performance of launchers are presented in Chapter 11.

10.1 SUBSYSTEMS

A telecommunication satellite incorporates various subsystems whose functions, in principle, are clearly distinct. It is customary to distinguish the payload, as discussed in the previous chapter, from the platform which supports and powers the payload (Figure 10.1). A list of the platform subsystems is given in Table 10.1 (those of the payload are included for reference). The functions to be provided and the most significant characteristics are indicated.

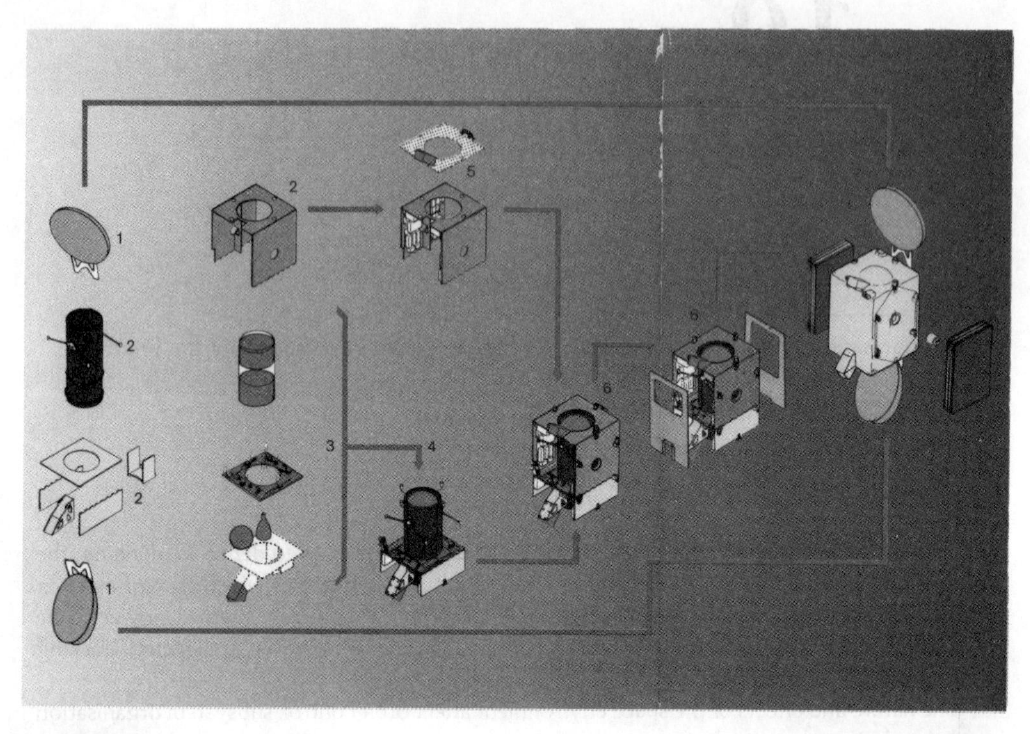

1 Antenna Reflector
2 Structure
3 Propulsion Equipment
4 Propulsion Integration
5 Payload Integration
6 Satellite Integration

Spacebus 100			
Max. GTO mass (kg)	1500	1900	2400
Central tube diameter	900	900	900
Length (mm)	2160	2160	2600
Body cross section (mm × mm)	1460 × 1640	1460 × 1640	1700 × 2300
Max. tank capacity (kg propellant)	750	950	1200
Tank type (2)	Spherical	Cylind. + Cassini	Cylind. + Cassini
Earth face position (on orbit)	Perpendicular to tube axis	Perpendicular to tube axis	Parallel or perpendicular to tube axis
Status	Existing	in development	Growth potential

SB 100 Major Characteristics Summary

Figure 10.1 Various subystems of a platform. (Reproduced courtesy of Aerospatiale.)

Table 10.1 Satellite subsystem

Subsystem	Principal functions	Characteristics
Attitude and orbit control (AOSC)	Attitude stabilisation Orbit determination	Accuracy
Propulsion	Provision of velocity increments	Specific impulse, Mass of propellant
Electric power supply	Provision of electrical energy	Power, Voltage stability
Telemetry, tracking and command (TTC)	Exchange of service information	Number of channels, Security of communication
Thermal control	Temperature maintenance	Dissipation capability
Structure	Equipment support	Rigidity, lightness
Antennas	Reception and transmission of radio-frequency signals	Coverage, Gain
Repeater	Signal amplification and frequency changing	Noise figure, power, linearity

Three common characteristics are not indicated but are essential and should be emphasised:

—minimum mass,
—minimum consumption,
—high reliability.

For the particular mission to be fulfilled, each subsystem is specified and designed taking account of these three criteria, the technology used and the characteristics of other subsystems. The performance and specification of a particular subsystem depend on the presence of other subsystems and this influences the interfaces between subsystems (Figure 10.2). Each interface is itself defined by numerous characteristics of which the most typical are given in Figure 10.2. Particular attention must be devoted to the problems of electromagnetic compatibility (EMC) and account taken of the numerous items of radio-frequency equipment operating in different frequency bands at different power levels.

In this chapter, the various subsystems are examined in order to clarify the essential characteristics.

10.2 ATTITUDE CONTROL

Movement of the satellite can be resolved into movement of the centre of mass in the earth's reference frame and movement of the body of the satellite about the centre of mass. The movement of the centre of mass is characteristic of the satellite orbit and its control has been discussed in Chapter 7. Movement of the body of the satellite about the centre of mass is determined by the evolution of the attitude.

The attitude of the satellite is represented with respect to the yaw, roll and pitch axes of a local orbital reference (Figure 10.3). This reference is centred on the centre of mass of the satellite; the yaw axis points in the direction of the centre of the earth, the roll axis is in the plane of the orbit, perpendicular to the first and oriented in the direction of the velocity;

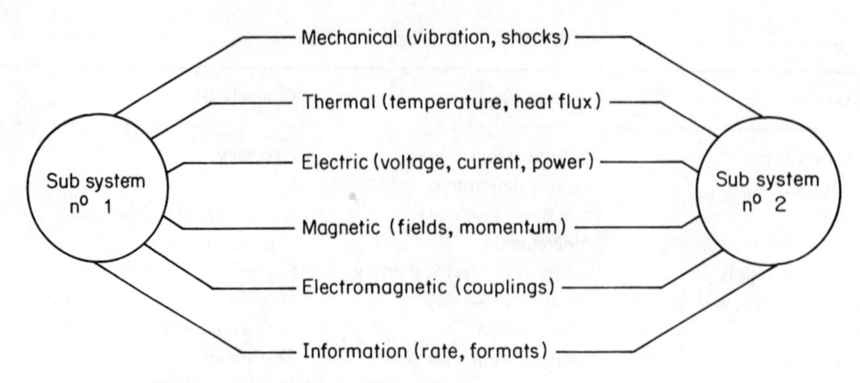

Figure 10.2 Interfaces between two subsystems.

the pitch axis is perpendicular to the two others (and hence perpendicular to the orbit) and oriented in such a way that the reference frame is direct (towards the south for a geostationary satellite). In the nominal attitude configuration, the axes of the mechanical reference of the satellite are, in principle, aligned with the axes of the local orbital reference. The attitude of the satellite is represented by the angles of rotation about the various axes between the orbital reference and the mechanical reference.

Maintaining attitude is fundamental for the satellite to fulfil its function. The accuracy and reliability of this subsystem determine the performance of most of the other subsystems; for example, narrow beam antennas and solar panels must be suitably oriented.

10.2.1 Attitude control functions

The role of attitude control usually consists of maintaining the mechanical axes in alignment with the orbital reference to an accuracy defined by the amplitude of rotation about each of the axes (the value of amplitude corresponds to a given probability of remaining within range). To

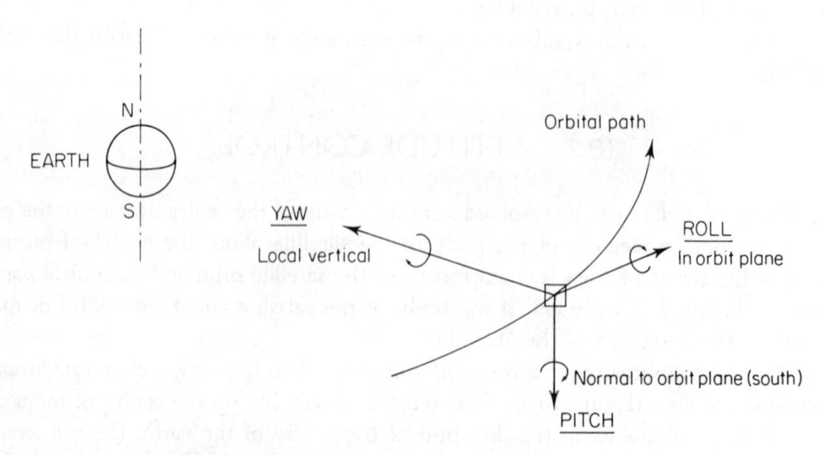

Figure 10.3 Reference axes for determination of satellite attitude.

be specific, typical ranges are $\pm 0.05°$ for roll and pitch and $\pm 0.2°$ for yaw for a geostationary satellite.

In certain cases a constant bias or a particular law of progression about one or more axes may be required in accordance with the requirements of the mission and the particular orbit concerned.

Maintaining attitude requires two functions of different kinds:

—A steering function which consists of causing the part of the satellite which must be oriented towards the earth to turn about the pitch axis in order to compensate for the apparent movement of the earth with respect to the satellite. For a geostationary satellite, this rotation is made at constant velocity equal to one revolution per day (0.25° per minute).
—A stabilisation function which involves compensating for the effects of attitude disturbing torques. The disturbing torques are created by gravitational forces, solar radiation pressure and interaction between current loops and the terrestrial magnetic field (see Chapter 12). These natural disturbing torques are very small (of the order of 10^{-4} to 10^{-5} N m). In contrast, disturbing torques created by lack of alignment of the thrust of the orbit control actuator with respect to the position of the centre of mass can be large (of the order of 10^{-2} to 10^{-3} N m).

In the past, passive attitude control has been considered. It involves using the effects of natural torques to maintain the required attitude. For example, use of the gravity gradient, which tends to align the axis of lowest intertia with the local vertical (see Section 12.2.1.1), has been investigated [MOB-68]. The accuracy obtained (at best a few degrees) is incompatible with the pointing requirements of telecommunication satellites which use narrow beam antennas requiring precise pointing.

Consequently, active attitude control is used. The process generally involves the following:

—measurement of the attitude of the satellite with respect to external references,
—determination of the attitude with respect to the defining reference,
—evalution of the actuator commands,
—execution of the corrections by means of actuators mounted on the satellite,
—evolution of the attitude in accordance with the satellite dynamics under the effect of actuating and disturbing torques.

The system can operate in closed loop on board the satellite, that is actuator control is directly generated by on-board equipment as a function of the outputs of attitude sensors. The attitude sensor outputs may also be transmitted to the ground on the telemetry channels and the actuators operated by the command channels to restore the attitude with evaluation on the ground of the required corrective actions. Closing of the control loop on the ground is possible only if dynamic progression of the satellite attitude is slow. In practice, according to the techniques used and the axes considered, a combination of these two principles is often used.

Although active attitude control is realised, it is useful to take advantage of natural effects as follows:

—use of the gyroscopic rigidity obtained by creation of angular momentum on board the satellite,
—creation of control torques using magnetic coils interacting with the terrestrial magnetic field,

—generation of active torques by taking advantage of solar radiation pressure.

Use of natural effects results in greater operational flexibility of attitude control and a reduction of the quantity of consumable propellant to be embarked on the satellite.

10.2.2 Attitude sensors

These sensors measure either the orientation of the satellite axes with respect to external references (such as the earth, the sun or the stars) or the progression of the orientation with time (gyrometers).

Their essential characteristic is accuracy. It depends not only on the procedure used but also on alignment errors of the detector with respect to the body of the satellite.

The sensors most used on board geostationary telecommunication satellites are solar detectors, terrestrial horizon detectors and gyrometers. For certain applications star detectors widen the range of possibilities. Finally, it is possible to use a radio-frequency beacon or a laser to obtain a measure of attitude.

10.2.2.1 Solar detectors

The solar detector uses photo-voltaic elements which produce a current when illuminated by the sun. Various arrangement are used (such as slit or digital sensors). The accuracy obtained for measurement of the angle between the direction of the sun and an axis related to the satellite is of the order of 0.005°.

10.2.2.2 Terrestrial horizon detectors

The earth surrounded by its atmosphere appears as a spherical black body at a temperature of 255 K when its radiation is measured in the infra-red absorption band of carbon dioxide (14–16 μm). As seen from space, the image of the earth contrasts strongly with respect to the background plane whose temperature is around 4 K. Measurement of infra-red radiation, whose emission is approximately uniform over the whole surface of the earth, by means of a thermosensitive element (such as a bolometer, thermocouple or thermopile) permits the contour of the terrestrial globe to be detected.

Scattering of reflected solar light (the earth's albedo) can also be used; this is detected by means of photoelectric cells or phototransistors. The measurements are corrupted due to the difficulty in separating it from the tropopause).

An accuracy of the order 0.05° is obtained.

10.2.2.3 Stellar sight

An image of a given portion of the sky provides a map of the stars whose relative positions are detected and compared with a reference map. The measurement accuracy is high (better than a thousandth of a degree); the penalty is the complexity of the sensor which is consequently delicate, costly and bulky. There is a danger of saturation by the sun, earth or other bright sources.

10.2.2.4 *Inertial detectors*

Inertial systems use gyroscopes to detect movement of the satellite or gyrometers to measure the angular velocity about one axis. They are of delicate construction, subject to drifts and their limited lifetime (about 10 000 h) prohibits their continuous use for conventional missions of 10 years.

10.2.2.5 *Radio-frequency detectors*

These detectors depend on measurement of the characteristics of radio waves transmitted to the satellite by ground radio beacons. These detectors permit measurement of the angle between the antenna axis on board the satellite and the direction of the beam. Rotation about the boresight (the yaw angle) is difficult to evaluate. By measuring the rotation of a polarised wave from a single radio beacon, a value of the yaw angle is obtained. Unfortunately, the orientation of the polarisation is affected by Faraday rotation and the accuracy of the yaw angle is of the order of 0.5°. On the other axes it could be as small as 0.01°.

10.2.2.6 *The laser detector*

The use of a laser beam has been considered for determining the orientation of the satellite. The expected accuracy is 0.006° for roll angles and 0.6° for the yaw angle. One of the major problems lies in attenuation of the beam by clouds.

10.2.3 Attitude determination

The purpose is to determine the orientation of the satellite in the local orbital reference as defined in Figure 10.3. For a rotating sensor the line of sight of a particular object defines a cone whose axis is the axis of rotation and whose vertex half angle is the angle between the direction of the object and the axis of rotation [WER-78]. Rotation of the sensor results either from rotation of the satellite on which the sensor is mounted or from a scanning device associated with the sensor.

The earth and sun are privileged objects. Determination of the direction of the centre of the earth may be achieved by using a detector with a narrow field of view which, in the course of rotation of the satellite, sweeps out a cone which intersects the terrestrial surface (Figure 10.4). Scanning the illuminated region of the earth produces a signal whose duration permits the nadir angle to be determined, that is the angle between the axis of rotation and the axis joining the satellite and the centre of the earth (the nadir axis).

A solar detector permits measurement of the vertex half angle of the second cone associated with the axis joining the satellite and the sun. The axis of rotation of the sensor is situated at one of the intersections of the two cones defined by the two observations (Figure 10.5).

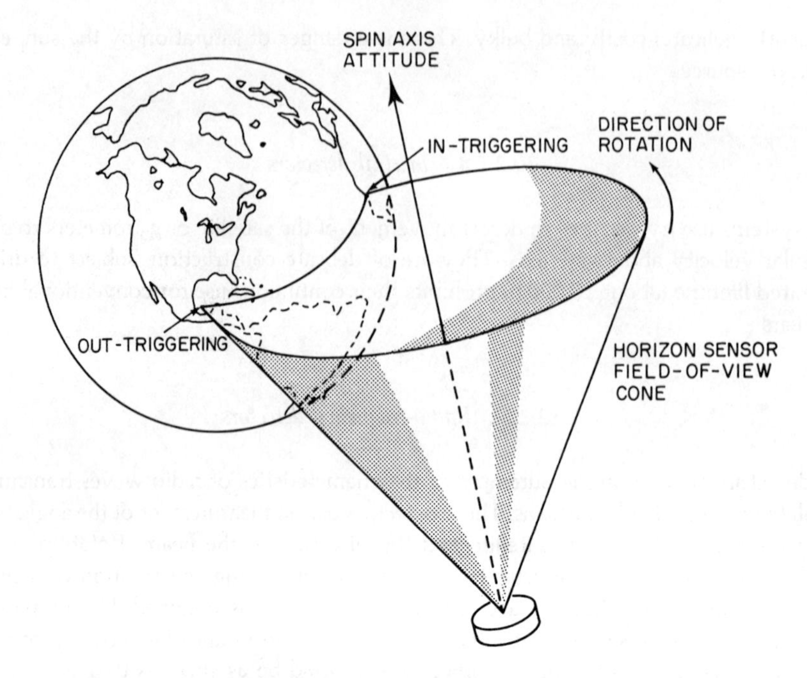

Figure 10.4 Earth contour detection [WER-78]. (Reproduced by permission of Kluwer Academic Publishers.)

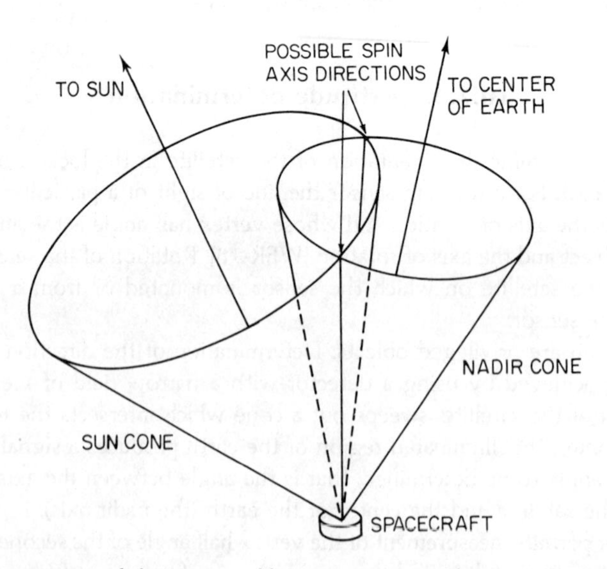

Figure 10.5 Determination of the two possible sensor spin axis directions [WER-78]. (Reproduced by permission of Kluwer Academic Publishers.)

Selection of one of the two intersections requires a third measurement or partial a priori knowledge of the orientation of the satellite.

The method presented above permits determination of the direction of the axis of rotation in space knowing the relative position of the satellite with respect to reference objects. This assumes, therefore, that the orbit of the satellite and the position of the satellite in the orbit have been exactly determined.

As far as a geostationary satellite is concerned, once the satellite is in position in its nominal orbit, the requirements of the mission (such as pointing the antennas towards the earth) no longer require determination of the attitude in space, only the orientation of the satellite with respect to the earth. An earth sensor thus readily permits attitude determination with respect to the roll and pitch axes. These measurements are obtained either by a mechanical scanning sensor or a static horizon detector using thermopiles; these provide a signal which is related to the separation between the direction of the centre of the earth and the direction of the optical axis of the sensor (Figure 10.6).

A scanning sensor can also be used (Figure 10.7). The advantage lies in the wider field of view which means that the sensor can be used for a large range of satellite altitudes.

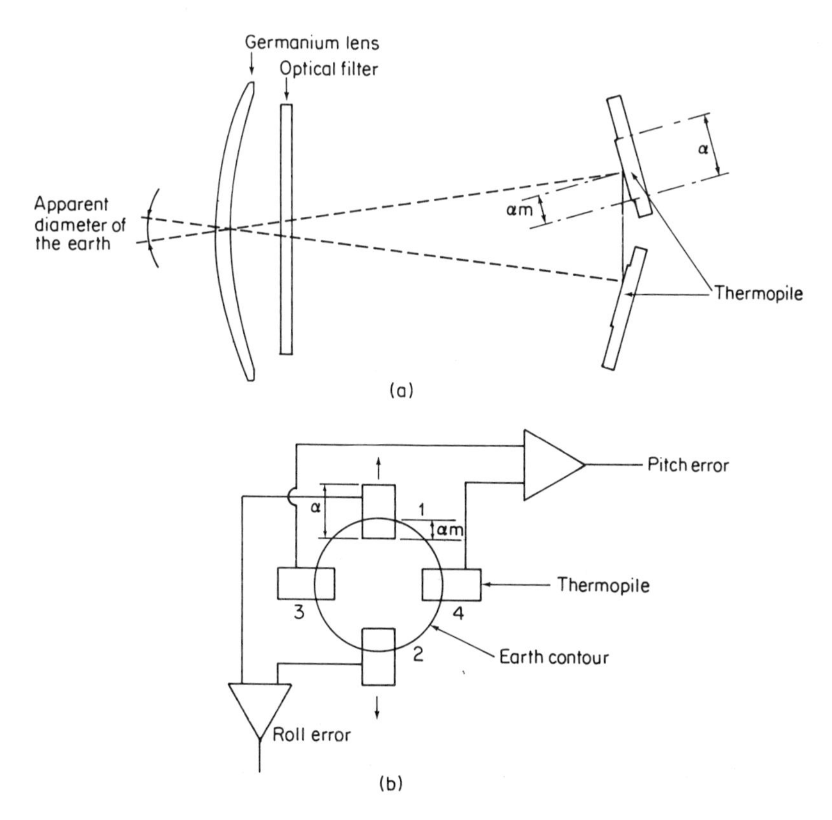

Figure 10.6 Determination of the direction of the centre of the Earth: (a) Earth image generation. (b) Wiring of thermopiles.

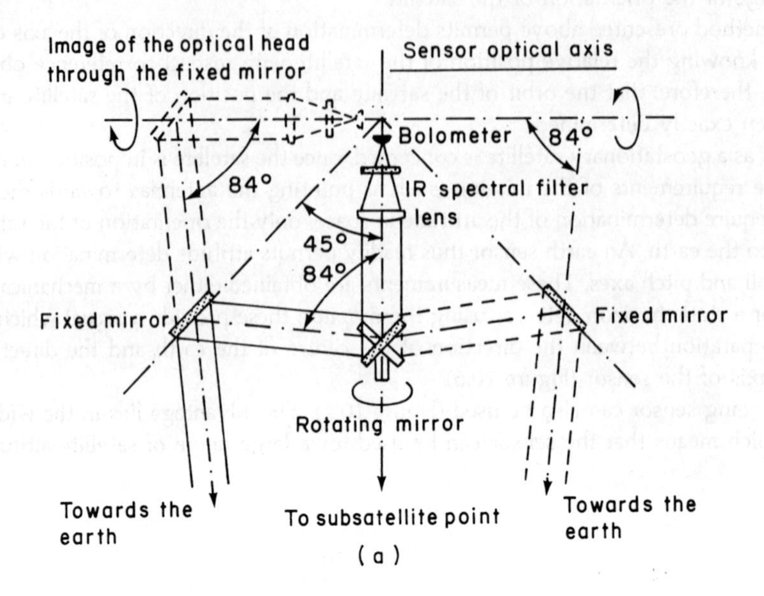

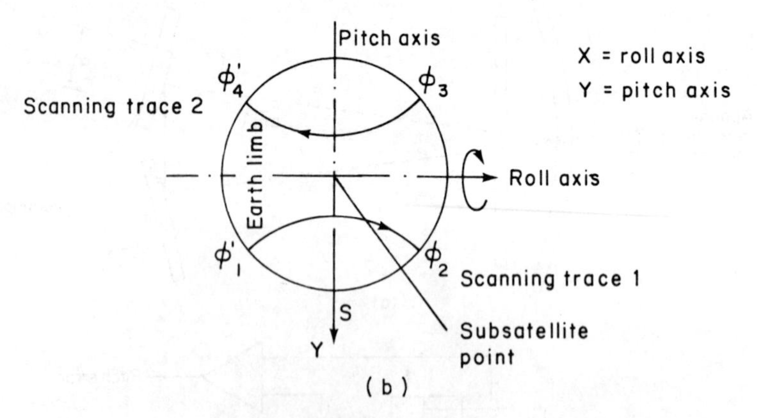

The satellite orientation is defined by the following equations:

$$- X \text{ deviation (pitch)} = \frac{(\phi_3+\phi_4)-(\phi_1+\phi_2)}{4} + \text{constant}$$

$$- Y \text{ deviation (roll)} = \frac{(\phi_4-\phi_3)-(\phi_2-\phi_1)}{4 \times \text{constant}}$$

Figure 10.7 Scanning sensor. (a) Principle. (b) Definition of measured parameters X and Y.

The circular form of the earth's image does not permit the pointing error about the yaw axis to be obtained directly. One approach is to combine a measurement made with a solar or stellar sensor. Otherwise, in the case where dynamic attitude progression is slow, knowledge of the orientation about the yaw axis can be estimated from a measurement of roll performed 6 h earlier, the directions in space of the roll and yaw axes interchange every 6 h as a consequence of the rotation of the satellite about the earth. The use of specific measurement is, therefore, not indispensable as long as the disturbing torques remain small. When large disturbing torques are generated (such as those caused by the use of thrusters for orbit control), a knowledge of the progression of the orientation in yaw is obtained by using an appropriate sensor (such as a gyrometer or solar sensor).

10.2.4 Actuators

Modification of the attitude is generally obtained by generating a torque which, taking account of the dynamics of the particular satellite, causes an angular acceleration, or velocity, about an axis. The attitude control actuators, therefore, have the purpose of generating torques. Various types are available as follows.

10.2.4.1 Angular momentum devices

These include reaction wheels and gyroscopes which exploit the principle of conservation of angular momentum.

Variation of the velocity of rotation ω of a flywheel (a reaction wheel), of moment of inertia I, causes the angular momentum $H = I\omega$ to be modified and a torque of moment T about the axis of the wheel is generated:

$$T = \mathrm{d}H/\mathrm{d}t = I\mathrm{d}\omega/\mathrm{d}t \qquad (\mathrm{N\,m}) \tag{10.1}$$

A steerable gyroscope consists of a flywheel rotating at constant velocity which is gimballed about one or two axes. Control of the orientation of the axis of the moment of inertia causes generation of a torque of moment T equal to the derivative of the variation with respect to time of the angular momentum vector. The limited lifetime of steering devices means that this type of device is very little used for active torque generation.

These devices are particularly suitable for maintaining attitude when the satellite is subject to cyclic disturbing torques (for example, the effect of the flattening of the earth). Disturbing torques with non-zero mean values (caused, for example, by the effect of solar radiation pressure), or disturbing torques of excessive amplitude can require a compensating angular momentum variation which exceeds the limits for the velocity of rotation of the flywheel or the orientation of the gyroscope; that is saturation occurs. It is then necessary to provide a desaturating torque by means of another power unit (a gas jet for example).

10.2.4.2 Thrusters

Thrusters produce reaction forces on the satellite by expelling material (propellant) through nozzles. The force obtained is a function of the quantity of material (mass) ejected per unit time

dm/dt and depends on the specific impulse I_{sp} of the propellant used (see Section 10.3):

$$F = gI_{sp}(dm/dt) \quad \text{(N)} \tag{10.2}$$

where g is the normalised terrestrial gravitational constant ($g = 9.807\,\text{m/s}^2$).

The torque obtained depends on the length d of the lever arm with respect to the centre of mass of the satellite:

$$T = Fd \quad \text{(N m)} \tag{10.3}$$

The torques to be applied are of the order of 10^{-4} to 10^{-1} N m. With a lever arm of 1 m, the thrusts are thus of the order of 10^{-4} to 10^{-1} N. In the interest of simplicity and reducing mass, these thrusters are generally part of the thruster assembly which provides orbit control after installation in orbit (see Chapter 7). These thrusters often provide thrusts greater than the values given above (from a few newtons to tens of newtons). Smaller thrusts, which can be modulated, are obtained as a mean value by using thrusters in an on-off mode of operation with a variable duty cycle.

10.2.4.3 *Magnetic coils*

Magnetic coils create a magnetic moment **M** when fed with a current I. This magnetic moment can generate a torque **T** by interaction with the terrestrial magnetic field **B**:

$$\mathbf{T} = \mathbf{M} \times \mathbf{B} \quad \text{(N m)} \tag{10.4}$$

As far as geostationary satellites are concerned, the value of the terrestrial magnetic field is very small (see Chapter 12). The torques obtained are equally small for realistic values of realisable magnetic moments. However, they can be sufficient to compensate for some of the disturbing torques exerted on the satellite.

10.2.4.4 *Solar Sails*

Solar radiation pressure (see Chapter 12) applied to a surface of sufficient size is capable of generating non-negligible torques. In general, these are disturbing torques which are opposed by designing the satellite so that the apparent surface in the direction of the sun is symmetrical with respect to the centre of mass. The most significant surfaces are undoubtedly those of the solar generators; a satellite stabilised in three axes also has two symmetrical panels which are aligned with the pitch axis.

By modifying the apparent surfaces of the two panels of the solar generator, it is possible to generate torques about the two axes (Figure 10.8). A torque about an axis in the direction of the sun is created by introducing a symmetrical bias between the normal to each panel and the direction of the sun (the windmill effect). A further torque can be obtained about an axis perpendicular to the direction of the sun in the plane of the orbit by means of a bias of one panel alone (asymmetry of the apparent surfaces).

The bias introduced must remain limited (to about ten degrees maximum) in order to avoid excessive reduction of the flux captured by the solar generator. The effect of the solar sails is increased by adding surfaces which are oblique with respect to the panels at the extremities of

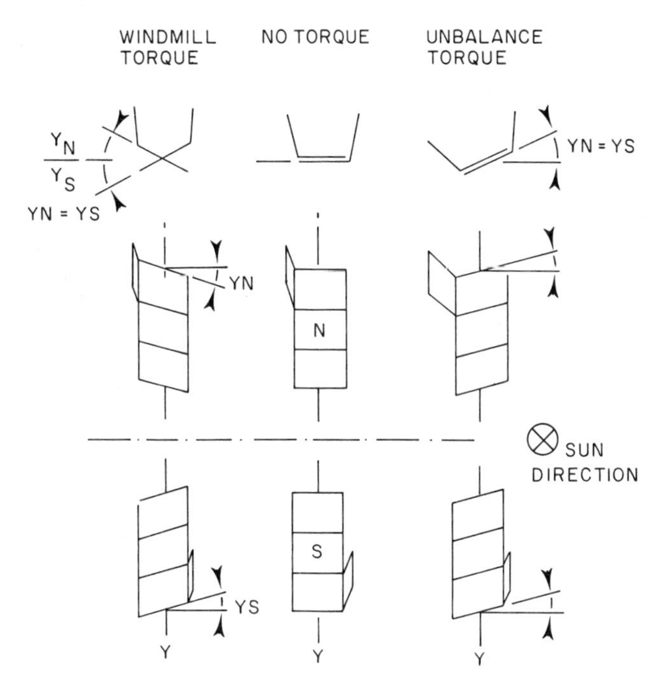

Figure 10.8 Solar sail.

the solar generators. The torques obtained are sufficient to compensate for most disturbing torques (except those induced during station-keeping manoeuvres).

10.2.5 The principle of gyroscopic stabilisation

Gyroscopic stabilisation is obtained by creating an angular momentum on board the satellite. By virtue of the principle of conservation of angular momentum, the orientation of the angular momentum tends to remain fixed in inertial space (gyroscopic rigidity). By choosing an angular momentum aligned with the pitch axis, which preserves a fixed orientation in space in spite of movement of the satellite in its orbit, the pitch axis benefits from gyroscopic rigidity and hence movements about the roll and yaw axes are limited.

The benefit of gyroscopic stabilisation is better appreciated if the effects of disturbing torques on a satellite are compared with and without angular momentum generation. Figure 10.9 shows the effect of a disturbing torque T_d which is exerted about the Z axis of a satellite whose mechanical axes x, y and z are initially aligned with the X, Y and Z axes of a reference frame.

—The satellite without angular momentum starts to rotate about the z axis with a constant angular acceleration $d\Omega/dt$ given by:

$$d\Omega/dt = T_d/I_z \qquad (\text{rad/s}^2) \tag{10.5}$$

where I_z is the moment of inertia of the satellite about the z axis.

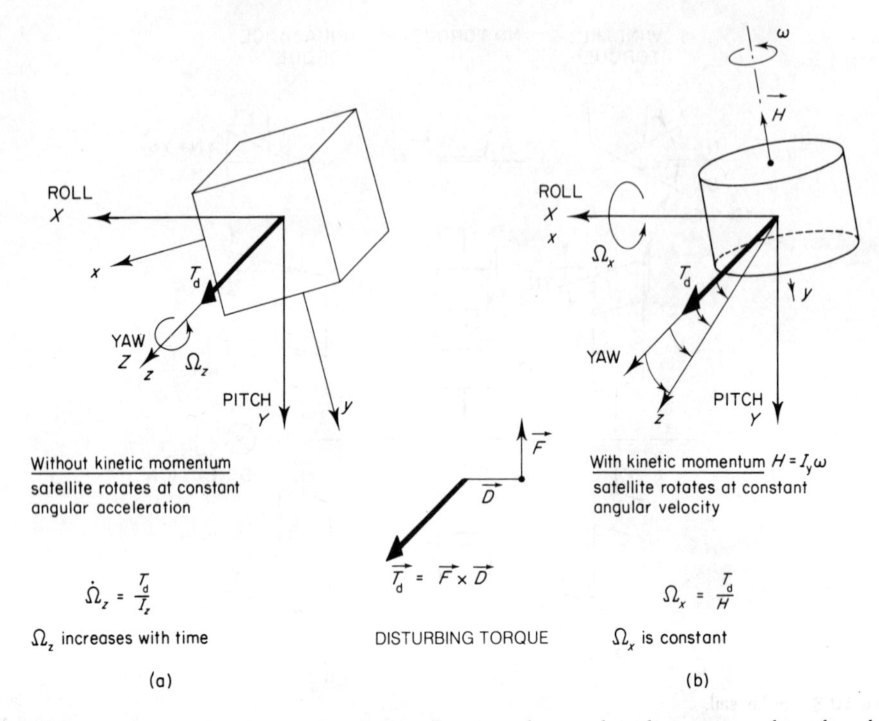

Figure 10.9 Action of a disturbing torque on a satellite: (a) without on-board momentum, (b) with on-board momentum.

—If the satellite has on-board angular momentum H about the x axis as a consequence of the gyroscopic effect, the rotation is about the x axis at a constant angular velocity Ω given by:

$$\Omega = T_d/H \qquad \text{(rad/s)} \tag{10.6}$$

Numerical example. The time required to obtain an attitude pointing error of 0.1° will be calculated assuming a disturbing torque of 1×10^{-5} N m and a moment of inertia about each of the axes 1000 m^2 kg.

In the first case (a satellite without angular momentum), the angular acceleration is:

$$d\Omega/dt = T_d/I_z = 1 \times 10^{-5}/1000 = 1 \times 10^{-8}\ \text{rad/s}^2$$
$$= (360/2\pi)10^{-8} = 5.73 \times 10^{-7}\ \text{degree/s}^2$$

The motion is at constant angular acceleration, hence $\theta = (1/2)(d\Omega/dt)t^2$ and therefore $t = [2\theta/(d\Omega/dt)]^{1/2} = [0.2/5.73 \times 10^{-7}]^{1/2} = 590\ \text{s} = 9.8\ \text{min}$. After 9.8 min the satellite, which was initially aligned with the reference frame, reaches the permitted pointing error limit and a corrective action must be performed.

If an angular momentum of 100 N m s now exists about the y axis, movement occurs at a constant angular velocity given by:

$$\Omega = T_d/H = 1 \times 10^{-5}/100 = 1 \times 10^{-7}\,\text{rad/s}$$
$$= (360/2\pi)10^{-7} = 5.73 \times 10^{-6}\,\text{degree/s}$$

Since the angular velocity is constant, $\theta = \Omega t$ and therefore $t = [\theta/\Omega] = [0.1/5.73 \times 10^{-6}] = 17\,452\,\text{s} = 290\,\text{min} = 4.8\,\text{h}$.

In this case 4.8 h are required for a satellite, which was initially aligned with the reference frame, to reach the permitted pointing error limit; a large time margin is, therefore, available before corrective action must be performed.

An angular momentum H of 100 N m s can be obtained either:

—by rotation of the entire satellite about the y axis with a velocity ω_y such that:

$$H = \omega_y I_y, \qquad \text{hence } \omega_y = H/I_y = 100/1000 = 0.1\,\text{rad/s} \cong 0.95\,\text{rev/min},$$

or

—by means of an inertia wheel mounted within the satellite. An inertia wheel is a heavy flywheel rotating at high speed and forming the rotor of an electric motor (Figure 10.10). According to the angular momentum to be obtained, the mass may be between 5 and 10 kg and the speed of rotation between 5000 and 20 000 revolutions per minute. To limit friction torques, the wheel may rotate in an evacuated enclosure and be suspended on magnetic bearings.

10.2.6 Stabilisation by rotation

Stabilisation by rotation was used on first generation telecommunication satellites and is still used (examples are numerous US national satellites, Brasilsat, INTELSAT VI, etc.). The satellite is given a rotating movement of several tens of revolutions per minute about one of the principal axes of inertia (spin stabilisation). This is a simple process which benefits from the properties of the gyroscope but has the disadvantage of either leading to a rotating, and hence low gain, antenna radiation pattern or necessitating contra-rotation of the antenna or the supporting platform (see Section 9.6). In the absence of disturbing torques, the angular momentum H maintains a fixed direction with respect to an absolute reference frame. For a geostationary satellite, the axis of rotation is thus always parallel to the axis of terrestrial rotation (the pitch axis).

Oscillations of the axis of rotation about the direction of the angular momentum H (nutation) arise when the moment of inertia about the axis of rotation is not sufficiently large with respect to that about the other perpendicular axes. These oscillations must be damped by internal dissipation of kinetic energy (nutation damping), or actively controlled by using thrusters in which case the system has a tendency to instability (when the moment of inertia about the axis of rotation is equal or smaller than that about the other axes). This situation arises with highly elongated satellites of which a large part (a contra-rotating platform) does not participate in the creation of angular momentum (dual spin stabilisation). This is the case, for example, with the INTELSAT VI satellite (Figure 10.11).

Disturbing torques have two effects; they reduce the velocity of rotation about the stabilised axis and cause a stabilised axis pointing error. It is, therefore, necessary to maintain the velocity of rotation (for example by means of thruster 1 of Figure 10.12) and correct the pointing error.

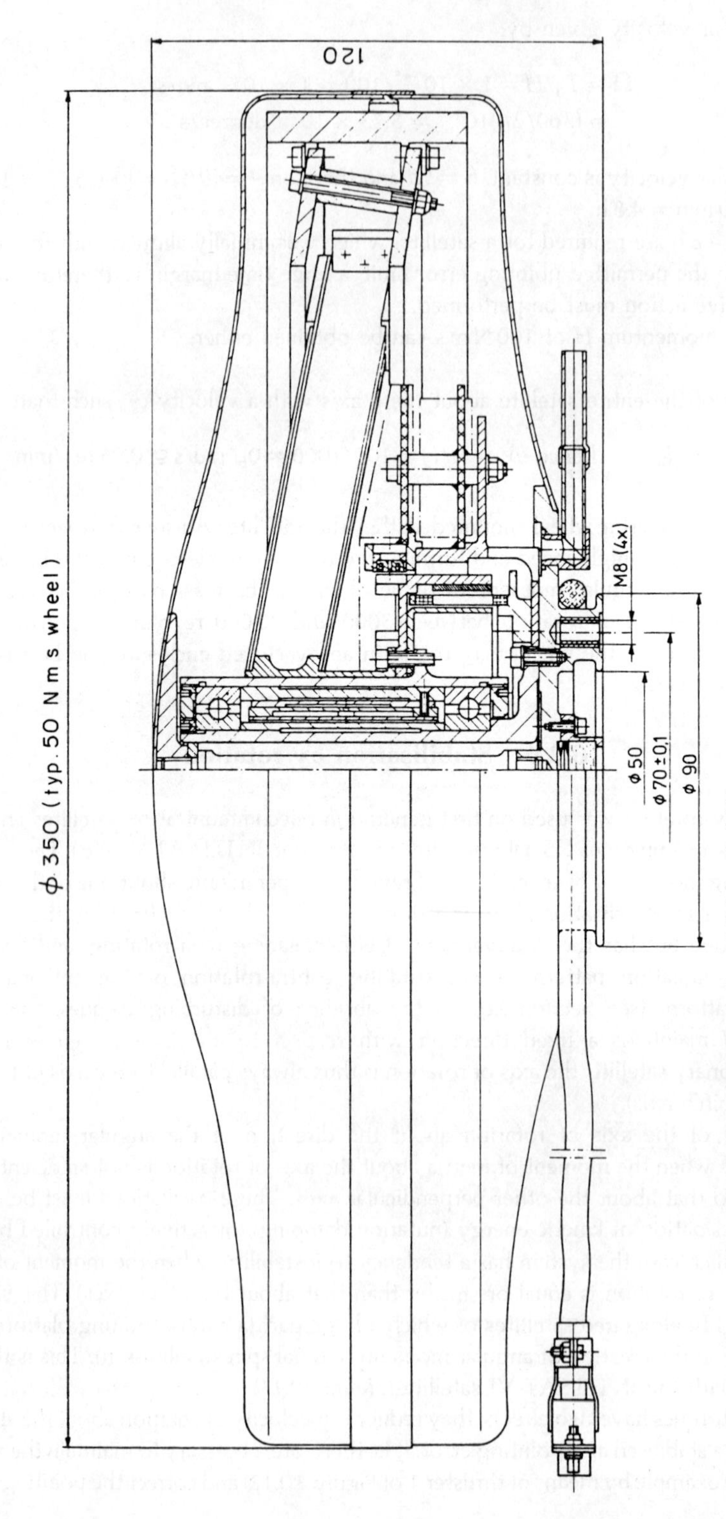

Figure 10.10 Inertia wheel. (Reproduced by permission of Teldix GmbH.)

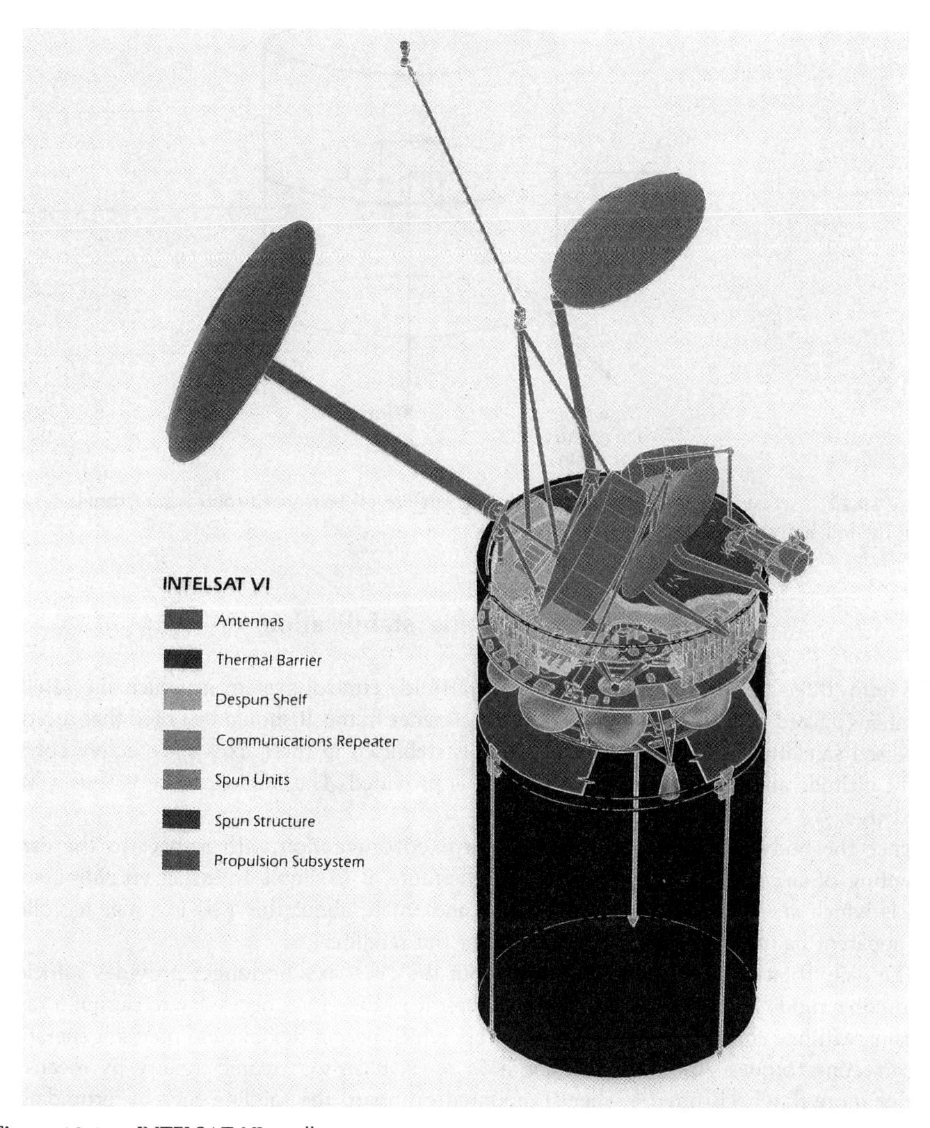

Figure 10.11 INTELSAT VI satellite.

When the component of the disturbing torque perpendicular to the axis of rotation is constant, the pointing error consists of a constant velocity drift about an axis perpendicular to the axis of the torque. Correction requires application of a torque which cancels the drift. In general, the correcting torque is applied periodically as soon as the pointing error reaches the maximum tolerated; a thruster such as 2 of Figure 10.12 is used. It operates by means of impulses in synchronism with the velocity of rotation of the satellite.

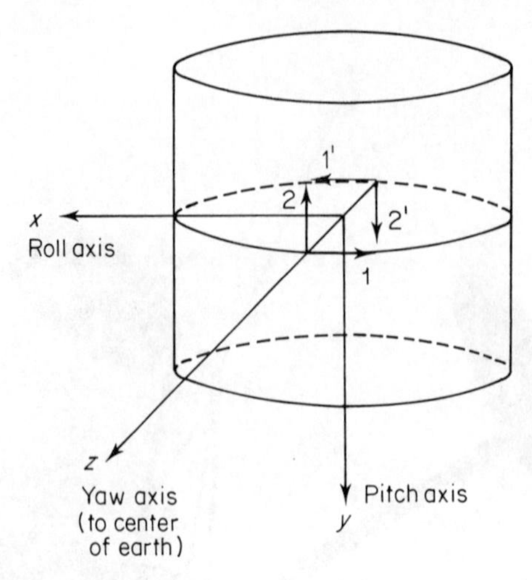

Figure 10.12 Actuator implementation on a spinning satellite. (1) Spin speed control. (2) Attitude control about the roll and yaw axes.

10.2.7 'Three-axis' stabilisation

The term 'three-axis stabilisation' denotes an attitude control system in which the satellite maintains a fixed orientation with respect to a reference frame. It should be noted that rotation stabilised satellites are also strictly described as stabilised in three axes since active control of the attitude about the three reference axes is provided. The nomenclature is thus a little restrictive.

Since the body of the satellite maintains a fixed orientation with respect to the earth, mounting of large antennas is facilitated. Furthermore, it is simple to install unfolding solar panels which are aligned with the pitch axis and rotate about this axis in order to follow the apparent daily movement of the sun about the satellite.

The daily rotation of the satellite body about the pitch axis no longer provides sufficient gyroscopic rigidity to combat disturbing torques. It is, therefore, necessary to design a rapid dynamic attitude control system using actuators which permit flexible and precise generation of correcting torques. Another technique is to re-establish gyroscopic rigidity by means of one or more flywheels (inertia wheels) mounted on board the satellite thereby providing a satellite with on-board angular momentum. This technique is the most widely used for communication satellites.

10.2.7.1 Satellite with on-board angular momentum (one momentum wheel)

The satellite (Figure 10.13) contains an inertia wheel whose axis is aligned with the pitch axis in the nominal attitude configuration (e.g. INTELSAT V or Telecom 2). The angular momentum generated about this axis provides gyroscopic rigidity which tends to keep the mechanical axis of the satellite on which the wheel is mounted and the pitch axis coincident with the orbital reference. Furthermore, by varying the velocity of the wheel slightly about its mean value, correcting torques along the pitch axis are easily generated.

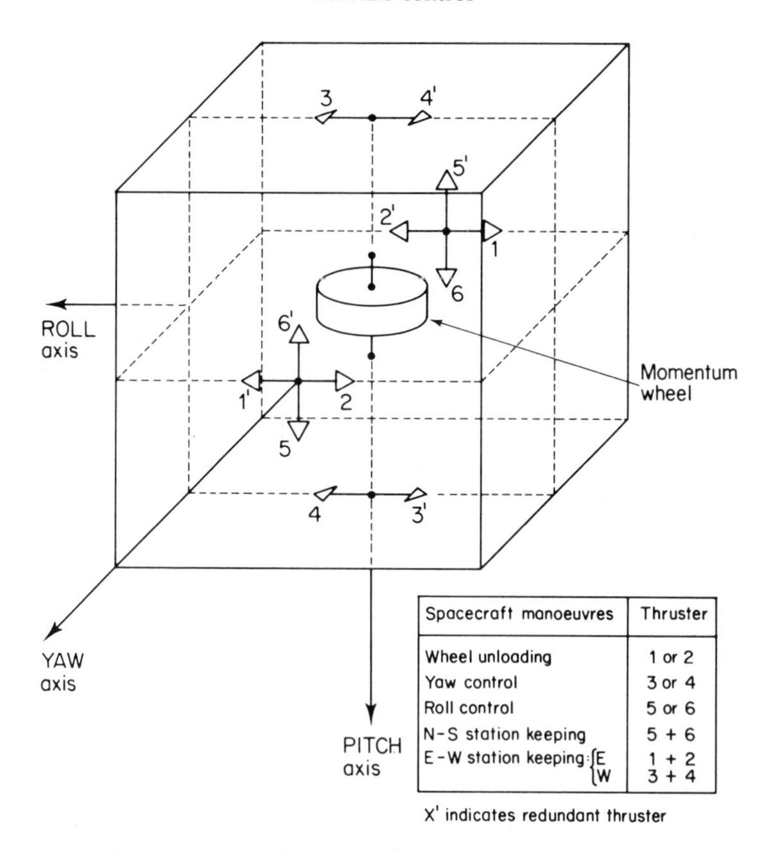

Spacecraft manoeuvres	Thruster
Wheel unloading	1 or 2
Yaw control	3 or 4
Roll control	5 or 6
N-S station keeping	5 + 6
E-W station keeping: E	1 + 2
W	3 + 4

Figure 10.13 Attitude control of a body-fixed satellite by momentum wheel.

Attitude control in normal mode is thus as follows. Pitch and roll angles are measured by one or more terrestrial sensors operating in the infra-red. The gyroscopic rigidity conveyed by the inertia wheel enables movement about the roll and yaw angles to be limited. Pitch control is realised by exchanging angular momentum between the wheel and the body of the vehicle (that is modulation of the velocity of the wheel). Roll control is obtained by using an actuator which generates a torque about this axis (such as a thruster, a magnetic coil or the use of solar panels as sails). The gyroscopic rigidity enables measurement of the yaw angle to be avoided. In the course of the orbit, there is an interchange between the roll axis and the yaw axis every 6 hours and this permits yaw control by measurement of roll.

During highly disturbed phases (for example during station-keeping operations) roll and yaw control are realised separately on each axis. Variations of yaw angle are then measured by a specific sensor such as an integrating gyrometer, solar sensor or stellar sight.

The pointing accuracy obtained with a stabilisation system of this type is of the order of $0.03°$ for the roll angle, $0.02°$ for the pitch angle and $0.3°$ for the yaw angle.

Pitch attitude control. Let T_d be a disturbing torque acting on the pitch axis. The moments of inertia of the satellite and the flywheel are I_S and I_W respectively and ω is the velocity of rotation of the wheel. If ϕ is the angle of the satellite with respect to the pitch axis, the following can be written using the angular momentum theorem:

$$I_S\ddot{\phi} + I_W\dot{\omega} = T_d \qquad \text{(N m s)} \tag{10.7}$$

As the satellite must rotate with a constant angular velocity of one revolution per day about the pitch axis, $\dot{\phi} = 0.25°/\text{min}$ and $\ddot{\phi} = 0$. Hence:

$$\dot{\omega} = T_d/I_W \qquad \text{(rad/s)}$$

If T_d is constant, the wheel must be accelerated by means of its drive motor. When the maximum or minimum velocity ω_M is reached, the wheel must be 'desaturated'; that is its velocity must be brought to a nominal value by applying a torque in opposition to the reaction torque. This torque is produced by two thrusters (one for each correction direction) oriented perpendicularly to the axis of the satellite (1 and 2 of Figure 10.13).

If, at time $t = 0$, the wheel is rotating with mean velocity ω_0, it will reach a velocity ω_M under the influence of torque T_d after a time t_1 such that:

$$(\omega_M - \omega_0)/t_1 = T_d/I_W, \quad \text{hence } t_1 = \Delta H/T_d \qquad \text{(s)} \tag{10.8}$$

where ΔH is the difference between the angular momentum H of the wheel at its mean velocity and at its extreme velocity ω_M.

The desaturation torque T_U must therefore be equal to the motor torque produced by the deceleration of the wheel in order not to disturb the attitude of the satellite. When this condition is fulfilled, the desaturation time t_U is such that

$$T_u t_u = T_d t_1, \qquad \text{hence } T_u = t_1(T_d/T_u) \qquad \text{(s)} \tag{10.9}$$

Control in roll and yaw. The control principles are the same for the two axes. Stabilisation about these axes is the gyroscopic stabilisation provided by the inertia wheel. As the angular momentum of the satellite is negligible compared with that of the wheel, a disturbing torque T_d on one axis (the yaw axis for example) causes a rotation at constant velocity Ω about this axis. If $H = I_W\omega_m$ is the angular momentum of the wheel (ω_m being the lowest value of the velocity of rotation of the wheel and therefore corresponding to the most unfavourable case), the drift velocity is $\Omega = T_d/H$. For a maximum pointing error ε, the time between two corrections will be:

$$t_2 = \varepsilon/\Omega \qquad \text{(s)} \tag{10.10}$$

The correcting torque, opposed to the disturbing torque, is generated by actuators. An example of pairs of thrusters (one per direction of correction) appears in Figure 10.13; thrusters 3 and 4 are for correction about the yaw axis and thrusters 5 and 6 are for correction about the roll axis. Each thruster acts for a time t_c such that:

$$T_t t_c = T_d t_2, \qquad \text{hence } t_c = t_2(T_d/T_t) \qquad \text{(s)} \tag{10.11}$$

where T_t is the torque exerted by the thruster.

Required mass of propellant. If F is the thrust, I_{SP} is the specific impulse and t_c is the total time of operation, the mass of propellant m is given by (see Section 10.3.2):

$$m = Ft_c/gI_{sp}$$

Over a period of one year, the number of thruster operations is $365 \times 24 \times 3600/t_c$ for unloading, $365 \times 24 \times 3600/t_2$ for corrections about the yaw axis and the same number for corrections about the roll axis. The total cumulative time of operation is:

$$t = 365 \times 24 \times 3600[(t_u/t_1) + 2(t_c/t_2)] \qquad \text{(s)}$$

The annual mass of propellant is:

$$m = 31.5 \times 10^6(F/gI_{sp})[(t_u/t_1) + 2(t_c/t_2)] \qquad \text{(kg)} \qquad (10.12)$$

Numerical example. Consider a satellite having the configuration of Figure 10.13 (thrusters on the periphery). Its characteristics are as follows:

Thrusters: thrust $F = 0.5$ N; specific impulse $I_{sp} = 290$ s; lever arm length $l = 0.75$ m. Flywheel: nominal velocity = 7500 rev/min; nominal angular momentum $H = 50$ N m s; permissible variation of angular momentum $\Delta = \pm 5$ N m s.

The disturbing torques considered are those which appear at times other than the station-keeping corrections; they are assumed to be constant and equal about each axis: $T_d = 5 \times 10^{-6}$ N m. Attitude control must be to within 0.1°.

Pitch control—The time t_1 between two unloading operations is:

$$t_1 = \Delta H/T_d = 1 \times 10^6 \text{ s} \qquad \text{(11.6 days)}$$

The unloading torque per pair of thrusters is $T_u = 2Fl = 0.75$ N m. The unloading time t_u is:

$$t_u = t_1(T_d/T_u) = 6.7 \text{ s}$$

Yaw (or roll) control—The drift velocity about the yaw axis is $\Omega = T_d/H = 1 \times 10^{-7}$ rad/s. The time t_2 between two corrections is $t_u = \varepsilon/\Omega = 1.7 \times 10^4$ s (4.7 h). The correcting torque per pair of thrusters is $T_t = 2Fl = 0.75$ N m. The operating time t_c is:

$$t_c = t_2(T_d/T_t) = 0.12 \text{ s}$$

Mass of propellant (10 years)—For a lifetime of 10 years, the mass of propellant is:

$$m = 31.5 \times 10^6(10F/gI_{sp})[(t_u/t_1) + 2(t_c/t_2)] = 2.3 \text{ kg}$$

A margin should be provided to take account of variations of thruster operation, ineffective propellant etc.

Steering by solar sails. Drift of the orientation of the angular momentum of the inertia wheel, under the effect of disturbing torques about the pitch and roll axes, can be continuously compensated by means of a torque generated by appropriate control of the orientation of solar generators to which flaps have been added in order to reinforce the solar sail effect (see Figure 10.8). This permits use of propulsion systems which eject mass to be limited and hence the quantity of propellant to be loaded on to the satellite and the operational constraints associated with the use of thrusters are reduced. By way of example, the Eurostar platform developed by Matra Space is designed with this type of attitude control about the roll and yaw axes. Control about the pitch axis is still realised by control of the velocity of the inertia wheel.

10.2.7.2 *Satellite with on-board angular momentum (several wheels)*

With a single inertia wheel, the axis of this wheel must be coincident with the pitch axis in the nominal attitude configuration in order to maintain a fixed orientation in space during rotation of the satellite in its orbit (control with zero degree of freedom (0 DOF)). It is not, therefore, feasible to introduce a bias on one of the axes in order to change, for example, the direction of the antenna boresight, as this requires us to modify the orientation of the angular momentum along the orbit which translates into propellant consumption.

Orientation of the angular momentum is facilitated by generating it using two or three wheels whose axes are inclined with respect to the pitch axis (Figure 10.14). The direction of the angular momentum depends on the relative velocities of rotation of the wheels and can thus be modified by adjusting their velocities. According to the number of wheels, one or two degrees of freedom are introduced into the attitude control (1 DOF or 2 DOF). An additional wheel for redundancy can also be added. An example of this configuration with two wheels in a V configuration (at angles of $\pm 20°$) about the satellite pitch axis in the pitch-yaw plane and one orthogonal to the pitch axis for redundancy is that of INTELSAT VII [NEY-90]. This additional degree of freedom allows operation of the satellite into an inclined orbit while maintaining the appropriate aiming of the antennas to provide the required coverage (the 'Comsat' manoeuver).

10.2.7.3 *Satellite without on-board angular momentum*

In spite of the constraints imposed by the rapidity of satellite attitude variations which require continuous control, it is feasible not to take advantage of the gyroscopic rigidity provided by installation of one or more inertia wheels on board the satellite. This permits considerable freedom in respect of satellite attitude and allows the orientation to be continuously modified, for example to shift the radio footprint of the antennas or to compensate for the effects of pointing displacement caused by a non-nominal orbit (e.g. non-zero inclination).

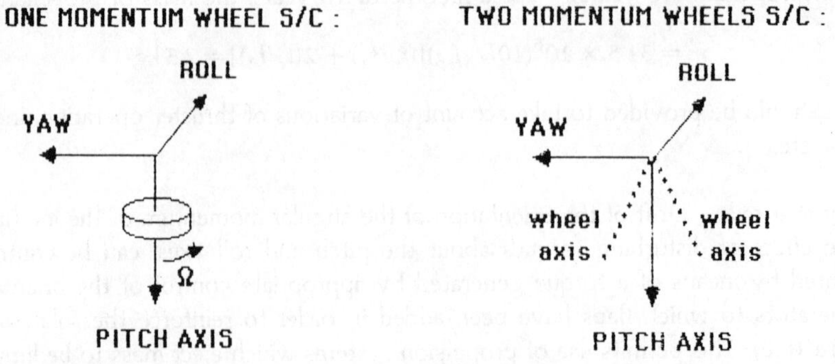

ONE MOMENTUM WHEEL S/C :
ROLL
YAW
Ω
PITCH AXIS

TWO MOMENTUM WHEELS S/C :
ROLL
YAW
wheel axis
wheel axis
PITCH AXIS

Figure 10.14 Angular momentum generated by one or several momentum wheels.

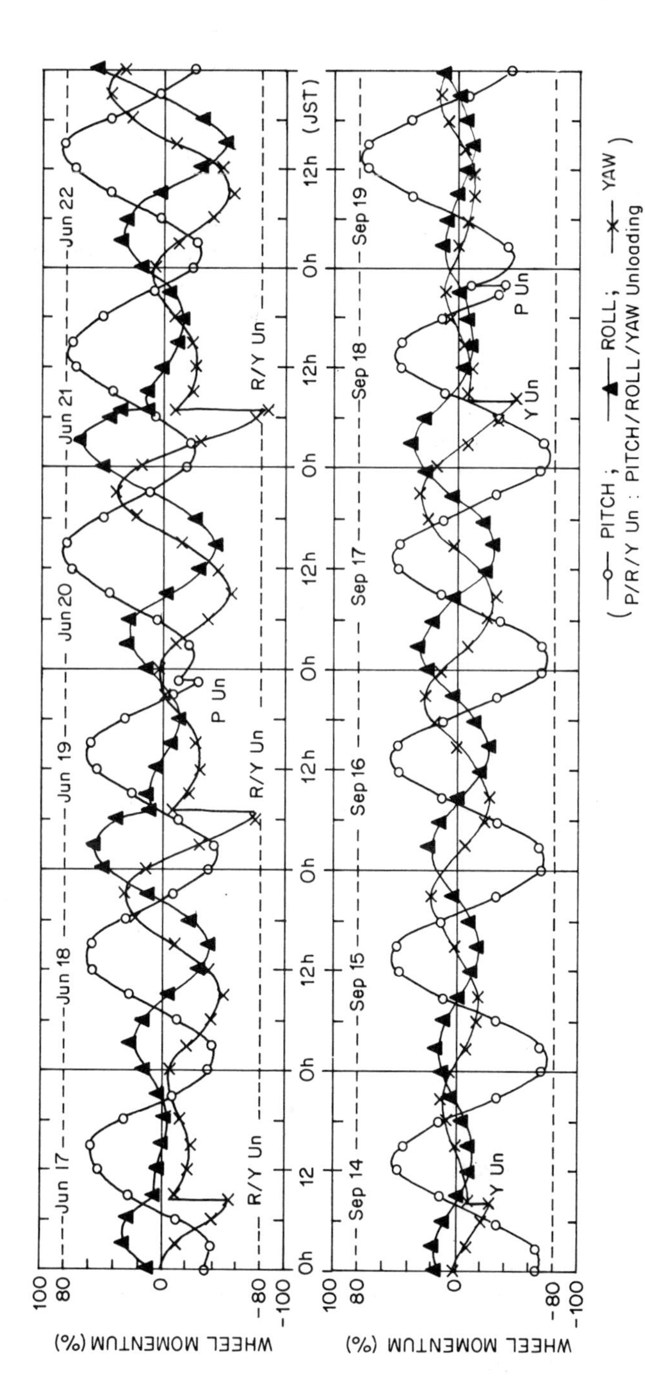

Figure 10.15 Variation of the angular momentum of the three reaction wheels (BSE satellite).

In this context it is useful to use actuators which are capable of generating finely modulated control torques. Three reaction wheels arranged along the three principal axes of the satellite allow disturbing torques to be compensated by exchanging angular momentum between the body of the vehicle and each of the three wheels. This exchange is obtained by continually controlling the velocity of rotation of the wheels from attitude information using an on-board computer (Figure 10.15). The gyroscopic rigidity introduced by the wheels is in this case a secondary, parasitic, effect which remains small on average; the mean velocity of rotation of the wheels is nominally close to zero.

The principle presented above for control about the pitch axis remains useful with reaction wheels on each of the three axes, but formulating complete equations for the system is much more complex as a consequence of gyroscopic coupling terms due to interaction of the movement of the satellite and the wheels. When the disturbing torque about a given axis has a non-zero mean value, compensation of the torque is accompanied by a continuous increase of the velocity of the wheel. When the maximum velocity is reached, it is necessary to unload the wheel by compensating the reaction torque generated by electrical braking by means of an external torque (such as a thruster). In Figure 10.15 it can be seen that unloading operations are performed when the angular momentum of the wheel exceeds $\pm 80\%$ of its maximum angular momentum. Unloading reduces the angular momentum of the wheel to $\pm 10\%$ of its maximum value. The example considered is that of the Japanese Broadcasting Satellite for Experimental Purposes (BSE), launched in 1978. The new Japanese Experimental Test Satellite (ETS-VI) is also of the zero-momentum type with three reaction wheels plus a redundant one.

Another example of a satellite using attitude control without on-board angular momentum is that of the Olympus satellite of the European Space Agency.

10.3 THE PROPULSION SUBSYSTEM

The role of the propulsion subsystem is mainly to generate forces which act on the centre of mass of the satellite. These forces modify the satellite orbit, either to ensure injection into a predetermined orbit or to control drift of the nominal orbit. The propulsion system also serves to produce torques to assist the attitude control system. The forces generated by the propulsion units are reaction forces resulting from the expulsion of material.

10.3.1 Characteristics of thrusters

There are two classes of thruster to be considered:

—Low power thrusters, from a few millinewtons to a few newtons, which are used for orbit control (reaction control system (RCS)).

—Medium and high power thrusters, from several hundreds of newtons to several tens of thousands of newtons, which are used for orbit changes during the launch phase. According to the type of launcher used, these thrusters form the apogee kick motor (AKM) or the perigee kick motor (PKM).

The specific characteristics of orbit control thrusters (RCS) are:

—low power levels (several tens of millinewtons to about ten newtons),
—a larger number of operating cycles of limited duration (a few hundreds of milliseconds to a few hours),
—a cumulative operating time of several hundreds or thousands of hours,
—a long lifetime of greater than ten years.

10.3.1.1 *Velocity increments*

The law of conservation of momentum can be written:

$$M\,\mathrm{d}V = v\,\mathrm{d}M \qquad \text{(N)} \tag{10.13}$$

It expresses the fact that between time t and time $t + \mathrm{d}t$, a satellite of initial mass M moving with velocity V has lost mass dM and increased its velocity by dV. The velocity of ejection of the mass dM with respect to the satellite is v. Integrating between time t_0 (satellite mass $= M + m$) and time t_1 (satellite mass $= M$) gives:

$$\Delta V = v \log[(M + m)/M] \qquad \text{(m/s)} \tag{10.14}$$

where m is the mass of material ejected and M is the mass of the satellite at the conclusion of the manoeuvre.

10.3.1.2 *Specific impulse*

The velocity increment obtained depends on the nature of the material ejected (the propellant) and the velocity of ejection v. The choice of propellant used is influenced by the ease of obtaining a high ejection velocity. Propellants are characterised by a parameter called the specific impulse I_{sp}. The specific impulse is the impulse (force × time) communicated during a time dt by unit weight of propellant consumed during this time interval:

$$I_{sp} = F\,\mathrm{d}t/g\,\mathrm{d}m = F/[g(\mathrm{d}m/\mathrm{d}t)] \qquad \text{(s)} \tag{10.15}$$

where $g = 9.807\,(\mathrm{m/s^2})$ is the terrestrial gravitational constant.

The specific impulse is thus also the thrust per unit weight of propellant consumed per second. As dm/dt is the mass flow rate ρ of propellant ejected:

$$I_{sp} = F/\rho g \qquad \text{(s)} \tag{10.16}$$

Expression (10.13) can also be written $M\,\mathrm{d}V/\mathrm{d}t = v(\mathrm{d}M/\mathrm{d}t)$, that is $F = v\rho$, hence:

$$I_{sp} = v/g \qquad \text{(s)} \tag{10.17}$$

The specific impulse is thus expressed in seconds. In certain cases, the specific impulse is defined as the impulse communicated per unit mass of propellant used and can thus be expressed in different units according to the system used: N s/kg or lbf s/lbm (1 lbm = 0.4536 kg, 1 lbf = 4.448 N, 9.807 lbf s/lbm = 1 N s/kg). The advantage of the definition of specific impulse in seconds is that the unit is universally used (I_{sp}(s) = I_{sp}(lbf s/lbm) = (1/9.807) I_{sp}(N s/kg)).

10.3.1.3 *Mass of propellant for a given velocity increment*

Combining equations (10.14) and (10.17) gives:

$$\Delta V = (gI_{sp}) \log[(M+m)/M] = gI_{sp} \log[M_i/M_f] \quad \text{(m/s)} \qquad (10.18)$$

where M_i is the initial mass and M_f is the final mass after combustion of the propellant.

The mass of propellant m necessary to provide a given ΔV to a satellite of mass M_f after combustion of propellant characterised by its specific impulse I_{sp} is obtained from:

$$m = M_f[\exp(\Delta V/gI_{sp}) - 1] \quad \text{(kg)} \qquad (10.19)$$

The mass of propellant m necessary to provide a given ΔV can also be expressed as a function of the initial mass M_i before combustion of the propellant:

$$m = M_i[1 - \exp(-\Delta V/gI_{sp})] \quad \text{(kg)} \qquad (10.20)$$

10.3.1.4 *Total impulse-time of operation*

The total impulse I_t communicated to the system by ejection of a quantity m of propellant is obtained by integrating the elementary impulse Fdt over the time of operation. Assuming the specific impulse to be constant over the time of operation, this gives:

$$I_t = gmI_{sp} \quad \text{(N s)} \qquad (10.21)$$

The time of operation T depends on the thrust F. Assuming the mass flow rate ρ to be constant, equations (10.16) and (10.17) lead to:

$$T = gmI_{sp}/F = I_t/F \quad \text{(s)} \qquad (10.22)$$

10.3.1.5 *Chemical and electrical propulsion*

Two classes of propulsion system exist:

—Chemical propulsion whose thrust level is between 0.5 newton and several hundreds of newtons for propulsion with liquid propellants and from tens to thousands of newtons for propulsion with solid propellants,

—Electrical propulsion which can deliver a thrust of the order of millinewtons up to a few tens of millinewtons.

The specific impulses depend on the propellant used and the type of thruster (Table 10.2).

Table 10.2 Specific impulses

Type of propellant	I_{sp}(s)
Cold gas (nitrogen)	70
Hydrazine	220
Heated hydrazine	300
Bipropellant	290[a]
	310[b]
Electric ions	1000 to 10 000
Solid	290

[a]Blowdown operation mode (i.e. pressurant gas is stored in the same tank as propellant and pressure decreases as propellant is consumed).
[b]Regulated pressure operation mode (i.e. a regulator maintains a constant gas pressure).

10.3.2 Chemical propulsion

The principle consists of generating gases at high temperature by chemical combustion of liquid or solid propellants. These gases are accelerated by the nozzle.

10.3.2.1 Solid propellants

Solid propellant motors are reserved for generating velocity increments for initial injection into orbit. These motors can be used once only and develop large thrusts (from tens to thousands of newtons). The specific impulse obtained is of the order of 295 s. A description of these motors, together with their characteristics, is given in Chapter 11 which is devoted to injection into orbit.

10.3.2.2 Cold gas

Cold gas propulsion consists of releasing a gas stored under pressure in a reservoir through a nozzle. The material used, according to its nature and the pressure can be in the liquid state (examples are freon, propane and ammonia) or gaseous state (such as nitrogen) in the reservoir. These systems are characterised by relative simplicity, low thrusts and small specific impulses (less than 100 s). They were used mainly on the first satellites and retain a benefit in the case of particular applications where problems of thermal control and pollution associated with hot gas systems arise.

10.3.2.3 Monopropellant hydrazine

A hot gas at a temperature of around 900 °C composed of ammonia, nitrogen and hydrogen is obtained by catalytic decomposition of hydrazine which is then released through a nozzle (Figure 10.16). The catalyst is a metal (iridium). The catalytic bed is designed is such a way that

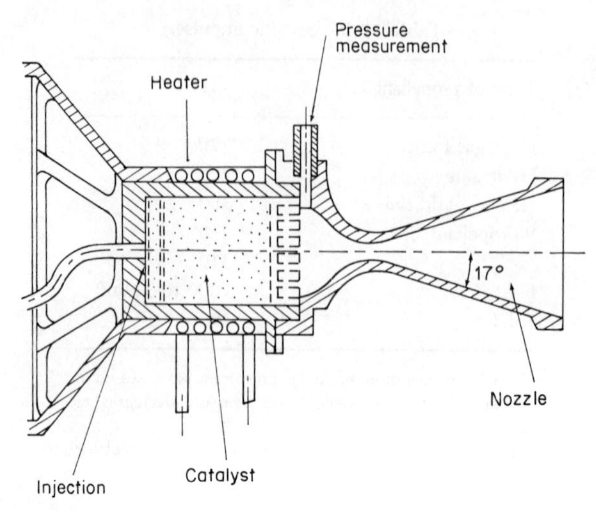

Figure 10.16 Hydrazine thruster.

the contact area is as large as possible within a small volume (small spherical granules are used). The performance of the propellant depends on the temperature of the catalyst and that of the hydrazine.

The thrust obtained is limited by the quantity of hydrazine which can be decomposed in unit time (a function of the area available to the reaction) and is typically of the order of 0.5 to 20 N.

The specific impulse is of the order of 220 s and depends on the operating conditions of the propellant (such as starting from cold or hot and operation in continuous or pulse mode). In particular, operation in pulse mode has a low performance as a consequence of the relatively long time required to establish the thrust.

10.3.2.4 *Electrothermal monopropellant hydrazine*

In order to increase the velocity of ejection and hence the specific impulse, the gas obtained after catalytic decomposition may be superheated to a temperature of the order of 2000 °C before release through the nozzle. The superheating is provided electrically in a heat exchanger.

The specific impulse obtained is of the order of 300 s which is more than 20% greater than that of hydrazine; this leads to an equivalent reduction in the quantity of propellant to be loaded on to the satellite to ensure provision of a given velocity increment. The disadvantages lie in the high electrical consumption of each motor (several hundreds of watts), the limited thrusts obtained (0.5 N), problems with the behaviour of materials at high temperature and, consequently, reliability.

10.3.2.5 *Bi-propellant propulsion*

Bi-propellant systems use an oxidant–fuel pair which has the property of spontaneous ignition (hypergolic propellants) when they are in contact in the combustion chamber in order to produce hot gases for release through the nozzle.

The most commonly used pair consists of nitrogen tetroxide (N_2O_4) as the oxidant and mono-methylhydrazine (CH_3NHNH_2 or MMH) as the fuel. The gas produced is a mixture of water, nitrogen, carbon dioxide, carbon monoxide and hydrogen.

The mixture ratio in the thruster is an important parameter on which the performance depends. It is defined as the ratio of the mass of oxidant to the mass of fuel flowing per unit time. For the mixture considered, the optimum ratio is of the order of 1.6. This value is also the ratio of the density of the two propellants. This property means that the volume of the reservoirs used to store the propellants on board the satellite is the same for the two propellants; this facilitates integration of the reservoirs and limits development costs.

Realisable thrusts range from tens of newtons for motors used for orbit control to several hundreds, even thousands, of newtons for the motors used for injection into orbit (see Section 11.1.4).

The specific impulse obtained is between 290 and 320 s; it depends on the mass flow rate, and hence the thrust concerned, and the supply pressure of the propellants in the combustion chamber (see Section 10.3.2.6). The useful performance obtained by virtue of the high value of specific impulse is affected by the greater dry mass of the propulsion system due to the duplication of valves, filters, reservoirs, pipework, etc. The overall balance becomes useful only for satellites with a sufficiently high mass in orbit (of the order of one tonne if only the orbit control propulsion system is considered). On the other hand, bi-propellant propulsion is particularly useful in connection with the unified propulsion concept (see Section 10.3.4.3).

10.3.2.6 Operation of liquid propellant propulsion

Liquid propellant propulsion systems contain reservoirs to store the propellants, a pressurisation system to drive the propellants from the reservoirs, pipework on which the filters, valves, pressure tappings, filling and draining orifices etc. are mounted and the thrusters themselves (see Figure 10.19, Section 10.3.4.3). Precise thermal control maintains the temperature of the various internal parts within a band which is often narrow; the difference between the freezing and boiling temperatures of propellants can be small. Thermal control is ensured by use of both insulating material and electrical heaters as far as the reservoirs, pipework and valves are concerned and by heatsinks which ensure extraction of the heat generated from the combustion chamber and the motor nozzles into the satellite structure.

The propellants are stored in one or, more often, several reservoirs arranged in such a way that the position of the centre of mass of the satellite varies as little as possible as the reservoirs become exhausted. Positioning is also influenced by consideration of the relative values of the moments of inertia about the various axes, particularly with satellites stabilised by rotation either in the transfer phase or in the normal mode. The problem of sloshing of the propellants always arises for satellites which contain a phase of attitude control by rotation; this can compromise control stability particularly when large quantities are embarked. The use of reservoirs of a particular form, with energy dissipating devices within the reservoirs, permits any oscillation which would otherwise arise to be rapidly damped.

Injection of propellants into the motors under a given pressure required for correct operation is ensured by a reservoir pressurising device. The simplest system is to incompletely fill the reservoirs (by r%) in order to reserve space for a pressurising gas which forces the propellants out of the reservoir. As the reservoir becomes exhausted, the initial pressure decreases (this is described as 'blow down'). The ratio of the initial pressure P_i to the final pressure P_f is equal to the ratio of the final volume V_f to the initial volume V_i available to the gas:

$$P_i/P_f = V_f/V_i = 1/(1-r) \qquad (10.23)$$

This pressure variation ratio (the 'blow down ratio'), equal to the inverse of the complement of the filling coefficient r, is limited to a maximum value which depends on the type of propulsion used. The limiting value is of the order of 4 for hydrazine monopropellant propulsion (a maximum filling coefficient of 3/4) and 2 for bi-propellant systems.

To increase the filling coefficient of the reservoir, it is possible to store the pressurising gas in a separate reservoir. The gas can be stored there either with the required initial pressure, which simply represents an increase of the total volume, or with a much higher pressure; in which case it feeds the reservoir by way of a pressure reducing valve. The pressure reducing valve provides regulation of the gas pressure on the propellants and this ensures that the thrusters operate under conditions for obtaining the best performance. The disadvantage is that the lifetime of the pressure regulator is limited and it is not feasible to operate the system continuously throughout the lifetime of the satellite. Operation under constant pressure in thus reserved for the orbit injection phase, particularly with systems with unified propulsion (see Section 11.1.4.2), and possible repressurising of reservoirs in the course of the life of the satellite when, during normal operation in 'blow down' mode, the pressure becomes too low.

Finally, the problem of separation of liquid and gas in the reservoirs arises. In fact, the pressurising gas and the propellant constitute an emulsion due to the absence of gravity on board the satellite. It must be ensured that only the liquid escapes through the duct which feeds the motors. For rotation stabilised satellites, positioning the duct on the periphery of the satellite permits separation of the gas and liquid by means of the artificial gravity caused by the rotation of the satellite. This gravity does not exist with three-axis stabilised satellites. It is necessary to separate the liquid and gaseous phases mechanically within the reservoir. A polymer membrane can be used with propellants which are not very corrosive (hydrazine). A metallic membrane (bellows) can be used for small reservoirs.

On the other hand, the use of a membrane is not possible with corrosive propellants (nitrogen peroxide) over long lifetimes. The use of surface tension forces which exist at the interface between a liquid and a solid surface can ensure that only the liquid is present within a network of fine cavities (a sieve or metallic sponge of porous material) which feeds the duct. A bubble trap blocks possible bubbles in the pipework which may form when the satellite is subjected to large accelerations.

10.3.2.7 *Location of thrusters*

As far as the thrusters used for attitude and orbit control are concerned, the number of thrusters and their location are dictated by various considerations. The forces to be generated are as follows:

—parallel to the orbit: thrust is exerted in the plane of the satellite orbit and serves to control the semi-major axis and eccentricity of the orbit (to maintain longitude) (i.e. thruster 1 controlled by pulses phased with the rotation of the satellite in Figure 10.12, thrusters 1 and 2 or 1′ and 2′ used simultaneously in Figure 10.13). The thrust also serves to desaturate the inertia wheel for satellites with on-board moment of inertia (1 and 2′ or 1′ and 2 used simultaneously) and to maintain the velocity of rotation of the satellite when it is stabilised by rotation (1 and 1′ in Figure 10.12).

—Perpendicular to the orbit: thrust is exerted along the pitch axis. It serves to correct the inclination (thrusters 2 and 2′ in Figure 10.12; 5 and 6 or 5′ and 6′ used simultaneously in Figure 10.13) and to modify the orientation of the north–south axis (2 and 2′ controlled by pulses phased with the rotation of the satellite in Figure 10.12; 5 and 5′ or 6 and 6′ in Figure 10.13).

The choice of thruster position must be made in accordance with the nature of the mechanical effects to be obtained (torques about the centre of mass or forces acting on the centre of mass).

The gas jet at the output of the nozzle is characterised by its angular width. This jet must not strike parts of the satellite since this would cause problems of deviation of the jet and hence deviation of the direction of the thrust in addition to thermal problems. One approach, when the thrust to be generated must be parallel to one surface of the satellite, is to mount the nozzle with a given inclination (10 to 15°) with respect to the surface. The interaction between the jet and the surface is thus reduced, to the detriment of the efficiency of the thruster in the required direction and the generation of an orthogonal thrust component. Furthermore, the location of the thrusters must take into account the problems of pollution of sensitive surrounding surfaces (such as solar cells, radiating surfaces, sensors etc.).

A compromise between the various requirements is sought by attempting to minimise the required number of thrusters. The ease of integration (which consists, for example, of locating several mutually pre-aligned thrusters on the same plate) also arises in the criteria to be considered.

10.3.3 Electric propulsion

Electric propulsion involves the use of an electrostatic or electromagnetic field to accelerate and eject ionised material. Electric propulsion is an advanced technology in comparison with chemical propulsion. It is characterised by low thrusts (less than 0.1 N) with a high specific impulse (1000 to 10 000s). A notable reduction can thus be achieved in the quantity of propellants to be embarked in comparison with conventional technologies. On the other hand, the operating times will be much greater in view of the low thrust. Electric propulsion, above all, requires a large amount of electrical power. In the specification of a system, therefore, not only the specific impulse but also the specific power must be considered; the latter is equal to the ratio of the electric power to the thrust. It is of the order of 25 to 50 W/mN according to the type of thruster.

Various electric propulsion techniques have been developed [PFE-76], [FEA-77], [CCIR, Rep 843]; these include plasma propulsion, ionic propulsion and Arcjet.

10.3.3.1 Plasma propulsion

The thruster is a form of capacitor using a rod of teflon placed between two electrodes. This capacitor is fed by an electric generator and charges until the high voltage causes a spark to flash across the surface of the rod. A layer of the material is ionised and the plasma is accelerated by the self-generated electromagnetic field [FRE-78]. Once the capacitor is discharged, it recharges until the next discharge occurs. Wear of the teflon rod is compensated by advancing the rod under spring pressure.

The technique is simple and does not require a neutralising device since the plasma ejected is electrically neutral. On the other hand, problems of pollution and electromagnetic compatibility are not negligible. This type of thruster has been used both experimentally and operationally on American satellites (LES-6, LES-8, LES-9). The specific impulses obtained are between 1000 and 5000 s.

10.3.3.2 Ionic propulsion

In an ion thruster, charged particles (ions) are accelerated by an electric field. Since the particles ejected are all of the same sign, it is necessary to make the electrically neutral beam by ejecting the same quantity of charge of the opposite sign in order to avoid raising the satellite to an excessive potential with respect to the surrounding medium. This is achieved by means of an electron gun (a neutraliser).

The ionised material is a heavy metal which is in the liquid state at the storage temperature in order to facilitate feeding to the thruster; examples are mercury, xenon and caesium. Various types of thruster have been developed; they differ in the technique used to obtain the ions from the metallic atoms as follows.

Prior ionisation. The electrons are first extracted from the atoms in an ionisation chamber after vaporisation of the propellant by electric heating. The ions created in this way are then accelerated by a grid raised to a high negative voltage (Figure 10.17). It is, of course, necessary to provide an electron gun to neutralise the beam.

The specific impulses obtained are of the order of 2000 to 3000 s and thrusts are of the order of 2 to 20 mN for a corresponding electric power consumption of 60 to 600 W.

The electrons are extracted either by means of electron bombardment of a cloud of atoms by an electron gun [FEA-78] or excitation by an induced radio-frequency field of several hundreds of kHz [BAS-79]. Motors using electronic bombardment are also called Kaufman motors and examples are the T5 and UK-18 motors developed by Marconi and 2 and 20 mN motors developed by Mitsubishi. Motors using radio-frequency fields include the RIT-10 and RIT-15 motors developed by MBB in Germany. Table 10.3 compares the characteristics of the UK-18 and RIT-15 motors recently developed in Europe.

Field emission. This process permits both ionisation and acceleration of the ions to be obtained. Two plates of which one side has a bevelled section with a very sharp edge are superposed in such a way that the extremities of the bevels coincide (Figure 10.18). The liquid metal progresses by capillarity between the strips to the extremity of the sharp edge. A slotted electrode raised to a high potential with respect to the strips (about 10 kV) creates an intense local electric field which is strengthened in the vicinity of the ridge by the point effect. The field on the ridge is sufficiently large to extract electrons from the propellant atoms and the ions generated in this way are directly accelerated by the electric field. This type of motor has been developed, in particular, by SEP under a European Space Agency contrast [BUG-82].

The specific impulses obtained are very high, between 8000 and 10 000 and thrusts of

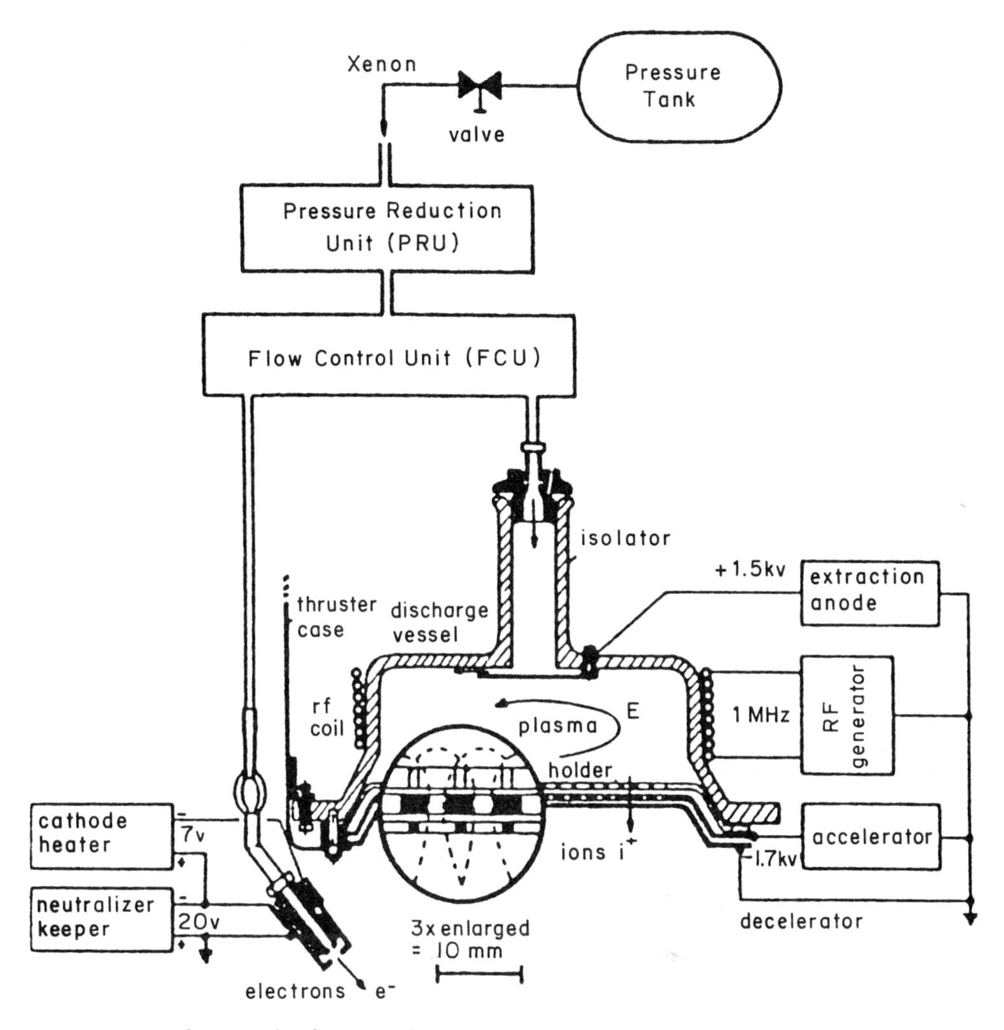

Figure 10.17 The principle of an ionic thruster.

Table 10.3 Ion thruster characteristics

	RITA-15	UK-18
Thrust level	15 mN	18 mN
Exhaust velocity	46 800 m/s	40 869 m/s
Beam current	236 mA	336 mA
Mass flow rate	0.46 mg/s	0.55 mg/s
Specific impulse	33 700 N s/kg	32 352 N s/kg
Total input power	552 W	559 W

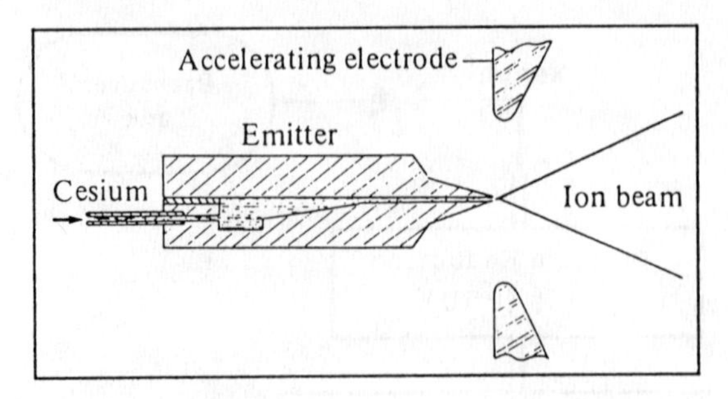

Figure 10.18 Field emission ionic thruster.

the order of 10 mN can be obtained by parallel operation. Electrical consumption, on the other hand, is relatively high, of the order of kW.

10.3.3.3 *Arcjet propulsion [GHI-90]*

A low power arcjet thruster consists basically of an anode, made out of high temperature materials such as pure tungsten or tungsten–rhenium alloy, which serves as chamber, throat and expansion nozzle.

The cathode is usually made from thoriated tungsten and has the shape of a rod with a conical tip.

The propellant gas (argon, ammonia or catalytically decomposed hydrazine), is fed into the plenum chamber and is heated by an arc discharge.

Low power arcjet thrusters operating at input powers of the order of 1 kW can offer considerable advantages, both in terms of increased payload capability and extended lifetime, over chemical thrusters in performing station-keeping manoeuvres of medium-small sized geostationary satellites.

Arcjet thrusters, besides their inherent simplicity from the system point of view, can use hydrazine as propellant allowing a high level of commonality with the other elements of the spacecraft propulsion subsystem.

This type of thruster is presently under development in the USA, Japan and Europe and promises to become a strong competitor in the market of medium sized geostationary satellites with beginning of life (BOL) mass around 1000 kg for mission durations between five and twelve years.

10.3.3.4 *Application specific characteristics*

High specific impulses enable the mass of propellant embarked to be reduced in comparison with chemical thrusters. On the other hand, electric thrusters require additional electrical energy which can lead to an increase of the mass of the solar generators. In the final analysis, they are advantageous only for satellites with a long lifetime (greater than seven years).

With ionic propulsion the width of the ion beam at the thruster output is large. To avoid interaction of the jet with the surface of the satellite when the force produced must be parallel to the surface, it can be necessary to incline the thrust axis with respect to the body and this leads to a loss of thrust efficiency. Moreover, the direction of thrust is not well defined ($\pm 30°$) and consequently disturbing torques are generated. The problem can be resolved by mounting the motor on a movable plate which permits the disturbing torques to be minimised after calibration in orbit [DHU-89]. Finally, it is necessary to take account of the low thrust in determining correction strategies [FRE-72] (see Section 10.3.4.4).

Arcjet propulsion suffers from a larger demand in power consumption and a limited lifetime compared with the increasing lifetime required by communication satellites.

10.3.4 Organisation of the propulsion subsystem

The general organisation of the propulsion subsystem varies in accordance with the types of propulsion used.

10.3.4.1 Solid apogee motor associated with hydrazine propulsion

This combination was widely used for communication satellites until the mid 1980s. It involves equipping the satellite with a solid propellant apogee motor, which is used only for injection into orbit, and a mono-propellant hydrazine propulsion system which is used for attitude and orbit control. The system benefits from relative simplicity and remains useful for small satellites from a mass balance point of view. There are several disadvantages:

—The very high thrust of the solid motor requires gyroscopic attitude stabilisation during the manoeuvre. This prohibits deployment of appendages (such as antennas and solar panels) which are unsuited to support the transmitted acceleration in a transfer orbit.
—The solid motor can be ignited only once and is not duplicated.
—The velocity increment provided is not adjustable once the motor is integrated into the satellite. Possible differences between the nominal transfer orbit and the transfer orbit into which the launcher has injected the satellite cannot be compensated.
—The specific impulses are not among the highest.
—It is necessary to integrate two different propulsion systems.

10.3.4.2 Solid apogee motor associated with bi-propellant propulsion

The use of bi-propellant in place of hydrazine permits a mass gain for large satellites (above 1200 kg). The bi-propellant propulsion system is, however, more complex and the usefulness of this combination remains limited. It is preferable to consider the unified propulsion concept.

10.3.4.3 Unified bi-propellant propulsion

All the propellants required for injection into orbit and attitude and orbit control are stored in a single set of reservoirs. The most used propellants remain monomethyl hydrazine and nitrogen peroxide.

A high thrust apogee motor (400 N for example) is used for injection into orbit. However, in view of the size of the velocity increment to be provided and the limited thrust, the manoeuvre must, in general, be performed on several occasions to avoid a loss of efficiency (see Section 11.1.3.5). A set of thrusters of lower thrust is used for attitude and orbit control.

The assembly is fed by one or more pairs of reservoirs which are pressurised by helium stored in a separate reservoir (Figure 10.19 and Table 10.4). Operation of the system is as follows:

—During the launch phase, the propellant reservoirs are isolated from the thrusters and the helium reservoir by closed pyrotechnic valves.
—Once in transfer orbit, these valves are opened and this permits the helium to pressurise the propellants under constant pressure by means of a regulator. A set of electric valves provides a feed to the apogee motor for the various manoeuvres; operation under constant pressure ensures a maximum specific impulse of the order of 320 s. When the satellite is in the definitive orbit, the apogee motor is totally isolated from the rest of the subsystem by a pyrotechnic valve. The helium reservoir is also isolated from the propellant reservoirs by closure of valves under electrical control. The supply to the attitude and orbit control thrusters is then realised in blow down mode, but the pressure variation remains small

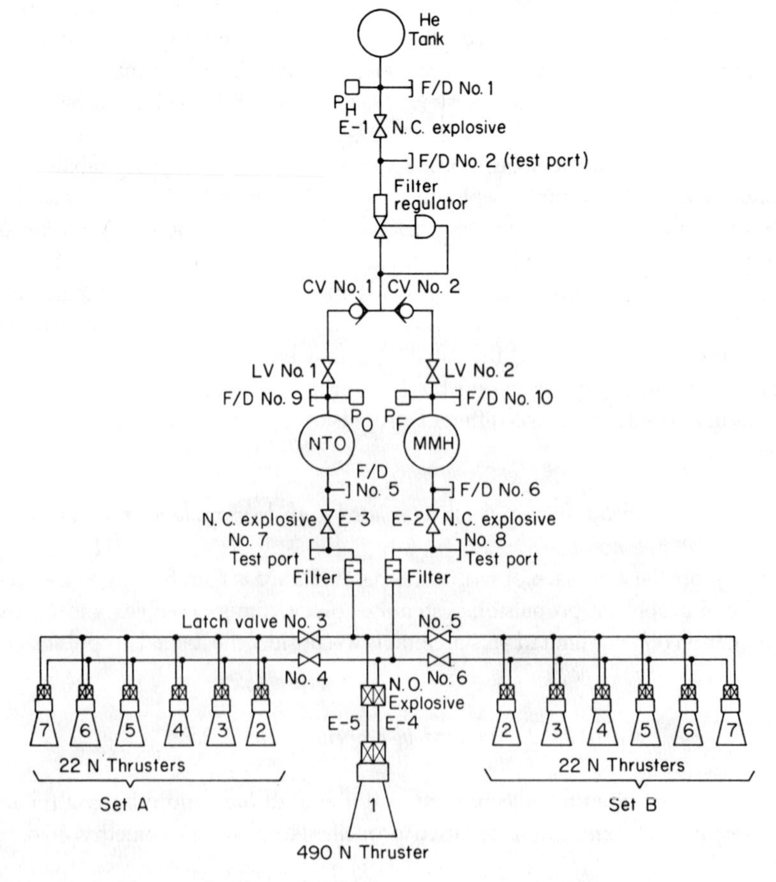

Figure 10.19 Unified bi-propellant propulsion.

Table 10.4

Item	Qty Per S/S	Nominal mass (kg) INSAT-1	Supplier
1. Propellent tanks	—	30.672	Lockheed
A. Propellent (MMH & NTO) tank	2	(29.443)	
B. Auxiliary fuel tank	—	(1.229)	
2. Pressurant (helium) tank	1	11.620	Brunswick
3. Apogee (490N) thruster	1	3.647	Marquardt
4. AOCS (22N) thruster	12	7.838	Marquardt
5. Pressurant (helium) module	1	2.840	Marquardt
A. Pyro valve (NC: normally closed)	1	(0.132)	Pyronetics
B. Regulator	1	(0.762)	Consolidated
C. Check valve (CV)	2	(0.308)	HTL
D. Latch valve (LV)	2	(0.572)	Marquardt
E. Pressure transducer LP	2	(0.379)	Paine
F. Mounting brackets/manifolds	A/R	(0.460)	FACC
G. Pressure transducer HP	—	(0.227)	Teledyne
6. Propellant (MMH & NTO) modules	2	2.832	Marquardt
A. Pyro valve (NC: normally closed)	2	(0.263)	Pyronetics
B. Filter	2	(0.282)	Wintec
C. Latch valve (LV)	4	(1.143)	Marquardt
D. Pyro value (NO: normally open)	2	(0.254)	Pyronetics
E. Mounting brackets/manifolds	A/R	(0.890)	FACC
7. Fill/drain valves (F/D)	8	0.768	Pyronetics
Total subsystem mass		60.217	

since a large proportion of the propellants is consumed during the apogee manoeuvres. The specific impulse obtained is of the order of 290 s.

There are many advantages:

—Dividing the apogee manoeuvre into several intervals permits accurate calibration and control of injection.
—When the transfer orbit is nominal, the propellants normally provided to allow for deviation from this orbit and not used by the apogee motor are available for attitude and orbit control and this enables the expected lifetime to be increased.
—Integration is facilitated by the presence of a single system.

The specific impulse is greater than that of other usable propellants but the dry mass of the propulsion system is greater. However, the benefit remains to the advantage of unified propulsion. For example, a satellite of 1240 kg in transfer orbit supports an additional 55 kg of payload with a unified system in comparison with a conventional solid apogee motor plus hydrazine (Table 10.5) [MOS-84].

Table 10.5 Bipropellant versus AKM + monopropellant AOC mass comparison (1240 kg spacecraft in transfer orbit)

	Bipropellant system (kg)	Solid AKM + monopropellant	Advantage (kg)
Expendables:			
Apogee manoeuvre (310 versus 285 I_{sp})	538	573	35
On-orbit needs (288 versus 220 I_{sp})	110	144	34
End-of-life mass:	592	523	69
Propulsion inerts:			
Helium	1.6	0.2	−1.4
AKM burn-out	—	34	34
Propulsion system	60	20	−40
Residuals	10	1.4	−8.5
Net S/C mass	520.4	467.4	53

10.3.4.4 *Unified bi-propellant propulsion combined with electric propulsion*

It is useful to consider the use of electric propulsion to provide north–south orbit control of the satellite. This control requires around 50 m/s for a geostationary satellite (see Section 7.3.3.4) and this represents more than 90% of the required velocity increments. Use of a high specific impulse propellant thus permits a large reduction of the global mass of the propulsion subsystem for missions with long lifetimes (greater than ten years). Figure 10.20 shows that, for a satellite of launch mass 2200 kg from Kourou, the use of electric propulsion is not beneficial as a consequence of the additional dry mass of the electric propulsion system for lifetimes of under six years; but, for a mission of ten years, combined use of unified bipropellant propulsion and electric propulsion enables a payload mass gain of 70 kg to be obtained [DUH-89].

The low thrust requires special strategies. Only correction of the secular component of drift of the inclination of the orbital plane is generally considered (see Section 7.3.4.5). The orientation of the secular drift is reasonably constant and manoeuvres are to be performed at two points of the orbit close to the perpendicular to the direction of the vernal point. The manoeuvres are, therefore, not performed during an eclipse.

An example of a possible strategy is as follows: when the inclination limit is reached, the manoeuvre is performed twice per day at the points mentioned above. Allowing for the modest thrust, several hours (3 or 4 hours) are necessary each time. The procedure is repeated on several days until the inclination reaches the limiting value in the opposite direction. The inclination then evolves freely for several days (up to 45 days) until the start of a new period of corrections.

The electric power required for operation of the electric thruster is readily available at the beginning of life without the need to overdimension the solar generators. The efficiency of solar cells decreases with time spent in orbit as a consequence of degradation due to high energy radiation (see Section 10.4). The generator is thus dimensioned to provide the nominal power at end of life, and an excess power (30%) is available at the beginning of life.

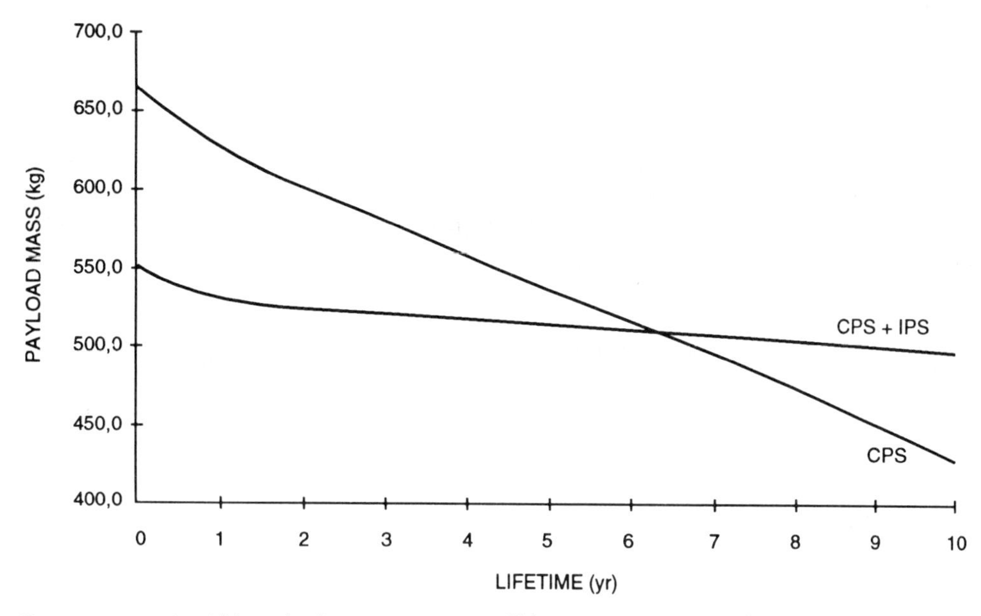

Figure 10.20 Available payload mass versus expected lifetime comparison with chemical propulsion system (CPS) only and use of electric propulsion for north–south station keeping.

Furthermore, it is possible to choose a strategy such that there are no north–south correction manoeuvres during the two periods of 45 days of eclipse around the equinoxes. Hence the part of the solar generator which is used to recharge the batteries during the eclipse periods (used only to provide trickle charge outside these periods) is available to provide the power required for operation of the electric thrusters. The use of electric propulsion, therefore, has only a limited influence on the dimensioning and mass of the electrical supply system.

The problem of úncertainty of the direction of the thrust vector of electric motors can be resolved by mounting the motor on a platform which permits orientation of the thrust direction. Calibration, however, requires accurate measurement of attitude variation during manoeuvres, particularly about the yaw axis. In view of the length (several hours) of the manoeuvres, use of a conventional integrating gyrometer is not possible due to the drift. The use of a solar sensor is one possible solution [DUH-89].

10.3.5 Example calculation

The mass m of propellant (specific impulse I_{sp}) required to impart a velocity impulse ΔV to a satellite of mass M is given equation (10.16). For a satellite of dry mass 500 kg equipped with a hydrazine propulsion system ($I_{sp} = 200$ s), north–south control during a 7 year period requires $\Delta V_I = 350$ m/s and hence a mass m_I of propellant of 88 kg. The longitude correction, for which ΔV_L is of the order of 7 m/s, leads to a mass $m_L = 1.6$ kg. Control of orbit drift and attitude require of the order of 10 kg in addition. The mass of the reservoirs, the pipework and the nozzles, which is of the order of 30 kg, must be added to the total mass of hydrazine required. In total, the global mass of the propulsion subsystem is of the order of 130 kg.

10.4 THE ELECTRIC POWER SUPPLY

In view of limitations in mass and volume, the electric power supply of a satellite poses one of the most restricting problems. The increase of equivalent radiated power necessary for the use of small earth stations means that the electric power required reaches several kilowatts for communication satellites, particularly those intended for direct broadcasting of television programmes or for links with mobiles. The electric power to be provided is directly related to the radio-frequency power of the amplifiers in the payload as a function of their efficiency.

The electric power supply subsystem consists of the following:

—A primary source of energy which supports conversion of energy available in another form into electrical energy (for civil applications it consists of a solar generator).
—A secondary source of energy which is substituted for the primary energy source when this cannot fulfil its function, for example in an eclipse period (such as a battery of electrochemical accumulators).
—Conditioning (regulation and distribution) and protection circuits.

10.4.1 Primary energy sources

The only external source is solar radiation. On-board sources of energy (nuclear piles or combustible materials) are not at the present technologically satisfactory for performing the mission of a geostationary communication satellite (CCIR Rep 673). However, during the first minutes following injection into the transfer orbit phase, on-board electrochemical accumulators, which will subsequently constitute the secondary energy source, play the role of a primary energy source.

10.4.1.1 Characteristics of solar radiation

The characteristics of solar radiation are presented in Chapter 12. The normalised solar flux at a distance of 1 astronomical unit is of the order of $1353\,W/m^2$. However, the value resulting from recent estimates is of the order of $1370\,W/m^2$. The solar flux captured by a surface perpendicular to the plane of the equator evolves in the course of the year as a function of variation of the earth–sun distance and the declination of the sun with respect to the equatorial plane (see Figure 12.2).

The sun can be considered to be a black body at 6000 K and spectral radiation is maximum around 0.5 micrometers; 90% of the power radiated is concentrated between 0.3 and 2.5 μm (see Figure 12.1).

10.4.1.2 Solar cells

Solar cells operate according to the principle of the voltaic effect (the appearance of a voltage at the connections to a p–n junction subjected to a photon flux).

Current–voltage characteristic. An example of the current–voltage characteristic as a function

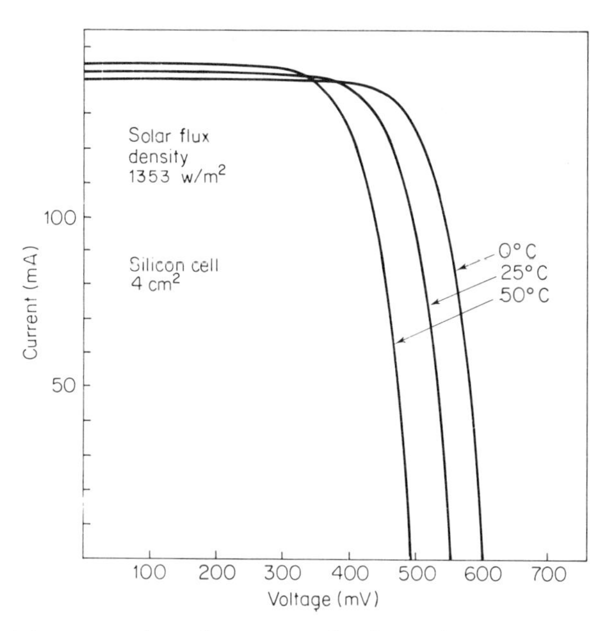

Figure 10.21 Typical current–voltage characteristic of a silicon solar cell.

of the load due to the circuit fed by the cell is represented in Figure 10.21 for a 2 cm by 2 cm silicon cell. The incident solar flux is assumed to be normal to the surface and equal to the normalised value (1353 W/m^2). It is of course necessary to take account of the angle between the normal to the surface and the direction of the sun; the flux actually captured is a function of the cosine of this angle (for angles which are not too large, i.e. less than 45°).

Maximum power is obtained when the product $V_c I_c$ is maximum, that is in the region of the knee of the characteristic. The maximum power and open circuit voltage depend on temperature (the open circuit voltage falls by 50% if the temperature rises from 27 to 150 °C).

Conversion efficiency. The mean efficiency at the point of maximum power of a conventional silicon cell subjected to solar radiation above the atmosphere at a temperature of 27 °C is from 10 to 15%. The efficiency decreases under the effect of radiation; a decrease of 30% in seven to ten years is typical for a satellite in geostationary orbit. The magnitude of the degradation depends on the type of orbit concerned, the mean solar activity during the period concerned and the occurrence of solar flares (see Chapter 12). Dimensioning of the solar generator must allow for degradation of the initial efficiency over the anticipated lifetime.

In order to limit degradation, the cell is protected by a covering which is transparent to the longer wavelengths for which the sensitivity of the cell is greatest but capable of attenuating the damaging part of the radiation. This window is realised in quartz or fused silica.

Technology. The material currently used for solar cells is silicon. Constant progress has enabled the efficiency of the cells to be increased and the mass decreased.

Current silicon cells are realised in a relatively thick monocrystalline chip of 200 to 280 μm. Thinner cells whose thickness is of the order of 50 to 80 μm have been developed in order to reduced the mass [TAK-86]. The increase of efficiency in the course of time (from less than

10% in the 1960s to around 15% at present has been obtained by optimisation of the doping of the semiconductor material, anti-reflecting surface treatment in order to favour penetration of solar light and the use of a reflecting deposit on the back face in order to make photons which have not given up their energy during initial passage through the cell pass through it again (back surface reflector (BSR)).

The use of gallium arsenide (GaAs) enables higher values of efficiency, which exceed 18%, to be obtained. Table 10.6(a) compares the characteristics of GaAs cells with those of a Si cell [TAK-86]. However, the use of these cells has remained experimental as a consequence of the relatively high mass of the cells (the density and thickness are greater); there is no advantage in reducing the number of cells, and hence the area, of the solar generator if it is heavier. Furthermore, gallium arsenide remains a high cost material so that mass production processes have not been brought into operation.

At the present time, mass production of GaAs cells comparable with that of silicon cells is possible. Table 10.6(b) illustrates the benefit of using GaAs cells with respect to silicon cells [YOS-88]. Furthermore, the resistance of the cells to radiation is greater than that of silicon cells. Overdimensioning of the solar generator at start of life is thus less (20%).

Table 10.6(a) Characteristics of GaAs solar cells

	MELCO [TAK-86]	MELCO [YOS-88]	Spectrolab [86]	Si
Size (cm^2)	2 × 2	2 × 4	2 × 4	2 × 2
Thickness (μm)	280	200	300	280
Mass (g)	0.8	1.15	1.6	0.48
Voc (V)	0.97	0.98	1.01	0.56
Isc (mA)	127.2	256.8	228	170
P_{max}(mW)	95.2	193.8	189.6	76
η (%, BOL)	17.5	18.9	17.5	14.1
Hardness*	0.78	0.78	0.78	0.68

*Radiation hardness: efficiency remaining factor (efficiency normalised to the initial value) after 1 MeV electron irradiation at a fluence of 1×10^{15} e/cm^2.

Table 10.6(b) Comparison of solar array performance and cost on GaAs and Si solar cells

	Case 1 4.3 ~ 5.1 kW (GEO, EOL/10 years)				Case 2 5.2 ~ 6.2 kW (LEO, EOL/15 years)			
Cell type	GaAs	GaAs	Si	Si	GaAs	GaAs	Si	Si
Efficiency	18.6%	20.1%	14.3%	13.3%	18.6	20.1%	14.3%	13.3%
Thickness	200 μm	200 μm	200 μm	50 μm	200 μm	200 μm	200 μm	50 μm
Specific weight (W/kg)	36.0	39.2	29.7	32.2	42.7	46.1	35.5	38.9
Power/area (W/m^2)	113.9	123.9	77.9	83.2	139.5	150.3	96.6	94.3
Power cost ratio (Cost/W)	1.1	1.0	1.0	1.3	1.1	1.0	1.0	1.4

The area of a cell in silicon or GaAs is of the order of 4 to 36 cm² (2 cm × 2 cm, 2 cm × 4 cm, 2 cm × 6 cm, 6 cm × 6 cm).

10.4.1.3 *Solar panels*

Several thousands of cells are interconnected in order to deliver the power P required. They are bonded to panels which provide the necessary rigidity and thermal regulation. The filling ratio f which characterises the ratio of the area occupied by the cells to the total area of the panel is of the order of 90%.

Interconnection of cells. The cells are connected in series and parallel in order to deliver a voltage of some tens of volts (e.g. 42 volts for INTELSAT V) and a current of several tens of amps. The potential difference V to be delivered at the generator terminals determines the number of cells to be connected in series; if V_c is the potential difference corresponding to the chosen operating point (of the order of 0.5 V), the number of cells in series is equal to V/V_c.

The number of branches in parallel depends on the current $I = P/V$ to be delivered; if I_c is the current corresponding to the chosen operating point (for example of the order of 0.14 A for a cell of 4 cm²), the number of branches to be connected in parallel is equal to I/I_c.

This basic organisation is modified in order to minimise the consequences of cell breakdown and the effect of shadow (due to the satellite body or antennas on the generator). An open circuit breakdown of one cell in a branch leads to loss of the whole branch since electrical continuity is no longer provided. This failing can be avoided by connecting groups of cells in parallel. In contrast, a cell in one branch becoming short circuit means that the electromotive force of this branch becomes less than that of all the others. The current distribution will, therefore, be unbalanced with a danger of insulation breakdown due to localised thermal dissipation. A diode in series with each branch of cells enables the defective branch to be isolated. The generator thus consists of small groups of cells arranged in parallel and series; the choice of series–parallel combination is such that it maximises the overall reliability, taking into account the relative failure rates associated with cell failure in a short- or open-circuit state (see Chapter 13).

A non-illuminated cell in a branch behaves as a load for the other cells. The current which passes through it can involve an excessive thermal dissipation which leads to insulation breakdown. Protection is provided by placing parallel diodes on one or more cells along the branch. Figure 10.22 shows the principles of solar cell arrangement.

Dimensioning. The power P_c delivered by a solar cell (of area s) is expressed by:

$$P_c = \phi e s(1 - l) \qquad \text{(W)} \tag{10.24}$$

where ϕ is the solar flux captured by the cell (W/m²), e is the efficiency of the cell (10 to 15% at start of life for a Si cell) s is the surface area of the cell (m²) and l are losses (%) due to cover, cabling etc. (a typical value is 10 to 15%).

The solar flux captured depends on the illumination conditions and is obtained from the nominal solar flux $W = 1370$ W/m² as a function of the distance d to the sun and the angle θ between the normal to the cell and the direction of the sun:

$$\phi = W(a^2/d^2)\cos\theta \tag{10.25}$$

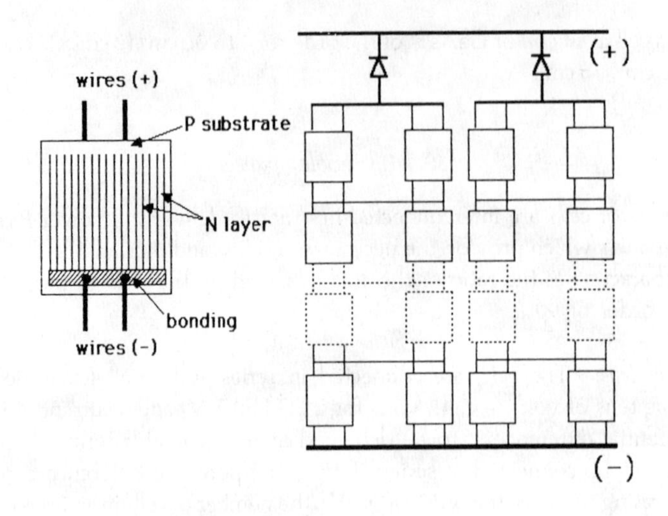

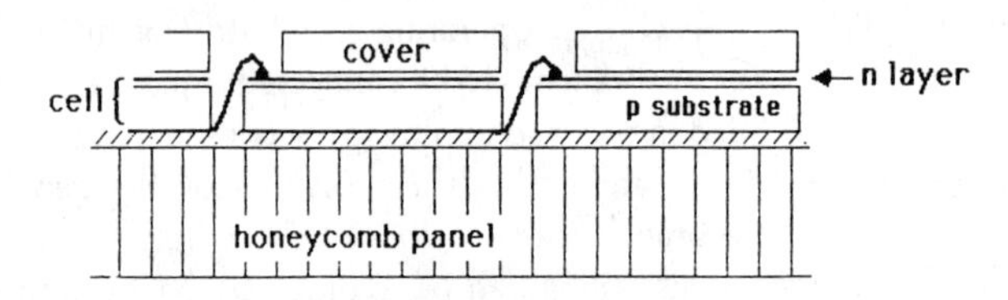

Figure 10.22 Arrangement of a solar generator.

where a is the mean sun–earth distance = 1 AU. Variation of the ratio a^2/d^2 as a function of the date in the year are given in Section 12.1.4. The efficiency of the cell varies as a function of the extent of the degradation caused by high energy radiation. The cell manufacturer provides values of the degradation of efficiency for various values of radiation dosage and this enables account to be taken of the estimated dosage for a particular mission to determine the magnitude of the degradation. In the absence of precise data, the degradation of efficiency can be modelled to a first approximation by an exponential law for a satellite in geostationary orbit. For example, the following is suggested:

$$e_{EOL} = e_{BOL}[\exp(-0.043T)] \tag{10.26}$$

where T is the time spent in orbit in years.

The surface area A of the solar panel required to generate a given power P is given by:

$$A = (P/P_c)s/f = ns/f \tag{10.27}$$

where P_c is the power delivered per cell (this depends on the illumination conditions), n is the number of cells required and f is the coefficient of filling (75 to 90%).

The power nP_c delivered by the solar generator varies with time. The requirements of the satellite also very with time. Dimensioning of the solar generator must, therefore, be performed for the least favourable conditions.

In the case of a geostationary satellite, the power delivered by the generator is lowest at the summer solstice. At the equinoxes, the power is greater but must recharge the battery since the satellite enters eclipse periods. It is necessary to check that the dimensioning does allow for the summer solstice.

Satellites stabilised by rotation. On spin-stabilised satellites, the solar cell panels form the exterior envelope of the body of the satellite. Additional cylindrical panels can be deployed after launching to increase the useful surface area (for example Intelsat VI, Figure 10.11).

The number of cells required is large since they are not all illuminated by the sun at the same time. By considering a cylinder with its axis normal to the direction of radiation, the variation of incidence between cells located on the illuminated side leads to a surface area which must be $\pi/2$ times greater than a plane perpendicular surface to obtain the same power. Including the surface in shadow, the number of cells to be installed is thus π times greater than if the generator were plane.

In practice, the different operating conditions of the cells limit this factor to a value between 2 and 2.5. In the course of rotation, following passage into shadow, the mean operating temperature of the cells is lower and the efficiency is higher. Furthermore, degradation due to solar radiation is appreciably less.

Three-axis stabilised satellites. Various types of solar cell panel can be used with three-axis stabilised satellites:

—Flexible panels which are rolled up during launching in a storage container and unrolled in orbit by a deployable mast.
—Semi-rigid hinged panels folded in concertina fashion during launching in a storage container and extended in orbit by a deployable mast (e.g. the Olympus satellite [BON-84]).
—Rigid panels of large dimensions (compatible with the capacity of the launcher) joined in groups of three or four to constitute a solar generator wing. The joint is realised by means of a hinge arrangement which permits folding for launching. Deployment in orbit is achieved by means of a set of springs, cables, pulleys and velocity regulators in order to ensure co-ordinated movement of the panels without shocks.

Once deployed, the solar generator wings are rotated in order to follow the apparent movement of the sun about the satellite. For a geostationary satellite, the generator wings are aligned with the pitch axis and rotation is daily.

Operation of an orientation device requires the following:

—solar sensors with electronic measurement and control circuits,
—a drive motor with sliding contacts to transfer the current to the satellite (a bearing and power transfer assembly (BAPTA), Figure 10.23).

Specific performance. Assuming best use of the cells, the mass balance of flat solar panels compared with panels mounted on the body of a rotation stabilised satellite remains in favour of flat solar arrays.

By way of indication, the panels of solar cells mounted on the body of a rotation stabilised satellite enable 30 to 35 W/m^2 and 8 to 12 W/kg to be obtained and the specific mass is of the order of 3 to 5 kg/m^2 (INTELSAT VI; 2 kW after 10 years, 59 m^2, 250 kg) [BRO-84]. For

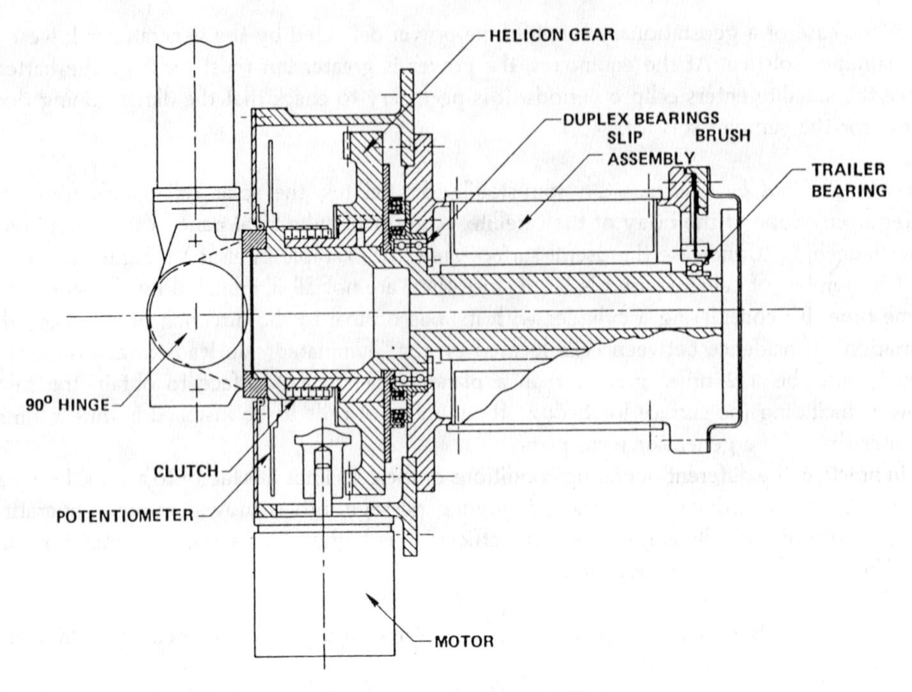

Figure 10.23 Solar generator bearing and power transfer assembly.

flat solar panels, the performance is of the order of 100 W/m^2 and 30 to 40 W/kg (Eutelsat II; 3 kW, 30 m^2, 90 kg) [KRA-84], [CHO-88].

Work in progress has the goal of increasing the permissible dimensions of deployable solar panels and reducing the weight. It is hoped to achieve specific powers of 100 W/kg and power levels of several tens of kW (using flexible deployable solar panels for satellites designed for mobile communications [FAY-86].

10.4.2 Secondary energy sources

The secondary energy source takes its energy from the primary energy source when this is operational (the storing function) and returns the stored energy when the primary energy source ceases to operate.

Batteries of electrochemical accumulators are the most appropriate means of providing this function. They play a particularly important role in the case of communication satellites for which operation during an eclipse is expected as a consequence of the availability objectives usually specified. Recall that, for a geostationary satellite, eclipses occur on 90 days per year and have a duration which can be as long as 70 min (see Chapter 7).

The following qualities are sought:

—Adequate lifetime which depends on the depth of discharge and the temperature.
—High specific energy in terms of Wh/kg.

10.4.2.1 *Characteristic parameters*

The characteristic parameters of an accumulator are as follows:

—Capacity C (A h), characteristic of the product of the current drawn and the time of use.
—Terminal voltage of the charged unit (volt).
—Specific energy (Wh/kg), the energy stored per unit mass.

In use other parameters arise:

—Mean discharge voltage V_d (volt), which depends on the intensity of the discharge current.
—Depth of discharge (DOD), which characterises the percentage of the stored energy which is effectively used at the end of the longest period of use without recharging.
—Charge efficiency η_{ch}, the ratio of the energy stored to the energy consumed for recharging.
—Discharge efficiency η_d, the ratio of recovered energy to that part of the stored energy which has been used.

The depth of discharge DOD is the only parameter directly defined by the user. The choice of the depth of discharge is dictated by the expected lifetime of the battery or more precisely by the number of charge and discharge cycles to be obtained. As the depth of discharge decreases, the number of charge and discharge cycles which the battery can support increases (Figure 10.24).

10.4.2.2 *Dimensioning*

Dimensioning the battery in terms of the energy to be provided is determined by the capacities of the components available on the market. The energy E_c recovered from an accumulator element of capacity C as a function of the parameters defined above is given by:

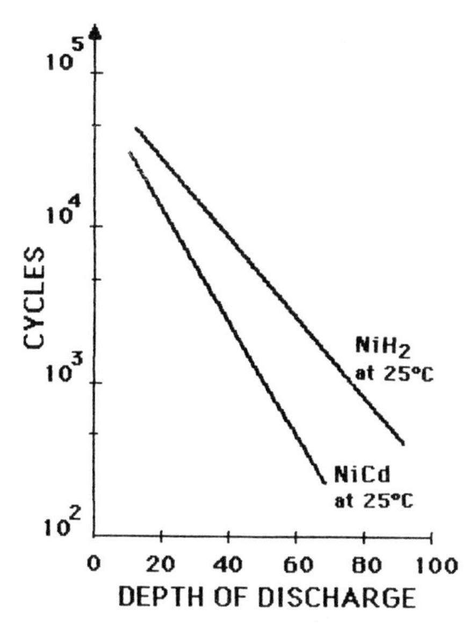

Figure 10.24 Number of cycles as a function of the depth of discharge.

$$E_c = CV_d \mathrm{DOD} \eta_d \qquad (\mathrm{W\,h}) \qquad (10.28)$$

The battery consists of n elements in series and the energy recovered is thus equal to nE_c. The number of elements n in series is chosen in such a way that nV_d is just greater than the voltage V to be obtained during discharge:

$$n = \text{integer} \geqslant V/V_d \qquad (10.29)$$

One element is often added to provide the proper voltage even in the case of short circuit breakdown of a battery element.

Let P be the power to be provided for the duration of an eclipse T_{ecl} (hour). The energy which the battery must have delivered after time T_{ecl} is thus given by:

$$E = PT_{ecl} \qquad (\mathrm{W\,h})$$

The capacity C of the battery elements under these conditions is given by:

$$C = PT_{ecl}/nV_d \mathrm{DOD}\, \eta_d \qquad (\mathrm{A\,h}) \qquad (10.30)$$

The calculation rarely leads to a capacity value corresponding to that of a component available on the market. It is, therefore, necessary to choose elements of a capacity just greater than that required. This leads to a battery which is overdimensioned with respect to the requirements thereby imposing a mass penalty. If the voltage to be delivered is not imposed, the number of elements n can be altered to provide the nominal power for an optimum battery mass. Note that the energy to be delivered during the eclipse is usually split between several batteries (typically two) to ensure some form of redundancy and facilitate integration in the satellite (see Section 10.4.3.6).

10.4.2.3 *Technologies*

Nickel-cadmium cells have been used since the advent of communication satellites for the storage of electrical energy. Used for the first time on the NTS 2 satellite in 1974, nickel-hydrogen cells are replacing nickel-cadmium on current satellites by reason of their higher specific energy and greater lifetime [JOY-86].

A higher specific energy permits the use of a lighter battery for the same electrical power delivered during an eclipse. This is obtained, in particular, as a consequence of the greater depth of discharge permitted with an NiH_2 battery than with a NiCd battery for the same number of cycles (Figure 10.24). For example, for a mission in geostationary orbit, the maximum depth of discharge for a NiCd battery is 50%; with an NiH_2 battery the depth of discharge can reach 75%.

Furthermore, NiCd batteries suffer from a performance limitation due to progressive deterioration of their constituents particularly the separators between electrodes. They are also as sensitive to overcharging as they are to excessive discharging. NiH_2 batteries are less subject to these limitations. On the other hand, the relatively high pressure which exists in the constituents requires a form of container which makes interconnection of the cells more difficult than in the case of parallelepiped shaped NiCd cells (Figure 10.25).

Other electrochemical combinations can be considered for particular applications; examples are the rechargeable silver-zinc batteries used on Ariane and silver-hydrogen batteries ($Ag{-}H_2$) which provide high specific energies, but a limited lifetime, for low orbit satellites. Table 10.7

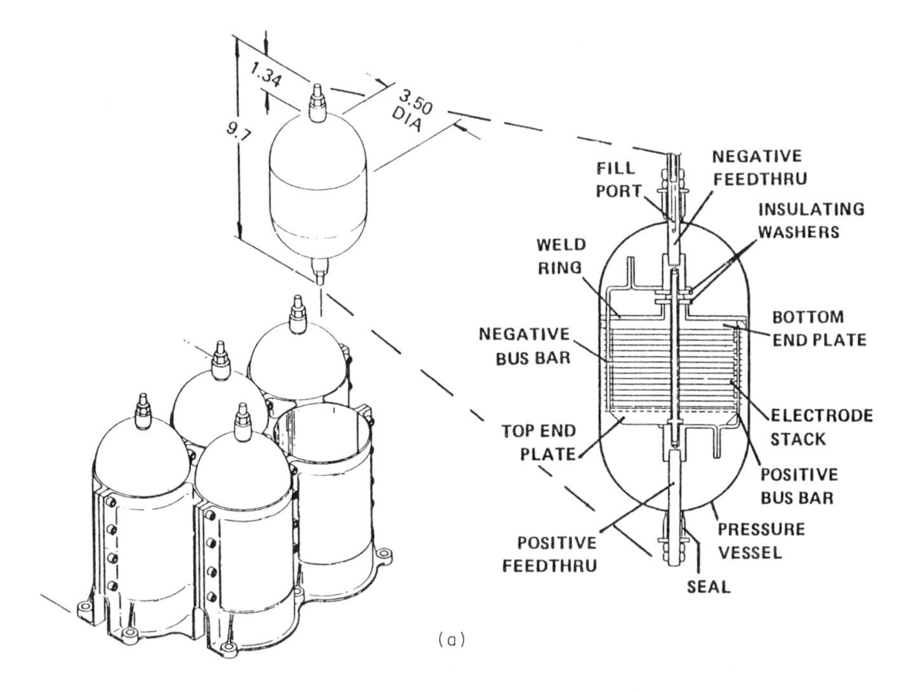

(a)

(b)

Figure 10.25 Assembly of NiH_2 batteries (courtesy of British Aerospace).

summarises the performance of various types of electrochemical cell; the performance indicated corresponds to a depth of discharge of 100%. Other types of cell are being examined; these include lithium and sodium, which require high operating temperatures (350 °C), hydrogen–oxygen and hydrogen–bromide [FAY-86].

10.4.2.4 Operation of the battery

Battery charging. Battery charging is generally performed at constant current, the current I_{ch} in amps being such that:

$$I_{ch} = C/10 \text{ to } C/15 \qquad (10.31)$$

where C is the capacity expressed in A h.

The recharge time T_{ch} is a function of the energy $E = PT_{ecl}$ (W h) provided during the eclipse, the battery terminal voltage V_{ch} during charging, the charging current I_{ch} and the charging efficiency η_{ch} (0.75 to 0.9):

$$T_{ch} = PT_{ecl}/I_{ch}V_{ch}\eta_{ch} = C\,\mathrm{DOD}/I_{ch}\eta_{ch} \quad \text{(h)} \qquad (10.32)$$

It must be verified that the charge time is less than the time between eclipses (22.8 h for a geostationary satellite). The power required for the recharge is given by:

$$P_{ch} = I_{ch}V_{ch}/\eta_{reg} \quad \text{(W)} \qquad (10.33)$$

where η_{reg} is the efficiency of the charging regulator. In the absence of a regulator (an unregulated bus), $\eta_{reg} = 1$.

Care must be taken not to overcharge the battery; protection against overcharging can be provided by limiting the terminal voltage at the end of charging; that is by terminating the charge at constant voltage. Another charging technique under examination consists of operating the end of charge at constant temperature [SPR-85]. This technique has the advantage of a better charging efficiency, better precision of the end of charge and an increase in the lifetime of the battery. Once the battery is charged, a continuous current of the order of $C/75$ to $C/50$ compensates for the phenomenon of self-discharge of the battery (trickle charge).

Battery discharge. The battery discharge current can be relatively large and is limited by heating problems associated with the internal resistance. A value of the order of $C/2$ to $C/5$ is a good compromise which enables an excessively rapid fall of voltage at the end of discharge to be avoided. The mean voltage during discharge depends on the current; it is of the order of 1.2 V per cell for NiCd and 1.3 V per cell for NiH_2 (Figure 10.26). At the end of the discharge, the voltage decreases rapidly. It is then necessary to stop the discharge (at a minimum voltage of the order of 0.7 V) in order to avoid the effects of polarity inversion.

Operating temperature; reconditioning. The temperature range within which the battery elements must be maintained is narrow, typically 0 to 15 °C. Too low a temperature affects the performance; a high temperature limits the lifetime. Furthermore the battery gives off heat during discharge and cools during charging. Thermal control is thus designed so that the battery can radiate heat and electric heaters limit the fall of temperature during recharging.

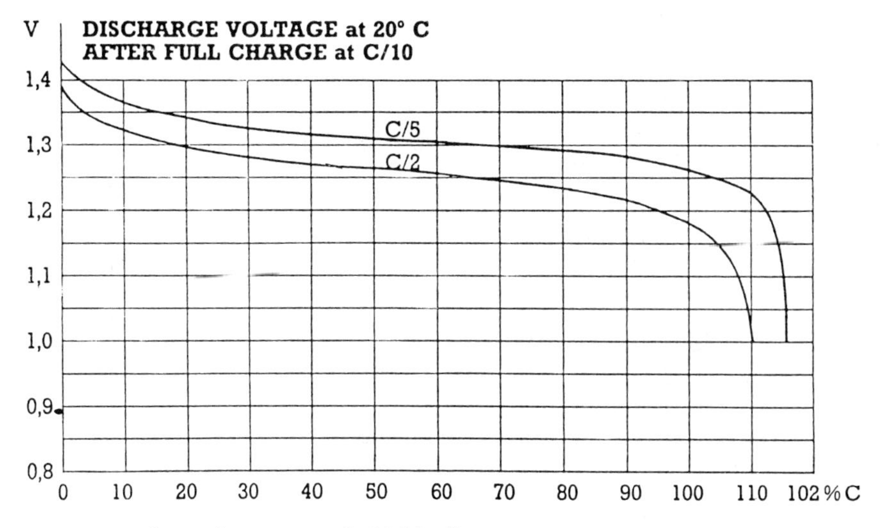

Figure 10.26 Discharge characteristic of a NiCd cell.

After a long period of non-use (with the battery fully charged but not used), it is necessary to perform reconditioning in order to restore the nominal performance of the battery. The battery is first completely discharged at low current. The battery is then fully recharged at low current; one or more charge–discharge cycles can be added.

10.4.3 Conditioning and protection circuits

The voltage delivered by the primary energy source (the solar generator) depends on the operating point. The power available depends on the incident solar flux and the degree of degradation of the solar cells due to irradiation.

During an eclipse, the temperature of the solar generator decreases and can fall to − 180 °C. After the eclipse the voltage delivered by the cells is around 2.5 times the nominal value which corresponds to the equilibrium temperature (of the order of a few tens of degrees in the illuminated phase). As far as the battery is concerned, the voltage delivered by a cell depends strongly on the state of charge. The battery terminal voltage can vary from 15 to 30% between the start and end of an eclipse according to the depth of discharge realised.

It is, therefore, necessary to provide conditioning circuits for the electrical energy intended for distribution to the equipment and to provide battery charging. The circuits will be associated with those for control and protection. Ohmic losses must, of course, be minimised in power distribution; evaluation of the losses associated with the various stages of regulation and distribution, to which ohmic losses are added, shows that, in certain cases, a third of the power generated by the solar generator is effectively used by the payload [CAP-85].

The system architecture results from a compromise between two extremes; the bus whose voltage is not regulated either in or out of eclipse ('night' or 'day'), and the bus whose voltage is regulated day and night.

Table 10.7 Characteristics of different storage cells

Type of cells	Electrolyte	Nominal voltage cell (V)	Energy Density (WH/kg)	Temperature (°C)	Cycle life at different depth of discharge levels			Whether space qualified
					25%	50%	75%	
Ni–Cd	Diluted potassium hydroxide (KOH) solution	1.25	25–30	−10 −40	21 000	3 000	800	Yes
Ni–H_2	KOH solution	1.30	50–80	−10 −40	> 15 000	> 10 000	> 4 000	Yes
Ag–Cd	KOH solution	1.10	60–70	0–40	3 500	750	100	Yes
Ag–Zn	KOH solution	1.50	120–130	10–40	2 000	400	75	Yes
Ag–H_2	KOH solution	1.15	80–100	10–40	> 18 000	—	—	No
Pb–Acid	Diluted sulphuric acid	2.10	30–35	10–40	1 000	700	250	—

10.4.3.1 *Unregulated bus*

The schematic arrangement is presented in Figure 10.27. The solar generator feeds the platform equipment directly and the payload by way of a distribution box. The operating point is defined by the intersection of the current–voltage characteristic of the generator and that representing the equipment load (a hyperbola representing operation at constant power P_E). Figure 10.28 illustrates these characteristics and the chosen operating (point N).

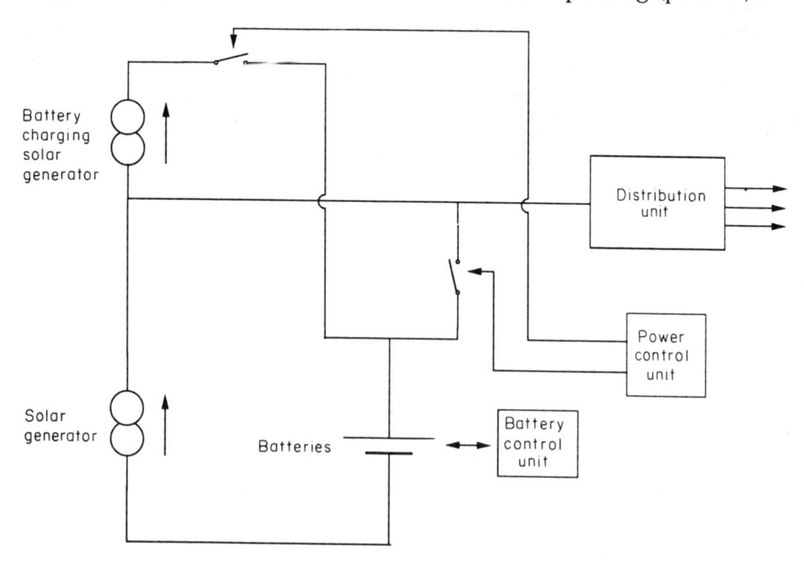

Figure 10.27 Unregulated bus power supply.

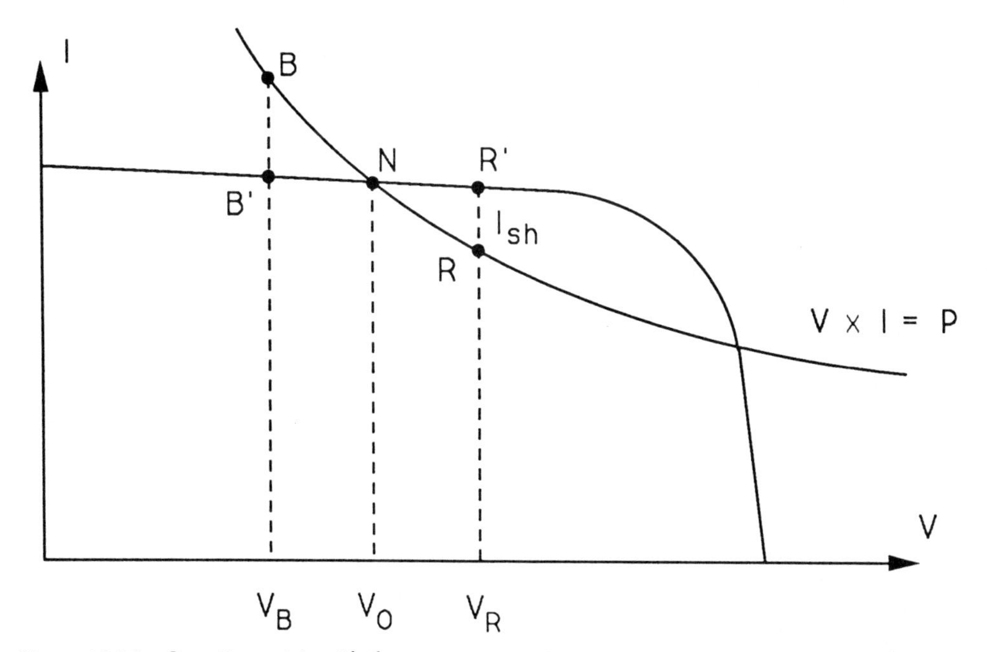

Figure 10.28 Operating points with the various types of voltage regulation buses.

The battery is connected through a switch either to a recharging solar generator in series with the main generator (out of eclipse) or to the supply bus during eclipse periods. Out of eclipse, the bus terminal voltage varies in accordance with the evolution of the characteristics of the solar generator and the power consumed.

When the satellite is in eclipse, the battery, which then provides the power, determines the potential V_B of the bus which feeds the equipment (point B; the battery has a characteristic of the voltage source type). The terminal voltage decreases with time during the discharge.

The unregulated bus has the advantage of simplicity and hence good reliability. On the other hand, the equipment is subjected to large voltage variations (10 to 40%). Some equipment may be designed to accept such variations of supply voltage (e.g. equipment to supply travelling wave tubes (TWT), others require a voltage converter and regulator at the point where the electrical energy is distributed to the equipment concerned (for both the payload and the platform).

10.4.3.2 *Sun regulated bus*

In order to limit voltage variations for most of the time, voltage regulation outside periods of eclipse can be considered. The organisation of the system is presented in Figure 10.29. The solar generator feeds the equipment at constant voltage by means of a voltage regulator. Outside eclipses, the equipment supply voltage is thus kept constant (equal to V_R) within a range of a few per cent depending on the performance of the regulator. Two operating points are now defined on Figure 10.28; these are that of the solar generator, point R, and that of the load, point R′. The segment RR′ represents the current I_{sh} shunted through the regulator.

A charge regulator connected to the bus provides recharging and maintains constant battery current outside eclipses (the bus voltage is greater than the battery voltage). When the satellite is in eclipse, the battery is directly connected to the terminals of the bus which feeds the equipment (point B). The bus voltage imposed by the battery voltage V_B then changes with the discharge of the battery.

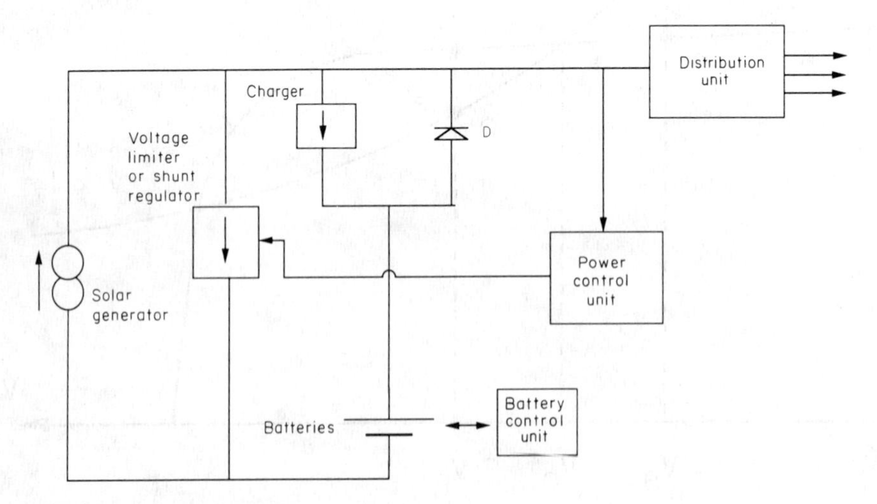

Figure 10.29 Sun regulated bus power supply.

The system daytime regulation remains relatively simple in concept and in operational use. The battery supply to the bus is provided without intervention (due to the diode) when the bus voltage falls below the battery voltage following a temporary increase of demand or entry of the satellite into eclipse. Voltage variations outside periods of eclipse are limited. On the other hand, the equipment is always subjected to variations of battery terminal voltage during discharge.

10.4.3.3 Regulated bus

Day and night voltage regulation is obtained by decoupling the battery from the bus by means of a discharge regulator (Figure 10.30). Out of eclipse, a regulating circuit fixes the potential of the solar generator and the supply bus to which the equipment is connected. During an eclipse, the battery provides the power to the load by way of the discharge regulator which keeps the bus terminal voltage constant and equal to V_R (point R on Figure 10.28).

The advantages of operating equipment at constant voltage are obtained to the detriment of system complexity and hence a potential reduction of reliability. Problems of electromagnetic compatibility can arise when the equipment is connected directly to the supply bus. Depending on the impedance of the system, variation of the current consumed by equipment may lead to voltage variations which can cause coupling between systems or noise; this applies particularly with operation in time division multiple access mode which involves current demands by the amplifiers at the frame rate.

10.4.3.4 Other architectures

The three architectures presented are not the only possible ones and other combinations can be conceived. Hence, it is possible to use one of more dedicated elements of the solar generator

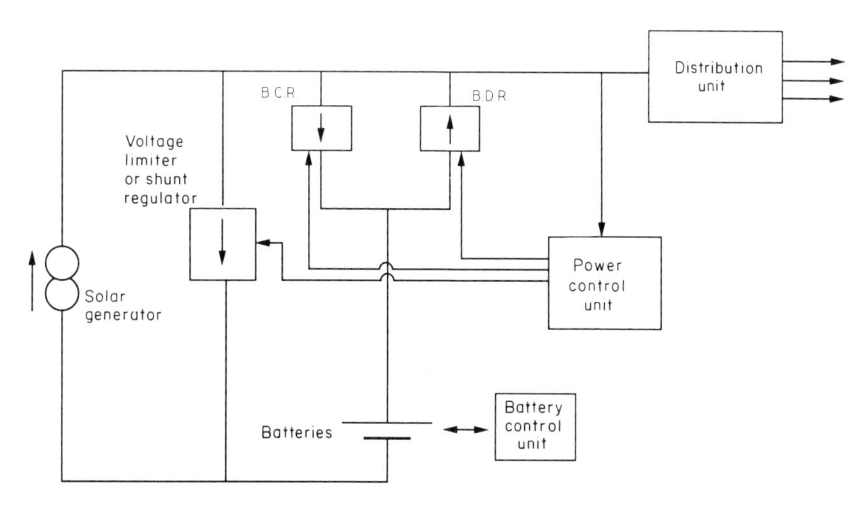

Figure 10.30 Regulated bus power supply.

for battery recharging instead of taking power from the main supply bus in the case of daytime, or day and night, regulated buses.

Another example (the TELECOM 1 satellite) involves use of the battery on exit from an eclipse as a buffer to limit the voltage increase with an unregulated bus; the voltage increase is due to the excess power delivered by the solar generator at low temperature.

It is also possible to define an architecture where the solar generator, the battery and the equipment are permanently in parallel. The battery thus feeds the equipment automatically on entry into an eclipse. It recharges itself on exit from the eclipse by imposing its voltage on the bus terminals. Once charged, the terminal voltage is relatively constant and the battery plays the role of a buffer. A shunt resistance connected in parallel with the battery by means of a switch enables any excess current to be absorbed.

A voltage regulator of the shunt type is usually used since, on the one hand, this type of regulator is well suited to the behaviour of a source of the current generator type and, on the other hand, it avoids any voltage drop between the generator and the equipment. However, due to the appearance of semiconductor components having a low voltage drop in the on state (such as HEXFET transistors), the use of regulators of the series type can be considered. The principle of the two types is illustrated in Figure 10.31.

10.4.3.5 *Comparison of the various architectures*

In addition to the advantages and weaknesses indicated previously, unregulated systems suffer during eclipse from a lock-out phenomenon on the battery voltage which necessitates over-dimensioning the solar generator in order to escape from the resulting stable state without interruption of the power supplied to the equipment [CAP-85].

For the first two architectures, on passing through an eclipse the solar generator characteristic disappears from Figure 10.28 and the operating point of the load moves from point P or R′ towards point B. At the end of the eclipse, the solar generator characteristic reappears and the operating point of the solar generator is then point B′. The solar generator thus provides a power $V_B I_G$; the balance of the load power P_E is borne by the battery (BB′V_B). This situation corresponds to a stable state with no exit (in order to be able to disconnect the battery) except by cancelling this balance of power. This means displacing the operating point of the generator from point B′ to point B and hence the availability of a generator capable of providing a larger power than that strictly necessary out of eclipse as defined by the operating points N and R.

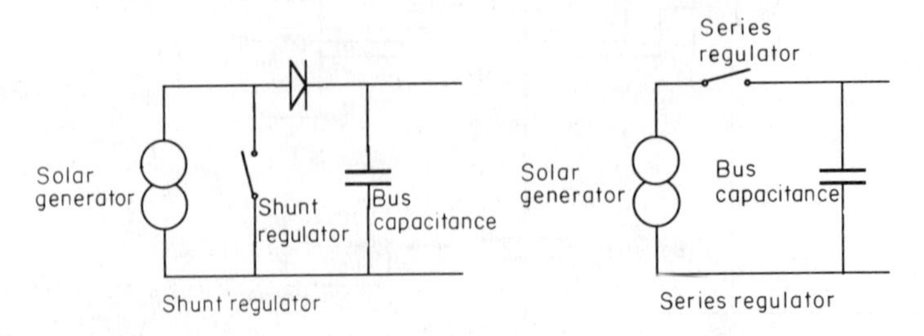

Figure 10.31 Architecture of shunt and series regulators.

One criterion in choosing between the various configurations is also the overall mass and efficiency of the power supply subsystem. A comparative study of the performance of the subsystem concerned for various satellites using different architectures indicates that the use of a regulated bus provides the best performance in terms of global mass and efficiency [CAP-85]. Even if the regulated bus requires additional equipment, its mass is compensated by the elimination of converters and regulators in the distribution of power to equipment. The operating point of the solar generator is optimised as a consequence of decoupling of the battery and operation at constant voltage.

Furthermore, with a regulated bus, the battery voltage is no longer directly determined by the bus supply voltage and can thus be chosen to optimise the number of elements in accordance with available capacities. The battery voltage can be chosen to be above or below the bus voltage. A parametric study of the various configurations by means of software simulation confirms that the day-and-night regulated bus architecture is the most useful in terms of mass for any power level up to 6 kW [LAC-89].

Results show that it is more useful to recharge the battery with a dedicated section of the solar generator than to extract the power required for recharging from the bus by means of a charging regulator as described in Section 10.4.3.3. Finally (below 4.5 kW), it is preferable to choose a battery voltage less than the bus voltage whatever the type of battery (NiCd or NiH_2). In contrast, to provide a power greater than 4.5 kW, a battery voltage higher than the bus permits a reduction of mass.

Finally, it is clear that the choice of a sufficiently high nominal bus voltage, which leads to lower cable currents for a given power, improves the system efficiency by reducing losses due to resistive dissipation. Voltages from 42 to 50 volts are traditional in Europe, while progression of the 28 V standard in the USA towards higher voltages is evident [FAY-86].

10.4.3.6 *Redundancy, protection and distribution circuits*

Redundancy of the main elements of the electric supply subsystem (the solar generator and the battery) is impossible on account of the mass and large bulk. To minimise the chance of catastrophic breakdown and loss of the whole subsystem due to failure of one element, the subsystem on a communication satellite is generally organised in two separate supply branches on which the satellite equipment is distributed. Each of these two branches is fed by one wing of the solar generator and associated with a separate battery. Interconnection facilities are provided for the various elements in case of breakdown of one of them. The secondary equipment (regulators, switches and control devices) are replicated.

The conditioning subsystem also contains circuits to protect operation (for example, limitation of battery discharge to avoid polarisation inversion of the cells) and to protect against breakdowns. In this way, certain precautions can be envisaged, for example allowance for failure of one cell of a battery (such as using reverse diodes or a parallel relay to ensure electrical continuity in case of open-circuit). These solutions are to be used with care; the addition of extra elements may lead to a larger overall probability of breakdown than that of the element which the device is intended to protect!

Finally, the subsystem contains regulation and conversion circuits which are intended to provide equipment with continuous regulated voltages. This equipment must have the highest possible efficiency, low mass and high reliability. The high efficiency is obtained by using chopping techniques of various kinds to raise or lower the voltage.

10.4.4 Example calculations

The power to be provided at the end of life (10 years) is 1200 W. This power is to be provided under the least favourable conditions, that is at the summer solstice when the solar flux is given by $\phi = 1353 \times 0.89\ \text{W/m}^2$ (see Figure 12.1).

10.4.4.1 *Spin-stabilised satellite*

A cylindrical satellite is covered on the external lateral surface with silicon solar cells of surface area $s = 4\ \text{cm}^2$. In the course of rotation, the cells successively face the sun and cold space. Hence their mean temperature remains low, of the order of 10°C. At 10°C, the efficiency of the cells is 14%. The degradation of efficiency at end of life is estimated at 22%. The various losses (due to cell protection windows, cabling, etc.) are included in a factor $l = 0.9$. The filling factor of the panels is $f = 0.85$.

As a consequence of the cylindrical form and rotation, the form factor F, which characterises the ratio of the actual surface on which the cells are mounted to the equivalent surface normal to the axis of the cylinder, is equal to π.

The power at the end of life is given by:

$$P = (1 - 0.22)\phi elns/F$$

The number of cells n required is thus:

$$n = 1200\pi/(1 - 0.22) \times 1353 \times 0.89 \times 0.14 \times 0.9 \times 4 \times 10^{-4} = 79\,750$$

The required surface A taking account of the filling factor f is equal to $A = ns/f = 79\,750 \times 4 \times 10^{-4}/0.85 = 37.5\ \text{m}^2$ which represents a cylinder of diameter 2.16 m and height 5.5 m.

By estimating the mean mass per cell (including cabling, mounting and the protecting window) at 0.8 g and a support of mass $1.6\ \text{kg/m}^2$, the total mass is $79\,750 \times 0.8 \times 10^{-3} + 37.5 \times 1.6 = 124$ kg.

The specific power is thus $1200/124 = 9.7$ W/kg.

10.4.4.2 *Three axis stabilised satellite*

The satellite contains two rectangular orientable solar panels covered with cells of $2 \times 4 = 8\ \text{cm}^2$. The panels are aligned with the pitch axis and can be oriented about this axis.

As the cells continuously face the sun, degradation is greater and equal to 28% in 10 years. The mean temperature is also higher and this leads to a lower efficiency at the start of life which thus has a value $e = 11\%$. The losses are slightly greater due, for example, to the shadow of the antenna tower on the panels, hence $l = 0.88$. The filling factor f is equal to 0.75.

The power at the end of life is given by:

$$P = (1 - 0.28)\phi\, elns$$

The number of cells n required is thus:

$$n = 1200/(1 - 0.28) \times 1353 \times 0.89 \times 0.11 \times 0.88 \times 8 \times 10^{-4} = 17\,870$$

The required surface area A taking account of the coefficient of filling f is equal to $79\,750 \times 8 \times 10^{-4}/0.75 = 19\,\text{m}^2$.

Estimating the mean mass per cell at 1.6 g and a support of mass $1.8\,\text{kg/m}^2$, the total mass is $17\,870 \times 1.6 \times 10^{-3} + 19 \times 1.8 = 63\,\text{kg}$.

The specific power is thus $1200/63 = 19\,\text{W/kg}$.

This example is similar to the generator of INTELSAT V [McK-78].

10.4.4.3 *The battery*

The power to be provided on board a geostationary satellite during the longest eclipse is $P = 800\,\text{W}$. NiCd batteries of capacity $C = 30\,\text{A h}$ are used with a depth of discharge DOD of 55% to guarantee a lifetime of 7 years; the mean voltage during discharge is $V_d = 1.2$ volt per cell and the discharge efficiency is $\eta_d = 0.9$.

Since the duration of the longest eclipse is 1.2h, the energy to be provided during the eclipse is $E = 1.2\,\text{P}$ (Wh). The energy E_c provided by a battery of capacity C is given by $E_c = CV_d\,\text{DOD}\,\eta_d$ (Wh). The number of cells required is thus:

$$n = 1.2\,P/CV_d\,\text{DOD}\,\eta_d = 1.2 \times 800/30 \times 1.2 \times 0.55 \times 0.9 = 54\ \text{cells}$$

To increase the reliability of the system, the battery is divided into two sections of 27 cells each. The nominal voltage during discharge is 32.4 V. To avoid loss of a cell by short circuit, one cell more than is necessary is often added to each half battery. Also, diodes are placed at the terminals of each battery to ensure electrical continuity in case of open circuit breakdown.

If the exact mass of a battery is not known, a rapid estimate of the mass can be obtained from typical values of specific energy per unit mass E_m of the technology used.

The energy to be stored is $E_s = 1.2\,P/\alpha\,\text{DOD}\,\eta_d = 1940\,\text{W h}$. The mass of the battery is thus given by:

$$M\,(\text{kg}) = E_s/E_m = 1.2\,P/\alpha\,\text{DOD}\,\eta_d\,E_m$$

Assuming a specific energy E_m equal to 30 W h/kg, the mass of the battery is estimated from $M = (1.2 \times 800/0.55 \times 0.9)\,(1/30) = 65\,\text{kg}$.

10.5 TELECONTROL, TRACKING AND COMMAND (TTC)

The TTC functions are as follows:

- —To receive control signals from the ground to initiate manoeuvres and to change the state or mode of operation of equipment.
- —To transmit results of measurements, information concerning satellite operation, the operation of equipment and verification of the execution of commands to the ground.
- —To measure the ground–satellite distance, and possibly the radial velocity, in order to permit location of the satellite.

The function described as 'on-board management' is often associated with the functions of the

TTC subsystem. On-board management includes all data processing and formatting operations together with data traffic and time management on-board the satellite.

The links concerned are service links and the information rates involved are low, a few kilobits per second at maximum. It is important to distinguish these from scientific satellite telemetry (such as observation of the earth) for which the data rates to be transmitted are much greater.

Finally, one of the major characteristics required of TTC links is availability. Ensuring availability of the TTC links is fundamental for diagnostics in case of breakdown and for performing corrective actions.

The necessary reliability is obtained by means of suitably replicated transmitting and receiving equipment (transponders). This equipment is associated with one or more antennas which have a radiation pattern such that the gain is as constant as possible, or at least greater than a minimum value, throughout most of the space around the satellite. This permits links to be established whatever the attitude of the satellite.

10.5.1 Frequencies used

The links involved constitute a satellite operation service and the frequencies used must be chosen from the bands allocated to this service. The frequencies normally used are in S band as follows:

—The uplink in the band 2025 to 2120 MHz.
—The downlink in the band 2200 to 2300 MHz.

Frequencies in the VHF band (140–150 MHz) have also been used.

It is clear that the available bandwidth (of the order of 100 MHz) is insufficient to accommodate all the modulated carriers from the various satellites in orbit. Also, this band is reserved for operations associated with putting the satellite into orbit and an emergency mode in case of a problem in normal mode in the operational phase. During these phases, the orientation of the satellite attitude with respect to the earth is arbitrary. An omnidirectional antenna is thus indispensable.

In order to liberate specific bands for particular operations, the TTC links in normal mode are routed through the satellite payload. They therefore use the frequency bands corresponding to the nominal satellite service, a fixed satellite service, for example.

The problem associated with the use of nominal frequency bands is that usually the payload antennas are directional and, in the case of a pointing error due to an attitude control problem, the TTC links are interrupted. It is therefore necessary for the TTC links to be rerouted via the transmitting and receiving system in S band; this should be automatic since access to the satellite from the ground is no longer possible. This function is realised by means of a selecting device which is activated by command from the ground once the satellite is on station in the nominal configuration and routes the service links via the satellite payload by means of a priority relay. The device then monitors the carrier level of the received command signal. As long as the level remains within predefined limits, the detector output signal keeps the relay energised. In the case of a pointing error, or a problem on the uplink of the nominal mission, the relay switches to the released position which corresponds to routing via the S band repeaters (Figure 10.32). Since the replicated repeaters are associated with one or more antennas which have a quasi-omnidirectional radiation pattern, the TTC links continue to be transmitted even if the attitude is highly perturbed. However, the nominal frequency band

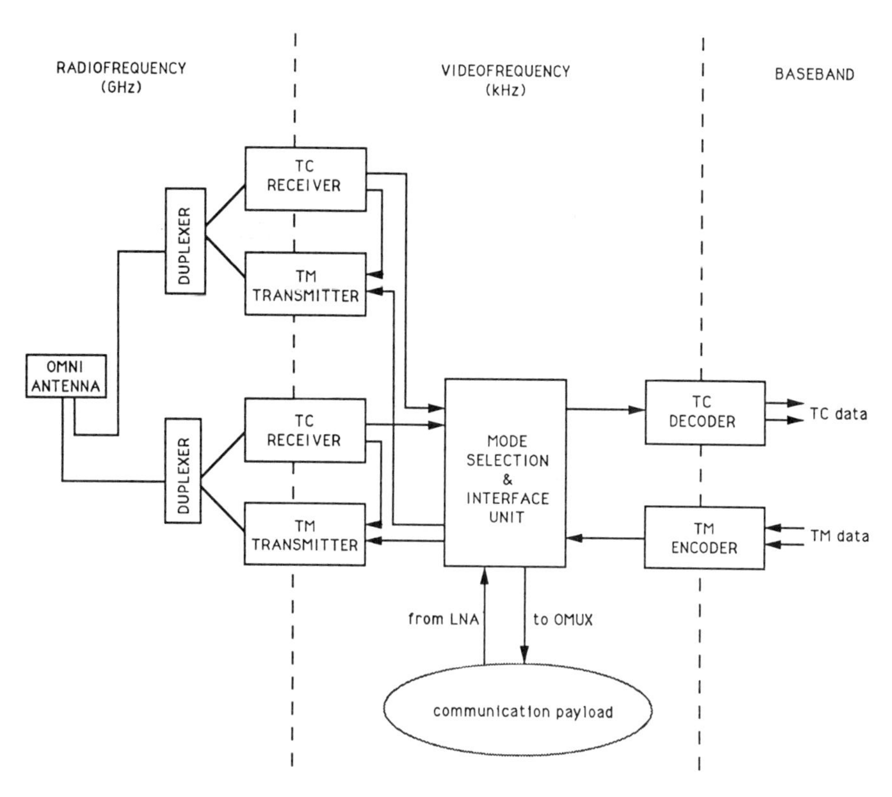

Figure 10.32 Block diagram of a centralised TTC subsystem.

of the communication payload can also be used to handle TTC carriers during the installation of the satellite in orbit. The same TC receiver and TM transmitter are then used during installation and operational life of the satellite. To obtain the omnidirectional coverage during installation, several wide beamwidth antennas on different sides of the satellite are usually used. Once on station, the input of the TC receiver and the output of the TM transmitter are then connected to the main antennas of the communication payload.

10.5.2 The command link

The command links are provided by a carrier whose frequency depends on the band used and is phase or frequency modulated by a subcarrier at a few kilohertz (e.g. 8 kHz). Digital phase modulation of this subcarrier by the data has previously been realised. The bit stream (often NRL-L formatted) has a data rate ranging from some hundreds of bit/s to several kilobit/s according to the application. On account of the low bit rate, the use of a subcarrier enables the useful spectrum (that of the data) to be separated from the carrier itself.

10.5.2.1 *Command message*

The command message is organised in lines preceded by a group of bits for acquisition of synchronism. Each line consists of words of several bits. The length of the line structure

Address and synchronisation word	Mode selection	Mode selection repeated	First data word	First data word repeated	Second data word	Second data word repeated	Third data word	Third data word repeated
16 bits	4 bits	4 bits	12 bits	12 bits	12 bits	12 bits	12 bits	12 bits

Figure 10.33 ESA standard.

depends on the standards used. For example, in the ESA standard, the line consists of 96 bits (Figure 10.33). The first word, of 16 bits, constitutes an address and synchronisation word specific to the destination decoder; it is followed by a mode selection word of 4 bits which is immediately repeated. The mode selection word is used to indicate the type of command transmitted in the three words of 12 bits which are repeated once and contain the data. The line terminates with a repetition of the address and selection word (80 bits between the two words). The 12 bit words used to carry the data can, for example, contain data coded in 8 bits which has been extended to 12 bits by means of an error correcting code.

The commands to be transmitted are either regulating commands to adjust a quantity on board the satellite to a particular value (for example, the helix current of a TWT) or to load registers in a computer or memory with binary system commands (for example, opening or closing a relay, 0 and 1 for open and closed).

Commands, according to the particular mode selected, can be:

—executed immediately after reception,
—stored in memory and executed on reception of a specific command,
—stored in memory and executed when activated at a given time determined by the time management system on board the satellite or when activated by a signal produced by one of the sub-systems on board the satellite.

10.5.2.2 Link security

Repetition enables the integrity of the received words to be verified before they are used. One of the important characteristics of the command link is security. It is fundamental for the survival of the satellite that it really is the required command which is executed.

Various precautions are taken such as error correcting coding of the data words, repetition for verification and detection of possible differences, deferred execution of commands etc. With deferred execution of commands, the command is detected on board the satellite, stored in memory, retransmitted to the ground by telemetry for verification of authenticity and executed only after authentication by an execution command sent on the command link.

Finally, precautions are also taken to make the system insensitive to signals transmitted by intruders; these include narrow band reception, input limiters, insensitivity to non-standard signals and possible encryption on the link.

Use of a spread spectrum link permits the problems of both interference between systems and protection from undesired signals to be solved and also makes efficient use of the frequency bands through the possibility of multiple access to the same band.

10.5.3 Telemetry links

Telemetry links are also provided by a carrier which is phase or frequency modulated by a sub-carrier at a few kilohertz (e.g. 40.96 kHz). The digital data phase modulates this sub-carrier. The data rate ranges from a few tens of bits/s to a few kilobits/s.

10.5.3.1 Telemetry format

The telemetry message is organised in lines; a group of lines (or frames) constitutes a format. Each line consists of 8-bit words and starts with a synchronising code; the first line contains a format identification word. Lines are identified by means of a counter. In the ESA standard, the format consists of 16 lines and each line contains 48 words. The 8-bit words constitute the data; this may be part of a data word for data whose resolution requires coding in more than 8 bits (the data word is then shared among several 8-bit words) or a block of data which requires only a single bit, for example, the state of a relay.

10.5.3.2 Data conditioning

The data to be transmitted may consist of analogue quantities, corresponding, for example, to the results of measurements, digital words (the value in a register or the output of an encoder) or binary system states (0 or 1, relay open or closed). Analogue quantities are sampled, quantised and encoded with a number of bits depending on the required resolution and the range of amplitude variation of the signal. A clock is necessary to discretise the analogue information. In accordance with the dynamic behaviour of the information transmitted, sampling is not performed at the same rate for all signals; with respect to a basic cycle, some signals are undersampled (sampled at multiples of the cycle) and others are oversampled (sampled several times per cycle).

The data are obtained either:

—directly from satellite equipment and conditioned (analogue-digital conversion, formatting etc.) in the telemetry encoder, or
—at the output of a processing unit on a network to which the various satellite equipment has access (see Section 10.5.5).

10.5.4 Location

10.5.4.1 Distance measurement

Distance measurement is performed by means of specific sub-carriers which modulate the telecommand carrier, are coherently demodulated in the receiver and are then used to modulate the telemetry carrier. Comparison of the initial phase of the signals with the phase of the demodulated signals on the ground enables the round-trip time to be obtained. This time, from which the precisely known time delay in the receiving equipment is deducted, permits the distance to the satellite to be calculated.

Various approaches are possible according to the nature of the sub-carrier; these include fixed frequency (tone), variable frequency, modulation by a pseudorandom (PN) sequence etc. The tone system is currently used. In this case the sub-carrier is a sinusoidal wave of fixed frequency f. Measurement of the phase shift between the transmitted and received tones, which is a function of the distance R from the station to the satellite (a round-trip trajectory of $2R$), enables this distance to be determined:

$$\Delta\phi = 2\pi f(2R)/c \tag{10.34}$$

where c is the velocity of light.

As phase shift is measured modulo 2π, the measurement will be the same for all values of R such that $2\pi f\,(2R)/c = K2\pi$. The distance ambiguity corresponds to the modulus of the distance obtained for $K = 1$. Table 10.8 gives the distance ambiguity as a function of the frequency of the tone used. Observe that, for a geostationary satellite, the frequency of the tone must be at most 8 Hz for the measurement to be made without ambiguity.

The choice of tone frequency is thus guided by incompatible considerations. A high frequency (100 kHz) is necessary to ensure accuracy of phase measurement; conversely the frequency must be low enough for the wavelength to be long enough with respect to the distance to be measured for there to be no ambiguity.

The difficulty is avoided by transmitting two tones simultaneously; these are a major tone at 100 kHz which permits good measurement accuracy and a minor tone, obtained by division of the major tone (and hence in phase with it), which enables the ambiguity to be resolved. The procedure is as follows; the major tone at 100 kHz is first transmitted with the first minor tone at 20 kHz (division by 5). On reception the minor tone is compared with five signals separated in phase by $2\pi/5$ which are obtained by division by five of the received major tone. Only one of these signals is in phase with the received minor tone and is selected (Figure 10.34a). A replica of the received minor tone at 20 kHz is obtained, but with the phase accuracy of a 100 kHz tone. This signal serves in turn to create five signals at 4 kHz also differing in phase by $2\pi/5$ which are compared in turn with the new minor tone at 4 kHz which has replaced that transmitted at 20 kHz (the major tone remains continuously transmitted to ensure continuity of the replica retrieved on reception). Only one of the five signals is in phase and is selected in its turn. The process is repeated with minor tones of 800 Hz, 160 Hz, 32 Hz and finally 8 Hz. In this way a signal at 8 Hz is obtained by successive division of the major tone at 100 kHz and whose phase relationship is known (Figure 10.34b). Comparison with the transmitted minor tone at 8 Hz thus permits determination of the distance with the accuracy obtained for a 100 kHz tone.

Figure 10.35 shows an example of the spectrum of the signal which modulates the telemetry carrier (sub-carrier modulated by the data at 40.96 kHz) or the telecontrol carrier (sub-carrier modulated by the data at 8 kHz). The minor tones are transposed in frequency between 16 and 20 kHz to reduce the required bandwidth (Table 10.9).

Table 10.8 Distance ambiguity ΔR as a function of tone frequency f

f	100 kHz	20 kHz	4 kHz	800 Hz	160 Hz	32 Hz	8 Hz
ΔR (km)	1.5	7.5	37.5	187.5	937.5	4687.5	18750

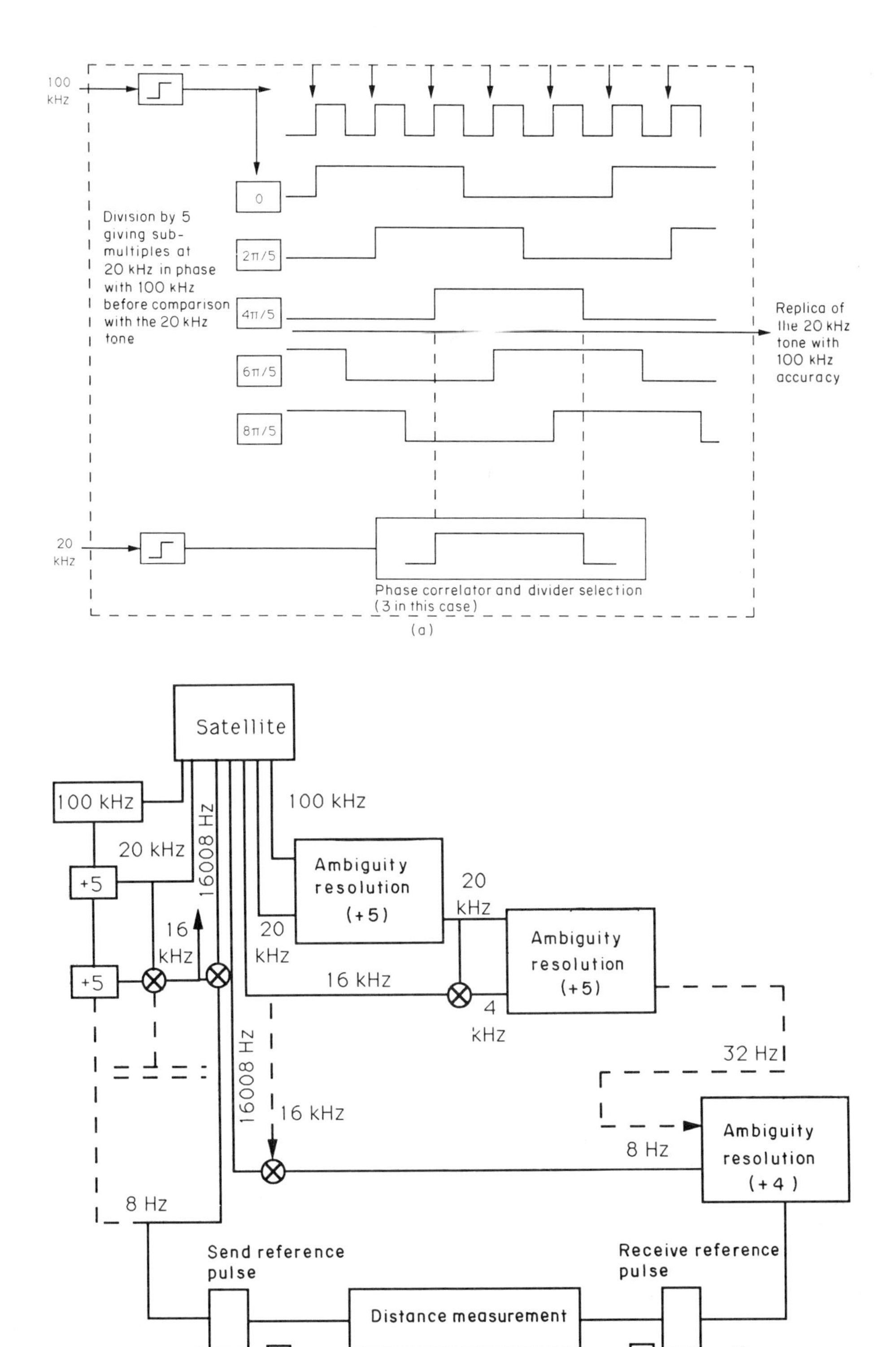

Figure 10.34 Principle of tone range measurement.

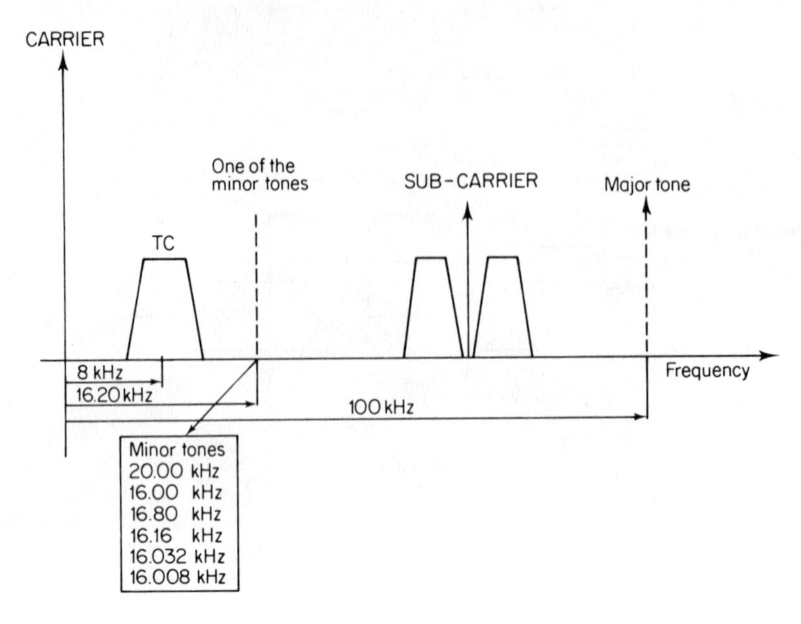

Figure 10.35 Spectrum of the telemetry carrier modulating signal.

Table 10.9 Major and minor tones

Major tone	Divider	Virtual minor tone (Hz)	Transmitted tone (Hz)
100 kHz	5	20 000	20 000
	5	4 000	16 000
	5	800	16 800
	5	160	16 160
	5	32	16 032
	4	8	16 008

Measurement accuracy depends on the tone frequency, the received signal-to-noise ratio, the stability of the transit time in the satellite equipment and variations of propagation time in the ionosphere. For a tone of frequency f, the quadratic mean value of the distance error is given by:

$$S = (c/4\pi f)[k/\sqrt{(S/N)}] \tag{10.35}$$

where $[k/\sqrt{(S/N)}]$ is the quadratic mean value of the phase error, a function of the constant k which depends on the structure of the receiver used, and S/N is the signal-to-noise ratio at the phase detector input. Variations of tropospheric propagation time can cause an error of between 0 and 300 m (at 2 GHz), but this can be estimated separately. Under good reception conditions, distance measurement can be performed with an error of the order of a few tens of metres.

10.5.4.2 *Measurement of radial velocity*

Radial velocity can be obtained by measurement of the Doppler effect. It is necessary to ensure frequency and phase coherence at the transponder between the downlink and uplink carriers. The nominal frequency f_d of the downlink carrier is such that:

$$f_d/f_u = 240/221 \tag{10.36}$$

where f_u is the nominal frequency of the uplink carrier. If the satellite is given a velocity V_r with respect to the control station, the frequency received on board is equal to:

$$f_u^* = f_u[1 + (V_r/c)] \quad \text{(Hz)} \tag{10.37}$$

where c is the velocity of light.

The frequency retransmitted on the downlink is obtained by multiplication by the ratio 240/221 and the frequency received by the control station is equal to:

$$f_d^* = (240/221)f_u[1 + (V_r/c)]^2 \quad \text{(Hz)} \tag{10.38}$$

Taking account of the very small value of V_r with respect to c, this gives:

$$f_d^* = (240/221)f_u[1 + (2V_r/c)] \tag{10.39}$$

The radial velocity is thus obtained as a function of the frequency difference Δf between the received frequency f_d^* and the nominal frequency f_d on the downlink ($f_d = (240/221)f_u$):

$$V_r = -(c/2)(221/240)\Delta f/f_u \quad \text{(m/s)} \tag{10.40}$$

Measurement of the radial velocity requires operation of the transponder in coherent mode which can be different from the normal non-coherent mode where the downlink carrier is obtained from an on-board oscillator. Mode selection is achieved by command.

10.5.5 On-board data handling

10.5.5.1 *Functions performed by on-board management*

—Command processing: decoding, validation, acknowledgement and execution (immediate or deferred) of command signals.
—Acquisition, compression and/or coding and formatting of telemetry information.
—Processing and storage: processing of data relating to the on-board management sub-system itself (such as time and configuration) and on demand from the satellite sub-systems, storage of telemetry data, modes and software.
—Synchronisation, data timing and traffic management: on-board time management, distribution of on-board timing and clock signals to sub-systems, dating of events and measurements, management of traffic between sub-systems.
—Monitoring and control: acquisition and analysis of monitored and diagnostic parameters, decision taking (e.g. changing into survival mode and reconfiguration), generation and execution of appropriate commands.

According to the type of satellite and its complexity, these functions may not all be required. In particular, on-board management of the first communication satellites was rudimentary in respect of the small number of TTC channels which communicated with the ground (a few tens). The number of channels was subsequently increased (to several hundreds at the end of the 1970s) but on-board management functions remained limited and could be satisfied by two specific pieces of equipment: the command decoder and the telemetry encoder. During the 1980s, the increase in size and complexity of satellites has led to an increase in the number of command and telemetry channels (4400 for INTELSAT VI). At the same time microprocessors and high performance memories, which can be used in a space environment, had been developed. These developments have led to organisation of the on-board management sub-system in association with the command and telemetry sub-systems in a modular form based on a data transfer bus.

10.5.5.2 Simplified architecture

The on-board management system can be restricted to decoding command signals and encoding telemetry signals. The interfaces with this equipment consist of a subcarrier modulated by a bit stream (a 'video' signal) from or to the respective radio-frequency equipment (the command receiver and the telemetry transmitter) and the command and telemetry signals to or from the satellite equipment on as many links as there are signals. The block diagram of this architecture is given in Figure 10.32.

The functions to be realised are limited mainly to processing of command signals (decoding, validation and execution) and telemetry signals (acquisition, formatting and dating):

- The command decoder detects bits from the noisy demodulated signal after recovery of the bit rate (primary synchronisation) and then separates the various format components (such as the address and the mode) from the data (secondary synchronisation), validates and transmits the execution commands after demultiplexing the data on the various equipment channels.
- The telemetry encoder performs analogue-to-digital conversion of analogue telemetry signals, multiplexes the different channels and generates the data format by adding identification and synchronisation bits, generates and modulates the subcarrier with the bit stream to obtain the 'video' signal destined for the telemetry transmitter.

The increase in the number of TTC channels leads to increased complexity of this equipment. Furthermore, it is necessary to route the various electrical signals separately from the satellite equipment to the TTC sub-system. This leads to bulky cabling with a high mass (13 km and 130 kg of cables on INTELSAT VI). This architecture is thus unsuitable for modern satellites which require a large number of TTC channels as a consequence of the complexity of their sub-systems.

10.5.5.3 Modular architecture

The system organisation uses a decentralised architecture with a communication bus between the various pieces of data processing and handling equipment (Figure 10.36). The various

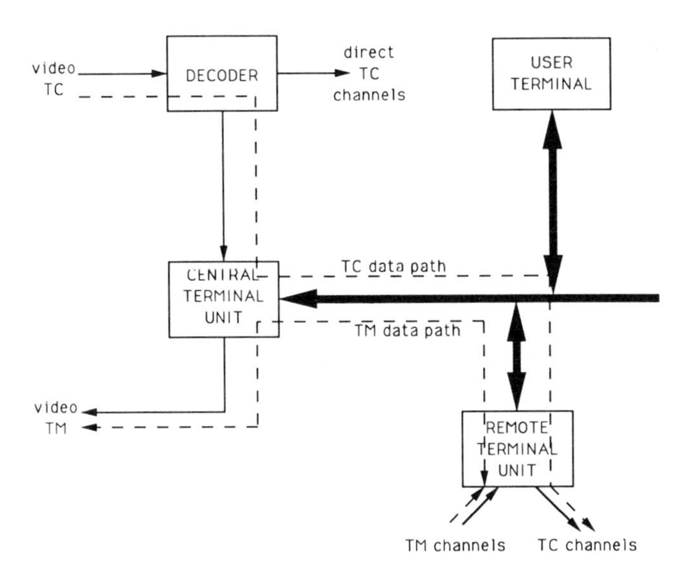

Figure 10.36 Modular architecture for data management.

modules are the command decoder, the main management equipment, the data bus and the remote terminals.

The command decoders. As with centralised architecture, this restores the bit stream and then separates the various components of the data format. Commands are divided into two categories: those which are demultiplexed in order of priority and transmitted directly on their respective channels to the equipment concerned, and those which are processed by the central terminal unit.

The central terminal unit (CTU). This equipment performs many functions as follows:

—handles the command data concerned and distributes it to the equipment via the bus,
—handles the data traffic on the bus, defines on-board time and distributes it to users,
—generates terminal interrogations for acquisition of telemetry data,
—multiplexes data from remote terminals via the bus and that which it generates directly and formats it to produce the telemetry bit stream,
—modulates the sub-carrier with the bit stream to obtain the 'video' signal destined for the telemetry transmitter.

If the need arises, this equipment can monitor critical parameters and take appropriate decisions (such as reconfiguration, changing to emergency mode and so on).

The main management equipment is associated with an on-board computer (OBC) which provides the required processing capacity. This computing power can be made available to the various satellite sub-systems to perform processing in time sharing on data relating to a sub-system (centralised processing).

Data bus. This carries data and clock signals between the central terminal unit and the various units; in certain cases, it also carries power signals to activate relays (to put equipment out of circuit). Date exchange management is governed by protocols.

Remote terminal unit (RTU). On activation by the central terminal unit, the RTU performs the following:

—acquires command data and interrogations on the bus,
—directs commands to the equipment concerned in the form of electrical signals which are user specific,
—delivers clock signals available on the bus to the users,
—acquires telemetry signals from users on the various channels, performs analogue-to-digital conversion and encoding if necessary, and transmits telemetry data on the bus (on request from the central terminal unit).

Remote terminal units are used to connect simple equipment, which does not have internal control capacity, to the bus (examples are relays and heaters for temperature control). More sophisticated equipment already has computing capability and is thus capable of controlling exchanges when connected directly to the data bus. Remote terminals and users can be provided with computing capacity (intelligent terminal unit (ITU) which permits data processing to be performed locally (decentralised processing as opposed to centralised processing where the computer is associated with the central terminal unit).

Some sub-systems which consist of several units dispersed about the satellite may have their own data bus which then accesses the main data bus by way of a remote interface unit.

10.5.5.4 *The OBDH standard*

In order to encourage development of standardised modular equipment, the interfaces between units have been the subject of standardisation known as the On-Board Data Handling (OBDH) standard. The OBDH standard defines the organisation of the data bus (an interrogation line and a response line on screened twisted pairs), the coding of the bits, the data structure, the

Table 10.10 Characteristics of the OBDH standard

Compatible with the PCM/TC ESA and PCM/TM ESA standards
Two wire full duplex bus protocol (interrogation and response)
Up to 31 users connected to the bus
Response time on the bus $< 140\,\mu s$
Data rate on the bus $\leqslant 500$ kbps
Bus length < 20 metres
Distribution of one to five clocks
No possibility of interruption of the central control unit (CTU) by users (CTU bus management is by polling)
No internal redundancy or possibility of standard redundancy
Up to 2 times 48 direct ON/OFF commands provided by the decoder
19 useful bits per 32 bit packet on the bus
On-board clock: 4 to 6 MHz, stability 10^{-6} s per year

protocols, the user interfaces, etc. The general characteristics of the standard are given in Table 10.10.

10.5.5.5 *Satellite control*

Satellite control operations consist, in some cases, of elaborating commands in order to execute actions in accordance with information available on board the satellite.

With the rudimentary on-board management architectures of first generation satellites (see Section 10.5.5.2), the information from telemetry channels is grouped in the control station where the context is analysed and the decisions taken are transmitted on the command channels. This mode of operation is quite suitable for geostationary communication satellites, which are continuously visible from the control station and whose management is simple (they have few TTC channels).

The existence of computing power on board the satellite enables information processing and direct generation of appropriate commands on board the satellite to be considered. This permits the load on the control station to be lightened and the availability of the satellite to be increased by automatic reconfiguration in case of breakdown; for orbiting satellites complex control actions can be contemplated even when they are invisible from earth stations.

A control hierarchy is thus always instituted; the first level corresponds to operations and major events which are still processed by the control station; the second level corresponds to events and operations directly processed on board the satellite. A third level can be introduced into the hierarchy when the sub-systems themselves are capable of controlling their mode of operation and reconfiguring in case of breakdown (a decentralised processing capacity in the intelligent remote terminals).

10.6 THERMAL CONTROL AND STRUCTURE

The purpose of thermal control is to maintain the satellite equipment within the temperature ranges which enable it to operate satisfactorily, by providing its nominal performance, and avoiding any irreversible deterioration when it is not operating.

This also applies to the structure of the satellite which must remain within a mean temperature range in order to minimise deformation and guarantee precise alignment of attitude stabilisation sensors and antennas.

10.6.1 Thermal control specifications

Thermal control must be optimised with respect to the constraints of both the operational and transfer phases. These constraints are very different (due to different orbits and attitudes, the state of the apogee motor etc.).

The objectives of thermal control are thus to maintain the equipment within specified temperature ranges; these differ for equipment when operating and when on stand-by. Furthermore, the behaviour of the equipment may differ according to whether it is operating or at rest; in operation, it will usually generate heat which the thermal control must remove; when at rest, the equipment must, in certain cases, be heated in order to avoid an excessively low temperature. Finally, the maximum values of temperature gradients (with respect to time) must also be considered.

10.6.1.1 *Specified temperature ranges*

The temperature ranges to be maintained differ greatly from one piece of satellite equipment to another.

Examples are:

—Antenna:	− 150 °C to + 80 °C
—Electronic equipment:	− 30 °C to + 55 °C (on stand-by)
	+ 10°C to + 45°C (operating)
—Solar generator:	− 160 °C to + 55 °C
—Battery:	− 10 °C to + 25 °C (on stand-by)
	+ 0 °C to + 10 °C (operating)
—Solar sensor:	− 30 °C to + 55°
—Propellant reservoir:	+ 10 °C to + 55 °C
—Pyrotechnic unit:	− 170 °C to + 55 °C

These temperature ranges are those which the equipment may be expected to encounter once in orbit. This implies that the equipment has been designed to operate at, or withstand, more extended temperature ranges than those stated. In particular, a specified range within which the nominal performance of the equipment must be maintained is defined by adding modelling errors to the limits of the estimated temperature range. A still wider range within which the equipment must not suffer irreversible degradation is also defined.

10.6.1.2 *Characteristics of the space environment*

The characteristics of the space environment are presented in Chapter 12. As far as thermal control is concerned, the most useful characteristics are recalled in Figure 10.37. It must be remembered that the satellite is subject to the effects of three radiation sources (the sun, the earth and the terrestrial albedo) which have different spectral distributions and geometric forms and are absorbed differently by the surface of the satellite. Eclipses and variations of attitude and distance modify the illumination conditions in the course of time. Cold space absorbs all radiation from the satellite. The ambient vacuum prevents convection.

10.6.1.3 *The principle of thermal control*

The mean temperature of a piece of equipment is the result of an equilibrium between the heat generated internally, the heat absorbed and radiated by the surfaces of the unit and the heat received or removed by conduction through the mechanical mounting of the equipment.

The mean temperature of the satellite is the result of equilibrium between the heat generated internally, the heat absorbed by the surfaces of the satellite and the heat radiated by the surfaces of the satellite.

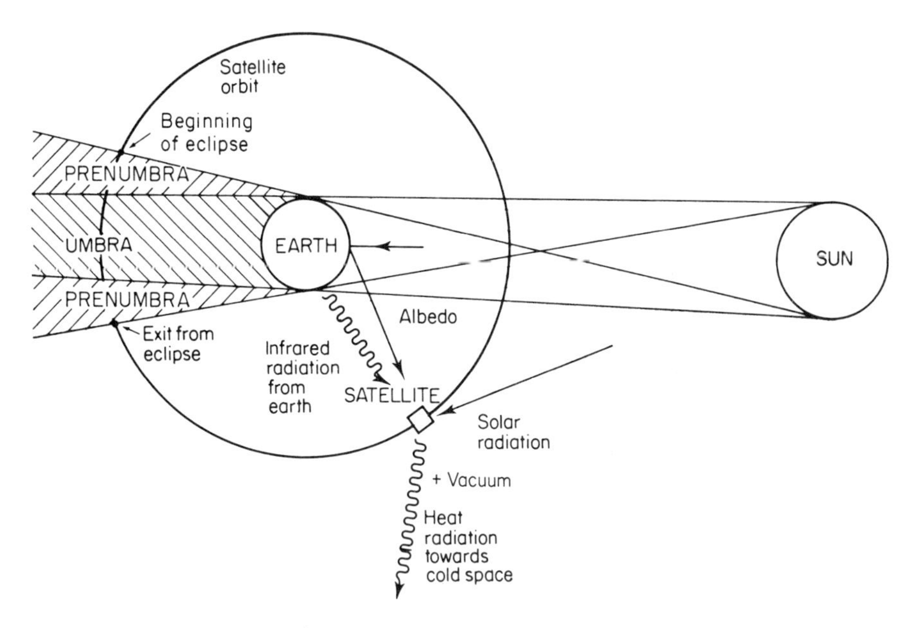

Figure 10.37 Space environment characteristics of importance to thermal control.

Thermal control, therefore, consists of:

—Adjusting the thermal conductivities between the various parts of the satellite, either to favour heat exchanges by conduction between one point and another (by choice of material type, surface and section, use of heat pipes etc.) or to limit exchanges (by use of insulating materials etc.).
—Making use of the thermo-optical properties (such as emissivity and absorption) of surfaces to favour, for example, the extraction of heat by radiation while minimising that captured (OSR radiators).
—Providing local sources of heat if necessary (electric heaters).
—Arranging for certain surfaces to be able to radiate to remote space without constraint in order to lower the temperature (infra-red detectors).

Thermal control is realised either passively or actively with a preference for passive control wherever possible in view of its simplicity, low cost and reliability.

10.6.2 Passive control

Passive control is based on the absorptive and emissive properties of surfaces. Absorptivity α is defined as the ratio of the power absorbed by unit surface area to the incident power. Emissivity ϵ is defined as the ratio of the power radiated per unit surface area to the power which would be radiated by unit surface area of a black body. Recall that the power radiated per unit surface area of an ideal black body (W/m^2) is σT^4, where T is the temperature of the black body (K) and $\sigma = 5.67 \times 10^{-8}\,W\,m^{-2}\,K^{-4}$ is the Stefan–Boltzmann constant (see Section 12.1.4).

According to the material used, the values of absorptivity α and emissivity ϵ vary between zero and unity. For a given material, the ratio α/ϵ is of prime importance in determining the mean temperature of a surface exposed to the sun.

10.6.2.1 *Types of cladding*

Various types of surface are used:

—White paint; this absorbs infra-red radiation (terrestrial flux) and reflects solar flux. It is a cold surface in sunlight ($-150\,^{\circ}$C to $-50\,^{\circ}$C) since the ratio α/ϵ is small (α can reach 0.9, ϵ is of the order of 0.17).
—Black paint; this absorbs all wavelengths but is also characterised by a high emissivity ($\epsilon = 0.89$). The absorptivity is very high ($\alpha = 0.97$). The temperature in sunlight is greater than $0\,^{\circ}$C.
—Aluminium paint; conversely, this absorbs little and emits little. The emissivity is low (ϵ of the order of 0.25) and so is the absorptivity (α of the order of 0.25). The equilibrium temperature in sunlight is close to $0\,^{\circ}$C. On the other hand, as the emissivity is much less than that of black paint, an aluminium covering is warmer in shadow than a black covering.
—Polished metal; this absorbs the visible part of the solar spectrum (solar absorbers) but reflects infra-red radiation. These surfaces are warm in sunlight ($50\,^{\circ}$C to $150\,^{\circ}$C) since the ratio α/ϵ is high (for example, for gold $\alpha = 0.04$ and $\varepsilon = 0.25$).

The values of the thermo-optical properties of claddings are affected by uncertainties and variations due to characterisation errors, problems of reproducibility during fabrication, sensitivity to contamination and degradation due to the effect of the space environment.

In order to limit the exchanges, a surface can be isolated by using superisolating padding or multilayer isolation (MLI). This padding consists of several sheets of plastic (mylar) aluminised on both faces and separated by a material of low conductivity (dacron mesh). The exterior layer is aluminised only on the internal face and consists of either kapton (with a golden appearance) for temperatures which do not exceed $150\,^{\circ}$C or titanium for high temperatures (up to $400\,^{\circ}$C). The conductance is of the order of $0.05\ \mathrm{W/m^2\,K}$.

10.6.2.2 *Radiating surfaces*

For communication equipment which dissipates heat (such as power amplifiers), radiators having a very low absorptivity to emissivity ratio are used. These surfaces are, therefore, capable of efficiently radiating the heat generated while absorbing the least possible solar radiation. These radiators are strips of silica silvered on the reverse side (optical solar reflectors (OSR)) or produced from sheets of plastic material (teflon, kapton or mylar) with a deposit of silver or aluminium on the reverse face (second surface mirror (SSM)).

A quick estimate of the surface area S required to extract the thermal power dissipated by the payload of a communication satellite can be obtained using equation (10.41). This equation states that when the equilibrium temperature T is reached, the sum of the thermal power P (heat) and the power absorbed from the sun is equal to the power radiated by the surface considered:

$$P + \alpha\phi S = \epsilon v S \sigma T^4 \tag{10.41}$$

where α and ϵ are the absorptivity and emissivity of the radiator, ϕ is the flux (W/m^2) received from the sun and depends on the earth–sun distance and the angle of incidence, S is the radiating surface area (m^2) and v is the view factor of the radiating surface.

The view factor is the complement of the percentage of the 2π steradians of space above the surface which is obstructed (occulted) by obstacles (such as the solar generator). For a three-axis stabilised satellite, the radiating surfaces are mounted on the north and south walls of the satellite (see Figure 10.1). The solar generators which are mounted in alignment with the north–south axis mask part of the space for the radiators. Typically, the coefficient v is between 0.85 and 0.9. This coefficient applies only to the corresponding term in the radiated power. For the absorbed power, having regard to the angle of incidence, almost all of the surface participates in capturing solar flux.

Taking an equilibrium temperature of several tens of degrees (30 °C to 40 °C), the surface area S is calculated for the most unfavourable case, that is at the end of life when α reaches its highest value as a consequence of degradation and at the time of year when the flux captured as a function of the orientation of the surface is maximum (see Section 12.1.4.1).

For a surface oriented perpendicularly to the equatorial plane,

$$\phi = \cos \partial \times d^{-2} \times 1370\,\text{W/m}^2$$

where ∂ is the declination and d the distance (in astronomical units AU) of the sun. The most unfavourable case occurs just before the spring equinox when, from Figure 12.2, $\partial = 4.3^\circ$ and $d^{-2} = 1.011$ from which $\phi = 1.008 \times 1370\,\text{W/m}^2$.

For a surface oriented parallel to the equatorial plane,

$$\phi = \cos(90^\circ - \partial) \times d^{-2} \times 1370\,\text{W/m}^2$$

The most unfavourable conditions arise:

—Just before the winter solstice for a surface situated on the south face of the satellite for which, from Figure 12.1, $\partial = 23.5^\circ$ and $d^{-2} = 1.033$ from which

$$\phi = 0.947 \times 1370\,\text{W/m}^2$$

—At the summer solstice for a surface situated on the north face of the satellite for which, from Figure 12.1, $\partial = 23.5^\circ$ and $d^{-2} = 0.965$ from which

$$\phi = 0.885 \times 1370\,\text{W/m}^2$$

It is clear that the values of ϕ are smaller for surfaces parallel to the equatorial plane. This justifies mounting radiating surfaces on the north and south walls of three axis stabilised satellites.

When the radiating surface is not illuminated by the sun (for example at the equinoxes and the summer solstice for a surface on the south face of the satellite), the equilibrium temperature of the surface (of a size already determined for the most unfavourable case) is obtained by putting $\phi = 0$ in equation (10.41). This temperature must not be too low in order to restrict thermal constraints.

10.6.2.3 *Example calculation for a three-axis stabilised satellite*

The following characteristics are suggested:

—Heat to be dissipated = 800 W (assumed to be equally shared between the north and south faces).
—α of the radiator after combustion of the apogee motor and 7 years in orbit = 0.17 (value before launch = 0.06, after combustion of the apogee motor = 0.1, degradation by 0.01 per year in orbit).
—ϵ of the radiator = 0.75.
—v is taken equal to 1.
—Equilibrium temperature of the radiator = 32 °C.

From equation (10.41):

$$S = P/(\epsilon v \sigma T^4 - \alpha\phi)$$

The area of the surface to be mounted on the south wall is:

$$S = 400/(0.75 \times 5.67 \times 10^{-8} \times 305^4 - 0.17 \times 0.412 \times 1370) = 1.46\,\mathrm{m}^2$$

For the north wall, a similar calculation leads to $1.43\,\mathrm{m}^2$.

The equilibrium temperature of the radiating surface of the south wall at the summer solstice is given by:

$$T = (P/S\epsilon\sigma)^{1/4} = (400/1.46 \times 0.75 \times 5.67 \times 10^{-8})^{1/4} = +10\,°\mathrm{C}$$

All this assumes that the heat dissipated by the radiating equipment is uniformly distributed on the surface and hence the temperature of the radiating surface is the same at all points. Since the heat is generated by devices (such as travelling wave tubes) which have a relatively small mounting area, it is necessary to provide a device (a piece of aluminium of trapezoidal form) to distribute the heat from the equipment. Heat pipes (see the following section) are also increasingly used.

10.6.3 Active control

This is used to complement passive control and may consist mainly of heaters, movable shutters and heat pipes which can also be classed as passive thermal control:

—electric resistance heaters, controlled by thermostats or by command,
—movable shutters (Maltese cross or louvres), cover more or less of the radiating surface and are controlled by a temperature transducer (a bimetallic strip) or by command,

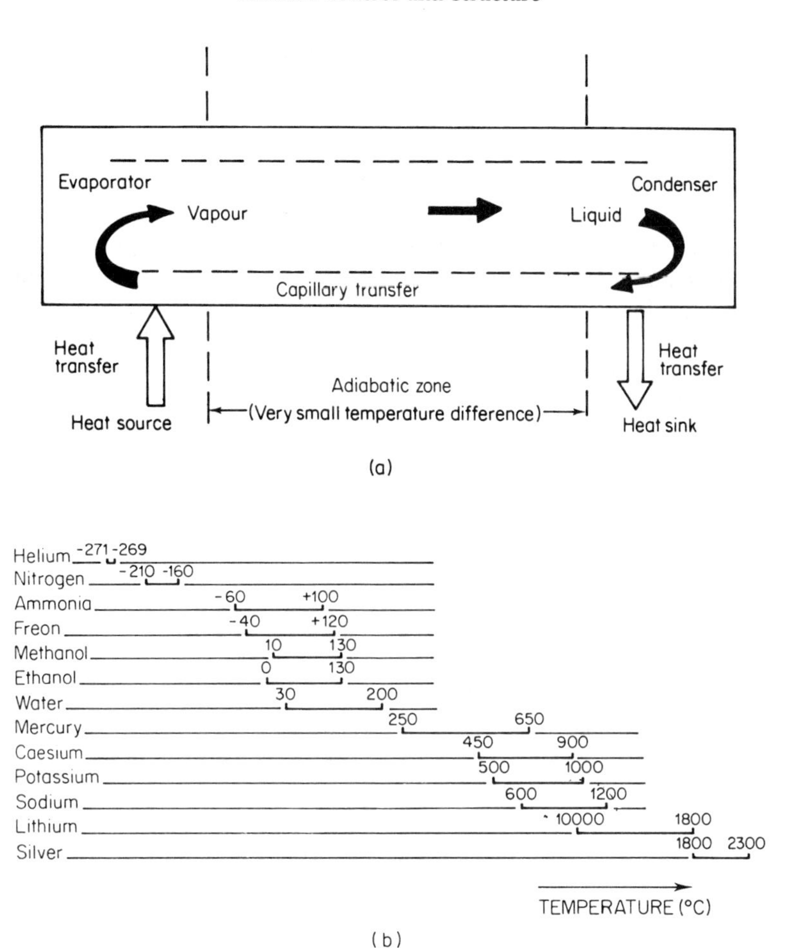

Figure 10.38 Heat pipe. (a) Principle of operation. (b) Temperature operating range.

—Heat pipes which transfer heat from hot points to radiators under a small temperature difference by means of successive vaporisation and condensation of a fluid at the two extremities of a tube (Figure 10.38). Heat pipes provide a large capacity for heat transfer due to the high values of latent heat of the fluids used.

Heat pipes are increasingly used to distribute the heat dissipated by equipment on the radiating surface. They can be buried in the honeycomb structure which forms the wall and are also used to connect the north and south walls to balance the wall temperatures at the solstices.

10.6.4 Structure

The functions of this sub-system can be classified as mechanical functions, geometric functions and others.

10.6.4.1 *Mechanical functions*

These consist of:

—supporting the on-board equipment, particularly during the launching phases when the mechanical constraints imposed by the launcher are highest (see Figure 12.5),
—permitting the various separations and deployments which change the satellite from its launch phase configuration to its operational configuration and accepting the forces acting during these operations (such as deployment of solar generators and antennas),
—providing the satellite with the required rigidity (launcher and appendices decoupling),
—permitting handling of the satellite on the ground.

10.6.4.2 *Geometric functions*

The geometric functions are related to the requirements of surface form and volume of the satellite. They are:

—to provide sufficient mounting surface for the satellite equipment (such as transponders and antennas),
—to reserve sufficient volume between the satellite and the fairing to accommodate folded appendices (such as antennas and the solar generator),
—to provide sufficient accessibility of equipment during integration of the satellite,
—to guarantee precise and stable location of equipment particularly the sensors and antennas.
—to provide sufficient space for the radiating surfaces, which must be conveniently situated on the satellite with respect to the mounting surfaces provided for power amplifiers,
—to provide a launcher and fairing interface.

10.6.4.3 *Other functions*

These functions are not directly mechanical or geometric:

—to provide a reference potential for the equipment,
—to gurantee the same potential at different parts of the satellite in order to avoid electrical discharge phenomena,
—to satisfy the requirements for thermal control (such as the value of thermal conductance between different points and support for isolating materials),
—to protect components from radiation and high energy particle flux.

10.6.4.4 *Materials used*

The essential qualities of this sub-system are contradictory—resistance to deformation and lightness. Since the forces along the principal axis of the satellite are, in principle, the greatest (this depends on the configuration of the satellite on the launcher and the type of apogee

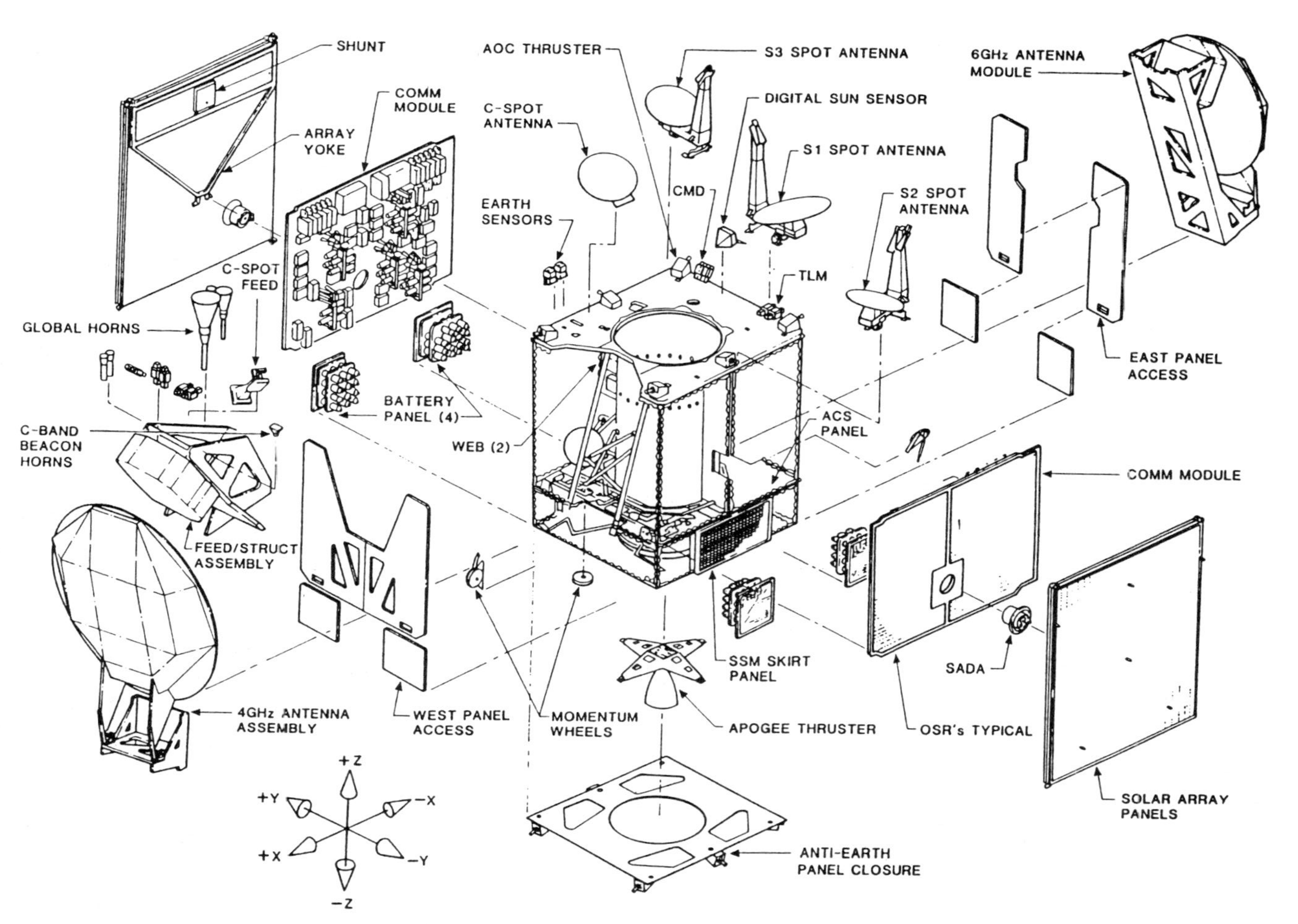

Figure 10.39 Exploded view of INTELSAT VII. (Reproduced by permission of Orbital Sciences Corporation.)

motor), an architecture based on a central tube which contains the solid apogee motor, or the propellant reservoirs in the case of unified propulsion, is often adopted (Figure 10.39).

Current techniques permit only about 6% of the total mass of the satellite to be devoted to the structure. This is achieved by using alloys of aluminium and magnesium, honeycomb panels, bonded assemblies and composite materials based on carbon fibres (for solar panels and antenna towers). Ths use of beryllium is limited on account of its prohibitive cost.

10.6.5 Conclusion

The preceding sections have defined the objectives and techniques of thermal control and its impact on the structure of the satellite. The impact of the choice of attitude stabilisation procedure on the design of the thermal control sub-system and the general organization of the structure must be emphasised. The type of propulsion system (powder or liquid propellant apogee motor) also influences the architecture.

The problems are very different for a spin stabilised satellite and a 'three-axis' stabilised satellite. For example, in the first case rotation of the satellite ensures uniform exposure conditions of the lateral faces of the satellite body to the various sources of radiation; this is not the case for three-axis stabilised satellites. Organisation of the structure is also influenced by rotational symmetry.

The area available for mounting radiating surfaces on a communication satellite is one of the factors which determines the maximum power which the satellite can radiate. Lack of space for installing these surfaces (which must be situated on the opposite face to that on which the equipment is mounted) is often a limit imposed on the electrical power. Use of amplifiers of higher efficiency permits this limit to be exceeded.

10.7 DEVELOPMENTS AND TRENDS

The development of satellite communications has been considerable during the last twenty years. This has resulted in an increase in the size and power of satellites. The Intelsat I satellite had a mass of 40 kg (beginning of life BOL) and the electrical power available on board was only 33 watts. At the end of the 1980s, the INTELSAT VI satellites weighed 2500 kg BOL) with a power of 2200 W. This corresponds to a GTO (geostationary transfer orbit) mass of 4170 kg [NEY-90]. At the start of the nineteen eighties, in response to the increasing communication requirements during the previous years, larger and larger communication satellites were considered. In this way, the satellites envisaged for the INTELSAT VII generation were much larger than the INTELSAT VI satellites. In fact, a certain reduction of communication requirements associated with the multiplicity of support available has led to a downward revision of the increasing size of satellites. Hence, the INTELSAT VII satellite has a GTO mass of 3610 kg and the installed electrical power, on the other hand, is 3970 W [NEY-90].

At the same time, technological progress has permitted better use of the mass and power installed in orbit; a more sophisticated payload has provided higher communication capacity to mass and power ratios, high efficiency electrical supply systems (solar cells and distribution), high specific energy (batteries), higher performance propulsion systems requiring less mass of propellent for a given velocity increment, and so on.

Several classes of communication satellite have developed:

—Satellites described as small, with a start of life mass of the order of 600 kg, which were able to satisfy national requirements during the 1980s.
—Medium satellites, with a mass in orbit at the start of life of the order of 1000 to 1500 kg, which are well suited to various requirements.
—Large satellites, of mass in orbit greater than 2000 kg, for specific applications requiring a high electrical power (such as direct television broadcasting, links with mobiles, VSAT, etc.).

The large majority of these are geostationary satellites and this orbit continues to be of major use for communication services. Non-geostationary orbits (of the Molnya and Tundra type) are however considered for mobile satellite services, for example (see Chapter 7.2).

In contrast, micro- (mass less than 50 kg) and mini- (mass of the order of 100 kg) satellites, may be able to offer useful facilities for some applications such as electronic mail and provision of communication services for users with limited requirements, etc.

As far as medium and large satellites are concerned, at the start of the 1990s the most useful compromise between organisation and technology for the various platform subsystems is as follows:

—Attitude control by three axis stabilisation, using angular momentum whose orientation may, or may not, be controllable, associated with compensation of disturbing torques by solar wing or magnetic coil.
—Unified bi-propellant propulsion system for injection into orbit and orbit control in the operational phase (with the potential use of ionic propulsion for north–south control).
—Power supply system with deployable solar generator (and potential use of GaAs cells) and NiH_2 batteries, power distribution by regulated bus (with the possible use of a series regulator in place of the conventional shunt regulator).
—Management of telecommand, telemetry and tracking according to a modular decentralised architecture based on a data exchange bus.
—Thermal control using networks of heat pipes to optimise the use of radiating surfaces.
—A structure using composite materials such as carbon fibre.

These technologies are used on new platforms developed by the various manufacturers such as Eurostar (Matra-British Aerospace) [BEN-86], HS 601 (Hughes Aircraft Company), Satcom (GE-Astro Space), Spacebus (Aerospatiale/MBB) [CHO 88], etc. Experimental platforms such as Olympus (ESA) [BON-86], ETS-VI (Japan) [KIT-89] may incorporate particular technologies (such as attitude stabilisation without on-board angular momentum).

REFERENCES

[ANA-84] F. Ananasso, E. Saggese, L. LoPresti, M. Pent (1984) Simulation-aided design of the ITALSAT satellite regenerative transmission channel, *AIAA 10th Communication Satellite Systems Conference*, pp. 263–269.

[ARI-88] Y. Arimoto, S. Yoshimoto, K. Ohmaru, T. Lida (1988) A multibeam antenna for 27/22 GHz band satellite broadcasting, *AIAA 12th International Communication Satellite Systems Conference*, pp. 670.

[ARN-84] R. Arnim (1984) The Franco-German DBS Program 'TV-SAT/TDF-1', *AIAA 10th Communication Satellite Systems Conference*, pp. 75.

[BAN-88] S. Bangara, A. Patatcchini (1988) EUTELSAT II—A second generation European regional satellite system, *AIAA 12th International Communication Satellite Systems Conference.*

[BAR-84] P.L. Bargellini (1984) A reassessment of satellite communications in the 20- and 30-GHz bands. *AIAA 10th Communication Satellite Systems Conference.*

[BAS-79] H. Bassner (1979) Development status and application of the electric propulsion system RII 10 used for station keeping. *XXXth IAF Congress*, Muncih, Paper 79–07.

[BEN-86] A.H. Bentley (1986) Eurostar, a new highly flexible platform, *AIAA 11th Communication Satellite Systems Conference, San Diego*, Paper 86–0712-CP, March.

[BHA-86] K.B. Bhasin, G. Anzic, R.R. Kunath, D.J. Connolly (1984) Optical techniques to feed and control gaAs MMIC modules for phased array antenna applications, *AIAA 11th Communication Satellite Systems Conference.*

[BIN-84] N. Bingham, A.D. Craig, L. Flook (1984) Evolution of European telecommunication satellite pointing performance, *AIAA 10th Communication Satellite Systems Conference*, pp. 336.

[BON-84] R. Bonhomme, B.L. Herdan, R. Steels (1984) Development and applications of new technologies in the ESA Olympus programme, *AIAA 10th Communication Satellite Systems Conference, Orlando*, pp. 249–262, March.

[BRO-84] H. Broderson, D. Pfefferkorn (1984) INTELSAT VI solar array design and performance, *AIAA 10th communication satellite systems conference.*

[BUG-82] M. Bugeat, D. Valentian (1982) Development status of a cesium field emission thruster, *Acta Astronautua*, **9**(9).

[CAP-85] A. Capel, D. O'Sullivan (1985) Influence of the bus regulation on telecommunication spacecraft power system and distribution, *Proc. ESA Sessions at 16th Annual IEEE PESC.*

[CLO-86] W. Clopp, T.A. Hawkes Jr., C.R. Bertles, B.A. Pontano, T. Kao (1986) Geostationary communications platform payload concepts, *AIAA 11th Communication Satellite Systems Conference*, pp. 577–587.

[CHO-88] M. Chognot, A. Jablonski (1988) The spacebus platforms, *AIAA 10th Communication Satellite Systems Conference.*

[CON-86] M.J. Conroy, R.J. Kerczewski (1986) Testing of 30-GHz low noise receivers, *AIAA 11th Communication Satellite Systems Conference*, pp. 326–340.

[DAY-88] J. Dayton (1988) High efficiency, longlife traveling wave tubes for future communications satellites, *AIAA 12th International Communication Satellite Systems Conference*, pp. 452.

[DRI-86] T.F. Driggers, E.M. Hunter (1986) Geostationary communications platform payload concepts, *AIAA 11th Communication Satellite Systems Conference*, pp. 566–576.

[DUH-89] T.G. Duhamel (1989) Implementation of electric propulsion for north–south station keeping on the EUROSTAR spacecraft, *AIAA 25th Joint Propulsion Conference, Monterey*, Paper 89–2274 (9 pages), July.

[DUR-83] G.W. Durling (1983) High power and pointing accuracy from body-spun spacecraft, *Space and Communication and Broadcasting*, **1**, No. 2, pp. 65–71.

[FAY-86] K.A. Faymon (1986) Spacecraft 2000, *AIAA 11th Communication Satellite Systems Conference, San Diego*, pp. 88–91, March.

[FEA-77] D.G. Feam (1977) A review of the UKTS electron bombardment mercury ion thruster, *ESTEC Conf. on Attitude and Orbit Control System*, Noorduijk.

[FOX-86] S.M. Fox (1986) Attitude control subsystem performance of the RCA series 3000 satellite, *AIAA 11th Communication Satellite Systems Conference*, pp. 79–87.

[FRE-78] B.A. Free, W.J. Guman, G. Herron, S. Zafran (1978) Electric propulsion for communications satellites, *AIAA, 7th CSSC*, San Diego, pp. 746–758.

[FRE-72] B.A. Free (1972) Chemical and electric propulsion tradeoffs for communications satellites, *Comsat Tech. Rev.*, **2**(1), 123–145.

[GEL-84] A. Gelly (1984) Planning of the 12 GHz broadcasting satellites, *AIAA 10th Communication Satellite Systems Conference*, pp. 1.

[GHI-90] L. Ghislanzoni, C. Petagna, A Trippi (1990) The application of ARCJET propulsion systems for geostationary satellites, *AIAA 21st International Electric Propulsion Conference*, 18–20 July, Orlando, Florida, USA.

[GOL-84] A.M. Goldman (1984) An electro-optical communications satellite transponder, *AIAA 10th Communication Satellite Systems Conference*, pp. 262–263.

[HAN-86] K. Handa, K. Honma, Y. Itoh, T. Mochizuki, W. Akinaga, S. Fukuda, I. Haga (1984) Trends and development of low noise amplifiers using new fet device, *AIAA 11th Communication Satellite Systems Conference*, pp. 308–312.

[HOE-86] C.F. Hoeber, C.F. Pollard, R.R. Nicholas (1986) Passive intermodulation product generation in high Power communications satellites, *AIAA 11th Communication Satellite Systems Conference*, pp. 361–374.

[HOO-84] M.L. Hoover, R.W. Lysak, J.R. Wright (1984) Operational aspects of AT&T's companded single sideband satellite system, *AIAA 10th Communication Satellite System Conference*, pp. 112.

[HUB-88] K. Hubner (1988) An advanced electronic power conditioner for 12 GHz 100–150 w traveling wave tube amplifiers, *AIAA 12th International Communication Satellite Systems Conference*, pp. 457.

[IZU-84] T. Izumisawa, S. Kato T. Kohri (1984) Regenerative SCPC satellite communications systems, *AIAA 10th Communication Satellite Systems Conference*, pp. 269.

[JAR-84] K. Jarett (1984) Operational aspects of INTELSAT VI satellite-switched TDMA communication system, *AIAA 10th Communication Satellite Systems Conference*, pp. 107.

[JOY-86] P.F. Joy, Goliazewski (1986) Advanced thermal and power systems for the satcom-ku satellites, *AIAA 11th Communication Satellite Systems Conference*, pp. 697–704.

[KAW-86] T. Kawashima, N. Takata, T. Mukai (1986) Graphite epoxy structure and GaAs solar array for CS-3 domestic communication satellite, *AIAA 11th Communication Satellite Systems Conference*, pp. 639–643.

[KEL-84] H. Kellermeier, D.E. Koelle, R. Barbera (1984) A standardized propulsion module for future communications satellite in the 2000 to 3000 kg class, *AIAA 10th Communication Satellite Systems Conference*.

[KEL-86] H. Kellermeier, H. Vorbrugg, K. Pontoppidan (1986) The MBB unfurlable mesh antenna (UMA) design and development, *AIAA 11th Communication Satellite Systems Conference*, pp. 417–425.

[KEI-86] J.E. Keigler, L. Muhlfelder (1986) Optimum antenna beam pointing for communication satellite, *AIAA 11th Communication Satellite Systems Conference*, pp. 70–78.

[KID-86] A.M. Kidd, H.J. Moody, H. Raine (1986) System design of the Canadian mobile communication satellite (M-SAT) space segment, *AIAA 11th Communication Satellite Systems Conference*, pp. 396–404.

[KRA-84] R. Krawczyk, L. Decramber (1984) Spot solar array, *Photovoltaic Generators in Space (ESA SP-210), Cannes*, pp. 191–199, Nov.

[KOB-88] Y. Kobayashi, S. Makino, T. Katagi (1988) Application of front fed offset Cassegrain antennas for communication satellites, *AIAA 12th International Communication Satellite Systems Conference*, pp. 678.

[LAC-89] B. Lacore (1989) Analysis of power topologies used in telecommunications satellites, *European Space Power Conference, Madrid*.

[LAP-86] N. LaPrade, H. Zelen, P. Caporossi, L. Dolan, Ku-band SSPA for communications satellites? *AIAA 11th Communication Satellite Systems Conference*, pp. 321–325.

[LOC-88] A. Lockwood (1988) The adaptive multiplexer—an exciting new multiplexing technique, *AIAA 12th International Communication Satellite Systems Conference*.

[LOS-88] G. Losquadro (1988) Multiple access system for a data relay satellite using a phased array antenna, *AIAA 12th International Communication Satellite Systems Conference*.

[MAR-84] E.R. Martin (1984) Satellite television corporation's DBS system an update, *AIAA 10th Communication Satellite Systems Conference*, pp. 84.

[MOB-68] F.L. Mobley (1968) Gravity gradient stabilization result from the Dodge Satellite, *AIAA, Paper 68–460*, San Francisco.

[MOR-88] G. Morelli, T. Mattiti (1988) The ITALSAT satellite program, *AIAA 12th International Communication Satellite Systems Conference.*

[MOS-84] V.A. Moseley (1984) Bipropellant propulsion systems for medium class satellites, *AIAA 10th Communication Satellite Systems Conference.*

[NEY-90] P. Neyrat, L. Dest, E. Hunter, L. Templeton (1990) The Intelsat VI spacecraft, *AIAA 13th Int. Communication Satellite Systems Conf.*, Los Angeles, May, pp. 95–110.

[ONO-88] T. Ono, H. Hirose, T. Imatani (1988) Ku-band solid state power amplifier, *AIAA 12th International Communication Satellite Systems Conference.*

[PEE-76] H.A. Pfeffer, E. Slachmuylders, C. Rosetti (1976) The future of European electric propulsion by ESA, *ESA Scientific and Technical Review*, **2**, 256–267.

[PEL-88] A. Pelletier, C. Stael (1988) Ku-band modern FWTA's for satellite transponders, *AIAA 12th International Communication Satellite Systems Conference*, pp. 438–444.

[RAA-86] A.R. Raab, C.E. Profera (1986) Wideband beam-forming networks for satellite antenna sysems a mass schedule and cost-effective route to better RF performance, *AIAA 11th Communication Satellite Systems Conference*, pp. 426–429.

[REN-84] U. Renner, J. Nauck (1984) Development trends in Europe on satellite clusters and geostationary platforms, *AIAA 10th Communication Satellite Systems Conference*, pp. 622.

[SAI-84] Saint-Aubert, D. Valentian, W. Berry, Utilization of electric propulsion for communication satellites, *AIAA 10th Communication Satellite Systems Conference*, pp. 354, 1984.

[SCH-88] R. Schreib (1988) Abstract—readiness appraisal: ion propulsion for communications satellites, *AIAA 12th International Communication Satellite Systems Conference.*

[SME-86] J. Smetana, T.J. Kascak, R.E. Alexovich (1986) MMIC antenna technology development in the 30/20 gigahertz band, *AIAA 11th Communication Satellite Systems Conference*, pp. 430–443.

[SMO-84] A.E. Smoll, T.E. Roberts, E.W. Matthews, E.A. Lee and C.C. Han (1984) A new multiple beam satellite antenna for 30/20 GHz communications coverage of conus—experiment evaluation, *AIAA 10th Communication Satellite Systems Conference*, pp. 33.

[SOR-88] R. Sorbello (1988) Advanced satellite antenna developments, *AIAA 12th International Communication Satellite Systems Conference*, pp. 652.

[TAK-86] N. Takata, S. Matsuda (1986) Development of new solar cells and their space applications, *Space Communications and Broadcasting*, **4**, No. 2, 101–113, Fall.

[TSU-88] H. Tsunoda, K. Nakajima, A. Miyasaka (1988) Thermal analysis method of high capacity communications satellite with heat pipes, *AIAA 12th International Communication Satellite Systems Conference.*

[TSU-88] N. Tsunoda, D. Collins (1988) Satellite communications in Japan via super bird, *AIAA 12th International Communications Satellite Systems Conference.*

[WAD-84] T.O. Wade, Approaches to optimization of SS/TDMA time slot assignement, *AIAA 10th Communication Satellite Systems Conference*, pp. 645.

[WEI-84] A.W. Weinrich, M.R. Freeling (1984) RCA advanced satcom—the first all-solid-state communications satellite, *AIAA 10th Communication Satellite Systems Conference*, pp. 581.

[WER-78] J.R. Wertz (1978) *Spacecraft Attitude Determination and Control*, Kluwer.

[YOS-88] S. Yoshida, K. Sato, H. Matsumoto, S. Hokuyo, K. Yamagami (1988) Evaluation of large area and thin GaAs solar cells for space use, *AIAA 12th International Communication Satellite Systems Conference, Arlington*, pp. 63–71, March.

11 SATELLITE INSTALLATION AND LAUNCHERS

In the preceding chapters, the system was in its nominal operating configuration with the satellite in its orbit. Telecommunications missions, the communication techniques used, orbits and system components have been successively examined. System installation, which determines successful commissioning of the system, remain to be described. The specific functions to be performed are presented in this chapter. Some characteristics of launchers will also be described.

11.1 INSTALLATION IN ORBIT

11.1.1 General principle

Installation into orbit consists of positioning the satellite in its nominal orbit from a launching base on the surface of the earth. A launch vehicle which may have various associated auxiliary propulsion systems is used to inject the satellite into an intermediate orbit called the transfer orbit. The procedure using a transfer orbit is based on the so-called Hohmann transfer which enables the satellite to move from a low altitude circular orbit to a higher altitude circular orbit with a minimum expenditure of energy [HOH-25]. The first velocity increment changes the circular orbit into an elliptical one whose perigee altitude is that of the circular orbit (the velocity vector before and just after the velocity increment is perpendicular to the radius vector of the orbit), and the altitude of the apogee depends on the magnitude of the applied velocity increment. A second velocity increment at the apogee of the transfer orbit enables a circular orbit to be obtained at the altitude of the apogee (the velocity vector just before and after the velocity increment is perpendicular to the radius vector of the corresponding orbit).

Figure 11.1 illustrates this procedure for the installation into orbit of a geostationary satellite which will serve as a reference in the following part of this chapter (most communication satellites are geostationary). The equatorial circular orbit at an altitude of 35 786 kilometres is reached by way of a transfer orbit into which the satellite has been injected by the launching system. Circularisation of the orbit is achieved by means of a velocity impulse provided at the apogee of the transfer orbit.

Additional details may be provided in accordance with the type of launcher; the procedure is covered by one of the following:

—From a low altitude earth orbit (LEO), the satellite may be injected into the transfer orbit

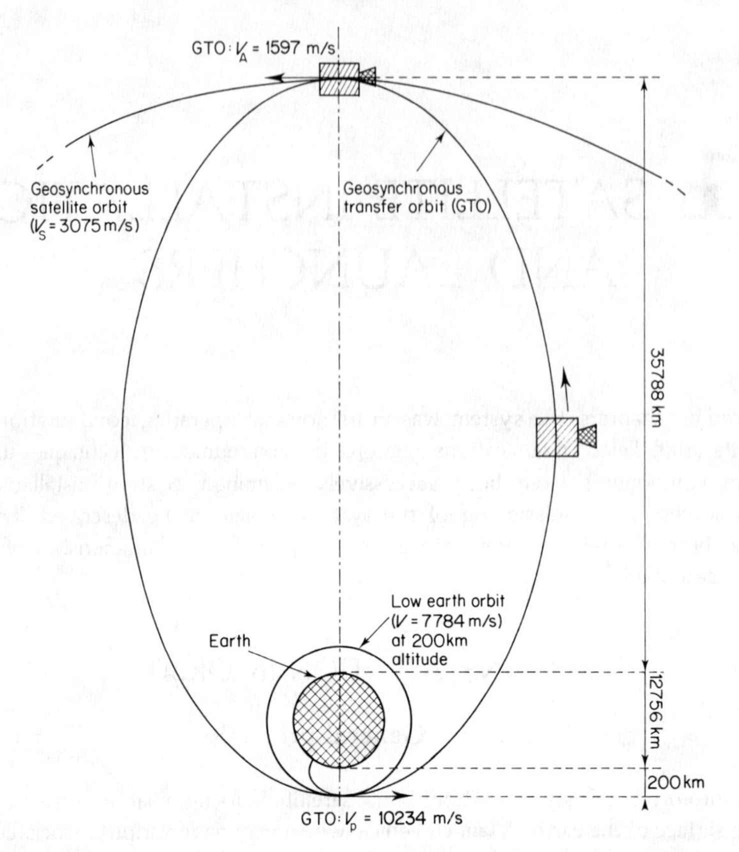

Figure 11.1 Geosynchronous transfer orbit (GTO) from a low earth orbit to a geosynchronous orbit.

by means of a specific propulsion system (the *perigee stage* or the *perigee motor* depending on whether the system is independent of the satellite or an integral part of it). This is the procedure when a satellite is launched using the American Space Shuttle or Space Transportation System (STS) or using the Commercial Titan launch vehicle. A second velocity increment must be provided at the apogee to circularise the orbit, either by an independent *transfer stage*, or by an *apogee motor* integrated into the satellite.

— The satellite may be directly injected into the geostationary transfer orbit (GTO). The launcher must communicate the appropriate velocity to the satellite at the perigee of an elliptic orbit whose perigee altitude is that of the injection point (injection at the perigee), and whose apogee altitude is that of the geostationary satellite orbit. This is the procedure used by most conventional launchers such as Ariane, Delta and Atlas Centaur. A velocity increment must be provided at the apogee of the transfer orbit by the satellite apogee motor to circularise the orbit.

— Finally, the launcher itself can inject the satellite into geostationary earth orbit (GEO). The launcher successively provides the velocity increments required to cause the satellite (and the last stage of the launcher) to move into the transfer orbit and the velocity increment to circularise the orbit. This procedure is that of several conventional launchers such as the US Titan IIIC and the Soviet Proton.

Precise determination of the transfer orbit parameters requires trajectory tracking on several successive orbits. In order to avoid excessive perturbations of successive orbits by atmospheric braking, the selected altitude of the perigee must not be below 150 kilometres. It is generally from 200 to 300 kilometres.

11.1.2 Calculation of the required velocity increments

11.1.2.1 Orbit velocity

The relation $V^2 = 2\mu/r - \mu/a$ permits the velocities at the perigee and apogee of the transfer orbit to be calculated (see Section 7.1.4.2) where:

— a is the semi-major axis of the ellipse,
— μ is the earth's gravitational constant ($\mu = 398,\ 603\,\text{km}^3/\text{s}^2$),
— r is the distance from the centre of the earth to the point concerned on the ellipse which moves with velocity V. The semi-major axis of the ellipse has a value given by:

$$a = [(h_P + h_A)/2] + R_E \qquad (11.1)$$

where h_P and h_A are the altitudes of the perigee and the apogee and R_E is the terrestrial radius equal to 6378 km. Taking 200 km for the altitude of the perigee:

$$a = (200 + 35\,786)/2 + 6378 = 24\,372\,\text{km}$$

Hence:
— At the perigee: $r_P = 6578\,\text{km}$ $V_P = 10\,234\,\text{m/s}$
— At the apogee: $r_A = 42\,166\,\text{km}$ $V_A = 1597\,\text{m/s}$

11.1.2.2 Direct injection of the satellite into the transfer orbit

Most conventional launchers inject the satellite at the perigee of the transfer orbit. The mission of the launcher is thus to convey the satellite to the required altitude h_P with a velocity vector parallel to the surface of the earth (that is perpendicular to the radius vector so that the injection point is the perigee) and having a modulus equal to V_P, the velocity at the perigee.

The required injection velocity is given by:

$$V_P = \sqrt{[(2\mu/(R_E + h_P)) - (\mu/a)]} \qquad \text{(m/s)} \qquad (11.2)$$

11.1.2.3 Coplanar velocity increments

If it is assumed that successive orbits are in the same plane, the velocity increment required to transfer from one orbit to another is equal to the difference between the velocity of the satellite in the final orbit and the velocity in the initial orbit.

For circularisation at the apogee of the transfer orbit, since the velocity of the geostationary satellite is 3075 m/s, the velocity increment to be provided is given by:

$$\Delta V = 3075 - 1597 \text{ m/s}$$
$$\Delta V = 1478 \text{ m/s} \tag{11.3}$$

The velocity increment at the perigee from a low altitude circular orbit is calculated in a similar manner.

11.1.3 Inclination correction and circularisation

In the previous discussion, all orbit changes have taken place in the same plane. If the final orbit is the geostationary satellite orbit, the initial orbits must be in the equatorial plane. What are the parameters which determine the inclination given to orbits by the launcher? What must be done if the inclination of the transfer orbit is not zero?

11.1.3.1 Minimum inclination of the initial orbit provided by the launcher

The launch vehicle takes off from a launch base M and follows a trajectory in a plane which contains the centre of the earth and is characterised by the angle A which forms a projection **U** on the horizontal plane of the velocity vector and the direction of the north (the launch azimuth). The components of the unit vectors along **OM** and **U** are (Figure 11.2):

	OM	**U**
Along Ox:	$\cos l$	—
Along Oy:	0	$\cos(90° - A)$
Along Oz:	$\sin l$	—

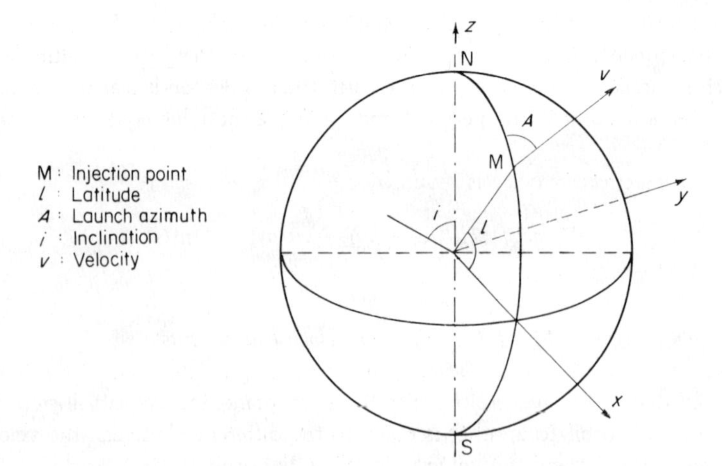

Figure 11.2 Launch azimuth A, launch-pad latitude l and transfer orbit inclination i.

where l is the latitude of the launching base. The unit vector in the direction of $\mathbf{OM} \wedge \mathbf{U}$ is perpendicular to the plane of the orbit and thus has a component along Oz given by:

$$\cos i = \sin A \cos l \tag{11.4}$$

The inclination i of the orbit obtained is thus greater than or equal to the latitude l of the launching base if the trajectory of the launch vehicle is planar; this is the usual procedure since every manoeuvre which changes the plane induces mechanical constraints and an additional expenditure of energy.

The minimum inclination equal to the latitude of the launch pad is obtained for a launch azimuth $A = 90°$ that is for a launch towards the east. A launch towards the east also enables the greatest benefit to be taken of the velocity introduced into the trajectory by the rotation of the earth. The velocities calculated for the orbits are absolute velocities in a reference fixed in space. At the instant of take-off, since the launch vehicle (and the satellite) are coupled to the rotating earth, they benefit in the plane of the trajectory from a velocity V_l induced by the rotation of the earth and equal to:

$$V_l = V_E \cos l \sin A \qquad \text{(m/s)} \tag{11.5}$$

where $V_E = \Omega_E R_E = 365$ m/s is the velocity of a point on the earth's equator, with Ω_E, the angular velocity of the rotation of the earth, equal to 360°/86 164 s (see Section 7.1.5) and $R_E = 6378$ km, the mean equatorial radius. It should be noticed that the velocity induced by the rotation of the earth can, in certain cases, be a disadvantage, particularly when polar orbits or an inclination greater than 90° (retrograde orbits) are to be obtained. It is then necessary to provide additional energy which becomes greater as the latitude of the launch base becomes less.

Without a manoeuvre to change the plane, zero inclination would therefore require a launch base situated on the equator. Furthermore, the velocity component induced by rotation of the earth would then be a maximum. If the latitude of the launching base is not zero, the inclination of the orbit obtained is no longer zero and an inclination correction manoeuvre must be performed.

For example, for a launch from the Kennedy Space Center (KSC) at Cape Canaveral in Florida (Eastern Test Range (ETR), latitude 28°, the orbit cannot be inclined at less than 28°. For a launch from the base at Kourou in Guyana (latitude 5.3°), the inclination cannot be less than 5.3°.

11.1.3.2 *The principle of the correction strategy*

Consider the transfer orbit into which the satellite is placed by a conventional launcher. The plane of the transfer orbit is defined by the centre of the earth and the velocity vector at a given instant; the inclination of this orbit is defined by the angle between the plane of the orbit and the equatorial plane (see Section 7.1.4).

Inclination correction, that is transferring the satellite from the plane of the transfer orbit to the plane of the equator (Figure 11.3a), requires a velocity increment to be applied as the satellite passes through one of the nodes of the orbit such that the resultant velocity vector V_S is in the plane of the equator (Figure 11.3b). As the final orbit should be within the equatorial plane, that is of zero inclination, the manoeuvre must be performed at the nodes.

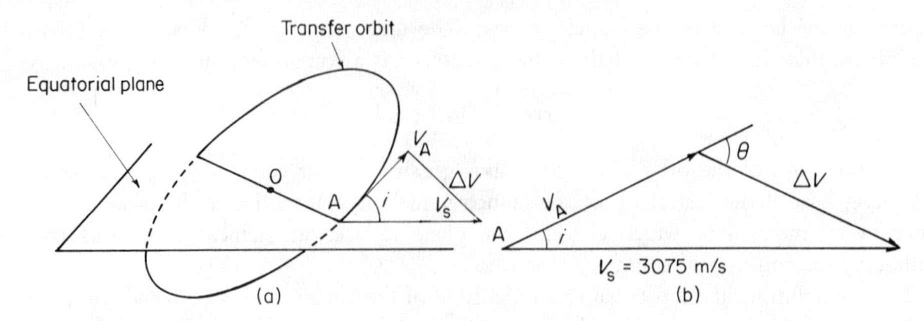

Figure 11.3 Inclination correction: (a) transfer orbit plane and equatorial plane; (b) required velocity increment (value and orientation) in a plane perpendicular to the line of nodes.

(This is not always the case when north–south station-keeping inclination corrections are performed (as described in Section 7.3.4.5); the latter can be performed away from the nodes since the orbit after correction is not necessarily of zero inclination.)

However, in some special cases (such as satellites without north–south control and satellites awaiting operational service in an inclined parking orbit, see Section 7.3.4.5), the intended orbit is not of zero inclination and the right ascension of the ascending node is imposed. It is then necessary to change the inclination of the initial orbit by a given amount while displacing the position of the ascending node; the velocity increment is no longer applied at the node and a special procedure is then used [SKI-86], [POC-87, pp. 34–38].

For a given inclination correction, the velocity impulse ΔV to be applied increases with the velocity of the satellite. The correcting operation is more economic when this velocity is low. The correction is thus performed at the apogee of the transfer orbit at the same time as circularisation. For this the following are required:

- The *perigee–apogee line* (the apsidal line) should be in the *equatorial plane,* that is coincident with the *line of nodes.* This implies that injection at the perigee of the transfer orbit occurs on crossing the equatorial plane,
- The apogee of the *transfer orbit* should be at the *altitude of the geostationary satellite orbit,*
- The *thrust direction of the apogee motor* should have *a correct orientation with respect to the satellite velocity vector* in the plane perpendicular to the local vertical. As the apogee motor is mounted rigidly along a mechanical axis of the satellite, the *orientation of this axis must be stabilised* during the manoeuvre.

Taking account of the fact that V_S is nearly twice V_A, the geometry of Figure 11.3b shows that θ is approximately $2i$. For Cape Canaveral θ is approximately 56°; for Kourou, since the nominal inclination of the Ariane transfer orbit is approximately 7°, θ has a value of around 14°. The exact value of θ is determined from:

$$\theta = \arcsin[(V_S \sin \Delta i)/\Delta V_A)] \quad \text{(rad)} \tag{11.6}$$

where ΔV_A is the total velocity increment to be applied for circularisation and inclination correction by an amount Δi. The value of this velocity increment is given by:

$$\Delta V_A = \sqrt{(V_S^2 + V_A^2 - 2V_A V_S \cos \Delta i)} \quad \text{(m/s)} \tag{11.7}$$

where V_S is the velocity of the satellite in the final circular orbit (equal to 3075 m/s for the geostationary satellite orbit) or by:

$$\Delta V_A = \sqrt{[\mu/(R_E + h_P)]}\sqrt{K}\sqrt{\{1 + [2K/(1 + K)] - 2\sqrt{[(2K\cos i)/(1 + K)]}\}} \quad (11.8)$$

where $K = (R_E + h_P)/(R_E + h_A)$, with:

h_P = altitude of the perigee
h_A = altitude of the apogee (equal to $R_0 = 35\,786$ km)
R_E = 6378 km, the mean equatorial radius
$\mu = 3.986 \times 10^{14}\,m^3/s^2$, the earth's gravitational constant.

By assuming that the altitude of the perigee is 200 km, the expression for ΔV reduces to:

$$\Delta V = \sqrt{[12.006 - 9.822\cos i]} \qquad (km/s) \quad (11.9)$$

It can be seen that, for i greater than 70°, the inclination correction requires an impulse greater than that required for circularisation.

11.1.3.3 *Procedures based on three impulses*

The Hohmann procedure is optimum for transfer between coplanar circular orbits using two velocity increments. When the ratio between the radius of the final and initial orbits is large (greater than 12), a procedure, called bi-elliptical, which use three velocity increments is more economical [MAR-79]. The same situation arises for smaller radius ratios (of the order of 6 which are characteristic of geostationary transfer orbits) when the inclination change to be achieved exceeds 40°. Then for launch bases at high latitude, such a procedure can be considered; this consists of injecting the satellite into a transfer orbit whose apogee altitude is greater than that of the geostationary satellite orbit (Figure 11.4) (supersynchronous transfer orbit). At the apogee of this transfer orbit, a manoeuvre which corrects the inclination and increases the altitude of the perigee to that of the geostationary satellite orbit is performed. This manoeuvre requires a velocity increment less than that required for the same operation at the geostationary satellite altitude since the velocity of the satellite is lower. A final velocity increment reduces the apogee altitude to that of geostationary satellites. This procedure is also of interest to reduce the propellant requirements for apogee boost to increase the lifetime of the satellite (more fuel available for station keeping) or the available payload mass, at the expense of an increase of the required launcher performance.

11.1.3.4 *Procedure from an initial inclined circular orbit*

When the launcher delivers the satellite into a low altitude circular orbit and if a change of inclination is necessary, two approaches are possible with a two impulse strategy:

— The satellite is first injected into a transfer orbit with an inclination equal to that of the initial orbit, then the orbit is circularised and the inclination is corrected at the same time at the apogee of the transfer orbit. This method may be preferable since the velocity is lower at the apogee than at the perigee and inclination correction at the former will be less costly. The velocity increment ΔV_P to be provided at the perigee will be given by:

$$\Delta V_P = V_P - V_I \qquad (m/s) \quad (11.10)$$

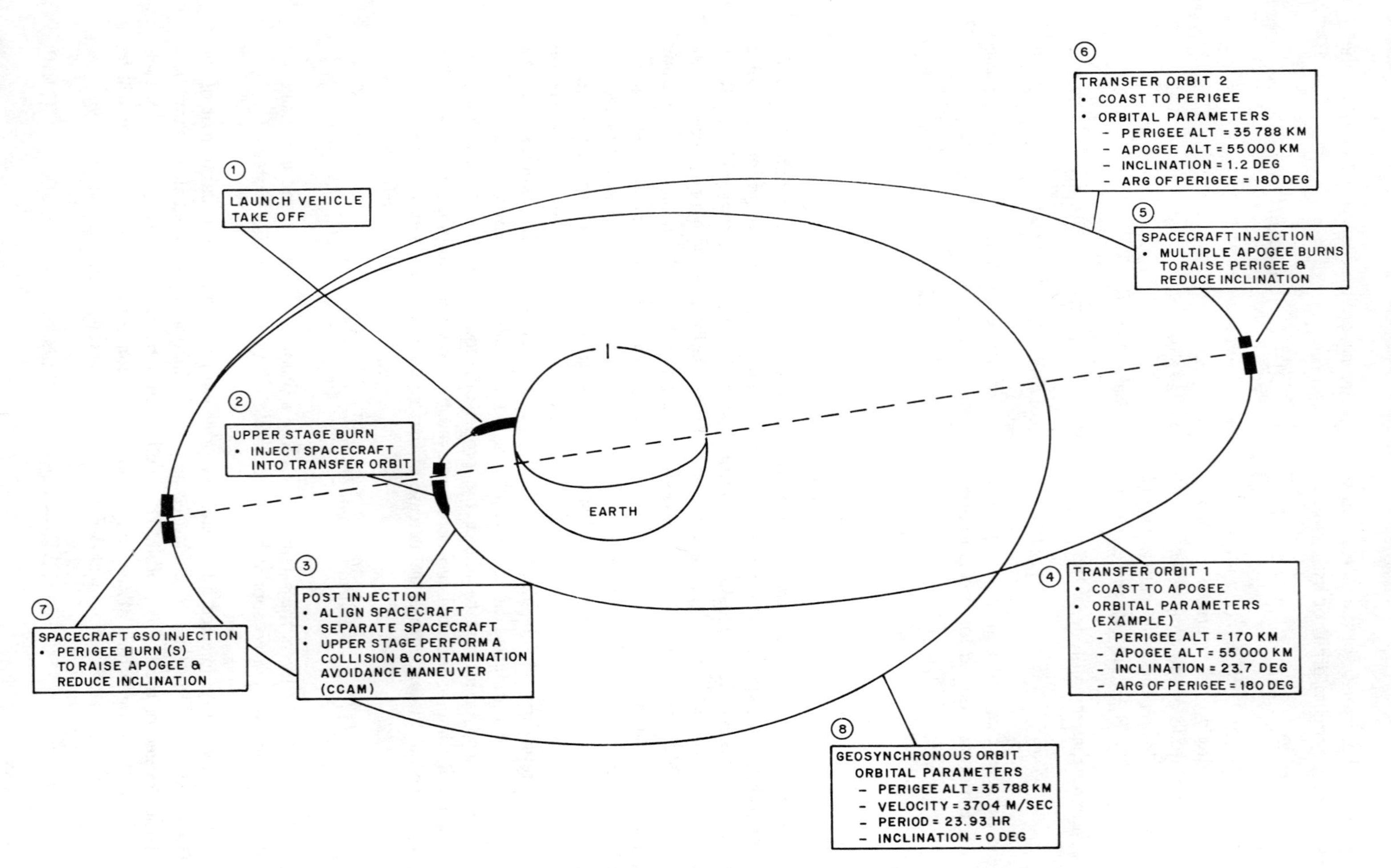

Figure 11.4 Supersynachronous transfer orbit ascent profile [WHI-90]. (Reproduced by permission of the American Institute of Aeronautics and Astronautics.)

where V_P is the velocity at the perigee of the transfer orbit (11.3) and V_l is the velocity in the initial circular orbit of altitude h_p. V_I is given by $\sqrt{[\mu/(R_E + h_P)]}$.

— Share the inclination correction between the velocity increment at the perigee and the velocity increment at the apogee. This procedure is called the generalised Hohmann method; it enables the total required velocity increment $\Delta V_A + \Delta V_P$ to be minimised by varying the magnitude of the inclination correction Δi_P obtained during the first velocity increment at the perigee. The velocity increment to be provided at the perigee then has a value given by:

$$\Delta V_P = \sqrt{(V_l^2 + V_P^2 - 2V_l V_P \cos \Delta i_P)} \quad \text{(m/s)} \tag{11.11}$$

The velocity increment to be provided at the apogee is evaluated using (11.7) where the inclination change will be given by $\Delta i_A = \Delta i_l - \Delta i_P$ (Δi_l is the inclination of the initial orbit).

In the case of an initial orbit of altitude 290 km and inclination 28.5°, optimisation shows that the total velocity increment is minimised by reducing the inclination by 2.2° on injection at the perigee of the transfer orbit which will thus have an inclination 26.3° (the altitude of the apogee is 35 786 km).

11.1.3.5 *Non-implusive velocity increments*

The procedures described above assume impulsive velocity increments are applied at specific points in the orbit (impulsive indicates a very short duration with respect to the period of the orbit). The example below will show that the quantity of propellant required for these manoeuvres is large and it is therefore necessary for the motor thrust to be high so that the combustion time is short (see Section 10.3 for the relations between the various magnitudes). With solid fuel motors (see Section 11.1.4.1), the impulse assumption is justified (the thrust is of several tens of thousands of newtons with a combustion time of only tens of seconds).

On the other hand, with bi-propellant motors (see Section 11.1.4.2), the thrust is limited to a few hundreds of newtons (typically 400 or 490 N) and the combustion time for an inclination correction and circularisation manoeuvre at the apogee of the transfer orbit can be of the order of a hundred minutes. During the thrust, the satellite moves significantly in the orbit and therefore does not remain in the vicinity of the apogee; this reduces the efficiency of the manoeuvre. This loss of efficiency causes an additional quantity of propellant to be consumed in comparison with the quantity which would be required by an impulsive manoeuvre [ROB-66].

A first step in reducing the loss of efficiency is obtained by igniting the motor before the satellite reaches the apogee in such a way that the combustion time extends over a section of the orbit which is symmetrical with respect to the apogee.

Two techniques allow the loss of efficiency to be further reduced in order to approach that of an impulsive manoeuvre:

— Control of the direction of the motor thrust in such a way that it always remains parallel to the orbit as does the velocity of the satellite.
— Subdividing the velocity increment into several manoeuvres.

In connection with thrust orientation, two techniques can be considered during apogee motor operation:

—Conservation, with inertial axes, of the thrust direction which remains fixed in space.
—Orientation of the thrust direction with displacement of the satellite.

Fixed orientation of the thrust direction in space is easily obtained by causing the satellite to rotate about its mechanical axis along which the motor is mounted. The axis will have been previously oriented in the required direction, as defined by the angle θ given by (11.6), in the plane perpendicular to the radius vector at the apogee. At some distance from the apogee, the orientation of the motor thrust is, therefore, not that of the satellite velocity vector and the efficiency of the thrust is reduced (the effective thrust is the actual thrust multiplied by the cosine of the angle between the thrust vector and the velocity vector).

Rotation of the thrust so that it remains aligned with the velocity vector requires active control of the satellite attitude in accordance with a particular control law. The efficiency of the manoeuvre is then increased and reaches around 99.5% with respect to an impulsive manoeuvre. This procedure has been used for the Olympus satellite of the European Space Agency [MUG-86].

Furthermore, an efficiency approaching that of an impulsive manoeuvre is obtained by subdividing the velocity increment required for circularisation and inclination correction into several manoeuvres. Passage from the transfer orbit to the geostationary satellite orbit is thus performed by means of several increases of perigee altitude using short duration thrusts as the satellite passes through the apogees of the resulting intermediate orbits. Figure 11.5 illustrates this procedure for installation of the TDF1 satellite.

The advantages of multiple thrusts are as follows:

—Since the portion of the orbit on which the thrust is exerted is reduced on each manoeuvre, the efficiency of the manoeuvre is higher.
—The motor thrust can be calibrated during the first operation thereby permitting more precise subsequent manoeuvres.
—Optimisation of successive manoeuvres can be performed by taking account of errors and variations of previous thrusts.
—By varying the amplitude and date of thrusts, it is possible to combine the operations of circularisation and inclination correction with those of positioning the satellite in orbit at its station longitude (see Section 11.1.7.2). This permits the quantity of propellant consumed for these operations to be minimised.

Between each manoeuvre, the satellite performs at least two revolutions in orbit so that the orbit parameters may be determined precisely.

The number of manoeuvres is chosen in such a way that loss of efficiency with respect to an impulsive manoeuvre is limited as far as possible. With a two manoeuvre strategy, the duration of each thrust remains long and efficiency is limited. A three manoeuvre strategy enables an efficiency of 99.75% to be achieved while allowing flexibility in the distribution of the total velocity increment among the three manoeuvres; this facilitates accommodation of various constraints, such as the solar aspect and visibility of stations, in the optimisation process [BOI-86], [POC-86]. For more than three manoeuvres the operational constraints of the control centre become prohibitive.

In distributing the total velocity increment among the manoeuvres, that of the third impulse is generally chosen to be the smallest in order to obtain small realisation errors; this enables deviation from the desired orbit to be minimised. In contrast, according to the optimisation

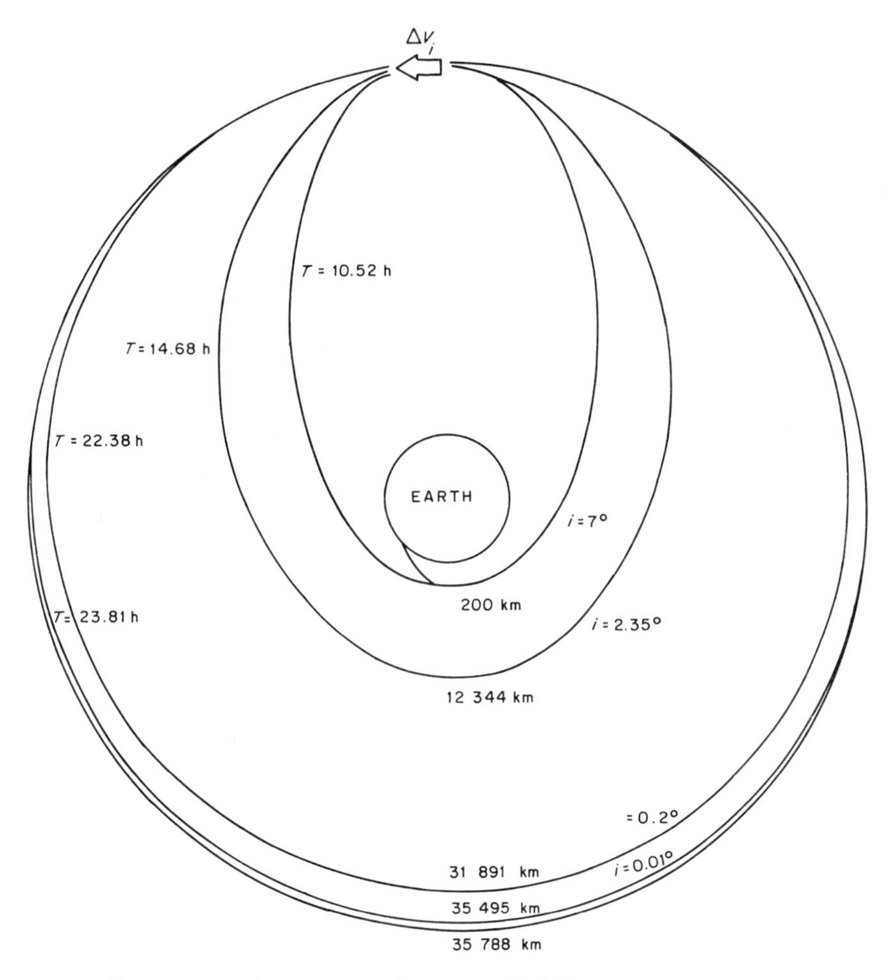

Figure 11.5 Geostationary orbit injection with apogee multiple burn strategy.

process used and the constraints taken into account, the values of the first two increments may be either decreasing (e.g. TVSAT: 57% and 36% [RAJ-86] and TDF1: 55% and 39% or increasing (e.g. 33% and 45% [POC-86]). Various approaches to the optimisation process and procedures are discussed in the references already cited in this section together with [ESC-86], [KAM-86], [POC-87, Chap 4] [WEI-85].

11.1.3.6 *Increase of velocity at the perigee*

The use of re-ignitable bi-propellant motors permits use, if necessary, of the motor for the first time on passing through the perigee to increase the velocity of the satellite (perigee velocity augmentation manoeuvre (PVA)).

This manoeuvre enables limited performance of the launcher with respect to the launch mass of the satellite to be alleviated. A launcher in a given configuration is capable of placing

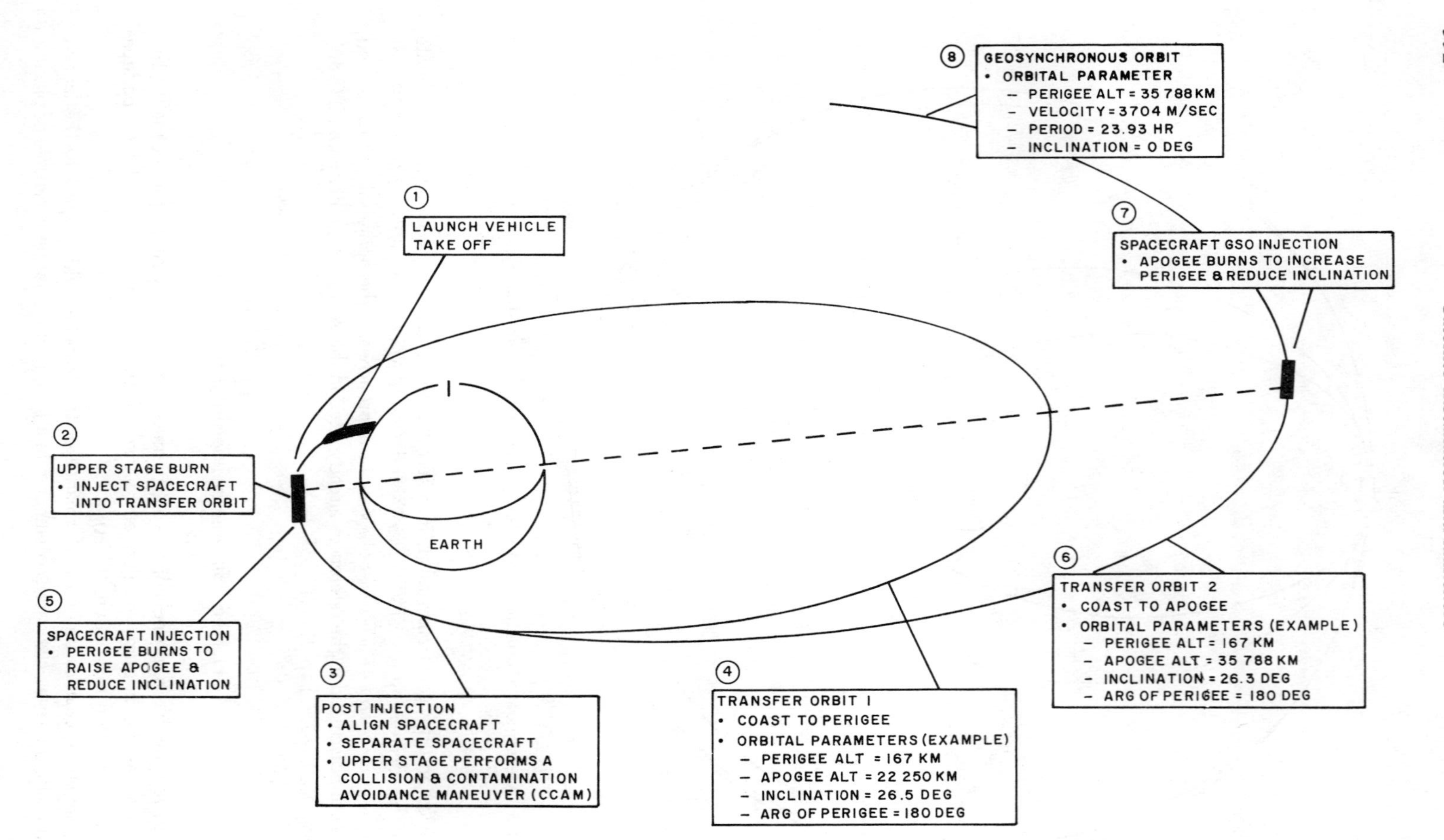

Figure 11.6 Geostationary transfer orbit ascent profile with perigee velocity augmentation [WHI-90]. (Reproduced by permission of the American Institute of Aeronautics and Astronautics).

a certain mass in an elliptical transfer orbit of nominal altitude at the apogee. If the mass is greater, the velocity obtained on injection at the perigee will be less than the nominal velocity and this will result in an altitude of the apogee which is lower than the nominal one. It is then possible to use the satellite motor to provide a velocity increment which compensates for the inadequate velocity at the perigee and hence restores the altitude of the apogee to its nominal value. This is clearly obtained to the detriment of the satellite propellant budget, but this approach can prove useful after overall optimisation to avoid the use of a more powerful (and hence more expensive) launcher when the mass of the satellite to be put into orbit slightly exceeds the nominal performance of the launcher. The performance of launchers is effectively discretised with steps which can be large (several hundreds of kilograms) according to the type or version. The procedure is illustrated in Figure 11.6 with the Delta launch vehicle [WHI-90].

11.1.4 The apogee (or perigee) motor

The velocity increments required for changes of orbit, when not provided by the launcher, are provided by either the so-called apogee or perigee motor according to the case. Since the velocity increments to be provided are large, the motor thrust must be high for the combustion time to be short. The technologies which permit high thrusts to be obtained are solid and bi-propellant propulsion. Definitions of the characteristic magnitudes of thrusters and the relations between the mass of propellant and the velocity increment provided have been explained in Section 10.3.

11.1.4.1 Solid propellant thrusters

Solid propulsion is a conventional technology for satellite apogee motors and the motors of transfer stages which are attached to them.

A solid propellant motor consists of a mixture of oxidiser and fuel in solid from in a case of titanium or composite material (an epoxy impregnated Kevlar wound shell) which communicates with the exterior via a nozzle (Figure 11.7). The nozzle is usually realised in

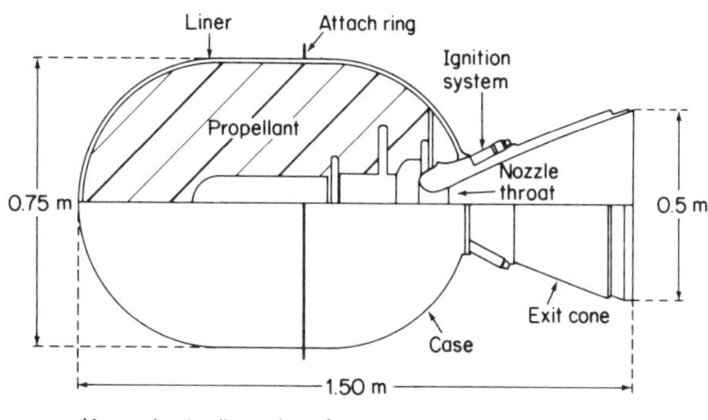

Figure 11.7 Construction of a solid propellant motor.

a carbon fibrous skeleton or substrate serving as a reinforcement and a carbon matrix which ensures binding of the fibres. This material can support very high temperatures (3500 °C), has good resistance to distortion and a low density [MEL-87].

The propellant grain uses a fuel with high percentages of polybutadiene carboxide or hydroxide to which additives are added. Aluminium powder is used as a combustant. Thermal protection is deposited on the inner surface of the case before melting the grain in order to protect the case during combustion. The block of powder after solidification is machined in such a way as to provide a pipe in which the combustion propagates. The form given to this pipe determines the variation with time of the surface of the propellant which is available for combustion and this in turn conditions the thrust profile of the motor. The mixture is ignited by an electrically controlled igniter located either at the extreme rear (opposite to the nozzle) or close to the throat of the nozzle.

For a case of given size, it is possible to vary the quantity of powder contained within limited proportions. This unloading factor, of the order of 10 to 15%, enables the velocity increment which the motor provides to be adjusted when manufacturing the motor. The specific impulse of solid propellant motors is of the order of 290 s.

Examples of these motors are the MAGE motors developed by SEP under contract to the European Space Agency [ASA-83], [ISO-79] and the STAR motors manufactured by THIOKOL in the USA. The performance of these motors, together with others which are also on the market, are given in Table 11.1. Most of these motors have been developed for use on missiles.

These motors are used as apogee or perigee motors according to their performance in

Table 11.1 Characteristics of various solid propellant thrusters which can be used as perigee and apogee motors

Name	Mass (kg)	Propellant (Max.) (kg)	Impulse (10 N s)	Max. thrust (empty) (N)	*Isp* (s)
MAGE 1	368	335	0.767	28 500	287.6
MAGE 1S	447	410	1.168	33 400	290.7
MAGE 2	529	490	1.410	46 700	293.8
STAR 30B	537	505	1.460	26 825	293.1
STAR 30C	621	585	1.645	31 730	285.2
STAR 30E	660	621	1.780	35 365	290.1
STAR 31	1398	1300	3.740	95 675	293.2
STAR 48	2115	1998	5.695	67 820	290.0
STAR 62	2890	2740	7.820	78 320	291.2
U.T. SRM-2	3020	2760	8.100	260 750	303.6
Aerojet 62	3605	3310	9.310	149 965	286.7
STAR 63E	4422	4059	11.866	133 485	298
STAR 75	4798	4563	13.265	143 690	296.3
Aerojet 66	7033	6256	17.596	268 335	286.7
Minuteman III	9085	8390	23.100	206 400	280.6
U.T. SRM-1	10390	9750	28.100	192 685	295.5

**Isp* is sometimes expressed in N s/kg or lbf s/lbm. The equivalences are (1 lbm = 0.4536 kg; 1 lbf = 4.448 N); *Isp* (s) = *Isp* (lbf sec/lbm) = *Isp* (N s/kg) × 1/9.807

accordance with the combined (satellite–motor) mass and the velocity increment to be imparted to it. These motors are used to propel transfer stages and even as the upper stage of certain conventional launchers (e.g. Delta).

11.1.4.2 Bi-propellant liquid motors

These motors use hypergolic (spontaneous ignition on contact) liquid propellants which generate hot gases by combustion and these expand in a convergent–divergent nozzle. Various fuel–oxidiser combinations can be used (Table 11.2, from [PRI-86]).

For satellite motors, a much used combination is nitrogen tetroxide (N_2O_4 or NTO) as the oxidiser and monomethylhydrazine ($CH_3.NH.NH_2$ or MMH) as the fuel; this enables combustion temperatures of the order of 3000 °C to be obtained. Thrusts are of the order of 400 N (the MBB motor) to 490 N (the Marquard motor).

The specific impulse is of the order of 310 to 320 s under nominal conditions of mixture ratio and propellant supply pressure to the motor. The mixture ratio is defined as the ratio of the mass of oxidiser to the mass of fuel injected into the motor combustion chamber in unit time. This ratio influences the specific impulse of the motor and must be optimised to obtain the maximum value of impulse as a function of the chemical composition of the propellants [GOR-71]. The stoichiometric value (to ensure combustion without residue) is of the order of 1.6. The supply of propellants under pressure is achieved by means of a pressuring gas (helium) stored under high pressure (200 bar) in a reservoir. This gas, after passing

Table 11.2 Oxidiser/fuel combinatons for bi-propellant propulsion

Fuel	Oxidiser	Specific impulse (s)
(C) Hydrogen (H_2)	(C) Oxygen (O_2)	430
(L) Kerosene (RP-1)	(C) Oxygen	328
(L) Hydrazine (N_2H_4)	(C) Oxygen	338
(L) UDMH†	(C) Oxygen	336
(C) Hydrogen	(C) Fluorine (F_2)	440
(L) Hydrazine	(C) Fluorine	388
(L) 0.5 UDMH–0.5 N_2H_4	(C) Fluorine	376
(L) Hydrazine	(L) Nitrogen tetroxide (N_2O_4)	314
(L) MMH* (CH_3NHNH_2)	(L) Nitrogen tetroxide	328
(L) Aerozine (AZ50)	(L) Nitrogen tetroxide	310
(L) UDMH†	(L) Nitrogen tetroxide	309
(L) 0.5 UDMH–0.5 N_2H_4	(L) Nitrogen tetroxide	312
(L) UH25‡	(L) Nitrogen tetroxide	320
(L) 0.5 UDMH–0.5 N_2H_4	(L) Nitric acid	297
(L) Pentaborane	(L) Nitric acid	321

(C), cryogenic; (L), liquid at ambient temperature.
*MMH: monomethylhydrazine.
†UDMH: unsymmetric dimethylhydrazine.
‡UH25: 25% hydrazine hydrate and 75% UDMH.

through a pressure regulator, forces the propellant from the satellite tanks into the motor under constant pressure (10 to 14 bar) during operation.

The use of electric pumps to feed the motor is being examined. Pumps enable the mass of the propulsion system to be reduced due to elimination of the pressurised helium reservoir and reduction of the mass of the propellant tanks. The latter need no longer support the motor supply pressure but only the pump feed pressure; this is obtained from helium which is stored under low pressure either directly in the propellant tanks or in a small auxiliary tank. The electric power required to operate the pumps can be provided by the solar generator and the satellite batteries or by separate batteries.

The use of bi-propellant propulsion for the apogee motor (and possibly the perigee motor) leads naturally to the concept of a unified propulsion system (see Section 10.3.4.3); this permits the mass to be reduced in comparison with the combination of a solid fuel apogee motor and a propulsion system for attitude and orbit control using catalytic decomposition of hydrazine. Furthermore, the excess propellant not used during a nominal apogee thrust can be used for orbit control thereby permitting an increase in the lifetime of the satellite.

For the transfer stages, it is advantageous to use motors of higher thrust [KEL-84]. The following can be cited among motors operating with an MMH/NTO combination: Mitsubishi RE10-300 of 2940 N and $Isp = 340$ s, MBB of 3000 N and $Isp = 320$ s, Rocketdyne RS-51 of 11790 N and $Isp = 320$ s.

For these transfer stages, cryogenic propulsion using a liquid oxygen (LOX)–liquid hydrogen (LH2) combination can also be considered and this enables much greater specific impulses, of the order of 470 s, to be obtained with the disadvantage of more sophisticated technology (Mitsubishi RE6 of 9810 N and $Isp = 470$ s, SEP HM7 of 63,000 N and $Isp = 444$ s, Rocketdyne RS-44 of 66720 N and $Isp = 480$ s).

11.1.4.3 *Hybrid propellant motors*

Hybird thrusters can be considered with a view to their use as re-ignitable motors for the transfer stage between a low altitude circular orbit and the geostationary satellite orbit. These consist of motors whose propellants are in different physical phases; the oxidiser is liquid but the fuel is solid. The advantages of such a system are relatively simple design, and hence moderate cost, flexibility which includes the possibility of extinguishing and reigniting the motor (a limited number of times) and high performance (Isp of the order of 295 s). Engineering difficulties lie mainly in the problems of thermal transfer (radiation cooling) due to the long combustion times [JAN-88].

11.1.4.4 *Example calculation: mass of propellant required for circularisation of the transfer orbit (bi-propellant apogee motor)*

The assumption of a transfer orbit with zero inclination will be made initially so that the effect of inclination on the operational mass of the satellite can be indicated in a subsequent section. The altitude of the perigee of the transfer orbit is 200 km and the velocity increment to be provided at the apogee is 1478 m/s (11.2). The satellite of mass $M_{GTO} = 1500$ kg in the transfer orbit (GTO) uses a bi-liquid apogee motor (MBB400) which delivers a specific impulse of 310 s and a thrust of 400 N. The mass m of propellant required for circularisation is deduced from equation (10.20) of Chapter 10:

$$m = M_{GTO}[1 - \exp(-\Delta V/g\,Isp)] = 1500[1 - \exp(-1478/9.81 \times 310)] = 577\,\text{kg}$$

Taking account of the dry mass of the reservoirs, the motor, the pipework and accessories, the mass of the propulsion system for the apogee thrust reaches 630 kg and the mass of the satellite in geostationary orbit is thus equal to 870 kg.

The thrust F of the apogee motor is 400 N. From equation (10.16) the mass flow rate of the propellant is given by:

$$\rho = (F/g\,Isp) = 400/310 \times 9.81 = 0.13\,\text{kg/s}$$

and the duration of combustion is:

$$t = m/\rho = 577/0.13 = 4440\,\text{s} = 74\,\text{min}.$$

During this period, which is long in comparison with the period of the orbit, the satellite moves with respect to the apogee and it is necessary to subdivide the manoeuvre into several thrusts so that its efficiency is not excessively reduced.

The maximum acceleration Γ to which the satellite is subjected is given by:

$$\Gamma = F/M = 400/870 = 0.46\,\text{m/s}^2$$

when M is the mass of the satellite at the end of combustion of the motor. With a solid fuel motor, the results would have been slightly different.

With a motor of specific impulse 293 s delivering a thrust of 35 000 N (e.g. STAR 30E), the mass of propellant required is 603 kg, the mass flow rate is 12.2 kg/s and the combustion time is 50 s. The thrust can thus be considered as an impulse. On the other hand, the mass of the satellite in geostationary orbit is only $1500 - (603 + 40) = 857$ kg (the dry mass of the motor is 40 kg) and the acceleration imparted to the satellite reaches $35\,000/857 = 40\,\text{m/s}^2$.

11.1.4.5 *Influence of the latitude of the launch base on the mass*

Assuming an inclination of the transfer orbit equal to the latitude of the launch base (launch towards the east), the curve representing ΔV as a function of the latitude of the launch pad is plotted in Figure 11.8a. Table 11.3 gives the required velocity increments and the corresponding mass of propellant for various launch bases and a satellite of mass 1500 kg equipped with a motor of specific impulse 293 s. Figure 11.8b illustrates the percentage mass reduction of the satellite in orbit as a function of the latitude of the launch base. In the case of a launch pad at 28.5° latitude (Cape Canaveral), the mass lost with respect to an equatorial launch pad is of the order of 12 per cent; this emphasises the advantage of a launch base situated close to the equator such as Kourou.

11.1.5 Injection into orbit with a conventional launcher

Installing a geostationary satellite into orbit using a launcher with several stages separates into the following three phases (Figure 11.9).

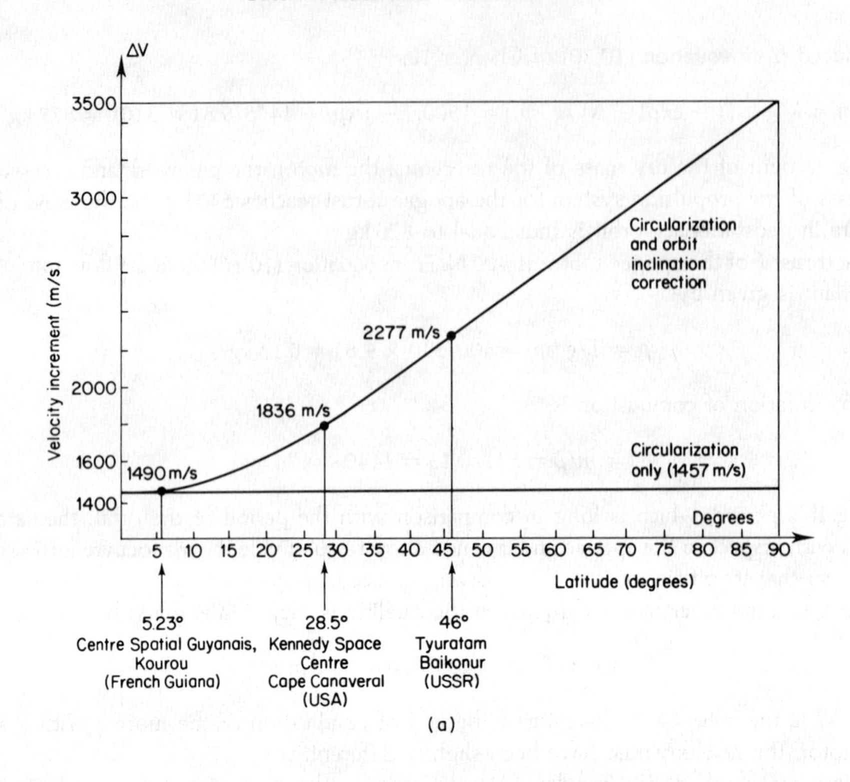

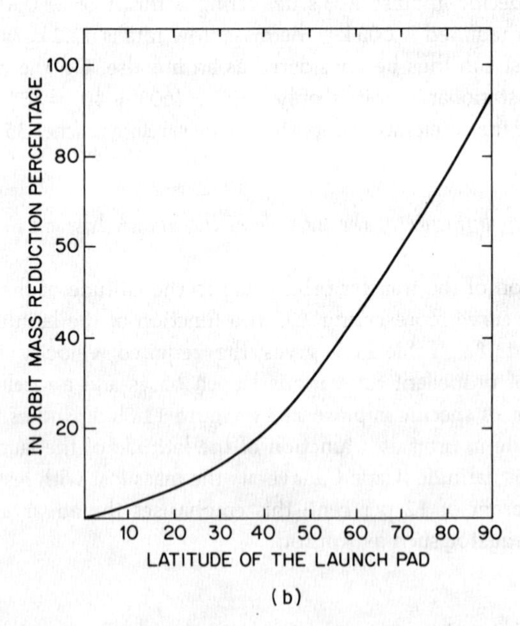

Figure 11.8 Influence of the launch-pad latitude. (a) Required velocity increment for circularisation and correction of transfer orbit inclination versus launch-pad latitude (perigee altitude 200 km); launch towards the east ensuring an orbit inclination equal to the latitude of the launch pad. (b) Influence of the latitude of the launch pad on the satellite mass at the beginning of life (mass at launch, 1000 kg; specific impulse 310 s).

Table 11.3 Influence of the latitude of the launching base

	Kourou (France)	Cape Canaveral (USA)	Tyuratam (USSR)
Latitude	5.23°	28.5°	46°
ΔV (m/s)	1490	1836	2277
Propellant mass (kg)	387	453	527
Loss with respect to Kourou (kg)	0	66	140
Usable sat. mass (kg)	613	547	473

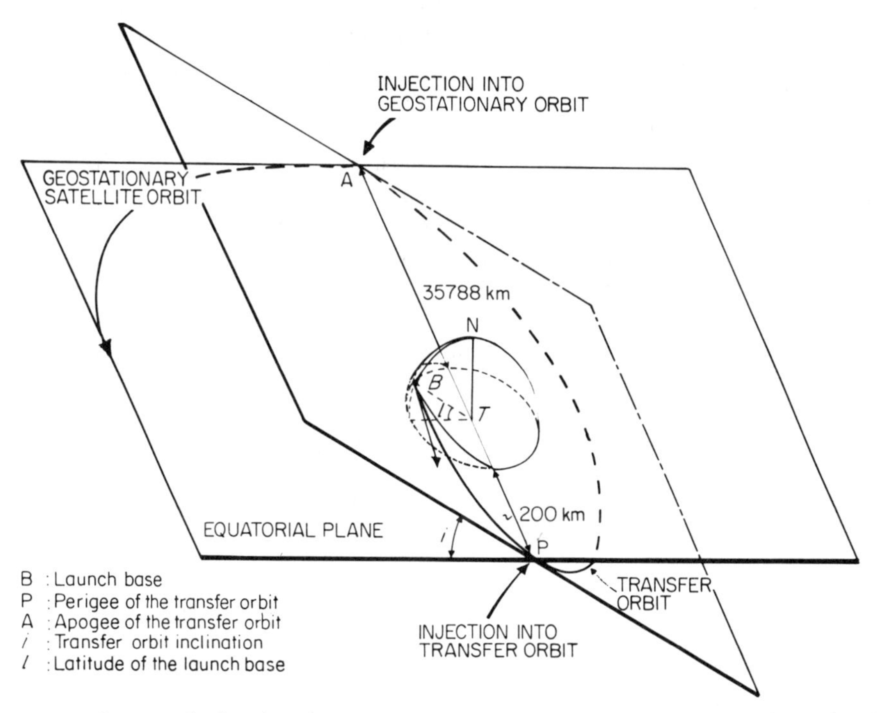

Figure 11.9 Sequence for launch and injection into transfer and geostationary orbit with an expendable launch vehicle.

11.1.5.1 *Launch phase*

From launcher take-off to injection at the perigee of the transfer orbit, the nominal functions to be provided are as follows:

— Increase the altitude in order to achieve the altitude of the perigee.
— Dispose of the fairing which protects the satellite as it passes through the dense layers of the atmosphere.
— Bring the last stage/satellite assembly onto a trajectory which intersects the equatorial

plane parallel to the surface of the earth with the required velocity on passing through the equatorial plane (the perigee of the transfer orbit).

Figures 11.10 and 11.11 illustrate two possible strategies.

The first (Delta and Atlas Centaur launchers) contain an intermediate ballistic phase during which propulsion is stopped. The last stage/satellite assembly is then orientated and caused to rotate in order to guarantee that this orientation is maintained during ignition of the last stage. This ballistic phase is made necessary by the long distance to be covered from the launch base (Cape Canaveral) to passing through the equatorial plane; this does not permit continuous thrust of the motors on the trajectory to be considered.

In the second (Ariane), the distance to be covered is shorter and the thrust is continuous (with the exception of the separation of the stages). The trajectory of the launcher is such that the intended point of injection is reached at an altitude of 200 km with the required velocity. As a consequence of the motor thrust on each phase of the flight, a given acceleration is conveyed to the launch vehicle at each instant as a function of the mass and a particular velocity will be reached at the end of a defined time associated with a related length of the trajectory. This trajectory should be contained between the launch pad and the crossing of the equatorial plane. Optimisation of the performance leads to a trajectory which rises above the required altitude, potential energy is thus gained and is then used later to provide an increase of velocity. Similarly, injection is performed at a point on the transfer orbit beyond the perigee (guidance having orientated the velocity vector in the appropriate direction).

It should also be noted that the argument of the perigee of the transfer orbit to be obtained is not 180° but about 178°. This is to take account of the drift of the argument of the perigee by 0.82°/day (see Section 7.3.2.3). It will cause the orbit to rotate in its plane in such a

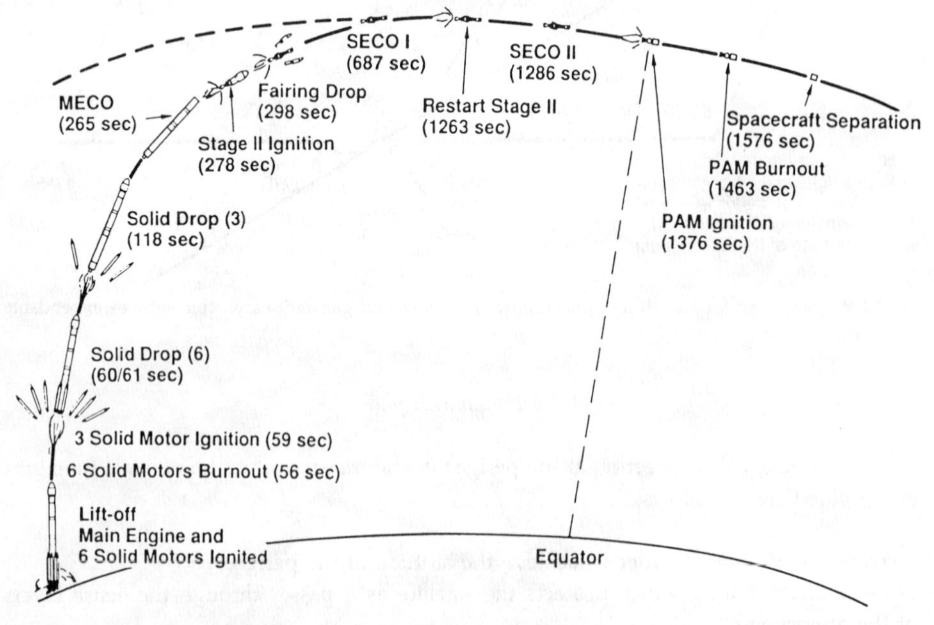

Figure 11.10 Mission profile of an expendable launch vehicle (Delta II) with a coasting phase for a typical GTO mission [POR-88]. (Reproduced by permision of American Institute for Aeronautics and Astronautics.)

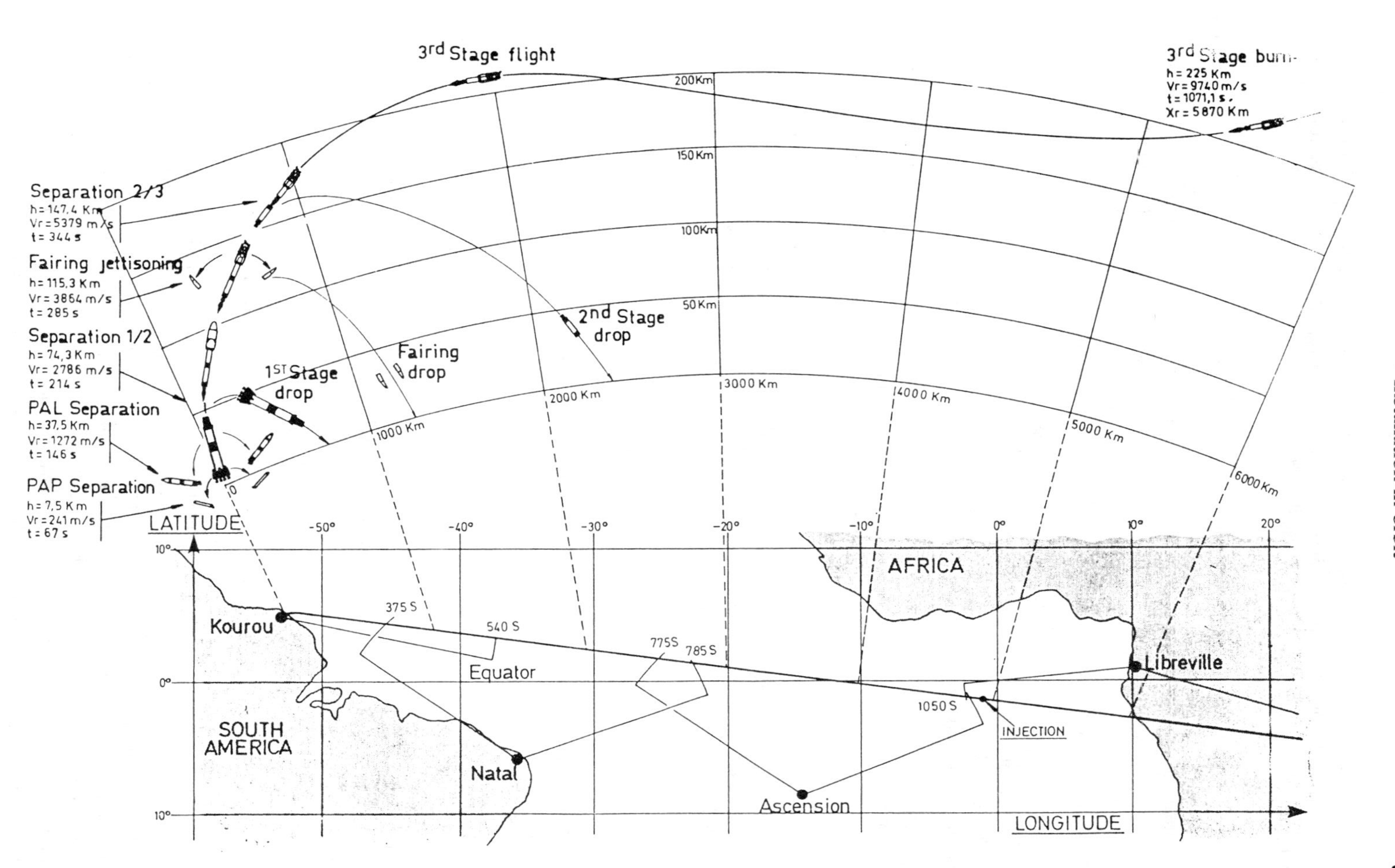

Figure 11.11 ARIANE 4 mission profile for a typical GTO mission. (Reproduced by permission of Arianespace.)

way that at the time of the nominal manoeuvre at apogee number 6 (after 5.5 orbits, that is 2.5 days) it will coincide with the ascending node and hence will be in the equatorial plane (the argument of the perigee is then equal to 180°).

11.1.5.2 The transfer phase

The transfer phase starts with injection of the composite launcher/satellite final stage and terminates with injection into the quasi-geostationary satellite orbit at the apogee of the transfer orbit.

In this phase the functions to be provided are as follows:

—separation of the satellite and the final stage,
—trajectory determination and restoration of the required transfer orbit,
—measurement of the satellite attitude,
—satellite orientation correction in view of the apogee manoeuvre.

This orientation may be maintained either by causing the satellite to rotate (gyroscopic stabilisation) or by active attitude control using sensors and actuators (three-axis stabilisation).

The importance of obtaining a transfer orbit whose parameters are close to those of the nominal orbit must be emphasised; the altitude of the apogee must be that of the geostationary satellite orbit and the apsidal line must be in the equatorial plane. If this is not the case, the result after injection into the geostationary satellite orbit will be non-zero eccentricity and inclination which the satellite must correct using its actuators, hence using on-board propellant which reduces the satellite lifetime.

11.1.5.3 The positioning phase

This phase starts with injection into the quasi-geostationary satellite orbit at the apogee of the transfer orbit and terminates with positioning of the satellite at the chosen station in the geostationary satellite orbit.

Other procedures for installation into orbit which permit the launch vehicle to move out of the plane defined by the launch azimuth and the latitude of the launch base ('dog leg' manoeuvres) can be envisaged.

In this way, by reigniting the last stage in flight, the Titan IIIC launcher permits direct injection of the satellite into the geostationary satellite orbit. This approach avoids the need for an apogee motor and thus achieves a gain in the useful mass of the satellite.

11.1.6 Injection into orbit from a quasi-circular low altitude orbit

The operations involved in putting a geostationary satellite into orbit from a quasi-circular low altitude orbit differ from those described above due to the inability of the launcher to inject the satellite directly into the transfer orbit. This is the case with launches using the Space Shuttle whose nominal orbit is circular, of altitude 290 km and inclination 28.5°. It is also the case with two-stage launchers, such as Commercial Titan III which delivers its payload into an elliptical orbit (148 × 259 km) with an inclination of 28.6°.

The changes with respect to the operations described above are as follows:

11.1.6.1 The launch phase

The launch phase does not lead directly to injection of the satellite at the perigee of the transfer orbit but places the launch vehicle in a circular or slightly elliptical orbit of non-zero inclination.

11.1.6.2 Injection into the transfer orbit

After separation from the launcher into a particular attitude configuration, a velocity increment is imparted to the satellite in order to inject it at the perigee of the transfer orbit. This increment is provided by the perigee motor on passing through the equatorial plane.

The perigee motor can be integrated into the satellite or can be part of an auxiliary stage to which the satellite is attached.

An integrated perigee motor can be implemented for this use (usually a solid fuel motor), with possible separation after use. This motor (in the case of a bi-propellant motor) can also be re-used to provide part or all of the velocity increment at the apogee (HS 399 satellite).

Table 11.4 Upper stages for injection into transfer and geostationary orbit

Name	Type (1)	Manu-facturer	Prop-ellant (2)	Mass (kg) Stage (3)	Mass (kg) With support (4)	Length (m) (5)	Performance (kg) GTO (6)	GEO (7)
IRIS	P/S	Aeritalia	S		2 700		950	—
PAM-D	P/S	McD. Douglas	S	2 150	3 320	1.9*	1250	—
PAM-D II	P/S	McD. Douglas	S	3 725	5 380	2.0*	1840	—
AMS	P/3A	OSC	L	5 130	6 430	1.65	2550	—
SCOTS	P/S	RCA/GE	S	4 560	5 800	2.5	2700	—
HPPM	TS/3A	Ford	L	5 970	6 400	1.5	—	1350
STV	P/3A		L		8 300		3400	—
LPM	TS	Aerojet	L	7 650		1.7	4400	1540
TOS	P/3A	OSC	S	10 925	12 950	3.3	6090	—
TRANSTAGE	TS/3A	M. Marietta	L	12 250		4.6		1900
IUS	TS/3A	Boeing	S	14 690	18 900	5.0	—	2270
TOS/AMS	TS/3A	OSC	S/L	16 075	18 320	4.8	8860	2960
CENTAUR	TS/3A	G. Dynamics	C	15 900	19 800	5.9	—	4540
CENTAUR G′	TS/3A	G. Dynamics	C		23 100	8.9	—	5910

(1) P = Perigee stage, TS = transfer stage (perigee + apogee), S = gyroscopic attitude stabilisation, 3A = three-axis attitude control.
(2) S = solid propellants, L = liquid propellants, C = cryogenic propellants.
(3) Stage with full propellant capacity without launching cradle and satellite.
(4) Stage with full propellant capacity and supporting cradle for use with the Space Shuttle.
(5) Length without satellite (*length in the Shuttle cargo bay: 2.2 m PAM-D and 2.4 m PAM-DII).
(6) From a circular orbit of 290 km altitude and 28.5° inclination. GTO = in geostationary transfer orbit, $i = 26.5°$, GEO = in geostationary orbit. (The mass in GEO with a perigee stage is about half the mass in GTO which includes the apogee motor required in this case.)
(7) The total mass in the Shuttle cargo bay is (4) + (6).

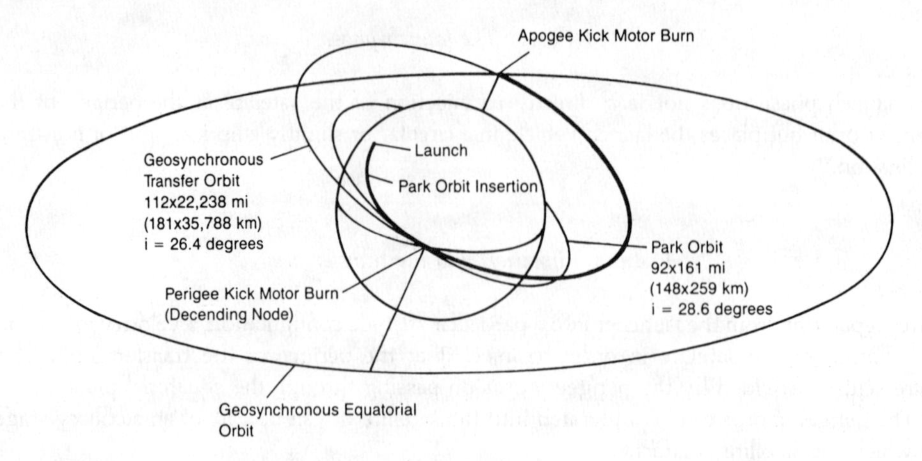

Figure 11.12 Geostationary orbit injection sequence from a low altitude parking orbit.

An auxiliary stage can merely fulfil the injection function at the perigee (the perigee stage, usually based on a solid fuel motor) or it can fulfil both the functions of injection at the perigee and injection at the apogee (the transfer stage). The transfer stage can use solid fuel motors (two are needed, the IUS stage for example). Alternatively it can use bi-propellant propulsion with a re-usable motor (for example the LPM or AMS stages) or dedicated motors with possible separation of used mass (empty tanks with or without the motor) after providing the first velocity increment. Finally, the transfer stage can be realised with a combination of two technologies by using a bi-propellant apogee stage associated with a separable solid fuel perigee stage (for example the TOS/AMS stage).

During injection into the transfer orbit, orientation of the perigee motor thrust is ensured either by causing the satellite to rotate (gyroscopic stabilisation) or by active attitude control using sensors and actuators (three-axis stabilisation). Attitude control of the satellite/stage combination can be ensured by the satellite attitude control sub-system (in the case of an integrated motor) or by a sub-system which is specific to the auxiliary stage (for example with the IUS). The characteristics of the various available perigee and transfer stages are given in Table 11.4.

11.1.6.3 *Transfer and positioning phases*

These phases are similar to those described above knowing that certain operations may possibly be realised by the transfer stage.

Figure 11.12 illustrates the various stages of the procedure during a geostationary satellite launch by the Titan launch vehicle.

11.1.7 Operations during installation (station acquisition)

Installation consists of transferring the satellite from the transfer orbit provided by the launch vehicle to the longitude of its station. Installation thus covers the transfer and positioning phases specified above. It also covers the operations of configuring the satellite to perform the mission for which it has been designed.

11.1.7.1 Choice of apogee where the manoeuvre is performed

The choice is influenced by various considerations:

— The satellite must perform at least one or two transfer orbits to permit orbit determination with adequate accuracy.
— The satellite must not remain too long in the transfer orbit since the space environment there is special and different from that of the final orbit for which the satellite has nominally been designed. In particular, the satellite is subjected to more numerous eclipses and on each orbit passes through a region of trapped particles (the Van Allen Belt, see Section 12.4.1). Finally, the electrical energy resources, which consist of batteries and possibly a partially deployed solar generator, are limited.
— Ignition of the apogee motor must be performed within the visibility of at least two control stations in order to ensure consistency of this difficult manoeuvre by means of redundancy.
— The apogee chosen must be close to the longitude of the desired station. After the apogee manoeuvre, the satellite will be quasi-geostationary close to the position of the apogee where the manoeuvre was performed.

Figure 11.13 shows the track of the satellite in the transfer orbit (for the case of a launch by Ariane) and indicates the position of successive apogees. In this way the number of the transfer orbit on which the manoeuvre must be performed in order to satisfy the constraints of visibility and proximity to the station is determined. For the launch of a satellite intended to be positioned in the vicinity of the Greenwich meridian, the nominal choice is apogee number four with a second chance at apogee number six in the case of a problem. These apogees are visible from the control stations at Toulouse and Kourou.

11.1.7.2 Drift orbit

Due to variations of the parameters of the transfer orbit and the apogee manoeuvre, and also in order to permit the satellite to achieve the longitude of its station, the orbit obtained after the apogee manoeuvre is never exactly the geostationary satellite orbit. A residual inclination and eccentricity exist and the semi-major axis differs from that of the synchronous orbit and this leads to drift of the satellite. The drift orbit attained in this way must be corrected by means of low thrust thrusters from the satellite orbit control system in such a way that after a certain time (several days) the satellite will reach the intended station longitude on an orbit of the desired eccentricity and inclination (not necessarily zero, see Section 7.3.5).

When a multiple thrust procedure is used, the drift orbit is generally included in the optimisation of the injection procedure which takes the satellite to its station longitude by taking account of inaccuracy of the transfer orbit provided by the launcher (see Section 11.1.3.5).

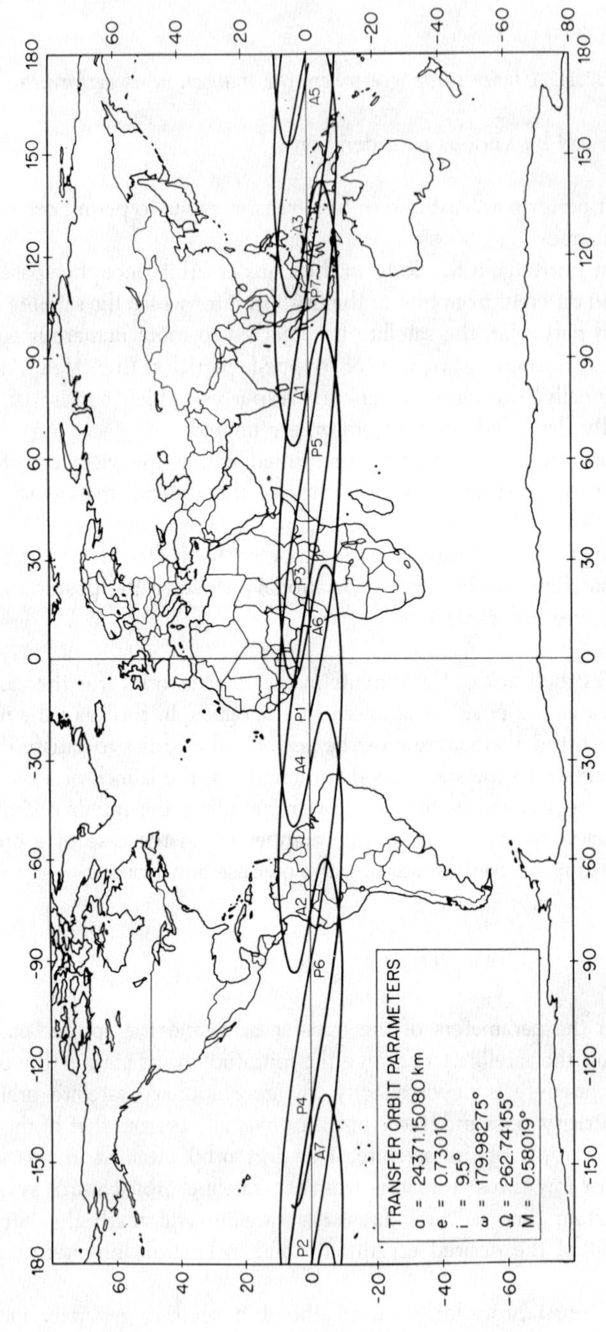

Figure 11.13 Satellite track during transfer orbit (Ariane launch) showing successive locations of the apogee (A) and perigee (P). (Reproduced by permission of Centre National d'Etudes Spatiales from Robert and Foliard (1980).)

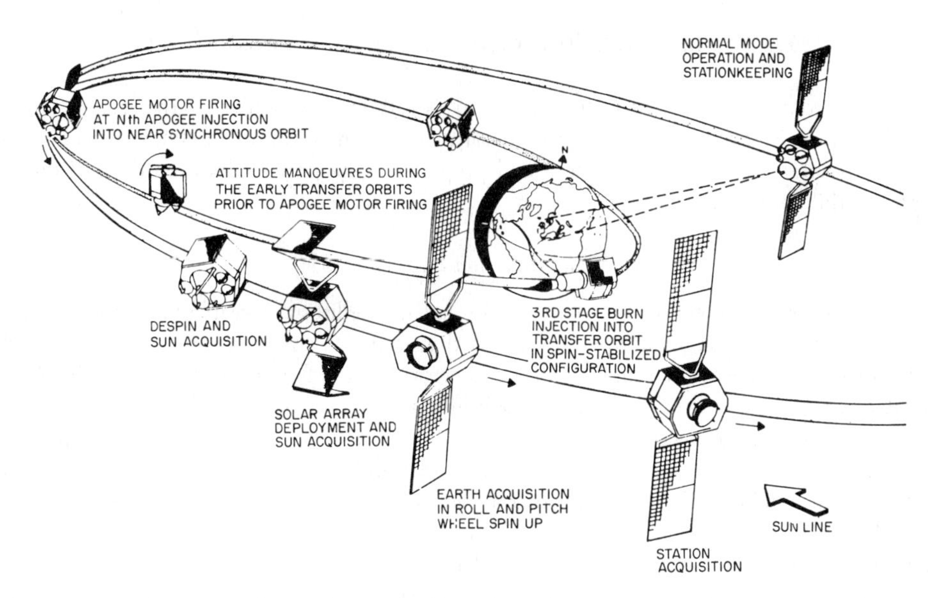

Figure 11.14 Operations during drift orbit. (Reproduced by permission of Centre National d'Etudes Spatiales from Robert and Foliard (1980).)

In addition to corrections of the orbit, the main operations to be performed are as follows (Figure 11.14):

— Reduction of the velocity of rotation in the case of attitude control by rotation during the apogee thrust.
— Acquisition of the sun by the solar sensors (already achieved during the transfer orbit with three axis control).
— Attitude control.
— Deployment of solar panels and activation of the panel rotation system to follow the apparent movement of the sun.
— Acquisition of the earth (using an infra-red sensor).
— Establishment of the desired attitude using information from the earth sensor.
— Activation of the momentum wheel(s).

The positioning phase lasts for several days after the apogee manoeuvre.

11.1.7.3 *Satellite test and acceptance*

Once the satellite is at the intended station longitude in its nominal orbit, station keeping is activated. The operational life of the satellite will start when various satellite sub-system tests have been performed and their performance has been evaluated. This test and acceptance phase lasts for several weeks.

11.1.8 Injection into orbits other than geostationary

In the preceding discussion, the final orbit to be attained was that of the geostationary satellite which is the orbit used by the majority of communication satellites. However, as seen in Chapter 7, other orbits have useful properties. The procedures used depend on the type of orbit to be attained.

11.1.8.1 *Injection into polar orbit*

Polar orbits are of interest for communications as the satellites are visible for a given period of time from any location at the surface of the earth. A constellation of satellites with appropriate phasing (several satellites on the same orbit and different orbit planes with evenly distributed right ascension values) allows a continuous coverage of the earth (e.g. the IRIDIUM system). Some applications of sun synchronous orbits have been proposed [PRI-88a]. Injection of the satellite into an orbit with the desired inclination is achieved by choosing the launch azimuth as a function of the latitude of the launch base according to equation (11.4). For low altitude orbits (several hundreds of kilometres), direct injection of the satellite into the final orbit is possible. For high altitude circular orbits (several thousands of kilometres), an intermediate transfer orbit must be used following the principle described in Section 11.1.1.

11.1.8.2 *Injection into inclined elliptic orbits*

The procedure used depends on the mass of the satellite to be placed in orbit and the capacity of the launcher.

For a satellite of high mass using the full capacity of the launcher, an orbit of the desired inclination will be obtained by adjusting the launch azimuth in the context of a dedicated launch. According to the characteristics of the final orbit, the satellite will be injected into either a transfer orbit or the final orbit.

For a satellite of limited mass which does not use the full capacity of the launcher, sharing of this capacity among several satellites permits the launching cost to be reduced. It is, however, unlikely that the intended orbit of the other satellite carried by the launcher will have the same desired inclination; most opportunities for multiple launching which arise concern missions in geostationary transfer orbit. It is thus necessary to consider procedures which permit the inclined elliptic orbit to be joined from the standard geostationary transfer orbit provided by the launcher, by modifying the inclination and raising the altitude of the perigee and the apogee.

For example, to obtain an orbit of the 'TUNDRA' type (see Section 7.2.1), procedures with two or three velocity increments can be considered as follows.

With the two velocity increment procedure, the first impulse modifies the inclination from 7° (Ariane launcher) to 63.4° and a second impulse raises the perigee from 200 km to 22 000 km.

In the three increment procedure, the first impulse at the perigee of the transfer orbit greatly increases the altitude of the apogee (for example to 100 000 km). At the apogee, the second impulse modifies the inclination at least cost due to the low velocity of the satellite (see Section 11.1.3.3). The third impulse enables the final orbit to be attained.

Optimisation shows that the total velocity increment to be provided is less with the three impulse procedure. To place a satellite in an orbit of the 'TUNDRA' type from the standard Ariane transfer orbit, the total increment to be provided is of the order of 2300 m/s instead of 2500 m/s with the two impulse procedure [ROU-88].

11.1.9 The launch window

The launch window specifies the time periods when launching a satellite is possible taking account of the following associated constraints:

— To permit determination of the attitude with the required accuracy (this relates to permissible ranges of the direction angles of the sun and the earth with respect to the satellite axes, considering that the earth, the satellite and the sun must not be aligned).
— To avoid saturating the sensors or disappearance of references during the apogee manoeuvre(s) (this relates to the position of the satellite with respect to the sun and the position and duration of eclipses).
— To ensure an electric power supply (this relates to the position of the satellite with respect to the sun and the position of eclipses).
— To guarantee thermal control (this relates to the position of the satellite with respect to the sun and the number and duration of eclipses).
— To be within radio visibility of the control station during the critical phases (this relates to radio disturbances at sunrise and sunset).

From a combination of the various constraints, possible time slots, or launch windows, for launching the satellite can be deduced for all days of the year. In the case of a double launch, the launch window must satisfy the constraints imposed by installation of both satellites. Figure 11.15 illustrates such a launch window for a double Ariane launch into the geostationary orbit [BOL-86], [MUA-87].

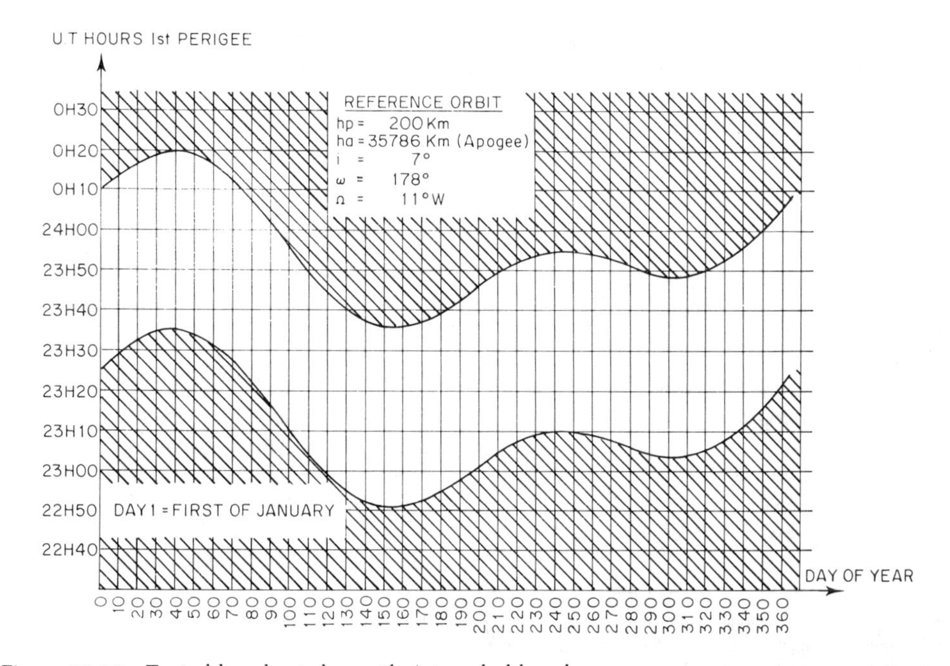

Figure 11.15 Typical launch window with Ariane dual launch.

11.2 LAUNCHERS

The economic risks represented by the installation of a satellite communication system have led to serious commercial competition in the launcher domain.

To be free of the constraints imposed by the only country selling launchers at the start of the 1980s (the United States with the Delta and Atlas Centaur launchers), some industrialised countries (China, Japan and Europe) have developed launcher programmes which enable satellites to be placed in geostationary orbit.

At the same time, the United States was engaged in a programme whose objective was to develop a recoverable and re-usable launcher with a view to reducing the cost of installation into orbit (the Space Shuttle) and with the idea of abandoning the production of

Table 11.5 Operational launchers at the start of the 1990s

Launcher	Country	Total Mass (tonnes)	Number d'étages of stages	Propellants (kg)	Performance: Low orbit (kg)	Performance: Geostationary transfer orbit. (kg)
Pegasus	USA	18.5	3	3S or 2S,1L	325	—
—ASLV	India	35	4	4 S	150	—
—Scout	USA	22	4	4S	230	55
—M3 SII	Japan	62	3+2 SB	3S+2S	770	60
—Lance	USSR		2	2L	700	—
Cosmos C1						
—N-II	Japan	135	3+9 SB	2L,1S+9S	2 000	720
—Delta 3920	USA	190	3+9 SB	2L,1S+9S	3 000	1 247
—Long March CZ-3	China	202	3	3L	3 800	1 400
—Delta II (6925)	USA	210	3+9 SB	2L,1S+9S	3 980	1 450
—Delta II (7925)	USA	220	3+9 SB	2L,1S+9S	5 040	1 820
—Atlas Centaur II	USA	163	21/2	1.5L,1C	6 100	2 300
—Atlas Centaur IIA	USA		21/2	1.5L,1C		2 900
—H-II	Japan	256	2+2 SB	1L,1C+2S	10 000	4 000
—Ariane	Europe	376	3+4 SB	2L,1C+ 2/4S/L	5 000 (A42 P)	1900 (A40) to 4400 (A44L)
—Titan III	USA	680	2+2 SB+US	2L+2S+S/L	14 800	5 000 with IUS
—Proton	USSR	700	4	4L	20 000	5 000 (2200 in GEO)
—STS	USA	2000	1+2 SB+US	1L+2S+P/L	29 000	8800 with TOS/AMS

S: Solid propellant, L: Liquid Propellant, C:Cryogenic propellant, SB: Strapon Booster,
S/L: Powder or liquid,, 2/4: 2 or 4 SB according to type,
US: acording to type Upper Stage.

conventional launchers. The Challenger catastrophe in January 1986, which led to immobilisation on the ground of the US civil and military satellites for more than two and a half years, caused a revision of this policy. The production and development of conventional launchers was consequently restarted. Furthermore, the Reagan administration decided that use of the Space Shuttle would be reserved for government missions. It also proposed that companies developing conventional launchers should commercialise their launching services and the installations at various launch bases which are government property (Air Force and NASA) were put at their disposal [STA-88].

In 1983, the Soviet Union decided to offer the service of its Proton launcher to the Western world to launch the Inmarsat II satellites. In 1985 a specific organisation, Licensintorg under the aegis of the Glavkosmos agency, was set up to commercialise the launcher; it offered guarantees with respect to problems of technology transfer with western countries.

Table 11.5 summarises the main characteristics of launchers which were operational, or nearly so, at the start of the 1990s. These launchers are mainly for commercial use with the exception of the Japanese type N, which is built under US licence, and the Space Shuttle.

Finally, the end of the 1980s has seen the beginning of an interest in small satellites (microsatellites—less than 50 kg; minisatellites—less than 500 kg). There seem to be many advantages which include reduced costs, short development times, a wide range of potential applications (in particular the LIGHTSTAT programme and the Defense Advanced Research Project Agency (DARPA) launched in 1988). Suitable launchers are necessary and these include existing low capacity launchers (SCOUT, MIIIS, Pegasus, etc.), adaptors for powerful launchers (ASAP, APEX with Ariane), the development of new launchers (Pegasus) and numerous projects (SCOUT II, Conestoga, Taurus, Amorc, Space Vector, etc.), the credibility of which is not certain.

11.2.1 China

In 1986 China decided to commercialise its launchers in a range called Long March. The characteristics of these launchers are given in Table 11.6. The Long March 3 or LM-3 launcher is a three-stage launcher which uses liquid propellants and cryogenic third stage (1st stage: diameter 3.35 m, length 20.3 m, 142 t of UDMH/N_2O_4; 2nd stage; length 9.7 m, 35 t of UDMH/N_2O_4; 3rd stage: length 7.48 m, 8.5 t of LH/LOX) it is capable of placing 1400 kg in a geostationary transfer orbit of inclination 31.1°. An improved version of the launcher (LM-3a) should be able to place up to 2.5 tonnes in geostationary transfer orbit in 1992, particularly due to the use of a new third stage.

An increase in the performance of the two-stage launcher LM-2c by the addition of four strap-on boosters (LM-2E) raises the launcher performance to 8.6 tonnes in low orbit and 3.3 tonnes in the geostationary transfer orbit. Launcher LM-4 differs from LM-3 in having a different third stage designed to inject 2.5 tonnes into a polar orbit.

The main launch base is situated at Liangshan near Xichang in the province of Sichuan (latitude 31.1°N) where the construction of a second launching assembly is planned. China also has an older base, the Chiu Chuan Space Center, near Juiquan (latitude 40.7°N) in the province of Gansu on the edge of the Gobi Desert. Construction of a third launch base is planned to the south of Beijing for launching satellites into polar orbit.

After long negotiations arising mainly from the problem of US satellite technology transfer into a non-Western country, China and the United States reached an agreement in December

Table 11.6 The Long March System. Family of the Chinese launch vehicles

Characteristics of the vehicle	LM-1 (Long March 1)	LM-2 (Long March 2)	LM-3 (Long March 3)
General configuration	3-Stage	2-Stage	3-Stage
Overall length	29.45 m	31.65 m	43.25 m
Lift-off mass	81.6 t	191 t	202 t
Diameter (1st stage)	2.25 m	3.35 m	3.35 m
Operational	in 1970	in 1974	in 1984
Payload dimensions:			
diameter	1 m?	3.17 m	232/272 m
length	2 m?	2.2 m/3 m	5 m
Orbital inclination	70°	42–63°	31.1°
Performance:			
400 km circular	300 kg	2 t	3.8 t
200 km/1500 km	—	1.8 t	4.8 t
geostationary transfer	—	—	1.4 t

1988 which provided for the launching by China of a maximum of nine satellites from 1989 to 1994 at world market cost, that is a turnover of 350 to 500 million dollars.

China was, therefore, able to launch the Australian Aussat B1 and B2 satellites built by Hughes Aircraft in 1991 and 1992. A long March 3 launcher launched the Asiasat satellite on 7 April 1990 on behalf of Asia Satellite Telecommunication Corporation (Hong Kong) (the Westar 6 satellite whose launch had failed due to maloperation of the perigee stage was recovered by the Space Shuttle in 1985 and resold by the insurance company which was the owner after recovery by Hughes Aircraft).

11.2.2 Europe (Ariane)

The Ariane family of launch vehicles (Figure 11.16a) has been developed by the European Space Agency under the management of the Centre National d'Etudes Spatiales (CNES). In July 1973, during the European Space Conference, it was decided to combine the European Launcher Development Organisation (ELDO) and the European Space Research Organisation (ESRO) into a single body called the European Space Agency (ESA). Among the objectives of the Agency was that of developing a range of three stage launchers to permit direct injection of the satellite/apogee motor combination from a transfer orbit into the geostationary satellite orbit without a ballistic phase, starting from the Kourou launch base in French Guyana. This orbit can be obtained accurately using guidance of the launch vehicle by an on-board computer which uses information provided by an inertial unit. Steering is provided by orientation of the jet of the main motors of the various stages. Furthermore, an attitude and roll control system (SCAR) enables the third stage and the satellite to be positioned in the desired attitude before separation with, if necessary, rotation at up to 10 revolutions per minute. An economy is thus realised in the quantity of propellant to be embarked for later corrections of the satellite orbit and attitude. This mass saving may

Table 11.7 Ariane launch vehicle performances

	Ariane 1 (max.)	Ariane 2 (max.)	Ariane 3	Ariane 4	Ariane 5
Year of operation	1981	1984	1984	1986	1995
Synchronous transfer orbit	1800 kg	2175 kg	2580 kg	1900–4200 kg	5200–8200 kg
Geostationary orbit[a]	1000 kg	1170 kg	1400 kg	1050–2300 kg	2800–4500 kg
Low Earth orbit (200 Km)	4900 kg	5100 kg	5900 kg	8700 kg	18 000 kg
Sun synchronous orbit[b]	2400 kg	2800 kg	3250 kg	4550 kg	10 000 kg
Escape	1100 kg	1330 kg	1550 kg	2580 kg	—

[a] Approximatively, as it depends on the performance of the apogee motor. Obtained by dividing the GTO capability by 1.8.
[b] 800 km, 98.6° inclination.

represent one to three years of extra satellite lifetime. The launchers can put several independent satellites into orbit simultaneously by means of a specific adaptation system (Sylda and Spelda).

The first firing of an Ariane 1 launcher took place on 24 December 1979. A programme of improvements by minor modification of the launcher (such as addition of extra solid boosters, modification of the motor combustion chamber pressures and increase of the mass of propellant) led to Ariane 3, whose first flight took place in August 1984. Table 11.7 and Figure 11.16a show the evolution of the performance and size of the Ariane launcher.

Commercial exploitation of the launchers is provided by the Arianespace company; it started with the ninth flight of the launcher after four test and four promotion firings.

11.2.2.1 Ariane 4

The Ariane 4 programme was developed concurrently as decided in January 1982. It is the latest of the family and the workhorse of the Arianespace company for the 1990s; the first flight of the Ariane 4 launcher took place on 15 June 1988. The principal characteristics of the launcher are given in Figure 11.16b and Table 11.8 [MUA-87].

The performance improvement is obtained by an increase of the capacity of the first stage and the use of additional, more powerful, solid or liquid propellant boosters [MEC-87]. As a consequence of the use of additional boosters (Figure 11.16a and Table 11.9), the nominal performance extended from 1.9 to 4.27 tonnes on a transfer orbit for the geostationary satellite orbit whose altitudes at the perigee and apogee are 200 km and 35 786 km; the inclination is 7°. The table also gives the performance in a typical sun synchronous orbit of altitude 800 km and inclination 98.6°.

Various types of fairing, of external diameter 4 metres (3.2 m for Ariane 3), are available for single and double launches (Figure 11.17a and b). Double launches use a different concept from SYLDA (the Ariane double-launch system) as used for Ariane 3 in which the supporting structure of the upper satellite is contained within the fairing thereby reducing the volume available for the lower satellite. SPELDA (external supporting structure for double Ariane launches) has the same external diameter as the fairing which covers the upper satellite and supports it. The volume available for the lower satellite is thus maximised. The sequence of

Kit DC Sept. 1990

THE ARIANE CONTINUITY

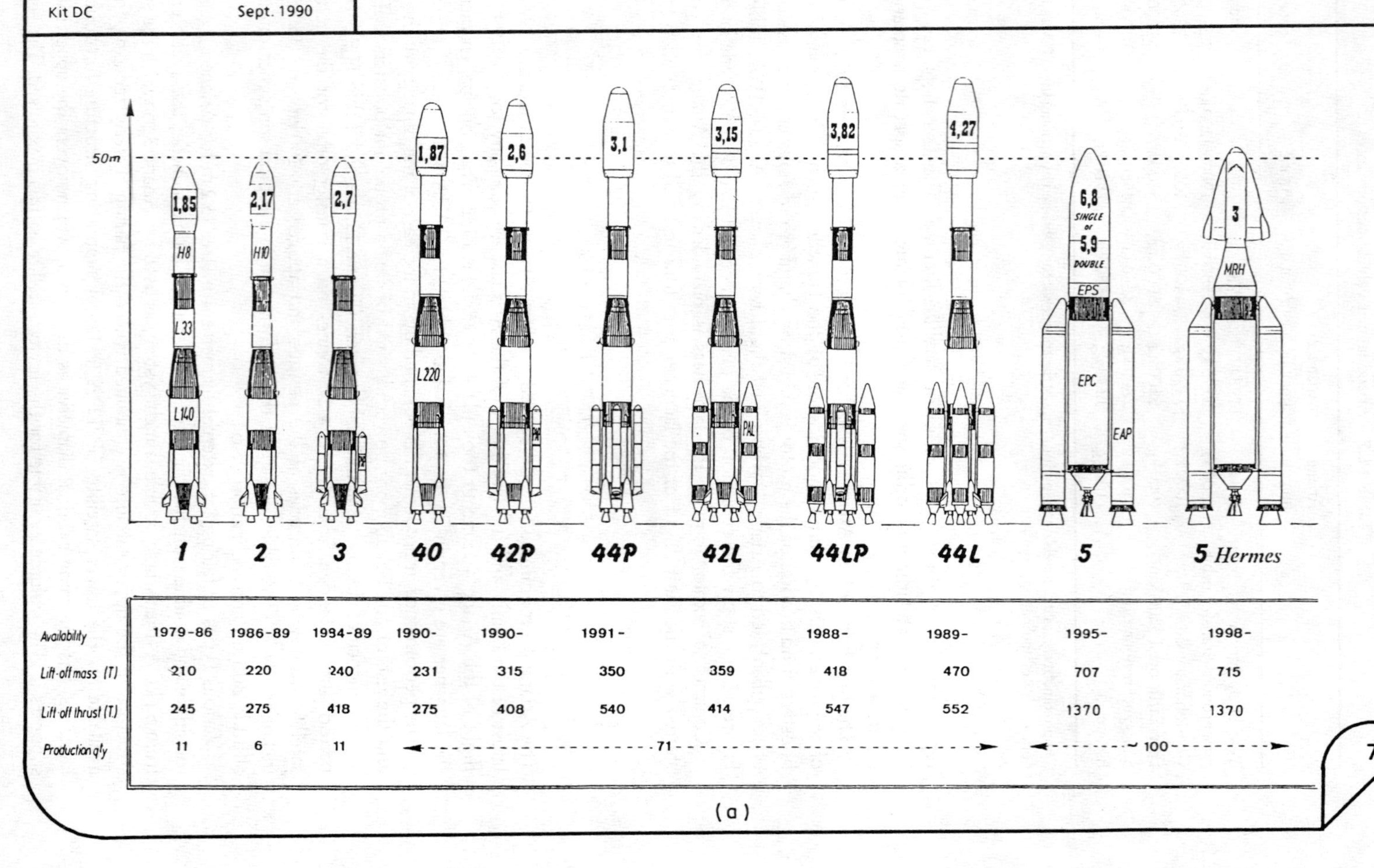

	1	2	3	40	42P	44P	42L	44LP	44L	5	5 Hermes
Availability	1979-86	1986-89	1984-89	1990-	1990-	1991-		1988-	1989-	1995-	1998-
Lift-off mass (T)	210	220	240	231	315	350	359	418	470	707	715
Lift-off thrust (T.)	245	275	418	275	408	540	414	547	552	1370	1370
Production q'ty	11	6	11	←			71		→	← ~ 100	→

(a)

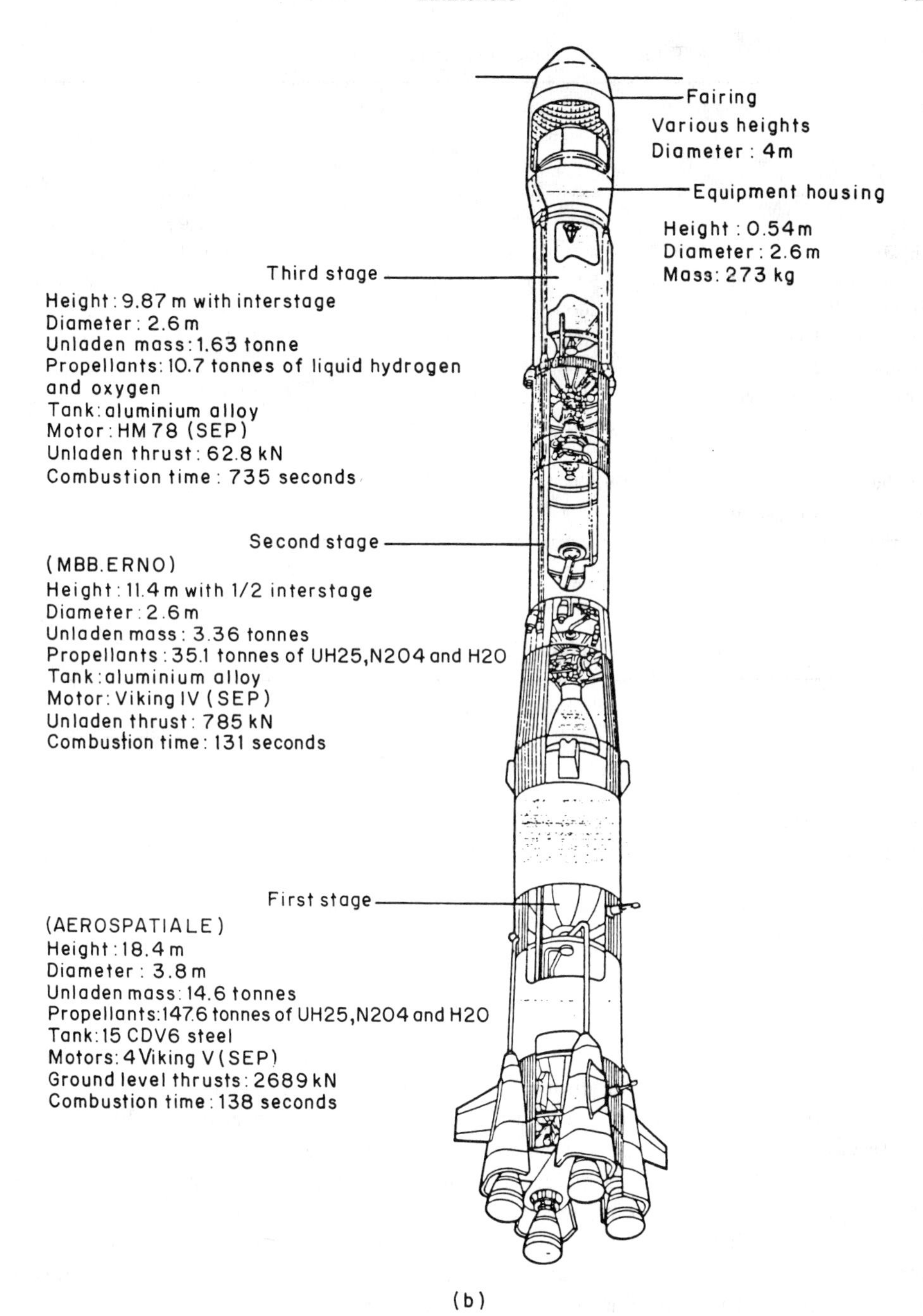

Figure 11.16 Ariane 4. (a) Family and different configurations. (Reproduced by permission of Arianespace) (b) Ariane 4 architecture.

Table 11.8 Dimensions and mass of Ariane 4

Item	Dry mass (kg)	Mass with propellants (kg)	Height (m)	Diameter (m)
Fairing	725 to 782	—	8.6 or 9.6	4.0
SPELDA	400 to 450	—	2.8 or 3.8	4.0
Vehicle equipment bay (VEB)	520	—	1	2.6 to 4.0
Third stage H10	1 200	10 700	9.9	2.6
Second stage L33	3 600	34 000	11.5	2.6
First stage L220	17 500	234 000	25.0	3.8
Liquid strap-on booster	4 500	39 000	19.0	2.2
Solid strap-on booster	3 200	9 500	12.0	1.1
Launcher (44LP, without payload)	39 300	375 000	59.8	

Table 11.9 Performance of the six versions of Ariane 4

Orbit	Payload capacity (kg) 40	42P	44P	42L	44LP	44L
Geostationary transfer (GTO)	1900	2600	3000	3200	3700	4200/4400*
Sun synchronous	2700	3400	4100	4500	5000	
Low altitude (LEO)	4600	5000				

*With H10$^+$ third stage.

the operations during a double launch is illustrated in Figure 11.17c. For double launches of low capacity, the use of SYLDA within the Ariane 4 fairing is possible.

A dedicated launching assembly called ELA-2 (Ariane launching ensemble) has been put into service; this permits the interval between two firings to be reduced to less than one month. The launch vehicle is assembled vertically in a special building and then moved on a mobile platform to the launching area which is remote from the assembly building. Mating of the fairing with pre-encapsulated satellites with the launch vehicle is performed in the launch area. During these last preparations, assembly of another launch vehicle on a second mobile platform can be started in the assembly building.

The launching programme starts approximately nine weeks before the firing with transportation of the launch vehicle components by boat from Le Havre. Assembly and launch vehicle testing in the assembly building require about four weeks and transfer to the launch area takes place two weeks before the firing. During this time the satellites arrive by air and are prepared and tested in special buildings before being enclosed in the SPELDA. The resulting combination is installed on the launcher five days before firing. The firing sequence lasts for

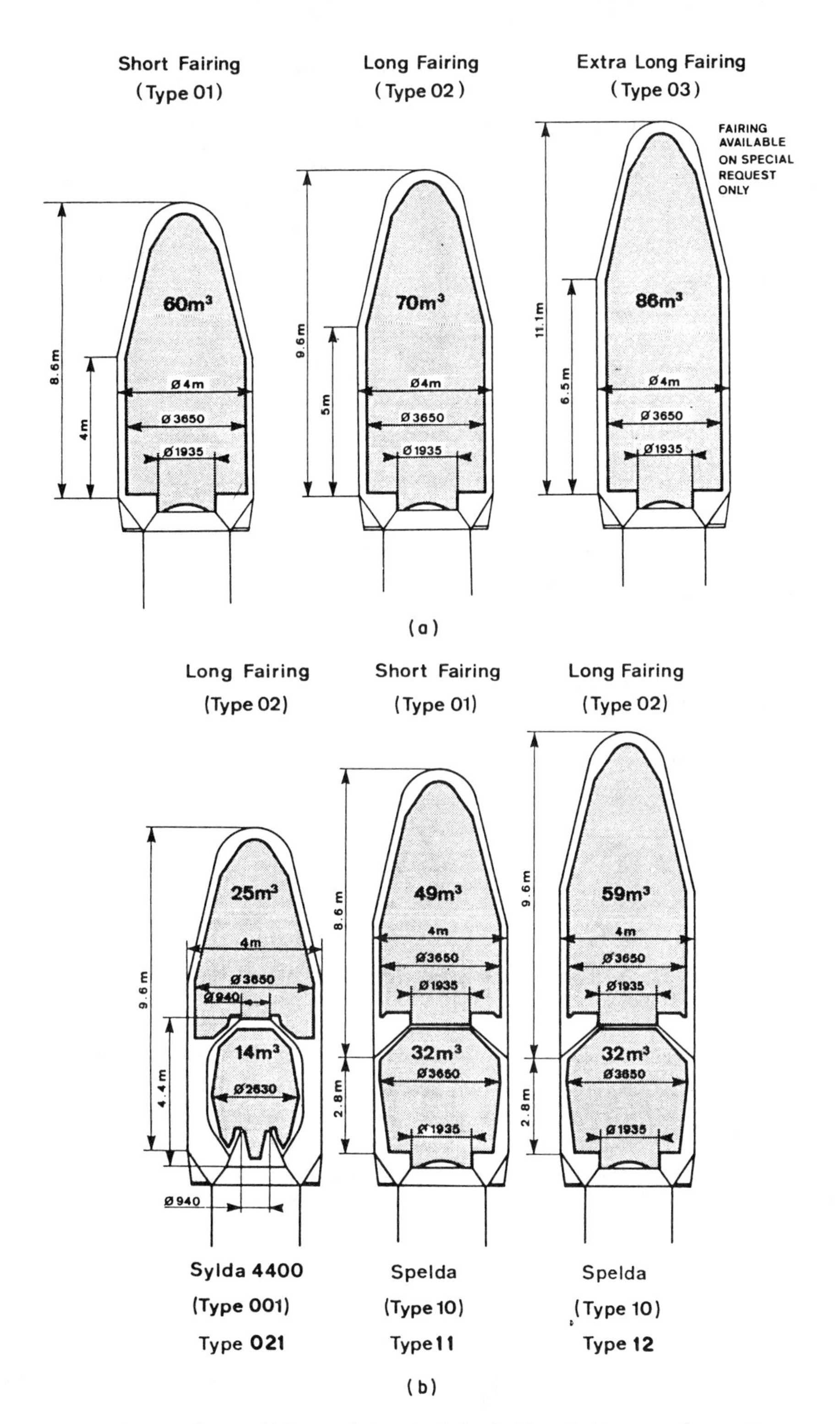

Figure 11.17 Ariane 4 fairings. (a) For single launch, (b) for dual launch, (c) spacecraft separation sequence with dual launch.

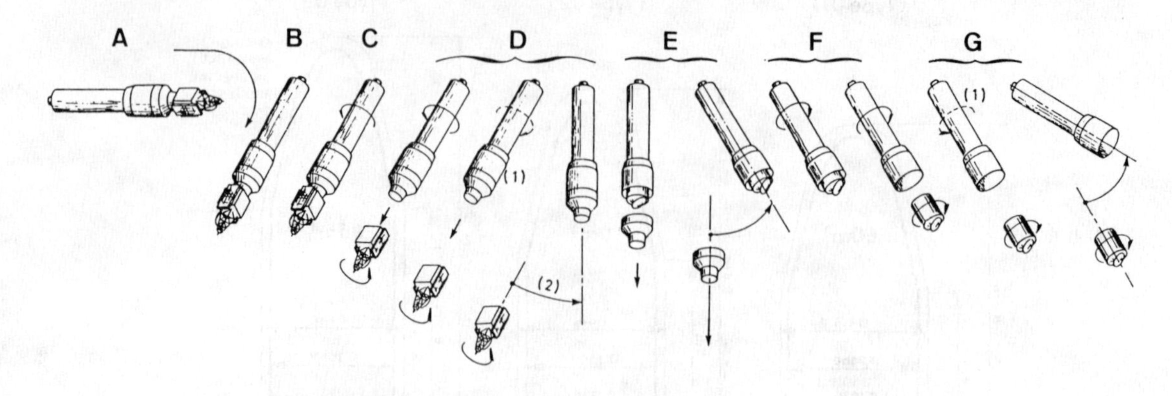

A and B : Orientation of composite (3rd stage + payload) by 3rd stage roll and attitude control system (SCAR)
C : Spin up by action of SCAR (2)
D : Separation of upper spacecraft. Then spin down (1) and reorientation by action of SCAR
E : Upper SPELDA jettisoning. Reorientation as requested by inner spacecraft.
F : Spin up and separation of inner spacecraft.
G : 3rd stage avoidance maneuver (spin down, reorientation of 3rd stage, spin up at 5 rpm and Lox valves opening).
Note : Spacecraft separations can also be accommodated under a 3 axis stabilized configuration.

(c)

Figure 11.17 (*cont.*)

38 hours, is spread over three days and covers the operations of filling the tanks of the various stages. Six minutes before firing, control of operations is taken over by computers which check the parameters of the launcher, control the separation of the cryogenic propellant filling arms of the third stage four seconds before the end of the sequence which is terminated by the command to ignite the first stage motors and the additional liquid propellant boosters, if any. The launch vehicle retaining clamps are freed when the motor parameters have been verified after 3.4 seconds; then the additional solid propellant thrusters are ignited. Separation of the additional solid propellant thrusters occurs after 67 seconds and the liquid propellant ones after 146 seconds. The flight of the launch vehicle up to extinguishing the third stage motor lasts approximately 18 minutes. the trajectory of the launcher and the sequence of events during the flight are illustrated in Figure 11.11 [MUA-87].

11.2.2.2 *Ariane 5*

The European Columbus programme provides for the development of an orbital structure containing modules and platforms which may be automatic or manned. The installation, maintenance and servicing of these elements may imply a new launching activity with the space aircraft Hermes.

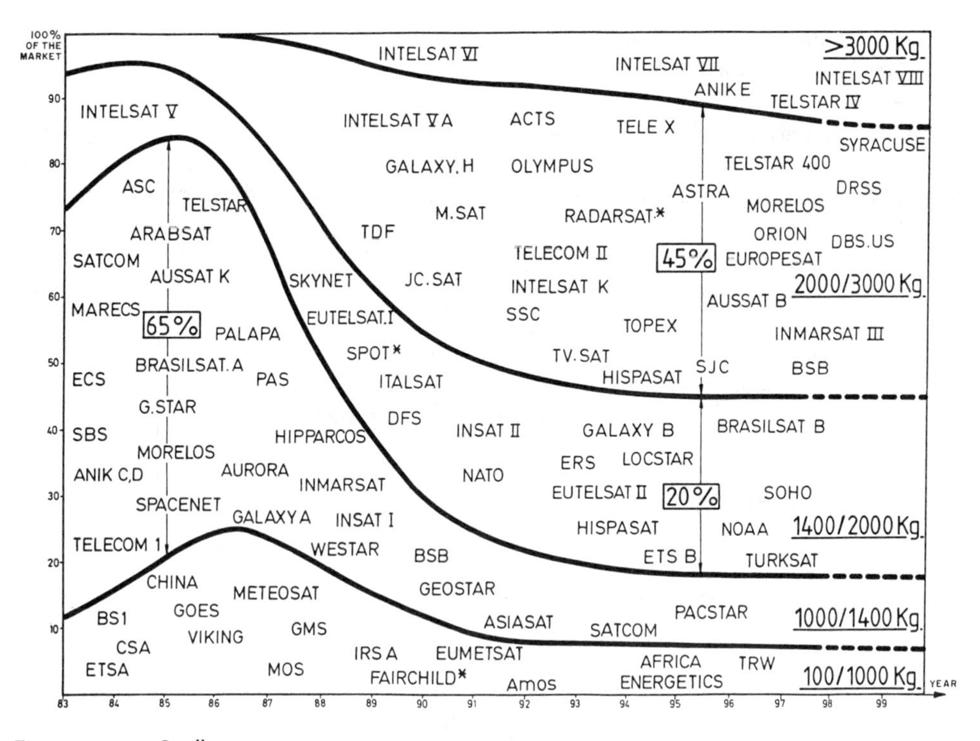

Figure 11.18 Satellite mass increase.

On the other hand, Figure 11.18 shows that during the last decade, a growing number of satellites with a large mass have taken an increasing part of the market. Future missions envisaged for the launchers lead to performance objectives of 2500 kg to 4000 kg in geostationary transfer orbit with an available diameter under the fairing of the order of 4.5 m (compared with 3.65 m for Ariane 4). To remain competitive, the cost of launching must be reduced as far as possible. One cost reduction factor involves the use of a sufficiently powerful launcher to provide simultaneous launching of several satellites. Finally, the highest possible reliability is required of a commercial launcher as the payloads become more and more costly.

The Ariane 5 launcher development programme, undertaken in 1985, permitted a response to these objectives by separating launching of the satellites, to be performed autonomously, and manned flights with Hermes; the latter constitute a particular payload and only the thrusters required for its orbital manoeuvres are provided.

The performance expected of the Ariane 5 launcher is as follows:

— A total capacity of 6800 kg in geostationary transfer orbit which may be for a single launch or shared among satellites and adapters for multiple launching (that is a total payload of the order of 5900 kg for a double launch).
— A capacity in low altitude circular orbit from 18 tonnes (550 km, inclination 28.5°) to 21 tonnes,
— Injection into orbit from the Hermes space plane which will have a mass of 21 tonnes in the final achieved circular orbit of altitude 500 km and inclination 28.5°.

—Useful fairing diameter 4.57 m.
—Reliability of 0.98 for the mission.

The architecture adopted for the Ariane 5 launcher is organised as two composite propulsive stages (Figure 11.19a). The lower composite is independent of the mission and consists of a principal cryogenic stage assisted at take-off by two additional solid propellant boosters. The

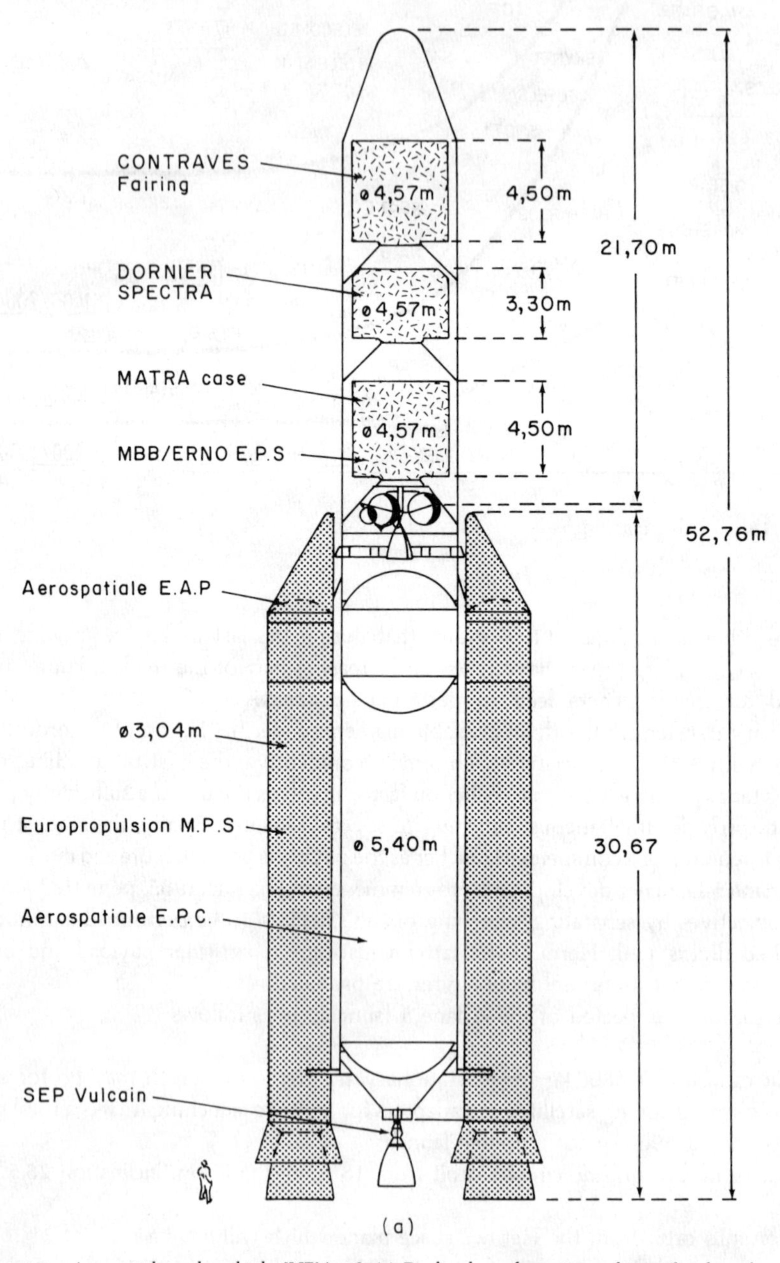

Figure 11.19 Ariane 5 launch vehicle [HEY-90]. (a) Payload configuration, (b) payload performance.

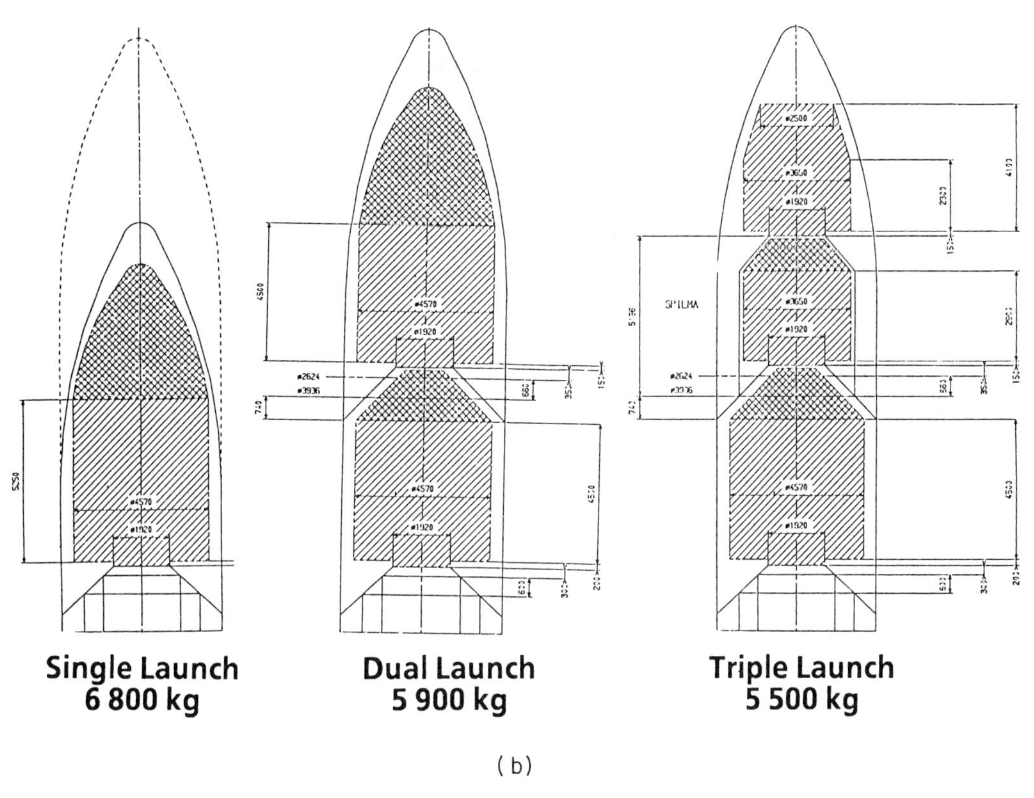

Figure 11.19 *(cont.)*

Table 11.10 General characteristics of Ariane 5

	P230	H155	L7	VEB	Fairing	Launcher
Diameter (m)	3.05	5.45	5.4	5.4	5.4	—
Height (m)	31.5	30	4.5	1.6	23/11	55.85
Total mass (t)	269/unit	170	9.0	1.1	2.4/1.4	728
Mass of propellant at take-off	230/unit	LOX:130 LH2:25	MMH:2.4 N_2O_4:4.8	Hydrazine 60 kg	—	622
Combustion (s)	125	615	1100	—	—	1715
Thrust (t)	600/unit (at take-off)	170 (empty)	2.8	—	—	1370 (at take-off)
(kN)	5890	1670	27.5	—	—	13450

principal stage (HI55) embarks 155 tonnes of liquid oxygen and hydrogen and is propelled from take-off for 615 seconds by the Vulcan HM60 motor which provides 60 tonnes thrust and is under development. The additional thrusters (P230) burn their 230 tonnes of propellant in 120 s while delivering a thrust of 750 tonnes. The upper composite could be the Hermes space plane for manned missions. For automatic launches, the upper composite contains a non-cryogenic liquid propellant propulsive stage (L7), the equipment bag, the fairing and, if necessary, one or two SPELTRA (external supporting structure for triple Ariane launches). A SPELTRA is used for double launches. Two superposed SPELTRAs permit simultaneous launching of three satellites. This architecture is thus different from that of previous launchers and provides different options (Figure 11.19b) [LAP-86] [HEY-90].

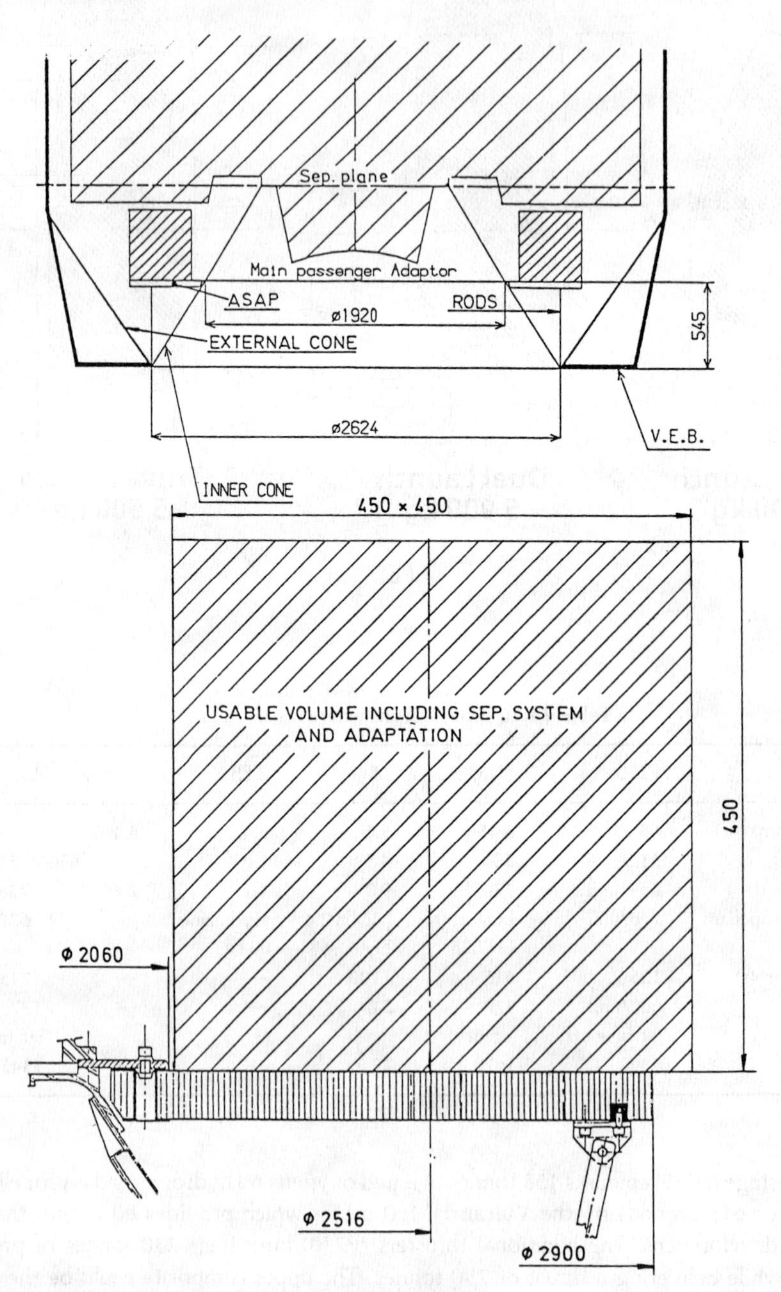

Figure 11.20 Ariane structure for auxiliary payloads (ASAP).

The general characteristics of Ariane 5 in its automatic version for injection from a transfer orbit into the geostationary satellite orbit are given in Table 11.10. Ariane 5 will be launched from a newly developed launching assembly (ELA 3) installed at the launching complex at Kourou.

11.2.2.3 Auxiliary payloads

Arianespace had developed various adaptors for use on Ariane 4 before the demand and market potential for launching small satellites (from a few tens to around a hundred kilograms) and scientific experiments arose [LAR-89].

The ASAP structure permits from one to four small auxiliary satellites of reduced mass to be carried in addition to the main payloads (Figure 11.20). These satellites use the volume which often remains available in the base of the fairing. APEX uses the space in the adaptor which supports the principal satellite and provides launcher accommodation for minisatellites. The ARTEP structure houses technological experiments in the upper part of the SELDA. These structures take advantage of the surplus capacity of the launcher compared with that actually required for the mission. The missions of auxiliary loads often require orbits with a high inclination; opportunities will arise during the six launches anticipated annually for satellites to be placed in sun synchronous orbit [LAR-83].

11.2.3 Europe other than Ariane

11.2.3.1 SCOUT-II/San Marco Scout

The SCOUT-II programme aims to develop a launcher capable of placing 450 kg in a low orbit of 550 km altitude from the Italian launch base of San Marco which is situated in the Indian Ocean and is close to the equator off the coast of Kenya [FAB-89]. This programme was proposed by the Italian company SNIA BDP in association with the US company LTV which produces the SCOUT launcher and provides the basis for development. The performance of the SCOUT (220 kg in a circular orbit of altitude 550 km) is increased by the adoption of two additional boosters (derived from those developed by SNIA-BDP for Ariane 3) and the use of a Mage motor as the last stage. The fairing is of larger dimensions.

11.2.3.2 Future transport systems

Access to space by means of vehicles which are between aircraft and rockets is envisaged; they would take-off from conventional runways and attain low altitude orbits where they would separate from their payload. The vehicles would then re-enter the atmosphere and land like an aircraft. Essentially reusable, these transport systems could provide much lower operational costs than those of conventional launchers.

Two projects are being researched in Europe:

— The Sänger project is a two-stage transport system (83 m long and 330 tonnes) proposed by MBB in the Federal Republic of Germany. The first stage is a hypersonic aircraft with atmospheric power units burning liquid hydrogen and capable of a cruising speed of Mach 4.4 at an altitude of 25 km. Having achieved the desired latitude, the vehicle accelerates to Mach 6.8 in order to separate the second stage at an altitude of 30 km. The first stage then lands while the second stage, propelled by its rocket motors, attains the intended orbit. The second stage is either a recoverable aircraft with a pilot (hypersonic orbital upper stage (HORUS)) capable of transporting four astronauts and 4 tonnes of payload or a transfer stage (cargo upper stage (CARGUS)) capable of injecting 2.5 tonnes into the geostationary satellite orbit. This stage would be derived from the central H155 stage of Ariane V [HOG-88], [KOE-88].

— HOTOL (horizontal take off and landing) is a single stage aircraft-rocket project researched by British Aerospace in Great Britain. The space plane takes off horizontally from a conventional 3000 m runway at a velocity of 290 knots. The propellants burn liquid hydrogen and oxygen from the air during the climb through the atmosphere at a gradient of 24°. Mach 5 is reached after 9 min at an altitude of 26 km. The use of atmospheric oxygen is no longer possible and is replaced by liquid oxygen. Propulsion continues to an altitude of 90 km where the required velocity is attained for a ballistic phase which will take the craft to its operational altitude of around 300 km. An auxiliary propulsion system is used for the orbit and attitude changes required by the mission. This system also permits the altitude of the perigee to be brought to around 70 km after the mission for the re-entry phase which is performed at high incidence (80°). Hypersonic gliding starts below an altitude of 25 km; the approach is made at a velocity of 250 knots at a gradient of 16°. The velocity at touch-down is 170 knots and the maximum coasting distance is 1800 m.

The principal characteristics are: length 76 m, wing-span 27 m, weight 200 tonnes at take-off, 42 tonnes on landing. The liquid hydrogen tank is housed in the forward part of the fuselage and the rear part contains the liquid oxygen tank required to feed the motors outside the atmosphere. The cargo is situated in the central part. The anticipated capacity is 7 tonnes in a circular equatorial orbit of 300 km altitude.

11.2.4 The United States

Prior to the start of the 1980s, the United States had various conventional launcher programmes of varying capacities [GIL-84], [MAH-84]. In parallel a different kind of space transport system, using recoverable units, was developed (the space transportation system (STS) or Space Shuttle). The decision by NASA in 1984 to abandon conventional launchers in favour of the Space Shuttle seemed to put an end to most programmes. The Challenger accident in January 1986 led, in the same year, to the USA taking the decision to continue conventional launcher development in order to provide a replacement for launching military satellites and to no longer use the Space Shuttle for launching commercial satellites. Hence, during 1987, American companies prepared themselves for commercial use of improved versions of their launchers with availability anticipated around 1989–1990 after restarting the production lines and launcher component supply channels.

11.2.4.1 Delta

The programme for the Delta launcher, developed by McDonnell Douglas Astronautics Company (MDAC), was initiated at the end of the 1950s. The first firing of a Delta launcher

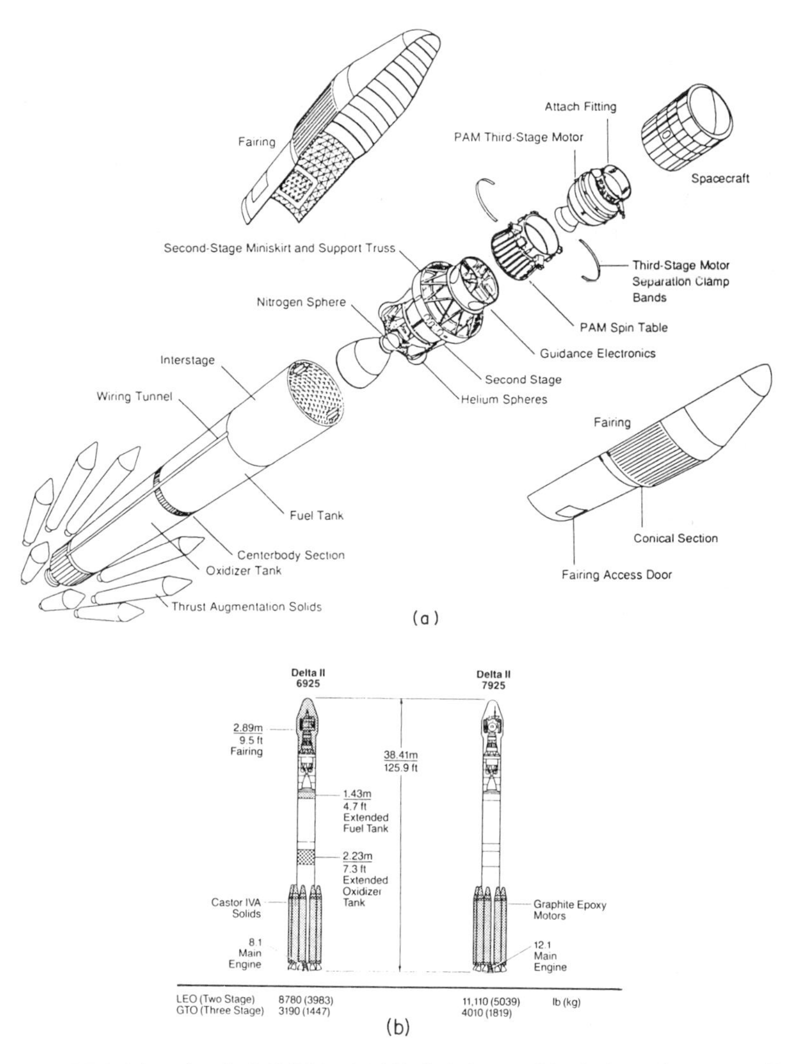

Figure 11.21 DELTA launch vehicle [MEY-90]. (a) Exploded view, (b) vehicle configurations. (Reproduced by permission of McDonnell Douglas Space Systems Company.)

took place in 1960; it had a capacity of 54 kg in geostationary transfer orbit. Successive modifications have led to the Delta 3920 PAM launcher which, in 1982, was capable of injecting 1270 kg into GTO due, in particular, to the use of the PAM upper stage (see Table 11.4) as the third stage. In 1984, NASA took the decision to abandon the programme but, following the unavailability of a means of launching during 1986, MDAC obtained a contract from the US Air Force in January 1987 for development of the Delta II programme (the MLVII launcher particularly for launching the Navstar/GPS satellites). At the same time, MDAC prepared to provide commercial launching services [POR-88].

The architecture of the Delta II launcher is presented in Figure 11.21 [MEY-90]. Two steps were anticipated in the development of the Delta II launcher. The first firing of launcher Delta 6925, capable of injecting 1447 kg into GTO, took place on 14 February 1989.

In comparison with the 3920 PAM-D launcher, the increase (164 kg) in performance of Delta 6925 is obtained by an increase, to 96.5 tonnes, in the capacity of the reservoirs of the first stage which burns kerosene and liquid oxygen and by replacement of the nine additional Castor IV solid boosters by Castor IVA solids of higher performance. The second stage (6 tonnes of AZ50 and N204 propellants) remains unchanged. The third stage consists of the upper PAM-D stage which uses the STAR 48 motor.

Guidance of the missile is provided by an inertia unit and computer located in the upper part of the second stage which controls the thrust orientation of the Rocketdyne RS-27 motor and the first and second stage Aerojet motors. During the ballistic phase of the second stage flight, orientation control is provided by a liquid gas system (nitrogen). For the third stage, rotation of the third stage-satellite assembly by a turntable before separation provides gyroscopic stabilisation of the attitude of the combined assembly during the 87 seconds of motor combustion.

With Delta 7925, the performance is increased to 1819 kg in geostationary transfer orbit by the use of additional Hercules boosters having a shell of composite material, which allows a gain in mass, and by modification of the settings of the RS-27 motor. In this way the total thrust of the first stage becomes 477 tonnes instead of 397 tonnes for the 6925.

Use of a fairing of 2.9 m diameter enables a greater volume to be provided for the satellites. An increase in the size of this fairing is possible by using sections of the fairings fabricated by MDAC for the Titan launchers. The diameter is then 3.05 m for a height of 9.4 m.

11.2.4.2 *Atlas/Centaur*

The Atlas/Centaur launcher developed by General Dynamics also builds up from a long history [BON-82]. An Atlas launched the first communication satellite SCORE in 1958. The Centaur stage was then developed to form the two stage Atlas/Centaur launcher.

The Atlas family includes, in particular, the Atlas H which is capable of placing 1960 kg in low orbit and the Atlas G which serves as the first stage of the commercial Atlas/Centaur or Atlas I [SCH-88] (Figure 11.22a) [WHI-90]. This launcher is built for commercial use by General Dynamics (the first commercial flight was on 15 July 1990); they provide launching and associated services within the agreement signed with NASA and the US Air Force for use of government installations such as the 36 B and 36 A launching ramps at Cape Canaveral. The Atlas/Centaur launcher also constitutes the MLVII launcher for the USA Department of Defense (for launching the DSCS III satellites).

The Atlas G with a diameter of 3.05 m and height of 23 m uses kerosene and liquid oxygen as propellants and develops a thrust of 1950 kN at launch thanks to three Rocketdyne motors. After about 2.5 minutes of flight, the launcher separates from the two motors in order to eliminate useless mass since sufficient thrust is then obtained with a single motor. The fairing is ejected shortly after. The propelled phase continues until the Atlas propellants are completely consumed after about 285 s. The Centaur then takes over after separation from the Atlas. The two Pratt & Whitney cryogenic motors develop a total of 147 kN of thrust and burn liquid hydrogen and oxygen. After 325 s of operation, the motors stop for a coasting phase which enables the launcher to reach the proximity of the point of injection. The motors are then reignited to inject the satellite into the transfer orbit.

Guidance is provided by an inertial system which allows good injection accuracy to be obtained. This system also enables the satellite to be delivered into the transfer orbit with the desired altitude by reorientation of the Centaur after the second extinction of the motors.

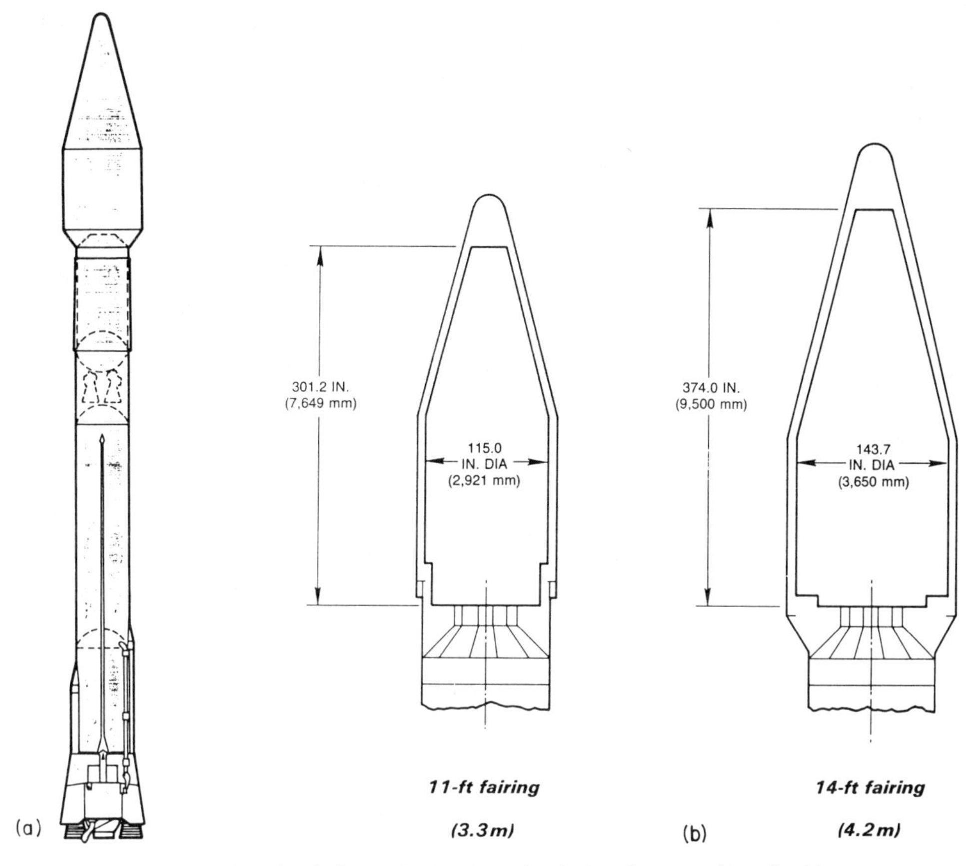

Figure 11.22 ATLAS launch vehicle [WHI-90]. (a) Launch vehicle architecture, (b) payload fairing options.

Two new fairings have been developed. One is called MPF (medium payload fairing) and has an external diameter of 3.3 m (useful diameter 2.92 m); the other is called LPF (large payload fairing) and has an external diameter of 4.2 m (useful diameter 3.65 m) (Figure 11.22b). The possibility of double launches has not been retained, According to the fairing used, the performance of the Atlas/Centaur I is 2340 kg with the MPF fairing and 2270 kg with the LPF fairing.

Atlas II differs from Atlas I in having elongated stages, increased first stage thrust and a modification of the cryogenic propellant mixture of the Centaur stage. The performance is increased to 2760 kg in GTO.

An increase of the capacity to 2900/2810 kg according to the fairing used is achieved with Atlas/Centaur IIA by increasing the quantity of propellant embarked and improvement of motor performance. A new inertial guidance system is also planned. By the addition of four extra Castor IV boosters, the capacity could rise to 3490 kg in GTO in 1993.

11.2.4.3 Titan

Titan III launchers have been used by the US Air Force for more than twenty years and originated from the Titan I and II intercontinental ballistic missiles. Titan II Geminis have been used on twelve occasions to put two men into earth orbit. Various versions of the Titan III have been developed from a central two-stage assembly using liquid propellants; they differ in the extra power units used and the type of upper stage [DUT-88]. For example the Titan IIIC, equipped with two additional five-segment solid propellant boosters, is capable of placing a 1900 kg satellite directly in geostationary orbit by using the re-ignitable Transtage upper stage. Titan IIID is similar to Titan IIIC with the exception of the Transtage stage.

Among the latest versions, Titan 34D (650 tonnes at take-off) is a more powerful version of Titan IIID which is capable of placing 15 tonnes in low orbit and 4 tonnes in geostationary transfer orbit. The most powerful launcher is the Titan IV capable of placing 20 tonnes in low orbit. This launcher differs from the Titan 34D by an increase in the thrust of the motors of the two central stages, an increase of the tank capacity of the central core and the use of seven-segment auxiliary boosters. A 5 m diameter fairing permits voluminous satellites to be housed.

Following the agreements signed with NASA and the US Air Force for the use of the Cape Canaveral government installations (which include 40 launching pads and a building for integration and assembly) Martin Marietta installed a structure for commercial use of the Titan Launcher [MAI-89]. The Commercial Titan III is derived from the Titan 34D. This launcher, which is 46 m high and weighs 680 tonnes, consists of a central core with two stages of 3 m diameter flanked by two large five-and-a-half segment solid propellant boosters each weighing 246 tonnes and developing a thrust of 6210 kN (Figure 11.23).

The auxiliary boosters on their own support take-off and propulsion of the launcher for about 108 s until the guidance system detects a fall in acceleration due to depletion of the propellants and commands ignition of the first central stage before separation of the auxiliary power units. Steering is provided during this first phase by direction of the jet from the thrusters.

The first and second stages use liquid propellants, nitrogen tetroxide N_2O_4 as oxidiser and Aerozine-50 as fuel. The first stage is 24 m high and delivers a thrust of 2437 kN by means of two Aerojet motors for around 152 s. When the guidance system detects a fall in acceleration, the second stage is ignited and then separated from the first by pyrotechnic decoupling; vent holes ensure the evacuation of gases from the interstage volume during this period. The second stage measures 9.4 m and provides a thrust of 467 kN for around 225 s to inject the composite payload into a parking orbit. A hydrazine orientation control system permits the required attitude to be established before separation of the composite with rotation of the assembly at up to 2 revolutions per minute. The capacity of the Commercial Titan III in low earth orbit (148 × 259 km) can attain 15 tonnes.

For injection of satellites into geostationary transfer orbit (GTO), Commercial Titan III must be assisted by a perigee stage capable of providing the velocity increment to transfer from the nominal 150 × 260 km parking orbit inclined at 28.6° to the 180 × 35 786 km transfer orbit inclined at 26.4°. Commercial Titan III is compatible with most of the stages cited in Table 11.4. The performance in the transfer orbit depends on the stage used and ranges from 1280 kg with PAM-D to 2700 kg with E-SCOTS for each satellite with a double launch. With a single launch, the use of IUS or TOS enables nearly 5 tonnes to be placed in GTO. The Centaur stage permits even higher performance.

The Transtage stage can also be used. This stage is equipped with a navigation, guidance and attitude control system which used two re-ignitable liquid propellant motors capable of delivering a total thrust of 71 170 kN. This stage can place around 4.3 tonnes in transfer orbit. Transtage is used firstly to increase the velocity produced by Titan III by initial ignition

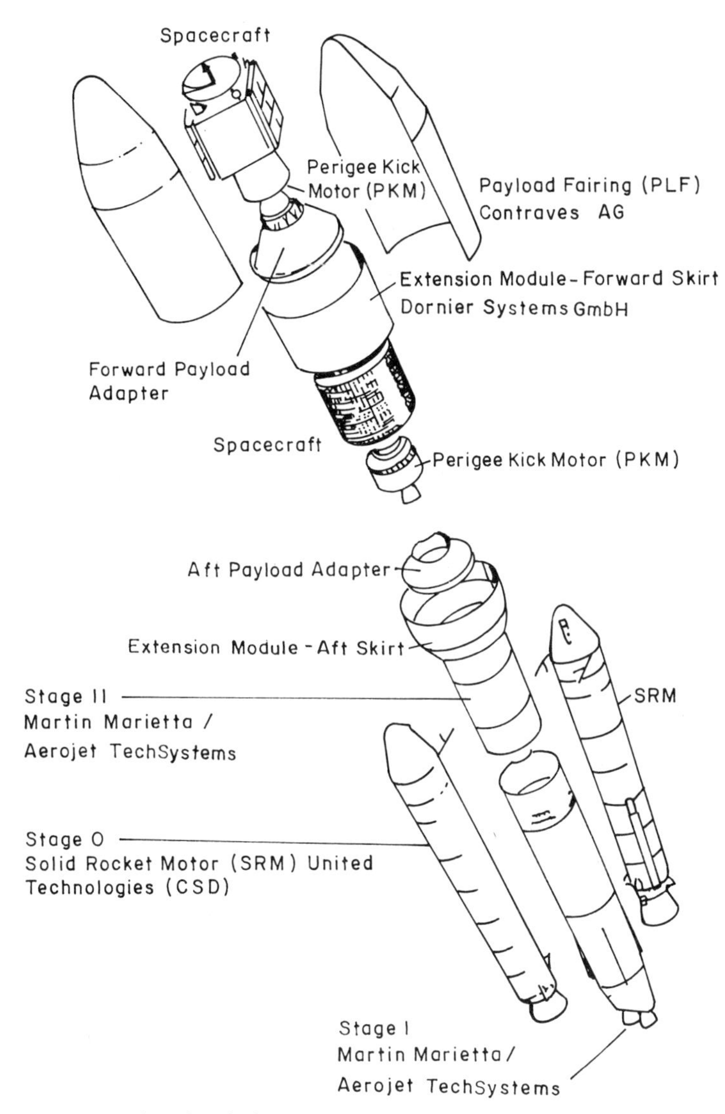

Figure 11.23 TITAN III launch vehicle.

at the end of combustion of the second stage and then, after extinction once the required velocity is achieved and a ballistic phase of around ten minutes, reignition of the motors for seven minutes injects the satellite into the transfer orbit. Transtage also permits the satellite to be placed directly into geostationary orbit by a third re-ignition at the apogee of the transfer orbit.

The Titan III launcher uses a fairing which is 4 m diameter and 10.4 m long, manufactured by Contraves and similar to that of Ariane 4. Simultaneous launching of two satellites is possible by means of an extension module 5.1 m long and 4 m diameter developed by Dornier. This module contains the lower satellite and supports the upper satellite together with the fairing which protects the latter.

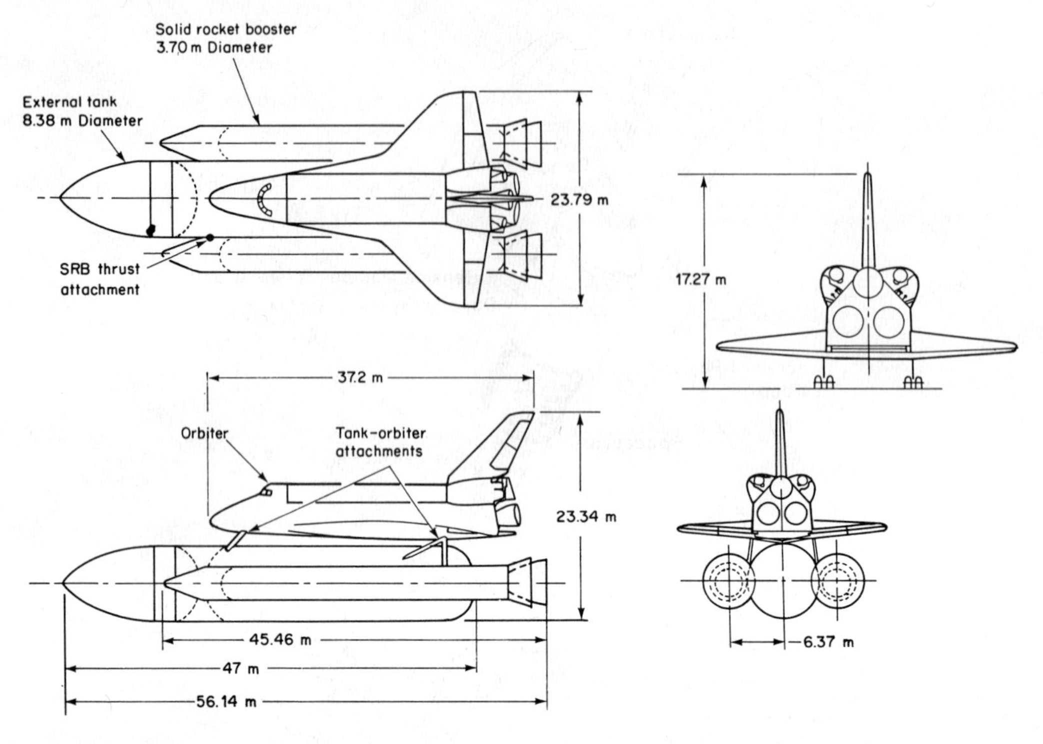

Figure 11.24 Space transportation system (STS).

11.2.4.4 Space Transportation System (Space Shuttle)

The Space Transportation System or Space Shuttle consists of three principal constituents:

- —The space aircraft or 'orbiter' which, after passing into orbit, is capable of descending into the dense layers of the atmosphere and gliding back to the surface of the earth.
- —An enormous releasable and non-recoverable external reservoir which contains the hydrogen–oxygen cryogenic propellants which feed the three motors each of 200 t thrust located on the aircraft.
- —Two additional recoverable solid propellant thrusters of 1225 t thrust (Figure 11.24).

The first flight in orbit of Columbia took place on 12 April 1981. Two other vehicles were built—Challenger and Discovery. A dramatic accident which cost the lives of seven members of the Challenger crew in January 1986 following a failure of a gasket on one of the additional boosters, kept the fleet immobile on the ground for redesign of the boosters and safety improvement. Flights restarted in October 1988 with Discovery.

The large dimensions of the cargo bay (18.3 m long and 4.6 m diameter) and the launching capacity (29 tonnes in circular orbit at 300 km) enable large satellites to be put into orbit and construction of space stations to be considered both for telecommunication purposes and for scientific missions.

On the other hand, the Shuttle can attain only a low circular orbit (nominally 300 km) and two large velocity increments are necessary to place geostationary satellites in their orbit in accordance with the procedure explained in Section 11.1.6 and Figure 11.12.

Although some operations have been successfully attempted in the past, missions to recover a satellite for repair seem to be limited economically to specific satellites circulating in low orbit. For satellites in higher altitude orbits, it is necessary to provide a towing vehicle (see later) and as far as communication satellites in geostationary orbit are concerned, the economic balance is not favourable [PRI-88b].

To place a satellite in orbit from the Shuttle, it is first necessary to eject the payload from the cargo bay. Ejection can use a spring device with prior establishment of rotation to ensure attitude stabilisation (e.g. PAM-D) (see Figure 11.25a). The 'frisbee' method, where a spring propels the satellite out of the cargo bay while imparting a rotating motion to it for attitude control, is also used. Finally, a manipulating arm can extract the satellite from the cargo bay and deposit it outside.

A suitable propulsion device subsequently enables the payload to join the final orbit. Various designs can be considered:

— *A perigee stage* which propels the satellite, equipped with an apogee motor, in the transfer orbit after ejection from the cargo bay. Examples of these upper stages are PAM-D and TOS (Table 11.4).

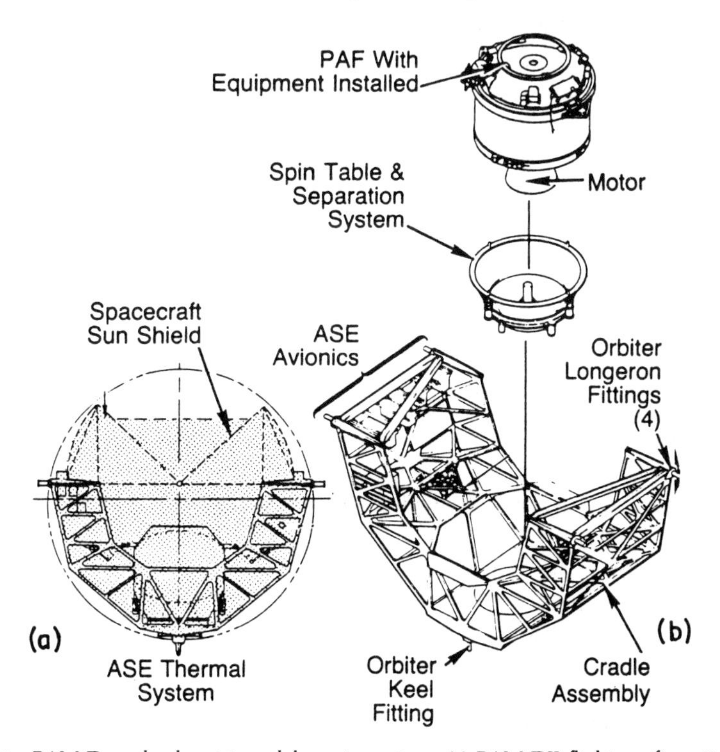

Figure 11.25 PAM-D payload assist module perigee stage. (a) PAM-DII flight configuration, (b) spring ejection from the cargo bay.

The PAM-D (Payload Assist Module Delta class) uses the Thiokol Star-48 motor to place 1247 kg in transfer orbit. The configuration includes the motor, an adaptor supporting the satellite and a supporting cradle mounted in the cargo bay of the Shuttle (ASE: airborne support equipment). This cradle is equipped with a turntable in order to cause the satellite-motor assembly to rotate at between 45 and 100 revolutions per minute before separation in order to ensure gyroscopic attitude stabilisation. Ejection is performed by springs which have previously been compressed during mounting of the motor in the cradle (Figure 11.25(a) and (b)). The total mass of the assembly in the Shuttle cargo bay (without the satellite) is 3319 kg [ORD-82].

After ejection of the combination, the Shuttle manoeuvres in order to place itself at a sufficient distance during ignition of the motor which occurs on crossing the equatorial plane [CAR-84]. Ignition is initiated 45 min after ejection by a sequence controller which is armed during separation. On ignition, the attitude of the combination in an inertial reference is that imposed by the Shuttle during separation due to gyroscopic stabilisation (which can require active control of nutation during the ballistic phase). Superior performance (1588 kg in GTO) is obtained with the PAM-D2 also manufactured by McDonnel Douglas.

The Transfer Orbit Stage (TOS) developed by Orbital Sciences Corporation (OSC) is a perigee stage capable of placing up to 6080 kg in GTO. The TOS uses the SRM-1 (United Technologies Corporation) solid propellant motor and has an inertial guidance system associated with a three-axis control system [WHI-86].

The IRIS developed by Italy [RUM-88], the E-SCOTS developed by RCA/General Electric [BAL-84], the AMS developed by OSC and others could also be cited.

— *A propulsion system integrated* into the satellite which provides the velocity increments at the perigee and apogee by means of its own propulsion system. The satellite may contain a specific perigee motor which could be separated after the manoeuvre and another motor for the apogee thrust (e.g. the Leasat satellite which uses a solid propellant perigee motor and a bipropellant apogee motor which completes the velocity increment at the perigee and subsequently provides that at the apogee). It can also be organised around a unified propulsion system which provides all the necessary velocity increments.

This procedure enables the volume occupied in the cargo bay to be optimised and the cost of launching to be reduced; the invoice presented to each user is established on the basis of his percentage occupation (in volume and mass) with respect to the capacity of the Shuttle.

— *A transfer stage* which is capable of placing a satellite without its own motor directly into the geostationary satellite orbit. The transfer stage can use two solid propellant motors, one for the impulse at the perigee and the other for the impulse at the apogee (e.g. Inertial Upper Stage (IUS)), a reignitable liquid propellant motor (e.g. Transtage, Centaur) or a combination of the two, a solid propellant motor for the perigee and a liquid propellant motor for the apogee (e.g. TOS/AMS).

IUS, developed by Boeing, is a transfer stage using SRM-1 and SRM-2 solid propellant motors and having a navigation and active attitude control system; it can place 2270 kg directly in geostationary orbit [HAN-80]. It has been used in particular for launching the TDRS data relay satellites.

TOS/AMS is a hybrid transfer stage consisting of a TOS perigee stage associated with a liquid propellant stage (apogee and manoeuvring system (AMS)) used to perform the manoeuvre at the apogee. The capacity in geostationary satellite orbit is 2958 kg [WHI-86].

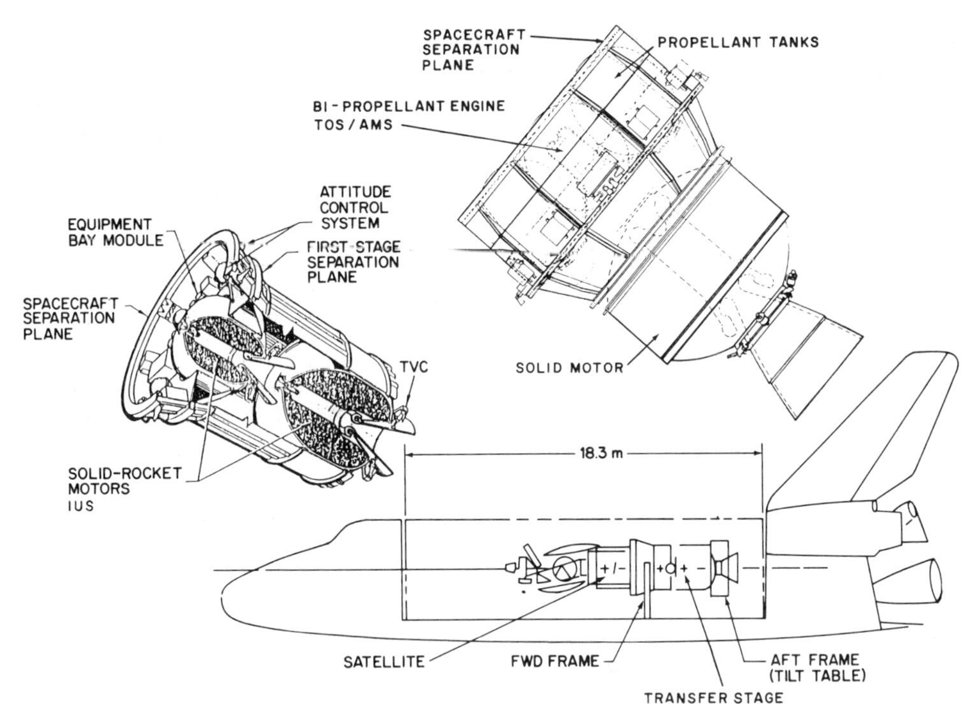

Figure 11.26 Transfer orbit stages: IUS and TOS/AMS. (Reproduced by permission of the European Space Agency.)

Figure 11.26 shows the IUS and TOS/ARS transfer stages and their arrangement within the cargo bay.

The Centaur developed by General Dynamics, the HPPM developed by Ford Aerospace [MOS-84], various propulsion system modules attached to satellites [MOS-84], [KEL-84] and others could also be cited.

—*A recoverable towing vehicle* (space tug) which can place the payload to which it is attached in the desired orbit (geostationary for example) as with a transfer stage but which can also dock with a satellite already in orbit and bring it to the vicinity of the Shuttle cargo bay for attention or recovery.

Among the tug projects, the Orbital Manoeuvring Vehicle (OMV) can be mentioned; TRW was chosen to develop this in 1986. The OMV is a tug with a limited operating radius designed to modify the orbit (injection or recovery) of satellites or service modules in connection with use of the Space Shuttle. The capacity of this tug enables it to modify the inclination of the orbit of a 2225 kg satellite by 8° or to increase its altitude by 2300 km.

A tug project of higher capacity, the Orbital Transfer Vehicle (OTV) is planned. Various designs have been suggested by different companies [BAN-88], [CHA-88], [HEN-88], [FES-88].

11.2.4.5 Future high capacity transport systems

The technology used for the Shuttle could be used to develop a high capacity space transport system. This non-recoverable pilotless vehicle, called Shuttle C, would enable more than 45 tonnes to be placed in low orbit. This launcher could take advantage of the production facilities established for the 24 flights per year anticipated before the Challenger accident. It is anticipated that non-commercial use agreed after this would require only 14 flights per year and this could enable advantage to be taken of the installed infrastructure to manufacture the components of Shuttle C. As well as using most of the principal components of the Shuttle (such as the external tank, the additional boosters and the main motors), an increase of capacity could be obtained by eliminating everything which enables the Shuttle to be a manned recoverable vehicle such as the wings, pressurised cabin, landing gear and so on. With the help of a high performance transfer stage such as the Centaur cryogenic stage developed by General Dynamics or the OTV, this launcher would have a high capacity (several tonnes) in the geostationary satellite orbit.

Furthermore, NASA has issued a tender for the development of the Advanced Launch System (ASL), a high capacity launcher of moderate cost [BRA-88]. Various manufacturers have made proposals such as General Dynamics for a launcher derived from the Atlas Centaur and Martin Marietta for two vehicles with cryogenic centre stages and extra liquid propellant boosters capable of placing 48 and 72 tonnes in low altitude circular orbit.

11.2.4.6 *Low capacity launchers*

Scout. The SCOUT is a four-stage solid propellant launcher capable of placing 220 kg in a circular orbit of 550 km altitude (21 tonnes at take off). It is built by the LTV company and can be launched from various US bases (including Cape Canaveral and Vandenberg) and also from the Italian platform at San Marco off the Kenya coast (see also SCOUT II in Section 11.2.3.1).

Pegasus. Pegasus is a novel launch vehicle launched from a B-52 aircraft flying at around 43 000 ft at Mach 0.8. The first flight took place on 5 April 1990. It consists of a three-stage solid propellant launcher developed by Orbital Sciences Corporation and Hercules Aerospace (Figure 11.27) [THO-90].

The Pegasus flight vehicle is 15 m long and 15.24 m in diameter and has a gross weight (excluding payload) of approximately 18 600 kg. A delta wing with a 6.7 m span and three 1.52 m span movable control fins are mounted on the first stage. Pegasus is carried aloft by

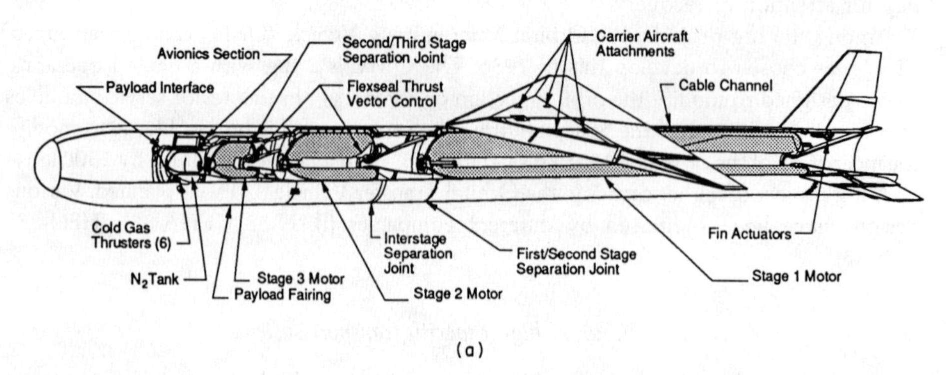

Figure 11.27 Pegasus architecture (a) and payload fairing dimensions (b). (Reproduced by permission of Orbital Sciences Corporation.)

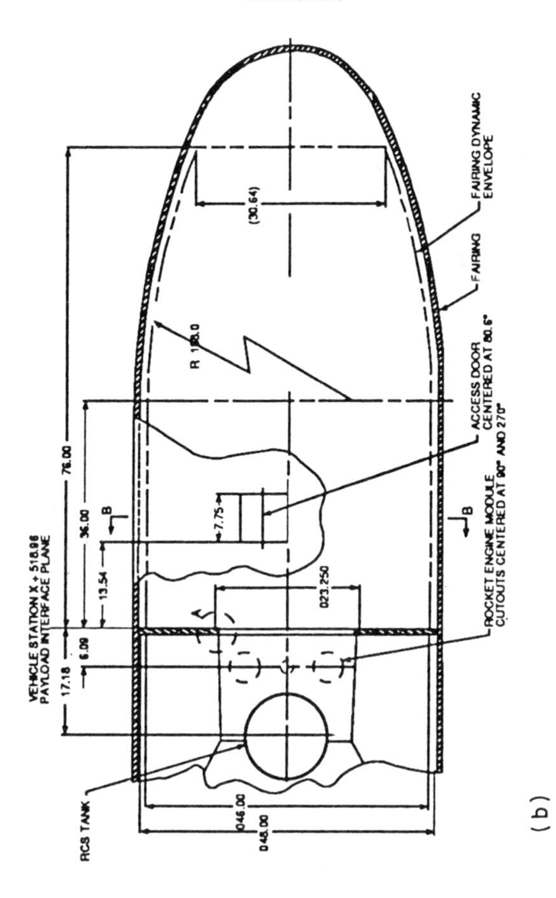

(b)

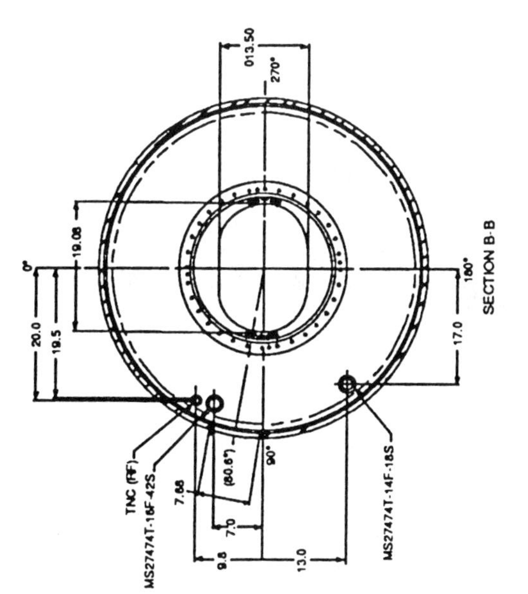

Figure 11.27 *(cont.)*

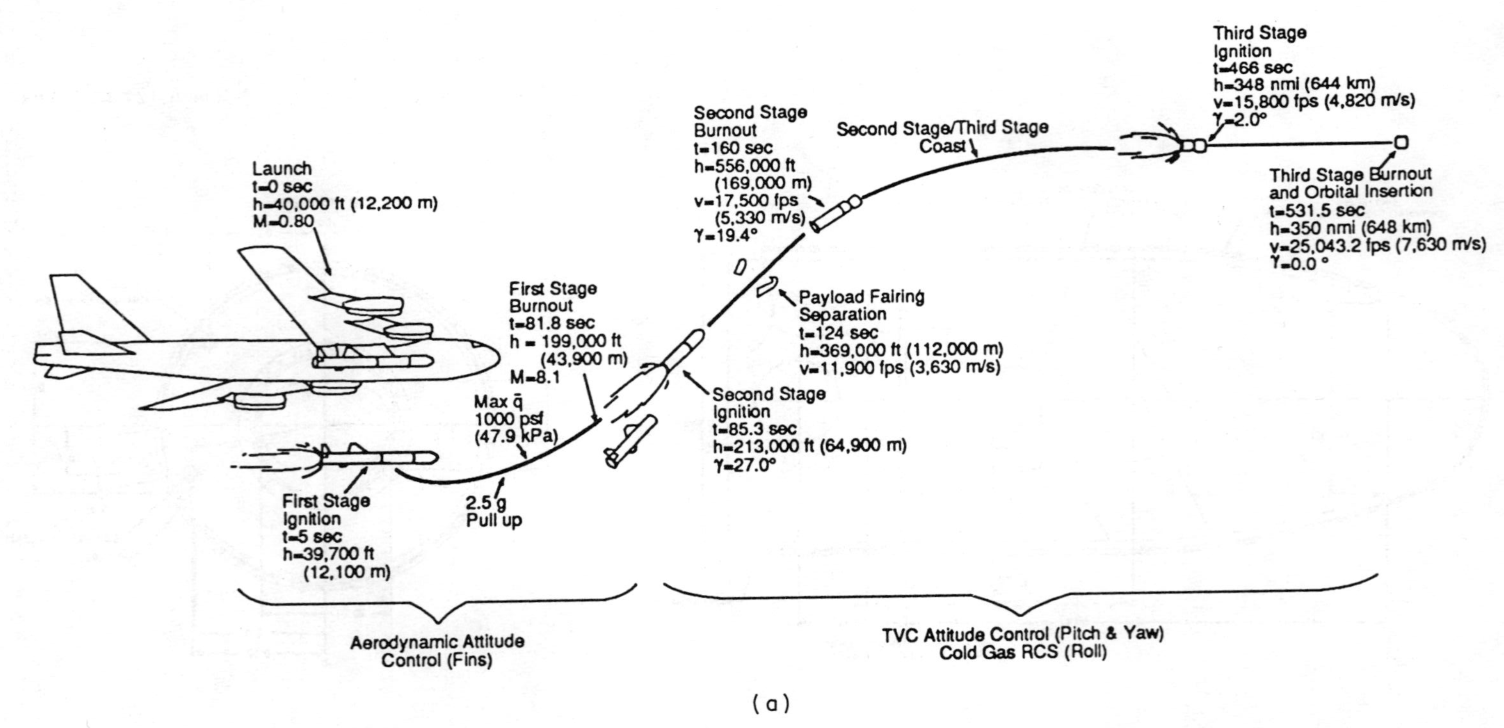

Figure 11.28 Pegasus baseline mission profile (a) and payload performance (b). (Reproduced by permission of Orbital Sciences Corporation.)

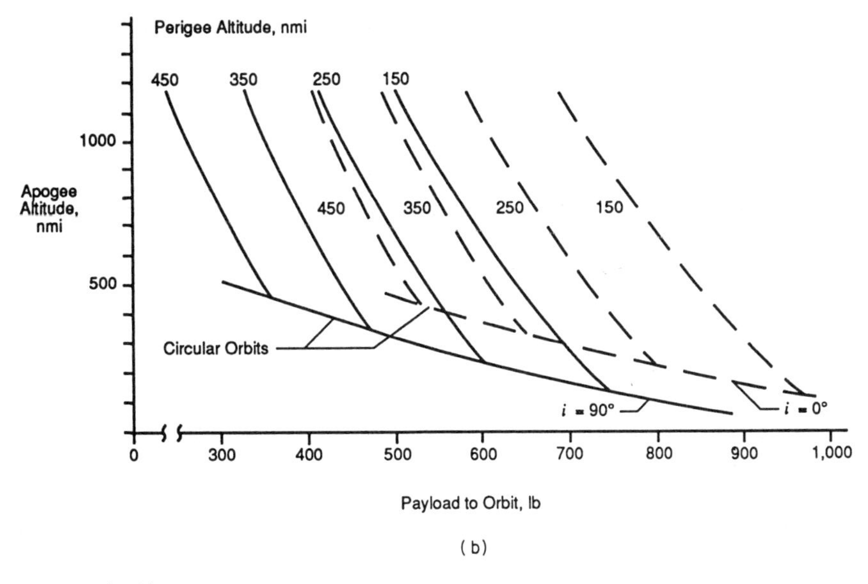

Figure 11.28 *(cont.)*

a conventional transport- or bomber-class aircraft to level flight conditions of 12.2 km and Mach 0.8 (Figure 11.27b). After release from the aircraft and ignition of its first stage motor, the vehicle's autonomous flight control system provides guidance through the required suborbital or orbital trajectory. Pegasus can deliver spacecraft weighing up to 408 kg into low Earth orbits (Figure 11.28a) or launch payloads up to 680 kg on suborbital, high Mach number cruise or ballistic flights. Spacecraft as large as 1.83 m long and 1.17 m in diameter can fit within the standard Pegasus payload fairing (Figure 11.27b). In addition, Pegasus can accommodate three-axis, gravity gradient or spin-stabilised spacecraft or multiple smaller satellites on a single launch.

An air-launched, lift-assisted trajectory enables Pegasus to deliver approximately twice the payload to orbit of an equivalent ground-launched vehicle. The superior performance of Pegasus is the result of several factors: the potential and kinetic energy contributed by the carrier aircraft, reduced drag due to lower air density flight profiles, improved propulsion efficiency due to higher nozzle expansion ratios, and reduced gravity losses due to its unique flat trajectory and wing-generated lift. Advantages over traditional pad-launched rockets include an increased range of orbital inclinations achievable without energy wasting 'dog-legs' or out-of-plane manoeuvring, and extended launch windows resulting from the flexibility in launch point selection. In addition, the Pegasus ascent profile generates lower accelerations, dynamic pressures, and structural and thermal stresses compared to ground-launched boosters, providing a more gentle ride to orbit for its satellite payloads.

Taurus. The launcher called Taurus or small satellite launch vehicle (SSLV) consists of a first stage derived from the first stage of the Peacekeeper missile and three stages of Pegasus. The capacity would be 1150 kg in polar orbit at 550 km. The volume available to the payload is 3 m^3 with a 1.27 m diameter.

Space Vector. This three-stage launch vehicle is organised as follows: the first stage consists of three Castor IV boosters; the second stage is a STAR 48 motor and the third stage is a STAR 27 motor. The capacity should be 340 kg in a 500 km polar orbit.

Conestoga. Space Services Inc. has developed a solid propellant launcher called Conestoga [SLA-89]. Several versions are proposed; they are organised around a first stage which consists of two (Conestoga 2) or 4 (Conestoga 4) Castor IVB (Thiokol) thrusters. The second stage consists of one (Conestoga 2) or two (Conestoga 4) Castor IVB thrusters and the third stage uses a Star 48 (Conestoga 2) or Castor IVB (Conestoga 4). Conestoga 4 also has a fourth stage equipped with a Star 48 motor.

The capacity of the launcher should be 280 kg (Conestoga 2) and 850 kg (Conestoga 4) in a circular polar orbit of altitude 500 km. The volume available to the payload is 5.36 m^2 with a diameter of 1.45 m.

Amroc. The American Rocket Company (AMROC) has proposed a launcher consisting of an assembly of standardised hybrid fuel (LOX/polybutadiene) modules of 315 kN thrust [KOO-89]. Various configurations are possible according to the number of modules; the basic configuration contains 22 and enables 1400 kg to be placed in a circular polar orbit.

Lockheed. This launch vehicle originates from the Peacekeeper missile. The first stage consists of one Castor 120 and four Castor IV. The second stage is a Castor 120 booster. The third stage is a STAR 75 motor. A fourth hydrazine stage allows accurate orbit injection. The capacity should be of the order of 2.5 tonnes in low Earth orbit. This launch vehicle has been proposed to launch four IRIDIUM satellites.

11.2.5 India

In the 1970s, India, under the administration of the Indian Space Research Organisation (ISRO), initiated the development of launchers for national scientific requirements. With the launching of a 35 kg satellite in 1980 using a four-stage solid propellant SLV-3 launcher, India became the sixth country to have launched a satellite by its own means.

A performance improvement programme led to the Augmented Satellite Launch Vehicle (ASLV) which has a capacity of 150 kg in low altitude orbit. The SRIHARITOKA launch base is situated 160 km to the north of Madras.

Concurrently the Polar SLV (PSLV) which is capable of placing 1000 kg in sun synchronous polar orbit was started; its second stage uses bi-liquid propulsion. A cryogenic motor development programme will enable the Geostationary Launch Vehicle (GSLV) to be produced; this will be capable of placing satellites of the INSAT II type (2 tonnes) in geostationary transfer orbit.

11.2.6 Israel

Israel launched its first satellite Offeq-1 on 19 September 1988 using the Shavit launcher. This two-stage solid propellant launcher injected the 150 kg satellite into a 1300 × 290 km elliptical orbit of inclination 140°. The volume under the fairing is tapered of length 2.3 m and diameters 0.7 m and 1.8 m.

11.2.7 Japan

11.2.7.1 NASDA programmes

The launchers for application programmes are developed in Japan by Mitsubishi Heavy Industries on behalf of the Japanese Space Agency (NASDA).

The N range of launchers consists of N1 and N2. Launchers N1 can place 150 kg in geostationary orbit under a fairing of 1.4 m useful diameter. The first two stages are developed by Mitsubishi, the first under licence from McDonnel Douglas. The third stage is manufactured by Nissan under licence from Thiokol. Model N2 is derived from the Delta 2914 launcher; the second stage is developed by Ishikawa-Harima Heavy Industries (IHI) under licence from Aérojet. This launcher is capable of placing 640 kg in geostationary transfer orbit and 360 kg in geostationary orbit (the useful diameter of the fairing is 2.2 m).

The H1 launcher is similar to the N2 launcher with the exception of the second stage which becomes cryogenic and uses the LE-5 motor of 10.5 tonnes thrust developed by IHI. This launcher, whose first firing took place in August 1986, can place 550 kg in geostationary orbit (the fairing diameter is 2.2 m).

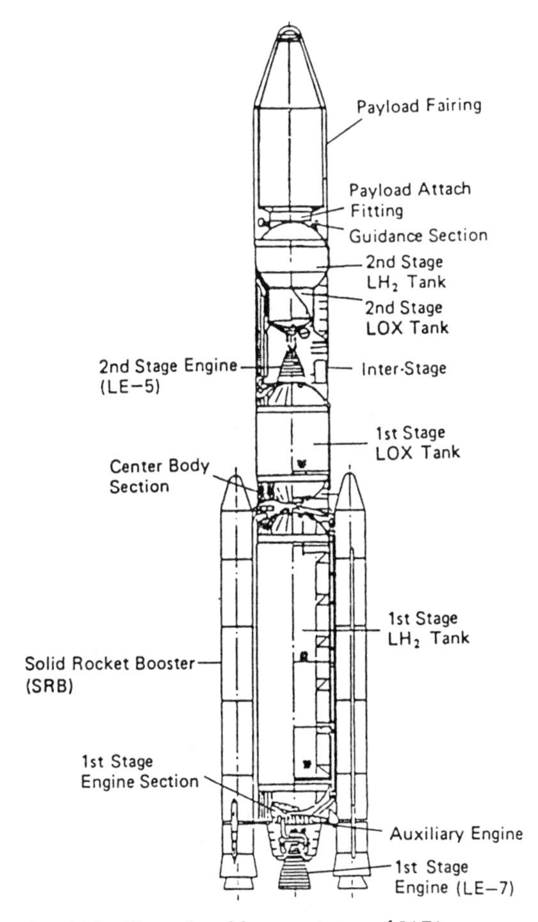

Figure 11.29 H-2 launch vehicle. (Reproduced by permission of IAF.)

Table 11.11 Characteristics of the H-II launcher.

Item	Specification		
Overall length	49.9 m		
Diameter	4 m		
Mass (including payload)	263.9 t		
	1st stage	Solid rocket boosters	2nd stage
Propellant	LOX/LH_2	Solid propellant	LOX/LH_2
Propellant weight	86.3 t	118 t (2 units)	16.7 t
Thrust	86 t (sea level)	318 t (2 units, sea level)	12.4 t (empty)
Total burning time	348 s	94 s	609 s (restart capability)
Isp	445.4 s (empty)	273 (empty)	452 s (empty)
Weight	98.1 t	140.9 t (2 units)	19.7 t
Fairing			
Total weight		1.4 t	
Diameter		4.1 (External diameter)	
Payload		3.7 m diameter × 10 m	
Guidance system		Strap-down inertial	

The H2 launcher has been designed with a new architecture containing a central body with two cryogenic stages and two large extra solid boosters (this architecture is similar to that used for Ariane 5). The characteristics of this launcher are given in Table 11.11 and Figure 11.29 [MOC-88], [ETO-88]. The launcher is 49 m high with a diameter of 4 m and a mass without payload of 256 tonnes. The first stage uses a new LE-7 cryogenic motor of around 100 tonnes thrust while the second stage motor is an improved version of the LE-5. The total propellant mass (LOX + LH2) is 99 tonnes. The additional solid boosters consist of four segments and are 23.4 m high with a diameter of 1.8 m; they contain 118 tonnes of a polybutadiene based mixture for a thrust of 320 tonnes.

Guidance is provided by a strap-down inertia unit which uses gyrolasers and allows for a ballistic phase. Steering is achieved by controlling the direction of the jet from the additional thrusters and the first and second motors. Hydrazine thrusters are used during the second stage ballistic phase.

Performance is of the order of 2 tonnes in geostationary orbit, 4 tonnes in a transfer orbit of inclination 30° and 10 tonnes in a circular orbit of 300 km altitude. The useful volume under the fairing is 3.7 m diameter and 12 m high. The first launch is planned for 1993.

The potential increase planned by NASDA is 4 tonnes in geostationary orbit which will be achieved by the addition of supplementary boosters. A fairing of 4.6 m useful diameter is also planned.

Japan is also developing an automatic recoverable space aircraft HOPE (H-II Orbiting Space Plane) of reduced size (11.5 m long, 8.8 tonnes at launch, useful capacity of the order of 1.2 tonnes) which will be placed in orbit by the H-II launcher.

Commercialisation of the N1, N2 and H1 launchers is impossible owing to the use of sections built under US licence. It is possible for the H2 but the launch base at Tanegashima (latitude 30.4°N) poses operational problems. This base is open only twice for 45 days per year on account of restrictions due to fishing.

11.2.7.2 The ISAS programme

The development of launchers for scientific programmes is provided by the Institute of Space Aeronautical Science (ISAS); manufacture is performed by Nissan. The latest version of the Mu series is the M-3SII launcher capable of placing a 770 kg scientific on a 250 km circular orbit. It consists of a three-stage solid propellant launcher plus two extra boosters which are 28 m high and weight 62 tonnes. The diameter of the fairing is less than 1.4 m.

A new vehicle in the series is under development (1994); it consists of the M-V launcher with four stages (height 30 m, mass 200 tonnes). The performance will be 2000 kg in an orbit of altitude 250 km with a fairing of diameter slightly less than 2.5 m.

11.2.8 USSR

The USSR has various series of launchers each containing diverse models of varying capacity and use.

Three launch pads are used. The oldest, Tuyratam (or Baikanur), is situated 370 km to the south-west of the town of Baikanur in the republic of Kazakhstan and around 2000 km south-east of Moscow. The latitude is 51.6°N. The complex extends to the town of Leninsk, 30 km from the launch ramp. It was used to launch the first artificial satellite, Sputnik 1. Manned flights and the Proton launcher continue to be fired from Tuyratam.

The most active complex is that of Plesetsk, near the frontier with Finland. A third complex Kapustir Yar/Volvograd Station is used mainly for firing missiles.

11.2.8.1 Series A

This series, which served to launch the first Sputniks, consists of two- and three-stage launchers, according to version, assisted by four additional boosters arranged obliquely. The propellants used are kerosene and liquid oxygen (hydrazine and liquid oxygen for the third stage). These launchers, assembled horizontally, are fired from Tuyratam or Plesetsk.

Version A2 or the Soyuz Launcher is much used, particularly for launching manned Soyuz and Cosmos reconnaissance vehicles. The performance is 7.5 tonnes in low orbit and 2.4 tonnes in geostationary transfer orbit.

11.2.8.2 Series C

This series, which followed a series of launchers of modest performance (Series B—250 to 500 kg in low orbit), is used to launch the Cosmos scientific satellites. They are two-stage launchers (Lance Vostok) which can be fired from the three launch bases. The performance is of the order of 1 to 1.5 tonne in low orbit.

11.2.8.3 Series D

The D series launchers are heavy launchers capable of placing from 17 to 27 tonnes in low orbit in accordance with the configuration. The stages (two to four according to version) use liquid propellants. The first stage consists of a central body containing the oxidant and six auxiliary reservoirs containing the fuel. Six motors (Glushko/GDL-OKB RD-253) are mounted at the end of the auxiliary reservoirs and develop a thrust of 150 tonnes for 130 seconds. The second stage is propelled for 140 s by four Kosberg/JRD motors each developing a thrust of the order of 650 kN. The third stage uses a single Kosberg/JRD motor and has four vernier motors for orientation control. The duration of operation is of the order of 140 seconds. These stages use hydrazine and nitrogen peroxide as propellants. The launchers, assembled horizontally, are fired from Tuyratam.

Version D1e or Proton Launch is used particularly for launching satellites into geostationary orbit. The Soviet Union decided in 1983 to offer the services of the Proton launcher to the Western world to launch the Inmarsat II satellites. A commercial organisation, Licensintorg, was established under the aegis of the Glavkosmos agency.

The Proton launcher has four stages and is 52 m high. Its mass at take-off is of the order of 700 tonnes [PIR-87]. The first three stages place the fourth stage and payload in a circular orbit of altitude 200 km and inclination 51.6°. The fourth stage is actually a transfer stage (Block D stage) which uses liquid oxygen and kerosene as propellants. This reignitable stage, of mass 17.3 tonnes, develops a thrust of 85 kN with a specific impulse of 350 s. The total duration of operation can exceed 600 seconds. Attitude control is provided by small thrusters (vernier motors) which use hydrazine and nitrogen peroxide. On passing through the node, the fourth stage places the combination in a transfer orbit prior to the geostationary satellite orbit. Re-ignition of the fourth stage at the first apogee of the transfer orbit then injects the satellite into the geostationary orbit.

The Proton launcher is thus capable of injecting a payload of the order of 2.5 tonnes under a fairing of 3.7 m diameter directly into the geostationary satellite orbit. The accuracy of positioning would not be very good; the inclination and residual eccentricity are estimated at 0.1° and 0.01 respectively and the error in the period could reach 20 minutes.

11.2.8.4 Series F

The series F launchers seem to be reserved for launching military satellites such as the Cosmos ocean surveillance satellites. The launchers have two or three stages according to version and the performance should be of the order of 4.5 tonnes in low orbit.

11.2.8.5 Energia

The Energia launcher has been developed to place heavy payloads in low orbits, particularly the recoverable space vehicle Buran. The first launching of Energia took place in May 1987. This launcher consists of an enormous central body on which four additional boosters are mounted [GUB-88]; it is 60 m high and weighs 2000 tonnes. The capacity in low orbit is of the order of 100 tonnes.

The Buran recoverable vehicle is 36.4 m long, 16.5 m high and has a wing-span of 24 m. Its mass at take-off is 105 tonnes without the payload and its carrying capacity is 30 tonnes. The hold is 4.7 m diameter and 18 m long. On its return the empty mass of Buran is 62 tonnes and landing is possible with 20 tonnes of payload. The first flight under automatic pilot took place on 15 November 1988.

11.2.9 Cost of installation in orbit

It is difficult to specify the cost of launching a satellite, since the cost depends on the type of service provided, the performance of the launcher, the commercial policy of the organisation which sells the service and so on. An order of magnitude is 100 to 120 M$ for a capacity of the order of 4000 kg at take-off, that is around 2200 kg in geostationary orbit or around 30 000$ per kilogramme at take-off. This cost is lower when the launcher used is more powerful.

It should be noted that cost comparison between one launcher and another is not easy, since it is not sufficient to compare the cost and capacity placed in orbit. For the same capacity, launchers differ in a large number of characteristics—inclination of the transfer orbit, accuracy with which the orbit is obtained, useful volume under the fairing, static and dynamic mechanical constraints (such as longitudinal and transverse acceleration, vibration, shock, noise spectrum), thermal constraints, interfaces etc. These characteristics can have a very significant impact on the design, lifetime and hence overall cost of the system.

With modern high capacity launchers which are capable of multiple launches, the problem of sharing the cost of launching among users also arises. Before the Challenger accident and the decision no longer to use the Shuttle commercially, NASA had established a tariff policy which took account of the capacity used. A dedicated launch using the full capacity was invoiced at 74 million dollars (the actual cost is estimated to be 300 million dollars). In the case of shared usage, the price took account of a filling rate of 75%, that is 74/0.75 M$, multiplied by the greater of the following two ratios—mass of the payload concerned to cargo capacity (29.5 tonnes) and length occupied to length of the hold (18.3 m). The price of the upper stage and the operating costs are added to this.

In the case of Ariane, the price is invoiced pro rata to the mass to be carried taking account of the constraints imposed by the available adaptors for multiple launches.

REFERENCES

[AMS-89] S. Amstrong, O.L. (1989) A commercial European small launcher, *Workshop on Flight Opportunities for Small Payloads (ESA SP-289), Esrin*, pp. 161–166, May.

[ASA-83] W. Asad (1963) The MAGE family of European solid-propellent apogee boost motors, *ESA Bulletin*, No. 33, February, pp. 6–11.

[BAL-84] D.E. Balser (1984) Shuttle-compatible orbit transfer subsystem, *35th International Congress of the IAF, Lausanne*, October, Paper 84–14.

[BAN-88] E.L. Bangsund (1988) Issues associated with a future orbit transfer vehicle (OTV), *39th International Congress of the IAF, Bangalore*, October, Paper 88–185.

[BOI-86] J. Boissieres (1986) Mise à Poste d'un Satellite Géostationnaire avec un Moteur Réallumable, *Mécanique Spatiale pour les Satellites Géostationnaires, Colloque CNES, Cepadues*, pp. 457–474.

[BOL-86] P. Boland (1986) Analyse du Problème des Fenêtres de Lancement Satellite pour les Lancements Doubles sur Ariane, *Mécanique Spatiale pour les Satellites Géostionnaires, Colloque CNES, Cepadues*, pp. 37–52.

[BON-82] M.M. Bonesteel (1982) Atlas and Centaur adaptation and evolution—27 years and counting, *IEEE International Conference on Communications, Philadelphia*, June, pp. 3F.2.1–3F.2.8.

[BRA-88] D.R. Branscome (1988) NASA launch vehicles—the next twenty years, *39th International Congress of the IAF, Bangalore*, October, Paper 88–189.

[BRE-89] J. Breton, P. Loire (1989) ARIANE structure for auxiliary payloads (ASAP), *Workshop on Flight Opportunities for Small Payloads (ESA SP-298), Esrin*, pp. 39–42, May.

[BRO-86] D.R. Brown (1986) Orbit geometry for calibration and launch operations of a three-axis controlled satellite with inertial reference, *Mécanique Spatiale pour les Satellites Géostationnaires, Colloque CNES, Cepadues* pp. 53–70.

[CAR-84] J. Caray (1984) The process of lauching communications satellites with the shuttle: an example using WESTAR VI, *AIAA 10th Communications Satellite Systems Conference, Orlando*, March, Paper 84–0759.

[CHA-88] J.B. Chambers, S.E. Doyle (1988) Propulsion requirements for orbital transfer and planetary mission support, *39th International Congress of the IAF, Bangalore*, Paper 88–183, October.

[DON-84] H. Donat (1984) Mise et Maintien à Poste des Satellites Géostationnaires, *Mathématiques Spatiales, Colloque CNES*, pp. 673–710, Cepadues.

[DUT-88] R. Dutton, S. Isakowitz (1988) Commercial TITAN: a proven derivative, *AIAA 12th Communications Satellite Systems Conference, Arlington*, Paper 86–0852, March.

[ESC-86] P. Escudier (1988) Définition d'une Structure pour les Poussée d'Apogée d'une Mise à Poste d'un Satellite Géostationnaire—Application à TDF1, *Mécanique Spatiale pour les Satellites Géostionnaires*, Colloque CNES, *Cepadues*, pp. 511–522.

[ETO-88] T. Eto *et al.* (1988) On the status report of the H-II rocket, *39th International Congress of the IAF, Bangalore*, October, Paper 88–167.

[FAB-88] A. Fabrizi, M. Green (1989) SCOUT II programme status, *Workshop on Flight Opportunities for Small Payloads (ESA SP-298), Esrin*, pp. 167–172, May.

[FES-88] D.A. Fester, B.A. Bicknell (1988) Short-length, high-performance cryogenic stage, *39th International Congress of the IAF, Bangalore*, October, Paper 88–181.

[GIL-84] M. Gilli (1984) Les Systèmes de Lancement de la Décennie 80–90, *Note Technique CNES No. 111*, Toulouse.

[GOR-71] S. Gordon, B.J. McBride (1971) Computer program for calculation of complex chemical equilibrium compositions, rocket performance, incident, and reflected shocks and Chapman-Jouguet detonations, *NASA SP-273*.

[GUB-88] B.I. Gubanov (1988) ENERGUIYA—new Soviet launch vehicle, *39th International Congress of the IAF, Bangalore*, October, Paper 88–172.

[HAN-80] D.R. Hanford (1989) IUS—A key transportation element for future communications satellites, *AIAA 8th Communications Satellite Systems Conference, Orlando*, March, Paper 80–0589.

[HEN-88] M.W. Henley (1988) Space-based orbital transfer system evolution to support lunar and Mars missions, *39th International Congress of the IAF, Bangalore*, Paper 88–184, October.

[HEY-90] D.A. Heydon (1990) ARIANE program plans and outlook for commercial launch services, *AIAA 13th International Communication Satellite Systems Conference, Los Angeles*, Paper 90–0890, March.

[HOG-88] E. Högenauer (1988) SÄNGER—European reusability, *Space*, **4**, No. 3, May, pp. 4–9.

[HOH-25] W. Hohmann (1925) *Die Errichbarkeit der Himmelskörper*, Odelbourg.

[ISO-79] R. Isopi, New solid propellant for European boost motor, *Journal on Spacecraft and Rockets*, **16**, No. 6, pp. 355–357.

[JAN-88] D.F. Jansen, Ph. Kletzkine (1988) Preliminary design for a 3 kN hybrid propellant engine, *ESA Journal*, **12**, No. 4, pp. 421–440.

[KAM-86] P.C. Kammeyer (1986) Targeting multiburn, long duration apogee boost maneuvers, *Mécanique Spatiale pour les Satellites Géostationnaires, Colloque CNES, Cepadues*, pp. 523–531.

[KEL-84] H. Kellermeir, D. Koelle, R. Barbera (1984) A standardized propulsion module for future communications satellites in the 2000 to 3000 kg class, *AIAA 10th International Communication Satellite Systems Conference, Orlando*, March pp. 345–353 (Paper 84-0727).

[KOE-88] D.E. Koelle, H. Kuczera (1988) SANGER space transportation system, *39th International Congress of the IAF, Bangalore, October*, Paper 88–192.

[KOO-89] G.A. Koopman (1989) The industrial launch vehicle family for commercial space transportation, *Workshop on Flight Opportunities for Small Payloads (ESA SP-298), Esrin*, pp. 149–154, May.

[LAP-86] H. Laporte Weywada (1986) ARIANE 5: configuration and performances for unmanned missions, *AIAA 11th Communications Satellite Systems Conference, San Diego*, March, Paper 86–0672.

[LAR-88] P. Larcher (1988) Ariane auxiliary payloads, *Présenté à Technospace 88*, Bordeaux, December.

[LAR-89] P. Larcher (1989) Arrangements for flying auxiliary payloads on ARIANE, *Workshop on Flight Opportunities for Small Payloads (ESA SP-289), Esrin*, pp. 173–176, May.

[MAH-84] J. Mahon, J. Wild (1984) Commercial launch vehicles and upper stages, *Space Communications and Broadcasting*, **2**, No. 4, pp. 339–362.

[MAI-89] B. Maikisch (1989) USAF goes commercial, *Space*, **5**, No. 2, March.

[MAR-79] J.P. Marec (1979) *Optimal Space Trajectories*, Elsevier.

[MEC-87] A. Mechkak (1987) Le Développement des Propulseurs d'Appoint à Liquides, *ESA Bulletin*, No. 49, February, pp. 27–32.

[MEL-87] A. Melchior, M.F. Pouliquen, E. Soler (1987) Thermostructural composite Materials for liquid-propellant rocket engines, *AIAA Paper 87–2119*.

[MEY-90] J. Meyers (1990) DELTA-II, reliable communications satellite launch services, *AIAA 13th Communication Satellite Systems Conference, Los Angeles*, Paper 90–0828, March.

[MOC-88] M. Mochizuki, E. Sogame, Y. Shibato (1988) Status report of the H-I and H-II vehicles, *AIAA 12th Communications Satellite Systems Conference, Arlington*, March, Paper 88–0853.

[MOS-84] V.A. Moseley (1984) The liquid bipropellant stage concept—filling the OTV gap, *JANNAF Propulsion Meeting, New Orleans*, February, 6 pp.

[MUA-87] ARIANESPACE (1987) *Manuel de l'utilisateur d'Ariane 4*, Edition Originale d'avril 1983 (Révisions périodiques), BP 177, 91006 Evry Cedex, France.

[MUG-86] R. Mugelli, W. Flury (1986) Olympus—apogee engine firing, *Mécanique Spatiale pour les Satellites Géostionnaires, Colloque CNES, Cepadues*, pp. 475–492.

[ORD-82] C.A. Ordahl (1982) The MDAC payload assist modules, *AIAA 9th Communications Satellite Systems Conference, San Diego*, March, Paper 82–0559.

[PAA-90] Y. Parker (1990) PEGASUS rides out, *Space*, **6**, No. 3, pp. 6–8, June.

[PIR-87] T. Pirard (1987) Marketing the Proton, *Satellite Communications*, June.

[POC-86] J.J. Pocha, M.C. Webber (1986) Operational strategies for multi-burn apogee manoeuvres of geostationary spacecraft, *Space Communication and Broadcasting*, **4**, No. 3 September, pp. 229–233.

[POC-87] J.J. Pocha (1987) *An Introduction to Mission Design for Geostationary Satellites*, D. Reidel.

[POR-88] J.P. Porter, W.C. Hampton (1988) DELTA—The space transportation workhorse for government and commerce, *AIAA 12th Communications Satellite Systems Conference, Arlington*, March, Paper 88-0855.

[POU-90] M. Pouliquen (1990) Recent developments in small launch systems, *Advanced Lecture 2.03. EN, International Space University, Toronto*, August.

[PRI-86] W.L. Pritchard, J.A. Sciulli, (1986) *Satellite Communications—Systems Engineering*, Prentice Hall.

[PRI-88a] K.M. Price et al. (1988) Communications satellites in non-geostationary orbits, *AIAA 12th Communications Satellite Systems Conference, Arlington*, March, Paper 88–0842.

[PRI-88b] K.M. Price, J.S. Greenberg (1988) The economics of satellite retrieval, *AIAA 12th Communications Satellite Systems Conference, Arlington*, March, Paper 88–0843.

[RAJ-86] C.K. Rajasingh, A.F. Leibold (1986) Optimal injection of TVSAT with multi-impluse apogee manoeuvres with mission constraints and thrust uncertainties, *Mécanique Spatiale pour les Satellites Géostationnaires, Colloque CNES, Cepadues*, pp. 493–510.

[ROB-66] H.M. Robbins (1966) An analytical study of the impulsive approximation, *AIAA Journal*, **4**, No. 8, August, pp. 1417–1423.

[ROU-88] D. Rouffet *et al.* (1988), SYCOMORES: a new concept for land mobile satellite communications, *IEE Conference on Mobile Satellite Communications, Brighton*, September, pp. 138–142.

[RUM-88] G. Rum (1988) IRIS system qualification program, *39th International Congress of the IAF, Bangalore*, October, Paper 88–178.

[SCH-88] L.R. Scherer, R.C. White, Commercial Atlas/Centaur update, *AIAA 12th Communications Satellite Systems Conference, Arlington*, March, Paper 88–0854.

[SKI-86] J.K. Skipper (1986) Optimal transfer to inclined geosynchronous orbits, *Mécanique Spatiale pour les Satellites Géostationnaires, Colloque CNES, Cepadues*, pp. 71–84.

[SLA-98] D.K. Slayton, P.J. Armitage (1989) Reliable, low-cost launch service (CONESTOGA), *Workshop on Flight Opportunities for Small Payloads (ESA SP-298), Esrin*, pp. 155–160, May.

[SOO-83] E.M. Soop (1983) Introduction to geostationary orbit, *ESA SP-1053*, ESTEC.

[STA-88] A. Stadd (1988) Status and issues in commercializing space transportation, *AIAA 12th Communication Satellite Systems Conference, Arlington*, March.

[THO-90] D. Thompson, C. Schade (1990) PEGASUS and TAURUS launch vehicles, *AIAA 13th Communication Satellite Systems Conference, Los Angeles*, Paper 90–0892, March.

[TRI-89] M. Trischberger, B. Lacoste (1989) ARIANE technology experiment platform (ARTEP), *Workshop on Flight Opportunities for Small Payloads (ESA SP-298), Esrin*, pp. 43–46, May.

[VAN-90] F. Van Rensselaer, E. Browne (1990) Commercial TITAN programme status and outlook, *AIAA 13th Communication Satellite Systems Conference, Los Angeles*, Paper 90–0891, March.

[WEI-85] J. Weiss (1985) manoeuvres with finite thrust, *ESA Journal*, **9**, No. 1, pp. 49–63.

[WHI-86] R.C. White (1986) Status and capability of the TOS and AMS upper stage family, *AIAA 11th Communication Satellite Systems Conference, San Diego*, March, Paper 86–0670.

[WHI-90] R. White, M. Platzer (1990), ATLAS family update, *AIAA 13th Communication Satellite Systems Conference, Los Angeles*, Paper 90–0827, March.

12 THE SPACE ENVIRONMENT

This chapter describes the main constituents of the space environment which affect the design and operation of the satellite during its lifetime in orbit. The special nature of the space environment includes the following:

—Gravitational and magnetic fields.
—Radiation sources and absorption sinks.
—Absence of atmosphere (vacuum).
—Meteorites and debris.

The particular environment during injection of the satellite into orbit (acceleration, vibration, noise and depressurisation) should also be considered.

The effects of the environment on the satellite are principally as follows:

—Mechanical, consisting of forces and torques which are exerted on the satellite and modify its orbit and attitude.
—Thermal, resulting from radiation from the sun and earth absorbed by the satellite and energy radiated towards distant space.
—Degradation of materials and surface states subjected to the action of radiation and high energy particles.

12.1 VACUUM

12.1.1 Characterisation

Vacuum is one of the essential characteristics of the space environment. The molecular density diminishes very rapidly with altitude (the variation is exponential); it depends on the latitude, time of day, solar activity etc. and has been represented by various models . At 36 000 km (the geostationary satellite altitude), the pressure is less than 10^{-13} Torr (millimetres of mercury).

12.1.2 Effects

12.1.2.1 Mechanical effects

The effect of atmospheric drag due to an imperfect vacuum has been considered in Section 7.3.1.4. The altitude of the apogee of an elliptic orbit tends to decrease as does the altitude of a circular orbit. Above 3000 km, atmospheric drag can be considered to be negligible.

12.1.2.2 *Effects on materials*

In vacuum, materials sublime and outgas; the corresponding loss of mass depends on the temperature (for example: 10^3 Å/year at 110 °C, 10^{-3} cm/year at 170 °C and 10^{-1} cm/year at 240 °C for magnesium). As temperatures greater than 200° are easy to avoid, and on condition that excessively thin skins are not used, these effects are not important. The possibility of condensation of gases on cold surfaces is more serious (it can cause short circuits on insulating surfaces and degrade thermo-optical properties); it is thus necessary to avoid the use of materials which sublime too easily such as zinc and caesium. Furthermore, polymers have a tendecy to decompose into volatile products.

On the other hand, a major advantage of vacuum is that metals are preserved from the effects of corrosion.

The surfaces of certain materials, particularly metals, when in contact under high pressure, have a tendency to diffuse into each other by a cold welding process; the result is a large frictional force on bearings and the moving mechanisms (for example the deployment of solar generators and antennas). It is therefore necessary to keep moving parts in sealed pressurised enclosures and use lubricants having a low rate of evaporation and sublimation. Special materials (e.g. ceramic and special alloys such as stellite) are also used for bearing manufacture.

12.2 THE MECHANICAL ENVIRONMENT

12.2.1 The gravitational field

12.2.1.1 *The nature of the gravitational field*

The satellite is, above all, subjected to the earth's gravitational field which primarily determines the movement of the centre of mass of the satellite. This gravitational field has asymmetries, due to the non-spherical and inhomogeneous nature of the earth, which cause perturbations of the orbit. Perturbations also result from the gravitational fields due to the attraction of the sun and moon. These gravitational fields have been described in Chapter 7.

12.2.1.2 *The effect on the orbit*

The asymmetry of the earth's gravitational field and the attraction of the sun and moon cause perturbations of the Keplerian orbit of the satellite as defined by the attraction of the earth when assumed to be spherical and homogeneous. These perturbations lead to variation with time of the orbital parameters which define the movement of the centre of mass of the satellite (see Section 7.3).

12.2.1.3 *Effect on the satellite attitude*

The strength of the earth's gravitational field varies with altitude so that parts of the satellite which are more distant from the centre of the earth are less attracted than the nearer parts.

Since the resultant of this gravity gradient does not pass through the centre of mass of the satellite, a torque is created.

The earth's gravity gradient has the effect of aligning the axis of lowest inertia of the satellite along the local vertical. Assuming that the z axis is an axis of symmetry of the satellite, the corresponding torque is given by:

$$T = 3(\mu/r^3)(I_z - I_x)\theta \qquad (12.1)$$

where μ is the attraction constant of the earth, r is the distance of the satellite from the centre of the earth, I_z is the moment of inertia about the z axis, I_x is the moment of inertia about an axis perpendicular to the z axis (smaller than I_z) and θ is the angle, assumed to be small, between the z axis and the plane of the orbit.

This torque, which can be used to stabilise satellites injected into low orbit, is difficult to use for stabilising geostationary satellites. It is thus easy for the latter to make its effects negligible. It is sufficient to make I_x and I_z not greatly different from each other. For example, with $I_z = 180\,\mathrm{m^2\,kg}$ and $I_x = 100\,\mathrm{m^2\,kg}$, the maximum torque is $T = 2.2 \times 10^{-7}\,\mathrm{N\,m}$ for θ less than $10°$.

12.2.1.4 *Lack of gravity*

As the terrestrial attraction is in equilibrium with the centrifugal force, the various parts of the satellite are not subject to gravity. This is particularly important for liquid propellants which cannot be extracted by gravity from the reservoirs in which they are stored. it is necessary to install a pressurising system with artificial separation of the liquid and gas either by means of a membrane or by using the properties of surface tension forces (see Section 10.3.2.6).

12.2.2 The earth's magnetic field

12.2.2.1 *Characterisation of the terrestrial magnetic field*

The terrestrial magnetic field H, at a great distance is that of a magnetic dipole of moment $M_E = 7.9 \times 10^{15}\,\mathrm{Wb\,m}$. This dipole makes an angle of $11.5°$ with the axis of rotation of the earth. It thus creates an induction B which has two components as follows:

—A normal component:

$$B_N = (M_E \sin\theta)/r^3 \qquad (\mathrm{Wb/m^2}) \qquad (12.2a)$$

—A radial component:

$$B_R = (2M_E \cos\theta)/r^3 \qquad (\mathrm{Wb/m^2}) \qquad (12.2b)$$

where r is the distance of the point concerned from the centre of the earth and θ is the angle between the radius vector and the axis of the dipole (using the polar coordinates of the point concerned in the reference system associated with the dipole).

For a geostationary satellite the normal component varies between 1.03×10^{-7} and 1.05×10^{-7} Wb/m^2 and the radial component between $\pm 0.42 \times 10^{-7}$ Wb/m^2. The component perpendicular to the equatorial plane is virtually constant and equal to 1.03×10^{-7} Wb/m^2.

12.2.2.2 *The influence of the terrestrial magnetic field*

The terrestrial magnetic induction **B** exerts a torque **C** on a satellite of magnetic moment **M** such that:

$$\mathbf{C} = \mathbf{M} \wedge \mathbf{B} \qquad (\mathrm{N\,m}) \tag{12.3}$$

For a geostationary satellite, the component of induction perpendicular to the equator, although the largest and constant, produces the smallest long-term effect. The corresponding torque is in the plane of the equator and since the satellite performs one complete rotation about its axis parallel to the axis of the poles per day, the sum of the torques cancels every 24 hours.

The overall magnetic moment of a satellite results from remanent moments, moments due to electric currents in the cabling and induced moments proportional to the earth's magnetic field. These moments can be reduced or compensated before launching so that the torque due to the earth's magnetic moment on the ground does not exceed 10^{-4} N m. As the magnetic field is inversely proportional to the cube of the distance from the centre of the earth, the torque in the geostationary satellite orbit is divided by $(42\,165/6378)^3 = 289$. It thus becomes equal to 3.5×10^{-7} N m. In practice, the launching conditions modify some of the settings made on the ground; it is prudent to introduce some margin and to consider a torque of moment $\mathbf{C} = 10^{-6}$ N m as the disturbing torque due to the earth's magnetic field when dimensioning the satellite attitude control system.

The earth's magnetic field can also be used in an active manner to generate satellite attitude control torques by using appropriate actuators (magnetic coils, see Section 10.2.2).

12.2.3 Solar radiation pressure

The solar radiation pressure on a surface element of area dS can be decomposed into a component normal to the surface and a tangential component which depend on the angle of incidence of the solar radiation on the surface, the coefficient of reflectivity of the surface ρ and the intensity of the solar flux W (see Section 7.3.1.3). The effect of these forces on the movement of the centre of mass has also been discussed in Chapter 7. The fact remains that the resultant of the forces exerted on all the surface elements dS does not in general coincide with its centre of mass. This results in a torque which perturbs the attitude.

Each elemental force is proportional to $W \cos \theta$; the torque applied to the satellite is thus proportional to $WS_a \cos \theta$, where S_a is the apparent surface of the satellite in the direction of the sun and is principally determined by the size of the solar generators.

The torque thus depends on the orientation of the sun with respect to the satellite. For geostationary satellites, the direction of the sun makes an angle between 66.5° and 113.5° with the axis perpendicular to the equatorial plane (the pitch axis). The torque causes a drift of the orientation of the north–south axis of the satellite. As described for magnetic torques, the torques due to solar radiation pressure, which are disturbing torques, can also be used in an active manner to participate in satellite attitude control (see Section 10.2.2).

12.2.4 Meteorites and material particles

The earth is surrounded by a cloud of meteorites (scrap material, rocks, pebbles etc.) whose density becomes lower as the altitude increases [LOF-88], [POT-88]. Their velocity varies from several kilometres per second to several tens of kilometres per second. The commonest meteorites have masses between 10^{-4} and 10^{-1} g. The flux N of particles of mass equal to or greater than m per square metre per second can be estimated from the following equations:

—For $10^{-6}\,\text{g} < m < 1\,\text{g}$

$$\log_{10} N(\geqslant m) = -14.37 - 1.213 \log_{10} m$$

—For $10^{-12}\,\text{g} < m < 10^{-6}\,\text{g}$

$$\log_{10} N(\geqslant m) = -14.34 - 1.534 \log_{10} m - 0.063 (\log_{10} m)^2 \qquad (12.4)$$

12.2.4.1 *Probability of impact*

The motion imparted to the satellite by impact with a meteorite can be evaluated in statistical terms, that is by the probability of meteorites of a given mass colliding with the satellite and by the resulting magnitude of the motion transferred. Collisions between the satellite and meteorites are assumed to occur randomly and to be modelled by a Poisson distribution. The probability of having n impacts with particles of mass between m_1 and m_2 on a surface S during time t is given by:

$$P(n) = [(Sft)^n \exp(-Sft)]/n! \qquad (12.5)$$

where f is the flux of particles of mass between m_1 and m_2 such that $f = N(>m_1) - N(>m_2)$ with $N(>m)$ given by (12.4), S is the exposed surface area (m^2) and t the exposure time (s).

12.2.4.2 *The effect on materials*

Meteorite impact causes an erosion of around 1 Å per year at the geostationary satellite altitude (200 Å at low altitudes). For the heaviest meteorites, these impacts can cause perforation of metal sheets, if too thin, which could be disastrous for the survival of the satellite. Protection is possible by using screens consisting of several superimposed sheets of metal. The outer sheets fragment the meteorites and subsequent ones halt the debris [KES-88], [COU-88].

12.2.5 Torques of internal origin

Relative movement of the antennas, the solar panels and the fuel causes torques which are exerted on the main body of the satellite. Furthermore, maintaining satellites in a stationary position requires periodic application of forces which act on the centre of mass of the satellite.

The satellite contains propellant reservoirs which empty in the course of the mission and it is impossible to have a centre of mass which is firmly fixed with respect to the satellite body and hence with respect to the jets. Also, during integration of the satellite, mounting and alignment of the jets are subject to some inaccuracy. The correcting forces required to maintain position will, therefore, not be applied exactly at the centre of mass; a torque which disturbs attitude maintenance will arise during these corrections.

By way of example, considering thrusters with thrust of 2 N and a maximum displacement of the centre of mass of 5 mm, the value of the disturbing torque is $C_p = 10^{-2}$ N m.

12.2.6 The effect of communication transmissions

The phenomenon is that of radiation pressure; electromagnetic radiation from the antennas creates a pressure which can be non-negligible if the transmitted power is high.

For a very directive antenna radiating a power P_T, the force F produced is:

$$F = -(dm/dt)c = -P_T/c \qquad \text{(N)} \tag{12.6}$$

where c is the speed of light (m/s).

For example, for a satellite transmitting 1 kW in a 1 degree beam (a direct television broadcasting satellite, for example) the force F is 0.3×10^{-5} N. If the lever arm is 1 metre, the torque is 3×10^{-6} N m.

The perturbation is large only in the case where the transmitted power is large and concentrated into a narrow beam; it is then necessary for the antenna axis to pass through the centre of mass or to provide two antennas whose axes are symmetrical with respect to the centre.

12.2.7 Conclusions

The satellite is subjected to perturbations which modify its nominal orbit and create torques which disturb the attitude. For a geostationary satellite, it was shown in Chapter 7 that the attraction of the sun and moon causes a variation of the inclination of the plane of the orbit of the order of 1° per year. The asymmetry of the terrestrial potential causes a longitude drift. Solar radiation pressure modifies the eccentricity of the orbit. Table 12.1 summarises the orders of magnitude of the disturbing torques in respect of attitude.

Table 12.1 Attitude disturbing torques

Origin	Moment of torque	Comments
Station keeping	10^{-2}	Only during corrections
Radiation pressure	5×10^{-6}	Continuous except during eclipses
Magnetic field	10^{-6}	Daily mean is less
Gravity gradient	10^{-7}	Continuous

12.3 RADIATION

The energy radiated by a body depends on its temperature T and its emittance ϵ. The Stefan–Boltzmann law defines the radiance M of a body, that is the flux radiated per surface element dS as follows:

$$M = \epsilon \sigma T^4 \qquad (\mathrm{W/m^2}) \tag{12.7}$$

where $\sigma = 5.67 \times 10^{-8}\,\mathrm{W\,m^{-2}\,K^{-4}}$ is Boltzmann's constant. For a black body $\epsilon = 1$. The emittance of a body is the ratio of the radiance of this body to the radiance of a black body at the same temperature.

Planck's law expresses the spectral radiance L_λ of a black body, that is the power per unit wavelength and per unit solid angle emitted in a given direction by a surface element of a black body, divided by the orthogonal projection of this surface element onto a plane perpendicular to the direction considered:

$$L_\lambda = C_{1L}\lambda^{-5}[\exp(C_2/\lambda T) - 1]^{-1} \qquad (\mathrm{W/m^3\,sr}) \tag{12.8}$$

where $C_{1L} = 1.19 \times 10^{-16}\,\mathrm{W\,m^2\,sr^{-1}}$ and $C_2 = 1.439 \times 10^{-2}\,\mathrm{m\,K}$.

A functional form of Planck's law, known as Wien's law, relates the wavelength λ_m for which the spectral radiance of the black body is a maximum to its temperature T:

$$\lambda_m T = b \tag{12.9}$$

$$L_m/T^5 = b'$$

where $b = 2.9 \times 10^{-3}\,\mathrm{m\,K}$ and $b' = 4.1 \times 10^{-6}\,\mathrm{W\,m^{-3}\,K^{-5}\,sr^{-1}}$.

Space radiates as a black body at a temperature of 5 K. It behaves like a cold sink with an absorbency of 1; all the thermal energy emitted is completely absorbed.

The radiation received by the satellite arrives principally from the sun and the earth.

12.3.1 Solar radiation

Figure 12.1 shows the spectral irradiance, that is the incident flux per unit wavelength and unit surface area perpendicular to the solar radiation as a function of wavelength for a surface located at a distance of 1 astronomical unit (AU), i.e. the mean sun–earth distance, from the sun. It can be seen that the sun behaves as a black body at a temperature of 6000 K. Of the power radiated, 90% is situated in the band 0.3 to 2.5 microns with a maximum in the region of 0.5 μm.

The incident flux is about 1353 W/m^2 on a surface normal to the radiation located at 1 AU from the sun. It is the flux to which an artificial satellite of the earth is subjected for about ten days of the year after the spring equinox (see Figure 7.5). The flux varies in the course of the year as a function of variation of the earth–sun distance (the earth–satellite distance being assumed to be negligible) and this variation is illustrated in Figure 12.2(b). The power received by a surface element of the satellite depends on the orientation of this surface with respect to the direction of the incident radiation which itself varies as a function of the declination of the sun. For a satellite in an orbit in the equatorial plane, if a surface perpendicular

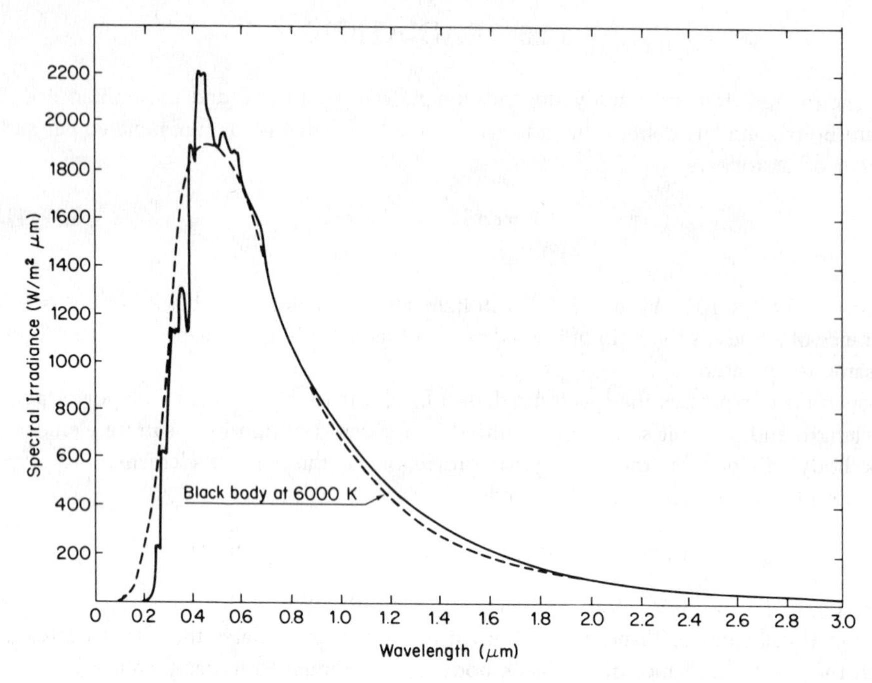

Figure 12.1 Spectral irradiance of the sun.

to the equatorial plane permanently orientated in the direction of the sun is considered, the incident flux is multiplied by the cosine of the declination; the result is presented in Figure 12.2(c). If a surface parallel to the plane of the equator is considered, the incident flux is multiplied by the sine of the declination and depends on the orientation (north or south) of the surface. The received power is zero at the equinoxes and during the six months of spring and summer for a surface oriented towards the north while it is zero during the six months of autumn and winter for a surface oriented towards the south.

The apparent diameter of the sun viewed from the earth is 32 min or around 0.5°.

12.3.2 Terrestrial radiation

Terrestrial radiation results from reflected solar radiation (the albedo) and its own radiation. The latter corresponds reasonably to that of a black body at 250 K, that is the irradiance is maximum in the infra-red band at 10–12μm. For a geostationary satellite, the total flux is less than 40 W/m^2 and is thus negligible compared with that provided by the sun.

12.3.3 Thermal effects

The satellite faces in view of the sun warm under the effect of solar radiation while its faces turned towards distant space become colder. Exchanges of heat by conduction and radiation thus occur (the vacuum prevents exchanges by convection). If the satellite retains a fixed orientation with respect to the sun, it establishes an equilibrium between the power absorbed

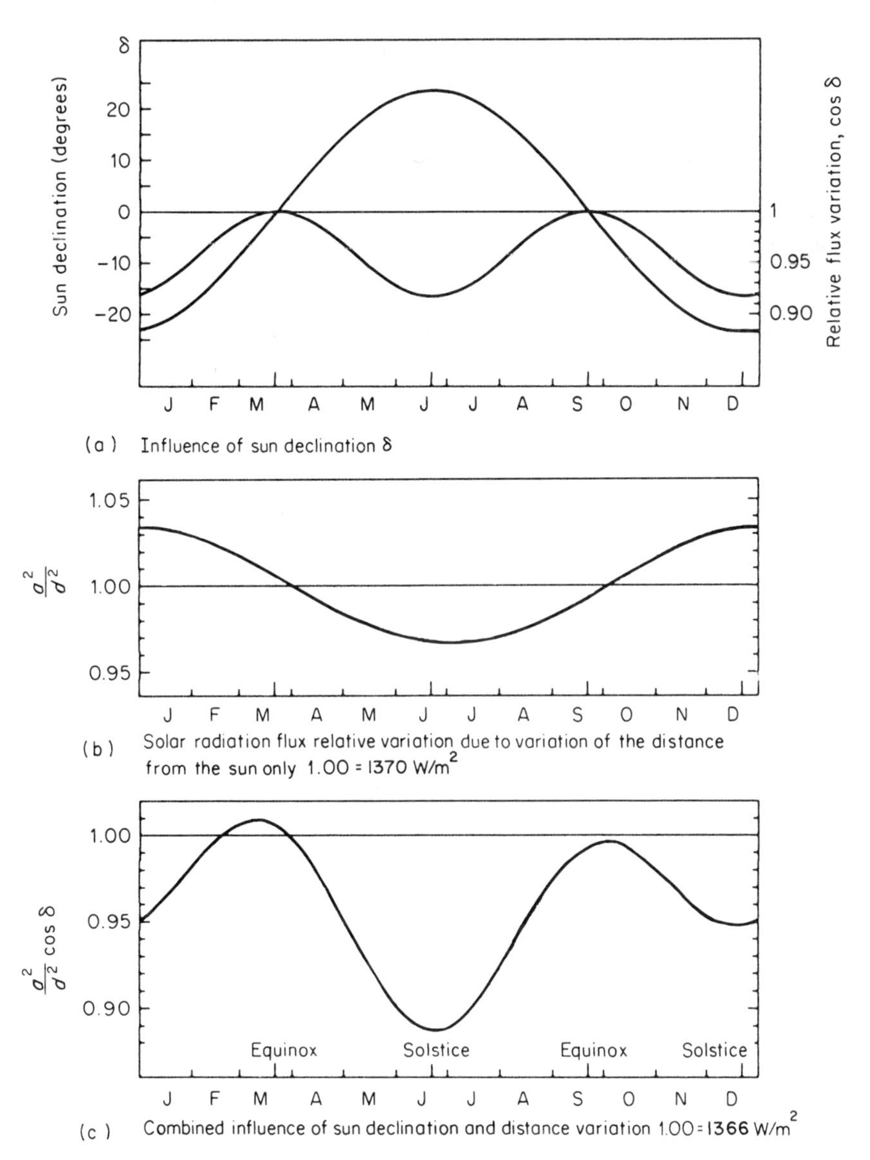

Figure 12.2 Solar radiation flux variations on a north—south sun-facing surface of a geostationary satellite. a = 1 IAU (semi-major axis of the earth orbit), d = sun to earth distance.

from the sun and the heat radiated. The mean temperature results from the following thermal balance:

$$P_S + P_I = P_R + P_A \tag{12.8}$$

where P_S is the power absorbed from the direct solar flux ($P_S = \alpha W S_a$, with W the solar flux, S_a the apparent surface and α the absorbency), P_I is the internal power dissipated, P_R is the thermal flux radiated and P_A is the power stored (or returned) by exchanges during temperature variation.

Table 12.2 Equilibrium temperature of a perfectly conducing inert sphere in space at a distance of 1 AU from the sun. (Absorbency α is the ratio of solar energy absorbed to solar energy received and emissivity ε is the ratio of thermal flux emitted to that which would be emitted by a black body raised to the same temperature.)

Cladding	Absorbency α	Emissivity ϵ	α/ϵ	T (°C)
Cold: white paint	0.20	0.80	0.25	−75
Medium: black paint	0.97	0.90	1.08	+12.5
Hot: bright gold	0.25	0.045	5.5	+155

For a perfectly conducting passive sphere of radius r at the equilibrium temperature T:

$$P_S = \alpha W \pi r^2$$

$$P_I = 0$$

$$P_A = 0$$

$$P_R = \epsilon \sigma T^4 4\pi r^2$$

(P_R is the power σT^4 radiated by the total surface $4\pi r^2$ of emittance ϵ). The equilibrium temperature is:

$$T = [(\alpha W)/(4\epsilon\sigma)]^{1/4} \tag{12.9}$$

where σ is the Stefan–Boltzmann constant (5.67×10^{-8} W m^{-2} K^{-4}). The equilibrium temperature of this spherical inert satellite ($P_R = 0$) depends only on the thermo-optical properties of the exterior surface, that is essentially its colour. Table 12.2 shows the equilibrium temperature for various claddings. It can vary from − 75 to + 155 °C.

The satellite equipment operates satisfactorily over a narrower range of temperatures; for example from 0 to +45 °C. The claddings will thus be chosen and combined judiciously to satisfy these conditions (see Section 10.6). As solar cells often cover the greatest part of the surfaces of the satellite, their thermo-optical properties are very important. Their absorbency is between 0.7 and 0.8 (in open circuit) and emissivity is between 0.80 and 0.85.

12.3.4 Effects on materials

Radiation in the ultra-violet whose spectrum extends from 100 to 1000 Å, causes ionisation in materials. This ionisation causes the following phenomena:

—Increase in the conductivity of insulators and modification of the absorbency and emissivity coefficients of thermo-optical claddings.
—Decrease of the conversion efficiency of solar cells with time spent in orbit (the order of magnitude is 30% decrease for silicon cells after 7 to 10 years).

At wavelengths greater than 1000 Å, solids can be excited; polymers are discoloured and their mechanical properties are weakened.

Above 3000 Å, the effects on metals and semiconductors are practically zero.

12.4 FLUX OF HIGH ENERGY PARTICLES

12.4.1 Cosmic particles

Cosmic particles are charged particles which consist mainly of high energy electrons and protons; they are emitted by the sun and various sources in space [STA-80].

The density and energy of these particles depend on the following:

—Altitude.
—Latitude.
—Solar activity.
—Time.

12.4.1.1 Cosmic radiation

Cosmic radiation consists mainly of protons (90%) and some alpha particles. The corresponding energies are in the gigaelectron-volt range, but the flux is low, of the order of 2.5 particles/cm^2 s.

12.4.1.2 Solar wind

Solar wind consists mainly of protons and electrons of lower energy. The mean density of protons during periods of low solar activity is of the order of 5 protons/cm^3 escaping from the sun at velocities around 400 km/s. The mean flux corresponding to the level of the earth's orbit is 2×10^8 protons/cm^2 with a mean energy of several kiloelectron-volts. According to solar activity, this flux can vary by a factor of 20. During periods of intense solar activity, solar eruptions occur more frequently and liberate proton fluxes with energies between several MeV and several hundreds of MeV. On rare occasions, with a periodicity of the order several years, the proton energy can reach GeV.

12.4.1.3 The Van Allen belts

As these particles are charged, they are subjected to the action of the terrestrial magnetic field and tend to form bands, called Van Allen belts, where the particles remain trapped.

For *electrons*, there are interior and exterior population zones. The boundary between the zones occurs at a distance of 2.8 earth radii. The energies of the electron population of the exterior zone range from several hundreds of keV to several MeV. The flux in this exterior zone can be evaluated from:

$N(>E) = 7 \times 10^{14} \exp(-4E)$ in periods of intense solar activity
$N(>E) = 1.3 \times 10^{13} E^{-1.5}$ in periods of minimum solar activity

where $N(>E)$ represents the flux of electrons per cm^2 per year with an energy expressed in MeV greater than E.

The high energy *protons* of the Van Allen belts are contained within a distance L equal to 4 earth radii with a maximum concentration around 1.5 and 2.2 earth radii for protons with energy ranging from 40 to 110 MeV (see Figure 12.3) [HES-68]. The energies range from MeV to several hundreds of MeV.

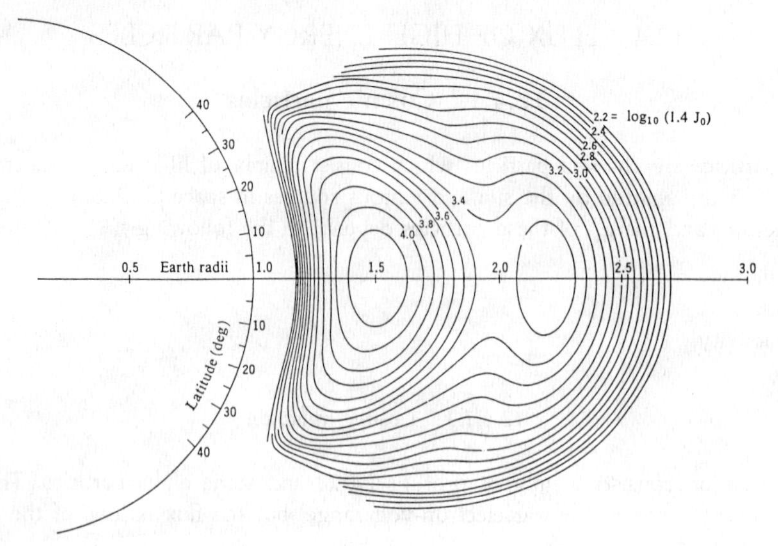

Figure 12.3 An R–λ map of experimentally measured proton fluxes of $40 < E < 110$ MEv. J_0 is the omnidirectional flux measured in protons/cm s [HES-63].

The geostationary satellite orbit (at 6.6 times the terrestrial radius) is outside the belt of trapped protons. Geostationary satellites are thus mainly affected by high energy protons generated by solar flares. The flux depends on the magnitude of solar activity which determines the occurrence of ordinary and extraordinary solar flares. It can be assumed that an extraordinary solar flare will almost certainly occur during the seven to ten years of the usual lifetime of satellites.

With respect to the electron flux, the geostationary satellite is in the exterior zone and the flux is evaluated from equations (12.10).

The various particle fluxes per cm^2 per year are given in Table 12.3 for the geostationary orbit. Figure 12.4 shows the total accumulated dose for a geostationary satellite after 12 years in orbit as a function of the orbital location considering a Si detector at the centre of a spherical shaped aluminium shielding 10 mm thick. The dose is the amount of energy absorbed per unit of mass of the considered matter (100 rad = 1 Gray = 1 J/kg).

12.4.2 Effects on materials

When subjected to charged particles, metals and semiconductors undergo excitation of the electron levels of the atoms [SRO-88], [PEA-88]. Plastics are ionised and insulating minerals undergo both effects.

Table 12.3 Total particle flux (number/cm^2 year) for the geostationary satellite orbit

Nature of particle	Low solar activity	Intense solar activity
Trapped electrons ($E > 0.5$ MeV)	3.5×10^{13}	9.5×10^{13}
Trapped protons	Negligible	Negligible
High energy solar protons ($E > 40$ MeV)	1×10^7	6.5×10^9

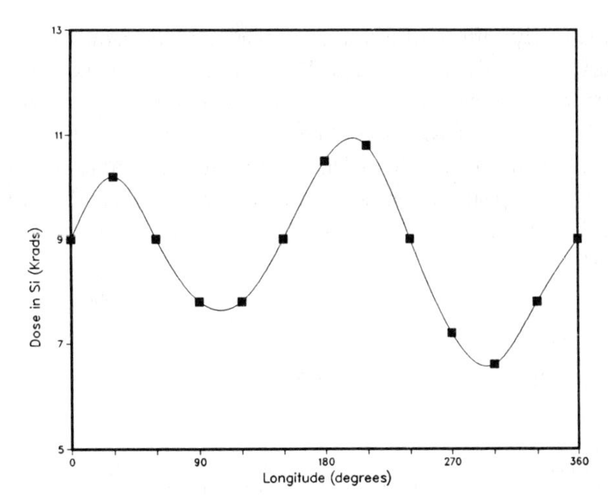

Figure 12.4 Total accumulated dose after 12 years for a geostationary satellite as a function of the station longitude (courtesy of Matra Marconi Space).

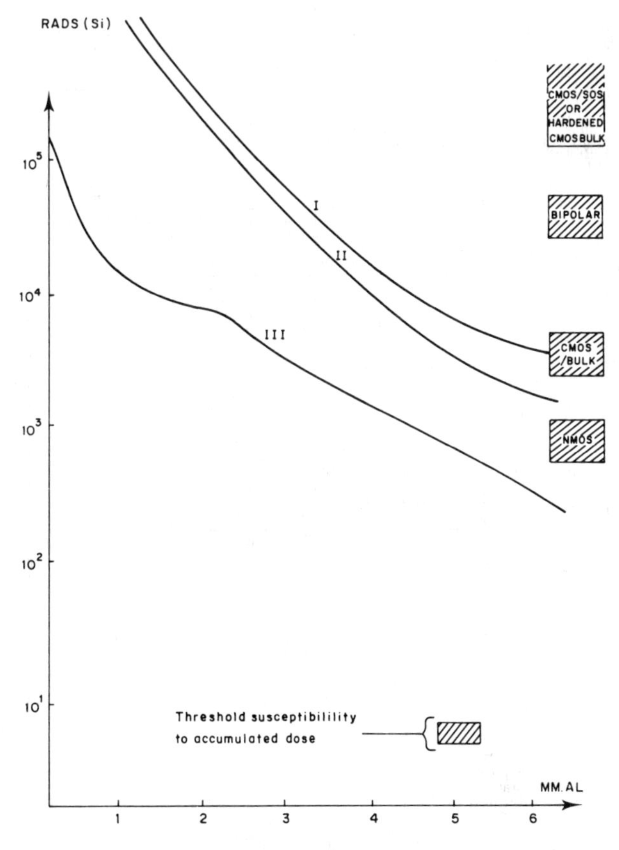

Figure 12.5 Radiation hardness of various integrated circuit technologies as a function of shield thickness. I Geostationary orbit during 7 years, II geostationary orbit during 3 years, III low polar orbit during 2 years.

Solar flares particularly affect the minority carriers in semiconductors, the optical transmission of glasses and certain polymers. The active components of electronic circuits in the satellite equipment can be protected against these effects by appropriate shielding (Figure 12.5) [LIN-81]. Equipment cases which contain sensitive components are produced in cast aluminium with wall thicknesses of the order of a centimetre [RAS-88]. The principal effects of high energy particles appear as degradation of the performance of solar cells which are directly exposed to the flux and modification of the thermo-optical characteristics of claddings which affect thermal control [WIL-83].

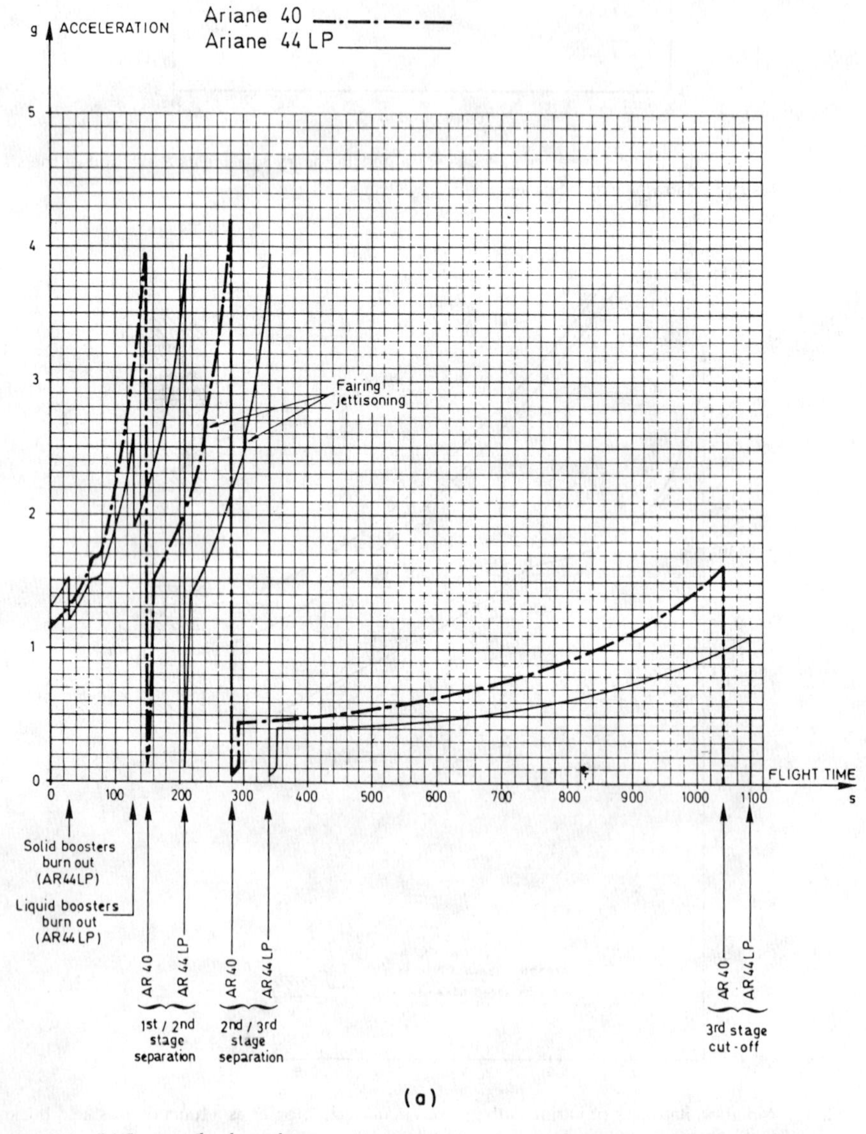

Figure 12.6 (a) Longitudinal acceleration (Ariane 4). (b) Noise spectrum under the fairing.

12.5 THE ENVIRONMENT DURING INSTALLATION

Installation, that is injection of the satellite into operational orbit, is preceded by two phases (see Chapter 11) during which the environment deviates somewhat from that described for the nominal orbit, particularly in the case of the orbit of geostationary satellites. These phases are as follows:

—The launch phase up to injection into the transfer orbit with a duration of tens of minutes.

—The transfer phase during which the satellite describes elliptical orbits whose apogee is at the altitude of the final orbit, for example a 200 km × 36,000 km orbit for the case of launching a geostationary satellite by Ariane. This transfer phase lasts for several tens of hours.

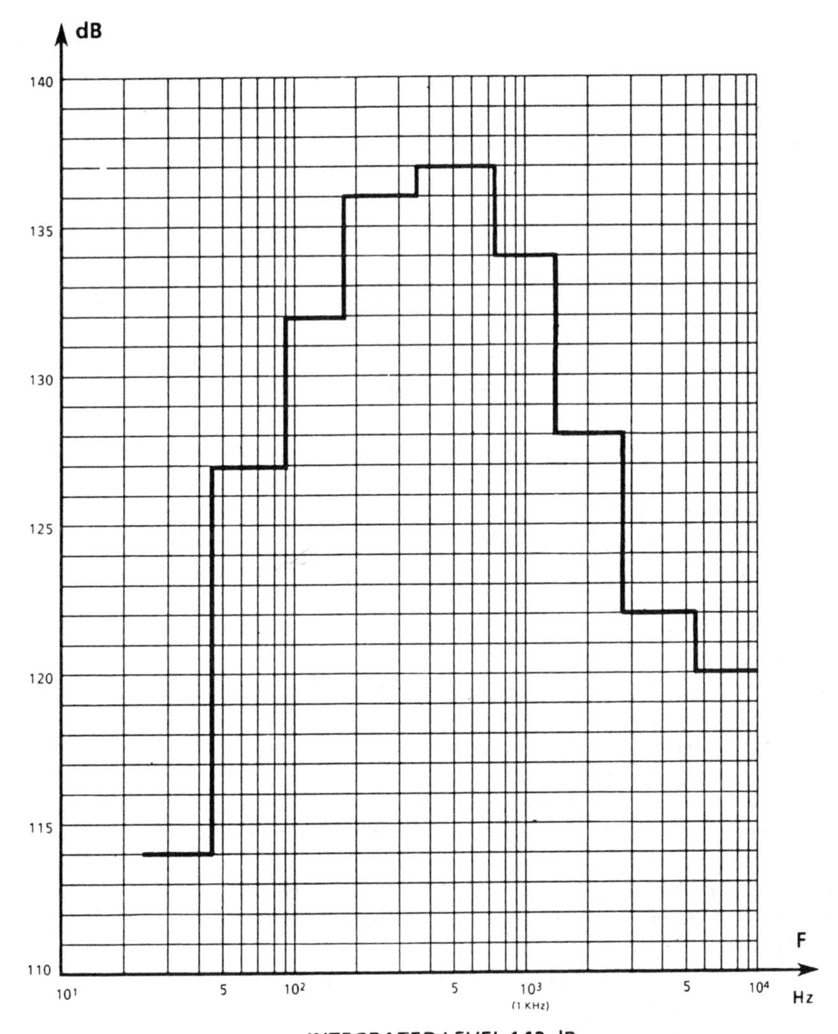

Figure 12.6 *(cont.)* (b)

12.5.1 The environment during launching

A fairing protects the satellite from aerodynamic heating as it passes through the dense layers of the atmosphere. Heating of the fairing has negligible effect on the satellite.

The most important constraints are longitudinal and transverse accelerations and vibrations, shocks communicated by the launcher during ignition of the motors and during propulsion phases. Acoustic noise under the fairing while passing through the atmosphere is also very high. The characteristics of these various excitations are given in the user manual for the launcher. An example is given in Figure 12.6 [MUA-87].

12.5.2 Environment in the transfer orbit

The satellite is usually stabilised by rotation and its configuration is different from its operational configuration since the apogee motor is full, the solar panels and antennas are folded and so on.

The environment and the effects discussed in the preceding sections are applicable with the following two differences:

—Thermal effects which depend on the earth, presence of the eclipses every 5 h, its own radiation and albedo,

—At the perigee, atmospheric drag is not negligible and the braking effect causes a reduction of altitude at the apogee.

REFERENCES

[COU-88] B.G. Cour-Palais, S.L. Avans, Shielding against debris, *Aerospace America,* June.

[ESA-81] European Space Agency (1981) *Ariane User Manual,* Paris.

[HES-68] W. Hess (1968) *The Radiation Belt and Magnetosphere.* Blaisdell.

[KESS-88] D.J. Kessler (1988) Predicting debris, *Aerospace America,* June.

[KIN-74] J.H. King (1974) Solar proton influences for 1977–1983 space missions, *J. Spacecraft,* **11**(3), pp. 401–408.

[LEG-80] P. Legendre (1980) Le maintien à posté des satellites geostationnaires. Evolution de l'orbite. *Le mouvement du véhicule spatial en orbite. Cours de technologie spatial.* CNES, pp. 583–607.

[LIN-81] F. Linder, Les Technologies d' Assemblage, *La Technologie des Expériences Scientifiques Spatiales, Cours de Technologie du CNES, Toulouse,* May pp. 659–688.

[LOF-88] J.P. Loftus *et al.* (1988) Decision time on orbit debris, *Aerospace America,* June.

[MuA-87] Arianespace (1987) *Manuel de l'utilisateur d' Ariane 4,* Edition Originale d' avid 1983 (Révisions Pérocdiques), BP 177, 91006 Everyledx, France.

[PEA-88] R.L. Pease *et al.* (1988) Radiation testing of semiconductor devices for space electronics, *Proceedings of the IEEE,* 76, No. 11, November.

[POT-88] A. Potter (1988) Measuring debris, *Aerospace America,* June.

[RAS-88] R.D. Rasmussen (1988) Spacecraft electronic design for radiation tolerance, *Proceedings of the IEEE,* 76, No. 11, November.

[SRO-88] J.R. Srour (1988) Radiation effects on microelectronics in space, *Proceedings of the IEEE,* **76**, No. 11, November.

[STAS-80] E.G. Stassinopoulos (1980) The geostationary radiation environment, *J. Spacecraft,* **17**(2), pp. 145–152.

[WIL-83] J.W. Wilson *et al.* A simple model of space radiation damage on GaAs solar cells, *NASA Technical Paper 2242,* December.

13 RELIABILITY OF SATELLITE COMMUNICATIONS SYSTEMS

The reliability of a system is defined by the probability of correct operation of the system during a given lifetime. The reliability of a complete satellite communication system depends on the reliability of its two principal constituents—the satellite and the ground stations.

The availability is the ratio of the actual period of correct operation of the system to the required period of correct operation. The availability of a complete satellite communication system depends not only on the reliability of the constituents of the system but also on the probability of successful launching, the replacement time and the number of operational and back-up satellites (in orbit and on the ground).

Availability of the ground stations depends not only on their reliability but also on their maintainability. For the satellite, availability depends only on reliability since maintenance is not envisaged with current techniques.

13.1 INTRODUCTION OF RELIABILITY

13.1.1 Failure rate

For complex equipment such as that of a satellite, two types of breakdown occur:

—Coincidental breakdown.
—Breakdowns resulting from usage (examples are wear of mechanical devices such as bearings and degradation of the cathodes of travelling wave tubes (TWT) and exhaustion of energy sources (such as the propellant reserves required for station keeping and attitude control).

The instantaneous failure rate $\lambda(t)$ of a given piece of equipment is defined as the limit, as the time interval tends to zero, of the ratio of the number of pieces of equipment which fail in the time interval concerned to the number of pieces of equipment in a correct operating state at the start of the time interval (a large number of identical pieces of equipment is assumed to operate at the same time).

The curve illustrating the variation of failure rate with time often has the form shown in Figure 13.1 (the 'bath-tub' curve), particularly for electronic equipment. Initially, the failure rate decreases rapidly with time. This is the period of early or infant failures. Subsequently, the failure rate is more or less constant. Finally, the failure rate increases rapidly with time, this is the wear-out period.

For space equipment, failures due to 'infant maladies' are eliminated before launching by

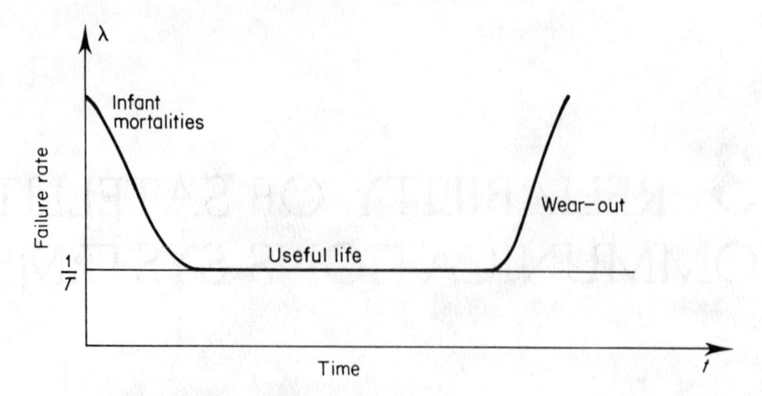

Figure 13.1 Failure rate versus time (bath-tub curve).

means of special preparation procedures (burn-in). Hence, during the period of useful life, most of the electronic and mechanical equipment has a constant failure rate λ. The instantaneous failure rate is thus often expressed in Fit (the number of failures in 10^9 h).

13.1.2 The probability of survival or reliability

If a piece of equipment has a failure rate $\lambda(t)$, its probability of survival from time 0 to t, or reliability $R(t)$, is given by:

$$R(t) = \exp\left[-\int_0^t \lambda(u)\,du\right] \tag{13.1}$$

This expression is of a general form which is independent of the law of variation of failure rate $\lambda(t)$ with time.

If the failure rate λ is constant, the expression for the reliability reduces to:

$$R(t) = e^{-\lambda t} \tag{13.2}$$

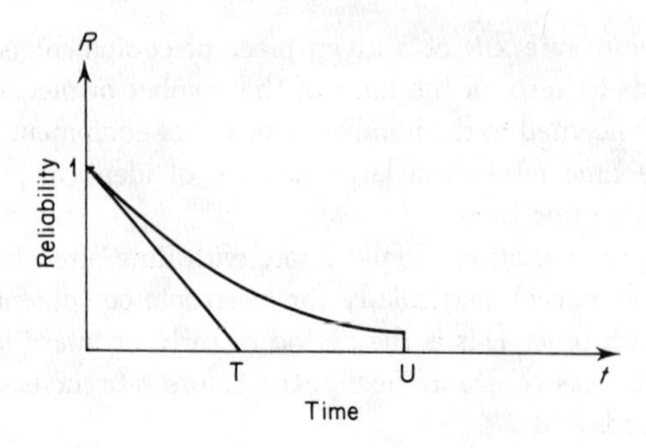

Figure 13.2 Reliability versus time.

For a satellite, the maximum mission life U can be defined at the end of which the service is no longer provided, usually due to exhaustion of the propellants. After time U, the probability of survival is zero. The curve of Figure 13.2 illustrates the variation of satellite reliability; the reliability is higher when λ is small.

13.1.3 Probability of death or unreliability

13.1.3.1 Unreliability D(t)

The unreliability or probability of having the system in a dead (failed) state at time t (failure has occurred between 0 and t) is the complement of the reliability:

$$R(t) + D(t) = 1 \tag{13.3}$$

13.1.3.2 Death probability density f(t)

The instantaneous probability of death is the derivative with respect to time of the unreliability. A death probability density $f(t)$ can thus be defined:

$$f(t) = \mathrm{d}D(t)/\mathrm{d}t = -\,\mathrm{d}R(t)/\mathrm{d}t \tag{13.4}$$

The probability of death occurring during a time interval t is thus:

$$D(t) = \int_0^t f(u)\,\mathrm{d}u \tag{13.5}$$

The failure rate $\lambda(t)$ is related to the death probability density $f(t)$ by:

$$\lambda(t) = f(t)/R(t) \tag{13.6}$$

If the failure rate λ is constant, $f(t) = \lambda \mathrm{e}^{-\lambda t}$.

For a satellite of maximum mission life U, the death probability density as a function of time is given in Figure 13.3

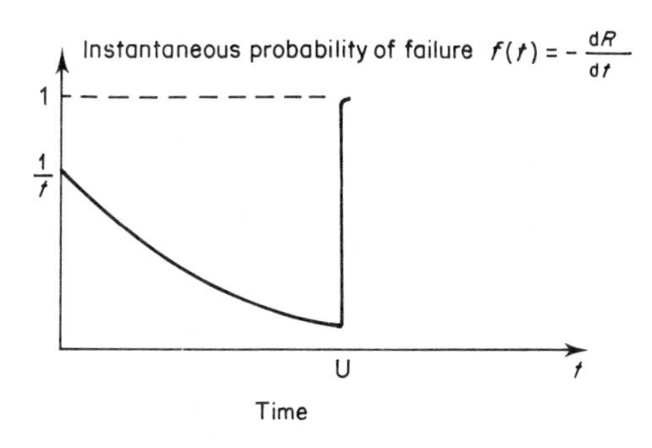

Figure 13.3 Instantaneous probability of failure versus time.

13.1.4 MTTF—mean lifetime

The mean lifetime or Mean Time To Failure (MTTF) is the mean time T of the occurrence of the first failure after entering service.

13.1.4.1 *Definition*

The mean time of occurrence of the first failure is obtained from the instantaneous probability of failure using:

$$T = \int_0^\infty t f(t)\,dt = \int_0^\infty R(t)\,dt \tag{13.7}$$

If the failure rate λ is constant, $T = 1/\lambda$.

For equipment which is repaired after the occurrence of a failure, the Mean Time Between Failures (MTBF) is defined in a similar manner.

13.1.4.2 *The case of a satellite of limited lifetime*

In the case of a satellite of maximum mission life U whose instantaneous probability of death is given by Figure 13.3, the mean lifetime τ can be considered as the sum of two integrals, the second is a delta function normalised so that the probability of death in a period of time of infinite duration is equal to 1. The mean lifetime τ can be written:

$$\tau = \int_0^U t\lambda e^{-\lambda t}\,dt + e^{-U/T} \int_U^\infty t\delta(t - U)\,dt \tag{13.8}$$

Hence:

$$\tau = T(1 - e^{-U/T}) \tag{13.9}$$

The mean lifetime τ depends on the mean time T of the occurrence of the first failure defined for a constant failure rate λ. The ratio τ/T is the probability of death during the maximum mission life U.

Table 13.1 gives the mean lifetime τ for an MTTF of 10 years as a function of the maximum mission life U.

Table 13.1 Average life for an MTTF of 10 years.

Max. Mission Life U (years)	Av. Life τ (years)
$U = T/3 = 3.3$	$\tau = 0.28\ T = 2.8$
$U = T/2 = 5$	$\tau = 0.39\ T = 3.9$
$U = T = 10$	$\tau = 0.63\ T = 6.3$
$U = 2T = 20$	$\tau = 0.86\ T = 8.6$
$U = 3T = 30$	$\tau = 0.95\ T = 9.5$

13.1.5 Reliability during the wear-out period

Components prone to wear-out, such as bearings, thrusters and vacuum tube cathodes, have failures at end of life whose probability density can be modelled by a normal distribution (the failure rate is no longer constant, failures are no longer accidental). The instantaneous probability of failure is thus of the form:

$$f(t) = \frac{\beta}{\alpha}\left(\frac{t-\gamma}{\alpha}\right)^{\beta-1} \exp\left[-\left(\frac{t-\gamma}{\alpha}\right)^{\beta}\right] \tag{13.10}$$

where μ is the mean lifetime and σ the standard deviation. The reliability becomes:

$$R(t) = 1 - \frac{1}{\sigma\sqrt{2\pi}} \int_t^{\infty} \exp\left[-\frac{1}{2}\left(\frac{t-\mu}{\sigma}\right)^2\right] dt \tag{13.11}$$

A hybrid reliability can be defined as the product of the reliability considering only wear-out and the reliability which characterises accidental failures. Equipment is generally designed in such a way that the lifetime determined by wear-out is long compared with the duration of the mission.

For components prone to wear-out, probability laws other than the normal distribution, such as the Weibull distribution for example, are also used to model the occurrence of failures. For the Weibull distribution, the expressions for death probability density $f(t)$ and reliability $R(t)$ are given by:

$$f(t) = \frac{1}{\sigma\sqrt{(2\pi)}} \exp\left[-\frac{1}{2}\left(\frac{t-\mu}{\sigma}\right)^2\right] \tag{13.12}$$

and

$$R = \exp\left[-\left(\frac{t-\gamma}{\alpha}\right)^{\beta}\right] \tag{13.13}$$

where α, β, γ are fitting parameters.

To model failure due to wear-out, the rate of which increases with time, the parameter β will be greater than 1 ($\beta = 1$ corresponds to a constant failure rate and $\beta < 1$ corresponds to a decreasing failure rate which can model early failures).

13.2 SATELLITE SYSTEM AVAILABILITY

Availability A is defined as $A =$ (required time $-$ down time)/(required time), where required time is the period of time for which the system is required to operate and down time is the cumulative time the system is out of order within the required time.

To provide a given system availability A for a given required time L, it is necessary to determine the number of satellites to be launched during the required time L. The number or satellites to be launched will affect the cost of the service.

The required number of satellites n and the availability A of the system will be evaluated for two typical cases for which t_R is the time required to replace a satellite in orbit and p is the probability of a successful launch [BAK-80].

13.2.1 No back-up satellite in orbit

13.2.1.1 *Number of satellites required*

As the mean lifetime of a satellite is τ, it will be necessary to put $S = L/\tau$ satellites into orbit on average during L years. As the probability of success of each launch is p, it will be necessary to attempt $n = S/p$ launches and the number of satellites n required is thus:

$$n = \frac{L}{pT[1 - \exp(-U/T)]} \tag{13.14}$$

13.2.1.2 *System availability*

If it is assumed that satellites close to their end of mission life U are replaced sufficiently in advance so that, even in the case of a launch failure, another launch can be attempted in time, the unavailability of the system at this time is small compared with the unavailability due to accidental failures.

During its lifetime U, the probability that a satellite fails in an accidental manner is $P_a = 1 - e^{-UT}$. In L years, there are S replacements to be performed of which $P_a \times S$ are for accidental failures. Each replacement requires a time t_r if it succeeds and, on average, a time t_r/p. The mean duration of unavailability during L years is $P_a S t_r/p = L t_r/pT$. The mean unavailability (breakdown) rate is:

$$B = t_r/pT \tag{13.15}$$

and the availability $A = 1 - B$ of the system is thus:

$$A = 1 - (t_r/pT) \tag{13.16}$$

13.2.2 Back-up satellite in orbit

By admitting, pessimistically but wisely, that a back-up satellite has a failure rate λ and a lifetime U equal to that of an active satellite, it is necessary to launch twice as many satellites during L years than in the previous case:

$$n = \frac{2L}{pT[1 - \exp(-U/T)]} \tag{13.17}$$

Taking account of the fact that t_r/T is small, the availability of the system becomes:

$$A = 1 - [(2t_r^2)/(pT)^2] \tag{13.18}$$

13.2.3 Conclusion

Table 13.2 provides three examples in which the time required for replacement t_r is 0.25 year and the probability p of a successful launch is 0.9.

Table 13.2 Examples of availability and number of satellites to be launched according to design life and MTTF

Design life U	5 years	7 years	10 years
MTTF T	10 years	20 years	20 years
Average lifetime τ	3.9 years	5.9 years	7.9 years
Probability of failure during life $P_f = \tau/T$	0.393	0.295	0.395
Time to replace T_R	0.25 year	0.25 year	0.25 year
Probability of launch p success	0.9	0.9	0.9
No spare			
Annual launch rate: n/L	0.28	0.19	0.14
Availability A	0.972	0.986	0.986
One in orbit spare			
Annual launch rate: n/L	0.56	0.38	0.28
Availability A	0.9985	0.9996	0.9996

To obtain a high availability A, it can be seen that the mean time of occurrence of the first failure (MTTF) of the satellite is significantly more important than the predicted lifetime U.

Without a back-up satellite, the service is not provided, in the examples in the table, for a mean of 3.4 or 1.7 months in 10 years according to the predicted lifetime. To limit the unavailability to one month implies an availability of at least 99.2% and this requires the presence of a back-up satellite in orbit, four to six launches and an MTTF of at least 10 years (10^5 h). Satellites must thus be designed with a failure rate less than 10^{-5} per hour (10^4 Fit).

13.3 SUB-SYSTEM RELIABILITY

Calculation of the reliability of a system is performed from the reliability of the elements which constitute the system. As far as the satellite is concerned, except in the special case where elements in parallel can independently fulfil a particular mission, most sub-systems are essentially in series from the point of view of reliability. This indicates that correct operation of each sub-system is indispensable for correct operation of the system.

13.3.1 Elements in series

13.3.1.1 Reliability

When elements are in series from the reliability point of view, the overall probability of correct operation is obtained by taking the product of the reliabilities of the elements. With n elements in series, the overall reliability R of the system can thus be written:

$$R = R_1 R_2 R_3, \ldots, R_n \tag{13.19}$$

The reliability of a system containing four elements in series, where each has a reliability of 0.98, is thus $0.98^4 = 0.922$.

13.3.1.2 Failure rate

The overall failure rate λ of the system is obtained by adding the failure rates λ_i of each of the constituents if these are constants. The overall failure rate is thus constant and the MTTF is $1/\lambda$.

A communication satellite includes about 10 subsystems (see Chapters 8, 9 and 10) and the mean failure rate per subsystem must be less than 10^{-7} h (10^2 Fit). To obtain this reliability, some equipment must be provided with partial or total redundancy.

13.3.2 Elements in parallel (static redundancy)

13.3.2.1 Reliability if one element out of n is sufficient

For elements in parallel in the reliability sense, the probability of death of the ensemble is the product of the probability of death of each of the elements:

$$D = D_1 D_2 D_3, \ldots, D_n \tag{13.20}$$

The reliability of the ensemble is given by: $R = 1 - D$. This relation is valid if correct operation of the ensemble is ensured with a single element out of the n.

13.3.2.2 Reliability if k out of n elements are necessary

If it is necessary to have k out of the n identical elements for correct operation, the various cases which correspond to correct operation must be analysed in order to evaluate the reliability. It can be shown that the probability p_k of having k elements out of the n in good order is given by the expansion of $(p+q)^n$ (the binomial rule), where p is the probability of correct operation of an element and q is that of not functioning ($p + q = 1$).

Example. A system consists of two elements of reliability R_i in parallel; correct operation is obtained if one of the two elements is in good order, the binomial rule gives $(R_i + D_i)^2 = R_i^2 + 2R_iD_i + D_i^2$.

Correct operation is obtained if both pieces of equipment are operating (reliability R_i^2) or if one has failed and the other is operational, or the inverse (reliability $2R_iD_i$)

The reliability R of the system is thus:

$$R = R_i^2 + 2R_iD_i = R_i^2 + 2R_i(1 - R_i) = 2R_i - R_i^2$$

If the failure rates are constant and equal to λ, this becomes:

$$R = 2\,\mathrm{e}^{-\lambda t} + \mathrm{e}^{-2\lambda t}$$

13.3.2.3 Failure rate

The overall failure rate is obtained from the ratio $f(t)/R$, where the death probability density is calculated from R using equation (13.4). Assuming the failure rates λ_i to be constant, it is found that the overall failure rate is a function of time and hence the overall failure rate is not constant.

Example. With two identical elements in parallel, it is found that:

$$\lambda = [2\lambda_i(1 - e^{-2\lambda_i t})/(2 - e^{-2\lambda_i t})]$$

As time tends to infinity, λ tends to λ_i.

13.3.2.4 Mean time of occurrence of a failure

The mean time of occurrence of the first failure is calculated from the reliability using equation (13.7). If the failure rates λ_i of the n elements in parallel are constant and identical and if correct operation is ensured with a single element, the MTTF of the overall system can be put in the form:

$$\text{MTTF} = \text{MTTF}_i + \text{MTTF}_i/2 + \text{MTTF}_i/3 + \cdots + \text{MTTF}_i/n \quad (13.22)$$

where $\text{MTTF}_i = 1/\lambda_i$ is characteristic of one element.

Example. With two identical elements in parallel, it is found that:

$$\text{MTTF} = 1/\lambda_i + 1/2\lambda_i = 3/2\lambda_i = 1.5\text{MTTF}_i$$

The mean time of occurrence of the failure is increased by 50% by the parallel connection of the two pieces of equipment.

13.3.3 Dynamic redundancy (with switching)

13.3.3.1 The Poisson distribution

Consider a system constituted, in the reliability sense, of m normally active elements in parallel, where n elements can, in turn, be placed in parallel to replace a failed main element. The failure rate λ_i of each of the elements is constant and the same for each. The reliability R of the system for m elements in a correct operational state is given by the Poisson distribution:

$$R = e^{-m\lambda_i t}[1 + m\lambda_i t + (m\lambda_i t)^2/2! + \cdots + (m\lambda_i t)^n/n!] \quad (13.23)$$

The mean time of occurrence of the first failure is given by:

$$\text{MTTF} = [(n + 1)/m]\text{MTTF}_i \quad (13.24)$$

where $\text{MTTF}_i = 1/\lambda_i$ is characteristic of one element,

The general expressions above assume failure rates which are constant and identical for the equipment while in service. They also assume that the back-up equipment is in good operational order at the time when it replaces a failed piece of principal equipment. The reliability of the switching devices is also assumed to be equal to 1.

It is useful to be able to consider equipment failure rates which differ and may be modified for back-up equipment which is in stand-by or operational mode. Since general expressions are either complex or impossible to establish, various special cases are presented below for particular examples.

13.3.3.2 *Redundancy with different failure rates which depend on the operational state*

A subsystem consists of a principal element and a back-up element which can replace it. The failure rate of the principal element is λ_p; the failure rate of the back-up element is λ_r when the element is inactive and λ_s when it is operating. The reliability of the subsystem is evaluated by considering the various probabilities of failure and correct operation.

The probability of correct operation of the principal element between time 0 and t is $e^{-\lambda_p t}$. The probability of failure of the principal element at time t_f (with $t_f < t$) is $\lambda_p e^{-\lambda_p tf}$. The probability of correct operation of the back-up element between 0 and t_f is $e^{-\lambda_r tf}$. The probability of correct operation of the back-up element between time t_f and time t is $e^{-\lambda(t-tf)}$.

Correct operation of the system at time t is thus ensured if the principal element is in good order at time t (reliability R_p) or if, after failure of the principal element at time t_f, the back-up element is in good order at time t (reliability R_s). For the back-up element to be in good order at time t, it must operate correctly between time 0 and time t_f and between t_f and time t.

The reliability R is thus equal to $R_p + R_s$ with:

$$R_p = e^{-\lambda_p t}$$

$$R_s = \int_0^t [\lambda_i e^{-\lambda_p tf})(e^{-\lambda_r tf})(e^{-\lambda_s(t-tf)})] dt_f$$

The reliability R of the system with redundancy is thus given by:

$$R = e^{-\lambda_p t} + [\lambda_p/(\lambda_p + \lambda_r + \lambda_s)][e^{-\lambda_s t} - e^{-(\lambda_p + \lambda_r)t}] \quad (13.25)$$

Calculation of the mean time T to the occurrence of the failure gives:

$$T = \text{MTTF} = (1/\lambda_p) + \lambda_p/\lambda_s(\lambda_p + \lambda_r) = T_p + [(T_s + T_r)/(T_p + T_r)] \quad (13.26)$$

where T_p, T_r, T_s are the mean times of occurrence of failure (MTTF) of the principal equipment, the back-up equipment when inactive and the back-up equipment when operating respectively.

A particular case. Consider a system having the same failure rate λ_i for the operational units and a zero failure rate for the inactive units ($\lambda_r = 0$). The expression for the reliability becomes:

$$R = e^{-\lambda_i t} + \lambda_i t e^{-\lambda_i}$$

This expression can be obtained directly from the Poisson distribution. The mean time of occurrence of the first failure under these conditions are:

$$T = \text{MTTF} = 2/\lambda_i = 2\text{MTTF}_i$$

The mean time of occurrence of the first failure is thus doubled by a redundancy of the 1/2 type (one active unit for two installed units).

13.3.3.3 *Equipment redundancy taking account of the reliability of the switching element*

The reliability will be evaluated for the example of a subsystem consisting of two units of which one is the principal and the other the back-up. The two cases to be considered are where the switching element is, and is not, necessary for operation of the principal element. The units have the same failure rate λ.

In the first case, the switch or switches are used to route the signals to the principal unit or to the back-up unit (Figure 13.4a). These switches are thus, from the reliability point of view, in series with the duplicated equipment. The reliability of the ensemble is thus equal to the product of the reliability of the switches (R_{sw} for a switch) and the reliability of the duplicated equipment which is obtained from the Poisson distribution:

$$R = R_{sw}^2[e^{-\lambda_i t}(1 + \lambda_i t)] \qquad (13.27)$$

In the second case, the principal element is accessible without passing through the switch which is used only to connect the back-up unit in parallel with the principal element at the time of failure (Figure 13.4b). Hence the reliability R_{sw} of the switch arises only in the back-up branch. The reliability of the system is thus obtained by considering that correct operation is obtained at time t if the principal equipment is operational at this time, or if, after failure of the principal equipment at time t_f, the back-up equipment operates correctly between t_f and t, on condition that the switch operates correctly. Calculation of the reliability gives:

$$R = e^{-\lambda_i t} + R_{sw}\lambda_i t e^{-\lambda_i t} \qquad (13.28)$$

If the mean failure rate of the switch λ_{sw} is constant, $R_{sw} = e^{-\lambda_{sw} t}$ and that reliability becomes $e^{-\lambda_i t} + \lambda t e^{-(\lambda_i + \lambda_{sw})t}$. The mean time T of occurrence of the failure is given by:

$$T = \text{MTTF} = (1/\lambda_i) + \lambda_i/(\lambda_i + \lambda_{sw})^2 \qquad (13.29)$$

If λ_{sw} is equal to 0, MTTF again has a value $2/\lambda_i = 2\,\text{MTTF}_i$.

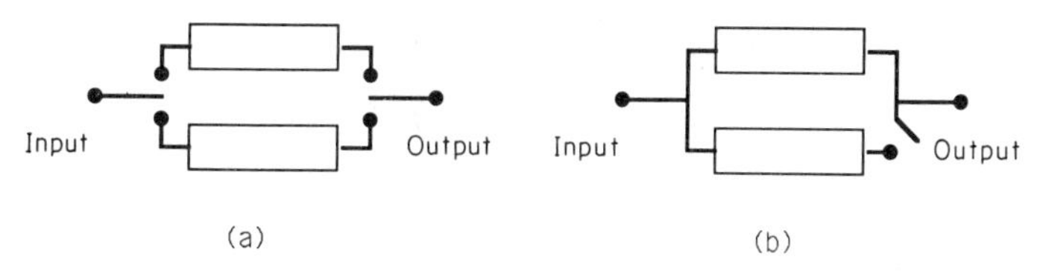

Figure 13.4 Equipment with 1/2 redundancy.

13.3.3.4 *Example: Redundancy of TWTs in the channelised part of a payload*

Consider the channelised part of the communication payload of a satellite, where two channels share three TWTs and the associated preamplifiers (2/3 redundancy). Access to the amplifiers is by way of two switches, a switch with two inputs and three outputs ($S_{2/3}$) for the input and a switch with three inputs and two outputs ($S_{3/2}$) for the output (Figure 13.5a). In normal operation, two of the amplifiers are active and the third is idle. The failure rate of the active equipment is λ_i. The failure rate of the idle equipment is λ_r (due to keeping it prewarmed for example).

The probability of correct operation of both channels at time t is obtained by considering the equivalent block diagram from the reliability point of view; this is given in Figure 13.5b. The reliability of the system is equal to the product of the reliability of the switches R_{sw} and that of the set of amplifiers R_A.

The reliability R_A of the set of amplifiers is evaluated by considering the various probabilities of failure and correct operation.

The probability of correct operation of both principal units between time 0 and t is $e^{-2\lambda_i t}$. The probability of failure of a principal unit at time t_f (with $t_f < t$) is $\lambda_i e^{-\lambda_i t_f}$. The probability of correct operation of the other principal equipment between time 0 and t is $e^{-\lambda_i t}$. The probability of correct operation of the back-up element between 0 and t_f is $e^{-\lambda_r t_f}$. The probability of correct operation of the back-up element between time t_f and t is $e^{-\lambda_i (t - t_f)}$.

Correct operation of the system at time t is thus ensured if:

—both principal units are in good order at time t,
—or if after failure of a principal unit at time t_f, the back-up element operates correctly between 0 and t_f and between t_f and t, knowing that the other principal unit continues to operate up to time t,
—or in the corresponding configuration to the previous one in case of failure of the other principal unit.

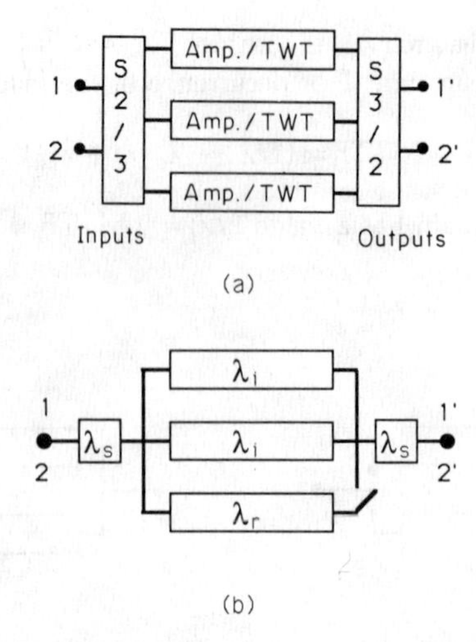

Figure 13.5 Equipment with 2/3 redundancy.

The reliability R_A is thus given by:

$$R_A = e^{-2\lambda_i t} + 2e^{-\lambda_i t} \int_0^t [(\lambda_i e^{-\lambda_i t_f})(e^{-\lambda_r t_f})(e^{-\lambda_i (t - t_f)})] dt_f$$

Hence:

$$R_A = e^{-2\lambda_i t}[1 + 2\lambda_i/\lambda_r)(1 - e^{-\lambda_r t})] \tag{13.30}$$

Notice that if the failure rate of the idle equipment is zero ($\lambda_r = 0$), the reliability is obtained directly from the Poisson distribution:

$$R_A = e^{-2\lambda t}(1 + 2\lambda t)$$

The probability of correct operation of both channels at time t including the reliability of the switches is thus given by:

$$R = R_{S2/3} R_A R_{S3/2}$$

Numerical example.

Amplifier failure rate $\lambda = 2300$ Fit (2300×10^{-9} per h)
Switch failure rate $\lambda_{sw} = 50$ Fit
The expected lifetime is 10 years.

After 10 years:

The reliability R_A of the set of amplifiers is $R_A = 0.937\,65$,
The reliability R_{sw} of a switch is $R_{sw} = e^{-\lambda_{sw} t} = 0.995\,63$,
The reliability R of the ensemble is thus $R_A R_{sw}^2 = 0.929\,47$.
In comparison, the reliability of a non-duplicated amplifier is $e^{-\lambda t} = 0.817\,52$.

13.3.4 Equipment having several failure modes

Some equipment and elements have several modes of failure, for example short circuit and open circuit for diodes, capacitors and so on. The consequences of a failure on the operation of the system concerned are not the same following a failure of one type or the other. The consequences also depend on the system architecture.

With a structure containing n elements in series, a failure of the open circuit type, characterized by a probability of death D_0 for one element, involves death of the ensemble. The probability of death of the ensemble is thus $1 - (1 - D_0)^n$. On the other hand, with a failure of the short-circuit type, characterised by a probability of death D_C, death of the ensemble requires failure of all the elements. The corresponding probability of death of the ensemble is thus $(D_C)^n$. The reliability R of the series structure is thus $R = (1 - D_0)^n - (D_C)^n$ and the series structure is robust with respect to failures of the short-circuit type.

When n elements in parallel are associated with failures of the open circuit type, characterised by a probability of death D_0 for one element, death of the ensemble requires failure of all the elements. The probability of death of the ensemble is thus $(D_0)^n$. On the other hand, failure of the short-circuit type, characterised by a probability of death D_C, involves that of the ensemble. The probability of death of the ensemble is thus $1-(1-D_C)^n$. The reliability R of the parallel structure is thus $R=(1-D_C)^n-(D_0)^n$ and the parallel structure is robust with respect to failures of the open circuit type.

For each of these structures, there is an optimum number of elements which enables maximum reliability to be achieved. If protection against both types of failure is required simultaneously, more complex structures must be used such as series-parallel or parallel-series types (COR-75). These procedures are used for wiring the solar cells of the power generator for example.

13.4 COMPONENT RELIABILITY

Certain sub-systems, such as the payload, contain several hundreds of components. To obtain failure rates of one or two per 100 000 h, each component must not exceed a failure rate of the order of 1 per 10 million hours.

During the design of the satellite, after the constraints have been analysed, provisional examination of the reliability enables redundancy arrangements to be defined together with the quality level of components and equipment.

13.4.1 Component reliability

Information on the failure rates of various types of component is available from the manufacturers who have the results of component tests under particular environmental conditions. The data provided in documents published by various organisations, such as report MIL-HDBK-217C [MIL-79] for example, can also be used. Table 13.3 gives the orders of magnitude of the failure rates of various components. The least reliable components are travelling wave tubes and components with moving parts (such as rotating bearings, relays and potentiometers). The most reliable components are passive ones such as resistors, capacitors, switching diodes and connectors.

The failure rate of a component can be greatly reduced by an appropriate choice of loading ratio (derating) . The mean power dissipated by the components is chosen to be a small percentage of the nominal power specified by the manufacturer. For example, a resistor capable of dissipating 1 W would be chosen for a resistance which must dissipate 300 mW; the loading ratio is thus 30%. In any case, operation should not be close to the specified parameter limits. The junction temperature of transistors must not exceed a specified value (typically 105°C).

The same principle is applied to the maximum values of voltages, currents, etc. which components and equipments must be capable of withstanding. Wear-out of elements is thus reduced according to a power law as a function of reduced loading. Applicable loading ratios appear in the preferential lists of components to be used in priority.

13.4.2 Component selection

Components are chosen after a functional examination of the equipment on preferential lists established by space agencies such as the European Space Agency (ESA/SCCG Space Component Coordination Group), CNES (CNES/QFT/IN-0500), NASA, etc. Special procedures

Table 13.3 Typical failure rates of components for space applications, expressed in fit (Failure rate at 75%—where applicable)

Component	Failure rate (fit)
Resistors	
Solid carbon	5
Metallic Film	5
Wirewound	10
Potentiometers	200
Capacitors	
Solid carbon	3
Polycarbonate	3
Mylar	5
Paper	20
Solid tantalum	20
Variable	20
High Voltage	100
Silicon diodes	
Switching	4
Standard	10
Power	20
Zener	50
Detector/mixer	100
Filter sections	
Hybrid	25
Passband	10
Couplers	10
Circulators	10
Connectors	1
Transistors (planar, silicon)	
Standard	10
Switching	10
HF	20
Power	50
Integrated circuits:	
Digital (bipolar)	10
Analog	20
FET IC	
1–10 gates	100
11–50 gates	500
TWT	1500
Transformers	200
Power	30
Signal	10
Inductors	
Power	20
Signal	10
Quartz crystals	80
Relays	400

are followed to ensure the manufacturing quality of the component chosen, the constancy of its properties from one sample to another and with time (these include purchase and acceptances specifications, qualifications of batches and component acceptance).

When a component which does not appear in the preferential lists is necessary, qualification of the component is performed using the same specifications as those which are supplied. It includes two main phases:

—The evaluation phase,
—The qualification phase.

13.4.2.1 *Evaluation*

This phase includes the following:

—Inspection of the manufacturing facilities.
—Detailed examination of the production line of the component concerned.
—Evaluation tests to the limits of the components.
—Examination of the manufacturing and monitoring documentation (the Process Identification Document (PID)).

The evaluation phase concludes with an evaluation report and a final review of the documentation. When this phase is completed in a satisfactory manner, the qualification phase is entered.

13.4.2.2. *Qualification*

This phase includes the following activities:

—Manufacture of the components which constitute the qualified batch.
—100% testing at the end of production and selection.
—Qualification testing of a sample of the batch.

If the results are satisfactory qualification is declared and a qualification certificate is delivered to the manufacturer. The qualified product is then entered in the preferential list such as the Qualified Products List (QPL) of the ESA.

Qualification is valid for a fixed period, after this period the validity could be extended on condition that batch tests are, or have been, performed and the process identification document has not been changed.

13.4.3 Manufacture

Having selected the components, the equipment manufacturing specifications must be defined. Technical design takes account of the constraints of performance, weight, volume, etc. and the constraints specific to the space environment [RAS-88]. The manufacturing specifications include the choice of wiring process (printed circuit or otherwise), the type of solder, the form of enclosure or protective cladding, etc.

Manufacturing quality control has the particular goal of verifying that manufacturing specifications are actually observed during the various stages and the components used are actually those which have been specified.

13.4.4 Quality assurance

Quality assurance is indispensable and complementary to security and reliability. More precisely, quality assurance ensures that a number of objectives and tasks relating to the requirements of the space project become facts.

The main elements of a quality assurance programme are [CHA-81] as follows.

13.4.4.1 *Quality of the pre-project studies and definition*

During the analysis phases, quality assurance consists of:

—Verifying the conformity of the design documentation (plans and specifications) with the requirements of the programme.

—Verifying the conformity of the definition with general quality rules and the particular requirements of the project.
—Ensuring that the requirements of reliability, security and quality are taken into account.

13.4.4.2 *Design quality*

At the design level, conformity of all designs, tests and detailed specifications to the quality rules and requirements of the projects must be verified. At the model testing level, conformity of the models to the design, conformity of the test conditions to the project requirements, the quality of the results and the standards must be verified.

13.4.4.3 *Supply quality*

The quality of supplies relies on:

—Definition of a supply specification which conforms to the reliability requirements and technical performance.
—Definition of the qualification and acceptance conditions.
—The choice of components and materials.
—Appraisal of defective components and materials.
—Definition of acceptance procedures for batches of components.

13.4.4.4 *Quality of manufacture*

Manufacturing quality depends on:

—Definition of an industrial document which conforms to the quality rules and the requirements of the project.
—Monitoring of the manufacturing, assembly, commissioning and repair procedures.
—Execution of the quality control plan during manufacture.

13.4.4.5 *Quality of testing*

As a general rule, the quality of testing is based on:

—Optimum definition of the test programme.
—A test procedure which conforms to the objective (qualification, acceptance or development testing) and is compatible with the requirements of the project (such as constraints encountered in the course of the mission and the duration of the mission).
—The quality, reliability and security of the test methods.
—The quality of the measuring equipment (ensured by periodic checking and suitable conditions of use).
—The quality of performance of the test.
—Utilisation of the results.

13.4.4.6 *Control of the configuration*

The quality of the whole project depends on thorough knowledge of the system at a given time and hence on subsequent control of the configuration. The organisation of the system, partitioning into assemblies, sub-assemblies, units, components and so on, the definition of basic documents, nomenclature, continuous updating, the availability and dissemination of documentation and information are the important factors for the quality of the configuration and its control.

13.4.4.7 *Non-conformity, failures, exemptions*

All recorded non-conformities and failures must be the object of a programme of treatment which includes analyses, expert appraisal, statistical evaluation, repairs, exemptions and modifications. This programme is particularly intended to identify the origin of the difficulty, the responsibility and the solution to be used to obtain conformity of the failed element to the reference models (this relates to the specification of the identification model, acceptance and qualification) and to avoid the occurrence of a further deviation in the subsequent part of the project.

13.4.4.8 *Development programme of models and mock-ups*

The quality of a development programme lies in the realisation of mock-ups and models (development mock-ups, mechanical models, thermal models, identifications models, qualifications models and flying models). Tests enable the feasibility, optimisation of structures and mass distribution, mechanical behaviour, thermal behaviour and adequacy of the hardware for the constraints which will be encountered to be ensured from one stage of the project to another. The combined results of the tests allow the design to be optimised and the adequacy of the model for the requirements of the mission to be verified.

13.4.4.9 *Storage, packaging, transport and handling*

Storage, packaging, handling and transport conditions are the subject of a set of rules specified with the intention of maintaining the quality of the hardware regardless of the level of integration. These rules contain a number of precautions which are taken so that the hardware is not weakened by constraints for which it is not designed and these must be considered in connection with the equipment used for packaging, handling and transport.

Application of these principles determines the validity of operations associated with the reliability and security of space programmes.

REFERENCES

[BAK-80] J.C. Baker, G.A. Baker (1980) Impact of the space environment on spacecraft lifetimes, *J. Spacecraft*, **17**(5), 479–480.

[CHA-81] R. Chauffriasse (1981) L'Assurance Produit dans les Programmes Spatiaux, *La Technologie des Expériences Scientifiques Spatiales, Cours de Technologie du CNES,* Toulouse, May, pp. 581–616.

[COR-75] M. Corazzo (1975) Techniques mathématiques de la fiabilité previsionnelle, Cépadues Editions, Toulouse.

[LIN-81] F. Linder (1981) Les Technologies d'Assemblage, *La Technologie des Expériences Scientifiques Spatiales, Cours de Technologie du CNES,* Toulouse, May, pp. 659–688.

[MIL-79] MIL-HDBK (1979) 217C—Reliability prediction of electronic equipment, Department of Defence, USA.

[RAS-88] R.D. Rasmussen (1988) Spacecraft electronic design for radiation tolerance, *Proceedings of the IEEE,* **76**, No. 11, November.

Index

A-MAC 73
ACl 131, 133, 405
Acknowledgement (ACK) 173
Acquisition 204
Active antenna 481
Active switching element 421
Active thermal control 572
ACTS satellite 4, 220, 429
Actuator 294, 507
Adaptive pulse code modulation (ADPCM) 381
Adaptivity 53
Adjacent channel 406
  interference (ACI) 131, 133, 405
Administrative organisation 56
ADPCM 381
Aerodynamic drag 279
Albedo 503
Allocation, frequency 57
ALOHA protocol 173
Alternate channel 406
Altitude 243
AM 89
AM/AM conversion 397
AM/PM conversion 397, 401
Amplification 369
Amplifier
  channel 416
  input 409
  low noise 359
  power 360
Amplitude modulation (AM) 89
Amplitude to phase modulation conversion 397, 401
Amroc 638
Analogue transmission 85, 120, 372
Angle, temporal 236
Angular beamwidth 17
Angular elongation, nodal 232
Angular measurement 306
Angular momentum 225, 507
Anomaly 229
Antenna 321, 329
  active 481
  alignment 26
  array 479
  Cassegrain 337
  characteristic 455
  coverage 437
  despun 473
  dual-grid 476
  gain 15, 352
  horn 329, 474
  lens 478
  multibeam 334
  multisource 476
  noise temperature 29, 33, 323, 326
  pointing 337
  pointing error 446
  reflecting 474
  type 483
  Vivaldi 483
Aperiodic coupling 365
Apeture blocking 330
Apogee motor 531, 582, 593
Apparent movement, satellite 344
Apsidal line 586
Arcjet propulsion 530
Argument of the perigee 232
Ariane 601, 609, 612
ARQ 117
Array antenna 479
Ascending node 231
Assignment 165
  packet 193
ASTRA 98
Asynchronous protocol 173
Asynchronous transfer mode (ATM) 218
Atlas 626
ATM 218
Atmospheric attenuation 50
Attenuation 25, 46, 50, 52
Attenuator noise temperature 30
Attitude control 500, 515
Attitude determination 503
Attitude sensor 502
Authenticity 103
Automatic repeat request (ARQ) 117
Automatic tracking 353
Availability 4, 55, 82, 667
Available power 313
Average talker 72
Axial ratio 19, 20
Axis, malor 228
Axis, semi-major 228
Azimuth 237
Azimuth angle 244, 267, 334
Azimuth-elevation mounting 337

B-MAC 75
Back-off 62, 137, 416
Back-up satellite 4, 668
Band, frequency 89
Bandwidth required 93, 97
Bandwidth-power interchange 118
Baseband processing 85, 434
Baseband switch 218, 435
Battery 546
Beam
  circular 458
  elliptical 461
  forming network 196
  interconnection 421
  lattice 470
  scanning 196, 219
  shaped 465
  shaping 466, 476, 481
  switching 189
Beamwidth 17, 345
BEP 111, 215
BER 78
Bi-propellant propulsion 524, 531, 595
Binary message 77
Binary phase shift keying (BPSK) 105, 110
Biphase coding 76
Bipolar coding 76
Bit 154
Bit error probability (BEP) 111, 215
Bit error rate (BER) 78
Bit rate 145
Bit stealing 379
Block cipher 103
Blocking probability 126
Boresight 15
BPSK 105, 110
Brightness temperature 29
Broadcast satellite service link 58
Burst assignment 147
Burst generation 141
Burst reception 144
Burst structure 142
Burst time plan 193
Bus, regulated 551
Bus, unregulated 549

C/N_0 32, 64, 67, 80, 122
C/T 32
C-MAC 75
Capacity 87
Capture effect 402
Carrier distribution 358
Carrier post-coupling 364
Carrier pre-coupling 363

Carrier to noise power spectral density ratio 136
Cassegrain antenna 337
Cassegrain mounting 331
CCI 186
CDMA 125, 129, 154. 164
Celestial latitude 237
Celestial longitude 237
Centaur 626
Central terminal unit (CTU) 565
CEPT hierarchy 101
Channel 77
amplifier 416
encoding 103
multiplication 101
telephone 71
Channelisation 405
Characteristic, antenna 455
China 611
Chip 154
Chrominance 72
Cipher 103
Circular beam 458
Circular orbit 7, 263, 265
Circular polarisation 19
Circularisation 584
Circulator 365
Civil time 239
Cladding 570
Clear sky 34, 38, 42, 46
Closed loop synchronisation 148
Closed loop tracking 347
Cluster, satellite 201
Co-channel interference (CCI) 186
Co-ordinates
ecliptic 237
equatorial 236
horizontal 237
temporal 236
terrestrial 236
topocentric 237
Co-polar polarisation 20
Code acquisition 161
Code division multiple access (CDMA) 125, 129, 154, 164
Code generation 160
Coding, downlink 216
Coding ratio 122
Coherent demodulation 110, 432
Cold gas propulsion 523
Collision 174
probability 305
Collision resolution algorithm (CRA) 177
Colour television signal 72
Command decoder 565
Command link 557
Common logic equipment 375
Communication services 57
Communication sub-system 366
Companding 86, 93
Component evaluation 677
Component reliability 676
Component selection 676
Compression 86
Computed tracking 347
Comsat manoeuvre 297
Confidentiality 103
Conditioning, battery 547
Conestoga 637
Configuration control 680
Conical scanning, tracking by 348
Conjunction, sun–satellite 247, 260, 274, 325
Constant width slot antenna (CWSA) 483
Constraints, operational 60
Contiguous beams 469
Continuous retransmission 117
Continuous visibility 254
Conventional satellite 4
Conversion efficiency 537
Conversion loss 409
Cooled amplifier 358
Correction cycle 304
Correction period 301
Correction strategy 585
Cosmic particles 657
Cosmic radiation 657
Coverage 8, 201, 245, 268, 44, 465
antenna 437
European 185, 472
geometric 441
radio-frequency 458, 461
CRA 177
Cross polarisation 49, 51
discrimination 20
isolation 20
Crosstalk 139
CTU 565
CWSA 483

D-MAC 75
D2-MAC 75
DAMA 178
Data 77, 81
Data bus 565
Data link interface (DLI) 380
Data transmission 84
DCME 380
DCU 189, 193
DE-BPSK 105, 110
De-emphasis 86, 97
DE-QPSK 105, 110
Death probability density 665
Declination 236
Decoding 114
Delay modulation 76
Delta 624
Demand assignment 171
Demand assignment multiple access (DAMA) 178
Demodulation 90, 109, 372
coherent 110, 432
differential 110, 432
gain 92
multicarrier 433
Demodulator 91, 96, 432
Demodulator threshold 92
Demultiplexing 88, 377
Depointing 26, 344, 446, 448, 451, 463
Descrambler 119
Despun antenna 473
Dichroic reflector 477
Differential demodulation 110, 432
Differential encoding 104, 107
Differential Manchester coding 76
Differentially encoded BPSK (DE-BPSK) 105, 110
Differentially encoded QPSK (DE-QPSK) 105, 110
Digital circuit multiplication equipment (DCME) 380
Digital modulation 104
Digital speech concentration 101
Digital speech interpolation (DSI) 101, 151, 378, 380
Digital transmission 80, 99, 122, 373
Digitisation 99
Direct encoding 107
Direct sequence transmission (DS-CDMA) 154, 160, 162
Distance ambiguity 560
Distance measurement 559
Distance, satellite 242, 266
Distribution circuit 553
Distribution control unit (DCU) 189, 193
Diversity 53
DLI 380
Doppler effect 146, 245
Down-converter 367
Downlink 24, 34, 42, 44, 61, 221, 556
Downlink coding 216
Drag, aerodynamic 279
Drift orbit 605
Drift, satellite 298
DS-CDMA 154, 160, 162
DSI 101, 151, 378, 380
Dual-grid antenna 476
Duobinary coding 76
Dwell area 219
Dynamic redundancy 671

Early Bird 9
Earth station 311
Earth's orbit 232
Earth–satellite geometry 240
Eccentric anomaly 229
Eccentricity 228, 282, 289, 291, 449
control of 302
Echo cancellation 383
Echo suppression 383
Echo suppressor 83
Eclipse 246, 260, 268, 271
Ecliptic co-ordinates 237
Efficiency 16
EIRP 21, 24, 45, 63, 183, 220, 313, 321, 393, 457, 463
Electric power supply 536
Electric propulsion 527
Electrical power 385
Electronic deviation, tracking by 349

Elevation 237
Elevation angle 8, 200, 243, 267, 334, 442
Elliptical beam 461
Elliptical orbit 5, 248, 259, 262
Elliptical polarisation 19
Encoding 100, 107
  channel 103
  differential 104
Encryption 102
End of life 305
Energy 228
Energy dispersion 98, 118
Energy source 536, 542
Equalisation 369
Equator, mean 235
Equatorial co-ordinates 236
Equatorial mounting 340
Equatorial orbit 8
Equatorial plane 586
Equivalent gain 314
Equivalent isotropic radiated power (EIRP) 21, 24, 45, 63, 183, 220, 313, 321, 393, 457, 463
Equivalent parabolic reflector antenna 332
Erlang 125
Error correction 114
Error probability 213
Europe 612
European Conference on Post and Telecommunications (CEPT) 101
European coverage 185, 472
EUTELSAT 142, 424, 467
EUTELAST standard 319
Evaluation, componenet 677
Expansion 86
Explicit reservation 178

Fabrication tolerance 16
Failure rate 663, 670
Faraday effect 51
Fast frequency shift keying (FFSK) 108
FDM 87, 378
FDM/CFM 95
FDM/FM 94
FDM/FM/FDMA 130
FDM/SSB-AM 96
FDMA 125, 130, 138
FFSK 108
FH-CDMA 157
Field 72
Field, traffic 142
Figure of merit 33, 45, 314, 321, 393, 457
Filtering 369
Fit 664, 669
Fixed antenna 346
Fixed assignment 165
Fixed mounting 352
Fixed satellite service link 58
Flux density, power 22
FM 90
Footprint, radio-frequency 457
Forward-acting error correction 115
Four-state modulation 106
Frame 72
  duration 151
  structure 142, 191
  throughput 195
France Telecom 168
Free space loss 22
Frequency allocation 57
Frequency band 89
Frequency conversion 358, 367, 370, 403, 409
Frequency division multiple access (FDMA) 125, 130, 138
Frequency division multiplexing (FDM) 87, 378
Frequency domain switching 427
Frequency hopping (FH-CDMA) 157
Frequency modulation (FM) 90
Frequency plan 424
Frequency re-use 183
Frequency translation 367, 369
G/T 33, 45, 314, 321, 393, 457
Gain 15, 63
  antenna 352
  demodulation 92
  diversity 53
  equivalent 314
  limiting value 465
  power 62, 396
  transmission 21, 313
Gamma filter 72
Geocentric latitude 236
Geographical latitude 236
Geographical longitude 236
Geometric coverage 441
Geometry, earth–satellite 240
Geostationary orbit 265
Geostationary satellite 9
Geosynchronous orbit 263
Geosynchronous transfer orbit (GTO) 582
Global coverage 442
Global efficiency 16
Global network 202
GMT 239
Gravitation constant 223
Gravitational field 648
Greenwich Mean Time (GMT) 239
Gregorian reflector 475
Ground, influence of 51
Ground radiation 35
Ground segment 2,4
Group delay 371
GTO 582
Guard time 146, 152
Guided-beam feed system 333
Gyroscope 507
Gyroscopic rigidity 514
Gyroscopic stabilisation 509

Handling 680
Header 142, 152
Hierarchy 101
Hopping, transponder 188
Horizontal co-ordinates 237
Horn antenna 329, 474
Hour angle 236, 341
Hour co-ordinates 236
Hybrid access 129
Hybrid coupler 365
Hybrid propellant motor 596
Hydrazine propulsion 531

IBO 62, 67, 135, 396, 399
Illumination efficiency 16
Implicit reservation 178
Improvement factor, diversity 53
Impulse-time 522
Inclination 231, 282, 287, 449
Inclination correction 584
Inclined plane strategy 300
India 638
Inertia wheel 512, 514
Inertial detector 503
Information storage 218
Injection into orbit 597, 602, 608
INMARSAT standard 320
Input amplifier 409
Input back-off (IBO) 62, 67, 135, 396, 399
Installation 581, 604, 660
Integrated Services Digital Network (ISDN) 81
Intelsat Network 443
INTELSAT 9, 97, 142, 371, 404, 513, 575
INTELSAT standard 315, 318
Intensity of precipitation 46
Interconnection, beam 421
Interference 8, 28, 59, 131, 157, 160, 186, 215, 326, 471
Interferometry 307
Intermediate frequency sub-system 375
Intermodulation 132, 135
  noise 135, 405
International meridian 237
Intersatellite link (ISL) 197
Ionic propulsion 528
Ionic thruster 529
IRIS 632
ISAS programme 640
ISDN 81
ISL 197
Israel 638
ITALSAT 4, 428

Japan 638
Julian century 239
Julian day 239

Kepler's laws 223

Laser detector 503
Latitude 241
  celestial 237
  geocentric 236
  geographical 236
Lattice converage 471

Launch azimuth 584
Launch base latitude 597
Launch phase 599
Launch window 609
Launcher 610
  low capacity 634
Launching 661
Legal time 239
Lens antenna 478
LEO 202, 581
Level diagram 40, 44
Linearly tapered slot antenna (LTSA) 483
Lifetime 4
Linear polarisation 19
Lineariser 363
Link
  satellite 58
  budget 211, 312
  cost 55
  intersatellite 197, 199
  optical 204
  radio frequency 203
  security 558
  station-to-station 61
Liquid propellant propulsion 525
LNA 38, 359, 409
Local sidereal time (LST) 237
Location 559
Lockheed 638
Long March 611
Longitude 241
  celestial 237
  geographical 236
  of the ascending node 231
  satellite 284
LOOPUS orbit 257
Loss 22, 25
Low capacity launcher 634
Low earth orbit (LEO) 202, 581
Low noise amplication 403
Low noise amplifier (LNA) 38, 359, 409
LST 237
LTSA 483
Luminance 72

MAC colour television signal 73
Magnetic coil 508
Magnetic field, earth's 649
MAILSTAR satellite 7
Major axis 228
Manufacture 678
Mean anomaly 230
Mean equator 235
Mean lifetime 666
Mean movement 230
Mean time to failure (MTTF) 666
Meridian, international 237
Message 77
Meteorites 651
Minimum shift keying (MSK) 108
Mobile satellite service link 58
Mode extraction monopulse 350
Modified polar mount 343
Modulation 372
  digital 104
  four-state 106
  index 90
  spurious 363
  two-state 105
MOLNYA orbit 7, 249, 255
Momentum wheel 515
Monitoring, alarms and control (MAC) 384
Monopropellant hydrazine 523
Monopulse technique 350
Movement, Mean 230
MSK 108
MTTF 666
Multi-clique operation 382
Multibeam antenna 334
Multibeam coverage 429
Multibeam satellite 181, 420
Multicarrier demodulation 433
Multicarrier operation 134, 398
Multidestination operation 382
Multipath effects 51
Multiple access 125, 128, 156, 159
Multiple beams 469
Multiple source monopulse 350
Multiple trajectories 157
Multiplexer 410
Multiplexer coupling 365
Multiplexing 87, 377
Multiplication factor 381
Multiservice systems 12
Multisource antenna 476

Nadir angle 244, 268, 274
NASDA programme 638
Network, global 202
Network, interface 377
Network synchronism 102
Newcomb's equation 239
Newton's law 223
Nodal angular elongation 232
Noise 27, 363
  figure 29
  power 27
  spectral density 91
  temperature 28, 30, 37, 61, 315
    antenna 29, 33, 323, 326
    sun 324
Non-linear characteristic, amplifier 394
NRZ coding 76, 154
NTSC colour television signal 72

OBDH 566
Oblateness 233
OBO 62, 67, 135, 396, 399
Official time 239
Offset mounting 330, 475
Offset QPSK (OQPSK) 108
On-board angular momentum 514
On-board data handling (OBDH) 566
On-board management 563
On-board processing 216
On-board regeneration 428
On-board switching 189, 422
On-demand assignment 165
One carrier per link 126, 141, 151
One carrier per station 127, 141, 151
Open loop synchronisation 149
Operational constraints 60
Optical link 204
Optical switching 423
OQPSK 108
Orbit
  circular 7, 263, 265
  correction 290, 294
  drift 605
  earth's 232
  elliptical 5, 248, 259, 262
  equatorial 8
  form of 230
  geostationary 265
  geosynchronous 263
  injection into 597, 602, 608
  installation in 581
  LOOPUS 257
  MOLNYA 7, 249, 255
  period 228
  perturbation 280
  restoration 306
  transfer 583, 586, 603, 662
  TUNDRA 7, 251, 255
  types of 5
  useful 248
  velocity 583
Orbital parameter 227, 280, 282
Orbital position 200
Orthogonal polarisation 19
Osculatory parameter 280
Output back-off (OBO) 62, 67, 135, 396, 399
Output power, saturation 62, 396

Packaging 680
Packet assignment 193
PAL colour television signal 72
PAM-D 631
Parabolic reflector 330
Parameter, orbital 227, 280, 282
Parameter, osculatory 280
Parameter, shape 227
Parametric amplifier 356
Passive thermal control 569
Path loss 27
Payload 3, 391
Payload Assist Module Delta class (PAM-D) 631
Pegasus 634
Perigee 591
  argument of 232
  motor 582, 593
  stage 582
Perigee velocity augmentation manoeuvre (PVA) 591
Perigee–apogee line 586
Period, orbit 228
Perturbation of orbit 276, 280
Phase shift keying (PSK) 105, 131
Phased array antenna 329, 467

Pitch attitude control 515
Plasma propulsion 527
Platform 3
  stabilised 474
  telecommunication 497
Plesiochronism 102
PN codes 129
Point coverage 444
Point of compression to 1 dB 397
Pointing error, antenna 446
Poisson distribution 671
Polar coding 76
Polar mounting 340, 343
Polarisation 18
  angle 336
  mismatch 26
Positioning phase 602, 604
Power amplifier 360, 419
Power, electrical 385
Power flux density 22, 63
Power gain 62, 39[illegible]
Power received [illegible]
Power spectral [illegible]
Power transfer ch[illegible] 63, 395, 398
Pre-emphasis 86, 97
Preamble 142
Precipitation 46
Primary energy source 536
Probability of death 665
Probability of survival 664
Programmed tracking 347, 353
Propagating medium 45
Propagation time 82, 245, 268, 362
Propulsion 520, 522
Protection, battery 547
Protection circuit 553
Protocol, asynchronous 173
Pseudo Noise (PN) codes 129
Pseudorandom sequence 119, 150, 159
PSK 205, 131
Psophometric weighting 80
PVA 591

QPSK 105, 110
Quadrature phase shift keying (QPSK) 105, 110
Quality assurance 678
Quantisation 100
Quasi-circular polarisation 20

Radial velocity measurement 563
Radiating surface 570
Radiation 260, 363, 653
  characteristic 321
  hardness 659
  pattern 16
  pressure, solar 278, 650
  resistance 437
Radio broadcast 76, 81
Radio-frequency coverage 458, 461
Radio-frequency detector 503
Radio-frequency footprint 457
Radio-frequency link 203
Radiocommunications regulations 56, 59
Rain 36, 41, 44, 46, 68
Rainfall rate 46
Random access 172
Range measurements 561
Re-utilisation factor 184
Reaction wheel 507
Received power 22
Receiver noise temperature 37
Receiving equipment 355
Reconfigurable interconnection 421
Recoverable towing vehicle 632
Reduced coverage 444
Redundancy 366, 407, 553, 671
Reference station 142
Reflecting antenna 474
Refraction 50
Regenerative satellite 4, 211
Regenerative transponder 212, 427, 432
Regulated bus 551
Regulations, radiocommunications 56, 59
Regulatory aspects 56
Reliability 4, 663, 667, 669, 676
Remote terminal unit (RTU) 566
Repeater 62
  architecture 404
  gain 66
Residual inclination 291
Retransmission 117
Return to centre strategy 301
Right ascension 236
Right ascension of the ascending node 231
Roll control 516
Rotation, stabilisation by 511
Routing, traffic 126
RTU 566
RZ coding 76

Sampling 100
Sandstorms, attenuation by 50
Satellite
  acceptance 607
  back-up 4, 668
  beam scanning 219
  cluster 201
  control 567
  conventional 4
  distance 242, 266
  drift 298
  geostationary 9
  link 58
  low orbit 202
  MAILSTAR 7
  multibeam 181, 420
  regenerative 4, 211
  service link 58
  SPOT 7
  subsystem 498
  switched TDMA (SS-TDMA) 189, 191
  test 607
  transparent 4
  track 240, 250, 606
  velocity 228
  visibility 253
Saturation output power 62, 396
Scanning sensor 506
SCOUT 623, 634
SCPC 92
SCPC/FDMA 131
SCPC/FM 93, 96
Scrambling 119
SECAM colour television signal 72
Secondary energy source 536, 542
Segment, ground 2, 4
Segment, space 2
Selective retransmission 117
Semi-major axis 228, 233
SEP 111
Series regulator 552
Service link, satellite 58
Shape parameter 227
Shaped beam 465
Shaped coverage 467
Shunt regulator 552
Side lobe radiation 321
Sidereal day 238
Sidereal time (ST) 237, 239
Signal 15
Signal-to-noise ratio 32, 45, 61, 78, 86, 90, 91, 92 96
Signature 164
Single channel per carrier (SCPC) 92
Site diversity 53
Sky noise 35
Slotted ALOHA protocol (S-ALOHA) 177
Solar
  array 538
  cell 536, 538
  day 238
  detector 502
  generator 540, 542
  panel 540
  radiation 536, 653
    pressure 278, 650
  sail 508, 517
  time 237
  wind 657
Solid propellant 523, 593
SORF 147
SOTF 147
Space communication service 57
Space segment 2
Space Shuttle 582, 629
Space Transportation System (STS) 582, 629
Space tug 632
Space Vector 637
SPADE system 167
Specific attenuation 48
Specific impulse 521
Spectral density 28
Spectral efficiency 109
Spectral occupation 87, 89, 91, 155, 158, 205

Speech 72
activation 85
concentration, digital 101
SPELTRA 621
Spill-over efficiency 16
Spillover 330
SPOT satellite 7
Spread spectrum multiple access (SSMA) 129
Spread spectrum transmission 164
Spreading ratio 155
Spurious modulation 363
SQPSK 108
SS/TDMA 189, 191
SSMA 129
ST 237, 239
Stabilisation by rotation 511
Stabilisation, three axis 514
Stabilised platform 474
Stable equilibrium point 285
Staggered QPSK (SQPSK) 108
Stand-by power 385
Standing wave ratio (SWR) 362
Start of receive frame (SORF) 147
Start of transmit frame (SOTF) 147
Station acquisition 604
Station keeping 290, 294, 297, 305
Station keeping window 146, 292
Station-to-station link 61
Stefan–Boltzman law 653
Stellar sight 503
Step-by-step tracking 348
Storage 680
Storage cell 539
Stream cipher 103
STS 582, 629
Subsystem reliability 669
Subsystem, satellite 498
Sun regulated bus 550
Sun–satellite conjunction 247, 274, 260, 325
Surface finish efficiency 16
Switching matrix 425
Switching, TST type 218, 434
SWR 362
Symbol error probability (SEP) 111
Symbol generator 104
Synchronisation 145, 148, 150, 160, 194
Synchronism, network 102
System availability 667

T-carrier hierarchy 101
Taurus 637
TDM 100, 378
TDM/PSL/FDMA 130
TDMA 125, 140, 151, 153, 374
TELECOM 168, 514
Telecommunication platform 497
Telecontrol, tracking and command (TTC) 555
Telemetry 559
Telephone 83
channel 71, 92
transmission 94, 96
Telephony 87
Television 72, 80
transmission 96
Temperature range 568
Temporal precipitation statistic 46
Terrestrial co-ordinates 236
Terrestrial horizon detector 502
Terrestrial radiation 654
Thermal control 567
Thermal effects 654
Third order intercept point 400
Third order intermodulation 399
Three axis stabilisation 514
Three axis stabilised satellite 541
Threshold, demodulator 92
Throughput 138, 151, 162, 179
frame 195
Thruster 507, 520, 526, 593
Time, sidereal 237, 239
Time, solar 237
Time division multiple access (TDMA) 125, 140, 151, 153, 374
Time division multiplexing (TDM) 100, 378
Time equation 237
Time reference 238
Time slot interchange (TSI) 380
Time–space–time structure (TST) 218, 434
Titan 627
Tone range measurement 561
Topocentric co-ordinates 237
Toroidal radiation pattern 473
TOS 632
Track, satellite 240, 250, 606
Tracking, 205, 344, 347, 352
code 161
Traffic density 125
Traffic field 142
Traffic routing 126
Traffic station 142
Trajectory 226
Transfer coefficient 401
TRANSDYN 168
Transfer characteristic 62, 134
Transfer orbit 583, 586, 603, 662
Transfer Orbit Stage (TOS) 632
Transfer phase 602, 604
Transfer stage 582, 632
Transistor
amplifier 357, 361, 419
analogue 120
delay 8
digital 80, 99, 122, 373
duration 8
equipment 360
gain 21, 313
time 152, 175
Transparent satellite 4
Transparent transponder 212
Transponder 63, 133, 211, 420
architecture 403
conventional 394
hopping 188
regenerative 212, 427, 432
Transport 680
Travelling wave tube amplifier 418
Tripod mounting 343
True anomaly 229
True view angle 438
True view representation 441
Trunking telephony 12
TSI 380
TST type switching 218, 434
TTC 555
Tube amplifier 361
TUNDRA orb[illegible] 7, 251, 255
Two reflecto[illegible]unting 475
Two-state m[illegible]tion 105

Unified bi-propellant propulsion 531
Uninterruptible power 385
Unique word (UW) 142, 145, 152
United States 624
Universal time (UT) 239
Unregulated 549
Unreliability 665
Unstable equilibrium point 285
Uplink 33, 38, 41, 61, 220, 556
USSR 641
UT 239
UW 142, 145, 152

Vacuum 647
Van Allen Belt 260, 657
Veis reference 232
Velocity impulse, annual 301
Velocity increment 294, 300, 521, 583
Velocity, satellite 228
Very small aperture terminal (VSAT) 4, 12
Visibility, continuous 254
Visibility, satellite 253
Vivaldi antenna 483
VSAT 4, 12

Wear-out 667
Window structure 192

X-Y mounting 339

Yaw control 516

Zenithal distance 237